Elementary Number Theory
and Its Applications

Fifth Edition

Elementary Number Theory

and Its Applications

Fifth Edition

Kenneth H. Rosen
AT&T Laboratories

PEARSON

Addison
Wesley

Boston San Francisco New York
London Toronto Sydney Tokyo Singapore Madrid
Mexico City Munich Paris Cape Town Hong Kong Montreal

Publisher: Greg Tobin
Senior Acquisitions Editor: William Hoffman
Editorial Assistants: Emily Portwood and Mary Reynolds
Marketing Manager: Yolanda Cossio
Marketing Coordinator: Heather Peck
Managing Editor: Karen Wernholm
Senior Production Supervisors: Jeffrey Holcomb and Julie LaChance
Project Management: Barbara Pendergast
Composition and Art Illustration: Windfall Software, using ZzTEX
Senior Manufacturing Buyer: Evelyn Beaton
Photo Research: Beth Anderson
Interior Design: Barbara T. Atkinson
Cover Design: Suzin Purney Osborne
Cover Image: © Jasper Johns / Licensed by VAGA, New York, NY / SuperStock

Photo Credits: Grateful acknowledgment is made to the copyright holders of the biographical photos, listed on page 721, which is hereby made part of this copyright page.

Many of the designations used by manufacturers and sellers to distinguish their products are claimed as trademarks. Where those designations appear in this book, and Addison-Wesley was aware of a trademark claim, the designations have been printed in initial caps or all caps.

Library of Congress Cataloging-in-Publication Data

Rosen, Kenneth H.
    Elementary number theory and its applications / Kenneth H. Rosen.—5th ed.
      p. cm.
    Includes bibliographical references and index.
    ISBN 0-321-23707-2 (alk. paper)
    1. Number theory.   I. Title.

QA241.R67 2005
512.7′2—dc22   2004057054

ISBN 0-321-23707-2
1 2 3 4 5 6 7 8 9 10-PHT-07 06 05 04

Preface

In olden times (well, before 1975) number theory had the reputation of being the purest part of mathematics. It was studied for its long and rich history, its wealth of easily accessible and fascinating questions, and its intellectual appeal. But, in the past few years, people have looked at number theory in a new way. Today, people study number theory both for the traditional reasons and for the compelling reason that number theory has become essential for cryptography. The first edition of this book was the first text to integrate the modern applications of elementary number theory with traditional topics. This fifth edition builds on the basic approach of the original text. No other number theory text presents elementary number theory and its applications in as thoughtful a fashion as this book does. Instructors will be pleasantly surprised to see how modern applications can be seamlessly woven into their number theory course when they use this text.

This book is designed as a text for an undergraduate number theory course at any level. No formal prerequisites are needed for most of the material, other than some level of mathematical maturity. This book is also designed to be a useful supplement for computer science courses and as a number theory primer for people interested in learning about new developments in number theory and cryptography.

This fifth edition has been designed to preserve the strengths of previous editions while providing substantial enhancements and improvements. Instructors familiar with previous editions will be comfortable with this new edition. Those examining this book for the first time will see an up-to-date text, which integrates gems of number theory dating back thousands of years with developments less than ten years old. Those familiar with previous editions will find that this book has become more flexible, easier to teach from, and more interesting and compelling. They will also find that additional emphasis has also been placed on the historical context of results and on the experimental side of number theory.

Changes in the Fifth Edition

This new edition incorporates many improvements made at the request of users and reviewers. The new edition should be easier to teach from, easier to read, and more interesting and informative. This edition more effectively conveys both the beauty and the utility of number theory. Noteworthy changes include:

- *More flexible organization*

The first section of the fourth edition has been divided into two shorter sections. The first covers types of numbers and sequences and introduces diophantine approximation. The second covers sums and products. Instructors can skip most of the material in these two sections if desired, although many will want to cover the material on diophantine approximation. Section 3.1 of the fourth edition has also been divided into two sections. The first of these sections introduces primes, establishes that there are infinitely many primes, and begins the discussion of how primes are found. The second section discusses the distribution of primes and introduces the prime number theorem and many conjectures about prime numbers.

- *Expanded coverage of cryptography*

Cryptanalysis of Vigènere ciphers has been added with the introduction of the Kasiski test and the index of coincidence. Recent developments in cryptography are mentioned, including the AES encryption standard. Attacks that have been devised on implementations on RSA are now described. One such attack is now developed in Chapter 12 using ideas from diophantine approximation using continued fractions. The weakness in a proposed zero-knowledge proof method is now included in an exercise.

- *Up-to-date discoveries*

The latest discoveries in number theory are reflected in the text, including a number of theoretical discoveries and discussions concerning the polynomial time algorithm for proving an integer in prime and the resolution of the Catalan conjecture. Computational discoveries, such as three new Mersenne primes, have been added. The Web site for the book will highlight the latest news in number theory and links will be provided that announce discoveries made subsequent to the publication of this book.

- *New and expanded topic coverage*

Dirichlet's theorem on approximation of real numbers by rational numbers has been added, introducing the subject of diophantine approximation to the first section of the text. A proof using the pigeonhole principle is provided. Many important topics whose full treatment is beyond the scope of an elementary number theory text are now discussed; the goal is to give the student a fuller appreciation of number theory. In a similar vein, the coverage of diophantine equations has been expanded. This edition includes brief discussions of Beal's conjecture, the Catalan conjecture and its recent resolution, and the Fermat-Catalan conjecture. The abc conjecture is also discussed, and how it can be used to prove results on diophantine equations is illustrated.

A new chapter on the Gaussian integers has been added. This chapter introduces Gaussian primes, the greatest common divisor of Gaussian integers, the Euclidean algorithm for Gaussian integers, and the unique factorization of Gaussian integers into Gaussian primes. This new chapter also explains how Gaussian integers can be used to find the number of ways to express a positive integer into the sum of two squares.

- *Improved examples and proofs*

Euclid's proof that there are infinitely many primes is now given in the text. A large number of other proofs of the infinitude of primes can be found in the exercises. Many proofs have been improved, either by simplification or by additional explanation.

- *Enhanced exercise sets*

This book has long been noted for its exceptional exercises; in this edition the exercises are even better. All exercises in the text have been reviewed and solved; exercises from the fourth edition found to be ambiguous or lacking assumptions have been clarified.

Several hundred new exercises have been added. Additional exercises involving Fibonacci identities have been inserted. New exercises also outline different proofs that there are infinitely many primes. There are many new exercises on cryptography, including many relating to the Vigènere cipher and the RSA cryptosystem. The newest proof of the law of quadratic reciprocity is outlined in an exercise. More exercises on nonlinear diophantine equations have been added, including exercises on Bachet's equation, Markov's equation, and congruent numbers.

- *Expanded historical context and biographies*

The history and status of the Riemann hypothesis are now covered. Skewes' constant, one of the largest numbers arising in a proof, is introduced. Also added is an account of the discovery by Thomas Nicely of the famous division flaw in the Pentium chip, found because two computations involving twin primes did not agree. This edition introduces many new biographies, including those of Bertrand, Farey, Waring, Bachet, Kronecker, Levi ben Gerson, and Catalan. Photographs have been also been added to many biographies.

- *Enhanced ancillaries and enhanced support for Maple® and Mathematica®*

The *Student's Solutions Manual* and the *Instructor's Manual* have been enhanced. They both now contain a comprehensive guide explaining how to use Maple for computations in number theory. Suggested syllabi for different courses are now contained in the *Instructor's Manual*. The *Instructor's Manual* and the Web site now both contain migration guides for the exercises showing where exercises in the fourth edition can be found in the fifth edition, and conversely, where exercises in the fifth edition were located in the fourth edition, if they were included in this previous edition.

Commands for carrying out computations with the Gaussian integers have been added to the appendix that describes number theory commands in Maple and *Mathematica*.

- *Extra focus on accuracy*

This edition benefits from extra resources devoted to ensure the accuracy of the text, as well as the exercises and their answers and solutions. Three accuracy checkers have spent long hours making sure that this book is as error-free as possible.

- *Expanded Web site*

The Web site for this text has been expanded and enhanced in several key ways. "Number Theory News" is a new feature highlighting recent discoveries in number theory. The extensive list of number theory Web sites keyed to the text has been expanded and all links have been updated. These links will be periodically updated during the life of this edition. The Web site now also supports an extensive collection of number theory and cryptography applets which can be used for computations and exploration, as well as a tutorial on PARI/GP, a computational system for fast computation in number theory upon which these applets are built. A collection of suggested group or individual student projects can also be found on the Web site.

Features

A Development of Classical Number Theory

The core of this book presents classical elementary number theory in a comprehensive and compelling manner. The historical context and importance of key results are noted. The basic material on each topic is developed carefully, followed by more sophisticated results on the same topic.

Applications

A key strength of this book is how applications of number theory are covered. Once the requisite theory has been developed, applications are woven into the text in a flexible way. These applications are designed to motivate the coverage of the theory and illustrate the usefulness of different aspects of elementary number theory. Extensive coverage is devoted to applications of number theory to cryptography. Classical ciphers, block and stream ciphers, public key cryptosystems, and cryptographic protocols are all covered. Other applications to computer science include fast multiplication of integers, pseudorandom numbers, and check digits. Applications to many other areas, such as scheduling, telephony, entomology, and zoology can also be found in the text.

Unifying Themes

Many concepts from elementary number theory are used in primality testing and factoring. Furthermore, primality testing and factoring play a key role in applications of number theory to cryptography. As such, these topics are used as unifying themes and are returned to repeatedly. Almost every chapter includes material on these topics.

Accessibility

This book has been designed with a minimum of prerequisites. The book is almost entirely self-contained, with only a knowledge of what is generally known as "college algebra" required. There are several places where knowledge of some concepts from calculus is needed (such as in the discussions of the distribution of primes and big-*O* notation). Concepts from discrete mathematics and linear algebra are needed in a few places. All material that depends on topics more advanced than college algebra is explicitly noted and is optional.

Accuracy

Great effort has been made to ensure the accuracy of this edition. Input from many users of the fourth edition, reviewers, and proofreaders has helped achieve this goal.

Extensive Exercise Sets

The best (and maybe the only) way to learn mathematics is by doing exercises. This text contains an extremely extensive and diverse collection of exercises. Many routine exercises are included to develop basic skills, with care taken so that both odd-numbered and even-numbered exercises of this type are included. A large number of intermediate-level exercises help students put several concepts together to form new results. Many other exercises and blocks of exercises are designed to develop new concepts. Challenging exercises are in ample supply and are marked with one star (∗) indicating a difficult exercise and two stars (∗∗) indicating an extremely difficult exercise. There are some exercises that contain results used later in the text; these are marked with a pointing-hand symbol (☞). These exercises should be assigned by instructors whenever possible.

An extensive collection of computer projects is also provided. Each section includes computations and explorations designed to be done with a computational program such as Maple or *Mathematica*, or using programs written by instructors and/or students. There are some routine exercises of this sort that students should do to learn how to apply basic commands from Maple or *Mathematica* (as described in Appendix D), as well as more open-ended questions designed for experimentation and creativity. Each section also includes a set of programming projects designed to be done by students using a programming language of their choice, such as the programming languages included with Maple and *Mathematica*, or another programming language of their choice.

Exercise Answers

The answers to all odd-numbered exercises are provided at the end of the text. More complete solutions to these exercises can be found in the *Student's Solutions Manual* that accompanies this text. All solutions have been carefully checked and rechecked to ensure accuracy.

Discovery via Empirical Evidence

In many places in the text numerical evidence is examined to help motivate key results. This gives an opportunity to students to come up with a conjecture much as the people who originally developed many of the results of number theory did.

Extensive Examples

This book includes examples that illustrate each important concept. These examples are designed to illustrate the definitions, algorithms, and proofs in the text. They are also designed to help students work many of the exercises found at the end of sections.

Carefully Motivated Proofs

Many proofs in this book are motivated with examples that precede the formal proof and illustrate the key ideas of the proof. The proofs themselves are presented in a careful, rigorous, and fully explained manner. The proofs are designed so that students can understand each step and the flow of logic. Numerical examples illustrating the steps of the proof are often provided following the formal proof as well.

Algorithmic Reasoning

The algorithmic aspects of elementary number theory are thoroughly covered in this text. Not only are many algorithms described, but their complexity is also analyzed. Among the algorithms described in this book are those for computing greatest common divisors in many different ways and for primality testing and factoring. The coverage of the complexity of algorithms has been included so that instructors can choose whether they want to include this material in their course.

Biographies and Historical Notes

More than 60 biographies of contributors to number theory are included in this edition. Contributors included lived in ancient times, the Middle Ages, the sixteenth through eighteenth centuries, the nineteenth century, and the twentieth century, and lived in the East and in the West. These biographies are designed to give students an appreciation of contributors as unique individuals who often led (or are leading) interesting lives.

Open Questions

Many open questions in number theory are described throughout the book. Some are described in the text itself and others are found in exercise sets. These questions show that the subject of number theory is a work in progress. Readers should be aware that attempting to solve such problems can often be time-consuming and futile. However, it would be surprising if some of these questions were not settled in the next few years.

Up-to-Date Content

The latest discoveries in number theory are included in this book. The current status of many open questions is described, as are new theoretical results. Discoveries of new primes and factorizations made as late as September 2004 are included with the first printing of this edition. These discoveries will help readers understand that number theory is an extremely active area of study. They may even see how they may participate in the search for new primes.

Bibliography

An extensive bibliography is provided for this book. This bibliography lists key printed number theory resources, including both books and papers. Many useful number texts are listed, as are books dealing with the history of number theory and particular aspects of the subject. Many original sources are included, as is material covering cryptography.

Maple and *Mathematica* Support

An appendix has been provided which lists the commands in both Maple and *Mathematica* for carrying out computations in number theory. These commands are listed according to the chapter of the text relevant to these commands.

Web Resources

The Web site for this book includes a Web guide to number theory that is keyed to this text, as well as an extensive collection of other resources. To access this site go to `www.awlonline.com/rosen`. For convenience, the most important number theory Web sites are highlighted in Appendix D.

Tables

A set of five tables is included to help students with their computations and experimentation. Looking at these tables can help students search for patterns and formulate conjectures. The use of a computational software package, such as Maple or *Mathematica* is recommended when these tables are insufficient.

List of Symbols

A list of symbols used in the text and where they are defined is included on the inside front cover of this book.

Ancillaries

Student's Solutions Manual (ISBN 0-321-26840-7)

The *Student's Solutions Manual* contains worked solutions to all the odd-numbered exercises in the text and other helpful material, including some tips on using Maple and

Mathematica to explore number theory. A tutorial for using Maple to do computations in number theory is provided.

Instructor's Manual (ISBN 0-321-26842-3)

The *Instructor's Manual* contains solutions to all exercises in the text. It also contains advice on planning which sections to cover. Sample tests are also provided.

Web Site

The Web site for this book contains a guide providing annotated links to a large number of Web sites relevant to number theory. These sites are keyed to the page in the book where relevant material is discussed. These locations are marked with an icon (✻) in the text. The Web site also contains a section highlighting the latest discoveries in number theory. An extensive collection of number theory and cryptography applets is also provided.

How to Use this Book

This text is designed to be extremely flexible. The essential, core material for a number theory course can be found in Section 1.4, which covers divisibility; Chapter 3, which covers primes, factoring, and greatest common divisors; Sections 4.1–4.3, which cover congruences; and Chapter 6, which covers important congruences including Fermat's little theorem. Instructors can design their own courses by supplementing core material with other content of their own choice. To help instructors decide which sections to cover, a brief description of the different parts of the book follows.

The material in Sections 1.1–1.4 is optional. Section 1.1 introduces different types of numbers, integer sequences, and countability. This section also introduces the notion of diophantine approximation. Section 1.2 reviews sums and products for students who need a review of these topics. Section 1.3 introduces mathematical induction, which students may already have studied elsewhere. (Material on integer axioms and the binomial theorem can be found in the appendices.) Section 1.4 introduces the Fibonacci numbers, a favorite topic of many instructors; students may have studied these numbers in a course in discrete mathematics. As stated previously, Section 1.5 presents core material on divisibility of integers and should be covered.

Chapter 2 is optional; it covers base b representations of integers, integer arithmetic, and the complexity of integer operations. Big-O notation is introduced in Section 2.3. This is important for students who have not seen this notation elsewhere, especially when the instructor wants to stress the complexity of computations in number theory.

As previously stated, Chapter 3 and Sections 4.1–4.3 present core material. Section 4.4, which deals with solving polynomial congruences modulo powers of primes is optional; it is important to development of p-adic number theory. Section 4.5 requires some background in linear algebra; the material in this section is used in Section 8.2; these sections may be omitted if desired. Section 4.6 introduces a particular factorization method (the Pollard rho method) and can be omitted.

Chapter 5 is optional. Instructors can pick and choose from a variety of applications of number theory. Section 5.1 introduces divisibility tests; Section 5.2 covers the perpetual calendar; Section 5.3 discusses scheduling round-robin tournaments; Section 5.4 shows how congruences can be used in hashing functions; and Section 5.5 describes how check digits are found and used. As mentioned previously, Chapter 6 presents core material.

Chapter 7 covers multiplicative functions. Section 7.1 should be covered; it introduces the basic concept of a multiplicative function and studies the Euler phi-function. The sum and number of divisors functions are studied in Section 7.2; this section is recommended for all instructors. All instructors will probably want to cover Section 7.3, which introduces the concept of a perfect number and describes the search for Mersenne primes.

Chapter 8 covers the applications of number theory to cryptology. It is highly recommended since this is such an important topic and one that students find extremely interesting. Section 8.1 introduces the basic terminology of this subject and some classical character ciphers; instructors who plan to cover cryptography in their course should be sure to include this section. Section 8.2 introduces block and stream ciphers, two important families of ciphers, and provides examples of these types of cipher that are based on number theory. Section 8.3 covers a particular type of block cipher based on modular exponentiation. Section 8.4 should be covered by all instructors. It introduces the fundamental concept of public key cryptography and illustrates this with the RSA cryptosystem. Section 8.5 discusses knapsack ciphers; it is an optional section. Section 8.6 provides an introduction to cryptographic protocols and is highly recommended for instructors interested in modern cryptographic applications. (Additional topics from cryptography are covered in Chapters 9, 10, and 11.)

Chapter 9 deals with the concept of the order of an integer, primitive roots, and index arithmetic. Sections 9.1–9.4 should be covered if possible. Section 9.5, which discusses how the concepts of this chapter are used in primality testing presents partial converses of Fermat's little theorem. Section 9.6, on universal exponents, is optional; it contains some interesting results about Carmichael numbers.

Chapter 10 introduces some applications that use the material from Chapter 9. The three sections that cover pseudorandom numbers, the ElGamal cryptosystem, and schemes for splicing telephone cable are optional. Instructors stressing cryptographic applications will especially want to cover Section 10.2.

Sections 11.1 and 11.2, which cover quadratic residues and quadratic reciprocity, a key result of number theory, should be covered whenever possible. Sections 11.3 and 11.4 deal with Jacobi symbols and Euler pseudoprimes and are optional. Section 11.5 covers zero-knowledge proofs; instructors interested in cryptography will want to cover this section if possible.

Section 12.1, which covers decimal fractions, will be covered by many instructors. Instructors with an interest in continued fractions will want to cover Sections 12.2–12.4, which establish the basic results about finite and periodic continued fractions. Section 12.5, which deals with factoring using continued fractions, is optional.

Most instructors will want to cover Sections 13.1 and 13.2, which deal with Pythagorean triples and Fermat's last theorem, respectively. Section 13.3, which covers sums of squares, and Section 13.4, which discusses the solution of Pell's equation and which uses continued fractions, are optional sections.

Chapter 14 is an optional chapter covering the Gaussian integers. Many of their properties analogous to those of the integers are developed in this chapter. In particular, Gaussian primes are introduced and the unique factorization of Gaussian integers is established. Finally, the number of ways a positive integer can be expressed as the sum of two squares is found using Gaussian integers.

The following figure showing the dependency of chapters will help instructors plan their course. Suggested syllabi for courses with different emphases are provided in the *Instructor's Resource Guide*. Although Chapter 2 may be omitted if desired, it does explain the big-*O* notation used throughout the text to describe the complexity of algorithms. Chapter 12 only depends on Chapter 1 as shown, except for Theorem 12.4, which depends on material from Chapter 9. Section 13.4 is the only part of Chapter 13 that depends on Chapter 12. Chapter 11 can be covered without covering Chapter 9 if the optional comments involving primitive roots in Section 9.1 are omitted. Section 14.3 should also be covered in conjunction with Section 13.3.

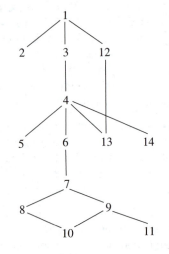

Acknowledgments

I wish to thank my management at AT&T Laboratories for their support in the preparation of this edition and for providing a stimulating professional environment. Special thanks go to Bart Goddard who has prepared the ancillaries to this book and to Douglas Eubert, Tom Wegleitner, and Steve Whalen for their help reviewing the manuscript for accuracy and for their assistance with the solution of the exercises and for checking and rechecking the answers and solutions to these exercises.

Thanks go to Bill Hoffman, the editor of this edition, for his support, and to all the other editors of previous editions of this book at Addison-Wesley, going back to Wayne

Yuhasz and Jeff Pepper who endorsed the original concept and recognized the potential appeal of this book at a time when other publishers considered number theory a dead course not worthy of new books. My appreciation also goes to the editorial, production, marketing, and media team behind this book, including Mary Reynolds, Julie LaChance, Jeffrey Holcomb, Barbara Atkinson, Beth Anderson, Barbara Pendergast, Paul Anagnostopoulos, Emily Portwood, Lynne Blaszak, Greg Tobin, and Phyllis Hubbard. I would also like to thank David Wright for his many contributions to the Web site for this book, including material on PARI/GP, number theory and cryptography applets, and suggested projects.

I have benefited from the thoughtful reviews and suggestions from users of previous editions of this book. Many of their ideas have been incorporated in this edition. My profound thanks go the following reviewers who helped me prepare this edition:

Reviewers

Ruth Berger, *Luther College*
Joel Cohen, *University of Maryland*
Michael Cullinane, *Keene State College*
Mark Dickinson, *University of Michigan*
George Greaves, *Cardiff University*
Kerry Jones, *Ball State University*
Slawomir Klimek, *Indiana University-Purdue University Indianapolis*
Stephen Kudla, *University of Maryland*
Jennifer McNulty, *The University of Montana*
Stephen Miller, *Rutgers University*
Michael Mossinghoff, *Davidson College*
Michael E. O'Sullivan, *San Diego State University*
Gary Towsley, *SUNY Geneseo*
David Wright, *Oklahoma State University*

I also wish to thank again the reviewers of previous editions of this book who have helped improve this book from edition to edition. Their affiliations at the time they reviewed the book are noted.

David Bressoud, *Pennsylvania State University*
Sydney Bulman-Fleming, *Wilfred Laurier University*
Richard Bumby, *Rutgers University*
Charles Cook, *University of South Carolina, Sumter*
Christopher Cotter, *University of Northern Colorado*
Euda Dean, *Tarleton State University*
Daniel Drucker, *Wayne State University*
Bob Gold, *Ohio State University*
Fernando Gouvea, *Colby College*
Jennifer Johnson, *University of Utah*
Roy Jordan, *Monmouth College*
Herbert Kasube, *Bradley University*
Neil Koblitz, *University of Washington*
Steven Leonhardi, *Winona State University*

Charles Lewis, *Monmouth College*
James McKay, *Oakland University*
John Mairhuber, *University of Maine–Orono*
Alexsandrs Mihailovs, *University of Pennsylvania*
Rudolf Najar, *California State University, Fresno*
Carl Pomerance, *University of Georgia*
Sinai Robins, *Temple University*
Tom Shemanske, *Dartmouth College*
Leslie Vaaler, *University of Texas, Austin*
Evelyn Bender Vaskas, *Clark University*
Samuel Wagstaff, *Purdue University*
Edward Wang, *Wilfred Laurier University*
Betsey Whitman, *Framingham State University*
David Wright, *Oklahoma State*
Paul Zwier, *Calvin College*

Finally, I thank in advance all those who send me suggestions and corrections in the future. You may send such material to me care of Addison-Wesley at `math@awl.com`.

Kenneth H. Rosen
Middletown, New Jersey

Contents

What Is Number Theory?

There is a buzz about number theory: Thousands of people work on communal number theory problems over the Internet . . . the solution of a famous problem in number theory is reported on the PBS television series NOVA . . . people study number theory to understand systems for making messages secret . . . What is this subject, and why are so many people interested in it today?

Number theory is the branch of mathematics that studies the properties of, and the relationships between, particular types of numbers. Of the sets of numbers studied in number theory, the most important is the set of positive integers. More specifically, the *primes,* those positive integers with no positive proper factors other than 1, are of special importance. A key result of number theory shows that the primes are the multiplicative building blocks of the positive integers. This result, called the *fundamental theorem of arithmetic,* tells us that every positive integer can be uniquely written as the product of primes in nondecreasing order. Interest in prime numbers goes back at least 2500 years, to the studies of ancient Greek mathematicians. Perhaps the first question about primes that comes to mind is whether there are infinitely many. In *The Elements,* the ancient Greek mathematician Euclid provided a proof that there are infinitely many primes. Interest in primes was rekindled in the seventeenth and eighteenth centuries, when mathematicians such as Pierre de Fermat and Leonhard Euler proved many important results, and conjectured approaches for generating primes. The study of primes progressed substantially in the nineteenth century; results included the infinitude of primes in arithmetic progressions, and sharp estimates for the number of primes not exceeding a positive number x. The twentieth century has seen the development of many powerful techniques for the study of primes, but even with these powerful techniques, many questions remain unresolved. An example of a notorious unsolved question is whether there are infinitely many twin primes, which are primes that differ by 2. New results will certainly follow in the coming decades, as researchers continue working on the many open questions involving primes.

The development of modern number theory was made possible by the German mathematician Carl Friedrich Gauss, one of the greatest mathematicians in history, who developed the language of *congruences* in the early nineteenth century. We say that two integers a and b are congruent modulo m, where m is a positive integer, if m divides $a - b$. This language makes it easy to work with divisibility relationships in much the same way that we work with equations. Gauss developed many important concepts in number theory; for example, he proved one of its most subtle and beautiful results, the *law of quadratic reciprocity*. This law relates whether a prime p is a perfect square modulo a second prime q to whether q is a perfect square modulo p. Gauss developed many different proofs of this law, some of which led to whole new areas of number theory.

Distinguishing primes from composite integers is a key problem of number theory. Work in this area has led to the development of an arsenal of *primality tests*. The simplest primality test is simply checking whether a positive integer is divisible by each prime not exceeding its square root. Unfortunately, this test is inefficient for extremely large positive integers. In the nineteenth century, Pierre de Fermat showed that p divides $2^p - 2$ whenever p is prime. Some mathematicians thought that the converse also was true (that is, that if n divides $2^n - 2$, then n must be prime). However, it is not; by the early nineteenth century, composite integers n, such as 341, were known for which n divides $2^n - 2$. Such integers are called *pseudoprimes*. Though pseudoprimes exist, primality tests based on the fact that most composite integers are not pseudoprimes are now used to quickly find extremely large primes.

Factoring a positive integer into primes is another central problem in number theory. The factorization of a positive integer can be found using trial division, but this method is extremely time-consuming. Fermat, Euler, and many other mathematicians devised imaginative factorization algorithms, which have been extended in the past 25 years into a wide array of factoring methods. Using the best-known techniques, we can easily find primes with hundreds of digits; factoring integers with the same number of digits, however, is beyond our most powerful computers.

The dichotomy between the time required to find large primes and the time required to factor large integers is the basis of an extremely important secrecy system, the *RSA cryptosystem*. The RSA system is a public-key cryptosystem, a security system in which each person has a public key and an associated private key. Messages can be encrypted by anyone using another person's public key, but these messages can be decrypted only by the owner of the private key. Concepts from number theory are essential to understanding the basic workings of the RSA cryptosystem, as well as many other parts of modern cryptography. The overwhelming importance of number theory in cryptography contradicts the earlier belief, held by many mathematicians, that number theory was unimportant for real-world applications. It is ironic that some famous mathematicians, such as G. H. Hardy, took pride in the notion that number theory would never be applied in the way that it is today.

The search for integer solutions of equations is another important part of number theory. An equation with the added proviso that only integer solutions are sought is called *diophantine,* after the ancient Greek mathematician Diophantus. Many different types of diophantine equations have been studied, but the most famous is the *Fermat equation*

$x^n + y^n = z^n$. *Fermat's last theorem* states that if n is an integer greater than 2, this equation has no solutions in integers x, y, and z, where $xyz \neq 0$. Fermat conjectured in the seventeenth century that this theorem was true, and mathematicians (and others) searched for proofs for more than three centuries, but it was not until 1995 that the first proof was given by Andrew Wiles.

As Wiles's proof shows, number theory is not a static subject! New discoveries continue steadily to be made, and researchers frequently establish significant theoretical results. The fantastic power available when today's computers are linked over the Internet yields a rapid pace of new computational discoveries in number theory. Everyone can participate in this quest; for instance, you can join the quest for the new *Mersenne primes,* primes of the form $2^p - 1$, where p itself is prime. In June 1999, the first prime with more than 1 million decimal digits was found: the Mersenne prime $2^{6,972,593} - 1$, and a concerted effort is under way to find a prime with more than 10 million digits. After learning about some of the topics covered in this text, you may decide to join the hunt yourself, putting your idle computing resources to good use.

What is elementary number theory? You may wonder why the word "elementary" is part of the title of this book. This book considers only that part of number theory called *elementary number theory,* which is the part not dependent on advanced mathematics, such as the theory of complex variables, abstract algebra, or algebraic geometry. Students who plan to continue the study of mathematics will learn about more advanced areas of number theory, such as analytic number theory (which takes advantage of the theory of complex variables), and algebraic number theory (which uses concepts from abstract algebra to prove interesting results about algebraic number fields).

Some words of advice. As you embark on your study, keep in mind that number theory is a classical subject with results dating back thousands of years, yet is also the most modern of subjects, with new discoveries being made at a rapid pace. It is pure mathematics with the greatest intellectual appeal, yet it is also applied mathematics, with crucial applications to cryptography and other aspects of computer science and electrical engineering. I hope that you find the many facets of number theory as captivating as aficionados who have preceded you, many of whom retained an interest in number theory long after their school days were over.

Experimentation and exploration form an indispensable part of the study of number theory. The results in this book were found by mathematicians who often examined large amounts of numerical evidence, looking for patterns and making conjectures. They worked diligently to prove their conjectures; some of these were proved and became theorems, others were rejected when counterexamples were found, and still others remain unresolved. As you study number theory, I recommend that you examine many examples, look for patterns, and formulate your own conjectures. This will help you to learn the subject—and you may even find some new results of your own!

1

The Integers

Introduction

In the most general sense, number theory deals with the properties of different sets of numbers. In this chapter, we will discuss some particularly important sets of numbers, including the integers, the rational numbers, and the algebraic numbers. We will briefly introduce the notion of approximating real numbers by rational numbers. We will also introduce the concept of a sequence, and particular sequences of integers, including some figurate numbers studied in ancient Greece. A common problem is the identification of a particular integer sequence from its initial terms; we will briefly discuss how to attack such problems.

Using the concept of a sequence, we will define countable sets and show that the set of rational numbers is countable. We will also introduce notations for sums and products, and establish some useful summation formulas.

One of the most important proof techniques in number theory (and in much of mathematics) is mathematical induction. We will discuss the two forms of mathematical induction, illustrate how they can be used to prove various results, and explain why mathematical induction is a valid proof technique.

Continuing, we will introduce the intriguing sequence of Fibonacci numbers, and describe the original problem from which they arose. We will establish some identities and inequalities involving the Fibonacci numbers, using mathematical induction for some of our proofs.

The final section of this chapter deals with a fundamental notion in number theory, that of divisibility. We will establish some of the basic properties of division of integers, including the "division algorithm." We will show how the quotient and remainder of a division of one integer by another can be expressed using values of the greatest integer function (we will describe a few of the many useful properties of this function, as well).

1.1 Numbers and Sequences

In this section, we introduce basic material that will be used throughout the text. In particular, we cover the important sets of numbers studied in number theory, the concept of integer sequences, and summations and products.

Numbers

To begin, we will introduce several different types of numbers. The *integers* are the numbers in the set

$$\{\ldots, -3, -2, -1, 0, 1, 2, 3, \ldots\}.$$

The integers play center stage in the study of number theory. One property of the positive integers deserves special mention.

The Well-Ordering Property   Every nonempty set of positive integers has a least element.

The well-ordering property may seem obvious, but it is the basic principle that allows us to prove many results about sets of integers, as we will see in Section 1.3.

The well-ordering property can be taken as one of the axioms defining the set of positive integers or it may be derived from a set of axioms in which it is not included. (See Appendix A for axioms for the set of integers.) We say that the set of positive integers is *well ordered*. However, the set of all integers is not well ordered, as there are sets of integers without a smallest element, such as the set of negative integers, the set of even integers less than 100, and the set of all integers itself.

Another important class of numbers in the study of number theory is the set of numbers that can be written as a ratio of integers.

Definition.   The real number r is *rational* if there are integers p and q, with $q \neq 0$, such that $r = p/q$. If r is not rational, it is said to be *irrational*.

Example 1.1.   The numbers $-22/7, 0 = 0/1, 2/17$, and $1111/41$ are rational numbers.

◄

Note that every integer n is a rational number, because $n = n/1$. Examples of irrational numbers are $\sqrt{2}$, π, and e. We can use the well-ordering property of the set of positive integers to show that $\sqrt{2}$ is irrational. The proof that we provide, although quite clever, is not the simplest proof that $\sqrt{2}$ is irrational. You may prefer the proof that we will give in Chapter 4, which depends on concepts developed in that chapter. (The proof that e is irrational is left as Exercise 44. We refer the reader to [HaWr79] for a proof that π is irrational. It is not easy.)

Theorem 1.1. $\sqrt{2}$ is irrational.

Proof.   Suppose that $\sqrt{2}$ were rational. Then there would exist positive integers a and b such that $\sqrt{2} = a/b$. Consequently, the set $S = \{k\sqrt{2} \mid k \text{ and } k\sqrt{2} \text{ are positive integers}\}$ is a nonempty set of positive integers (it is nonempty because $a = b\sqrt{2}$ is a member of S). Therefore, by the well-ordering property, S has a smallest element, say $s = t\sqrt{2}$.

We have $s\sqrt{2} - s = s\sqrt{2} - t\sqrt{2} = (s - t)\sqrt{2}$. Because $s\sqrt{2} = 2t$ and s are both integers, $s\sqrt{2} - s = s\sqrt{2} - t\sqrt{2} = (s - t)\sqrt{2}$ must also be an integer. Furthermore, it is positive, because $s\sqrt{2} - s = s(\sqrt{2} - 1)$ and $\sqrt{2} > 1$. It is less than s, because $s = t\sqrt{2}, s\sqrt{2} = 2t$ and $\sqrt{2} < 2$. This contradicts the choice of s as the smallest positive integer in S. It follows that $\sqrt{2}$ is irrational. ∎

The sets of integers, positive integers, rational numbers, and real numbers are traditionally denoted by $\mathbf{Z}$, $\mathbf{Z}^+$, $\mathbf{Q}$, and $\mathbf{R}$, respectively. Also, we write $x \in S$ to indicate that x belongs to the set S. Such notation will be used occasionally in this book.

We briefly mention several other types of numbers here, though we do not return to them until Chapter 12.

Definition.   A number α is *algebraic* if it is the root of a polynomial with integer coefficients; that is, α is algebraic if there exist integers $a_0, a_1, \ldots, a_n$ such that $a_n\alpha^n + a_{n-1}\alpha^{n-1} + \cdots + a_0 = 0$. The number α is called *transcendental* if it is not algebraic.

Example 1.2.   The irrational number $\sqrt{2}$ is algebraic, because it is a root of the polynomial $x^2 - 2$.   ◀

Note that every rational number is algebraic. This follows from the fact that the number a/b, where a and b are integers and $b \neq 0$, is the root of $bx - a$. In Chapter 12, we will give an example of a transcendental number. The numbers e and π are also transcendental, but the proofs of these facts (which can be found in [HaWr79]) are beyond the scope of this book.

The Greatest Integer Function

In number theory a special notation is used for the largest integer that is less than or equal to a particular real number.

Definition.   The *greatest integer* in a real number x, denoted by $[x]$, is the largest integer less than or equal to x. That is, $[x]$ is the integer satisfying

$$[x] \leq x < [x] + 1.$$

Example 1.3.   We have $[5/2] = 2, [-5/2] = -3, [\pi] = 3, [-2] = -2$, and $[0] = 0$.   ◀

Remark.   The greatest integer function is also known as the *floor function*. Instead of using the notation $[x]$ for this function, computer scientists usually use the notation $\lfloor x \rfloor$. The *ceiling function* is a related function often used by computer scientists. The ceiling function of a real number x, denoted by $\lceil x \rceil$, is the smallest integer greater than or equal to x. For example, $\lceil 5/2 \rceil = 3$ and $\lceil -5/2 \rceil = -2$.

The greatest integer function arises in many contexts. Besides being important in number theory, as we will see throughout this book, it plays an important role in the analysis of algorithms, a branch of computer science. The following example establishes a useful property of this function. Additional properties of the greatest integer function are found in the exercises at the end of this section and in [GrKnPa94].

Example 1.4.   Show that if n is an integer, then $[x + n] = [x] + n$ whenever x is a real number. To show that this property holds, let $[x] = m$, so that m is an integer. This implies that $m \leq x < m + 1$. We can add n to this inequality to obtain $m + n \leq x + n < m + n + 1$. This shows that $m + n = [x] + n$ is the greatest integer less than or equal to $x + n$. Hence $[x + n] = [x] + n$.   ◄

Definition.   The *fractional part* of a real number x, denoted by $\{x\}$, is the difference between x and the largest integer less than or equal to x, namely $[x]$. That is, $\{x\} = x - [x]$.

Because $[x] \leq x < [x] + 1$, it follows that $0 \leq \{x\} = x - [x] < 1$ for every real number x. The greatest integer in x is also called the *integral part* of x because $x = [x] + \{x\}$.

Example 1.5.   We have $\{5/4\} = 5/4 - [5/4] = 5/4 - 1 = 1/4$ and $\{-2/3\} = -2/3 - [-2/3] = -2/3 - (-1) = 1/3$.   ◄

Diophantine Approximation

We know that the distance of a real number to the integer closest to it is at most $1/2$. But can we show that one of the first k multiples of a real number must be much closer to an integer? An important part of number theory called *diophantine approximation* studies questions such as this. In particular, it concentrates on questions that involve the approximation of real numbers by rational numbers. (The adjective *diophantine* comes from the Greek mathematician Diophantus, whose biography can be found in Section 13.1.)

Here we will show that among the first n multiples of a real number α, there must be at least one at a distance less than $1/n$ from the integer nearest it. The proof will depend on the famous *pigeonhole principle,* introduced by the German mathematician Dirichlet.[1] Informally, this principle tells us if we have more objects than boxes, when these objects are placed in the boxes, at least two must end up in the same box. Although this seems like a particularly simple idea, it turns out to be extremely useful in number theory and combinatorics. We now state and prove this important fact, which is known as the pigeonhole principle because if you have more pigeons than roosts, two pigeons must end up in the same roost.

---

[1] Instead of calling Theorem 1.2 the pigeonhole principle, Dirichlet called it the *Schubfachprinzip* in German, which translates to the *drawer principle* in English. A biography of Dirichlet can be found in Section 3.1.

Theorem 1.2. *The Pigeonhole Principle.* If $k + 1$ or more objects are placed into k boxes, then at least one box contains two or more of the objects.

Proof. If none of the k boxes contains more than one object, then the total number of objects would be at most k. This contradiction shows that one of the boxes contains at least two or more of the objects. ∎

We now state and prove the approximation theorem, which guarantees that one of the first n multiples of a real number must be within $1/n$ of an integer. The proof we give illustrates the utility of the pigeonhole principle. (See [Ro03] for more applications of the pigeonhole principle.) (Note that in the proof we make use of the *absolute value function.* Recall that $|x|$, the absolute value of x, equals x if $x \geq 0$ and $-x$ if $x < 0$. Also recall that $|x - y|$ gives the distance between x and y.)

Theorem 1.3. *Dirichlet's Approximation Theorem.* If α is a real number and n is a positive integer, then there exist integers a and b with $1 \leq a \leq n$ such that $|a\alpha - b| < 1/n$.

Proof. Consider the $n + 1$ numbers $0, \{\alpha\}, \{2\alpha\}, \ldots, \{n\alpha\}$. These $n + 1$ numbers are the fractional parts of the numbers $j\alpha$, $j = 0, 1, \ldots, n$, so that $0 \leq \{j\alpha\} < 1$ for $j = 0, 1, \ldots, n$. Each of these $n + 1$ numbers lies in one of the n disjoint intervals $0 \leq x < 1/n, 1/n \leq x < 2/n, \ldots, (j - 1)/n \leq x < j/n, \ldots, (n - 1)/n \leq x < 1$. Because there are $n + 1$ numbers under consideration, but only n intervals, the pigeonhole principle tells us that at least two of these numbers lie in the same interval. Because each of these intervals has length $1/n$ and does not include its right endpoint, we know that the distance between two numbers that lie in the same interval is less than $1/n$. It follows that there exist integers j and k with $0 \leq j < k \leq n$ such that $|\{k\alpha\} - \{j\alpha\}| < 1/n$. Now let $a = k - j$ and $b = [k\alpha] - [j\alpha]$. Because $0 \leq j < k \leq n$, we see that $1 \leq a \leq n$. Moreover,

$$\begin{aligned}|a\alpha - b| &= |(k - j)\alpha - ([k\alpha] - [j\alpha])| \\ &= |(k\alpha - [k\alpha]) - (j\alpha - [j\alpha])| \\ &= |\{k\alpha\} - \{j\alpha\}| < 1/n.\end{aligned}$$

Consequently, we have found integers a and b with $1 \leq a \leq n$ and $|a\alpha - b| < 1/n$, as desired. ∎

Example 1.6. Suppose that $\alpha = \sqrt{2}$ and $n = 6$. We find that $1 \cdot \sqrt{2} \approx 1.414, 2 \cdot \sqrt{2} \approx 2.828, 3 \cdot \sqrt{2} \approx 4.243, 4 \cdot \sqrt{2} \approx 5.657, 5 \cdot \sqrt{2} \approx 7.071$, and $6 \cdot \sqrt{2} \approx 8.485$. Among these numbers $5 \cdot \sqrt{2}$ has the smallest fractional part. We see that $|5 \cdot \sqrt{2} - 7| \approx |7.071 - 7| = 0.071 \leq 1/6$. It follows that when $\alpha = \sqrt{2}$ and $n = 6$, we can take $a = 5$ and $b = 7$ to make $|a\alpha - b| < 1/n$. ◀

Our proof of Theorem 1.3 follows Dirichlet's original 1834 proof. Proving a stronger version of Theorem 1.3 with $1/(n + 1)$ replacing $1/n$ in the approximation is not difficult (see Exercise 32). Furthermore, in Exercise 34 we show how to use the Dirichlet approximation theorem to show that, given an irrational number α, there are infinitely many different rational numbers p/q such that $|\alpha - p/q| < 1/q^2$, and important result in the theory of diophantine approximation. We will return to this topic in Chapter 12.

Sequences

A *sequence* $\{a_n\}$ is a list of numbers $a_1, a_2, a_3, \ldots$. The terms of a sequence can be put into a one-to-one correspondence with the set of positive integers using the mapping $f(i) = a_i$. (Recall that a *one-to-one correspondence*, also called a *bijection*, is a function that is both one-to-one and onto.) We will consider many particular integer sequences in our study of number theory. We introduce several useful sequences in the following examples.

Example 1.7.   The sequence $\{a_n\}$, where $a_n = n^2$, begins with the terms $1, 4, 9, 16, 25, 36, 49, 64, \ldots$. This is the sequence of the squares of integers. The sequence $\{b_n\}$, where $b_n = 2^n$, begins with the terms $2, 4, 8, 16, 32, 64, 128, 256, \ldots$. This is the sequence of powers of 2. The sequence $\{c_n\}$, where $c_n = 0$ if n is odd and $c_n = 1$ if n is even, begins with the terms $0, 1, 0, 1, 0, 1, 0, 1, \ldots$. ◀

There are many sequences in which each successive term is obtained from the previous term by multiplying by a common factor. For example, each term in the sequence of powers of 2 is 2 times the previous term. This leads to the following definition.

Definition.   A *geometric progression* is a sequence of the form a, ar, ar^2, $ar^3, \ldots$, $ar^k, \ldots$, where a, the *initial term*, and r, the *common ratio*, are real numbers.

Example 1.8.   The sequence $\{a_n\}$, where $a_n = 3 \cdot 5^n$, $n = 0, 1, 2, \ldots$, is a geometric sequence with initial term 3 and common ratio 5. (Note that we have started the sequence with the term a_0. We can start the index of the terms of a sequence with 0 or any other integer that we choose.) ◀

A common problem in number theory is finding a formula or rule for constructing the terms of a sequence, even when only a few terms are known (such as trying to find a formula for the nth triangular number $1 + 2 + 3 + \cdots + n$). Even though the initial terms of a sequence do not determine the sequence, knowing the first few terms can lead to a conjecture for a formula or rule for the terms. Consider the following examples.

Example 1.9.   Conjecture a formula for a_n, where the first eight terms of $\{a_n\}$ are $4, 11, 18, 25, 32, 39, 46, 53$. We note that each term, starting with the second, is obtained by adding 7 to the previous term. Consequently, the nth term could be the initial term plus $7(n - 1)$. A reasonable conjecture is that $a_n = 4 + 7(n - 1) = 7n - 3$. ◀

The sequence proposed in Example 1.9 is an *arithmetic progression*, that is, a sequence of the form $a, a + d, a + 2d, \ldots, a + nd, \ldots$. The particular sequence in Example 1.9 has $a = 4$ and $d = 7$.

Example 1.10.   Conjecture a formula for a_n, where the first eight terms of the sequence $\{a_n\}$ are $5, 11, 29, 83, 245, 731, 2189, 6563$. We note that each term is approximately 3 times the previous term, suggesting a formula for a_n in terms of 3^n. The integers 3^n for

$n = 1, 2, 3, \ldots$ are $3, 9, 27, 81, 243, 729, 2187, 6561$. Looking at these two sequences together, we find that the formula $a_n = 3^n + 2$ produces these terms. ◄

Example 1.11. Conjecture a formula for a_n, where the first ten terms of the sequence $\{a_n\}$ are $1, 1, 2, 3, 5, 8, 13, 21, 34, 55$. After examining this sequence from different perspectives, we notice that each term of this sequence, after the first two terms, is the sum of the two preceding terms. That is, we see that $a_n = a_{n-1} + a_{n-2}$ for $3 \le n \le 10$. This is an example of a recursive definition of a sequence, discussed in Section 1.3. The terms listed in this example are the initial terms of the Fibonacci sequence, which is discussed in Section 1.4. ◄

Integer sequences arise in many contexts in number theory. Among the sequences we will study are the Fibonacci numbers, the prime numbers (covered in Chapter 3), and the perfect numbers (introduced in Section 7.3). Integer sequences appear in an amazing range of subjects besides number theory. A fantastically diverse collection of more than 8000 integer sequences has been amassed by Neil Sloane, who created *The Encyclopedia of Integer Sequences* ([SlPl95]) with Simon Plouffe. An extended version of this list, and a program for finding sequences that match initial terms provided as input, can be found on the Web. You may find this a valuable resource as you continue your study of number theory (as well as other subjects).

We now define what it means for a set to be countable, and show that a set is countable if and only if its elements can be listed as the terms of a sequence.

Definition. A set is *countable* if it is finite or it is infinite and there exists a one-to-one correspondence between the set of positive integers and the set. A set that is not countable is called *uncountable*.

An infinite set is countable if and only if its elements can be listed as the terms of a sequence indexed by the set of positive integers. To see this, simply note that a one-to-one correspondence f from the set of positive integers to a set S is exactly the same as a listing of the elements of the set in a sequence $a_1, a_2, \ldots, a_n, \ldots$, where $a_i = f(i)$.

Example 1.12. The set of integers is countable, because the integers can be listed starting with 0, followed by 1 and -1, followed by 2 and -2, and so on. This produces the sequence $0, 1, -1, 2, -2, 3, -3, \ldots$, where $a_1 = 0$, $a_{2n} = n$, and $a_{2n+1} = -n$ for $n = 1, 2, \ldots$. ◄

Is the set of rational numbers countable? At first glance, it may seem unlikely that there would be a one-to-one correspondence between the set of positive integers and the set of all rational numbers. However, there is such a correspondence, as the following theorem shows.

Theorem 1.4. The set of rational numbers is countable.

Proof. We can list the rational numbers as the terms of a sequence, as follows. First, we arrange all the rational numbers in a two-dimensional array, as shown in Figure 1.1. We put all fractions with a denominator of 1 in the first row. We arrange these by placing the fraction with a particular numerator in the position this numerator occupies in the list of all integers given in Example 1.12. Next, we list all fractions on successive diagonals, following the order shown in Figure 1.1. Finally, we delete from the list all fractions that represent rational numbers that have already been listed. (For example, we do not list 2/2, because we have already listed 1/1.)

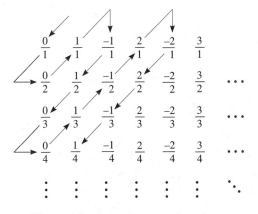

Figure 1.1 *Listing the rational numbers.*

The initial terms of the sequence are $0/1 = 0$, $1/1 = 1$, $-1/1 = -1$, $1/2$, $1/3$, $-1/2$, $2/1 = 2$, $-2/1 = -2$, $-1/3$, $1/4$, and so on.) We leave it to the reader to fill in the details, to see that this procedure lists all rational numbers as the terms of a sequence. ∎

1.1 Exercises

1. Determine whether each of the following sets is well ordered. Either give a proof using the well-ordering property of the set of positive integers, or give an example of a subset of the set that has no smallest element.

   a) the set of integers greater than 3

   b) the set of even positive integers

   c) the set of positive rational numbers

   d) the set of positive rational numbers that can be written in the form $a/2$, where a is a positive integer

   e) the set of nonnegative rational numbers

☞ 2. Show that if a and b are positive integers, then there is a smallest positive integer of the form $a - bk$, $k \in \mathbf{Z}$.

3. Prove that both the sum and the product of two rational numbers are rational.

4. Prove or disprove each of the following statements.

   a) The sum of a rational and an irrational number is irrational.

   b) The sum of two irrational numbers is irrational.

   c) The product of a rational number and an irrational number is irrational.

   d) The product of two irrational numbers is irrational.

* **5.** Use the well-ordering property to show that $\sqrt{3}$ is irrational.

6. Show that every nonempty set of negative integers has a greatest element.

7. Find the following values of the greatest integer function.

    a) [1/4]   d) [−2]
    b) [−3/4]   e) [[1/2] + [1/2]]
    c) [22/7]   f) [−3 + [−1/2]]

8. Find the following values of the greatest integer function.

    a) [−1/4]   d) [[1/2]]
    b) [−22/7]   e) [[3/2] + [−3/2]]
    c) [5/4]   f) [3 − [1/2]]

9. Find the fractional part of each of the following numbers.

    a) 8/5   c) −11/4
    b) 1/7   d) 7

10. Find the fractional part of each of the following numbers.

    a) −8/5   c) −1
    b) 22/7   d) −1/3

11. What is the value of $[x] + [-x]$ where x is a real number?

12. Show that $[x] + [x + 1/2] = [2x]$ whenever x is a real number.

13. Show that $[x + y] \geq [x] + [y]$ for all real numbers x and y.

14. Show that $[2x] + [2y] \geq [x] + [y] + [x + y]$ whenever x and y are real numbers.

15. Show that if x and y are positive real numbers, then $[xy] \geq [x][y]$. What is the situation when both x and y are negative? When one of x and y is negative and the other positive?

16. Show that $-[-x]$ is the least integer greater than or equal to x when x is a real number.

17. Show that $[x + 1/2]$ is the integer nearest to x (when there are two integers equidistant from x, it is the larger of the two).

18. Show that if m and n are integers, then $[(x + n)/m] = [([x] + n)/m]$ whenever x is a real number.

* **19.** Show that $\left[\sqrt{[x]}\right] = \left[\sqrt{x}\right]$ whenever x is a nonnegative real number.

* **20.** Show that if m is a positive integer, then

$$[mx] = [x] + [x + (1/m)] + [x + (2/m)] + \cdots + [x + (m - 1)/m]$$

whenever x is a real number.

21. Conjecture a formula for the nth term of $\{a_n\}$, if the first ten terms of this sequence are as follows.

    a) $3, 11, 19, 27, 35, 43, 51, 59, 67, 75$
    b) $5, 7, 11, 19, 35, 67, 131, 259, 515, 1027$
    c) $1, 0, 0, 1, 0, 0, 0, 0, 1, 0$
    d) $1, 3, 4, 7, 11, 18, 29, 47, 76, 123$

22. Conjecture a formula for the nth term of $\{a_n\}$, if the first ten terms of this sequence are as follows.

  a) $2, 6, 18, 54, 162, 486, 1458, 4374, 13122, 39366$

  b) $1, 1, 0, 1, 1, 0, 1, 1, 0, 1$

  c) $1, 2, 3, 5, 7, 10, 13, 17, 21, 26$

  d) $3, 5, 11, 21, 43, 85, 171, 341, 683, 1365$

23. Find three different formulas or rules for the terms of a sequence $\{a_n\}$, if the first three terms of this sequence are $1, 2, 4$.

24. Find three different formulas or rules for the terms of a sequence $\{a_n\}$, if the first three terms of this sequence are $2, 3, 6$.

25. Show that the set of all integers greater than -100 is countable.

26. Show that the set of all rational numbers of the form $n/5$, where n is an integer, is countable.

27. Show that the set of all numbers of the form $a + b\sqrt{2}$, where a and b are integers, is countable.

$*$ **28.** Show that the union of two countable sets is countable.

$*$ **29.** Show that the union of a countable number of countable sets is countable.

30. Using a computational aid, if needed, find integers a and b such that $1 \le a \le 8$ and $|a\alpha - b| < 1/8$, where α is

  a) $\sqrt{2}$.   c) π.

  b) $\sqrt[3]{2}$.   d) e.

31. Using a computational aid, if needed, find integers a and b such that $1 \le a \le 10$ and $|a\alpha - b| < 1/10$, where α is

  a) $\sqrt{3}$.   c) π^2.

  b) $\sqrt[3]{3}$.   d) e^3.

32. Prove the following stronger version of Dirichlet's approximation. If α is a real number and n is a positive integer, there are integers a and b such that $1 \le a \le n$ and $|a\alpha - b| \le 1/(n+1)$. (*Hint:* Consider the $n + 2$ numbers $0, \ldots, \{j\alpha\}, \ldots, 1$ and the $n + 1$ intervals $(k-1)/(n+1) \le x < k/(n+1)$ for $k = 1, \ldots, n+1$.)

33. Show that if α is a real number and n is a positive integer, then there is an integer k such that $|\alpha - u/k| \le 1/2k$.

34. Use Dirichlet's approximation theorem to show that if α is an irrational number, then there are infinitely many positive integers q for which there is an integer p such that $|\alpha - p/q| \le 1/q^2$.

35. Find four rational numbers p/q with $|\sqrt{2} - p/q| \le 1/q^2$.

36. Find five rational numbers p/q with $|\sqrt[3]{5} - p/q| \le 1/q^2$.

37. Show that if $\alpha = a/b$ is a rational number, then there are only finitely many rational numbers p/q such that $|p/q - a/b| < 1/q^2$.

The *spectrum sequence* of a real number α is the sequence that has $[n\alpha]$ as its nth term.

38. Find the first ten terms of the spectrum sequence of each of the following numbers.

  a) 2
  b) $\sqrt{2}$

  c) $2 + \sqrt{2}$
  d) e

  e) $(1 + \sqrt{5})/2$

39. Find the first ten terms of the spectrum sequence of each of the following numbers.

  a) 3
  b) $\sqrt{3}$

  c) $(3 + \sqrt{3})/2$
  d) π

40. Prove that if $\alpha \neq \beta$, then the spectrum sequence of α is different from the spectrum sequence of β.

** **41.** Show that every positive integer occurs exactly once in the spectrum sequence of α or in the spectrum sequence of β if and only if α and β are positive irrational numbers such that $1/\alpha + 1/\beta = 1$.

The *Ulam numbers* $u_n, n = 1, 2, 3, \ldots$ are defined as follows. We specify that $u_1 = 1$ and $u_2 = 2$. For each successive integer $m, m > 2$, this integer is an Ulam number if and only if it can be written uniquely as the sum of two distinct Ulam numbers. These numbers are named for *Stanislaw Ulam,* who first described them in 1964.

STANISLAW M. ULAM (1909–1984) was born in Lvov, Poland. He became interested in astronomy and physics at age 12, after receiving a telescope from his uncle. He decided to learn the mathematics required to understand relativity theory, and at the age of 14 he used textbooks to learn calculus and other mathematics.

Ulam received his Ph.D. from the Polytechnic Institute in Lvov in 1933, completing his degree under the mathematician Banach, in the area of real analysis. In 1935, he was invited to spend several months at the Institute for Advanced Study; in 1936, he joined Harvard University as a member of the Society of Fellows, remaining in this position until 1940. During these years he returned each summer to Poland where he spent time in cafes, such as the Scottish Cafe, intensely doing mathematics with his fellow Polish mathematians.

Luckily for Ulam, he left Poland in 1939, just one month before the outbreak of World War II. In 1940, he was appointed to a position as an assistant professor at the University of Wisconsin, and in 1943, he was enlisted to work in Los Alamos on the development of the first atomic bomb, as part of the Manhattan Project. Ulam made several key contributions that led to the creation of thermonuclear bombs. At Los Alamos, Ulam also developed the Monte Carlo method, which uses a sampling technique with random numbers to find solutions of mathematical problems.

Ulam remained at Los Alamos after the war until 1965. He served on the faculties of the University of Southern California, the University of Colorado, and the University of Florida. Ulam had a fabulous memory and was an extremely verbal person. His mind was a repository of stories, jokes, puzzles, quotations, formulas, problems, and many other types of information. He wrote several books, including *Sets, Numbers, and Universes* and *Adventures of a Mathematician.* He was interested in and contributed to many areas of mathematics, including number theory, real analysis, probability theory, and mathematical biology.

42. Find the first ten Ulam numbers.

* **43.** Show that there are infinitely many Ulam numbers.

* **44.** Prove that e is irrational. (*Hint:* Use the fact that $e = 1 + 1/1! + 1/2! + 1/3! + \cdots$.)

1.1 Computational and Programming Exercises

Computations and Explorations

Using a computation program such as Maple or *Mathematica,* or programs you have written, carry out the following computations and explorations.

1. Find 10 rational numbers p/q such that $|\pi - p/q| \leq 1/q^2$.

2. Find 20 rational numbers p/q such that $|e - p/q| \leq 1/q^2$.

3. Find as many terms as you can of the spectrum sequence of $\sqrt{2}$. (See the preamble to Exercise 38 for the definition of spectrum.)

4. Find as many terms as you can of the spectrum sequence of π. (See the preamble to Exercise 38 for the definition of spectrum.)

5. Find the first 1000 Ulam numbers.

6. How many pairs of consecutive integers can you find, where both are Ulam numbers?

7. Can the sum of any two consecutive Ulam numbers, other than 1 and 2, be another Ulam number? If so, how many examples can you find?

8. How large are the gaps between consecutive Ulam numbers? Do you think that these gaps can be arbitrarily long?

9. What conjectures can you make about the number of Ulam numbers less than an integer n? Do your computations support these conjectures?

Programming Projects

Write programs using Maple, *Mathematica,* or a language of your choice to do the following.

1. Given a number α, find rational numbers p/q such that $|\alpha - p/q| \leq 1/q^2$.

2. Given a number α, find its spectrum sequence.

3. Find the first n Ulam numbers, where n is a positive integer.

1.2 Sums and Products

Because summations and products arise so often in the study of number theory, we now introduce notation for summations and products. The following notation represents the sum of the numbers $a_1, a_2, \ldots, a_n$:

$$\sum_{k=1}^{n} a_k = a_1 + a_2 + \cdots + a_n.$$

The letter k, the *index of summation,* is a "dummy variable" and can be replaced by any letter. For instance,

$$\sum_{k=1}^{n} a_k = \sum_{j=1}^{n} a_j = \sum_{i=1}^{n} a_i, \text{ and so forth.}$$

Example 1.13. We see that $\sum_{j=1}^{5} j = 1 + 2 + 3 + 4 + 5 = 15$, $\sum_{j=1}^{5} 2 = 2 + 2 + 2 + 2 + 2 = 10$, and $\sum_{j=1}^{5} 2^j = 2 + 2^2 + 2^3 + 2^4 + 2^5 = 62$.

We also note that, in summation notation, the index of summation may range between any two integers, as long as the lower limit does not exceed the upper limit. If m and n are integers such that $m \le n$, then $\sum_{k=m}^{n} a_k = a_m + a_{m+1} + \cdots + a_n$. For instance, we have $\sum_{k=3}^{5} k^2 = 3^2 + 4^2 + 5^2 = 50$, $\sum_{k=0}^{2} 3^k = 3^0 + 3^1 + 3^2 = 13$, and $\sum_{k=-2}^{1} k^3 = (-2)^3 + (-1)^3 + 0^3 + 1^3 = -8$. ◀

We will often need to consider sums in which the index of summation ranges over all those integers that possess a particular property. We can use summation notation to specify the particular property or properties the index must have for a term with that index to be included in the sum. This use of notation is illustrated in the following example.

Example 1.14. We see that

$$\sum_{\substack{j \le 10 \\ j \in \{n^2 \mid n \in \mathbf{Z}\}}} 1/(j+1) = 1/1 + 1/2 + 1/5 + 1/10 = 9/5,$$

because the terms in the sum are all those for which j is an integer not exceeding 10 that is a perfect square. ◀

The following three properties for summations are often useful. We leave their proofs to the reader.

(1.1)
$$\sum_{j=m}^{n} k a_j = k \sum_{j=m}^{n} a_j$$

(1.2)
$$\sum_{j=m}^{n} (a_j + b_j) = \sum_{j=m}^{n} a_j + \sum_{j=m}^{n} b_j$$

(1.3)
$$\sum_{i=m}^{n} \sum_{j=p}^{q} a_i b_j = \left(\sum_{i=m}^{n} a_i \right) \left(\sum_{j=p}^{q} b_j \right) = \sum_{j=p}^{q} \sum_{i=m}^{n} a_i b_j$$

Next, we develop several useful summation formulas. We often need to evaluate sums of consecutive terms of a geometric series. The following example shows how a formula for such sums can be derived.

Example 1.15.   To evaluate

$$S = \sum_{j=0}^{n} ar^j,$$

the sum of the first $n + 1$ terms of the geometric series $a, ar, \ldots, ar^k, \ldots$, we multiply both sides by r and manipulate the resulting sum to find:

$$rS = r \sum_{j=0}^{n} ar^j$$

$$= \sum_{j=0}^{n} ar^{j+1}$$

$$= \sum_{k=1}^{n+1} ar^k \qquad \qquad \text{(shifting the index of summation)}$$

$$= \sum_{k=0}^{n} ar^k + (ar^{n+1} - a) \quad \text{(removing the term with } k = n + 1 \\ \qquad \qquad \qquad \qquad \qquad \text{from the set and adding the term with } k = 0)$$

$$= S + (ar^{n+1} - a).$$

It follows that

$$rS - S = (ar^{n+1} - a).$$

Solving for S shows that when $r \neq 1$,

$$S = \frac{ar^{n+1} - a}{r - 1}.$$

Note that when $r = 1$, we have $\sum_{j=0}^{n} ar^j = \sum_{j=0}^{n} a = (n + 1)a$.   ◀

Example 1.16.   Taking $a = 3$, $r = -5$, and $n = 6$ in the formula found in Example 1.15, we see that $\sum_{i=0}^{6} 3(-5)^j = \frac{3(-5)^7 - 3}{-5 - 1} = 39{,}063$.   ◀

The following example shows that the sum of the first n consecutive powers of 2 is one less than the next power of 2.

Example 1.17.   Let n be a positive integer. To find the sum

$$\sum_{k=0}^{n} 2^k = 1 + 2 + 2^2 + \cdots + 2^n,$$

we use Example 1.15, with $a = 1$ and $r = 2$, to obtain

$$1 + 2 + 2^2 + \cdots + 2^n = \frac{2^{n+1} - 1}{2 - 1} = 2^{n+1} - 1.$$ ◀

A summation of the form $\sum_{j=1}^{n}(a_j - a_{j-1})$, where $a_0, a_1, a_2, \ldots, a_n$ is a sequence of numbers, is said to be *telescoping*. Telescoping sums are easily evaluated because

$$\sum_{j=1}^{n} a_j - a_{j-1} = (a_1 - a_0) + (a_2 - a_1) + \cdots + (a_n - a_{n-1})$$
$$= a_n - a_0.$$

The ancient Greeks were interested in sequences of numbers that can be represented by regular arrangements of equally spaced points. The following example illustrates one such sequence of numbers.

Example 1.18. The *triangular numbers* $t_1, t_2, t_3, \ldots, t_k, \ldots$ is the sequence where t_k is the number of dots in the triangular array of k rows with j dots in the jth row. ◀

Figure 1.2 illustrates that t_k counts the dots in successively larger regular triangles for $k = 1, 2, 3, 4,$ and 5.

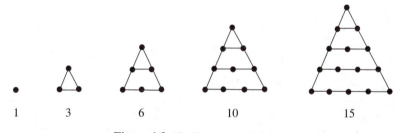

1   3   6   10   15

Figure 1.2 *The Triangular Numbers.*

Next, we will determine an explicit formula for the kth triangular number t_k.

Example 1.19. How can we find a formula for the nth triangular number? One approach is to use the identity $(k + 1)^2 - k^2 = 2k + 1$. When we isolate the factor k, we find that $k = ((k + 1)^2 - k^2)/2 - 1/2$. When we sum this expression for k over the values $k = 1, 2, \ldots, n$, we obtain

$$t_n = \sum_{k=1}^{n} k$$

$$= \left(\sum_{k=1}^{n}((k + 1)^2 - k^2)/2\right) - \sum_{k=1}^{n} 1/2 \quad \textit{(replacing k with } ((k + 1)^2 - k^2)/2\textit{)}$$

$$= ((n + 1)^2/2 - 1/2) - n/2 \quad\quad\quad \textit{(simplifying a telescoping sum)}$$

$$= (n^2 + 2n)/2 - n/2$$

$$= (n^2 + n)/2$$

$$= n(n + 1)/2.$$

The second equality here follows by the formula for the sum of a telescoping series with $a_k = (k+1)^2 - k^2$. We conclude that the nth triangular number $t_n = n(n+1)/2$. (See Exercise 7 for another way to find t_n.) ◀

We also define a notation for products, analogous to that for summations. The product of the numbers $a_1, a_2, \ldots, a_n$ is denoted by

$$\prod_{j=1}^{n} a_j = a_1 a_2 \cdots a_n.$$

The letter j above is a "dummy variable," and can be replaced arbitrarily.

Example 1.20.   To illustrate the notation for products, we have

$$\prod_{j=1}^{5} j = 1 \cdot 2 \cdot 3 \cdot 4 \cdot 5 = 120,$$

$$\prod_{j=1}^{5} 2 = 2 \cdot 2 \cdot 2 \cdot 2 \cdot 2 = 2^5 = 32, \text{ and}$$

$$\prod_{j=1}^{5} 2^j = 2 \cdot 2^2 \cdot 2^3 \cdot 2^4 \cdot 2^5 = 2^{15}.$$

◀

The *factorial function* arises throughout number theory.

Definition.   Let n be a positive integer. Then $n!$ (read as "n factorial") is the product of the integers $1, 2, \ldots, n$. We also specify that $0! = 1$. In terms of product notation, we have $n! = \prod_{j=1}^{n} j$.

Example 1.21.   We have $1! = 1$, $4! = 1 \cdot 2 \cdot 3 \cdot 4 = 24$, and $12! = 1 \cdot 2 \cdot 3 \cdot 4 \cdot 5 \cdot 6 \cdot 7 \cdot 8 \cdot 9 \cdot 10 \cdot 11 \cdot 12 = 479,001,600$. ◀

1.2 Exercises

1. Find each of the following sums.

   a) $\sum_{j=1}^{5} j^2$ b) $\sum_{j=1}^{5}(-3)$ c) $\sum_{j=1}^{5} 1/(j+1)$

2. Find each of the following sums.

   a) $\sum_{j=0}^{4} 3$ b) $\sum_{j=0}^{4}(j-3)$ c) $\sum_{j=0}^{4}(j+1)/(j+2)$

3. Find each of the following sums.

   a) $\sum_{j=1}^{8} 2^j$ b) $\sum_{j=1}^{8} 5(-3)^j$ c) $\sum_{j=1}^{8} 3(-1/2)^j$

4. Find each of the following sums.

   a) $\sum_{j=0}^{10} 8 \cdot 3^j$ b) $\sum_{j=0}^{10}(-2)^{j+1}$ c) $\sum_{j=0}^{10}(1/3)^j$

* **5.** Find and prove a formula for $\sum_{k=1}^{n}[\sqrt{k}]$ in terms of n and $[\sqrt{n}]$.

6. By putting together two triangular arrays, one with n rows and one with $n - 1$ rows, to form a square (as illustrated for $n = 4$), show that $t_{n-1} + t_n = n^2$, where t_n is the nth triangular number.

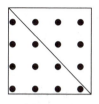

7. By putting together two triangular arrays, each with n rows, to form a rectangular array of dots of size n by $n + 1$ (as illustrated for $n = 4$), show that $2t_n = n(n + 1)$. From this, conclude that $t_n = n(n + 1)/2$.

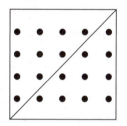

The *pentagonal numbers* $p_1, p_2, p_3, \ldots, p_k, \ldots$, are the integers that count the number of dots in k nested pentagons, as shown in the following figure.

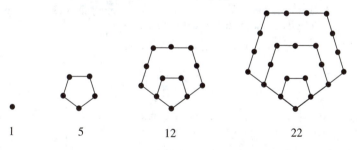

1   5   12   22

8. Show that $p_1 = 1$ and $p_k = p_{k-1} + (3k - 2)$ for $k \geq 2$. Conclude that $p_n = \sum_{k=1}^{n}(3k - 2)$.

9. Prove that the sum of the $(n - 1)$st triangular number and the nth square number is the nth pentagonal number.

10. a) Define the hexagonal numbers in a manner analogous to the definitions of triangular, square, and pentagonal numbers. (Recall that a hexagon is a six-sided polygon.)

   b) Find a closed formula for hexagonal numbers.

11. a) Define the heptagonal numbers in a manner analogous to the definitions of triangular, square, and pentagonal numbers. (Recall that a heptagon is a seven-sided polygon.)

   b) Find a closed formula for heptagonal numbers.

The *tetrahedral numbers* $T_1, T_2, T_3, \ldots, T_k, \ldots$, are the integers that count the number of dots on the faces of k nested tetrahedra, as shown in the following figure.

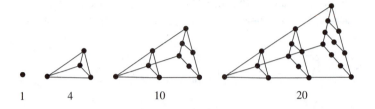

     1   4   10   20

12. Show that the nth tetrahedral number is the sum of the first n triangular numbers.

13. Find and prove a closed formula for the nth tetrahedral number.

14. Find $n!$ for n equal to each of the first ten positive integers.

15. List the integers $100!$, 100^{100}, 2^{100}, and $(50!)^2$ in order of increasing size. Justify your answer.

16. Express each of the following products in terms of $\prod_{i=1}^{n} a_i$, where k is a constant.

    a) $\prod_{i=1}^{n} k a_i$ b) $\prod_{i=1}^{n} i a_i$ c) $\prod_{i=1}^{n} a_i^k$

17. Use the identity $\frac{1}{k(k+1)} = \frac{1}{k} - \frac{1}{k+1}$ to evaluate $\sum_{k=1}^{n} \frac{1}{k(k+1)}$.

18. Use the identity $\frac{1}{k^2-1} = \frac{1}{2}\left(\frac{1}{k-1} - \frac{1}{k+1}\right)$ to evaluate $\sum_{k=2}^{n} \frac{1}{k^2-1}$.

19. Find a formula for $\sum_{k=1}^{n} k^2$ using a technique analogous to that in Example 1.19 and the formula found there.

20. Find a formula for $\sum_{k=1}^{n} k^3$ using a technique analogous to that in Example 1.19, and the results of that example and Exercise 19.

21. Without multiplying all the terms, show that

    a) $10! = 6!\, 7!$.   c) $16! = 14!\, 5!\, 2!$.

    b) $10! = 7!\, 5!\, 3!$.   d) $9! = 7!\, 3!\, 3!\, 2!$.

22. Let $a_1, a_2, \ldots, a_n$ be positive integers. Let $b = (a_1!\, a_2! \ldots a_n!) - 1$, and $c = a_1!\, a_2! \ldots a_n!$. Show that $c! = a_1!\, a_2! \cdots a_n!\, b!$.

23. Find all positive integers x, y, and z such that $x! + y! = z!$.

24. Find the values of the following products.

    a) $\prod_{j=2}^{n}(1 - 1/j)$ b) $\prod_{j=2}^{n}(1 - 1/j^2)$

1.2 Computational and Programming Exercises

Computations and Explorations

Using a computation program such as Maple or *Mathematica,* or programs you have written, carry out the following computations and explorations.

1. What are the largest values of n for which $n!$ has fewer than 100 decimal digits, fewer than 1000 decimal digits, and fewer than 10,000 decimal digits?

Programming Projects

Write programs using Maple, *Mathematica,* or a language of your choice to do the following.

1. Given the terms of a sequence $a_1, a_2, \ldots, a_n$, compute $\sum_{j=1}^{n} a_j$ and $\prod_{j=1}^{n} a_j$.

2. Given the terms of a geometric progression, find the sum of its terms.

1.3 Mathematical Induction

By examining the sums of the first n odd positive integers for small values of n, we can conjecture a formula for this sum. We have

$$1 = 1,$$
$$1 + 3 = 4,$$
$$1 + 3 + 5 = 9,$$
$$1 + 3 + 5 + 7 = 16,$$
$$1 + 3 + 5 + 7 + 9 = 25,$$
$$1 + 3 + 5 + 7 + 9 + 11 = 36.$$

From these values, we conjecture that $\sum_{j=1}^{n}(2j - 1) = 1 + 3 + 5 + 7 + \cdots + 2n - 1 = n^2$ for every positive integer n.

How can we prove that this formula holds for all positive integers n?

The *principle of mathematical induction* is a valuable tool for proving results about the integers—such as the formula just conjectured for the sum of the first n odd positive integers. First, we will state this principle, and then we will show how it is used. Subsequently, we will use the well-ordering principle to show that mathematical induction is a valid proof technique. We will use the principle of mathematical induction, and the well-ordering property, many times in our study of number theory.

We must accomplish two things to prove by mathematical induction that a particular statement holds for every positive integer. Letting S be the set of positive integers for which we claim the statement to be true, we must show that 1 belongs to S; that is, that the statement is true for the integer 1. This is called the *basis step*.

Second, we must show, for each positive integer n, that $n + 1$ belongs to S if n does; that is, that the statement is true for $n + 1$ if it is true for n. This is called the *inductive step*. Once these two steps are completed, we can conclude by the principle of mathematical induction that the statement is true for all positive integers.

Theorem 1.5. *The Principle of Mathematical Induction.*   A set of positive integers that contains the integer 1, and that has the property that, if it contains the integer k, then it also contains $k + 1$, must be the set of all positive integers.

We illustrate the use of mathematical induction by several examples; first, we prove the conjecture made at the start of this section.

Example 1.22.   We will use mathematical induction to show that

$$\sum_{j=1}^{n}(2j-1)=1+3+\cdots+(2n-1)=n^2$$

for every positive integer n. (By the way, if our conjecture for the value of this sum was incorrect, mathematical induction would fail to produce a proof!)

We begin with the basis step, which follows because

$$\sum_{j=1}^{1}(2j-1)=2\cdot1-1=1=1^2.$$

For the inductive step, we assume the inductive hypothesis that the formula holds for n; that is, we assume that $\sum_{j=1}^{n}(2j-1)=n^2$. Using the inductive hypothesis, we have

$$\sum_{j=1}^{n+1}(2j-1)=\sum_{j=1}^{n}(2j-1)+(2(n+1)-1) \quad \textit{(splitting off the term with } j=n+1\textit{)}$$

$$=n^2+2(n+1)-1 \qquad\qquad \textit{(using the inductive hypothesis)}$$

$$=n^2+2n+1$$

$$=(n+1)^2.$$

Because both the basis and the inductive steps have been completed, we know that the result holds.   ◄

Next, we prove an inequality via mathematical induction.

Example 1.23.   We can show by mathematical induction that $n!\le n^n$ for every positive integer n. The basis step, namely the case where $n=1$, holds since $1!=1\le 1^1=1$. Now, assume that $n!\le n^n$; this is the inductive hypothesis. To complete the proof, we must show, under the assumption that the inductive hypothesis is true, that

---

The Origin of Mathematical Induction

The first known use of mathematical induction appears in the work of the sixteenth-century mathematician Francesco Maurolico (1494–1575). In his book *Arithmeticorum Libri Duo,* Maurolico presented various properties of the integers, together with proofs. He devised the method of mathematical induction so that he could complete some of the proofs. The first use of mathematical induction in his book was in the proof that the sum of the first n odd positive integers equals n^2.

$(n + 1)! \le (n + 1)^{n+1}$. Using the inductive hypothesis, we have

$$(n + 1)! = (n + 1) \cdot n!$$
$$\le (n + 1)n^n$$
$$< (n + 1)(n + 1)^n$$
$$\le (n + 1)^{n+1}.$$

This completes both the inductive step and the proof.   ◄

We now show that the principle of mathematical induction follows from the well-ordering principle.

Proof.   Let S be a set of positive integers containing the integer 1, and the integer $n + 1$ whenever it contains n. Assume (for the sake of contradiction) that S is not the set of all positive integers. Therefore, there are some positive integers not contained in S. By the well-ordering property, because the set of positive integers not contained in S is nonempty, there is a least positive integer n that is not in S. Note that $n \ne 1$, since 1 is in S.

Now, because $n > 1$ (as there is no positive integer n with $n < 1$), the integer $n - 1$ is a positive integer smaller than n, and hence must be in S. But because S contains $n - 1$, it must also contain $(n - 1) + 1 = n$, which is a contradiction, as n is supposedly the smallest positive integer not in S. This shows that S must be the set of all positive integers.   ∎

A slight variant of the principle of mathematical induction is also sometimes useful in proofs.

Theorem 1.6. *The Second Principle of Mathematical Induction.*   A set of positive integers that contains the integer 1, and that has the property that, for every positive integer n, if it contains all the positive integers $1, 2, \ldots, n$, then it also contains the integer $n + 1$, must be the set of all positive integers.

The second principle of mathematical induction is sometimes called *strong induction* to distinguish it from the principle of mathematical induction, which is also called *weak induction*.

Before proving that the second principle of mathematical induction is valid, we will give an example to illustrate its use.

Example 1.24.   We will show that any amount of postage more than one cent can be formed using just two-cent and three-cent stamps. For the basis step, note that postage of two cents can be formed using one two-cent stamp and postage of three cents can be formed using one three-cent stamp.

For the inductive step, assume that every amount of postage not exceeding n cents, $n \ge 3$, can be formed using two-cent and three-cent stamps. Then a postage amount of $n + 1$ cents can be formed by taking stamps of $n - 1$ cents together with a two-cent stamp. This completes the proof.   ◄

We will now show that the second principle of mathematical induction is a valid technique.

Proof. Let T be a set of integers containing 1 and such that for every positive integer n, if it contains $1, 2, \ldots, n$, it also contains $n + 1$. Let S be the set of all positive integers n such that all the positive integers less than or equal to n are in T. Then 1 is in S, and by the hypotheses, we see that if n is in S, then $n + 1$ is in S. Hence, by the principle of mathematical induction, S must be the set of all positive integers, so clearly T is also the set of all positive integers, since S is a subset of T. ∎

Recursive Definitions

The principle of mathematical induction provides a method for defining the values of functions at positive integers. Instead of explicitly specifying the value of the function at n, we give the value of the function at 1 and give a rule for finding, for each positive integer n, the value of the function at $n + 1$ from the value of the function at n.

Definition.   We say that the function f is *defined recursively* if the value of f at 1 is specified and if for each positive integer n a rule is provided for determining $f(n + 1)$ from $f(n)$.

The principle of mathematical induction can be used to show that a function that is defined recursively is defined uniquely at each positive integer (see Exercise 25 at the end of this section). We illustrate how to define a function recursively with the following definition.

Example 1.25.   We will recursively define the *factorial function* $f(n) = n!$. First, we specify that

$$f(1) = 1.$$

Then we give a rule for finding $f(n + 1)$ from $f(n)$ for each positive integer, namely

$$f(n + 1) = (n + 1) \cdot f(n).$$

These two statements uniquely define $n!$ for the set of positive integers.

To find the value of $f(6) = 6!$ from the recursive definition, use the second property successively, as follows:

$$f(6) = 6 \cdot f(5) = 6 \cdot 5 \cdot f(4) = 6 \cdot 5 \cdot 4 \cdot f(3) = 6 \cdot 5 \cdot 4 \cdot 3 \cdot f(2) = 6 \cdot 5 \cdot 4 \cdot 3 \cdot 2 \cdot f(1).$$

Then use the first statement of the definition to replace $f(1)$ by its stated value 1, to conclude that

$$6! = 6 \cdot 5 \cdot 4 \cdot 3 \cdot 2 \cdot 1 = 720. \qquad \blacktriangleleft$$

The second principle of mathematical induction also serves as a basis for recursive definitions. We can define a function whose domain is the set of positive integers by specifying its value at 1 and giving a rule, for each positive integer n, for finding $f(n)$

from the values $f(j)$ for each integer j with $1 \leq j \leq n - 1$. This will be the basis for the definition of the sequence of Fibonacci numbers discussed in Section 1.4.

1.3 Exercises

1. Use mathematical induction to prove that $n < 2^n$ whenever n is a positive integer.

2. Conjecture a formula for the sum of the first n even positive integers. Prove your result using mathematical induction.

3. Use mathematical induction to prove that $\sum_{k=1}^{n} \frac{1}{k^2} = \frac{1}{1^2} + \frac{1}{2^2} + \cdots + \frac{1}{n^2} \leq 2 - \frac{1}{n}$ whenever n is a positive integer.

4. Conjecture a formula for $\sum_{k=1}^{n} \frac{1}{k(k+1)} = \frac{1}{1 \cdot 2} + \frac{1}{2 \cdot 3} + \cdots + \frac{1}{n(n+1)}$ from the value of this sum for small integers n. Prove that your conjecture is correct using mathematical induction. (Compare this to Exercise 17 in Section 1.2.)

5. Conjecture a formula for $\mathbf{A}^n$ where $\mathbf{A} = \begin{pmatrix} 1 & 1 \\ 0 & 1 \end{pmatrix}$. Prove your conjecture using mathematical induction.

6. Use mathematical induction to prove that $\sum_{j=1}^{n} j = 1 + 2 + 3 + \cdots + n = n(n+1)/2$ for every positive integer n. (Compare this to Example 1.19 in Section 1.2.)

7. Use mathematical induction to prove that $\sum_{j=1}^{n} j^2 = 1^2 + 2^2 + 3^2 + \cdots + n^2 = n(n+1)(2n+1)/6$ for every positive integer n.

8. Use mathematical induction to prove that $\sum_{j=1}^{n} j^3 = 1^3 + 2^3 + 3^3 + \cdots + n^3 = [n(n+1)/2]^2$ for every positive integer n.

9. Use mathematical induction to prove that $\sum_{j=1}^{n} j(j+1) = 1 \cdot 2 + 2 \cdot 3 + \cdots + n \cdot (n+1) = n(n+1)(n+2)/3$ for every positive integer n.

10. Use mathematical induction to prove that $\sum_{j=1}^{n} (-1)^{j-1} j^2 = 1^2 - 2^2 + 3^2 - \cdots + (-1)^{n-1} n^2 = (-1)^{n-1} n(n+1)/2$ for every positive integer n.

11. Find a formula for $\prod_{j=1}^{n} 2^j$.

12. Show that $\sum_{j=1}^{n} j \cdot j! = 1 \cdot 1! + 2 \cdot 2! + \cdots + n \cdot n! = (n+1)! - 1$ for every positive integer n.

13. Show that any amount of postage that is an integer number of cents greater than 11 cents can be formed using just 4-cent and 5-cent stamps.

14. Show that any amount of postage that is an integer number of cents greater than 53 cents can be formed using just 7-cent and 10-cent stamps.

Let H_n be the nth partial sum of the harmonic series, that is, $H_n = \sum_{j=1}^{n} 1/j$.

* 15. Use mathematical induction to show that $H_{2^n} \geq 1 + n/2$.

* 16. Use mathematical induction to show that $H_{2^n} \leq 1 + n$.

17. Show by mathematical induction that if n is a positive integer, then $(2n)! < 2^{2n}(n!)^2$.

18. Use mathematical induction to prove that $x - y$ is a factor of $x^n - y^n$, where x and y are variables.

☞ **19.** Use the principle of mathematical induction to show that a set of integers that contains the integer k, such that this set contains $n + 1$ whenever it contains n, contains the set of integers that are greater than or equal to k.

20. Use mathematical induction to prove that $2^n < n!$ for $n \geq 4$.

21. Use mathematical induction to prove that $n^2 < n!$ for $n \geq 4$.

22. Show by mathematical induction that if $h \geq -1$, then $1 + nh \leq (1 + h)^n$ for all nonnegative integers n.

23. A jigsaw puzzle is solved by putting its pieces together in the correct way. Show that exactly $n - 1$ moves are required to solve a jigsaw puzzle with n pieces, where a move consists of putting together two blocks of pieces, with a block consisting of one or more assembled pieces. (*Hint:* Use the second principle of mathematical induction.)

24. Explain what is wrong with the following proof by mathematical induction that all horses are the same color: Clearly all horses in any set of 1 horse are all the same color. This completes the basis step. Now assume that all horses in any set of n horses are the same color. Consider a set of $n + 1$ horses, labeled with the integers $1, 2, \ldots, n + 1$. By the induction hypothesis, horses $1, 2, \ldots, n$ are all the same color, as are horses $2, 3, \ldots, n, n + 1$. Because these two sets of horses have common members, namely horses $2, 3, 4, \ldots, n$, all $n + 1$ horses must be the same color. This completes the induction argument.

25. Use the principle of mathematical induction to show that the value at each positive integer of a function defined recursively is uniquely determined.

26. What function $f(n)$ is defined recursively by $f(1) = 2$ and $f(n + 1) = 2f(n)$ for $n \geq 1$? Prove your answer using mathematical induction.

27. If g is defined recursively by $g(1) = 2$ and $g(n) = 2^{g(n-1)}$ for $n \geq 2$, what is $g(4)$?

28. Use the second principle of mathematical induction to show that if $f(1)$ is specified and a rule for finding $f(n + 1)$ from the values of f at the first n positive integers is given, then $f(n)$ is uniquely determined for every positive integer n.

29. We define a function recursively for all positive integers n by $f(1) = 1$, $f(2) = 5$, and for $n > 2$, $f(n + 1) = f(n) + 2f(n - 1)$. Show that $f(n) = 2^n + (-1)^n$, using the second principle of mathematical induction.

30. Show that $2^n > n^2$ whenever n is an integer greater than 4.

31. Suppose that $a_0 = 1$, $a_1 = 3$, $a_2 = 9$, and $a_n = a_{n-1} + a_{n-2} + a_{n-3}$ for $n \geq 3$. Show that $a_n \leq 3^n$ for every nonnegative integer n.

✿ **32.** The tower of Hanoi was a popular puzzle of the late nineteenth century. The puzzle includes three pegs and eight rings of different sizes placed in order of size, with the largest on the bottom, on one of the pegs. The goal of the puzzle is to move all of the rings, one at a time, without ever placing a larger ring on top of a smaller ring, from the first peg to the second, using the third as an auxiliary peg.

a) Use mathematical induction to show that the minimum number of moves to transfer n rings from one peg to another, with the rules we have described, is $2^n - 1$.

b) An ancient legend tells of the monks in a tower with 64 gold rings and 3 diamond pegs. They started moving the rings, one move per second, when the world was created. When they finish transferring the rings to the second peg, the world will end. How long will the world last?

∗ **33.** The *arithmetic mean* and the *geometric mean* of the positive real numbers $a_1, a_2, \ldots,$ a_n are $A = (a_1 + a_2 + \cdots + a_n)/n$ and $G = (a_1 a_2 \cdots a_n)^{1/n}$, respectively. Use mathematical induction to prove that $A \geq G$ for every finite sequence of positive real numbers. When does equality hold?

34. Use mathematical induction to show that a $2^n \times 2^n$ chessboard with one square missing can be covered with L-shaped pieces, where each L-shaped piece covers three squares.

∗ **35.** A *unit fraction* is a fraction of the form $1/n$, where n is a positive integer. Because the ancient Egyptians represented fractions as sums of distinct unit fractions, such sums are called *Egyptian fractions*. Show that every rational number p/q, where p and q are integers with $0 < p < q$, can be written as a sum of distinct unit fractions, that is, as an Egyptian fraction. (*Hint:* Use strong induction on the numerator p to show that the algorithm that adds the largest possible unit fraction at each stage always terminates. For example, running this algorithm shows that $5/7 = 1/2 + 1/5 + 1/70$.)

36. Using the algorithm in Exercise 35, write each of these numbers as Egyptian fractions.

a) 2/3   c) 11/17

b) 5/8   d) 44/101

1.3 Computational and Programming Exercises

Computations and Explorations

Using a computation program such as Maple or *Mathematica,* or programs you have written, carry out the following computations and explorations.

1. Complete the basis and inductive steps, using both numerical and symbolic computation, to prove that $\sum_{j=1}^{n} j = n(n+1)/2$ for all positive integers n.

2. Complete the basis and inductive steps, using both numerical and symbolic computation, to prove that $\sum_{j=1}^{n} j^2 = n(n+1)(2n+1)/6$ for all positive integers n.

3. Complete the basis and inductive steps, using both numerical and symbolic computation, to prove that $\sum_{j=1}^{n} j^3 = (n(n+1)/2)^2$ for all positive integers n.

4. Use the values $\sum_{j=1}^{n} j^4$ for $n = 1, 2, 3, 4, 5, 6$ to conjecture a formula for this sum that is a polynomial of degree 5 in n. Attempt to prove your conjecture via mathematical induction using numerical and symbolic computation.

5. Paul Erdős and E. Strauss have conjectured that the fraction $4/n$ can be written as the sum of three unit fractions, that is, $4/n = 1/x + 1/y + 1/z$, where x, y, and z are distinct positive integers for all integers n with $n > 1$. Find such representation for as many positive integers n as you can.

6. It is conjectured that the rational number p/q, where p and q are integers with $0 < p < q$ and q is odd, can be expressed as an Egyptian fraction which is the sum of unit fractions with odd denominators. Explore this conjecture using the algorithm that successively adds the unit fraction with the least positive odd denominator q at each stage. (For example, $2/7 = 1/5 + 1/13 + 1/115 + 1/10{,}465$.)

Programming Projects

Write programs using Maple, *Mathematica,* or a language of your choice to do the following.

* **1.** List the moves in the tower of Hanoi puzzle (see Exercise 32). If you can, animate these moves.

** **2.** Cover a $2^n \times 2^n$ chessboard that is missing one square using L-shaped pieces (see Exercise 34).

3. Given a rational number p/q, express p/q as an Egyptian fraction using the algorithm described in Exercise 35.

1.4   The Fibonacci Numbers

In his book *Liber Abaci,* written in 1202, the mathematician *Fibonacci* posed a problem concerning the growth of the number of rabbits in a certain area. This problem can be phrased as follows: A young pair of rabbits, one of each sex, is placed on an island. Assuming that rabbits do not breed until they are two months old and after they are two months old, each pair of rabbits produces another pair each month, how many pairs are there after n months?

Let f_n be the number of pairs of rabbits after n months. We have $f_1 = 1$ because only the original pair is on the island after one month. As this pair does not breed during the second month, $f_2 = 1$. To find the number of pairs after n months, add the number on the island the previous month, f_{n-1}, to the number of newborn pairs, which equals f_{n-2}, because each newborn pair comes from a pair at least two months old. This leads to the following definition.

Definition.   The *Fibonacci sequence* is defined recursively by $f_1 = 1$, $f_2 = 1$, and $f_n = f_{n-1} + f_{n-2}$ for $n \geq 3$. The terms of this sequence are called the *Fibonacci numbers.*

The mathematician Edouard Lucas named this sequence after Fibonacci in the nineteenth century when he established many of its properties. The answer to Fibonacci's question is that there are f_n rabbits on the island after n months.

FIBONACCI (c. 1180–1228) (short for *filus Bonacci,* son of Bonacci), also known as Leonardo of Pisa, was born in the Italian commercial center of Pisa. Fibonacci was a merchant who traveled extensively throughout the Mideast, where he came into contact with mathematical works from the Arabic world. In his *Liber Abaci* Fibonacci introduced Arabic notation for numerals and their algorithms for arithmetic into the European world. It was in this book that his famous rabbit problem appeared. Fibonacci also wrote *Practica geometriae,* a treatise on geometry and trigonometry, and *Liber quadratorum,* a book on diophantine equations.

Examining the initial terms of the Fibonacci sequence will be useful as we study their properties.

Example 1.26. We compute the first ten Fibonacci numbers as follows:

$$f_3 = f_2 + f_1 = 1 + 1 = 2$$
$$f_4 = f_3 + f_2 = 2 + 1 = 3$$
$$f_5 = f_4 + f_3 = 3 + 2 = 5$$
$$f_6 = f_5 + f_4 = 5 + 3 = 8$$
$$f_7 = f_6 + f_5 = 8 + 5 = 13$$
$$f_8 = f_7 + f_6 = 13 + 8 = 21$$
$$f_9 = f_8 + f_7 = 21 + 13 = 34$$
$$f_{10} = f_9 + f_8 = 34 + 21 = 55.$$ ◀

We can define the value of $f_0 = 0$, so that $f_2 = f_1 + f_0$. We can also define f_n where n is a negative number so that the equality in the recursive definition is satisfied (see Exercise 37.)

The Fibonacci numbers occur in an amazing variety of applications. For example, in botany the number of spirals in plants with a pattern known as phyllotaxis is always a Fibonacci number. They occur in the solution of a tremendous variety of counting problems, such as counting the number of bit strings with no two consecutive 1s (see [Ro03]).

The Fibonacci numbers also satisfy an extremely large number of identities. For example, we can easily find an identity for the sum of the first n consecutive Fibonacci numbers.

Example 1.27. The sum of the first n Fibonacci numbers for $3 \le n \le 8$ equals 1, 2, 4, 7, 12, 20, 33, and 54. Looking at these numbers, we see that they are all just 1 less than the Fibonacci number f_{n+2}. This leads us to the conjecture that

$$\sum_{k=1}^{n} f_k = f_{n+2} - 1.$$

Can we prove this identity for all positive integers n?

We will show, in two different ways, that this identity does hold for all integers n. We provide two different demonstrations, to show that there is often more than one way to prove that an identity is true.

First, we use the fact that $f_n = f_{n-1} + f_{n-2}$ for $n = 2, 3, \ldots$ to see that $f_k = f_{k+2} - f_{k+1}$ for $k = 1, 2, 3, \ldots$. This means that

$$\sum_{k=1}^{n} f_k = \sum_{k=1}^{n} (f_{k+2} - f_{k+1}).$$

We can easily evaluate this sum because it is telescoping. Using the formula for a telescoping sum found in Section 1.2, we have

$$\sum_{k=1}^{n} f_k = f_{n+2} - f_2 = f_{n+2} - 1.$$

This proves the result.

We can also prove this identity using mathematical induction. The basis step holds because $\sum_{k=1}^{1} f_k = 1$ and this equals $f_{1+2} - 1 = f_3 - 1 = 2 - 1 = 1$. The inductive hypothesis is

$$\sum_{k=1}^{n} f_k = f_{n+2} - 1.$$

We must show that, under this assumption,

$$\sum_{k=1}^{n+1} f_k = f_{n+3} - 1.$$

To prove this, note that by the inductive hypothesis we have

$$\sum_{k=1}^{n+1} f_k = \left(\sum_{k=1}^{n} f_k \right) + f_{n+1}$$
$$= (f_{n+2} - 1) + f_{n+1}$$
$$= (f_{n+1} + f_{n+2}) - 1$$
$$= f_{n+3} - 1. \qquad \blacktriangleleft$$

The exercise set at the end of this section asks you to prove many other identities of the Fibonacci numbers.

How Fast Do the Fibonacci Numbers Grow?

The following inequality, which shows that the Fibonacci numbers grow faster than a geometric series with common ratio $\alpha = (1 + \sqrt{5})/2$, will be used in Chapter 3.

Example 1.28.   We can use the second principle of mathematical induction to prove that $f_n > \alpha^{n-2}$ for $n \geq 3$ where $\alpha = (1 + \sqrt{5})/2$. The basis step consists of verifying this inequality for $n = 3$ and $n = 4$. We have $\alpha < 2 = f_3$, so the theorem is true for $n = 3$. Since $\alpha^2 = (3 + \sqrt{5})/2 < 3 = f_4$, the theorem is true for $n = 4$.

The inductive hypothesis consists of assuming that $\alpha^{k-2} < f_k$ for all integers k with $k \leq n$. Because $\alpha = (1 + \sqrt{5})/2$ is a solution of $x^2 - x - 1 = 0$, we have $\alpha^2 = \alpha + 1$. Hence

$$\alpha^{n-1} = \alpha^2 \cdot \alpha^{n-3} = (\alpha + 1) \cdot \alpha^{n-3} = \alpha^{n-2} + \alpha^{n-3}.$$

By the inductive hypothesis, we have the inequalities

$$\alpha^{n-2} < f_n, \quad \alpha^{n-3} < f_{n-1}.$$

By adding these two inequalities, we conclude that

$$\alpha^{n-1} < f_n + f_{n-1} = f_{n+1}.$$

This finishes the proof.   ◄

We conclude this section with an explicit formula for the nth Fibonacci number. We will not provide a proof in the text, but Exercises 41 and 42 at the end of this section outline how this formula can be found using linear homogeneous recurrence relations and generating functions, respectively. Furthermore, Exercise 40 asks that you prove this identity by showing that the terms satisfy the same recursive definition as the Fibonacci numbers do, and Exercise 45 asks for a proof via mathematical induction. The advantage of the first two approaches is that they can be used to find the formula, while the second two approaches cannot.

Theorem 1.7.   Let n be a positive integer and let $\alpha = \frac{1+\sqrt{5}}{2}$ and $\beta = \frac{1-\sqrt{5}}{2}$. Then the nth Fibonacci number f_n is given by

$$f_n = \frac{1}{\sqrt{5}}(\alpha^n - \beta^n).$$

We have presented a few important results involving the Fibonacci numbers. There is a vast literature concerning these numbers and their many applications to botany, computer science, geography, physics, and other areas (see [Va89]). There is even a scholarly journal, *The Fibonacci Quarterly,* devoted to their study.

1.4 Exercises

1. Find the following Fibonacci numbers.

   a) f_{10} c) f_{15} e) f_{20}

   b) f_{13} d) f_{18} f) f_{25}

2. Find each of the following Fibonacci numbers.

   a) f_{12} c) f_{24} e) f_{32}

   b) f_{16} d) f_{30} f) f_{36}

3. Prove that $f_{n+3} + f_n = 2f_{n+2}$ whenever n is a positive integer.

4. Prove that $f_{n+3} - f_n = 2f_{n+1}$ whenever n is a positive integer.

5. Prove that $f_{2n} = f_n^2 + 2f_{n-1}f_n$ whenever n is a positive integer. (Recall that $f_0 = 0$.)

6. Prove that $f_{n-2} + f_{n+2} = 3f_n$ whenever n is an integer with $n \geq 2$. (Recall that $f_0 = 0$.)

7. Find and prove a simple formula for the sum of the first n Fibonacci numbers with odd indices when n is a positive integer. That is, find a simple formula for $f_1 + f_3 + \cdots + f_{2n-1}$.

8. Find and prove a simple formula for the sum of the first n Fibonacci numbers with even indices when n is a positive integer. That is, find a simple formula for $f_2 + f_4 + \cdots + f_{2n}$.

9. Find and prove a simple formula for the expression $f_n - f_{n-1} + f_{n-2} - \cdots + (-1)^{n+1}f_1$ when n is a positive integer.

10. Prove that $f_{2n+1} = f_{n+1}^2 + f_n^2$ whenever n is a positive integer.

11. Prove that $f_{2n} = f_{n+1}^2 - f_{n-1}^2$ whenever n is a positive integer. (Recall that $f_0 = 0$.)

12. Prove that $f_n + f_{n-1} + f_{n-2} + 2f_{n-3} + 4f_{n-4} + 8f_{n-5} + \cdots + 2^{n-3} = 2^{n-1}$ whenever n is an integer with $n \geq 3$.

13. Prove that $\sum_{j=1}^{n} f_j^2 = f_1^2 + f_2^2 + \cdots + f_n^2 = f_n f_{n+1}$ for every positive integer n.

14. Prove that $f_{n+1}f_{n-1} - f_n^2 = (-1)^n$ for every positive integer n.

15. Prove that $f_{n+1}f_n - f_{n-1}f_{n-2} = f_{2n-1}$ for every positive integer n, $n > 2$.

16. Prove that $f_1 f_2 + f_2 f_3 + \cdots + f_{2n-1}f_{2n} = f_{2n}^2$ if n is a positive integer.

17. Prove that $f_{m+n} = f_m f_{n+1} + f_n f_{m-1}$ whenever m and n are positive integers.

The *Lucas numbers,* named after *François-Eduoard-Anatole Lucas* (see Chapter 7 for a biography), are defined recursively by

$$L_n = L_{n-1} + L_{n-2}, \quad n \geq 3$$

with $L_1 = 1$ and $L_2 = 3$. They satisfy the same recurrence relation as the Fibonacci numbers, but the two initial values are different.

18. Find the first 12 Lucas numbers.

19. Find and prove a formula for the sum of the first n Lucas numbers when n is a positive integer.

20. Find and prove a formula for the sum of the first n Lucas numbers with odd indices when n is a positive integer.

21. Find and prove a formula for the sum of the first n Lucas numbers with even indices when n is a positive integer.

22. Prove that $L_n^2 - L_{n+1}L_{n-1} = 5(-1)^n$ when n is an integer with $n \geq 2$.

23. Prove that $L_1^2 + L_2^2 + \cdots + L_n^2 = L_n L_{n+1} - 2$ when n is an integer with $n \geq 1$.

24. Show that the nth Lucas number L_n is the sum of the $(n+1)$st and $(n-1)$st Fibonacci numbers, f_{n+1} and f_{n-1}, respectively.

25. Show that $f_{2n} = f_n L_n$ for all integers n with $n \geq 1$, where f_n is the nth Fibonacci number and L_n is the nth Lucas number.

26. Prove that $5f_{n+1} = L_n + L_{n+2}$ whenever n is a positive integer, f_n is the nth Fibonacci number, and L_n is the nth Lucas number.

* 27. Prove that $L_{m+n} = f_{m+1}L_n + f_m L_{n-1}$ whenever m and n are positive integers with $n > 1$, f_n is the nth Fibonacci number, and L_n is the nth Lucas number.

28. Show that L_n, the nth Lucas number, is given by

$$L_n = \alpha^n + \beta^n,$$

where $\alpha = (1 + \sqrt{5})/2$ and $\beta = (1 - \sqrt{5})/2$.

The *Zeckendorf representation* of a positive integer is the unique expression of this integer as the sum of distinct Fibonacci numbers, where no two of these Fibonacci numbers are consecutive terms in the Fibonacci sequence and where the term $f_1 = 1$ is not used (but the term $f_2 = 1$ may be used.)

29. Find the Zeckendorf representation of each of the integers 50, 85, 110, and 200.

* **30.** Show that every positive integer has a unique Zeckendorf representation.

31. Show that $f_n \le \alpha^{n-1}$ for every integer n with $n \ge 2$, where $\alpha = (1 + \sqrt{5})/2$.

32. Show that

$$\binom{n}{0} + \binom{n-1}{1} + \binom{n-2}{2} + \cdots = f_{n+1},$$

where n is a nonnegative integer and f_{n+1} is the $(n + 1)$st Fibonacci number. (See Appendix B for a review of binomial coefficients. Here, the sum ends with the term $\binom{1}{n-1}$.)

33. Prove that whenever n is a nonnegative integer, $\sum_{j=1}^{n} \binom{n}{j} f_j = f_{2n}$, where f_j is the jth Fibonacci number.

34. Let $\mathbf{F} = \begin{pmatrix} 1 & 1 \\ 1 & 0 \end{pmatrix}$. Show that $\mathbf{F}^n = \begin{pmatrix} f_{n+1} & f_n \\ f_n & f_{n-1} \end{pmatrix}$ when $n \in \mathbf{Z}^+$.

35. By taking determinants of both sides of the result of Exercise 34, prove the identity in Exercise 14.

36. Define the *generalized Fibonacci numbers* recursively by $g_1 = a$, $g_2 = b$, and $g_n = g_{n-1} + g_{n-2}$ for $n \ge 3$. Show that $g_n = af_{n-2} + bf_{n-1}$ for $n \ge 3$.

37. Give a recursive definition of the Fibonacci number f_n when n is a negative integer. Use your definition to find f_n for $n = -1, -2, -3, \ldots, -10$.

38. Use the results of Exercise 37 to formulate a conjecture that relates the values of f_{-n} and f_n when n is a positive integer. Prove this conjecture using mathematical induction.

39. What is wrong with the claim that an 8×8 square can be broken into pieces that can be reassembled to form a 5×13 rectangle as shown?

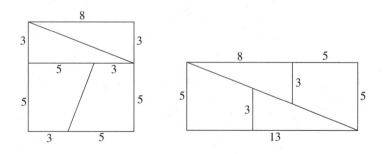

(*Hint:* Look at the identity in Exercise 14. Where is the extra square unit?)

40. Show that if $a_n = \frac{1}{\sqrt{5}}(\alpha^n - \beta^n)$, where $\alpha = (1 + \sqrt{5})/2$ and $\beta = (1 - \sqrt{5})/2$, then $a_n = a_{n-1} + a_{n-2}$ and $a_1 = a_2 = 1$. Conclude that $f_n = a_n$, where f_n is the nth Fibonacci number.

A *linear homogeneous recurrence relation of degree 2 with constant coefficients* is an equation of the form

$$a_n = c_1 a_{n-1} + c_2 a_{n-2},$$

where c_1 and c_2 are real numbers with $c_2 \neq 0$. It is not difficult to show (see [Ro03]) that if the equation $r^2 - c_1 r - c_2 = 0$ has two distinct roots r_1 and r_2, then the sequence $\{a_n\}$ is a solution of the linear homogeneous recurrence relation $a_n = c_1 a_{n-1} + c_2 a_{n-2}$ if and only if $a_n = C_1 r_1^n + C_2 r_2^n$ for $n = 0, 1, 2, \ldots$, where C_1 and C_2 are constants. The values of these constants can be found using the two initial terms of the sequence.

41. Find an explicit formula for f_n, proving Theorem 1.7, by solving the recurrence relation $f_n = f_{n-1} + f_{n-2}$ for $n = 2, 3, \ldots$ with initial conditions $f_0 = 0$ and $f_1 = 1$.

The *generating function* for the sequence $a_0, a_1, \ldots, a_k, \ldots$ is the infinite series

$$G(x) = \sum_{k=0}^{\infty} a_k x^k.$$

42. Use the generating function $G(x) = \sum_{k=0}^{\infty} f_k x^k$ where f_k is the kth Fibonacci number to find an explicit formula for f_k, proving Theorem 1.7. (*Hint:* Use the fact that $f_k = f_{k-1} + f_{k-2}$ for $k = 2, 3, \ldots$ to show that $G(x) - xG(x) - x^2 G(x) = x$. Solve this to show that $G(x) = x/(1 - x - x^2)$ and then write $G(x)$ in terms of partial fractions, as is done in calculus.) (See [Ro03] for information on using generating functions.)

43. Find an explicit formula for the Lucas numbers using the technique of Exercise 41.

44. Find an explicit formula for the Lucas numbers using the technique of Exercise 42.

45. Use mathematical induction to prove Theorem 1.7.

1.4 Computational and Programming Exercises

Computations and Explorations

Using a computation program such as Maple or *Mathematica,* or programs you have written, carry out the following computations and explorations.

1. Find the Fibonacci numbers f_{100}, f_{200}, and f_{500}.

2. Find the Lucas numbers L_{100}, L_{200}, and L_{500}.

3. A surprising theorem states that the Fibonacci numbers are the positive values of the polynomial $2xy^4 + x^2 y^3 - 2x^3 y^2 - y^5 - x^4 y + 2y$ as x and y range over all nonnegative integers. Verify this conjecture for the values of x and y where x and y are nonnegative integers with $x + y \leq 100$.

Programming Projects

Write programs using Maple, *Mathematica,* or a language of your choice to do the following.

1. Given a positive integer n, find the first n terms of the Fibonacci sequence.

2. Given a positive integer n, find the first n terms of the Lucas sequence.

1.5 Divisibility

The concept of the divisibility of one integer by another is central in number theory.

Definition.   If a and b are integers with $a \neq 0$, we say that a *divides* b if there is an integer c such that $b = ac$. If a divides b, we also say that a is a *divisor* or *factor* of b and that b is a *multiple* of a.

If a divides b we write $a \mid b$, and if a does not divide b we write $a \nmid b$. (Be careful not to confuse the notations $a \mid b$, which denotes that a divides b, and a/b, which is the quotient obtained when a is divided by b.)

Example 1.29.   The following statements illustrate the concept of the divisibility of integers: $13 \mid 182$, $-5 \mid 30$, $17 \mid 289$, $6 \nmid 44$, $7 \nmid 50$, $-3 \mid 33$, and $17 \mid 0$.   ◄

Example 1.30.   The divisors of 6 are $\pm 1, \pm 2, \pm 3, \pm 6$. The divisors of 17 are $\pm 1, \pm 17$. The divisors of 100 are $\pm 1, \pm 2, \pm 4, \pm 5, \pm 10, \pm 20, \pm 25, \pm 50, \pm 100$.   ◄

In subsequent chapters, we will need some simple properties of divisibility, which we now state and prove.

Theorem 1.8.   If a, b, and c are integers with $a \mid b$ and $b \mid c$, then $a \mid c$.

Proof.   Because $a \mid b$ and $b \mid c$, there are integers e and f such that $ae = b$ and $bf = c$. Hence, $c = bf = (ae)f = a(ef)$, and we conclude that $a \mid c$.   ∎

Example 1.31.   Because $11 \mid 66$ and $66 \mid 198$, Theorem 1.8 tells us that $11 \mid 198$.   ◄

Theorem 1.9.   If a, b, m, and n are integers, and if $c \mid a$ and $c \mid b$, then $c \mid (ma + nb)$.

Proof.   Because $c \mid a$ and $c \mid b$, there are integers e and f such that $a = ce$ and $b = cf$. Hence, $ma + nb = mce + ncf = c(me + nf)$. Consequently, we see that $c \mid (ma + nb)$.   ∎

Example 1.32.   As $3 \mid 21$ and $3 \mid 33$, Theorem 1.9 tells us that 3 divides

$$5 \cdot 21 - 3 \cdot 33 = 105 - 99 = 6.$$ ◄

The following theorem states an important fact about division.

Theorem 1.10. *The Division Algorithm.*   If a and b are integers such that $b > 0$, then there are unique integers q and r such that $a = bq + r$ with $0 \leq r < b$.   ∎

In the equation given in the division algorithm, we call q the *quotient* and r the *remainder.* We also call a the *dividend* and b the *divisor.* (*Note:* We use the traditional name for this theorem even though the division algorithm is not actually an algorithm. We discuss algorithms in Section 2.2.)

We note that a is divisible by b if and only if the remainder in the division algorithm is 0. Before we prove the division algorithm, consider the following examples.

Example 1.33. If $a = 133$ and $b = 21$, then $q = 6$ and $r = 7$, because $133 = 21 \cdot 6 + 7$. Likewise, if $a = -50$ and $b = 8$, then $q = -7$ and $r = 6$, because $-50 = 8(-7) + 6$.   ◄

We now prove the division algorithm using the well-ordering property.

Proof.   Consider the set S of all integers of the form $a - bk$ where k is an integer, that is, $S = \{a - bk \mid k \in \mathbf{Z}\}$. Let T be the set of all nonnegative integers in S. T is nonempty, because $a - bk$ is positive whenever k is an integer with $k < a/b$.

By the well-ordering property, T has a least element $r = a - bq$. (These are the values for q and r specified in the theorem.) We know that $r \geq 0$ by construction, and it is easy to see that $r < b$. If $r \geq b$ then $r > r - b = a - bq - b = a - b(q + 1) \geq 0$, which contradicts the choice of $r = a - bq$ as the least nonnegative integer of the form $a - bk$. Hence $0 \leq r < b$.

To show that these values for q and r are unique, assume that we have two equations $a = bq_1 + r_1$ and $a = bq_2 + r_2$, with $0 \leq r_1 < b$ and $0 \leq r_2 < b$. By subtracting the second of these equations from the first, we find that

$$0 = b(q_1 - q_2) + (r_1 - r_2).$$

Hence, we see that

$$r_2 - r_1 = b(q_1 - q_2).$$

This tells us that b divides $r_2 - r_1$. Because $0 \leq r_1 < b$ and $0 \leq r_2 < b$, we have $-b < r_2 - r_1 < b$. Hence, b can divide $r_2 - r_1$ only if $r_2 - r_1 = 0$ or, in other words, if $r_1 = r_2$. Because $bq_1 + r_1 = bq_2 + r_2$ and $r_1 = r_2$, we also see that $q_1 = q_2$. This shows that the quotient q and the remainder r are unique.   ■

We now use the greatest integer function (defined in Section 1.1) to give explicit formulas for the quotient and remainder in the division algorithm. Because the quotient q is the largest integer such that $bq \leq a$, and $r = a - bq$, it follows that

(1.4) $q = [a/b], \quad r = a - b[a/b].$

The following examples display the quotient and remainder of a division.

Example 1.34. Let $a = 1028$ and $b = 34$. Then $a = bq + r$ with $0 \leq r < b$, where $q = [1028/34] = 30$ and $r = 1028 - [1028/34] \cdot 34 = 1028 - 30 \cdot 34 = 8$.   ◄

Example 1.35. Let $a = -380$ and $b = 75$. Then $a = bq + r$ with $0 \leq r < b$, where $q = [-380/75] = -6$ and $r = -380 - [-380/75] \cdot 75 = -380 - (-6)75 = 70$.   ◄

We can use Equation (1.4) to prove a useful property of the greatest integer function.

Example 1.36. Show that if n is a positive integer, then $[x/n] = [[x]/n]$ whenever x is a real number. To prove this identity, suppose that $[x] = m$. By the division algorithm, we have integers q and r such that $m = nq + r$, where $0 \leq r \leq n - 1$. By Equation (1.4), we have $q = [[x]/n]$. Because $[x] \leq x < [x] + 1$, it follows that $x = [x] + \epsilon$, where $0 \leq \epsilon < 1$. We see that $[x/n] = [([x] + \epsilon)/n] = [(m + \epsilon)/n] = [((nq + r) + \epsilon)/n] = [q + (r + \epsilon)/n]$. Because $0 \leq \epsilon < 1$, we have $0 \leq r + \epsilon < (n - 1) + 1 = n$. It follows that $[x/n] = [q]$.   ◀

Given a positive integer d, we can classify integers according to their remainders when divided by d. For example, with $d = 2$, we see from the division algorithm that every integer when divided by 2 leaves a remainder of either 0 or 1. This leads to the following definition of some common terminology.

Definition.   If the remainder when n is divided by 2 is 0, then $n = 2k$ for some integer k, and we say that n is *even*, whereas if the remainder when n is divided by 2 is 1, then $n = 2k + 1$ for some integer k, and we say that n is *odd*.

Similarly, when $d = 4$, we see from the division algorithm that when an integer n is divided by 4, the remainder is either 0, 1, 2, or 3. Hence, every integer is of the form $4k$, $4k + 1$, $4k + 2$, or $4k + 3$, where k is a positive integer.

We will pursue these matters further in Chapter 4.

1.5 Exercises

1. Show that $3 \mid 99$, $5 \mid 145$, $7 \mid 343$, and $888 \mid 0$.

2. Show that 1001 is divisible by 7, by 11, and by 13.

3. Decide which of the following integers are divisible by 7.

   a) 0   d) 123321
   b) 707   e) −285714
   c) 1717   f) −430597

4. Decide which of the following integers are divisible by 22.

   a) 0   d) 192544
   b) 444   e) −32516
   c) 1716   f) −195518

5. Find the quotient and remainder in the division algorithm, with divisor 17 and dividend

   a) 100.   c) −44.
   b) 289.   d) −100.

6. What can you conclude if a and b are nonzero integers such that $a \mid b$ and $b \mid a$?

7. Show that if a, b, c, and d are integers with a and c nonzero, such that $a \mid b$ and $c \mid d$, then $ac \mid bd$.

8. Are there integers a, b, and c such that $a \mid bc$, but $a \nmid b$ and $a \nmid c$?

9. Show that if a, b, and $c \neq 0$ are integers, then $a \mid b$ if and only if $ac \mid bc$.

10. Show that if a and b are positive integers and $a \mid b$, then $a \le b$.

11. Show that if a and b are integers such that $a \mid b$, then $a^k \mid b^k$ for every positive integer k.

12. Show that the sum of two even or of two odd integers is even, whereas the sum of an odd and an even integer is odd.

13. Show that the product of two odd integers is odd, whereas the product of two integers is even if either of the integers is even.

14. Show that if a and b are odd positive integers and $b \nmid a$, then there are integers s and t such that $a = bs + t$, where t is odd and $\mid t \mid < b$.

15. When the integer a is divided by the integer b, where $b > 0$, the division algorithm gives a quotient of q and a remainder of r. Show that if $b \nmid a$, when $-a$ is divided by b, the division algorithm gives a quotient of $-(q + 1)$ and a remainder of $b - r_j$, whereas if $b \mid a$, the quotient is $-q$ and the remainder is 0.

16. Show that if a, b, and c are integers with $b > 0$ and $c > 0$, such that when a is divided by b the quotient is q and the remainder is r, and when q is divided by c the quotient is t and the remainder is s, then when a is divided by bc, the quotient is t and the remainder is $bs + r$.

17. a) Extend the division algorithm by allowing negative divisors. In particular, show that whenever a and $b \ne 0$ are integers, there are unique integers q and r such that $a = bq + r$, where $0 \le r < \mid b \mid$.

    b) Find the remainder when 17 is divided by -7.

18. Show that if a and b are positive integers, then there are unique integers q and r such that $a = bq + r$, where $-b/2 < r \le b/2$.

19. Show that if m and $n > 0$ are integers, then

$$\left[\frac{m + 1}{n}\right] = \begin{cases} \left[\frac{m}{n}\right] & \text{if } m \ne kn - 1 \text{ for some integer } k; \\ \left[\frac{m}{n}\right] + 1 & \text{if } m = kn - 1 \text{ for some integer } k. \end{cases}$$

20. Show that the integer n is even if and only if $n - 2[n/2] = 0$.

21. Show that the number of positive integers less than or equal to x, where x is a positive real number, that are divisible by the positive integer d equals $[x/d]$.

22. Find the number of positive integers not exceeding 1000 that are divisible by 5, by 25, by 125, and by 625.

23. How many integers between 100 and 1000 are divisible by 7? by 49?

24. Find the number of positive integers not exceeding 1000 that are not divisible by 3 or 5.

25. Find the number of positive integers not exceeding 1000 that are not divisible by 3, 5, or 7.

26. Find the number of positive integers not exceeding 1000 that are divisible by 3 but not by 4.

27. In 1999, to mail a first-class letter in the United States of America it cost 33 cents for the first ounce and 22 cents for each additional ounce or fraction thereof. Find a formula involving the greatest integer function for the cost of mailing a letter in 1999. Could it possibly have cost \$1.45 or \$2.31 to mail a first-class letter in the United States of America in 1999?

28. Show that if a is an integer, then 3 divides $a^3 - a$.

29. Show that the product of two integers of the form $4k + 1$ is again of this form, whereas the product of two integers of the form $4k + 3$ is of the form $4k + 1$.

30. Show that the square of every odd integer is of the form $8k + 1$.

31. Show that the fourth power of every odd integer is of the form $16k + 1$.

32. Show that the product of two integers of the form $6k + 5$ is of the form $6k + 1$.

33. Show that the product of any three consecutive integers is divisible by 6.

34. Use mathematical induction to show that $n^5 - n$ is divisible by 5 for every positive integer n.

35. Use mathematical induction to show that the sum of the cubes of three consecutive integers is divisible by 9.

In Exercises 36–40, let f_n denote the nth Fibonacci number.

36. Show that f_n is even if and only if n is divisible by 3.

37. Show that f_n is divisible by 3 if and only if n is divisible by 4.

38. Show that f_n is divisible by 4 if and only if n is divisible by 6.

39. Show that $f_n = 5f_{n-4} + 3f_{n-5}$ whenever n is a positive integer with $n > 5$. Use this result to show that f_n is divisible by 5 whenever n is divisible by 5.

* **40.** Show that $f_{n+m} = f_m f_{n+1} + f_{m-1} f_n$ whenever m and n are positive integers with $m > 1$. Use this result to show that $f_n \mid f_m$ when m and n are positive integers with $n \mid m$.

Let n be a positive integer. We define

$$T(n) = \begin{cases} n/2 & \text{if } n \text{ is even;} \\ (3n+1)/2 & \text{if } n \text{ is odd.} \end{cases}$$

We then form the sequence obtained by iterating T: n, $T(n)$, $T(T(n))$, $T(T(T(n)))$, ... For instance, starting with $n = 7$, we have $7, 11, 17, 26, 13, 20, 10, 5, 8, 4, 2, 1, 2, 1, 2, 1, \ldots$. A well-known conjecture, sometimes called the *Collatz conjecture*, asserts that the sequence obtained by iterating T always reaches the integer 1 no matter which positive integer n begins the sequence.

41. Find the sequence obtained by iterating T starting with $n = 39$.

42. Show that the sequence obtained by iterating T starting with $n = (2^{2k} - 1)/3$, where k is a positive integer greater than 1, always reaches the integer 1.

43. Show that the Collatz conjecture is true if it can be shown that for every positive integer n with $n \geq 2$ there is a term in the sequence obtained by iterating T that is less than n.

44. Verify that there is a term in the sequence obtained by iterating T, starting with the positive integer n, that is less than n for all positive integers n with $2 \leq n \leq 100$. (*Hint:* Begin by considering sets of positive integers for which it is easy to show that this is true.)

* **45.** Show that $[(2 + \sqrt{3})^n]$ is odd whenever n is a nonnegative integer.

1.5 Computational and Programming Exercises

Computations and Explorations

Using a computation program such as Maple or *Mathematica,* or programs you have written, carry out the following computations and explorations.

1. Verify the Collatz conjecture described in the preamble to Exercise 41 for all integers n not exceeding 10,000.

2. Using numerical evidence, what sort of conjectures can you make concerning the number of iterations needed before the sequence of iterations $T(n)$ reaches 1, where n is a given positive integer?

3. Using numerical evidence, make conjectures about the divisibility of Fibonacci numbers by 7, by 8, by 9, by 11, by 13, and so on.

Programming Projects

Write programs using Maple, *Mathematica,* or a language of your choice to do the following.

1. Decide whether an integer is divisible by a given integer.

2. Find the quotient and remainder in the division algorithm.

3. Find the quotient, remainder, and sign in the modified division algorithm given in Exercise 18.

4. Compute the terms of the sequence n, $T(n)$, $T(T(n))$, $T(T(T(n)))$, $\ldots$ for a given positive integer n, as defined in the preamble to Exercise 41.

2

Integer Representations and Operations

Introduction

The way in which integers are represented has a major impact on how easily people and computers can do arithmetic with these integers. The purpose of this chapter is to explain how integers are represented using base b expansions, and how basic arithmetic operations can be carried out using these expansions. In particular, we will show that when b is a positive integer, every positive integer has a unique base b expansion. For example, when b is 10, we have the decimal expansion of an integer; when b is 2, we have the binary expansion of this integer; and when b is 16, we have the hexadecimal expansion. We will describe a procedure for finding the base b expansion of an integer, and describe the basic algorithms used to carry out integer arithmetic with base b expansions. Finally, after introducing big-O notation, we will analyze the computational complexity of these basic operations in terms of big-O estimates of the number of bit operations that they use.

2.1 Representations of Integers

In daily life, we use decimal notation to represent integers. We write out numbers using digits to represent powers of ten. For instance, when we write out the integer 37465, we mean

$$3 \cdot 10^4 + 7 \cdot 10^3 + 4 \cdot 10^2 + 6 \cdot 10 + 5.$$

Decimal notation is an example of a *positional number system,* in which the position a digit occupies in a representation determines the quantity it represents. Throughout ancient and modern history, many other notations for integers have been used. For example, Babylonian mathematicians who lived more than 3000 years ago expressed integers using sixty as a base. The Romans employed Roman numerals, which are used even today to represent years. The ancient Mayans used a positional notation with twenty as a base. Many other systems of integer notation have been invented and used over time.

There is no special reason for using ten as the base in a fixed positional number system, other than that we have ten fingers. As we will see, any positive integer can be used as a base. With the invention and proliferation of computers, bases other than ten have become increasingly important. In particular, base 2, base 8, and base 16 representations of integers are used extensively by computers for various purposes.

In this section, we will demonstrate that no matter which positive integer b is chosen as a base, every positive integer can be expressed uniquely in base b notation. In Section 2.2, we will show how these expansions can be used to do arithmetic with integers. (See the exercise set at the end of this section to learn about one's and two's complement notations, which are used by computers to represent both positive and negative integers.)

For more information about the fascinating history of positional number systems, the reader is referred to [Or88] or [Kn97], where extensive surveys and numerous references may be found.

We now show that every positive integer greater than 1 may be used as a base.

Theorem 2.1.   Let b be a positive integer with $b > 1$. Then every positive integer n can be written uniquely in the form

$$n = a_k b^k + a_{k-1} b^{k-1} + \cdots + a_1 b + a_0,$$

where k is a nonnegative integer, a_j is an integer with $0 \le a_j \le b - 1$ for $j = 0, 1, \ldots, k$, and the initial coefficient $a_k \ne 0$.

Proof.   We obtain an expression of the desired type by successively applying the division algorithm in the following way. We first divide n by b to obtain

$$n = bq_0 + a_0, \quad 0 \le a_0 \le b - 1.$$

If $q_0 \ne 0$, we continue by dividing q_0 by b to find that

$$q_0 = bq_1 + a_1, \quad 0 \le a_1 \le b - 1.$$

We continue this process to obtain

$$q_1 = bq_2 + a_2, \quad 0 \le a_2 \le b - 1,$$
$$q_2 = bq_3 + a_3, \quad 0 \le a_3 \le b - 1,$$

$$\vdots$$

$$q_{k-2} = bq_{k-1} + a_{k-1}, \quad 0 \le a_{k-1} \le b - 1,$$
$$q_{k-1} = b \cdot 0 + a_k, \quad 0 \le a_k \le b - 1.$$

The last step of the process occurs when a quotient of 0 is obtained. To see this, first note that the sequence of quotients satisfies

$$n > q_0 > q_1 > q_2 > \cdots \ge 0.$$

Because the sequence $q_0, q_1, q_2, \ldots$ is a decreasing sequence of nonnegative integers that continues as long as its terms are positive, there are at most q_0 terms in this sequence, and the last term equals 0.

From the first equation above, we find that

$$n = bq_0 + a_0.$$

We next replace q_0 using the second equation, to obtain

$$n = b(bq_1 + a_1) + a_0 = b^2 q_1 + a_1 b + a_0.$$

Successively substituting for $q_1, q_2, \ldots, q_{k-1}$, we have

$$n = b^3 q_2 + a_2 b^2 + a_1 b + a_0,$$

$$\vdots$$

$$= b^{k-1} q_{k-2} + a_{k-2} b^{k-2} + \cdots + a_1 b + a_0,$$
$$= b^k q_{k-1} + a_{k-1} b^{k-1} + \cdots + a_1 b + a_0$$
$$= a_k b^k + a_{k-1} b^{k-1} + \cdots + a_1 b + a_0,$$

where $0 \le a_j \le b - 1$ for $j = 0, 1, \ldots, k$ and $a_k \ne 0$, given that $a_k = q_{k-1}$ is the last nonzero quotient. Consequently, we have found an expansion of the desired type.

To see that the expansion is unique, assume that we have two such expansions equal to n, that is,

$$n = a_k b^k + a_{k-1} b^{k-1} + \cdots + a_1 b + a_0$$
$$= c_k b^k + c_{k-1} b^{k-1} + \cdots + c_1 b + c_0,$$

where $0 \le a_k < b$ and $0 \le c_k < b$ (and where, if necessary, we have added initial terms with zero coefficients to one of the expansions to have the number of terms agree). Subtracting one expansion from the other, we have

$$(a_k - c_k)b^k + (a_{k-1} - c_{k-1})b^{k-1} + \cdots + (a_1 - c_1)b + (a_0 - c_0) = 0.$$

If the two expansions are different, there is a smallest integer j, $0 \le j \le k$, such that $a_j \ne c_j$. Hence,

$$b^j \left((a_k - c_k)b^{k-j} + \cdots + (a_{j+1} - c_{j+1})b + (a_j - c_j) \right) = 0,$$

so that

$$(a_k - c_k)b^{k-j} + \cdots + (a_{j+1} - c_{j+1})b + (a_j - c_j) = 0.$$

Solving for $a_j - c_j$, we obtain

$$a_j - c_j = (c_k - a_k)b^{k-j} + \cdots + (c_{j+1} - a_{j+1})b$$
$$= b \left((c_k - a_k)b^{k-j-1} + \cdots + (c_{j+1} - a_{j+1}) \right).$$

Hence, we see that

$$b \mid (a_j - c_j).$$

But because $0 \le a_j < b$ and $0 \le c_j < b$, we know that $-b < a_j - c_j < b$. Consequently, $b \mid (a_j - c_j)$ implies that $a_j = c_j$. This contradicts the assumption that the two expansions are different. We conclude that our base b expansion of n is unique. ∎

For $b = 2$, we see by Theorem 2.1 that the following corollary holds.

Corollary 2.1.1. Every positive integer may be represented as the sum of distinct powers of 2. ∎

Proof. Let n be a positive integer. From Theorem 2.1 with $b = 2$, we know that $n = a_k 2^k + a_{k-1} 2^{k-1} + \cdots + a_1 2 + a_0$, where each a_j is either 0 or 1. Hence, every positive integer is the sum of distinct powers of 2. ∎

In the expansions described in Theorem 2.1, b is called the *base* or *radix* of the expansion. We call base 10 notation, our conventional way of writing integers, *decimal* notation. Base 2 expansions are called *binary* expansions, base 8 expansions are called *octal* expansions, and base 16 expansions are called *hexadecimal*, or *hex* for short. The coefficients a_j are called the *digits* of the expansion. Binary digits are called *bits* (*bi*nary dig*its*) in computer terminology.

To distinguish representations of integers with different bases, we use a special notation. We write $(a_k a_{k-1} \ldots a_1 a_0)_b$ to represent the number $a_k b^k + a_{k-1} b^{k-1} + \cdots + a_1 b + a_0$.

Example 2.1. To illustrate base b notation, note that $(236)_7 = 2 \cdot 7^2 + 3 \cdot 7 + 6 = 125$ and $(10010011)_2 = 1 \cdot 2^7 + 1 \cdot 2^4 + 1 \cdot 2^1 + 1 = 147$. ◀

The proof of Theorem 2.1 provides a method of finding the base b expansion $(a_k a_{k-1} \ldots a_1 a_0)_b$ of any positive integer n. Specifically, to find the base b expansion of n, we first divide n by b. The remainder is the digit a_0. Then, we divide the quotient $[n/b] = q_0$ by b. The remainder is the digit a_1. We continue this process, successively dividing the quotient obtained by b, to obtain the digits in the base b expansion of n. The process stops once a quotient of 0 is obtained. In other words, to find the base b expansion of n, we perform the division algorithm repeatedly, replacing the dividend each time with the quotient, and stop when we come to a quotient that is 0. We then read up the list of remainders to find the base b expansion. We illustrate this procedure in Example 2.2.

Example 2.2. To find the base 2 expansion of 1864, we use the division algorithm successively:

$$1864 = 2 \cdot 932 + 0,$$
$$932 = 2 \cdot 466 + 0,$$
$$466 = 2 \cdot 233 + 0,$$
$$233 = 2 \cdot 116 + 1,$$
$$116 = 2 \cdot 58 + 0,$$
$$58 = 2 \cdot 29 + 0,$$
$$29 = 2 \cdot 14 + 1,$$
$$14 = 2 \cdot 7 + 0,$$
$$7 = 2 \cdot 3 + 1,$$
$$3 = 2 \cdot 1 + 1,$$
$$1 = 2 \cdot 0 + 1.$$

To obtain the base 2 expansion of 1864, we simply take the remainders of these divisions. This shows that $(1864)_{10} = (11101001000)_2$. ◀

Computers represent numbers internally by using a series of "switches" that may be either "on" or "off." (This may be done mechanically, using magnetic tape, electrical switches, or by other means.) Hence, we have two possible states for each switch. We can use "on" to represent the digit 1 and "off" to represent the digit 0; this is why computers use binary expansions to represent integers internally.

Computers use base 8 or base 16 for display purposes. In base 16 (hexadecimal) notation there are 16 digits, usually denoted by 0, 1, 2, 3, 4, 5, 6, 7, 8, 9, A, B, C, D, E, F. The letters A, B, C, D, E, and F are used to represent the digits that correspond to 10, 11, 12, 13, 14, and 15 (written in decimal notation). The following example demonstrates the conversion from hexadecimal to decimal notation.

Example 2.3. To convert $(A35B0F)_{16}$ from hexadecimal to decimal notation, we write

$$(A35B0F)_{16} = 10 \cdot 16^5 + 3 \cdot 16^4 + 5 \cdot 16^3 + 11 \cdot 16^2 + 0 \cdot 16 + 15$$
$$= (10705679)_{10}.$$ ◀

A simple conversion is possible between binary and hexadecimal notation. We can write each hex digit as a block of four binary digits according to the correspondences given in Table 2.1.

Example 2.4. An example of conversion from hex to binary is $(2FB3)_{16} = (10111110110011)_2$. Each hex digit is converted to a block of four binary digits (the initial zeros in the initial block $(0010)_2$ corresponding to the digit $(2)_{16}$ are omitted).

To convert from binary to hex, consider $(11110111101001)_2$. We break this into blocks of four, starting from the right. The blocks are, from right to left, 1001, 1110, 1101, and 0011 (with two initial zeros added). Translating each block to hex, we obtain $(3DE9)_{16}$. ◀

Hex Digit	Binary Digits	Hex Digit	Binary Digits
0	0000	8	1000
1	0001	9	1001
2	0010	A	1010
3	0011	B	1011
4	0100	C	1100
5	0101	D	1101
6	0110	E	1110
7	0111	F	1111

Table 2.1 *Conversion from hex digits to blocks of binary digits.*

We note that a conversion between two different bases is as easy as binary–hex conversion whenever one of the bases is a power of the other.

2.1 Exercises

1. Convert $(1999)_{10}$ from decimal to base 7 notation. Convert $(6105)_7$ from base 7 to decimal notation.

2. Convert $(89156)_{10}$ from decimal to base 8 notation. Convert $(706113)_8$ from base 8 to decimal notation.

3. Convert $(10101111)_2$ from binary to decimal notation and $(999)_{10}$ from decimal to binary notation.

4. Convert $(101001000)_2$ from binary to decimal notation and $(1984)_{10}$ from decimal to binary notation.

5. Convert $(100011110101)_2$ and $(11101001110)_2$ from binary to hexadecimal.

6. Convert $(ABCDEF)_{16}$, $(DEFACED)_{16}$, and $(9A0B)_{16}$ from hexadecimal to binary.

7. Explain why we really are using base 1000 notation when we break large decimal integers into blocks of three digits, separated by commas.

8. Show that if b is a negative integer less than -1, then every nonzero integer n can be uniquely written in the form
$$n = a_k b^k + a_{k-1} b^{k-1} + \cdots + a_1 b + a_0,$$
where $a_k \neq 0$ and $0 \leq a_j < |b|$ for $j = 0, 1, 2, \ldots, k$. We write $n = (a_k a_{k-1} \ldots a_1 a_0)_b$, just as we do for positive bases.

9. Find the decimal representation of $(101001)_{-2}$ and $(12012)_{-3}$.

10. Find the base -2 representations of the decimal numbers -7, -17, and 61.

11. Show that any weight not exceeding $2^k - 1$ may be measured using weights of $1, 2, 2^2, \ldots, 2^{k-1}$, when all the weights are placed in one pan.

12. Show that every nonzero integer can be uniquely represented in the form

$$e_k 3^k + e_{k-1} 3^{k-1} + \cdots + e_1 3 + e_0,$$

where $e_j = -1, 0$, or 1 for $j = 0, 1, 2, \ldots, k$ and $e_k \neq 0$. This expansion is called a *balanced ternary expansion*.

13. Use Exercise 12 to show that any weight not exceeding $(3^k - 1)/2$ may be measured using weights of $1, 3, 3^2, \ldots, 3^{k-1}$, when the weights may be placed in either pan.

14. Explain how to convert from base 3 to base 9 notation, and from base 9 to base 3 notation.

15. Explain how to convert from base r to base r^n notation, and from base r^n to base r notation, when $r > 1$ and n are positive integers.

16. Show that if $n = (a_k a_{k-1} \ldots a_1 a_0)_b$, then the quotient and remainder when n is divided by b^j are $q = (a_k a_{k-1} \ldots a_j)_b$ and $r = (a_{j-1} \ldots a_1 a_0)_b$, respectively.

17. If the base b expansion of n is $n = (a_k a_{k-1} \ldots a_1 a_0)_b$, what is the base b expansion of $b^m n$?

One's complement representations of integers are used to simplify computer arithmetic. To represent positive and negative integers with absolute value less than 2^n, a total of $n + 1$ bits is used.

The leftmost bit is used to represent the sign. A 0 in this position is used for positive integers and a 1 in this position is used for negative integers.

For positive integers, the remaining bits are identical to the binary expansion of the integer. For negative integers, the remaining bits are obtained by first finding the binary expansion of the absolute value of the integer, and then taking the complement of each of these bits, where the complement of a 1 is a 0 and the complement of a 0 is a 1.

18. Find the one's complement representations, using bit strings of length six, of the following integers.

a) 22   c) −7
b) 31   d) −19

19. What integer does each of the following one's complement representations of length five represent?

a) 11001   c) 10001
b) 01101   d) 11111

20. How is the one's complement representation of $-m$ obtained from the one's complement of m, when bit strings of length n are used?

21. Show that if m is an integer with one's complement representation $a_{n-1} a_{n-2} \ldots a_1 a_0$, then $m = -a_{n-1}(2^{n-1} - 1) + \sum_{i=0}^{n-2} a_i 2^i$.

Two's complement representations of integers also are used to simplify computer arithmetic (in fact, they are used much more commonly than one's complement representations). To represent an integer x with $-2^{n-1} \leq x \leq 2^{n-1} - 1$, n bits are used.

The leftmost bit represents the sign, with a 0 used for positive integers and a 1 for negative integers.

For a positive integer, the remaining $n - 1$ bits are identical to the binary expansion of the integer. For a negative integer, the remaining bits are the bits of the binary expansion of $2^{n-1} - |x|$.

22. Find the two's complement representations, using bit strings of length six, of the integers in Exercise 18.

23. What integers do the representations in Exercise 19 represent if each is the two's complement representation of an integer?

24. Show that if m is an integer with two's complement representation $a_{n-1}a_{n-2} \cdots a_1 a_0$, then $m = -a_{n-1} \cdot 2^{n-1} + \sum_{i=0}^{n-2} a_i 2^i$.

25. How is the two's complement representation of $-m$ obtained from the two's complement representation of m, when bit strings of length n are used?

26. How can the two's complement representation of an integer be found from its one's complement representation?

27. Sometimes integers are encoded by using four-digit binary expansions to represent each decimal digit. This produces the *binary coded decimal* form of the integer. For instance, 791 is encoded in this way by 011110010001. How many bits are required to represent a number with n decimal digits using this type of encoding?

A *Cantor expansion* of a positive integer n is a sum

$$n = a_m m! + a_{m-1}(m-1)! + \cdots + a_2 2! + a_1 1!,$$

where each a_j is an integer with $0 \le a_j \le j$ and $a_m \ne 0$.

28. Find Cantor expansions of $14, 56,$ and 384.

∗ **29.** Show that every positive integer has a unique Cantor expansion. (*Hint:* For each positive integer n there is a positive integer m such that $m! \le n < (m+1)!$. For a_m, take the quotient from the division algorithm when n is divided by $m!$, then iterate.)

The Chinese game of *nim* is played as follows. There are several piles of matches, each containing an arbitrary number of matches at the start of the game. To make a move a player removes one or more matches from one of the piles. The players take turns, and the player who removes the last match wins the game.

A *winning position* is an arrangement of matches in piles such that if a player can move to this position, then (no matter what the second player does) the first player can continue to play in a way that will win the game. An example is the position where there are two piles, each containing one match; this is a winning position, because the second player must remove a match, leaving the first player the opportunity to win by removing the last match.

30. Show that the position in nim where there are two piles, each with two matches, is a winning position.

31. For each arrangement of matches into piles, write the number of matches in each pile in binary notation, and then line up the digits of these numbers into columns (adding initial zeros where necessary). Show that a position is a winning one if and only if the number of 1s in each column is even. (For example: Three piles of 3, 4, and 7 give

$$\begin{array}{ccc} 0 & 1 & 1 \\ 1 & 0 & 0 \\ 1 & 1 & 1 \end{array}$$

where each column has exactly two 1s.) (*Hint:* Show that any move from a winning position produces a nonwinning one. Show that there is a move from any nonwinning position to a winning one.)

Let a be an integer with a four-digit decimal expansion, where not all digits are the same. Let a' be the integer with a decimal expansion obtained by writing the digits of a in descending order, and let a'' be the integer with a decimal expansion obtained by writing the digits of a in ascending order. Define $T(a) = a' - a''$. For instance, $T(7318) = 8731 - 1378 = 7353$.

* **32.** Show that the only integer with a four-digit decimal expansion (where not all digits are the same) such that $T(a) = a$ is $a = 6174$. The integer 6174 is called *Kaprekar's constant,* after the Indian mathematician *D. R. Kaprekar,* because it is the only integer with this property.

** **33.** a) Show that if a is a positive integer with a four-digit decimal expansion where not all digits are the same, then the sequence $a, T(a), T(T(a)), T(T(T(a))), \ldots$, obtained by iterating T, eventually reaches the integer 6174.

b) Determine the maximum number of steps required for the sequence defined in part (a) to reach 6174.

Let b be a positive integer and let a be an integer with a four-digit base b expansion, with not all digits the same. Define $T_b(a) = a' - a''$, where a' is the integer with base b expansion obtained by writing the base b digits of a in descending order, and a'' is the integer with base b expansion obtained by writing the base b digits of a in ascending order.

** **34.** Let $b = 5$. Find the unique integer a_0 with a four-digit base 5 expansion such that $T_5(a_0) = a_0$. Show that this integer a_0 is a Kaprekar constant for base 5; in other words, that $a, T(a), T(T(a)), T(T(T(a))), \ldots$ eventually reaches a_0, whenever a is an integer with a four-digit base 5 expansion where not all digits are the same.

* **35.** Show that no Kaprekar constant exists for four-digit numbers to the base 6.

* **36.** Determine whether there is a Kaprekar constant for three-digit integers to the base 10. Prove that your answer is correct.

D. R. KAPREKAR (1905–1986) was born in Dahanu, India, and was interested in numbers even as a small child. He received his secondary school education in Thana and studied at Ferguson College in Poona. Kaprekar attended the University of Bombay, receiving his bachelor's degree in 1929. From 1930 until his retirement in 1962, he worked as a schoolteacher in Devlali, India. Kaprekar discovered many interesting properties in recreational number theory. He published extensively, writing about such topics as recurring decimals, magic squares, and integers with special properties.

2.1 Computational and Programming Exercises

Computations and Explorations

Using a computation program such as Maple or *Mathematica,* or programs you have written, carry out the following computations and explorations.

1. Find the binary, octal, and hexadecimal expansions of each of the following integers.

   a) 9876543210   b) 1111111111   c) 10000000001

2. Find the decimal expansion of each of the following integers.

   a) $(1010101010101)_2$ b) $(765432101234567)_8$ c) $(ABBAFADACABA)_{16}$

3. Evaluate each of the following sums, expressing your answer in the same base used to represent the summands.
   a) $(1101101101101101)_2 + (1001001001001001001001)_2$
   b) $(12345670123456)_8 + (765432107654321)_8$
   c) $(123456789ABCD)_{16} + (BABACACADADA)_{16}$

4. Find the Cantor expansions of the integers 100,000, 10,000,000, and 1,000,000,000. (See the preamble to Exercise 28 for the definition of Cantor expansions.)

5. Verify the result described in Exercise 33 for several different four-digit integers, in which not all digits are the same.

6. Use numerical evidence to make conjectures about the behavior of the sequence $a, T(a), T(T(a)), \ldots$ where a is a five-digit integer in base 10 notation in which not all digits are the same, and $T(a)$ is defined as in the preamble to Exercise 32.

7. Explore the behavior for different bases b of the sequence $a, T(a), T(T(a)), \ldots$ where a is a three-digit integer in base b notation. What conjectures can you make? Repeat your exploration using four-digit and then five-digit integers in base b notation.

Programming Projects

Write programs using Maple, *Mathematica,* or a language of your choice to do the following.

1. Find the binary expansion of an integer from the decimal expansion of this integer, and vice versa.

2. Convert from base b_1 notation to base b_2 notation, where b_1 and b_2 are arbitrary positive integers greater than 1.

3. Convert from binary notation to hexadecimal notation, and vice versa.

4. Find the base (-2) notation of an integer from its decimal notation (see Exercise 8).

5. Find the balanced ternary expansion of an integer from its decimal expansion (see Exercise 12).

6. Find the Cantor expansion of an integer from its decimal expansion (see the preamble to Exercise 28).

7. Play a winning strategy in the game of nim (see the preamble to Exercise 30).

* 8. Investigate the sequence $a, T(a), T(T(a)), T(T(T(a))), \ldots$ (defined in the preamble to Exercise 32), where a is a positive integer, to discover the minimum number of iterations required to reach 6174.

2.2 Computer Operations with Integers

Before computers were invented, mathematicians did computations either by hand or by using mechanical devices. Either way, they were only able to work with integers of rather limited size. Many number theoretic problems, such as factoring and primality testing, require computations with integers of as many as 100 or even 200 digits. In this section, we will study some of the basic algorithms for doing computer arithmetic. In the following section, we will study the number of basic computer operations required to carry out these algorithms.

We have mentioned that computers internally represent numbers using bits, or binary digits. Computers have a built-in limit on the size of integers that can be used in machine arithmetic. This upper limit is called the *word size,* which we denote by w. The word size is usually a power of 2, such as 2^{32} for Pentium machines or 2^{35}, although sometimes the word size is a power of 10.

To do arithmetic with integers larger than the word size, it is necessary to devote more than one word to each integer. To store an integer $n > w$, we express n in base w notation, and for each digit of this expansion we use one computer word. For instance, if the word size is 2^{35}, using ten computer words we can store integers as large as $2^{350} - 1$, since integers less than 2^{350} have no more than ten digits in their base 2^{35} expansions. Also note that to find the base 2^{35} expansion of an integer, we need only group together blocks of 35 bits.

The first step in discussing computer arithmetic with large integers is to describe how the basic arithmetic operations are methodically performed.

We will describe the classical methods for performing the basic arithmetic operations with integers in base r notation, where $r > 1$ is an integer. These methods are examples of *algorithms*.

Definition.   An *algorithm* is a finite set of precise instructions for performing a computation or for solving a problem.

We will describe algorithms for performing addition, subtraction, and multiplication of two n-digit integers $a = (a_{n-1}a_{n-2} \ldots a_1a_0)_r$ and $b = (b_{n-1}b_{n-2} \ldots b_1b_0)_r$, where initial digits of zero are added if necessary to make both expansions the same length. The algorithms described are used for both binary arithmetic with integers less than the word size of a computer, and *multiple precision* arithmetic with integers larger than the word size w, using w as the base.

Addition   When we add a and b, we obtain the sum

$$a + b = \sum_{j=0}^{n-1} a_j r^j + \sum_{j=0}^{n-1} b_j r^j = \sum_{j=0}^{n-1} (a_j + b_j) r^j.$$

To find the base r expansion of $a + b$, first note that by the division algorithm, there are integers C_0 and s_0 such that

$$a_0 + b_0 = C_0 r + s_0, \quad 0 \leq s_0 < r.$$

Because a_0 and b_0 are positive integers not exceeding r, we know that $0 \le a_0 + b_0 \le 2r - 2$, so that $C_0 = 0$ or 1; here, C_0 is the *carry* to the next place. Next, we find that there are integers C_1 and s_1 such that

$$a_1 + b_1 + C_0 = C_1 r + s_1, \quad 0 \le s_1 < r.$$

Since $0 \le a_1 + b_1 + C_0 \le 2r - 1$, we know that $C_1 = 0$ or 1. Proceeding inductively, we find integers C_i and s_i for $1 \le i \le n - 1$ by

$$a_i + b_i + C_{i-1} = C_i r + s_i, \quad 0 \le s_i < r,$$

with $C_i = 0$ or 1. Finally, we let $s_n = C_{n-1}$, since the sum of two integers with n digits has $n + 1$ digits when there is a carry in the nth place. We conclude that the base r expansion for the sum is $a + b = (s_n s_{n-1} \ldots s_1 s_0)_r$.

When performing base r addition by hand, we can use the same familiar technique as is used in decimal addition.

Example 2.5.   To add $(1101)_2$ and $(1001)_2$, we write

$$
\begin{array}{ccccc}
\mathit{1} & & & \mathit{1} & \\
& 1 & 1 & 0 & 1 \\
+ & 1 & 0 & 0 & 1 \\
\hline
1 & 0 & 1 & 1 & 0 \\
\end{array}
$$

where we have indicated carries by 1s in italics written above the appropriate column. We found the binary digits of the sum by noting that $1 + 1 = 1 \cdot 2 + 0, 0 + 0 + 1 = 0 \cdot 2 + 1, 1 + 0 + 0 = 0 \cdot 2 + 1$, and $1 + 1 + 0 = 1 \cdot 2 + 0$. ◀

Subtraction   Assume that $a > b$. Consider

$$a - b = \sum_{j=0}^{n-1} a_j r^j - \sum_{j=0}^{n-1} b_j r^j = \sum_{j=0}^{n-1} (a_j - b_j) r^j.$$

Note that by the division algorithm, there are integers B_0 and d_0 such that

$$a_0 - b_0 = B_0 r + d_0, \quad 0 \le d_0 < r,$$

Where the Word "Algorithm" Comes from

"Algorithm" is a corruption of the original term "algorism," which originally comes from the name of the author of the ninth-century book *Kitab al-jabr w'al-muqabala (Rules of Restoration and Reduction)*, *Abu Ja'far Mohammed ibn Mûsâ al-Khwârizmî* (see his biography included on the next page). The word "algorism" originally referred only to the rules of performing arithmetic using Hindu-Arabic numerals, but evolved into "algorithm" by the eighteenth century. With growing interest in computing machines, the concept of an algorithm became more general, to include all definite procedures for solving problems, not just the procedures for performing arithmetic with integers expressed in Arabic notation.

and because a_0 and b_0 are positive integers less than r, we have

$$-(r-1) \leq a_0 - b_0 \leq r - 1.$$

When $a_0 - b_0 \geq 0$, we have $B_0 = 0$. Otherwise, when $a_0 - b_0 < 0$, we have $B_0 = -1$; B_0 is the *borrow* from the next place of the base r expansion of a. We use the division algorithm again to find integers B_1 and d_1 such that

$$a_1 - b_1 + B_0 = B_1 r + d_1, \quad 0 \leq d_1 < r.$$

From this equation, we see that the borrow $B_1 = 0$ as long as $a_1 - b_1 + B_0 \geq 0$, and that $B_1 = -1$ otherwise, because $-r \leq a_1 - b_1 + B_0 \leq r - 1$. We proceed inductively to find integers B_i and d_i, such that

$$a_i - b_i + B_{i-1} = B_i r + d_i, \quad 0 \leq d_i < r$$

with $B_i = 0$ or -1, for $1 \leq i \leq n - 1$. We see that $B_{n-1} = 0$, because $a > b$. We can conclude that

$$a - b = (d_{n-1} d_{n-2} \ldots d_1 d_0)_r.$$

When performing base r subtraction by hand, we use the familiar technique used in decimal subtraction.

Example 2.6. To subtract $(10110)_2$ from $(11011)_2$, we have

$$
\begin{array}{rccccc}
& & \mathit{-1} & & & \\
& 1 & 1 & 0 & 1 & 1 \\
- & 1 & 0 & 1 & 1 & 0 \\
\hline
& & & 1 & 0 & 1
\end{array}
$$

where the -1 in italics above a column indicates a borrow. We found the binary digits of the difference by noting that $1 - 0 = 0 \cdot 2 + 1$, $1 - 1 + 0 = 0 \cdot 2 + 0$, $0 - 1 + 0 = -1 \cdot 2 + 1$, $1 - 0 - 1 = 0 \cdot 2 + 0$, and $1 - 1 + 0 = 0 \cdot 2 + 0$. ◄

ABU JA'FAR MOHAMMED IBN MÛSÂ AL-KHWÂRIZMÎ (c. 780–c. 850), an astronomer and mathematician, was a member of the House of Wisdom, an academy of scientists in Baghdad. The name al-Khwârizmî means "from the town of Kowarzizm," now known as Khiva in modern Uzbekistan. Al-Khwârizmî was the author of books on mathematics, astronomy, and geography. People in the West first learned about algebra from his works; the word "algebra" comes from *al-jabr*, part of the title of his book *Kitab al-jabr w'al muqabala*, which was translated into Latin and widely used as a text. Another book describes procedures for arithmetic operations using Hindu-Arabic numerals.

Multiplication   Before discussing multiplication, we describe *shifting*. To multiply $(a_{n-1} \ldots a_1 a_0)_r$ by r^m, we need only shift the expansion left m places, appending the expansion with m zero digits.

Example 2.7.   To multiply $(101101)_2$ by 2^5, we shift the digits to the left five places and append the expansion with five zeros, obtaining $(10110100000)_2$.   ◄

We first discuss the multiplication of an n-place integer by a one-digit integer. To multiply $(a_{n-1} \ldots a_1 a_0)_r$ by $(b)_r$, we first note that

$$a_0 b = q_0 r + p_0, \quad 0 \le p_0 < r,$$

and $0 \le q_0 \le r - 2$, because $0 \le a_0 b \le (r-1)^2$. Next, we have

$$a_1 b + q_0 = q_1 r + p_1, \quad 0 \le p_1 < r,$$

and $0 \le q_1 \le r - 1$. In general, we have

$$a_i b + q_{i-1} = q_i r + p_i, \quad 0 \le p_i < r,$$

and $0 \le q_i \le r - 1$. Furthermore, we have $p_n = q_{n-1}$. This yields $(a_{n-1} \ldots a_1 a_0)_r (b)_r = (p_n p_{n-1} \cdots p_1 p_0)_r$.

To perform a multiplication of two n-place integers, we write

$$ab = a \left(\sum_{j=0}^{n-1} b_j r^j \right) = \sum_{j=0}^{n-1} (a b_j) r^j.$$

For each j, we first multiply a by the digit b_j, then shift j places to the left, and finally add all of the n integers we have obtained to find the product.

When multiplying two integers with base r expansions, we use the familiar method of multiplying decimal integers by hand.

Example 2.8.   To multiply $(1101)_2$ and $(1110)_2$, we write

```
            1  1  0  1
         ×  1  1  1  0
         ─────────────
            0  0  0  0
         1  1  0  1
      1  1  0  1
   1  1  0  1
   ──────────────────
   1  0  1  1  0  1  1  0
```

Note that we first multiplied $(1101)_2$ by each digit of $(1110)_2$, shifting each time by the appropriate number of places, and then we added the appropriate integers to find our product.   ◄

Division   We wish to find the quotient q in the division algorithm

$$a = bq + R, \quad 0 \leq R < b.$$

If the base r expansion of q is $q = (q_{n-1}q_{n-2} \cdots q_1q_0)_r$, then we have

$$a = b\left(\sum_{j=0}^{n-1} q_j r^j\right) + R, \quad 0 \leq R < b.$$

To determine the first digit q_{n-1} of q, notice that

$$a - bq_{n-1}r^{n-1} = b\left(\sum_{j=0}^{n-2} q_j r^j\right) + R.$$

The right-hand side of this equation is not only positive, but also less than br^{n-1}, because $\sum_{j=0}^{n-2} q_j r^j \leq \sum_{j=0}^{n-2}(r-1)r^j = \sum_{j=1}^{n-1} r^j - \sum_{j=0}^{n-2} r^j = r^{n-1} - 1$. Therefore, we know that

$$0 \leq a - bq_{n-1}r^{n-1} < br^{n-1}.$$

This tells us that

$$q_{n-1} = \left[\frac{a}{br^{n-1}}\right].$$

We can obtain q_{n-1} by successively subtracting br^{n-1} from a until we obtain a negative result; q_{n-1} is then one less than the number of subtractions.

To find the other digits of q, we define the sequence of *partial remainders* R_i by

$$R_0 = a$$

and

$$R_i = R_{i-1} - bq_{n-i}r^{n-i}$$

for $i = 1, 2, \ldots, n$. By mathematical induction, we show that

(2.1)
$$R_i = \left(\sum_{j=0}^{n-i-1} q_j r^j\right)b + R.$$

For $i = 0$, this is clearly correct, because $R_0 = a = qb + R$. Now, assume that

$$R_k = \left(\sum_{j=0}^{n-k-1} q_j r^j\right)b + R.$$

Then

$$R_{k+1} = R_k - bq_{n-k-1}r^{n-k-1}$$

$$= \left(\sum_{j=0}^{n-k-1} q_j r^j \right) b + R - bq_{n-k-1}r^{n-k-1}$$

$$= \left(\sum_{j=0}^{n-(k+1)-1} q_j r^j \right) b + R,$$

establishing (2.1).

By (2.1) we see that $0 \le R_i < r^{n-i}b$, for $i = 1, 2, \ldots, n$, because $\sum_{j=0}^{n-i-1} q_j r^j \le r_{n-i} - 1$. Consequently, because $R_i = R_{i-1} - bq_{n-i}r^{n-i}$ and $0 \le R_i < r^{n-1}b$, we see that the digit q_{n-i} is given by $[R_{i-1}/(br^{n-i})]$ and can be obtained by successively subtracting br^{n-i} from R_{i-1} until a negative result is obtained, and then q_{n-i} is one less than the number of subtractions. This is how we find the digits of q.

Example 2.9. To divide $(11101)_2$ by $(111)_2$, we let $q = (q_2 q_1 q_0)_2$. We subtract $2^2(111)_2 = (11100)_2$ once from $(11101)_2$ to obtain $(1)_2$, and once more to obtain a negative result, so that $q_2 = 1$. Now, $R_1 = (11101)_2 - (11100)_2 = (1)_2$. We find that $q_1 = 0$, because $R_1 - 2(111)_2$ is less than zero, and likewise $q_0 = 0$. Hence, the quotient of the division is $(100)_2$ and the remainder is $(1)_2$. ◀

2.2 Exercises

1. Add $(101111011)_2$ and $(1100111011)_2$.

2. Add $(10001000111101)_2$ and $(11111101011111)_2$.

3. Subtract $(11010111)_2$ from $(1111000011)_2$.

4. Subtract $(101110101)_2$ from $(1101101100)_2$.

5. Multiply $(11101)_2$ and $(110001)_2$.

6. Multiply $(1110111)_2$ and $(10011011)_2$.

7. Find the quotient and remainder when $(110011111)_2$ is divided by $(1101)_2$.

8. Find the quotient and remainder when $(110100111)_2$ is divided by $(11101)_2$.

9. Add $(1234321)_5$ and $(2030104)_5$.

10. Subtract $(434421)_5$ from $(4434201)_5$.

11. Multiply $(1234)_5$ and $(3002)_5$.

12. Find the quotient and remainder when $(14321)_5$ is divided by $(334)_5$.

13. Add $(ABAB)_{16}$ and $(BABA)_{16}$.

14. Subtract $(CAFE)_{16}$ from $(FEED)_{16}$.

15. Multiply $(FACE)_{16}$ and $(BAD)_{16}$.

16. Find the quotient and remainder when $(BEADED)_{16}$ is divided by $(ABBA)_{16}$.

17. Explain how to add, subtract, and multiply the integers 18235187 and 22135674 on a computer with word size 1000.

18. Write algorithms for the basic operations with integers in base (-2) notation (see Exercise 8 of Section 2.1).

19. How is the one's complement representation of the sum of two integers obtained from the one's complement representations of those integers?

20. How is the one's complement representation of the difference of two integers obtained from the one's complement representations of those integers?

21. Give an algorithm for adding and an algorithm for subtracting Cantor expansions (see the preamble to Exercise 28 of Section 2.1).

22. A *dozen* equals 12, and a *gross* equals 12^2. Using base 12, or *duodecimal* arithmetic, answer the following questions.
    a) If 3 gross, 7 dozen, and 4 eggs are removed from a total of 11 gross and 3 dozen eggs, how many eggs are left?
    b) If 5 truckloads of 2 gross, 3 dozen, and 7 eggs each are delivered to the supermarket, how many eggs are delivered?
    c) If 11 gross, 10 dozen, and 6 eggs are divided in 3 groups of equal size, how many eggs are in each group?

23. A well-known rule used to find the square of an integer with decimal expansion $(a_n a_{n-1} \ldots a_1 a_0)_{10}$ and final digit $a_0 = 5$ is to find the decimal expansion of the product $(a_n a_{n-1} \ldots a_1)_{10}[(a_n a_{n-1} \ldots a_1)_{10} + 1]$, and append this with the digits $(25)_{10}$. For instance, we see that the decimal expansion of $(165)^2$ begins with $16 \cdot 17 = 272$, so that $(165)^2 = 27225$. Show that this rule is valid.

24. In this exercise, we generalize the rule given in Exercise 23 to find the squares of integers with final base $2B$ digit B, where B is a positive integer. Show that the base $2B$ expansion of the integer $(a_n a_{n-1} \ldots a_1 a_0)_{2B}$ starts with the digits of the base $2B$ expansion of the integer $(a_n a_{n-1} \ldots a_1)_{2B} [(a_n a_{n-1} \ldots a_1)_{2B} + 1]$ and ends with the digits $B/2$ and 0 when B is even, and the digits $(B-1)/2$ and B when B is odd.

2.2 Computational and Programming Exercises

Computations and Explorations

Using a computation program such as Maple or *Mathematica,* or programs you have written, carry out the following computations and explorations.

1. Verify the rules given in Exercises 23 and 24 for examples of your choice.

Programming Projects

Write programs using Maple, *Mathematica,* or a language of your choice to do the following.

1. Perform addition with arbitrarily large integers.

2. Perform subtraction with arbitrarily large integers.

3. Multiply two arbitrarily large integers using the conventional algorithm.

4. Divide arbitrarily large integers, finding the quotient and remainder.

2.3 Complexity of Integer Operations

Once an algorithm has been specified for an operation, we can consider the amount of time required to perform this algorithm on a computer. We will measure the amount of time in terms of *bit operations*. By a bit operation we mean the addition, subtraction, or multiplication of two binary digits, the division of a two-bit by a one-bit integer (obtaining a quotient and a remainder), or the shifting of a binary integer one place. (The actual amount of time required to carry out a bit operation on a computer varies depending on the computer architecture and capacity.) When we describe the number of bit operations needed to perform an algorithm, we are describing the *computational complexity* of this algorithm.

In describing the number of bit operations needed to perform calculations, we will use *big-O* notation. Big-O notation provides an upper bound on the size of a function in terms of a particular well-known reference function whose size at large values is easily understood.

To motivate the definition of this notation, consider the following situation. Suppose that to perform a specified operation on an integer n requires at most $n^3 + 8n^2 \log n$ bit operations. Since $8n^2 \log n < 8n^3$ for every positive integer, less than $9n^3$ bit operations are required for this operation for every integer n. Since the number of bit operations required is always less than a constant times n^3, namely $9n^3$, we say that $O(n^3)$ bit operations are needed. In general, we have the following definition.

Definition. If f and g are functions taking positive values, defined for all $x \in S$, where S is a specified set of real numbers, then f is $O(g)$ on S if there is a positive constant K such that $f(x) < Kg(x)$ for all sufficiently large $x \in S$. (Normally, we take S to be the set of positive integers, and we drop all reference to S.)

Big-O notation is used extensively throughout number theory and in the analysis of algorithms. *Paul Bachmann* introduced big-O notation in 1892 ([Ba94]). The big-O notation is sometimes called a Landau symbol, after *Edmund Landau*, who used this notation throughout his work in the estimation of various functions in number theory. The use of big-O notation in the analysis of algorithms was popularized by renowned computer scientist *Donald Knuth*.

We illustrate this concept of big-O notation with several examples.

Example 2.10. We can show on the set of positive integers that $n^4 + 2n^3 + 5$ is $O(n^4)$. To do this, note that $n^4 + 2n^3 + 5 \leq n^4 + 2n^4 + 5n^4 = 8n^4$ for all positive integers. (We take $K = 8$ in the definition.) The reader should also note that n^4 is $O(n^4 + 2n^3 + 5)$. ◀

Example 2.11. We can easily give a big-O estimate for $\sum_{j=1}^{n} j$. Noting that each summand is less than n tells us that $\sum_{j=1}^{n} j \leq \sum_{j=1}^{n} n = n \cdot n = n^2$. Note that we could also derive this estimate easily from the formula $\sum_{j=1}^{n} j = n(n+1)/2$. ◀

We now will give some useful results for working with big-O estimates for combinations of functions.

Theorem 2.2.   If f is $O(g)$ and c is a positive constant, then cf is $O(g)$.

Proof. If f is $O(g)$, then there is a constant K with $f(x) < Kg(x)$ for all x under consideration. Hence $cf(x) < (cK)g(x)$, so cf is $O(g)$.   ■

Theorem 2.3.   If f_1 is $O(g_1)$ and f_2 is $O(g_2)$, then $f_1 + f_2$ is $O(g_1 + g_2)$, and $f_1 f_2$ is $O(g_1 g_2)$.

Proof. If f is $O(g_1)$ and f_2 is $O(g_2)$, then there are constants K_1 and K_2 such that $f_1(x) < K_1 g_1(x)$ and $f_2(x) < K_2 g_2(x)$ for all x under consideration. Hence,

$$f_1(x) + f_2(x) < K_1 g_1(x) + K_2 g_2(x)$$
$$\leq K(g_1(x) + g_2(x)),$$

where K is the maximum of K_1 and K_2. Hence, $f_1 + f_2$ is $O(g_1 + g_2)$.

Also,

$$f_1(x) f_2(x) < K_1 g_1(x) K_2 g_2(x)$$
$$= (K_1 K_2)(g_1(x) g_2(x)),$$

so $f_1 f_2$ is $O(g_1 g_2)$.   ■

PAUL GUSTAV HEINRICH BACHMANN (1837–1920), the son of a pastor, shared his father's pious lifestyle, as well as his love of music. His talent for mathematics was discovered by one of his early teachers. After recovering from tuberculosis, he studied at the University of Berlin and later in Göttingen, where he attended lectures presented by Dirichlet. In 1862, he received his doctorate under the supervision of the number theorist Kummer. Bachmann became a professor at Breslau and later at Münster. After retiring, he continued mathematical research, played the piano, and served as a music critic for newspapers. His writings include a five-volume survey of number theory, a two-volume work on elementary number theory, a book on irrational numbers, and a book on Fermat's last theorem (this theorem is discussed in Chapter 13). Bachmann introduced big-O notation in 1892.

EDMUND LANDAU (1877–1938) was the son of a Berlin gynecologist, and attended high school in Berlin. He received his doctorate in 1899 under the direction of Frobenius. Landau first taught at the University of Berlin and then moved to Göttingen, where he was full professor until the Nazis forced him to stop teaching. His main contributions to mathematics were in the field of analytic number theory; he established several important results concerning the distribution of primes. He authored a three-volume work on number theory and many other books on mathematical analysis and analytic number theory.

Corollary 2.3.1.   If f_1 and f_2 are $O(g)$, then $f_1 + f_2$ is $O(g)$.

Proof.   Theorem 2.3 tells us that $f_1 + f_2$ is $O(2g)$. But if $f_1 + f_2 < K(2g)$, then $f_1 + f_2 < (2K)g$, so $f_1 + f_2$ is $O(g)$.   ∎

The goal in using big-O estimates is to give the best big-O estimate possible while using the simplest reference function possible. Well-known reference functions used in big-O estimates include $1, \log n, n, n \log n, n \log n \log \log n, n^2$, and 2^n, as well as some other important functions. Calculus can be used to show that each function in this list is smaller than the next function in the list, in the sense that the ratio of the function and the next function tends to 0 as n grows without bound. Note that more complicated functions than these occur in big-O estimates, as you will see in later chapters.

We illustrate how to use theorems for working with big-O estimates with the following example.

Example 2.12.   To give a big-O estimate for $(n + 8 \log n)(10n \log n + 17n^2)$, first note that $n + 8 \log n$ is $O(n)$ and $10n \log n + 17n^2$ is $O(n^2)$ (because $\log n$ is $O(n)$ and $n \log n$ is $O(n^2)$) by Theorems 2.2 and 2.3 and Corollary 2.3.1. By Theorem 2.3, we see that $(n + 8 \log n)(10n \log n + 17n^2)$ is $O(n^3)$.   ◄

Using big-O notation, we can see that to add or subtract two n-bit integers takes $O(n)$ bit operations, whereas to multiply two n-bit integers in the conventional way takes $O(n^2)$ bit operations (see Exercises 12 and 13 at the end of this section). Surprisingly,

DONALD KNUTH (b. 1938) grew up in Milwaukee where his father owned a small printing business and taught bookkeeping. He was an excellent student who also applied his intelligence in unconventional ways, such as finding more than 4500 words that could be spelled from the letters in "Ziegler's Giant Bar," winning a television set for his school and candy bars for everyone in his class.

Knuth graduated from Case Institute of Technology in 1960 with B.S. and M.S. degrees in mathematics, by special award of the faculty who considered his work outstanding. At Case he managed the basketball team and applied his mathematical talents by evaluating each player using a formula he developed (receiving coverage on CBS television and in *Newsweek*). Knuth received his doctorate in 1963 from the California Institute of Technology.

Knuth taught at the California Institute of Technology and Stanford University, retiring in 1992 to concentrate on writing. He is especially interested in updating and adding to his famous series, *The Art of Computer Programming*. This series has had a profound influence on the development of computer science. Knuth is the founder of the modern study of computational complexity and has made fundamental contributions to the theory of compilers. Knuth has also invented the widely used TeX and Metafont systems used for mathematical (and general) typography. TeX played an important role in the development of HTML and the Internet. He popularized the big-O notation in his work on the analysis of algorithms.

Knuth has written for a wide range of professional journals in computer science and mathematics. However, his first publication, in 1957, when he was a college freshman, was the "The Potrzebie System of Weights and Measures," a parody of the metric system, which appeared in *MAD Magazine*.

there are faster algorithms for multiplying large integers. To develop one such algorithm, we first consider the multiplication of two $2n$-bit integers, say $a = (a_{2n-1}a_{2n-2}\ldots a_1a_0)_2$ and $b = (b_{2n-1}b_{2n-2}\ldots b_1b_0)_2$. We write

$$a = 2^n A_1 + A_0 \quad b = 2^n B_1 + B_0,$$

where

$$A_1 = (a_{2n-1}a_{2n-2}\ldots a_{n+1}a_n)_2 \quad A_0 = (a_{n-1}a_{n-2}\ldots a_1a_0)_2$$
$$B_1 = (b_{2n-1}b_{2n-2}\ldots b_{n+1}b_n)_2 \quad B_0 = (b_{n-1}b_{n-2}\ldots b_1b_0)_2.$$

We will use the identity

(2.2) $ab = (2^{2n} + 2^n)A_1B_1 + 2^n(A_1 - A_0)(B_0 - B_1) + (2^n + 1)A_0B_0.$

To find the product of a and b using (2.2) requires that we perform three multiplications of n-bit integers (namely, A_1B_1, $(A_1 - A_0)(B_0 - B_1)$, and A_0B_0), as well as a number of additions and shifts. This is illustrated by the following example.

Example 2.13.   We can use (2.2) to multiply $(1101)_2$ and $(1011)_2$. We have $(1101)_2 = 2^2(11)_2 + (01)_2$ and $(1011)_2 = 2^2(10)_2 + (11)_2$. Using (2.2), we find that

$$(1101)_2(1011)_2 = (2^4 + 2^2)(11)_2(10)_2 + 2^2((11)_2 - (01)_2) \cdot ((11)_2 - (10)_2) +$$
$$(2^2 + 1)(01)_2(11)_2$$
$$= (2^4 + 2^2)(110)_2 + 2^2(10)_2(01)_2 + (2^2 + 1)(11)_2$$
$$= (1100000)_2 + (11000)_2 + (1000)_2 + (1100)_2 + (11)_2$$
$$= (10001111)_2. \qquad \blacktriangleleft$$

We will now estimate the number of bit operations required to multiply two n-bit integers by using (2.2) repeatedly. If we let $M(n)$ denote the number of bit operations needed to multiply two n-bit integers, we find from (2.2) that

(2.3) $$M(2n) \leq 3M(n) + Cn,$$

where C is a constant, because each of the three multiplications of n-bit integers takes $M(n)$ bit operations, whereas the number of additions and shifts needed to compute ab via (2.2) does not depend on n, and each of these operations takes $O(n)$ bit operations.

From (2.3), using mathematical induction, we can show that

(2.4) $$M(2^k) \leq c(3^k - 2^k),$$

where c is the maximum of the quantities $M(2)$ and C (the constant in (2.3)). To carry out the induction argument, we first note that with $k = 1$, we have $M(2) \leq c(3^1 - 2^1) = c$, because c is the maximum of $M(2)$ and C.

As the induction hypothesis, we assume that

$$M(2^k) \leq c(3^k - 2^k).$$

Then, using (2.3), we have

$$
\begin{aligned}
M(2^{k+1}) &\leq 3M(2^k) + C2^k \\
&\leq 3c(3^k - 2^k) + C2^k \\
&\leq c3^{k+1} - c \cdot 3 \cdot 2^k + c2^k \\
&\leq c(3^{k+1} - 2^{k+1}).
\end{aligned}
$$

This establishes that (2.4) is valid for all positive integers k.

Using inequality (2.4), we can prove the following theorem.

Theorem 2.4.   Multiplication of two n-bit integers can be performed using $O(n^{\log_2 3})$ bit operations. (*Note:* $\log_2 3$ is approximately 1.585, which is considerably less than the exponent 2 that occurs in the estimate of the number of bit operations needed for the conventional multiplication algorithm.)

Proof.   From (2.4), we have

$$
\begin{aligned}
M(n) = M(2^{\log_2 n}) &\leq M(2^{[\log_2 n]+1}) \\
&\leq c(3^{[\log_2 n]+1} - 2^{[\log_2 n]+1}) \\
&\leq 3c \cdot 3^{[\log_2 n]} \leq 3c \cdot 3^{\log_2 n} = 3cn^{\log_2 3} \quad (\textit{because } 3^{\log_2 n} = n^{\log_2 3}).
\end{aligned}
$$

Hence, $M(n)$ is $O(n^{\log_2 3})$.   ■

We now state, without proof, two pertinent theorems. Proofs may be found in [Kn97] or [Kr79].

Theorem 2.5.   Given a positive number $\epsilon > 0$, there is an algorithm for multiplication of two n-bit integers using $O(n^{1+\epsilon})$ bit operations.

Note that Theorem 2.4 is a special case of Theorem 2.5 with $\epsilon = \log_2 3 - 1$, which is approximately 0.585.

Theorem 2.6.   There is an algorithm to multiply two n-bit integers using $O(n \log_2 n \log_2 \log_2 n)$ bit operations.

Since $\log_2 n$ and $\log_2 \log_2 n$ are much smaller than n^ϵ for large numbers n, Theorem 2.6 is an improvement over Theorem 2.5. Although we know that $M(n)$ is $O(n \log_2 n \log_2 \log_2 n)$, for simplicity we will use the obvious fact that $M(n)$ is $O(n^2)$ in our subsequent discussions.

The conventional algorithm described in Section 2.2 performs a division of a $2n$-bit integer by an n-bit integer with $O(n^2)$ bit operations. However, the number of bit operations needed for integer division can be related to the number of bit operations needed for integer multiplication. We state the following theorem, which is based on an algorithm discussed in [Kn97].

Theorem 2.7. There is an algorithm to find the quotient $q = [a/b]$, when the $2n$-bit integer a is divided by the integer b (having no more than n bits), using $O(M(n))$ bit operations, where $M(n)$ is the number of bit operations needed to multiply two n-bit integers.

2.3 Exercises

1. Determine whether each of the following functions is $O(n)$ on the set of positive integers.

   a) $2n + 7$ d) $\log(n^2 + 1)$
   b) $n^2/3$ e) $\sqrt{n^2 + 1}$
   c) 10 f) $(n^2 + 1)/(n + 1)$

2. Show that $2n^4 + 3n^3 + 17$ is $O(n^4)$ on the set of positive integers.

3. Show that $(n^3 + 4n^2 \log n + 101n^2)(14n \log n + 8n)$ is $O(n^4 \log n)$.

4. Show that $n!$ is $O(n^n)$ on the set of positive integers.

5. Show that $(n! + 1)(n + \log n) + (n^3 + n^n)((\log n)^3 + n + 7)$ is $O(n^{n+1})$.

6. Suppose that m is a positive real number. Show that $\sum_{j=1}^{n} j^m$ is $O(n^{m+1})$.

* 7. Show that $n \log n$ is $O(\log n!)$ on the set of positive integers.

8. Show that if f_1 and f_2 are $O(g_1)$ and $O(g_2)$, respectively, and c_1 and c_2 are constants, then $c_1 f_1 + c_2 f_2$ is $O(g_1 + g_2)$.

9. Show that if f is $O(g)$, then f^k is $O(g^k)$ for all positive integers k.

10. Let r be a positive real number greater than 1. Show that a function f is $O(\log_2 n)$ if and only if f is $O(\log_r n)$. (*Hint:* Recall that $\log_a n / \log_b n = \log_a b$.)

11. Show that the base b expansion of a positive integer n has $[\log_b n] + 1$ digits.

12. Analyzing the conventional algorithms for subtraction and addition, show that these operations require $O(n)$ bit operations with n-bit integers.

13. Show that to multiply an n-bit and an m-bit integer in the conventional manner requires $O(nm)$ bit operations.

14. Estimate the number of bit operations needed to find $1 + 2 + \cdots + n$,
    a) by performing all the additions;
    b) by using the identity $1 + 2 + \cdots + n = n(n + 1)/2$, and multiplying and shifting.

15. Give an estimate for the number of bit operations needed to find each of the following quantities.

    a) $n!$ b) $\binom{n}{k}$

16. Give an estimate of the number of bit operations needed to find the binary expansion of an integer from its decimal expansion.

17. Use identity (2.2) with $n = 2$ to multiply $(1001)_2$ and $(1011)_2$.

18. Use identity (2.2) with $n = 4$, and then with $n = 2$, to multiply $(10010011)_2$ and $(11001001)_2$.

19. a) Show there is an identity analogous to (2.2) for decimal expansions.

b) Using part (a), multiply 73 and 87 performing only three multiplications of one-digit integers, plus shifts and additions.

c) Using part (a), reduce the multiplication of 4216 and 2733 to three multiplications of two-digit integers, plus shifts and additions; then, using part (a) again, reduce each of the multiplications of two-digit integers into three multiplications of one-digit integers, plus shifts and additions. Complete the multiplication using only nine multiplications of one-digit integers, and shifts and additions.

20. If $\mathbf{A}$ and $\mathbf{B}$ are $n \times n$ matrices, with entries a_{ij} and b_{ij} for $1 \leq i \leq n$, $1 \leq j \leq n$, then $\mathbf{AB}$ is the $n \times n$ matrix with entries $c_{ij} = \sum_{k=1}^{n} a_{ik}b_{kj}$. Show that n^3 multiplications of integers are used to find $\mathbf{AB}$ directly from its definition.

21. Show that it is possible to multiply two 2×2 matrices using only seven multiplications of integers, by using the identity

$$
\begin{pmatrix} a_{11} & a_{12} \\ a_{21} & a_{22} \end{pmatrix} \begin{pmatrix} b_{11} & b_{12} \\ b_{21} & b_{22} \end{pmatrix}
$$

$$
= \begin{pmatrix} a_{11}b_{11} + a_{12}b_{21} & \begin{aligned} & x + (a_{21} + a_{22})(b_{12} - b_{11}) \\ & \quad + (a_{11} + a_{12} - a_{21} - a_{22})b_{22} \end{aligned} \\ \begin{aligned} & x + (a_{11} - a_{21})(b_{22} - b_{12}) \\ & \quad - a_{22}(b_{11} - b_{21} - b_{12} + b_{22}) \end{aligned} & \begin{aligned} & x + (a_{11} - a_{21})(b_{22} - b_{12}) \\ & \quad + (a_{21} + a_{22})(b_{12} - b_{11} \end{aligned} \end{pmatrix},
$$

where $x = a_{11}b_{11} - (a_{11} - a_{21} - a_{22})(b_{11} - b_{12} + b_{22})$.

* **22.** Using an inductive argument, and splitting $(2n) \times (2n)$ matrices into four $n \times n$ matrices, use Exercise 21 to show that it is possible to multiply two $2^k \times 2^k$ matrices using only 7^k multiplications, and less than 7^{k+1} additions.

23. Conclude from Exercise 22 that two $n \times n$ matrices can be multiplied using $O(n^{\log_2 7})$ bit operations when all entries of the matrices have less than c bits, where c is a constant.

2.3 Computational and Programming Exercises

Computations and Explorations

Using a computation program such as Maple or *Mathematica,* or programs you have written, carry out the following computations and explorations.

1. Multiply 81,873,569 and 41,458,892 by using identity (2.2) with these eight-digit integers, with the resulting four-digit integers, and with the resulting two-digit integers.

2. Multiply two 8×8 matrices of your choice, by using the identity in Exercise 21 with these matrices and then again for the multiplication of the resulting 4×4 matrices.

Programming Projects

Write programs using Maple, *Mathematica,* or a language of your choice to do the following.

* **1.** Multiply two arbitrarily large integers using identity (2.2).

** **2.** Multiply two $n \times n$ matrices using the algorithm discussed in Exercises 21–23.

3

Primes and Greatest Common Divisors

Introduction

This chapter introduces a central concept of number theory, namely that of a prime number. A prime is an integer with precisely two positive integer divisors. Prime numbers were studied extensively by the ancient Greeks, who discovered many of their basic properties. In the past three centuries, mathematicians have devoted countless hours to exploring the world of primes. They have discovered many fascinating properties, formulated diverse conjectures, and proved interesting and surprising results. Research into questions involving primes continues today, partly driven by the importance of primes in modern cryptography. Open questions about primes stimulate new research. There are also hordes of people trying to enter the record books by finding the largest prime yet known.

In this chapter, we will show that there are infinitely many primes. The proof we will give dates back to ancient times. We will also show how to find all the primes not exceeding a given integer, using the sieve of Eratosthenes, also dating back to antiquity. We will discuss the distribution of primes, and state the famous prime number theorem that was proved at the end of the nineteenth century. This theorem provides an accurate estimate for the number of primes not exceeding a given integer. Many questions about primes remain open despite attention from mathematicians over hundreds of years; we will discuss two of the best known, the twin prime conjecture and Goldbach's conjecture.

This chapter also shows that every positive integer can be written uniquely as the product of primes (when the primes are written in increasing order of size). This result is known as the *fundamental theorem of arithmetic*. To prove this theorem, we will use the concept of the greatest common divisor of two integers. We will establish many important properties of the greatest common divisor in this chapter, such as the fact that it is the smallest linear combination of these integers. We will describe the Euclidean algorithm that can be used for finding the greatest common divisor of two integers, and analyze its computational complexity. We will discuss methods used to find the factorization of integers into products of primes, and discuss the complexity of these methods. Numbers of special form are often studied in number theory; in this chapter, we will introduce the Fermat numbers, which are integers of the form $2^{2^n} + 1$. (Fermat conjectured that they are all prime but this turns out not to be true.)

Finally, we will introduce the concept of a diophantine equation, which is an equation where only solutions in integers are sought. We will show how greatest common divisors can be used to help solve linear diophantine equations. Unlike many other diophantine equations, linear diophantine equations can be solved easily and systematically.

3.1 Prime Numbers

The positive integer 1 has just one positive divisor. Every other positive integer has at least two positive divisors, because it is divisible by 1 and by itself. Integers with exactly two positive divisors are of great importance in number theory; they are called *primes*.

Definition.   A *prime* is a positive integer greater than 1 that is divisible by no positive integers other than 1 and itself.

Example 3.1.   The integers 2, 3, 5, 13, 101, and 163 are primes.   ◄

Definition.   A positive integer greater than 1 that is not prime is called *composite*.

Example 3.2.   The integers $4 = 2 \cdot 2$, $8 = 4 \cdot 2$, $33 = 3 \cdot 11$, $111 = 3 \cdot 37$, and $1001 = 7 \cdot 11 \cdot 13$ are composite.   ◄

The primes are the multiplicative building blocks of the integers. Later, we will show that every positive integer can be written uniquely as the product of primes.

In this section, we will discuss the distribution of prime numbers among the set of positive integers, and prove some elementary properties about this distribution. We will also discuss more powerful results about the distribution of primes. The theorems we will introduce include some of the most famous results in number theory.

You can find all primes less than 10,000 in Table E.1 at the end of the book.

The Infinitude of Primes   We start by showing that there are infinitely many primes, for which the following lemma is needed.

Lemma 3.1.   Every positive integer greater than 1 has a prime divisor.

Proof.   We prove the lemma by contradiction; we assume that there is a positive integer greater than 1 having no prime divisors. Then, since the set of positive integers greater than 1 with no prime divisors is nonempty, the well-ordering property tells us that there is a least positive integer n greater than 1 with no prime divisors. Since n has no prime divisors and n divides n, we see that n is not prime. Hence, we can write $n = ab$ with $1 < a < n$ and $1 < b < n$. Because $a < n$, a must have a prime divisor. By Theorem 1.8, any divisor of a is also a divisor of n, so n must have a prime divisor, contradicting the fact that n has no prime divisors. We can conclude that every positive integer greater than 1 has at least one prime divisor. ∎

We now show that there are infinitely many primes, a wondrous result known by the ancient Greeks. This is one of the key theorems in number theory that can be proved in a variety of ways. The proof we will provide was presented by Euclid in his book the *Elements* (Book IX, 20). This simple, yet elegant proof is considered by many to be particularly beautiful. It is not surprising that the very first proof found in the book *Proofs from THE BOOK* [AiZi03], a collection of particularly insightful and clever proofs, begins with this proof found in Euclid. Moreover, this book presents six quite different proofs of the infinitude of primes. (Here, *THE BOOK* refers to the imagined collection of perfect proofs that Paul Erdős claimed is maintained by God.) We will introduce a variety of different proofs that there are infinitely many primes later in this chapter. (See Exercise 8 at the end of this section, the exercise sets in Sections 3.3 and 3.5, and Section 3.6.)

Theorem 3.1. There are infinitely many primes.

Proof. Suppose that there are only finitely many primes, p_1, p_2, ... p_n, where n is a positive integer. Consider the integer Q_n, obtained by multiplying these primes together and adding one, that is,

$$Q_n = p_1 p_2 \cdots p_n + 1.$$

By Lemma 3.1, Q has at least one prime divisor, say q. We obtain a contradiction by showing that q is not one of the primes listed. (These supposedly formed a complete list of all primes.) If $q = p_j$ for some integer j with $1 \leq j \leq n$, then since $Q_n - p_1 p_2 \cdots p_n = 1$, because q divides both terms on the left-hand side of this equation, by Theorem 1.9 it follows that $q | 1$. This is impossible because no prime divides 1. Consequently, q must be a prime we have not listed. This contradiction shows that there are infinity many primes. ∎

The proof of Theorem 3.1 is nonconstructive because the integer we have constructed in the proof, Q_n, which is one more than the product of the first n primes, may or may not be prime (see Exercise 11). Consequently, in the proof we have not found a new prime, but we know that one exists.

Finding Primes In later chapters, we will be interested in finding and using extremely large primes. Tests distinguishing between primes and composite integers will be crucial; such tests are called *primality tests*. The most basic primality test is *trial division,* which tells us that the integer n is prime if and only if it is not divisible by any prime not exceeding $\sqrt{n}$. We now prove that this test can be used to determine whether n is prime.

Theorem 3.2. If n is a composite integer, then n has a prime factor not exceeding $\sqrt{n}$.

Proof. Since n is composite, we can write $n = ab$, where a and b are integers with $1 < a \leq b < n$. We must have $a \leq \sqrt{n}$, since otherwise $b \geq a > \sqrt{n}$ and $ab > \sqrt{n} \cdot \sqrt{n} = n$. Now, by Lemma 3.1, a must have a prime divisor, which by Theorem 1.8 is also a divisor of n and which is clearly less than or equal to $\sqrt{n}$. ∎

1	2	3	~~4~~	5	~~6~~	7	~~8~~	9	~~10~~
11	~~12~~	13	~~14~~	~~15~~	~~16~~	17	~~18~~	19	~~20~~
~~21~~	~~22~~	23	~~24~~	~~25~~	~~26~~	~~27~~	~~28~~	29	~~30~~
31	~~32~~	~~33~~	~~34~~	~~35~~	~~36~~	37	~~38~~	~~39~~	~~40~~
41	~~42~~	43	~~44~~	~~45~~	~~46~~	47	~~48~~	~~49~~	~~50~~
~~51~~	~~52~~	53	~~54~~	~~55~~	~~56~~	~~57~~	~~58~~	59	~~60~~
61	~~62~~	~~63~~	~~64~~	~~65~~	~~66~~	67	~~68~~	~~69~~	~~70~~
71	~~72~~	73	~~74~~	~~75~~	~~76~~	~~77~~	~~78~~	79	~~80~~
~~81~~	~~82~~	83	~~84~~	~~85~~	~~86~~	~~87~~	~~88~~	89	~~90~~
~~91~~	~~92~~	~~93~~	~~94~~	~~95~~	~~96~~	97	~~98~~	~~99~~	~~100~~

Figure 3.1 *Using the sieve of Eratosthenes to find the primes less than 100.*

 We can use Theorem 3.2 to find all the primes less than or equal to a given positive integer n. This procedure is called the *sieve of Eratosthenes,* since it was invented by the ancient Greek mathematician *Eratosthenes.* We illustrate its use in Figure 3.1 by finding all primes less than 100. We first note that every composite integer less than 100 must have a prime factor less than $\sqrt{100} = 10$. Since the only primes less than 10 are 2, 3, 5, and 7, we only need to check each integer less than 100 for divisibility by these primes. We first cross out, with a horizontal slash (—), all multiples of 2 greater than 2. Next, we cross out with a slash (/) those integers remaining that are multiples of 3, other than 3 itself. Then all multiples of 5, other than 5, that remain are crossed out with a backslash (\). Finally, all multiples of 7, other than 7, that are left are crossed out with a vertical slash (|). All remaining integers (other than 1) must be prime.

Although the sieve of Eratosthenes produces all primes less than or equal to a fixed integer, to determine in this manner whether a particular integer n is prime it is necessary to check n for divisibility by all primes not exceeding $\sqrt{n}$. This is quite inefficient; later, we will give better methods for deciding whether or not an integer is prime.

We now introduce a function that counts the primes not exceeding a specified number.

Definition. The function $\pi(x)$, where x is a positive real number, denotes the number of primes not exceeding x.

Example 3.3. From our illustration of the sieve of Eratosthenes, we see that $\pi(10) = 4$ and $\pi(100) = 25$. ◄

Primes in Arithmetic Progressions Every odd integer is either of the form $4n + 1$ or the form $4n + 3$. Are there infinitely many primes in both these forms? The primes $5, 13, 17, 29, 37, 41, \ldots$ are of the form $4n + 1$ and the primes $3, 7, 11, 19, 23, 31, 43, \ldots$ are of the form $4n + 3$. Looking at this evidence hints that there are infinitely many primes in both these progressions. What about other arithmetic progressions such as $3n + 1, 7n + 4, 8n + 7$, and so on? Does each of these contain infinitely many primes? German mathematician *G. Lejeune Dirichlet* settled this question in 1837, when he used methods from complex analysis to prove the following theorem.

Theorem 3.3. *Dirichlet's Theorem on Primes in Arithmetic Progressions.* Suppose that a and b are positive integers not divisible by the same prime. Then the arithmetic progression $an + b, n = 1, 2, 3, \ldots$, contains infinitely many primes.

No simple proof of Dirichlet's theorem on primes in arithmetic progressions is known. (Dirichlet's original proof used complex variables. In the 1950s an elementary but complicated proof was found by Selberg.) However, special cases of Dirichlet's theorem can be proved quite easily. We will illustrate this in Section 3.5, by showing that there are infinitely many primes of the form $4n + 3$.

The Largest Known Primes For hundreds if not thousands of years, professional and amateur mathematicians have been motivated to find a prime larger than any currently known. The person who discovers such a prime becomes famous, at least for a time,

G. LEJEUNE DIRICHLET (1805–1859) was born into a French family living in the vicinity of Cologne, Germany. He studied at the University of Paris when this was an important world center of mathematics. He held positions at the University of Breslau and the University of Berlin, and in 1855 was chosen to succeed Gauss at the University of Göttingen. Dirichlet is said to be the first person to master Gauss's *Disquisitiones Arithmeticae,* which had appeared 20 years earlier. He is said to have kept a copy of this book at his side even when he traveled. His book on number theory, *Vorlesungen über Zahlentheorie,* helped make Gauss's discoveries accessible to other mathematicians. Besides his fundamental work in number theory, Dirichlet made many important contributions to analysis. His famous "drawer principle," also called the pigeonhole principle, is used extensively in combinatorics and in number theory.

and has his or her name entered into the record books. Because there are infinitely many prime numbers, there is always a prime larger than the current record. Looking for new primes is done somewhat systematically; rather than checking randomly, people examine numbers that have a special form. For example, in Chapter 7 we will discuss primes of the form $2^p - 1$, where p is prime; such numbers are called *Mersenne primes*. We will see that there is a special test that makes it possible to determine whether $2^p - 1$ is prime, without performing trial divisions. The largest known prime number has been a Mersenne prime for most of the past hundred years. Currently, the world record for the largest prime known is $2^{24,036,583} - 1$.

Formulas for Primes   Is there a formula that generates only primes? This is another question that has interested mathematicians for many years. No polynomial in one variable has this property, as Exercise 23 demonstrates. It is also the case that no polynomial in n variables, where n is a positive integer, generates only primes (a result that is beyond the scope of this book). There are several impractical formulas that generate only primes. For example, Mills has shown that there is a constant Θ such that the function $f(n) = [\Theta^{3^n}]$ generates only primes. Here the value of Θ is known only approximately, with $\Theta \approx 1.3064$. This formula is impractical for generating primes not only because the exact value of Θ is not known, but also because to compute Θ you must know the primes that $f(n)$ generates (see [Mi47] for details).

If no useful formula can be used to generate large primes, how can they be generated? In Chapter 6, we will learn how to generate large primes using what are known as probabilistic primality tests.

Primality Proofs

If someone presents you with a positive integer n and claims that n is prime, how can you be sure that n really is prime? We already know that we can determine whether n is prime by performing trial divisions of n by the primes not exceeding $\sqrt{n}$. If n is not divisible by any of these primes, it itself is prime. Consequently, once we have determined that n is not divisible by any prime not exceeding its square root, we have produced a proof that n is prime. Such a proof is also known as a *certificate of primality*.

Unfortunately, using trial division to produce a certificate of primality is extremely inefficient. To see this, we estimate the number of bit operations used by this test. Using the prime number theorem, we can estimate the number of bit operations needed to show that an integer n is prime by trial divisions of n by all primes not exceeding $\sqrt{n}$. The prime number theorem tells us that there are approximately $\sqrt{n}/\log \sqrt{n} = 2\sqrt{n}/\log n$ primes not exceeding $\sqrt{n}$. To divide n by an integer m takes $O(\log_2 n \cdot \log_2 m)$ bit operations. Therefore, the number of bit operations needed to show that n is prime by this method is at least $(2\sqrt{n}/\log n)(c \log_2 n) = c\sqrt{n}$ (where we have ignored the $\log_2 m$ term because it is at least 1, even though it sometimes is as large as $(\log_2 n)/2$). This method of showing that an integer n is prime is very inefficient, for it is necessary not only to know all the primes not larger than $\sqrt{n}$, but to do at least a constant multiple of $\sqrt{n}$ bit operations.

To input an integer into a computer program, we input the binary digits of the integer. Consequently, the computational complexity of algorithms for determining whether an integer is prime is measured in terms of the number of binary digits in the integer. By Exercise 11 in Section 2.3 we know that a positive integer n has $[\log_2 n] + 1$ binary digits. Consequently, a big-O estimate for the computational complexity of an algorithm in terms of number of binary digits of n translates to the same big-O estimate in terms of $\log_2 n$, and vice versa. Note that the algorithm using trial divisions to determine whether an integer n is prime is exponential in terms of the number of binary digits of n, or in terms of $\log_2 n$, because $\sqrt{n} = 2^{\log_2 n/2}$. That is, this algorithm has exponential time complexity, measured in terms of the number of binary digits in n. As n gets large, an algorithm with exponential complexity quickly becomes impractical. Determining whether a number with 200 digits is prime using trial division still takes billions of years on the fastest computers.

Mathematicians have looked for efficient primality tests for many years. In particular, they have searched for an algorithm that produces a certificate of primality in polynomial time, measured in terms of the number of binary digits of the integer input. In 1975, G. L. Miller developed an algorithm that can prove that an integer is prime using $O((\log n)^5)$ bit operations, assuming the validity of a hypothesis called the generalized Riemann hypothesis. Unfortunately, the generalized Riemann hypothesis remains an open conjecture. In 1983, Leonard Adleman, Carl Pomerance, and Robert Rumely developed an algorithm that can prove an integer is prime using $(\log n)^{c \log \log \log n}$ bit operations, where c is a constant. Although their algorithm does not run in polynomial time, it runs in close to polynomial time because the function $\log \log \log n$ grows so slowly. To use their algorithm with an up-to-date PC to determine whether a 100-digit integer is prime requires just a few milliseconds, determining whether a 400-digit integer is prime requires less than a second, and determining whether a 1000-digit integer is prime takes less than an hour. (For more information about their test, see [AdPoRu83] and [Ru83].)

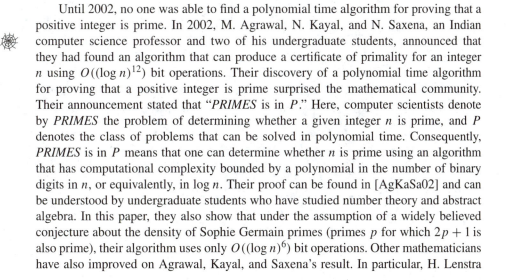

Until 2002, no one was able to find a polynomial time algorithm for proving that a positive integer is prime. In 2002, M. Agrawal, N. Kayal, and N. Saxena, an Indian computer science professor and two of his undergraduate students, announced that they had found an algorithm that can produce a certificate of primality for an integer n using $O((\log n)^{12})$ bit operations. Their discovery of a polynomial time algorithm for proving that a positive integer is prime surprised the mathematical community. Their announcement stated that "*PRIMES* is in P." Here, computer scientists denote by *PRIMES* the problem of determining whether a given integer n is prime, and P denotes the class of problems that can be solved in polynomial time. Consequently, *PRIMES* is in P means that one can determine whether n is prime using an algorithm that has computational complexity bounded by a polynomial in the number of binary digits in n, or equivalently, in $\log n$. Their proof can be found in [AgKaSa02] and can be understood by undergraduate students who have studied number theory and abstract algebra. In this paper, they also show that under the assumption of a widely believed conjecture about the density of Sophie Germain primes (primes p for which $2p + 1$ is also prime), their algorithm uses only $O((\log n)^6)$ bit operations. Other mathematicians have also improved on Agrawal, Kayal, and Saxena's result. In particular, H. Lenstra and C. Pomerance have reduced the exponent 12 in the original estimate to $6 + \epsilon$, where ϵ is any positive real number.

It is important to note that in our discussion of primality tests, we have only addressed *deterministic* algorithms, that is, algorithms that decide with certainty whether an integer is prime. In Chapter 6, we will introduce the notion of probabilistic primality tests, that is, tests that tell us that there is a high probability, but not a certainty, that an integer is prime.

3.1 Exercises

1. Determine which of the following integers are primes.

   a) 101   c) 107   e) 113
   b) 103   d) 111   f) 121

2. Determine which of the following integers are primes.

   a) 201   c) 207   e) 213
   b) 203   d) 211   f) 221

3. Use the sieve of Eratosthenes to find all primes less than 150.

4. Use the sieve of Eratosthenes to find all primes less than 200.

5. Find all primes that are the difference of the fourth powers of two integers.

6. Show that no integer of the form $n^3 + 1$ is a prime, other than $2 = 1^3 + 1$.

7. Show that if a and n are positive integers with $n > 1$ and $a^n - 1$ is prime, then $a = 2$ and n is prime. (*Hint:* Use the identity $a^{kl} - 1 = (a^k - 1)(a^{k(l-1)} + a^{k(l-2)} + \cdots + a^k + 1)$.)

8. (This exercise constructs another proof of the infinitude of primes.) Show that the integer $Q_n = n! + 1$, where n is a positive integer, has a prime divisor greater than n. Conclude that there are infinitely many primes.

9. Can you show that there are infinitely many primes by looking at the integers $S_n = n! - 1$, where n is a positive integer?

10. Using Euclid's proof that there are infinitely many primes, show that the nth prime p_n does not exceed $2^{2^{n-1}}$ whenever n is a positive integer. Conclude that when n is a positive integer, there are at least $n + 1$ primes less than 2^{2^n}.

11. Let $Q_n = p_1 p_2 \ldots p_n + 1$, where $p_1, p_2, \ldots, p_n$ are the n smallest primes. Determine the smallest prime factor of Q_n for $n = 1, 2, 3, 4, 5$, and 6. Do you think that Q_n is prime infinitely often? (*Note:* This is an unresolved question.)

12. Show that if p_k is the kth prime, where k is a positive integer, then $p_n \le p_1 p_2 \cdots p_{n-1} + 1$ for all integers n with $n \ge 3$.

13. Show that if the smallest prime factor p of the positive integer n exceeds $\sqrt[3]{n}$, then n/p must be prime or 1.

14. Show that if p is a prime in the arithmetic progression $3n + 1, n = 1, 2, 3, \ldots$, then it is also in the arithmetic progression $6n + 1, n = 1, 2, 3, \ldots$.

15. Find the smallest prime in the arithmetic progression $an + b$, where

a) $a = 3, b = 1$.   b) $a = 5, b = 4$.   c) $a = 11, b = 16$.

16. Find the smallest prime in the arithmetic progression $an + b$, where

a) $a = 5, b = 1$.   b) $a = 7, b = 2$.   c) $a = 23, b = 13$.

17. Use the second principle of mathematical induction to prove that every integer greater than 1 is either prime or the product of two or more primes.

∗ 18. Use the principle of inclusion–exclusion (Exercise 16 of Appendix B) to show that

$$\pi(n) = (\pi(\sqrt{n}) - 1) + n - \left(\left[\frac{n}{p_1} \right] + \left[\frac{n}{p_2} \right] + \cdots + \left[\frac{n}{p_r} \right] \right)$$

$$+ \left(\left[\frac{n}{p_1 p_2} \right] + \left[\frac{n}{p_1 p_3} \right] + \cdots + \left[\frac{n}{p_{r-1} p_r} \right] \right)$$

$$- \left(\left[\frac{n}{p_1 p_2 p_3} \right] + \left[\frac{n}{p_1 p_2 p_4} \right] + \cdots + \left[\frac{n}{p_{r-2} p_{r-1} p_r} \right] \right) + \cdots,$$

where $p_1, p_2, \ldots, p_r$ are the primes less than or equal to $\sqrt{n}$ (with $r = \pi(\sqrt{n})$). (*Hint:* Let property P_i be the property that an integer is divisible by p_i.)

19. Use Exercise 18 to find $\pi(250)$.

20. Show that $x^2 - x + 41$ is prime for all integers x with $0 \le x \le 40$. Show, however, that it is composite for $x = 41$.

21. Show that $2n^2 + 11$ is prime for all integers n with $0 \le n \le 10$, but is composite for $n = 11$.

22. Show that $2n^2 + 29$ is prime for all integers n with $0 \le n \le 28$, but is composite for $n = 29$.

∗ 23. Show that if $f(x) = a_n x^n + a_{n-1} x^{n-1} + \cdots + a_1 x + a_0$, where the coefficients are integers, then there is an integer y such that $f(y)$ is composite. (*Hint:* Assume that $f(x) = p$ is prime, and show that p divides $f(x + kp)$ for all integers k. Conclude that there is an integer y such that $f(y)$ is composite from the fact that a polynomial of degree $n, n > 1$, takes on each value at most n times.)

The *lucky numbers* are generated by the following sieving process: Start with the positive integers. Begin the process by crossing out every second integer in the list, starting your count with the integer 1. Other than 1, the smallest integer not crossed out is 3, so we continue by crossing out every third integer left, starting the count with the integer 1. The next integer left is 7, so we cross out every seventh integer left. Continue this process, where at each stage we cross out every kth integer left, where k is the smallest integer not crossed out, other than 1, not yet used in the sieving process. The integers that remain are the lucky numbers.

24. Find all lucky numbers less than 100.

25. Show that there are infinitely many lucky numbers.

26. Suppose that t_k is the smallest prime greater than $Q_k = p_1 p_2 \cdots p_k + 1$, where p_j is the jth prime number.
    a) Show that $t_k - Q_k + 1$ is not divisible by p_j for $j = 1, 2, \ldots, k$.
    b) R. F. Fortune conjectured that $t_k - Q_k + 1$ is prime for all positive integers k. Show that this conjecture is true for all positive integers k with $k \leq 5$.

3.1 Computational and Programming Exercises

Computations and Explorations

Using a computation program such as Maple or *Mathematica*, or programs you have written, carry out the following computations and explorations.

1. Find the nth prime, where n is each of the following integers.

   a) 1,000,000   b) 333,333,333   c) 1,000,000,000

2. Find the smallest prime greater than each of the following integers.

   a) 1,000,000   b) 100,000,000   c) 100,000,000,000

3. Plot the nth prime as a function of n for $1 \leq n \leq 100$.

4. Plot $\pi(x)$ for $1 \leq x \leq 500$.

5. Find the smallest prime factor of $n! + 1$ for all positive integers n not exceeding 20.

6. Find the smallest prime factor of $p_1 p_2 \cdots p_k + 1$, where $p_1, p_2, \ldots, p_k$ are the kth smallest primes for all positive integers k not exceeding 50.

7. Use the sieve of Eratosthenes to find all primes less than 10,000.

8. Use the result given in Exercise 18 to find $\pi(10,000)$, the number of primes not exceeding 10,000.

9. Verify R. F. Fortune's conjecture that $t_k - Q_k + 1$ is prime for all positive integers k, where t_k is the smallest prime greater than $Q_k = \prod_{j=1}^{k} p_j + 1$ for as many k as you can.

10. Find all lucky numbers (as defined in the preamble to Exercise 24) not exceeding 10,000.

Programming Projects

Write programs using Maple, *Mathematica*, or a language of your choice to do the following.

1. Decide whether a given positive integer is prime, using trial division of the integer by all primes not exceeding its square root.

* 2. Use the sieve of Eratosthenes to find all primes less than n, where n is a given positive integer.

** 3. Find $\pi(n)$, the number of primes less than or equal to n, using Exercise 18.

4. Given positive integers a and b not divisible by the same prime, find the smallest prime number in the arithmetic progression $an + b$, where n is a positive integer.

* 5. Find the lucky numbers less than n, where n is a given integer (see the preamble to Exercise 24).

3.2 The Distribution of Primes

We know that there are infinitely many primes, but can we estimate how many primes there are less than a positive real number x? One of the most famous theorems of number theory, and of all mathematics, is the *prime number theorem,* which answers this question.

Mathematicians in the late eighteenth century examined tables of prime numbers created using hand calculations. Using these values, they looked for functions that estimated $\pi(x)$. In 1798, French mathematician Adrien-Marie Legendre (see Chapter 11 for a biography) used tables of primes up to 400,031, computed by Jurij Vega, to note that $\pi(x)$ could be approximated by the function

$$\frac{x}{\log x - 1.08366}.$$

The great German mathematician Karl Friedrich Gauss (see Chapter 4 for a biography) conjectured that $\pi(x)$ increases at the same rate as the functions

$$x/\log x \quad \text{and} \quad \text{Li}(x) = \int_2^x \frac{dt}{\log t}$$

(where $\int_2^x \frac{dt}{\log t}$ represents the area under the curve $y = 1/\log t$ and above the t-axis from $t = 2$ to $t = x$). (The name *Li* is an abbreviation of *logarithmic integral.*)

Neither Legendre nor Gauss managed to prove that these functions approximated $\pi(x)$ closely for large values of x. By 1811, a table of all primes up to 1,020,000 had been produced (by Chernac), which could be used to provide evidence for these conjectures.

The first substantial result showing that $\pi(x)$ could be approximated by $x/\log x$ was established in 1850 by Russian mathematician *Pafnuty Lvovich Chebyshev.* He showed that there are positive real numbers C_1 and C_2, with $C_1 < 1 < C_2$, such that

$$C_1(x/\log x) < \pi(x) < C_2(x/\log x)$$

PAFNUTY LVOVICH CHEBYSHEV (1821–1894) was born on the estate of his parents in Okatovo, Russia. His father was a retired army officer. In 1832, Chebyshev's family moved to Moscow, where he completed his secondary education with study at home. In 1837, Chebyshev entered Moscow University, graduating in 1841. While still an undergraduate, he made his first original contribution, a new method for approximating roots of equations. Chebyshev joined the faculty of St. Petersburg University in 1843, where he remained until 1882. His doctoral thesis, written in 1849, was long used as a number theory textbook at Russian universities. Chebyshev made contributions to many areas of mathematics besides number theory, including probability theory, numerical analysis, and real analysis. He worked in theoretical and applied mechanics, and had a bent for constructing mechanisms, including linkages and hinges. He was a popular teacher, and had a strong influence on the development of Russian mathematics.

for sufficiently large values of x. (In particular, he showed that this result holds with $C_1 = 0.929$ and $C_2 = 1.1$.) He also demonstrated that if the ratio of $\pi(x)$ and $x/\log x$ approaches a limit as x increases, then this limit must be 1.

The prime number theorem, which states that the ratio of $\pi(x)$ and $x/\log x$ approaches 1 as x grows without bound, was finally proved in 1896, when French mathematician *Jacques Hadamard* and Belgian mathematician *Charles-Jean-Gustave-Nicholas de la Vallée-Poussin* produced independent proofs. Their proofs were based on results from the theory of complex analysis. They used ideas developed in 1859 by German mathematician Bernhard Riemann, which related $\pi(x)$ to the behavior of the function

$$\zeta(s) = \sum_{n=1}^{\infty} \frac{1}{n^s}$$

in the complex plane. (The function $\zeta(s)$ is known as the *Riemann zeta function.*) The connection between the Riemann zeta function and the prime numbers comes from the identity

$$\zeta(s) = \sum_{n=1}^{\infty} \frac{1}{n^s} = \prod_{p}(1 - \frac{1}{p^s})^{-1},$$

JACQUES HADAMARD (1865–1963) was born in Versailles, France. His father was a Latin teacher and his mother a distinguished piano teacher. After completing his undergraduate studies, he taught at a Paris secondary school. After receiving his doctorate in 1892, he became lecturer at the Faculté des Sciences of Bordeaux. He subsequently served on the faculties of the Sorbonne, the Collège de France, the École Polytechnique, and the École Centrale des Arts et Manufactures. Hadamard made important contributions to complex analysis, functional analysis, and mathematical physics. His proof of the prime number theorem was based on his work in complex analysis. Hadamard was a famous teacher; he wrote numerous articles about elementary mathematics that were used in French schools, and his text on elementary geometry was used for many years.

CHARLES-JEAN-GUSTAVE-NICHOLAS DE LA VALLEÉ-POUSSIN (1866–1962), the son of a geology professor, was born at Louvain, Belgium. He studied at the Jesuit College at Mons, first studying philosophy, later turning to engineering. After receiving his degree, instead of pursuing a career in engineering, he devoted himself to mathematics. De la Valleé-Poussin's most significant contribution to mathematics was his proof of the prime number theorem. Extending this work, he established results about the distribution of primes in arithmetic progression and the distribution of primes represented by quadratic forms. Furthermore, he refined the prime number theorem to include error estimates. He made important contributions to differential equations, approximation theory, and analysis. His textbook, *Cours d'analyse,* had a strong impact on mathematical thought in the first half of the twentieth century.

where the product on the right-hand side of the equation extends over all primes p. We will explain why this identity is true in Section 3.5.

In addition to proving the prime number theorem, de la Vallée-Poussin showed that the function $\mathrm{Li}(x)$ is a closer approximation to $\pi(x)$ than $x/(\log x - a)$ for all values of the constant a.

The proofs of the prime number theorem found by Hadamard and de la Valleé-Poussin depend on complex analysis, though the theorem itself does not involve complex numbers. This left open the challenge of finding a proof that did not use the theory of complex variables. It surprised the mathematical community when, in 1949, Norwegian mathematician *Atle Selberg* and Hungarian mathematician *Paul Erdős* independently found elementary proofs of the prime number theorem. Their proofs, though elementary (meaning that they do not use the theory of complex variables), are quite complicated and difficult.

We now formally state the prime number theorem.

Theorem 3.4. *The Prime Number Theorem.*   The ratio of $\pi(x)$ to $x/\log x$ approaches 1 as x grows without bound. (Here, $\log x$ denotes the natural logarithm of x and in the language of limits, we have $\lim_{x \to \infty} \pi(x)/(x/\log x) = 1$.)

Remark.   A concise way to state the prime number theorem is to write $\pi(x) \sim x/\log x$. Here the symbol $\sim$ denotes "is asymptotic to." We write $a(x) \sim b(x)$ to denote that $\lim_{x \to \infty} a(x)/b(x) = 1$, and we say that $a(x)$ is asymptotic to $b(x)$.

The prime number theorem tells us that the ratio between $x/\log x$ and $\pi(x)$ is close to 1 when x is large. However, there are functions for which the ratio between these functions and $\pi(x)$ approaches 1 more rapidly than it does for $x/\log x$. In particular, it

ATLE SELBERG (b. 1917), born in Langesund, Norway, became interested in mathematics as a schoolboy. He was inspired by Ramanujan's writing, both by the mathematics and the "air of mystery" surrounding Ramanujan's personality. Selberg received his doctorate in 1943 from the University of Oslo. He remained at the university until 1947, when he married and took a position at the Institute for Advanced Study in Princeton. After a brief stay at Syracuse University, he returned to the Institute for Advanced Study, where he was appointed a permanent member in 1949; he became a professor at Princeton University in 1951. Selberg received the Fields Medal, the most prestigious award in mathematics, for his work on sieve methods and on the properties of the set of zeros of the Riemann zeta function. He is also well known for his elementary proofs of the prime number theorem (also done by Paul Erdős), Dirichlet's theorem on primes in arithmetic progressions, and the generalization of the prime number theorem for primes in arithmetic progressions.

x	$\pi(x)$	$x/\log x$	$\pi(x)/\frac{x}{\log x}$	$Li(x)$	$\pi(x)/Li(x)$
10^3	168	144.8	1.160	178	0.9438202
10^4	1229	1085.7	1.132	1246	0.9863563
10^5	9592	8685.9	1.104	9630	0.9960540
10^6	78498	72382.4	1.085	78628	0.9983466
10^7	664579	620420.7	1.071	664918	0.9998944
10^8	5761455	5428681.0	1.061	5762209	0.9998691
10^9	50847534	48254942.4	1.054	50849235	0.9999665
10^{10}	455052512	434294481.9	1.048	455055614	0.9999932
10^{11}	4118054813	3948131663.7	1.043	4118165401	0.9999731
10^{12}	37607912018	36191206825.3	1.039	37607950281	0.9999990
10^{13}	346065536839	334072678387.1	1.036	346065645810	0.9999997
10^{14}	3204941750802	3102103442166.0	1.033	3204942065692	0.9999999

Table 3.1 *Approximations to $\pi(x)$.*

has been shown that $Li(x)$ is an even better approximation. In Table 3.1, we see evidence for the prime number theorem and that $Li(x)$ is an excellent approximation of $\pi(x)$. (Note that the values of $Li(x)$ have been rounded to the nearest integer.)

It is not necessary to find all primes not exceeding x to compute $\pi(x)$. One way to evaluate $\pi(x)$ without finding all the primes less than x is to use a counting ar-

PAUL ERDŐS (1913-1996), born in Budapest, Hungary, was the son of high-school mathematics teachers. When he was three years old, he could multiply three-digit numbers in his head, and when he was four, he discovered negative numbers on his own. At 17 he entered Eőtvős University, graduating in four years with a Ph.D. in mathematics. After graduating, he spent four years at Manchester University, England, as a postdoctoral fellow. In 1938 he came to the United States because of the difficult political situation in Hungary, especially for Jews.

Erdős made many significant contributions to combinatorics and to number theory. One of the discoveries of which he was most proud was his elementary proof of the prime number theorem. He also participated in the modern development of Ramsey theory, a part of combinatorics. Erdős traveled extensively throughout the world to work with other mathematicians. He traveled from one mathematician or group of mathematicians to the next, proclaiming, "My brain is open." Erdős wrote more than 1500 papers, with almost 500 coauthors. Erdős offered monetary rewards for the solutions of problems he found particularly interesting. Two recently published biographies ([Sc98] and [Ho99]) give further details on his life and work.

gument based on the sieve of Eratosthenes (see Exercise 18 in Section 3.1). Efficient ways of computing $\pi(x)$ requiring only $O(x^{(3/5)+\epsilon})$ bit operations have been devised by Lagarias and Odlyzko [LaOd82]. The current world record is $\pi(4 \cdot 10^{22}) = 783,964,159,847,056,303,858$, found as part of a distributed computing effort on the Internet. (Efforts to extend these computations to larger values of x have temporarily hit a snag.)

The Riemann Hypothesis

Many mathematicians consider the *Riemann hypothesis*, a conjecture about the zeros of the zeta function, the most important open problem in pure mathematics. For more than 100 years, number theorists have struggled to solve this problem. Interest in it has spread, perhaps because a prize of one million dollars for a proof (if it is indeed true) has been offered by the Clay Mathematics Institute. Recently, many general-interest books about the Riemann hypothesis, such as [De03], [Sa03a], and [Sa03b], have appeared, even though the hypothesis involves sophisticated notions from complex analysis. We will briefly describe the Riemann hypothesis for the benefit of readers familiar with complex analysis, as well as for the general appreciation of others.

We have defined the Riemann zeta function as $\zeta(s) = \sum_{n=1}^{\infty} \frac{1}{n^s}$. This definition is valid for all complex numbers s with $Re(s) > 1$, where $Re(s)$ is the real part of the complex number s. Riemann was able to extend the function defined by the infinite series to a function in the entire complex plane with a pole at $s = 1$. In his famous 1859 paper [Ri59], Riemann connected the zeta function with the distribution of prime numbers. He derived a formula for $\pi(x)$ in terms of the zeros of $\zeta(s)$. The more we understand about the location of the zeros of the zeta function, the more we know about the distribution of the primes. The Riemann hypothesis is a statement about the location of the zeros of this function. Before stating the hypothesis, we first note that the zeta function has zeros at the negative even integers $-2, -4, -6, \ldots$, called the *trivial zeros*. The Riemann hypothesis is the assertion that the nontrivial zeros of $\zeta(s)$ all have real part equal to 1/2. Note that there is an equivalent formulation of the Riemann hypothesis in terms of the error introduced when $\mathrm{Li}(x)$ is used to estimate $\pi(x)$; this alternative formulation does not involve complex variables. In 1901, von Koch showed that the Riemann hypothesis is equivalent to the statement that the error that occurs when $\pi(x)$ is estimated by $\mathrm{Li}(x)$ is $O(x^{1/2} \log x)$.

Many mathematicians believe the Riemann hypothesis is true, particularly because of the wealth of evidence supporting it. First, a vast amount of numerical evidence has been found. We now know that the first 2.5×10^{11} zeros (in order of increasing imaginary parts) have real part equal to 1/2. (These computations were done by Sebastian Wedeniwski, who has set up a distributed computing project to carry them out called ZetaGrid). Second, we know that at least 40% of the nontrivial zeros of the zeta function are simple and have real part equal to 1/2. Third, we know that if there are exceptions to the Riemann hypothesis, they must be rare as we move away from the line $Re(s) = 1/2$. Of course, it is still possible that this evidence is misleading us and that the Riemann hypothesis is not true. Perhaps this famous problem will be resolved in the next few years, or maybe it will resist all attacks for hundreds of years into the future. For more technical information about the Riemann hypothesis, consult the article by Enrico Bomberi on the Web and [Ed01].

How big is the nth prime? The prime number theorem has the following corollary, which can be proved using calculus (see page 10 of [HaWr79]).

Corollary 3.4.1. Let p_n be the nth prime, where n is a positive integer. Then $p_n \sim n \log n$.

What is the probability that a randomly selected positive integer is prime? Given that there are approximately $x/\log x$ primes not exceeding x, the probability that x is prime is approximately $(x/\log x)/x = 1/\log x$. For example, the probability that an integer near 10^{1000} is prime is approximately $1/\log 10^{1000} \approx 1/2302$. Suppose that you want to find a prime with 1000 digits; what is the expected number of integers you must select before you find a prime? The answer is that you must select roughly $1/(1/2302) = 2302$ integers of this size before one of them will be a prime. Of course, you will need to check each one to determine whether it is prime. In Chapter 6, we will discuss how this can be done efficiently.

Gaps in the Distribution of Primes   We have shown that there are infinitely many primes and we have discussed the abundance of primes below a given bound x, but we have yet to discuss how regularly primes are distributed throughout the positive integers. We first give a result that shows that there are arbitrarily long runs of integers containing no primes.

Theorem 3.5. For any positive integer n, there are at least n consecutive composite positive integers.

Proof. Consider the n consecutive positive integers

$$(n+1)! + 2, \quad (n+1)! + 3, \quad \ldots, \quad (n+1)! + n + 1.$$

When $2 \le j \le n + 1$, we know that $j \mid (n+1)!$. By Theorem 1.9 it follows that $j \mid (n+1)! + j$. Hence, these n consecutive integers are all composite. ∎

One of the Largest Numbers Ever Appearing Naturally in a Mathematical Proof
Using the data in Table 3.1, we can show that for all x in the table, the difference $\mathrm{Li}(x) - \pi(x)$ is positive and increases as x grows. Gauss, who only had access to the data in the first few rows of this table, believed this trend held for all positive integers x. However, in 1914, the English mathematician J. E. Littlewood showed that $\mathrm{Li}(x) - \pi(x)$ changes sign infinitely many times. In his proof, Littlewood did not establish a lower bound for the first time that $\mathrm{Li}(x) - \pi(x)$ changes from positive to negative. This was done in 1933 by Samuel Skewes, a student of Littlewood's, who managed to show that $\mathrm{Li}(x) - \pi(x)$ changes signs for at least one x with $x < 10^{10^{10^{34}}}$, a humongous number. This number, known as *Skewes' constant*, became famous as the largest number to appear naturally in a mathematical proof. Fortunately, in the past seven decades, considerable progress has been made in reducing this bound. The best current results show that $\mathrm{Li}(x) - \pi(x)$ changes sign near $x = 1.39822 \times 10^{316}$.

Example 3.4. The seven consecutive integers beginning with $8! + 2 = 40,322$ are all composite. (However, these are much larger than the smallest seven consecutive composites, 90, 91, 92, 93, 94, 95, and 96.)   ◄

Conjectures About Primes

Professional and amateur mathematicians alike find the prime numbers fascinating. It is not surprising that a tremendous variety of conjectures have been formulated concerning prime numbers. Some of these conjectures have been settled, but many still elude resolution. We will describe some of the best known of these conjectures here.

Looking at tables of primes led mathematicians in the first half of the nineteenth century to make conjectures that the distribution of primes satisfies some basic properties, such as this following conjecture.

Bertrand's Conjecture. In 1845, the French mathematician Joseph Bertrand conjectured that for every positive integer n with $n > 1$, there is a prime p such that $n < p < 2n$. Bertrand verified this conjecture for all n not exceeding 3,000,000, but he could not produce a proof. The first proof of this conjecture was found by Pafnuty Lvovich Chebyshev in 1852. Because this conjecture has been proved, it is often called *Bertrand's postulate*. (See Exercises 22–24 for an outline of a proof.)

Theorem 3.5 shows that the gap between consecutive primes is arbitrarily long. On the other hand, primes may often be close together. The only consecutive primes are 2 and 3, because 2 is the only even prime. However, many pairs of primes differ by two; these pairs of primes are called *twin primes*. Examples are the pairs 3, 5 and 7, 11 and 13, 101 and 103, and 4967 and 4969.

Evidence seems to indicate that there are infinitely many pairs of twin primes. There are 35 pairs of twin primes less than 10^3; 8169 pairs less than 10^6; 3,424,506 pairs less than 10^9; and 1,870,585,220 pairs less than 10^{12}. This leads to the following conjecture.

Twin Prime Conjecture. There are infinitely many pairs of primes p and $p + 2$.

JOSEPH LOUIS FRANÇOIS BERTRAND (1822–1900) was born in Paris. He studied at the École Polytechnique from 1839 until 1841 and at the École des Mines from 1841 to 1844. Instead of becoming a mining engineer, he decided to become a mathematician. Bertrand was appointed to a position at the École Polytechnique in 1856 and, in 1862, he also became professor at the Collège de France. In 1845, on the basis of extensive numerical evidence in tables of primes, Bertrand conjectured that there is at least one prime between n and $2n$ for every integer n with $n > 1$. This result was first proved by Chebyshev in 1852. Besides working in number theory, Bertrand worked on probability theory and differential geometry. He wrote several brief volumes on the theory of probability and on analyzing data from observations. His book *Calcul des probabilitiés*, written in 1888, contains a paradox on continuous probabilities now known as Bertrand's paradox. Bertrand was considered to be kind at heart, extremely clever, and full of spirit.

In 1966, Chinese mathematician J. R. Chen showed, using sophisticated sieve methods, that there are infinitely many primes p such that $p + 2$ has at most two prime factors. An active competition is under way to produce new largest pairs of twin primes. The current record for the largest pair of twin primes is $33,218,925 \cdot 2^{169,690} \pm 1$, a pair of primes with 51,090 digits each, discovered by Daniel Papp and Yves Gallot in 2002.

Viggo Brun showed that the sum $\sum_{\text{primes } p \text{ with } p+2 \text{ prime}} \frac{1}{p} = (1/3 + 1/5) + (1/5 + 1/7) + (1/11 + 1/13) + \cdots$ converges to a constant called *Brun's constant*, which is approximately equal to 1.9021605824. Surprisingly, the computation of Brun's constant has played a role in discovering flaws in Intel's original Pentium chip. In 1994, Thomas Nicely at Lynchburg College in Virginia computed Brun's constant in two different ways using different methods on a Pentium PC and came up with different answers. He traced the error back to a flaw in the Pentium chip and he alerted Intel to this problem. (See page 85 for more information about Nicely's discovery.)

We now discuss perhaps the most notorious conjecture about primes.

Goldbach's Conjecture. Every even positive integer greater than 2 can be written as the sum of two primes.

Example 3.5.   The integers 10, 24, and 100 can be written as the sum of two primes in the following ways:

$$10 = 3 + 7 = 5 + 5,$$
$$24 = 5 + 19 = 7 + 17 = 11 + 13,$$
$$100 = 3 + 97 = 11 + 89 = 17 + 83$$
$$= 29 + 71 = 41 + 59 = 47 + 53.$$
◀

This conjecture was stated by *Christian Goldbach* in a letter to Leonhard Euler in 1742. It has been verified for all even integers less than $4 \cdot 10^{14}$, with this limit increasing as computers become more powerful. Usually, there are many ways to write a particular even integer as the sum of primes, as Example 3.5 illustrates. However, a proof that there is always at least one way has not yet been found. The best result known to date is due to J. R. Chen, who showed (in 1966), using powerful sieve methods, that all sufficiently large integers are the sum of a prime and the product of at most two primes.

Goldbach's conjecture asserts that infinitely many primes occur as pairs of consecutive odd numbers. However, consecutive primes may be far apart. A consequence of

JING RUN CHEN (1933–1996) was a student of the prominent Chinese number theorist Loo Keng Hua. Chen was almost entirely devoted to mathematical research. During the Cultural Revolution in China, he continued his research, working almost all day and night in a tiny room with no electric lights, no table or chairs, only a small bed and his books and papers. It was during this period that he made his most important discoveries concerning twin primes and Goldbach's conjecture. Although he was a mathematical prodigy, Chen was considered to be next to hopeless in other aspects of life. He died in 1996 after a long illness.

the prime number theorem is that as n grows, the average gap between the consecutive primes p_n and p_{n+1} is $\log n$. Number theorists have worked hard to prove results that show that the gaps between consecutive primes are much smaller than average for infinitely many primes. For example, it has been shown that $p_{n+1} - p_n < 0.2486 \log n$ for infinitely many positive integers n. Showing that for every positive real number ϵ, there are infinitely many positive integers n such that $(p_{n+1} - p_n)/\log n < \epsilon$ remains an elusive goal on the way toward the proof of Goldbach's conjecture.

There are many conjectures concerning the number of primes of various forms, such as the following conjecture.

The $n^2 + 1$ Conjecture. There are infinitely many primes of the form $n^2 + 1$, where n is a positive integer.

The smallest primes of the form $n^2 + 1$ are $5 = 2^2 + 1$, $17 = 4^2 + 1$, $37 = 6^2 + 1$, $101 = 10^2 + 1$, $197 = 14^2 + 1$, $257 = 16^2 + 1$, and $401 = 20^2 + 1$. The best result known

---

Pentium Chip Flaw

The story behind the Pentium chip flaw encountered by Thomas Nicely shows that answers produced by computers should not always be trusted. A surprising number of hardware and software problems arise that lead to incorrect computational results. This story also shows that companies risk serious problems when they hide errors in their products. In June 1994, testers at Intel discovered that Pentium chips did not always carry out computations correctly. However, Intel decided not to make public information about this problem. Instead, they concluded that because the error would not affect many users, it was unnecessary to alert the millions of owners of Pentium computers. The Pentium flaw involved an incorrect implementation of an algorithm for floating-point division. Although the probability is low that divisions of numbers affected by this error come up in a computation, such divisions arise in many computations in mathematics, science, and engineering, and even in spreadsheets running business applications.

Later in that same month, Nicely came up with two different results when he used a Pentium computer to compute Brun's constant in different ways. In October 1994, after checking all possible sources of computational error, Nicely contacted Intel customer support. They duplicated his computations and verified the existence of an error. Furthermore, they told him that this error had not been previously reported. After not hearing any additional information from Intel, Nicely sent e-mail to a few people telling them about this. These people forwarded the message to other interested parties, and within a few days, information about the bug was posted on an Internet newsgroup. By late November, this story was reported by CNN, the *New York Times*, and the Associated Press.

Surprised by the bad publicity, Intel offered to replace Pentium chips, but only for users running applications determined by Intel to be vulnerable to the Pentium division flaw. This offer did not mollify the Pentium user community. All the bad publicity drove Intel stock down several dollars a share and Intel became the object of many jokes, such as: "At Intel, quality is job 0.999999998." Finally, in December 1994, Intel decided to offer a replacement Pentium chip upon request. They set aside almost half a billion dollars to cover costs, and they hired hundreds of extra employees to handle customer requests. Nevertheless, this story does have a happy ending for Intel. Their corrected and improved version of the Pentium chip was extremely successful.

to date is that there are infinitely many integers n for which $n^2 + 1$ is either a prime or the product of two primes. This was shown by Henryk Iwaniec in 1973. Conjectures such as the $n^2 + 1$ conjecture may be easy to state, but are sometimes extremely difficult to resolve (see [Ri96] for more information).

3.2 Exercises

1. Find the smallest five consecutive composite integers.

2. Find one million consecutive composite integers.

3. Show that there are no "prime triplets," that is, primes p, $p + 2$, and $p + 4$, other than 3, 5 , and 7.

4. Find the smallest four sets of prime triplets of the form p, $p + 2$, $p + 6$.

5. Find the smallest four sets of prime triplets of the form p, $p + 4$, $p + 6$.

6. Find the smallest prime between n and $2n$ when n is

    a) 3.   c) 19.

    b) 5.   d) 31.

7. Find the smallest prime between n and $2n$ when n is

    a) 4.   c) 23.

    b) 6.   d) 47.

An unsettled conjecture asserts that for every positive integer n there is a prime between n^2 and $(n + 1)^2$.

8. Find the smallest prime between n^2 and $(n + 1)^2$ for all positive integers n with $n \leq 10$.

9. Find the smallest prime between n^2 and $(n + 1)^2$ for all positive integers n with $11 \leq n \leq 20$.

10. Verify Goldbach's conjecture for each of the following values of n.

    a) 50   c) 102   e) 200

    b) 98   d) 144   f) 222

CHRISTIAN GOLDBACH (1690–1764) was born in Königsberg, Prussia (the city noted in mathematical circles for its famous bridge problem). He became professor of mathematics at the Imperial Academy of St. Petersburg in 1725. In 1728, Goldbach went to Moscow to tutor Tsarevich Peter II. In 1742, he entered the Russian Ministry of Foreign Affairs as a staff member. Goldbach is most noted for his correspondence with eminent mathematicians, in particular Leonhard Euler and Daniel Bernoulli. Besides his well-known conjectures that every even positive integer greater than 2 is the sum of two primes and that every odd positive integer greater than 5 is the sum of three primes, Goldbach made several notable contributions to analysis.

11. Goldbach also conjectured that every odd positive integer greater than 5 is the sum of three primes. Verify this conjecture for each of the following odd integers.

a) 7   c) 27   e) 101
b) 17   d) 97   f) 199

12. Show that every integer greater than 11 is the sum of two composite integers.

13. Show that Goldbach's conjecture that every even integer greater than 2 is the sum of two primes is equivalent to the conjecture that every integer greater than 5 is the sum of three primes.

14. Let $G(n)$ denote the number of ways to write the even integer n as the sum $p + q$, where p and q are primes with $p \leq q$. Goldbach's conjecture asserts that $G(n) \geq 1$ for all even integers n with $n > 2$. A stronger conjecture asserts that $G(n)$ tends to infinity as the even integer n grows without bound.

a) Find $G(n)$ for all even integers n with $4 \leq n \leq 30$.
b) Find $G(158)$.
c) Find $G(188)$.

* 15. Show that if n and k are positive integers with $n > 1$ and all n positive integers $a, a + k, \ldots, a + (n-1)k$ are odd primes, then k is divisible by every prime less than n.

Use Exercise 15 to help you solve Exercises 16–19.

16. Find an arithmetic progression of length six that begins with the integer 7 and where every term is a prime.

17. Find the smallest possible minimum difference for an arithmetic progression that contains four terms and where every term is a prime.

18. Find the smallest possible minimum difference for an arithmetic progression that contains five terms and where every term is a prime.

* 19. Find the smallest possible minimum difference for an arithmetic progression that contains six terms and where every term is a prime.

20. a) In 1848, A. de Polignac conjectured that every odd positive integer is the sum of a prime and a power of two. Show that this conjecture is false by showing that 509 is a counterexample.
    b) Find the next smallest counterexample after 509.

* 21. A *prime power* is an integer of the form p^n, where p is prime and n is a positive integer greater than 1. Find all pairs of prime powers that differ by 1. Prove that your answer is correct.

22. Let n be a positive integer greater than 1 and let $p_1, p_2, \ldots, p_t$ be the primes not exceeding n. Show that $p_1 p_2 \cdots p_t < 4^n$.

* 23. Let n be a positive integer greater than 3 and let p be a prime such that $2n/3 < p \leq n$. Show that p does not divide the binomial coefficient $\binom{2n}{n}$.

** 24. Use Exercises 22 and 23 to show that if n is a positive integer, then there exists a prime p such that $n < p < 2n$. (This is *Bertrand's conjecture*.)

25. Use Exercise 24 to show that if p_n is the nth prime, then $p_n \leq 2^n$.

26. Use Bertrand's conjecture to show that every positive integer n with $n \geq 7$ is the sum of distinct primes.

27. Use Bertrand's postulate to show that $\frac{1}{n} + \frac{1}{n+1} + \cdots + \frac{1}{n+m}$ does not equal an integer when n and m are positive integers.

* 28. In this exercise, we show that if n is an integer with $n \geq 4$, then $p_{n+1} < p_1 p_2 \cdots p_n$, where p_k is the kth prime. This result is known as *Bonse's inequality*.

   a) Let k be a positive integer. Show that none of the integers $p_1 p_2 \cdots p_{k-1} \cdot 1 - 1$, $p_1 p_2 \cdots p_{k-1} \cdot 2 - 1$, $\ldots$, $p_1 p_2 \cdots p_{k-1} \cdot p_k - 1$ is divisible by one of the first $k - 1$ primes and that if a prime p divides one of these integers, it cannot divide another of these integers.

   b) Conclude from part (a) that if $n - k + 1 < p_k$, then there is an integer among those listed in part (a) not divisible by p_j for $j = 1, \ldots, n$. (*Hint:* Use the pigeonhole principle.)

   c) Use part (b) to show that if $n - k + 1 < p_k$, then $p_{n+1} < p_1 p_2 \cdots p_k$. Fix n and suppose that k is the least positive integer such that $n - k + 1 < p_k$. Show that $n - k \geq p_{k-1} - 2$ and that $p_{k-1} - 2 \geq k$ when $k \geq 5$ and that if $n \geq 10$, then $k \geq 5$. Conclude that if $n \geq 20$, then $p_{(n+1)} < p_2 p_2 \cdots p_k$ for some k with $n - k \geq k$. Use this to derive Bonse's inequality when $n \geq 10$.

   d) Check the cases when $4 \leq n < 10$ to finish the proof.

29. Show that 30 is the largest integer n with the property that if $k < n$ and there is no prime p that divides both k and n, then k is prime. (*Hint:* Show that if n has this property and $n \geq p^2$ where p is prime, then $p \mid n$. Conclude that if $n \geq 7^2$, then n must be divisible by 2, 3, 5, and 7. Apply Bonse's inequality to show that such an n must be divisible by every prime, a contradiction. Show that 30 has the desired property, but no n with $30 < n < 49$ does.)

* 30. Show that $p_{n+1} p_{n+2} < p_1 \cdot p_2 \cdots p_n$, where p_k is the kth prime whenever n is an integer with $n \geq 4$. (*Hint:* Use Bertrand's postulate and the work done in part (c) of the proof of Bonse's inequality.

31. Show that $p_n^2 < p_{n-1} p_{n-2} p_{n-3}$, where p_k is the kth prime number and $n \geq 6$. Also, show that inequality does not hold when $n = 3, 4$, or 5. (*Hint:* Use Bertrand's postulate to obtain $p_n < 2 p_{n-1}$ and $p_{n-1} < 2 p_{n-2}$.)

32. Show that for every positive integer N there is an even number K so that there are more than N pairs of successive primes such that K is the difference between these successive primes. (*Hint:* Use the prime number theorem.)

3.2 Computational and Programming Exercises

Computations and Explorations

Using a computation program such as Maple or *Mathematica,* or programs you have written, carry out the following computations and explorations.

1. Verify as much of the information given in Table 3.1 as you can.

2. Find as many tuples of primes of the form p, $p + 2$, and $p + 6$ as you can.

3. Verify Goldbach's conjecture for all even positive integers less than 10,000.

4. Find all twin primes less than 10,000.

5. Find the first pair of twin primes greater than each of the integers in Computation 1.

6. Plot $\pi_2(x)$, the number of twin primes not exceeding x, for $1 \leq x \leq 1000$ and $1 \leq x \leq 10,000$.

7. Hardy and Littlewood conjectured that $\pi_2(x)$, the number of twin primes not exceeding x, is asymptotic to $2C_2 x/(\log x)^2$ where $C_2 = \prod_{p>2}\left(1 - \frac{1}{(p-1)^2}\right)$. The constant C_2 is approximately equal to 0.66016. Determine how accurate this asymptotic formula for $\pi_2(x)$ is for values of x as large as you can compute.

8. Compute Brun's constant with as much accuracy as possible.

9. Explore the conjecture that $G(n)$, the number of ways to write the even integer n as the sum $p + q$, where p and q are primes with $p \leq q$, satisfies $G(n) \geq 10$ for all even integers n with $n \geq 188$.

10. An unsettled conjecture asserts that for every positive integer n, there is an arithmetic progression of length n comprised of n consecutive prime numbers. The longest such arithmetic progression currently known consists of 22 consecutive primes. Find arithmetic progressions consisting of three consecutive primes with all primes less than 100 and four consecutive primes with all primes less than 500.

11. Show that all terms of the arithmetic progression of length five that begins with 1464481 and has common difference 210 are prime.

12. Show that all terms of the arithmetic progression of length twelve that begins with 23143 and has common difference 30030 are prime.

13. Find an arithmetic progression containing ten primes that begins with 199.

14. An unsettled conjecture asserts that for all positive integers n, there is a prime p such that $n^2 < p < (n + 1)^2$. Verify this conjecture for as many positive integers n as you can.

15. Explore the conjecture that every even integer is the sum of two, not necessarily distinct, lucky numbers. Continue by exploring the conjecture that given a positive integer k, there is a positive integer n that can be expressed as the sum of two lucky numbers in exactly k ways.

Programming Projects

Write programs using Maple, *Mathematica,* or a language of your choice to do the following.

1. Verify Goldbach's conjecture for all even integers less than n, where n is a given positive integer.

2. Find all twin primes less than n, where n is a given positive integer.

3. Find the first m primes of the form $n^2 + 1$, where n is a positive integer and m is a given positive integer.

4. Find $G(n)$, the number of ways to write the even integer n as the sum $p + q$, where p and q are primes with $p \leq q$.

5. Given a positive integer n, find as many arithmetic progressions of length n, where every term is a prime.

3.3 Greatest Common Divisors

If a and b are integers, not both 0, then the set of common divisors of a and b is a finite set of integers, always containing the integers $+1$ and -1. We are interested in the largest integer among the common divisors of the two integers.

Definition.   The *greatest common divisor* of two integers a and b, which are not both 0, is the largest integer that divides both a and b.

The greatest common divisor of a and b is written as (a, b). (Note that the notation $\gcd(a, b)$ is also used, especially outside of number theory. We will use the traditional notation (a, b) here, even though it is the same notation used for ordered pairs.) We also define $(0, 0) = 0$.

Even though every positive integer divides 0, we define $(0, 0) = 0$. This is done to ensure that the results we prove about greatest common divisors hold in all cases.

Example 3.6.   The common divisors of 24 and 84 are ± 1, ± 2, ± 3, ± 4, ± 6, and ± 12. Hence, $(24, 84) = 12$. Similarly, looking at sets of common divisors, we find that $(15, 81) = 3$, $(100, 5) = 5$, $(17, 25) = 1$, $(0, 44) = 44$, $(-6, -15) = 3$, and $(-17, 289) = 17$. ◀

We are particularly interested in pairs of integers sharing no common divisors greater than 1. Such pairs of integers are called *relatively prime*.

Definition.   The integers a and b are *relatively prime* if a and b have greatest common divisor $(a, b) = 1$.

Example 3.7.   Since $(25, 42) = 1$, 25 and 42 are relatively prime. ◀

Note that since the divisors of $-a$ are the same as the divisors of a, it follows that $(a, b) = (|a|, |b|)$ (where $|a|$ denotes the absolute value of a, which equals a if $a \geq 0$ and $-a$ if $a < 0$). Hence, we can restrict our attention to the greatest common divisors of pairs of positive integers.

In Example 3.6, we noted that $(15, 81) = 3$. If we divide 15 and 81 by $(15, 81) = 3$, we obtain two relatively prime integers, 5 and 27. This is no surprise, because we have removed all common factors. This illustrates the following theorem, which tells us that we obtain two relatively prime integers when we divide each of two original integers by their greatest common divisor.

Theorem 3.6.   Let a and b be integers with $(a, b) = d$. Then $(a/d, b/d) = 1$.

Proof.   Let a and b be integers with $(a, b) = d$. We will show that a/d and b/d have no common positive divisors other than 1. Assume that e is a positive integer such that $e \mid (a/d)$ and $e \mid (b/d)$. Then, there are integers k and l with $a/d = ke$ and $b/d = le$, so that $a = dek$ and $b = del$. Hence, de is a common divisor of a and b. Since d is the greatest common divisor of a and b, $de \leq d$, so that e must be 1. Consequently, $(a/d, b/d) = 1$. ∎

We do not change the greatest common divisor of two integers when we add a multiple of one of the integers to the other. In Example 3.6, we showed that $(24, 84) = 12$. When we add any multiple of 24 to 84, the greatest common divisor of 24 and the resulting number is still 12. For example, since $2 \cdot 24 = 48$ and $(-3) \cdot 24 = -72$, we see that $(24, 84 + 48) = (24, 132) = 12$ and $(24, 84 + (-72)) = (24, 12) = 12$. The reason for this is that the common divisors of 24 and 84 are the same as the common divisors of 24 and the integer that results when a multiple of 24 is added to 84. The proof of the following theorem justifies this reasoning.

Theorem 3.7. Let a, b, and c be integers. Then $(a + cb, b) = (a, b)$.

Proof. Let a, b, and c be integers. We will show that the common divisors of a and b are exactly the same as the common divisors of $a + cb$ and b. This will show that $(a + cb, b) = (a, b)$. Let e be a common divisor of a and b. By Theorem 1.9, we see that $e \mid (a + cb)$, so that e is a common divisor of $a + cb$ and b. If f is a common divisor of $a + cb$ and b, then by Theorem 1.9, we see that f divides $(a + cb) - cb = a$, so that f is a common divisor of a and b. Hence, $(a + cb, b) = (a, b)$. ∎

We will show that the greatest common divisor of the integers a and b, not both 0, can be written as a sum of multiples of a and b. To phrase this more succinctly, we use the following definition.

Definition. If a and b are integers, then a *linear combination* of a and b is a sum of the form $ma + nb$, where both m and n are integers.

Example 3.8. What are the linear combinations $9m + 15n$, where m and n are both integers? Among these combinations are $-6 = 1 \cdot 9 + (-1) \cdot 15$; $-3 = (-2)9 + 1 \cdot 15$; $0 = 0 \cdot 9 + 0 \cdot 15$; $3 = 2 \cdot 9 + (-1) \cdot 15$; $6 = (-1) \cdot 9 + 1 \cdot 15$; and so on. It can be shown that the set of all linear combinations of 9 and 15 is the set $\{\ldots, -12, -9, -6, -3, 0, 3, 6, 9, 12, \ldots\}$, as the reader should verify after reading the proofs of the following two theorems. ◀

In Example 3.8, we found that $(9, 15) = 3$ appears as the smallest positive linear combination with integer coefficients of 9 and 15. This is no accident, as the following theorem demonstrates.

Theorem 3.8. The greatest common divisor of the integers a and b, not both 0, is the least positive integer that is a linear combination of a and b.

Proof. Let d be the least positive integer that is a linear combination of a and b. (There is a *least* such positive integer, using the well-ordering property, since at least one of two linear combinations $1 \cdot a + 0 \cdot b$ and $(-1)a + 0 \cdot b$, where $a \neq 0$, is positive.) We write

(3.1) $$d = ma + nb,$$

where m and n are integers. We will show that $d \mid a$ and $d \mid b$.

By the division algorithm, we have

$$a = dq + r, \quad 0 \le r < d.$$

From this equation and (3.1), we see that

$$r = a - dq = a - q(ma + nb) = (1 - qm)a - qnb.$$

This shows that the integer r is a linear combination of a and b. Since $0 \le r < d$, and d is the least positive linear combination of a and b, we conclude that $r = 0$, and hence $d \mid a$. In a similar manner, we can show that $d \mid b$.

We have shown that d, the least positive integer that is a linear combination of a and b, is a common divisor of a and b. What remains to be shown is that it is the *greatest common divisor* of a and b. To show this, all we need show is that any common divisor c of a and b must divide d, since any proper positive divisor of d is less than d. Since $d = ma + nb$, if $c \mid a$ and $c \mid b$, Theorem 1.9 tells us that $c \mid d$, so that $d \ge c$. This concludes the proof.   ∎

Because we will often need to apply Theorem 3.8 in the case where a and b are relatively prime integers, we state the following corollary.

Corollary 3.8.1.   If a and b are relatively prime integers, then there are integers m and n such that $ma + nb = 1$.

Proof.   To prove this corollary, we note that if a and b are relatively prime, then $(a, b) = 1$. Consequently, by Theorem 3.8, 1 is the least positive integer that is a linear combination of a and b. It follows that there are integers m and n such that $ma + nb = 1$.   ∎

Theorem 3.8 is valuable: We can obtain results about the greatest common divisor of two integers using the fact that the greatest common divisor is the least positive linear combination of these integers. Having different representations of the greatest common divisor of two integers allows us to choose the one that is most useful for a particular purpose. This is illustrated in the proof of the following theorem.

Theorem 3.9.   If a and b are positive integers, then the set of linear combinations of a and b is the set of integer multiples of (a, b).

Proof.   Suppose that $d = (a, b)$. We first show that every linear combination of a and b must also be a multiple of d. First note that by the definition of greatest common divisor, we know that $d \mid a$ and $d \mid b$. Now every linear combination of a and b is of the form $ma + nb$, where m and n are integers. By Theorem 1.9, it follows that whenever m and m are integers, d divides $ma + nb$. That is, $ma + nb$ is a multiple of d.

We now show that every multiple of d is also a linear combination of a and b. By Theorem 3.8 we know that there are integers r and s such that $(a, b) = ra + sb$. The multiples of d are the integers of the form jd, where j is an integer. Multiplying both sides of the equation $d = ra + sb$ by j, we see that $jd = (jr)a + (js)b$. Consequently, every multiple of d is a linear combination of a and b. This completes the proof.   ∎

We have defined greatest common divisors using the notion that the integers are ordered. That is, given two distinct integers, one is larger than the other. However, we can define the greatest common divisor of two integers without relying on this notion of order, as we do in Theorem 3.10. This characterization of the greatest common divisor of two integers not depending on ordering is generalized in the study of algebraic number theory to apply to what are known as algebraic number fields.

Theorem 3.10. If a and b are integers, not both 0, then a positive integer d is the greatest common divisor of a and b if and only if:

   (i) $d \mid a$ and $d \mid b$
   (ii) if c is an integer with $c \mid a$ and $c \mid b$, then $c \mid d$.

Proof.   We will first show that the greatest common divisor of a and b has these two properties. Suppose that $d = (a, b)$. By the definition of common divisor, we know that $d \mid a$ and $d \mid b$. By Theorem 3.8, we know that $d = ma + nb$, where m and n are integers. Consequently, if $c \mid a$ and $c \mid b$, then by Theorem 1.9, $c \mid d = ma + nb$. We have now shown that if $d = (a, b)$, then properties (i) and (ii) hold.

Now assume that properties (i) and (ii) hold. Then we know that d is a common divisor of a and b. Furthermore, by property (ii), we know that if c is a common divisor of a and b, then $c \mid d$, so that $d = ck$ for some integer k. Hence, $c = d/k \leq d$. (We have used the fact that a positive integer divided by any nonzero integer is less than that integer.) This shows that a positive integer satisfying (i) and (ii) must be the greatest common divisor of a and b. ■

We have shown that the greatest common divisor of a and b, not both 0, is a linear combination of a and b. However, we have not explained how to find a particular linear combination of a and b that equals (a, b). In the next section, we will provide an algorithm that finds a particular linear combination of a and b that equals (a, b).

We can also define the greatest common divisor of more than two integers.

Definition.   Let $a_1, a_2, \ldots, a_n$ be integers, not all 0. The *greatest common divisor* of these integers is the largest integer that is a divisor of all of the integers in the set. The greatest common divisor of $a_1, a_2, \ldots, a_n$ is denoted by $(a_1, a_2, \ldots, a_n)$. (Note that the order in which the a_i's appear does affect the result.)

Example 3.9.   We easily see that $(12, 18, 30) = 6$ and $(10, 15, 25) = 5$. ◄

We can use the following lemma to find the greatest common divisor of a set of more than two integers.

Lemma 3.2.   If $a_1, a_2, \ldots, a_n$ are integers, not all 0, then $(a_1, a_2, \ldots, a_{n-1}, a_n) = (a_1, a_2, \ldots, a_{n-2}, (a_{n-1}, a_n))$.

Proof.   Any common divisor of the n integers $a_1, a_2, \ldots, a_{n-1}, a_n$ is, in particular, a divisor of a_{n-1} and a_n, and therefore a divisor of (a_{n-1}, a_n). Also, any common divisor of the $n-1$ integers $a_1, a_2, \ldots, a_{n-2}$, and (a_{n-1}, a_n) must be a common divisor of all n integers, for if it divides (a_{n-1}, a_n), it must divide both a_{n-1} and a_n. Since the set of n integers and the set of the first $n-2$ integers together with the greatest common divisor of the last two integers have exactly the same divisors, their greatest common divisors are equal. ∎

Example 3.10.   To find the greatest common divisor of the three integers 105, 140, and 350, we use Lemma 3.2 to see that $(105, 140, 350) = (105, (140, 350)) = (105, 70) = 35.$ ◄

Example 3.11.   Consider the integers 15, 21, and 35. We find that the greatest common divisor of these three integers is 1 using the following steps:

$$(15, 21, 35) = (15, (21, 35)) = (15, 7) = 1.$$

Each pair among these integers has a common factor greater than 1, since $(15, 21) = 3$, $(15, 35) = 5$, and $(21, 35) = 7$. ◄

Example 3.11 motivates the following definition.

Definition.   We say that the integers $a_1, a_2, \ldots, a_n$ are *mutually relatively prime* if $(a_1, a_2, \ldots, a_n) = 1$. These integers are called *pairwise relatively prime* if, for each pair of integers a_i and a_j with $i \neq j$ from the set, $(a_i, a_j) = 1$; that is, if each pair of integers from the set is relatively prime.

The concept of pairwise relatively prime is used much more often than the concept of mutually relatively prime. Also, note that pairwise relatively prime integers must be mutually relatively prime, but that the converse is false (as the integers 15, 21, and 35 in Example 3.11 show).

3.3 Exercises

1. Find the greatest common divisor of each of the following pairs of integers.

   a) 15, 35   d) 99, 100
   b) 0, 111   e) 11, 121
   c) −12, 18   f) 100, 102

2. Find the greatest common divisor of each of the following pairs of integers.

   a) 5, 15   d) −90, 100
   b) 0, 100   e) 100, 121
   c) −27, −45   f) 1001, 289

3. Let a be a positive integer. What is the greatest common divisor of a and $2a$?

4. Let a be a positive integer. What is the greatest common divisor of a and a^2?

5. Let a be a positive integer. What is the greatest common divisor of a and $a + 1$?

6. Let a be a positive integer. What is the greatest common divisor of a and $a + 2$?

7. Show that if a and b are integers, not both 0, and c is a nonzero integer, then $(ca, cb) = |c|(a, b)$.

8. Show that if a and b are integers with $(a, b) = 1$, then $(a + b, a - b) = 1$ or 2.

9. What is $(a^2 + b^2, a + b)$, where a and b are relatively prime integers that are not both 0?

10. Show that if a and b are both even integers that are not both 0, then $(a, b) = 2(a/2, b/2)$.

11. Show that if a is an even integer and b is an odd integer, then $(a, b) = (a/2, b)$.

12. Show that if a, b, and c are integers such that $(a, b) = 1$ and $c \mid (a + b)$, then $(c, a) = (c, b) = 1$.

13. Show that if a, b, and c are mutually relatively prime nonzero integers, then $(a, bc) = (a, b)(a, c)$.

☞ 14. a) Show that if a, b, and c are integers with $(a, b) = (a, c) = 1$, then $(a, bc) = 1$.

    b) Use mathematical induction to show that if $a_1, a_2, \ldots, a_n$ are integers, and b is another integer such that $(a_1, b) = (a_2, b) = \cdots = (a_n, b) = 1$, then $(a_1 a_2 \cdots a_n, b) = 1$.

15. Find a set of three integers that are mutually relatively prime, but any two of which are not relatively prime. Do not use examples from the text.

16. Find four integers that are mutually relatively prime such that any three of these integers are not mutually relatively prime.

17. Find the greatest common divisor of each of the following sets of integers.

    a) 8, 10, 12   d) 6, 15, 21
    b) 5, 25, 75   e) $-7, 28, -35$
    c) 99, 9999, 0   f) 0, 0, 1001

18. Find three mutually relatively prime integers from among the integers 66, 105, 42, 70, and 165.

19. Show that if $a_1, a_2, \ldots, a_n$ are integers that are not all 0 and c is a positive integer, then $(ca_1, ca_2, \ldots, ca_n) = c(a_1, a_2 \ldots, a_n)$.

20. Show that the greatest common divisor of the integers $a_1, a_2, \ldots, a_n$, not all 0, is the least positive integer that is a linear combination of $a_1, a_2, \ldots, a_n$.

21. Show that if k is an integer, then the integers $6k - 1, 6k + 1, 6k + 2, 6k + 3$, and $6k + 5$ are pairwise relatively prime.

22. Show that if k is a positive integer, then $3k + 2$ and $5k + 3$ are relatively prime.

23. Show that $8a + 3$ and $5a + 2$ are relatively prime for all integers a.

24. Show that if a and b are relatively prime integers, then $(a + 2b, 2a + b) = 1$ or 3.

25. Show that every positive integer greater than 6 is the sum of two relatively prime integers greater than 1.

The *Farey series* $\mathcal{F}_n$ *of order* n is the set of fractions h/k, where h and k are integers, $0 \le h \le k \le n$, and $(h, k) = 1$, in ascending order. We include 0 and 1 in the forms $0/1$ and $1/1$, respectively. For instance, the Farey series of order 4 is

$$\frac{0}{1}, \frac{1}{4}, \frac{1}{3}, \frac{1}{2}, \frac{2}{3}, \frac{3}{4}, \frac{1}{1}.$$

Exercises 26–29 deal with Farey series.

26. Find the Farey series of order 7.

∗ **27.** Show that if a/b, c/d, and e/f are successive terms of a Farey series, then

$$\frac{c}{d} = \frac{a+e}{b+f}.$$

∗ **28.** Show that if a/b and c/d are successive terms of a Farey series, then $ad - bc = -1$.

∗ **29.** Show that if a/b and c/d are successive terms of the Farey series of order n, then $b + d > n$.

JOHN FAREY (1766–1826) attended school in Woburn, England, until the age of 16. In 1782, he entered a school in Halifax, Yorkshire, where he studied mathematics, drawing, and surveying. In 1790, he married and his first son was born the following year. In 1792, the Duke of Bedford appointed Farey as land steward for his Woburn estates. Farey held this post until 1802, developing expertise in geology. When the duke died suddenly, the duke's brother dismissed Farey, who went to London and established an extensive practice as a surveyor and geologist.

Farey's geologic work included studies of soils and strata in Derbyshire. He also produced a map of the strata visible between London and Brighton. Farey also produced extensive scientific writings, publishing around 60 articles in philosophical and scientific magazines. These articles address a wide range of topics, including geology, forestry, physics, and many other areas.

Although he achieved moderate fame as a geologist, ironically Farey is remembered for a contribution to mathematics. In his four-paragraph 1816 article, "On a curious property of vulgar fractions," Farey noted that a reduced fraction p/q with $0 < p/q < 1$ and $q < n$ equals the fraction whose numerator and denominator are the sum of the numerators and the sum of the denominators, respectively, of the fractions on either side of p/q when all reduced fractions between 0 and 1 with denominators not exceeding n are written in increasing order (see Exercise 27). Farey said he was unaware whether this property was already known. He also wrote that he did not have a proof. The French mathematician Cauchy read Farey's article and proved this property in the book *Exercises de mathématique*, published in 1816. It was Cauchy who coined the name *Farey series* because he thought Farey was the first person to notice this property.

Not surprisingly, Farey was not the first person to notice the property for which he became famous. In 1802, C. Haros wrote an article in which he approximates decimal fractions using common fractions, constructing the Farey sequence for $n = 99$ employing this curious property in his construction.

* **30.** a) Show that if a and b are positive integers, then $((a^n - b^n)/(a - b), a - b) = (n(a, b)^{n-1}, a - b)$.

   b) Show that if a and b are relatively prime positive integers, then
   $((a^n - b^n)/(a - b), a - b) = (n, a - b)$.

31. Show that if a, b, c, and d are integers such that b and d are positive, $(a, b) = (c, d) = 1$, and $\frac{a}{b} + \frac{c}{d}$ is an integer, then $b = d$.

32. What can you conclude if a, b, and c are positive integers such that $(a, b) = (b, c) = 1$ and $\frac{1}{a} + \frac{1}{b} + \frac{1}{c}$ is an integer?

33. Show that if a and b are positive integers, then $(a, b) = 2 \sum_{i=1}^{a-1} [bi/a] + a + b - ab$. (*Hint:* Count the number of lattice points, that is, points with integer coordinates, inside or on the triangle with vertices $(0, 0)$, $(0, b)$, and $(a, 0)$ in two different ways.)

34. Show that if n is a positive integer and i and j are integers with $1 \leq i < j \leq n$, then $(n! \cdot i + 1, n! \cdot j + 1) = 1$.

35. Use Exercise 34 to show that there are infinitely many primes. (*Hint:* Assume that there are exactly r primes and consider the $r + 1$ numbers $(r + 1)! \cdot i + 1$. This proof was discovered by P. Schorn.)

36. Show that if c and d are relatively prime positive integers, then the integers $a_j, j = 0, 1, 2, \ldots$, defined by $a_0 = c$ and $a_n = a_0 a_1 \cdots a_{n-1} + d$ for $n = 1, 2, \ldots$, are pairwise relatively prime.

3.3 Computational and Programming Exercises

Computations and Explorations

Using a computation program such as Maple or *Mathematica,* or programs you have written, carry out the following computations and explorations.

1. Construct the Farey series of order 100.

2. Verify the properties of the Farey series given in Exercises 27, 28, and 29 for successive terms of your choice in the Farey series of order 100.

Programming Projects

Write programs using Maple, *Mathematica,* or a language of your choice to do the following.

1. Find the greatest common divisor of two integers from the lists of their divisors.

2. Print out the Farey series of order n for a given positive integer n.

3.4 The Euclidean Algorithm

We are going to develop a systematic method, or algorithm, to find the greatest common divisor of two positive integers. This method is called the *Euclidean algorithm.* It is named after the ancient Greek mathematician *Euclid*, who describes this algorithm in his *Elements.* (The same method for finding greatest common divisors was also described in the sixth century by the Indian mathematician *Aryabhata*, who called it "the pulverizer.")

Before we discuss the algorithm in general, we demonstrate its use with an example. We find the greatest common divisor of 30 and 72. First, we use the division algorithm to write $72 = 30 \cdot 2 + 12$, and we use Theorem 3.7 to note that $(30, 72) = (30, 72 - 2 \cdot 30) = (30, 12)$. Note that we have replaced 72 by the smaller number 12 in our computations because $(72, 30) = (30, 12)$. Next, we use the division algorithm again to write $30 = 2 \cdot 12 + 6$. Using the same reasoning as before, we see that $(30, 12) = (12, 6)$. Because $12 = 6 \cdot 2 + 0$, we now see that $(12, 6) = (6, 0) = 6$. Consequently, we can conclude that $(72, 30) = 6$, without finding all the common divisors of 30 and 72.

We now present the general form of the Euclidean algorithm for computing the greatest common divisor of two positive integers.

Theorem 3.11. *The Euclidean Algorithm.*   Let $r_0 = a$ and $r_1 = b$ be integers such that $a \geq b > 0$. If the division algorithm is successively applied to obtain $r_j = r_{j+1} q_{j+1} + r_{j+2}$, with $0 < r_{j+2} < r_{j+1}$ for $j = 0, 1, 2, \ldots, n - 2$ and $r_{n+1} = 0$, then $(a, b) = r_n$, the last nonzero remainder.   ∎

From this theorem, we see that the greatest common divisor of a and b is the last nonzero remainder in the sequence of equations generated by successively applying the division algorithm and continuing until a remainder is 0—where, at each step, the dividend and divisor are replaced by smaller numbers, namely the divisor and remainder.

To prove that the Euclidean algorithm produces greatest common divisors, the following lemma will be helpful.

Lemma 3.3.   If e and d are integers and $e = dq + r$, where q and r are integers, then $(e, d) = (d, r)$.

Proof.   This lemma follows directly from Theorem 3.7, taking $a = r$, $b = d$, and $c = q$.   ∎

We now prove that the Euclidean algorithm produces the greatest common divisor of two integers.

EUCLID (c. 350 B.C.E) was the author of the most successful mathematics textbook ever written, namely his *Elements,* which has appeared in over a thousand editions from ancient to modern times. Very little is known about Euclid's life, other than that he taught at the famed academy at Alexandria. Evidently he did not stress the applications of mathematics, for it is reputed that when asked by a student for the use of geometry, Euclid had his slave give the student some coins, "since he must needs make gain of what he learns." Euclid's *Elements* provides an introduction to plane and solid geometry, and to number theory. The Euclidean algorithm is found in Book VII of the thirteen books in the *Elements,* and his proof of the infinitude of primes is found in Book IX. Euclid also wrote books on a variety of other topics, including astronomy, optics, music, and mechanics.

Proof. Let $r_0 = a$ and $r_1 = b$ be positive integers with $a \geq b$. By successively applying the division algorithm, we find that

$$
\begin{aligned}
r_0 &= r_1 q_1 + r_2 & 0 \leq r_2 < r_1, \\
r_1 &= r_2 q_2 + r_3 & 0 \leq r_3 < r_2, \\
&\ \ \vdots \\
r_{j-2} &= r_{j-1} q_{j-1} + r_j & 0 \leq r_j < r_{j-1}, \\
&\ \ \vdots \\
r_{n-4} &= r_{n-3} q_{n-3} + r_{n-2} & 0 \leq r_{n-2} < r_{n-3}, \\
r_{n-3} &= r_{n-2} q_{n-2} + r_{n-1} & 0 \leq r_{n-1} < r_{n-2}, \\
r_{n-2} &= r_{n-1} q_{n-1} + r_n & 0 \leq r_n < r_{n-1}, \\
r_{n-1} &= r_n q_n.
\end{aligned}
$$

We can assume that we eventually obtain a remainder of zero, because the sequence of remainders $a = r_0 \geq r_1 > r_2 > \cdots \geq 0$ cannot contain more than a terms (because each remainder is an integer). By Lemma 3.3, we see that $(a, b) = (r_0, r_1) = (r_1, r_2) = (r_2, r_3) = \cdots = (r_{n-3}, r_{n-2}) = (r_{n-2}, r_{n-1}) = (r_{n-1}, r_n) = (r_n, 0) = r_n$. Hence, $(a, b) = r_n$, the last nonzero remainder. ∎

We illustrate the use of the Euclidean algorithm with the following example.

Example 3.12. The steps used by the Euclidean algorithm to find $(252, 198)$ are

$$
\begin{aligned}
252 &= 1 \cdot 198 + 54 \\
198 &= 3 \cdot 54 + 36 \\
54 &= 1 \cdot 36 + 18 \\
36 &= 2 \cdot 18.
\end{aligned}
$$

ARYABHATA (476–550) was born in Kusumapura (now Patna), India. He is the author of the *Aryabhatiya*, a summary of Hindu mathematics written entirely in verse. This book covers astronomy, geometry, plane and spherical trigonometry, arithmetic, and algebra. Topics studied include formulas for areas and volumes, continued fractions, sums of power series, an approximation for π, and tables of sines. Aryabhata also described a method for finding greatest common divisors which is the same as the method described by Euclid. His formulas for the areas of triangles and circles are correct, but those for the volumes of spheres and pyramids are wrong. Aryabhata also produced an astronomy text, *Siddhanta*, which includes a number of remarkably accurate statements (as well as other statements that are not correct). For example, he states that the orbits of the planets are ellipses, and he correctly describes the causes of solar and lunar eclipses. India named its first satellite, launched in 1975 by the Russians, *Aryabhata*, in recognition of his fundamental contributions to astronomy and mathematics.

We summarize these steps in the following table:

j	r_j	r_{j+1}	q_{j+1}	r_{j+2}
0	252	198	1	54
1	198	54	3	36
2	54	36	1	18
3	36	18	2	0

The last nonzero remainder (found in the next-to-last row in the last column) is the greatest common divisor of 252 and 198. Hence, $(252, 198) = 18$. ◄

The Euclidean algorithm is an extremely fast way to find greatest common divisors.

Later, we will see this when we estimate the maximum number of divisions used by the Euclidean algorithm to find the greatest common divisor of two positive integers. However, we first show that, given any positive integer n, there are integers a and b such that exactly n divisions are required to find (a, b) using the Euclidean algorithm. We can find such numbers by taking successive terms of the Fibonacci sequence.

The reason that the Euclidean algorithm operates so slowly when it finds the greatest common divisor of successive Fibonacci numbers is that the quotient in all but the last step is 1, as illustrated in the following example.

Example 3.13. We apply the Euclidean algorithm to find $(34, 55)$. Note that $f_9 = 34$ and $f_{10} = 55$. We have

$$55 = 34 \cdot 1 + 21$$
$$34 = 21 \cdot 1 + 13$$
$$21 = 13 \cdot 1 + 8$$
$$13 = 8 \cdot 1 + 5$$
$$8 = 5 \cdot 1 + 3$$
$$5 = 3 \cdot 1 + 2$$
$$3 = 2 \cdot 1 + 1$$
$$2 = 1 \cdot 2.$$

Observe that when the Euclidean algorithm is used to find the greatest common divisor of $f_9 = 34$ and $f_{10} = 55$, a total of eight divisions are required. Furthermore, $(34, 55) = 1$, since 1 is the last nonzero remainder. ◄

The following theorem tells us how many divisions are used by the Euclidean algorithm to find the greatest common divisor of successive Fibonacci numbers.

Theorem 3.12. Let f_{n+1} and f_{n+2} be successive terms of the Fibonacci sequence, with $n > 1$. Then the Euclidean algorithm takes exactly n divisions to show that $(f_{n+1}, f_{n+2}) = 1$.

Proof. Applying the Euclidean algorithm, and using the defining relation for the Fibonacci numbers $f_j = f_{j-1} + f_{j-2}$ in each step, we see that

$$f_{n+2} = f_{n+1} \cdot 1 + f_n,$$
$$f_{n+1} = f_n \cdot 1 + f_{n-1},$$
$$\vdots$$
$$f_4 = f_3 \cdot 1 + f_2,$$
$$f_3 = f_2 \cdot 2.$$

Hence, the Euclidean algorithm takes exactly n divisions, to show that $(f_{n+2}, f_{n+1}) = f_2 = 1$. ∎

The Complexity of the Euclidean Algorithm We can now prove a theorem first proved by *Gabriel Lamé*, a French mathematician of the nineteenth century, which gives an estimate for the number of divisions needed to find the greatest common divisor using the Euclidean algorithm.

Theorem 3.13. ***Lamé's Theorem.*** The number of divisions needed to find the greatest common divisor of two positive integers using the Euclidean algorithm does not exceed five times the number of decimal digits in the smaller of the two integers.

Proof. When we apply the Euclidean algorithm to find the greatest common divisor of $a = r_0$ and $b = r_1$ with $a > b$, we obtain the following sequence of equations:

$$r_0 = r_1 q_1 + r_2, \qquad 0 \le r_2 < r_1,$$
$$r_1 = r_2 q_2 + r_3, \qquad 0 \le r_3 < r_2,$$
$$\vdots$$
$$r_{n-2} = r_{n-1} q_{n-1} + r_n, \qquad 0 \le r_n < r_{n-1},$$
$$r_{n-1} = r_n q_n.$$

We have used n divisions. We note that each of the quotients $q_1, q_2, \ldots, q_{n-1} \ge 1$, and $q_n \ge 2$, because $r_n < r_{n-1}$. Therefore,

GABRIEL LAMÉ (1795–1870) was a graduate of the École Polytechnique. A civil and railway engineer, he advanced the mathematical theory of elasticity and invented curvilinear coordinates. Although his main contributions were to mathematical physics, he made several discoveries in number theory, including the estimate of the number of steps required by the Euclidean algorithm, and the proof that Fermat's last theorem holds for $n = 7$ (see Section 13.2). It is interesting to note that Gauss considered Lamé to be the foremost French mathematician of his time.

$$r_n \geq 1 = f_2,$$
$$r_{n-1} \geq 2r_n \geq 2f_2 = f_3,$$
$$r_{n-2} \geq r_{n-1} + r_n \geq f_3 + f_2 = f_4,$$
$$r_{n-3} \geq r_{n-2} + r_{n-1} \geq f_4 + f_3 = f_5,$$
$$\vdots$$
$$r_2 \geq r_3 + r_4 \geq f_{n-1} + f_{n-2} = f_n,$$
$$b = r_1 \geq r_2 + r_3 \geq f_n + f_{n-1} = f_{n+1}.$$

Thus, for there to be n divisions used in the Euclidean algorithm, we must have $b \geq f_{n+1}$. By Example 1.28, we know that $f_{n+1} > \alpha^{n-1}$ for $n > 2$, where $\alpha = (1 + \sqrt{5})/2$. Hence, $b > \alpha^{n-1}$. Now, since $\log_{10} \alpha > 1/5$, we see that

$$\log_{10} b > (n - 1) \log_{10} \alpha > (n - 1)/5.$$

Consequently,

$$n - 1 < 5 \cdot \log_{10} b.$$

Let b have k decimal digits, so that $b < 10^k$ and $\log_{10} b < k$. Hence, we see that $n - 1 < 5k$, and because k is an integer, we can conclude that $n \leq 5k$. This establishes Lamé's theorem. ∎

The following result is a consequence of Lamé's theorem. It tells us that the Euclidean algorithm is very efficient.

Corollary 3.13.1. The greatest common divisor of two positive integers a and b with $a > b$ can be found using $O((\log_2 a)^3)$ bit operations.

Proof. We know from Lamé's theorem that $O(\log_2 a)$ divisions, each taking $O((\log_2 a)^2)$ bit operations, are needed to find (a, b). Hence, by Theorem 2.3, (a, b) may be found using a total of $O((\log_2 a)^3)$ bit operations. ∎

Expressing Greatest Common Divisors—As Linear Combinations   The Euclidean algorithm can be used to express the greatest common divisor of two integers as a linear combination of these integers. We illustrate this by expressing $(252, 198) = 18$ as a linear combination of 252 and 198. Referring to the steps of the Euclidean algorithm used to find $(252, 198)$, by the next to the last step we see that

$$18 = 54 - 1 \cdot 36.$$

By the preceding step, it follows that

$$36 = 198 - 3 \cdot 54,$$

which implies that

$$18 = 54 - 1 \cdot (198 - 3 \cdot 54) = 4 \cdot 54 - 1 \cdot 198.$$

Likewise, by the first step, we have

$$54 = 252 - 1 \cdot 198,$$

so that

$$18 = 4(252 - 1 \cdot 198) - 1 \cdot 198 = 4 \cdot 252 - 5 \cdot 198.$$

This last equation exhibits $18 = (252, 198)$ as a linear combination of 252 and 198.

In general, to see how $d = (a, b)$ may be expressed as a linear combination of a and b, refer to the series of equations that is generated by the Euclidean algorithm. By the penultimate equation, we have

$$r_n = (a, b) = r_{n-2} - r_{n-1}q_{n-1}.$$

This expresses (a, b) as a linear combination of r_{n-2} and r_{n-1}. The second to the last equation can be used to express r_{n-1} as $r_{n-3} - r_{n-2}q_{n-2}$. Using this last equation to eliminate r_{n-1} in the previous expression for (a, b), we find that

$$r_n = r_{n-3} - r_{n-2}q_{n-2},$$

so that

$$(a, b) = r_{n-2} - (r_{n-3} - r_{n-2}q_{n-2})q_{n-1}$$
$$= (1 + q_{n-1}q_{n-2})r_{n-2} - q_{n-1}r_{n-3},$$

which expresses (a, b) as a linear combination of r_{n-2} and r_{n-3}. We continue working backward through the steps of the Euclidean algorithm to express (a, b) as a linear combination of each preceding pair of remainders, until we have found (a, b) as a linear combination of $r_0 = a$ and $r_1 = b$. Specifically, if we have found at a particular stage that

$$(a, b) = sr_j + tr_{j-1},$$

then, since

$$r_j = r_{j-2} - r_{j-1}q_{j-1},$$

we have

$$(a, b) = s(r_{j-2} - r_{j-1}q_{j-1}) + tr_{j-1}$$
$$= (t - sq_{j-1})r_{j-1} + sr_{j-2}.$$

This shows how to move up through the equations that are generated by the Euclidean algorithm so that, at each step, the greatest common divisor of a and b may be expressed as a linear combination of a and b.

This method for expressing (a, b) as a linear combination of a and b is somewhat inconvenient for calculation, because it is necessary to work out the steps of the Euclidean algorithm, save all these steps, and then proceed backward through the steps to write (a, b) as a linear combination of each successive pair of remainders. There is another method for finding (a, b) which requires working through the steps of the Euclidean algorithm only once. The following theorem gives this method, which is called the *extended Euclidean algorithm*.

Theorem 3.14. Let a and b be positive integers. Then

$$(a, b) = s_n a + t_n b,$$

where s_n and t_n are the nth terms of the sequences defined recursively by

$$s_0 = 1, \quad t_0 = 0,$$
$$s_1 = 0, \quad t_1 = 1,$$

and

$$s_j = s_{j-2} - q_{j-1} s_{j-1}, \quad t_j = t_{j-2} - q_{j-1} t_{j-1}$$

for $j = 2, 3, \ldots, n$, where the q_j are the quotients in the divisions of the Euclidean algorithm when it is used to find (a, b).

Proof. We will prove that

(3.2) $$r_j = s_j a + t_j b$$

for $j = 0, 1, \ldots, n$. Since $(a, b) = r_n$, once we have established (3.2), we will know that

$$(a, b) = s_n a + t_n b.$$

We prove (3.2) using the second principle of mathematical induction. For $j = 0$, we have $a = r_0 = 1 \cdot a + 0 \cdot b = s_0 a + t_0 b$. Hence, (3.2) is valid for $j = 0$. Likewise, $b = r_1 = 0 \cdot a + 1 \cdot b = s_1 a + t_1 b$, so that (3.2) is valid for $j = 1$.

Now, we assume that

$$r_j = s_j a + t_j b$$

for $j = 1, 2, \ldots, k - 1$. Then, from the kth step of the Euclidean algorithm, we have

$$r_k = r_{k-2} - r_{k-1} q_{k-1}.$$

Using the induction hypothesis, we find that

$$
\begin{aligned}
r_k &= (s_{k-2} a + t_{k-2} b) - (s_{k-1} a + t_{k-1} b) q_{k-1} \\
&= (s_{k-2} - s_{k-1} q_{k-1}) a + (t_{k-2} - t_{k-1} q_{k-1}) b \\
&= s_k a + t_k b.
\end{aligned}
$$

This finishes the proof. ∎

The following example illustrates the use of this algorithm for expressing (a, b) as a linear combination of a and b.

Example 3.14. We summarize the steps used by the extended Euclidean algorithm to express $(252, 198)$ as a linear combination of 252 and 198 in the following table.

j	r_j	r_{j+1}	q_{j+1}	r_{j+2}	s_j	t_j
0	252	198	1	54	1	0
1	198	54	3	36	0	1
2	54	36	1	18	1	-1
3	36	18	2	0	-3	4
4					4	-5

The values of s_j and t_j, $j = 0, 1, 2, 3, 4$, are computed as follows:

$$s_0 = 1, \qquad\qquad\qquad\qquad\qquad t_0 = 0,$$
$$s_1 = 0, \qquad\qquad\qquad\qquad\qquad t_1 = 1,$$
$$s_2 = s_0 - s_1 q_1 = 1 - 0 \cdot 1 = 1, \qquad t_2 = t_0 - t_1 q_1 = 0 - 1 \cdot 1 = -1,$$
$$s_3 = s_1 - s_2 q_2 = 0 - 1 \cdot 3 = -3, \qquad t_3 = t_1 - t_2 q_2 = 1 - (-1)3 = 4,$$
$$s_4 = s_2 - s_3 q_3 = 1 - (-3) \cdot 1 = 4, \qquad t_4 = t_2 - t_3 q_3 = -1 - 4 \cdot 1 = -5.$$

Because $r_4 = 18 = (252, 198)$ and $r_4 = s_4 a + t_4 b$, we have

$$18 = (252, 198) = 4 \cdot 252 - 5 \cdot 198. \qquad \blacktriangleleft$$

Note that the greatest common divisor of two integers may be expressed as a linear combination of these integers in an infinite number of ways. To see this, let $d = (a, b)$ and let $d = sa + tb$ be one way to write d as a linear combination of a and b, guaranteed to exist by the previous discussion. Then for all integers k,

$$d = (s + k(b/d))a + (t - k(a/d))b.$$

Example 3.15.   With $a = 252$ and $b = 198$, we have $18 = (252, 198) = (4 + 11k)252 + (-5 - 14k)198$ for any integer k. $\qquad \blacktriangleleft$

3.4 Exercises

1. Use the Euclidean algorithm to find each of the following greatest common divisors.

   a) $(45, 75)$ c) $(666, 1414)$
   b) $(102, 222)$ d) $(20785, 44350)$

2. Use the Euclidean algorithm to find each of the following greatest common divisors.

   a) $(51, 87)$ c) $(981, 1234)$
   b) $(105, 300)$ d) $(34709, 100313)$

3. For each pair of integers in Exercise 1, express the greatest common divisor of the integers as a linear combination of these integers.

4. For each pair of integers in Exercise 2, express the greatest common divisor of the integers as a linear combination of these integers.

5. Find the greatest common divisor of each of the following sets of integers.

   a) 6, 10, 15   b) 70, 98, 105   c) 280, 330, 405, 490

6. Find the greatest common divisor of each of the following sets of integers.

a) 15, 35, 90   b) 300, 2160, 5040   c) 1240, 6660, 15540, 19980

The greatest common divisor of the n integers $a_1, a_2, \ldots, a_n$ can be expressed as a linear combination of these integers. To do this, first express (a_1, a_2) as a linear combination of a_1 and a_2. Then express $(a_1, a_2, a_3) = ((a_1, a_2), a_3)$ as a linear combination of a_1, a_2, and a_3. Repeat this until $(a_1, a_2, \ldots, a_n)$ is expressed as a linear combination of $a_1, a_2, \ldots, a_n$. Use this procedure in Exercises 7 and 8.

7. Express the greatest common divisor of each set of numbers in Exercise 5 as a linear combination of the numbers in that set.

8. Express the greatest common divisor of each set of numbers in Exercise 6 as a linear combination of the numbers in that set.

The greatest common divisor of two positive integers can be found by an algorithm that uses only subtractions, parity checks, and shifts of binary expansions, without using any divisions. The algorithm proceeds recursively using the following reduction:

$$(a, b) = \begin{cases} a & \text{if } a = b; \\ 2(a/2, b/2) & \text{if } a \text{ and } b \text{ are even;} \\ (a/2, b) & \text{if } a \text{ is even and } b \text{ is odd;} \\ (a - b, b) & \text{if } a \text{ and } b \text{ are odd, where } a > b. \end{cases}$$

(*Note:* Reverse the roles of a and b when necessary.) Exercises 9–13 refer to this algorithm.

9. Find $(2106, 8318)$ using this algorithm.

10. Show that this algorithm always produces the greatest common divisor of a pair of positive integers.

* **11.** How many steps does this algorithm use to find (a, b) if $a = (2^n - (-1)^n)/3$ and $b = 2(2^{n-1} - (-1)^{n-1})/3$, when n is a positive integer?

* **12.** Show that to find (a, b) this algorithm uses the subtraction step in the reduction no more than $1 + [\log_2 \max(a, b)]$ times.

* **13.** Devise an algorithm for finding the greatest common divisor of two positive integers using their balanced ternary expansions.

In Exercise 18 of Section 1.5, a modified division algorithm is given, which states that if a and $b > 0$ are integers, then there exist unique integers q, r, and e such that $a = bq + er$, where $e = \pm 1$, $r \geq 0$, and $-b/2 < er \leq b/2$. We can set up an algorithm, analogous to the Euclidean algorithm, based on this modified division algorithm, called the *least-remainder algorithm*. It works as follows: Let $r_0 = a$ and $r_1 = b$, where $a > b > 0$. Using the modified division algorithm repeatedly, obtain the greatest common divisor of a and b as the last nonzero remainder r_n in the sequence of divisions

$$r_0 = r_1 q_1 + e_2 r_2, \qquad\qquad -r_1/2 < e_2 r_2 \leq r_1/2$$
$$\vdots$$
$$r_{n-2} = r_{n-1} q_{n-1} + e_n r_n, \qquad -r_{n-1}/2 < e_n r_n \leq r_{n-1}/2$$
$$r_{n-1} = r_n q_n.$$

14. Use the least-remainder algorithm to find $(384, 226)$.

15. Show that the least-remainder algorithm always produces the greatest common divisor of two integers.

** **16.** Show that the least-remainder algorithm is always at least as fast as the Euclidean algorithm. (*Hint:* First show that if a and b are positive integers with $2b < a$, then the least-remainder algorithm can find (a, b) with no more steps than it uses to find $(a, a - b)$.)

* **17.** Find a sequence of integers $v_0, v_1, v_2, \ldots$, such that the least-remainder algorithm takes exactly n divisions to find (v_{n+1}, v_{n+2}).

* **18.** Show that the number of divisions needed to find the greatest common divisor of two positive integers using the least-remainder algorithm is less than $8/3$ times the number of digits in the smaller of the two numbers, plus $4/3$.

* **19.** Let m and n be positive integers and let a be an integer greater than 1. Show that $(a^m - 1, a^n - 1) = a^{(m,n)} - 1$.

* **20.** Show that if m and n are positive integers, then $(f_m, f_n) = f_{(m,n)}$.

The next two exercises deal with the *game of Euclid.* Two players begin with a pair of positive integers and take turns making moves of the following type. A player can move from the pair of positive integers $\{x, y\}$ with $x \geq y$, to any of the pairs $\{x - ty, y\}$, where t is a positive integer and $x - ty \geq 0$. A *winning move* consists of moving to a pair with one element equal to 0.

21. Show that every sequence of moves starting with the pair $\{a, b\}$ must eventually end with the pair $\{0, (a, b)\}$.

* **22.** Show that in a game beginning with the pair $\{a, b\}$, the first player may play a winning strategy if $a = b$ or if $a > b(1 + \sqrt{5})/2$; otherwise, the second player may play a winning strategy. (*Hint:* First show that if $y < x \leq y(1 + \sqrt{5})/2$, then there is a unique move from $\{x, y\}$ that goes to a pair $\{z, y\}$ with $y > z(1 + \sqrt{5})/2$.)

* **23.** Show that the number of bit operations needed to use the Euclidean algorithm to find the greatest common divisor of two positive integers a and b with $a > b$ is $O((\log_2 a)^2)$. (*Hint:* First show that the complexity of division of the positive integer q by the positive integer d is $O(\log d \log q)$.)

* **24.** Let a and b be positive integers and let r_j and q_j, $j = 1, 2, \ldots, n$ be the remainders and quotients of the steps of the Euclidean algorithm as defined in this section.
   a) Find the value of $\sum_{j=1}^{n} r_j q_j$.
   b) Find the value of $\sum_{j=1}^{n} r_j^2 q_j$.

25. Suppose that a and b are positive integers with $a \geq b$. Let q_i and r_i be the quotients and remainders in the steps of the Euclidean algorithm for $i = 1, 2, \ldots, n$, where r_n is the last nonzero remainder. Let $Q_i = \begin{pmatrix} q_i & 1 \\ 1 & 0 \end{pmatrix}$ and $Q = \prod_{i=0}^{n} Q_i$. Show that $\begin{pmatrix} a \\ b \end{pmatrix} = Q \begin{pmatrix} r_n \\ 0 \end{pmatrix}$.

3.4 Computational and Programming Exercises

Computations and Explorations

Using a computation program such as Maple or *Mathematica,* or programs you have written, carry out the following computations and explorations.

1. Find (9876543210, 123456789), (11111111111, 1000000001) and (45666020043321, 73433510078091009).

2. Verify Lamé's theorem for several different pairs of large positive integers of your choice.

3. Compare the number of steps required to find the greatest common divisor of different pairs of large positive integers of your choice using the Euclidean algorithm, the algorithm described in the preamble to Exercise 9, and the least-remainder algorithm described in the preamble to Exercise 14.

4. Estimate the proportion of pairs of positive integers (a, b) that are relatively prime, where a and b are positive integers not exceeding 1000, not exceeding 10,000, not exceeding 100,000, and not exceeding 1,000,000. To do so, you may want to test a random selection of a small number of such pairs (see Section 10.1 for material on pseudorandom numbers). Can you make any conjectures from this evidence?

Programming Projects

Write programs using Maple, *Mathematica,* or a language of your choice to do the following.

1. Find the greatest common divisor of two integers using the Euclidean algorithm.

2. Find the greatest common divisor of two integers using the modified Euclidean algorithm given in the preamble to Exercise 14.

3. Find the greatest common divisor of two integers using no divisions (see the preamble to Exercise 9).

4. Find the greatest common divisor of a set of more than two integers.

5. Express the greatest common divisor of two integers as a linear combination of these integers.

6. Express the greatest common divisor of a set of more than two integers as a linear combination of these integers.

* 7. Play the game of Euclid described in the preamble to Exercise 21.

3.5 The Fundamental Theorem of Arithmetic

The fundamental theorem of arithmetic is an important result that shows that the primes are the multiplicative building blocks of the integers.

Theorem 3.15. *The Fundamental Theorem of Arithmetic.* Every positive integer greater than 1 can be written uniquely as a product of primes, with the prime factors in the product written in nondecreasing order.

Sometimes, the fundamental theorem of arithmetic is extended to apply to the integer 1. That is, 1 is considered to be written uniquely as the empty product of primes.

Example 3.16.   The factorizations of some positive integers are given by

$$240 = 2 \cdot 2 \cdot 2 \cdot 2 \cdot 3 \cdot 5 = 2^4 \cdot 3 \cdot 5, \quad 289 = 17 \cdot 17 = 17^2, \quad 1001 = 7 \cdot 11 \cdot 13. \quad \blacktriangleleft$$

Note that it is convenient to combine all the factors of a particular prime into a power of this prime, such as in the previous example: For the factorization of 240, all the factors of 2 were combined to form 2^4. Factorizations of integers in which the factors of primes are combined to form powers are called *prime-power factorizations*.

To prove the fundamental theorem of arithmetic, we need the following lemma concerning divisibility. This lemma turns out to be a crucial part of the proof.

Lemma 3.4.   If a, b, and c are positive integers such that $(a, b) = 1$ and $a \mid bc$, then $a \mid c$.

Proof.   Since $(a, b) = 1$, there are integers x and y such that $ax + by = 1$. Multiplying both sides of this equation by c, we have $acx + bcy = c$. By Theorem 1.9, a divides $acx + bcy$, because this is a linear combination of a and bc, both of which are divisible by a. Hence, $a \mid c$.   ∎

The following consequence of this lemma will be needed in the proof of the fundamental theorem of arithmetic.

Lemma 3.5.   If p divides $a_1 a_2 \cdots a_n$, where p is a prime and $a_1, a_2, \ldots, a_n$ are positive integers, then there is an integer i with $1 \le i \le n$ such that p divides a_i.

Proof.   We prove this result by induction. The case where $n = 1$ is trivial. Assume that the result is true for n. Consider a product of $n + 1$ integers $a_1 a_2 \cdots a_{n+1}$ that is divisible by the prime p. We know that either $(p, a_1 a_2 \cdots a_n) = 1$ or $(p, a_1 a_2 \cdots a_n) = p$. If $(p, a_1 a_2 \cdots a_n) = 1$, then by Lemma 3.4, $p \mid a_{n+1}$. On the other hand, if $p \mid a_1 a_2 \cdots a_n$, using the induction hypothesis, there is an integer i with $1 \le i \le n$ such that $p \mid a_i$. Consequently, $p \mid a_i$ for some i with $1 \le i \le n + 1$. This proves the result.   ∎

We now begin the proof of the fundamental theorem of arithmetic. First, we will show that every positive integer greater than 1 can be written as the product of primes in at least one way. Then we will show that this product is unique up to the order of primes that appear.

Proof.   We use proof by contradiction. Assume that some positive integer cannot be written as the product of primes. Let n be the smallest such integer (such an integer must exist, from the well-ordering property). If n is prime, it is obviously the product of a set of primes, namely the one prime n. So n must be composite. Let $n = ab$, with $1 < a < n$ and $1 < b < n$. But since a and b are smaller than n, they must be the product of primes. Then, since $n = ab$, we conclude that n is also a product of primes. This contradiction shows that every positive integer can be written as the product of primes.

We now finish the proof of the fundamental theorem of arithmetic by showing that the factorization is unique. Suppose that there is an integer n that has two different factorizations into primes:

$$n = p_1 p_2 \cdots p_s = q_1 q_2 \cdots q_t,$$

where $p_1, p_2, \ldots, p_s$, and $q_1, q_2, \ldots, q_t$ are all primes, with $p_1 \leq p_2 \leq \cdots \leq p_s$ and $q_1 \leq q_2 \leq \cdots \leq q_t$.

Remove all common primes from the two factorizations to obtain

$$p_{i_1} p_{i_2} \cdots p_{i_u} = q_{j_1} q_{j_2} \cdots q_{j_v}$$

where the primes on the left-hand side of this equation differ from those on the right-hand side, $u \geq 1$, and $v \geq 1$ (because the two original factorizations were presumed to differ). However, this leads to a contradiction of Lemma 3.5; by this lemma, p_{i_1} must divide q_{j_k} for some k, which is impossible, since each q_{j_k} is prime and is different from p_{i_1}. Hence, the prime factorization of a positive integer n is unique. ∎

Where Unique Factorization Fails   The fact that every positive integer has a unique factorization into primes is a special property of the set of integers that is shared by some, but not all, systems of numbers. In Chapter 13, we will study the diophantine equation $x^n + y^n = z^n$. In the nineteenth century, mathematicians thought they could prove that this equation has no solutions in nonzero integers when n is an integer with $n \geq 3$ (a result known as Fermat's last theorem), using a form of unique factorization for certain types of algebraic numbers. It turned out that these numbers do not enjoy the property of unique factorization. The supposed proofs were incorrect, a problem that escaped the notice of many eminent mathematicians.

Although we do not want to go too far afield (by introducing algebraic number theory, for instance), we can provide an example showing that unique factorization fails for certain types of numbers. Consider the set of numbers of the form $a + b\sqrt{-5}$, where a and b are integers. This set contains every integer (taking $b = 0$), as well as other numbers such as $3\sqrt{-5}$, $-1 + 4\sqrt{-5}$, $7 - 5\sqrt{-5}$, and so on. A number of this form is prime (in this context) if it cannot be written as the product of two other numbers of this form both different than ± 1. Note that $6 = 2 \cdot 3 = (1 + \sqrt{-5})(1 - \sqrt{-5})$. Each of the numbers $2, 3, 1 + \sqrt{-5}$, and $1 - \sqrt{-5}$ is a prime (see Exercises 19–22 at the end of this section to see how this can be established). It follows that the set of numbers of the form $a + b\sqrt{-5}$ does not enjoy the property of unique factorization into primes. On the other hand, numbers of the form $a + b\sqrt{-1}$, where a and b are integers, do have unique factorization, as we will show in Chapter 14.

Using Prime Factorizations

The prime-power factorization of a positive integer n encodes essential information about n. Given this factorization, we can immediately deduce whether a prime p divides n since p divides n if and only if it appears in this factorization. (We can obtain a contradiction of the uniqueness of the prime-power factorization of n if a prime q divided n, but did not appear in the prime-power factorization of n. The reader should fill in the other parts of the proof.) For instance, since $168 = 2^3 \cdot 3 \cdot 7$, each of the primes 2, 3, and 7 divides 120, but none of the primes 5, 11, and 13 do. Furthermore, the highest power of a prime p that divides n is the power of this prime in the prime-power factorization of n. For instance, each of 2^3, 3, and 7 divides 168, but none of 2^4, 3^2, and 7^2 do. Moreover, an integer d divides n if and only if all the primes in the prime-power factorization of d appear in the prime-power factorization of n to powers at least as large as they do in the prime-power factorization of d. (The reader should also verify that this follows from the fundamental theorem of arithmetic.) The following example illustrates how we can find all the positive divisors of a positive integer using this observation.

Example 3.17. The positive divisors of $120 = 2^3 \cdot 3 \cdot 5$ are those positive integers with prime-power factorizations containing only the primes 2, 3, and 5, to powers less than or equal to 3, 1, and 1, respectively. These divisors are

1	3	5	$3 \cdot 5 = 15$
2	$2 \cdot 3 = 6$	$2 \cdot 5 = 10$	$2 \cdot 3 \cdot 5 = 30$
$2^2 = 4$	$2^2 \cdot 3 = 12$	$2^2 \cdot 5 = 20$	$2^2 \cdot 3 \cdot 5 = 60$
$2^3 = 8$	$2^3 \cdot 3 = 24$	$2^3 \cdot 5 = 40$	$2^3 \cdot 3 \cdot 5 = 120.$

◄

Another way in which we can use prime factorizations is to find greatest common divisors, as illustrated in the following example.

Example 3.18. To be a common divisor of $720 = 2^4 \cdot 3^2 \cdot 5$ and $2100 = 2^2 \cdot 3 \cdot 5^2 \cdot 7$, a positive integer can contain only the primes 2, 3, and 5 in its prime-power factorization, and the power to which one of these primes appears cannot be larger than either of the powers of that prime in the factorizations of 720 and 2100. Consequently, to be a common divisor of 720 and 2100, a positive integer can contain only the primes 2, 3, and 5 to powers no larger than 2, 1, and 1, respectively. Therefore, the greatest common divisor of 720 and 2100 is $2^2 \cdot 3 \cdot 5 = 60$. ◄

To describe, in general, how prime factorizations can be used to find greatest common divisors, let $\min(a, b)$ denote the smaller, or minimum, of the two numbers a and b. Now, let the prime factorizations of a and b be

$$a = p_1^{a_1} p_2^{a_2} \cdots p_n^{a_n}, \quad b = p_1^{b_1} p_2^{b_2} \cdots p_n^{b_n},$$

where each exponent is a nonnegative integer, and where all primes occurring in the prime factorizations of a and of b are included in both products, perhaps with 0 exponents. We note that

$$(a, b) = p_1^{\min(a_1, b_1)} p_2^{\min(a_2, b_2)} \cdots p_n^{\min(a_n, b_n)},$$

because for each prime p_i, a and b share exactly $\min(a_i, b_i)$ factors of p_i.

Prime factorizations can also be used to find the smallest integer that is a multiple of each of two positive integers. The problem of finding this integer arises when fractions are added.

Definition.   The *least common multiple* of two nonzero integers a and b is the smallest positive integer that is divisible by a and b.

The least common multiple of a and b is denoted by $[a, b]$. (*Note:* The notation $\text{lcm}(a, b)$ is also commonly used to denote the least common multiple of a and b.)

Example 3.19.   We have the following least common multiples: $[15, 21] = 105$, $[24, 36] = 72$, $[2, 20] = 20$, and $[7, 11] = 77$.   ◀

Once the prime factorizations of a and b are known, it is easy to find $[a, b]$. If $a = p_1^{a_1} p_2^{a_2} \cdots p_n^{a_n}$ and $b = p_1^{b_1} p_2^{b_2} \cdots p_n^{b_n}$, where $p_1, p_2, \ldots, p_n$ are the primes occurring in the prime-power factorizations of a and b (where we might have $a_i = 0$ or $b_i = 0$ for some i), then for an integer to be divisible by both a and b, it is necessary that in the factorization of the integer, each p_j occurs with a power at least as large as a_j and b_j. Hence, $[a, b]$, the smallest positive integer divisible by both a and b, is

$$[a, b] = p_1^{\max(a_1, b_1)} p_2^{\max(a_2, b_2)} \cdots p_n^{\max(a_n, b_n)}$$

where $\max(x, y)$ denotes the larger, or maximum, of x and y.

Finding the prime factorization of large integers is time-consuming. Therefore, we would prefer a method for finding the least common multiple of two integers without using the prime factorizations of these integers. We will show that we can find the least common multiple of two positive integers once we know the greatest common divisor of these integers. The latter can be found via the Euclidean algorithm. First, we prove the following lemma.

Lemma 3.6.   If x and y are real numbers, then $\max(x, y) + \min(x, y) = x + y$.

Proof.   If $x \geq y$, then $\min(x, y) = y$ and $\max(x, y) = x$, so that $\max(x, y) + \min(x, y) = x + y$. If $x < y$, then $\min(x, y) = x$ and $\max(x, y) = y$, and again we find that $\max(x, y) + \min(x, y) = x + y$.   ∎

We use the following theorem to find $[a, b]$ once (a, b) is known.

Theorem 3.16.   If a and b are positive integers, then $[a, b] = ab/(a, b)$, where $[a, b]$ and (a, b) are the least common multiple and greatest common divisor of a and b, respectively.

Proof.   Let a and b have prime-power factorizations $a = p_1^{a_1} p_2^{a_2} \cdots p_n^{a_n}$ and $b = p_1^{b_1} p_2^{b_2} \cdots p_n^{b_n}$, where the exponents are nonnegative integers and all primes occurring in either factorization occur in both, perhaps with 0 exponents. Now let $M_j = \max(a_j, b_j)$

and $m_j = \min(a_j, b_j)$. Then, we have

$$
\begin{aligned}
[a, b](a, b) &= p_1^{M_1} p_2^{M_2} \cdots p_n^{M_n} p_1^{m_1} p_2^{m_2} \cdots p_n^{m_n} \\
&= p_1^{M_1+m_1} p_2^{M_2+m_2} \cdots p_n^{M_n+m_n} \\
&= p_1^{a_1+b_1} p_2^{a_2+b_2} \cdots p_n^{a_n+b_n} \\
&= p_1^{a_1} p_2^{a_2} \cdots p_n^{a_n} p_1^{b_1} \cdots p_n^{b_n} \\
&= ab,
\end{aligned}
$$

since $M_j + m_j = \max(a_j, b_j) + \min(a_j, b_j) = a_j + b_j$ by Lemma 3.6. ∎

The following consequence of the fundamental theorem of arithmetic will be needed later.

Lemma 3.7. Let m and n be relatively prime positive integers. Then, if d is a positive divisor of mn, there is a unique pair of positive divisors d_1 of m and d_2 of n such that $d = d_1 d_2$. Conversely, if d_1 and d_2 are positive divisors of m and n, respectively, then $d = d_1 d_2$ is a positive divisor of mn.

Proof. Let the prime-power factorizations of m and n be $m = p_1^{m_1} p_2^{m_2} \cdots p_s^{m_s}$ and $n = q_1^{n_1} q_2^{n_2} \cdots q_t^{n_t}$. Since $(m, n) = 1$, the set of primes $p_1, p_2, \ldots p_s$ and the set of primes $q_1, q_2, \ldots, q_t$ have no common elements. Therefore, the prime-power factorization of mn is

$$
mn = p_1^{m_1} p_2^{m_2} \cdots p_s^{m_s} q_1^{n_1} q_2^{n_2} \cdots q_t^{n_t}.
$$

Hence, if d is a positive divisor of mn, then

$$
d = p_1^{e_1} p_2^{e_2} \cdots p_s^{e_s} q_1^{f_1} q_2^{f_2} \cdots q_t^{f_t},
$$

where $0 \le e_i \le m_i$ for $i = 1, 2, \ldots, s$ and $0 \le f_j \le n_j$ for $j = 1, 2, \ldots, t$. Now, let $d_1 = (d, m)$ and $d_2 = (d, n)$, so that

$$
d_1 = p_1^{e_1} p_2^{e_2} \cdots p_s^{e_s} \quad \text{and} \quad d_2 = q_1^{f_1} q_2^{f_2} \cdots q_t^{f_t}.
$$

Clearly, $d = d_1 d_2$ and $(d_1, d_2) = 1$. This is the decomposition of d that we desire. Furthermore, this decomposition is unique. To see this, note that every prime power in the factorization of d must occur in either d_1 or d_2, that prime powers in the factorization of d that are powers of primes dividing m must appear in d_1, and that prime powers in the factorization of d that are powers of primes dividing n must appear in d_2. It follows that d_1 must be (d, m) and d_2 must be (d, n).

Conversely, let d_1 and d_2 be positive divisors of m and n, respectively. Then

$$
d_1 = p_1^{e_1} p_2^{e_2} \cdots p_s^{e_s},
$$

where $0 \le e_i \le m_i$ for $i = 1, 2, \ldots, s$, and

$$
d_2 = q_1^{f_1} q_2^{f_2} \cdots q_t^{f_t},
$$

where $0 \le f_j \le n_j$ for $j = 1, 2, \ldots, t$. The integer

$$d = d_1 d_2 = p_1^{e_1} p_2^{e_2} \cdots p_s^{e_s} q_1^{f_1} q_2^{f_2} \cdots q_t^{f_t}$$

is clearly a divisor of

$$mn = p_1^{m_1} p_2^{m_2} \cdots p_s^{m_s} q_1^{n_1} q_2^{n_2} \cdots q_t^{n_t},$$

because the power of each prime occurring in the prime-power factorization of d is less than or equal to the power of that prime in the prime-power factorization of mn.   ∎

A Proof of a Special Case of Dirichlet's Theorem   Unique factorization can be used to prove special cases of Dirichlet's theorem, which states that the arithmetic progression $an + b$ contains infinitely many primes whenever a and b are relatively prime positive integers. We will illustrate this with a proof of Dirichlet's theorem for the progression $4n + 3$.

Theorem 3.17.   There are infinitely many primes of the form $4n + 3$, where n is a positive integer.

Before we prove this result, we prove a useful lemma.

Lemma 3.8.   If a and b are integers, both of the form $4n + 1$, then the product ab is also of this form.

Proof.   Since a and b are both of the form $4n + 1$, there exist integers r and s such that $a = 4r + 1$ and $b = 4s + 1$. Hence,

$$ab = (4r + 1)(4s + 1) = 16rs + 4r + 4s + 1 = 4(4rs + r + s) + 1,$$

which is again of the form $4n + 1$.   ∎

We now prove the desired result.

Proof.   Let us assume that there are only a finite number of primes of the form $4n + 3$, say $p_0 = 3, p_1, p_2, \ldots, p_r$. Let

$$Q = 4p_1 p_2 \cdots p_r + 3.$$

Then, there is at least one prime in the factorization of Q of the form $4n + 3$. Otherwise, all of these primes would be of the form $4n + 1$, and by Lemma 3.8, this would imply that Q would also be of this form, which is a contradiction. However, none of the primes $p_0, p_1, \ldots, p_n$ divides Q. The prime 3 does not divide Q, for if $3 \mid Q$, then $3 \mid (Q - 3) = 4p_1 p_2 \cdots p_r$, which is a contradiction. Likewise, none of the primes p_j can divide Q, because $p_j \mid Q$ implies $p_j \mid (Q - 4p_1 p_2 \cdots p_r) = 3$, which is absurd. Hence, there are infinitely many primes of the form $4n + 3$.   ∎

Results About Irrational Numbers   We conclude this section by proving some results about irrational numbers. If α is a rational number, then we may write α as the quotient of two integers in infinitely many ways, for if $\alpha = a/b$, where a and b are integers with $b \neq 0$, then $\alpha = ka/kb$ whenever k is a nonzero integer. It is easy to see that a positive rational number may be written uniquely as the quotient of two relatively prime positive integers; when this is done we say that the rational number is in *lowest terms*. We note that the rational number 11/21 is in lowest terms. We also see that

$$\cdots = -33/-63 = -22/-42 = -11/-21 = 11/21 = 22/42 = 33/63 = \cdots.$$

The next two results show that certain numbers are irrational. We start by giving another proof that $\sqrt{2}$ is irrational (we proved this originally in Section 1.1).

Example 3.20. Suppose that $\sqrt{2}$ is rational. Then $\sqrt{2} = a/b$, where a and b are relatively prime integers with $b \neq 0$. It follows that $2 = a^2/b^2$, so that $2b^2 = a^2$. Since $2 \mid a^2$, it follows (see Exercise 40 at the end of this section) that $2 \mid a$. Let $a = 2c$, so that $b^2 = 2c^2$. Hence, $2 \mid b^2$, and by Exercise 40, 2 also divides b. However, since $(a, b) = 1$, we know that 2 cannot divide both a and b. This contradiction shows that $\sqrt{2}$ is irrational. ◀

We can also use the following more general result to show that $\sqrt{2}$ is irrational.

Theorem 3.18. Let α be a root of the polynomial $x^n + c_{n-1}x^{n-1} + \cdots + c_1x + c_0$, where the coefficients $c_0, c_1, \ldots, c_{n-1}$ are integers. Then α is either an integer or an irrational number.

Proof. Suppose that α is rational. Then we can write $\alpha = a/b$, where a and b are relatively prime integers with $b \neq 0$. Because α is a root of $x^n + c_{n-1}x^{n-1} + \cdots + c_1x + c_0$, we have

$$(a/b)^n + c_{n-1}(a/b)^{n-1} + \cdots + c_1(a/b) + c_0 = 0.$$

Multiplying by b^n, we find that

$$a^n + c_{n-1}a^{n-1}b + \cdots + c_1ab^{n-1} + c_0b^n = 0.$$

Since

$$a^n = b(-c_{n-1}a^{n-1} - \cdots - c_1ab^{n-2} - c_0b^{n-1}),$$

we see that $b \mid a^n$. Assume that $b \neq \pm 1$. Then, b has a prime divisor p. Since $p \mid b$ and $b \mid a^n$, we know that $p \mid a^n$. Hence, by Exercise 41, we see that $p \mid a$. However, since $(a, b) = 1$, this is a contradiction, which shows that $b = \pm 1$. Consequently, if α is rational then $\alpha = \pm a$, so that α must be an integer. ∎

We illustrate the use of Theorem 3.18 with the following example.

Example 3.21. Let a be a positive integer that is not the mth power of an integer, so that $\sqrt[m]{a}$ is not an integer. Then $\sqrt[m]{a}$ is irrational by Theorem 3.18, since $\sqrt[m]{a}$ is a root of $x^m - a$. Consequently, such numbers as $\sqrt{2}$, $\sqrt[3]{5}$, $\sqrt[10]{17}$, etc., are irrational. ◀

The fundamental theorem of arithmetic can be used to prove the following result, which relates the famous Riemann zeta function to the prime numbers.

Theorem 3.19. If s is a real number with $s > 1$, then

$$\zeta(s) = \sum_{n=1}^{\infty} \frac{1}{n^s} = \prod_{p \text{ prime}} \left(1 - \frac{1}{p^s}\right)^{-1}.$$

Not surprisingly, we will not prove Theorem 3.19 because its proof depends on results from analysis. We note here that the proof uses the fundamental theorem of arithmetic to show that the term $1/n^s$, where n is a positive integer, appears exactly once when the terms of the product on the right-hand side are expanded. To see this, we use the fact that

$$\frac{1}{1 - p_j^{-s}} = \sum_{k=0}^{\infty} \frac{1}{p_j^{k_s}}$$

and then we multiply these sums together, obtaining the term

$$\frac{1}{p_1^{k_1} p_2^{k_2} \cdots p_r^{k_r}}$$

when the denominator is the prime-power factorization of n exactly once. The details of the proof can be found in [HaWr79].

3.5 Exercises

1. Find the prime factorizations of each of the following integers.

   a) 36   e) 222   i) 5040
   b) 39   f) 256   j) 8000
   c) 100   g) 515   k) 9555
   d) 289   h) 989   l) 9999

2. Find the prime factorization of 111,111.

3. Find the prime factorization of 4,849,845.

4. Find all of the prime factors of each of the following integers.

   a) 100,000   b) 10,500,000   c) 10!   d) $\binom{30}{10}$

5. Find all of the prime factors of each of the following integers.

   a) 196,608   b) 7,290,000   c) 20!   d) $\binom{50}{25}$

6. Show that all of the powers in the prime-power factorization of an integer n are even if and only if n is a perfect square.

7. Which positive integers have exactly three positive divisors? Which have exactly four positive divisors?

8. Show that every positive integer can be written as the product of possibly a square and a square-free integer. A *square-free integer* is an integer that is not divisible by any perfect squares other than 1.

9. An integer n is called *powerful* if, whenever a prime p divides n, p^2 divides n. Show that every powerful number can be written as the product of a perfect square and a perfect cube.

10. Show that if a and b are positive integers and $a^3 \mid b^2$, then $a \mid b$.

Let p be a prime and n a positive integer. If $p^a \mid n$, but $p^{a+1} \nmid n$, we say that p^a *exactly divides* n, and we write $p^a \parallel n$.

11. Show that if $p^a \parallel m$ and $p^b \parallel n$, then $p^{a+b} \parallel mn$.

12. Show that if $p^a \parallel m$, then $p^{ka} \parallel m^k$.

13. Show that if $p^a \parallel m$ and $p^b \parallel n$ with $a \neq b$, then $p^{\min(a,b)} \parallel (m+n)$.

14. Let n be a positive integer. Show that the power of the prime p occurring in the prime-power factorization of $n!$ is

$$[n/p] + [n/p^2] + [n/p^3] + \cdots.$$

15. Use Exercise 14 to find the prime-power factorization of $20!$.

16. How many zeros are there at the end of $1000!$ in decimal notation? How many in base 8 notation?

17. Find all positive integers n such that $n!$ ends with exactly 74 zeros in decimal notation.

18. Show that if n is a positive integer, it is impossible for $n!$ to end with exactly 153, 154, or 155 zeros when it is written in decimal notation.

Let $\alpha = a + b\sqrt{-5}$, where a and b are integers. Define the *norm* of α, denoted by $N(\alpha)$, as $N(\alpha) = a^2 + 5b^2$.

19. Show that if $\alpha = a + b\sqrt{-5}$ and $\beta = c + d\sqrt{-5}$, where $a, b, c,$ and d are integers, then $N(\alpha\beta) = N(\alpha)N(\beta)$.

20. A number of the form $a + b\sqrt{-5}$ is *prime* if it cannot be written as the product of numbers α and β, where neither α nor β equals ± 1. Show that the number 2 is a prime number of the form $a + b\sqrt{-5}$. (*Hint:* Start with $N(2) = N(\alpha\beta)$, and use Exercise 19.)

21. Use an argument similar to that in Exercise 20 to show that 3 is a prime number of the form $a + b\sqrt{-5}$.

22. Use arguments similar to that in Exercise 20 to show that both $1 \pm \sqrt{-5}$ are prime numbers of the form $a + b\sqrt{-5}$.

23. Find two different factorizations of the number 21 into primes of the form $a + b\sqrt{-5}$, where a and b are integers.

* 24. Show that the set of all numbers of the form $a + b\sqrt{-6}$, where a and b are integers, does not enjoy the property of unique factorization.

The next four exercises present another example of a system where unique factorization into primes fails. Let H be the set of all positive integers of the form $4k + 1$, where k is a nonnegative integer.

25. Show that the product of two elements of H is also in H.

✳ **26.** An element $h \neq 1$ in H is called a *Hilbert prime* (named after famous German mathematician *David Hilbert*) if the only way it can be written as the product of two integers in H is $h = h \cdot 1 = 1 \cdot h$. Find the 20 smallest Hilbert primes.

27. Show that every element of H can be factored into Hilbert primes.

28. Show that factorization of elements of H into Hilbert primes is not necessarily unique, by finding two different factorizations of 693 into Hilbert primes.

29. Which positive integers n are divisible by all integers not exceeding $\sqrt{n}$?

30. Find the least common multiple of each of the following pairs of integers.

   a) 8, 12   d) 111, 303
   b) 14, 15   e) 256, 5040
   c) 28, 35   f) 343, 999

31. Find the least common multiple of each of the following pairs of integers.

   a) 7, 11   d) 101, 333
   b) 12, 18   e) 1331, 5005
   c) 25, 30   f) 5040, 7700

32. Find the greatest common divisor and least common multiple of the following pairs of integers.
   a) $2 \cdot 3^2 5^3$, $2^2 3^3 7^2$
   b) $2 \cdot 3 \cdot 5 \cdot 7$, $7 \cdot 11 \cdot 13$
   c) $2^8 3^6 5^4 11^{13}$, $2 \cdot 3 \cdot 5 \cdot 11 \cdot 13$
   d) $41^{101} 47^{43} 103^{1001}$, $41^{11} 43^{47} 83^{111}$

DAVID HILBERT (1862–1943), born in Königsberg, the city famous in mathematics for its seven bridges, was the son of a judge. During his tenure at Göttingen University, from 1892 to 1930, Hilbert made many fundamental contributions to a wide range of mathematical subjects. He almost always worked on one area of mathematics at a time, making important contributions, then moving to a new mathematical subject. Some areas in which Hilbert worked are the calculus of variations, geometry, algebra, number theory, logic, and mathematical physics. Besides his many outstanding original contributions, Hilbert is remembered for his famous list of 23 difficult problems. He described these problems at the 1900 International Congress of Mathematicians, as a challenge to mathematicians at the birth of the twentieth century. Since that time, they have spurred a tremendous amount and variety of research. Although many of these problems have now been solved, several remain open, including the Riemann hypothesis, which is part of Problem 8 on Hilbert's list. Hilbert was also the author of several important textbooks in number theory and geometry.

33. Find the greatest common divisor and least common multiple of the following pairs of integers.

    a) $2^2 3^3 5^5 7^7$, $2^7 3^5 5^3 7^2$

    b) $2 \cdot 3 \cdot 5 \cdot 7 \cdot 11 \cdot 13$, $17 \cdot 19 \cdot 23 \cdot 29$

    c) $2^3 5^7 11^{13}$, $2 \cdot 3 \cdot 5 \cdot 7 \cdot 11 \cdot 13$

    d) $47^{11} 79^{111} 101^{1001}$, $41^{11} 83^{111} 101^{1000}$

34. Show that every common multiple of the positive integers a and b is divisible by the least common multiple of a and b.

35. Periodical cicadas are insects with very long larval periods and brief adult lives. For each species of periodical cicada with a larval period of 17 years, there is a similar species with a larval period of 13 years. If both the 17-year and 13-year species emerged in a particular location in 1900, when will they next both emerge in that location?

36. Which pairs of integers a and b have greatest common divisor 18 and least common multiple 540?

37. Show that if a and b are positive integers, then $(a, b) \mid [a, b]$. When does $(a, b) = [a, b]$?

38. Show that if a and b are positive integers, then there are divisors c of a and d of b with $(c, d) = 1$ and $cd = [a, b]$.

☞ **39.** Show that if $a, b,$ and c are integers, then $[a, b] \mid c$ if and only if $a \mid c$ and $b \mid c$.

☞ **40.** Use Lemma 3.4 to show that if p is a prime and a is an integer with $p \mid a^2$, then $p \mid a$.

☞ **41.** Show that if p is a prime, a is an integer, and n is a positive integer such that $p \mid a^n$, then $p \mid a$.

42. Show that if $a, b,$ and c are integers with $c \mid ab$, then $c \mid (a, c)(b, c)$.

43. a) Show that if a and b are positive integers with $(a, b) = 1$, then $(a^n, b^n) = 1$ for all positive integers n.

    b) Use part (a) to prove that if a and b are integers such that $a^n \mid b^n$, where n is a positive integer, then $a \mid b$.

44. Show that $\sqrt[3]{5}$ is irrational:

    a) by an argument similar to that given in Example 3.20;

    b) using Theorem 3.18.

45. Show that $\sqrt{2} + \sqrt{3}$ is irrational.

46. Show that $\log_2 3$ is irrational.

47. Show that $\log_p b$ is irrational, where p is a prime and b is a positive integer that is not the second or higher power of p.

* **48.** Let n be a positive integer greater than 1. Show that $1 + \frac{1}{2} + \frac{1}{3} + \cdots + \frac{1}{n}$ is not an integer.

49. Show that if a and b are positive integers, then $(a, b) = (a + b, [a, b])$.

50. Find the two positive integers with sum 798 and least common multiple 10,780. (*Hint:* Use Exercise 49.)

51. Show that if $a, b,$ and c are positive integers, then $([a, b], c) = [(a, c), (b, c)]$ and $[(a, b), c] = ([a, c], [b, c])$.

The *least common multiple* of the integers $a_1, a_2, \ldots, a_n$, which are not all zero, is the smallest positive integer that is divisible by all the integers $a_1, a_2, \ldots, a_n$; it is denoted by $[a_1, a_2, \ldots, a_n]$.

52. Find $[6, 10, 15]$ and $[7, 11, 13]$.

53. Show that $[a_1, a_2, \ldots, a_{n-1}, a_n] = [[a_1, a_2, \ldots, a_{n-1}], a_n]$.

54. Let n be a positive integer. How many pairs of positive integers satisfy $[a, b] = n$? (*Hint:* Consider the prime factorization of n.)

55. a) Show that if a, b, and c are positive integers, then
$$\max(a, b, c) = a + b + c - \min(a, b) - \min(a, c) - \min(b, c) + \min(a, b, c).$$

   b) Use part (a) to show that
$$[a, b, c] = \frac{abc(a, b, c)}{(a, b)(a, c)(b, c)}.$$

56. Generalize Exercise 55 to find a formula relating $(a_1, a_2, \ldots, a_n)$ and $[a_1, a_2, \ldots, a_n]$, where $a_1, a_2, \ldots, a_n$ are positive integers.

57. Show that if a, b, and c are positive integers, then $(a, b, c)[ab, ac, bc] = abc$.

58. Show that if a, b, and c are positive integers, then $[a, b, c](ab, ac, bc) = abc$.

59. Show that if a, b, and c are positive integers, then $([a, b], [a, c], [b, c]) = [(a, b), (a, c), (b, c)]$.

60. Prove that there are infinitely many primes of the form $6k + 5$, where k is a positive integer.

* 61. Show that if a and b are positive integers, then the arithmetic progression a, $a + b$, $a + 2b, \ldots$, contains an arbitrary number of consecutive composite terms.

62. Find the prime factorizations of each of the following integers.

   a) $10^6 - 1$ d) $2^{24} - 1$

   b) $10^8 - 1$ e) $2^{30} - 1$

   c) $2^{15} - 1$ f) $2^{36} - 1$

63. A discount store sells a camera at a price less than its usual retail price of $99 but more than $1. If they sell $8137 worth of this camera and the discounted dollar price is an integer, how many cameras did they sell?

64. A publishing company sells $375,961 worth of a particular book. How many copies of the book did they sell if their price is an exact dollar amount which is more than $1?

65. If a store sells $139,499 worth of electronic organizers at a sale price which is an exact dollar amount less than $300 and more than $1, how many electronic organizers did they sell?

66. Show that if a and b are positive integers, then $a^2 \mid b^2$ implies that $a \mid b$.

67. Show that if a, b, and c are positive integers with $(a, b) = 1$ and $ab = c^n$, then there are positive integers d and e such that $a = d^n$ and $b = e^n$.

☞ 68. Show that if $a_1, a_2, \ldots, a_n$ are pairwise relatively prime integers, then $[a_1, a_2, \ldots, a_n] = a_1 a_2 \cdots a_n$.

69. Show that among any set of $n + 1$ positive integers not exceeding $2n$, there is an integer that divides a different integer in the set.

70. Show that $(m + n)!/m!n!$ is an integer whenever m and n are positive integers.

* **71.** Find all solutions of the equation $m^n = n^m$, where m and n are integers.

72. Let $p_1, p_2, \ldots, p_n$ be the first n primes and let m be an integer with $1 < m < n$. Let Q be the product of a set of m primes in the list and let R be the product of the remaining primes. Show that $Q + R$ is not divisible by any primes in the list, and hence must have a prime factor not in the list. Conclude that there are infinitely many primes.

73. This exercise presents another proof that there are infinitely many primes. Assume that there are exactly r primes $p_1, p_2, \ldots, p_r$. Let $Q_k = \left(\prod_{j=1}^{r} p_j \right) / p_k$ for $k = 1, 2, \ldots, r$. Let $S = \sum_{j=1}^{r} Q_j$. Show that S must have a prime factor not among the r primes listed. Conclude that there are infinitely many primes. (This proof was published by G. Métrod in 1917.)

74. Show that if p is prime and $1 \leq k < p$, then the binomial coefficient $\binom{p}{k}$ is divisible by p.

75. Prove that in the prime factorization of $n!$, where n is an integer with $n > 1$, there is at least one prime factor with 1 as its exponent. (*Hint:* Use Bertrand's postulate.)

Exercises 76 and 77 outline two additional proofs that there are infinitely many primes.

76. Suppose that $p_1, \ldots, p_j$ are the first j primes, listed in increasing order. Denote by $N(x)$ the number of integers n not exceeding the integer x that are not divisible by any prime exceeding p_j.
   a) Show that every integer n not divisible by any prime exceeding p_j can be written in the form $n = r^2 s$, where s is square-free.
   b) Show there are only 2^j possible values of s in part (a) by looking at the prime factorization of such an integer n, which is a product of terms $p_k^{e_k}$, where $0 \leq k \leq j$ and e_k is 0 or 1.
   c) Show that if $n \leq x$, then $r \leq \sqrt{n} \leq \sqrt{x}$, where r is in part (a). Conclude that there are no more than $\sqrt{x}$ different values possible for r. Conclude that $N(x) \leq 2^j \sqrt{x}$.
   d) Show that if the number of primes is finite and p_j is the largest prime, then $N(x) = x$ for all integers x.
   e) Show from parts (c) and (d) that $x \leq 2^j \sqrt{x}$, so that $x \leq 2^{2j}$ for all x, leading to a contradiction. Conclude that there must be infinitely many primes.

* **77.** This exercise develops a proof that there are infinitely many primes based on the fundamental theorem of arithmetic published by A. Auric in 1915. Assume that there are exactly r primes, $p_1 < p_2 < \cdots < p_r$. Suppose that n is a positive integer and let $Q = p_r^n$.
   a) Show that an integer m with $1 \leq m \leq Q$ can be written uniquely as $m = p_1^{e_1} p_2^{e_2} \cdots p_r^{e_r}$, where $e_i \geq 0$ for $i = 1, 2, \ldots, r$. Furthermore, show that for the integer m with this factorization, $p_1^{e_1} \leq m \leq Q = p_r^n$.
   b) Let $C = (\log p_r)/(\log p_1)$. Show that $e_i \leq nC$ for $i = 1, 2, \ldots, r$ and that Q does not exceed the number of r-tuples $(e_1, e_2, \ldots, e_r)$ of exponents in the prime-power factorizations of integers m with $1 \leq m \leq Q$.

c) Conclude from part (b) that $Q = p_r^n \leq (Cn + 1)^r \leq n^r(C + 1)^r$.

d) Show that the inequality in part (c) cannot hold for sufficiently large values of n. Conclude that there must be infinitely many primes.

Suppose that n is a positive integer. We define the *Smarandache function* $S(n)$ by specifying that $S(n)$ is the least positive integer for which n divides $S(n)!$. For example, $S(8) = 4$ since 8 does not divide $1! = 1$, $2! = 2$, and $3! = 6$, but it does divide $4! = 24$.

78. Find $S(n)$ for all positive integers n not exceeding 12.

79. Find $S(n)$ for $n = 40, 41$, and 43.

80. Show that $S(p) = p$ whenever p is prime.

Let $a(n)$ be the least inverse of the Smarandache function, that is, the least positive integer for m for which $S(m) = n$. In other words, $a(n)$ is the position of the first occurrence of the integer n in the sequence $S(1), S(2), \ldots, S(k), \ldots$.

81. Find $a(n)$ for all positive integers n not exceeding 11.

∗ **82.** Find $a(12)$.

83. Show that $a(p) = p$ whenever p is prime.

Let $\mathrm{rad}(n)$ be the product of the primes that occur in the prime-power factorization of n. For example, $\mathrm{rad}(360) = \mathrm{rad}(2^3 \cdot 3^2 \cdot 5) = 2 \cdot 3 \cdot 5 = 60$.

84. Find $\mathrm{rad}(n)$ for each of these values of n.

    a) 300   c) 44004

    b) 44   d) 128128

85. Show that $\mathrm{rad}(n) = n$ when n is a positive integer if and only if n is square-free.

86. What is the value of $\mathrm{rad}(n!)$ when n is a positive integer?

87. Show that $\mathrm{rad}(nm) \leq \mathrm{rad}(n)\mathrm{rad}(m)$ for all positive integers m and n. For which positive integers m and n does equality hold?

The next six exercises establish some estimates for the size of $\pi(x)$, the number of primes less than or equal to x. These results were originally proved in the nineteenth century by Chebyshev.

88. Let p be a prime and let n be a positive integer. Show that p divides $\binom{2n}{n}$ exactly

$$([2n/p] - 2[n/p]) + ([2n/p^2] - 2[n/p^2]) + \cdots + ([2n/p^t] - 2[n/p^t])$$

times, where $t = [\log_p 2n]$. Conclude that if p^r divides $\binom{2n}{n}$, then $p^r \leq 2n$.

89. Use Exercise 88 to show that

$$\binom{2n}{n} \leq (2n)^{\pi(2n)}.$$

90. Show that the product of all primes between n and $2n$ is between $\binom{2n}{n}$ and $n^{\pi(2n)-\pi(n)}$. (*Hint:* Use the fact that every prime between n and $2n$ divides $(2n)!$ but not $(n!)^2$.)

91. Use Exercises 89 and 90 to show that

$$\pi(2n) - \pi(n) < n \log 4 / \log n.$$

* **92.** Use Exercise 91 to show that
$$\pi(2n) = (\pi(2n) - \pi(n)) + (\pi(n) - \pi(n/2)) + (\pi(n/2) - \pi(n/4))$$
$$+ \cdots \leq n \log 64 / \log n.$$

* **93.** Use Exercises 89 and 92 to show that there are positive constants c_1 and c_2 such that

$$c_1 x / \log x < \pi(x) < c_2 x / \log x$$

for all $x \geq 2$. (Compare this to the strong statement given in the prime number theorem, stated as Theorem 3.4 in Section 3.2.)

3.5 Computational and Programming Exercises

Computations and Explorations

Using a computation program such as Maple or *Mathematica,* or programs you have written, carry out the following computations and explorations.

1. Find the prime factorizations of 8,616,460,799; 1,234,567,890; 111,111,111,111; and 43,854,532,213,873.

2. Compare the number of primes of the form $4n + 1$ and the number of primes of the form $4n + 3$ for a range of values of n. Can you make any conjectures about the relationship between these numbers?

3. Find the smallest prime of the form $an + b$, given integers a and b, for a range of values of a and b. Can you make any conjectures about such primes?

Programming Projects

Write programs using Maple, *Mathematica,* or a language of your choice to do the following.

1. Find all of the positive divisors of a positive integer from its prime factorization.

2. Find the greatest common divisor of two positive integers from their prime factorizations.

3. Find the least common multiple of two positive integers from their prime factorizations.

4. Find the number of zeros at the end of the decimal expansion of $n!$, where n is a positive integer.

5. Find the prime factorization of $n!$, where n is a positive integer.

3.6 Factorization Methods and the Fermat Numbers

By the fundamental theorem of arithmetic, we know that every positive integer can be written uniquely as the product of primes. In this section, we discuss the problem of determining this factorization, and we introduce several simple factoring methods. Factoring integers is an extremely active area of mathematical research, especially because it is important in cryptography, as we will see in Chapter 8. In that chapter, we will learn that the security of the RSA public-key cryptosystem is based on the observation that factoring integers is much, much harder than finding large primes.

Before we discuss the current status of factoring algorithms, we will consider the most direct way to factor integers, called *trial division*. We will explain why it is not very efficient. Recall from Theorem 3.2 that n either is prime, or has a prime factor not exceeding $\sqrt{n}$. Consequently, when we divide n successively by the primes $2, 3, 5, \ldots$, not exceeding $\sqrt{n}$, either we find a prime factor p_1 of n or we conclude that n is prime. If we have located a prime factor p_1 of n, we next look for a prime factor of $n_1 = n/p_1$, beginning our search with the prime p_1, as n_1 has no prime factor less than p_1, and any factor of n_1 is also a factor of n. We continue, if necessary, determining whether any of the primes not exceeding $\sqrt{n_1}$ divide n_1. We continue in this manner, proceeding iteratively, to find the prime factorization of n.

Example 3.22.   Let $n = 42,833$. We note that n is not divisible by 2, 3, or 5, but that $7 \mid n$. We have

$$42,833 = 7 \cdot 6119.$$

Trial divisions show that 6119 is not divisible by any of the primes 7, 11, 13, 17, 19, or 23. However, we see that

$$6119 = 29 \cdot 211.$$

Since $29 > \sqrt{211}$, we know that 211 is prime. We conclude that the prime factorization of 42,833 is $42,833 = 7 \cdot 29 \cdot 211$.   ◄

Unfortunately, this method for finding the prime factorization of an integer is quite inefficient. To factor an integer N, it may be necessary to perform as many as $\pi(\sqrt{N})$ divisions (assuming that we already have a list of the primes not exceeding $\sqrt{N}$), altogether requiring on the order of $\sqrt{N} \log N$ bit operations because, from the prime number theorem, $\pi(\sqrt{N})$ is approximately $\sqrt{N}/\log \sqrt{N} = 2\sqrt{N}/\log N$, and from Theorem 2.7, these divisions take $O(\log^2 N)$ bit operations each.

Modern Factorization Methods

Mathematicians have long been fascinated with the problem of factoring integers. In the seventeenth century, *Pierre de Fermat* invented a factorization method based on the idea of representing a composite integer as the difference of two squares. This method is of theoretical and some practical importance, but is not very efficient in itself. We will discuss Fermat's factorization method later in this section.

Since 1970, many new factorization methods have been invented that make it possible, using powerful modern computers, to factor integers that had previously seemed impervious. We will describe several of the simplest of these newer methods. However, the most powerful factorization methods currently known are extremely complicated. Their description is beyond the scope of this book, but we will discuss the size of the integers that they can factor.

Among recent factorization methods (developed in the past twenty-five years) are several invented by J. M. Pollard, including the Pollard rho method (discussed in Section 4.6) and the Pollard $p - 1$ method (discussed in Section 6.1). These two methods are generally too slow for difficult factoring problems, unless the numbers being factored have special properties. In Section 12.5, we will introduce another method for factoring that uses continued fractions. A variation of this method, introduced by Morrison and Brillhart, was the major method used to factor large integers during the 1970s. This algorithm was the first factoring algorithm to run in *subexponential time,* which means that the number of bit operations required to factor an integer n could be written in the form $n^{\alpha(n)}$ where $\alpha(n)$ decreases as n increases. A useful notation for describing the number of bit operations required to factor a number by an algorithm running in subexponential time is $L(a, b)$, which implies that the number of bit operations used by the algorithm is $O(\exp(b(\log n)^a(\log\log n)^{1-a}))$. (The precise definition of $L(a, b)$ is somewhat more complicated.) The variation of the continued fraction algorithm invented by Morrison and Brillhart uses $L(1/2, \sqrt{3/2})$ bit operations. Its greatest success was the factorization of a 63-digit number in 1970.

The *quadratic sieve,* described by Carl Pomerance in 1981, made it possible for the first time to factor numbers having more than one hundred digits not of a special form. This method, with many enhancements added after its original invention, uses $L(1/2, 1)$ bit operations. Its great success was in factoring a 129-digit integer known as RSA-129, whose factorization was posed as a challenge by the inventors of the RSA cryptosystem discussed in Chapter 8. Currently, the best general-purpose factoring algorithm for integers with more than 115 digits is the *number field sieve,* originally suggested by Pollard and improved by Buhler, Lenstra, and Pomerance, which uses $L(1/3, (64/9)^{1/3})$ bit operations. Its greatest success has been the factorization of a 160-digit integer known as RSA-160 in early 2003. For factoring numbers with fewer than 115 digits, the quadratic sieve still seems to be quicker than the number field sieve.

An important feature of the number field and quadratic sieves (as well as other methods) is that these algorithms can be run in parallel on many computers (or processors) at the same time. This makes it possible for large teams of people to work on factoring the

---

PIERRE DE FERMAT (1601–1665) was a lawyer by profession. He was a noted jurist at the provincial parliament in the French city of Toulouse. Fermat was probably the most famous amateur mathematician in history. He published almost none of his mathematical discoveries, but did correspond with contemporary mathematicians about them. From his correspondents, especially the French monk Mersenne (discussed in Chapter 6), the world learned about his many contributions to mathematics. Fermat was one of the inventors of analytic geometry. Furthermore, he laid the foundations of calculus. Fermat, along with Pascal, gave a mathematical basis to the concept of probability. Some of Fermat's discoveries come to us only because he made notes in the margins of his copy of the work of Diophantus. His son found his copy with these notes, and published them so that other mathematicians would be aware of Fermat's results and claims.

Number of Decimal Digits	Approximate MIPS–Years Required
150	10^4
225	10^8
300	10^{11}
450	10^{16}
600	10^{20}

Table 3.2 *Computing power required to factor integers using the number field sieve.*

same integer. (See the historical note on factoring RSA-129 and other RSA challenge numbers, at the end of this subsection.)

How big will the numbers be that can be factored in the future? The answer depends on whether (or, more likely, how soon) more efficient algorithms are invented, as well as how quickly computing power advances. A useful and commonly used measure for estimating the amount of computing required to factor integers of a certain size is millions of instructions per second–years, or MIPS–years. (One MIPS–year represents the computing power of the classical DEC VAX 11/780 during one year. It is still used as a reference point even though this computer is obsolete. Pentium PCs operate at hundreds of MIPS.) Table 3.2 (adapted from information in [Od95]) displays the computing power (in terms of MIPS–years, rounded to the nearest power of ten) required to factor integers of a given size using the number field sieve. Teams of people can work together, dedicating thousands or even millions of MIPS–years to factor particular numbers. Consequently, even without the development of new algorithms, it might not be surprising to see the factorization, within the next ten years, of integers (not of a special form) with 200, or perhaps as many as 250 decimal digits.

For further information on factoring algorithms, we refer the reader to [Br89], [Br00], [Di84], [Gu75], [Od95], [Po84], [Po90], [Ri94], [Ru83], [WaSm87], and [Wi84].

Fermat Factorization   We now describe a factorization technique that is interesting, although it is not always efficient. This technique, discovered by Fermat, is known as *Fermat factorization,* and is based on the following lemma.

Lemma 3.9. If n is an odd positive integer, then there is a one-to-one correspondence between factorizations of n into two positive integers and differences of two squares that equal n.

Proof.   Let n be an odd positive integer and let $n = ab$ be a factorization of n into two positive integers. Then n can be written as the difference of two squares, because

$$n = ab = s^2 - t^2,$$

where $s = (a + b)/2$ and $t = (a - b)/2$ are both integers because a and b are both odd.

Conversely, if n is the difference of two squares, say $n = s^2 - t^2$, then we can factor n by noting that $n = (s - t)(s + t)$.

We leave it to the reader to show that this is a one-to-one correspondence.   ∎

To carry out the method of Fermat factorization, we look for solutions of the equation $n = x^2 - y^2$ by searching for perfect squares of the form $x^2 - n$. Hence, to find factorizations of n, we search for a square among the sequence of integers

$$t^2 - n, (t + 1)^2 - n, (t + 2)^2 - n, \ldots$$

where t is the smallest integer greater than $\sqrt{n}$. This procedure is guaranteed to terminate, since the trivial factorization $n = n \cdot 1$ leads to the equation

$$n = \left(\frac{n + 1}{2}\right)^2 - \left(\frac{n - 1}{2}\right)^2.$$

Example 3.23.   We factor 6077 using the method of Fermat factorization. Since $77 < \sqrt{6077} < 78$, we look for a perfect square in the sequence

$$78^2 - 6077 = 7$$
$$79^2 - 6077 = 164$$
$$80^2 - 6077 = 323$$
$$81^2 - 6077 = 484 = 22^2.$$

Since $6077 = 81^2 - 22^2$, we see that $6077 = (81 - 22)(81 + 22) = 59 \cdot 103.$ ◄

Unfortunately, Fermat factorization can be very inefficient. To factor n using this technique, it may be necessary to check as many as $(n + 1)/2 - [\sqrt{n}]$ integers to

---

The RSA Factoring Challenge

The RSA Factoring Challenge is an ongoing contest that challenges mathematicians to factor certain large integers. The first RSA challenge, posed in 1977 in Martin Gardner's column in *Scientific American*, was to factor a 129-digit integer, known as RSA-129. A $100 prize was offered for the decryption of a message; the message could be decrypted easily when this 129-digit number was factored, but not otherwise. Seventeen years passed before this challenge was met in 1994. The factorization of RSA-129 using the quadratic sieve method took approximately 5000 MIPS–years, and was carried out in eight months by more than 600 people working together. RSA Labs, a part of RSA Data Security (the company that holds the patents for the RSA cryptosystem discussed in Chapter 8), sponsors the challenge, offering cash prizes for the factorization of integers on challenge lists. So far, they have awarded more than $40,000 for successful factorizations. Factorizations of numbers on their list have led to world records. For example, in 1996, a team led by Arjen Lenstra used the number field sieve to factor RSA-130. This took approximately 750 MIPS–years. In 1999, the number field sieve was used to factor RSA-140 and RSA-155, using 2000 and 8000 MIPS–years, respectively. The factorization of RSA-160 in April 2003 is the current world record for the factorization of a number not of a special form.

determine whether they are perfect squares. Fermat factorization works best when it is used to factor integers having two factors of similar size. Although Fermat factorization is rarely used to factor large integers, its basic idea is the basis for many more powerful factorization algorithms used extensively in computer calculations.

The Fermat Numbers

The integers $F_n = 2^{2^n} + 1$ are called the *Fermat numbers*. Fermat conjectured that these integers are all primes. Indeed, the first few are primes, namely $F_0 = 3$, $F_1 = 5$, $F_2 = 17$, $F_3 = 257$, and $F_4 = 65,537$. Unfortunately, $F_5 = 2^{2^5} + 1$ is composite, as we will now demonstrate.

Example 3.24.   The Fermat number $F_5 = 2^{2^5} + 1$ is divisible by 641. We can show that $641 \mid F_5$ without actually performing the division, using several not-so-obvious observations. Note that

$$641 = 5 \cdot 2^7 + 1 = 2^4 + 5^4.$$

Hence,

$$2^{2^5} + 1 = 2^{32} + 1 = 2^4 \cdot 2^{28} + 1 = (641 - 5^4)2^{28} + 1$$
$$= 641 \cdot 2^{28} - (5 \cdot 2^7)^4 + 1 = 641 \cdot 2^{28} - (641 - 1)^4 + 1$$
$$= 641(2^{28} - 641^3 + 4 \cdot 641^2 - 6 \cdot 641 + 4).$$

Therefore, we see that $641 \mid F_5$.   ◀

The following result is a valuable aid in the factorization of Fermat numbers.

Theorem 3.20.   Every prime divisor of the Fermat number $F_n = 2^{2^n} + 1$ is of the form $2^{n+2}k + 1$.

The proof of Theorem 3.20 is presented as an exercise in Chapter 11. Here, we indicate how Theorem 3.20 is useful in determining the factorization of Fermat numbers.

Example 3.25.   From Theorem 3.20, we know that every prime divisor of $F_3 = 2^{2^3} + 1 = 257$ must be of the form $2^5 k + 1 = 32 \cdot k + 1$. Because there are no primes of this form less than or equal to $\sqrt{257}$, we can conclude that $F_3 = 257$ is prime.   ◀

Example 3.26.   When factoring $F_6 = 2^{2^6} + 1$, we use Theorem 3.20 to see that all of its prime factors are of the form $2^8 k + 1 = 256 \cdot k + 1$. Hence, we need only perform trial divisions of F_6 by primes of the form $256 \cdot k + 1$ that do not exceed $\sqrt{F_6}$. After considerable computation, we find that a prime divisor is obtained with $k = 1071$, that is, $274{,}177 = (256 \cdot 1071 + 1) \mid F_6$.   ◀

The Factorization of Fermat Numbers   A tremendous amount of effort has been devoted to the factorization of Fermat numbers. As yet, no new Fermat primes (beyond

F_4) have been found. Many mathematicians believe that no additional Fermat primes exist. We will develop a primality test for Fermat numbers in Chapter 11, which has been used to show that many Fermat numbers are composite. (When such a test is used, it is not necessary to use trial division to show that a number is not divisible by a prime not exceeding its square root.)

As of this writing (2004), a total of 214 Fermat numbers are known to be composite, but the complete factorizations are known for only seven composite Fermat numbers: F_5, F_6, F_7, F_8, F_9, F_{10}, and F_{11}. The Fermat number F_9, a number with 155 decimal digits, was factored in 1990 by Mark Manasse and Arjen Lenstra, using the number field sieve, which breaks the problem of factoring an integer into a large number of smaller factoring problems that can be done in parallel. Though Manasse and Lenstra farmed out computations for the factorization of F_9 to hundreds of mathematicians and computer scientists, it still took about two months to complete the computations. (For details of the factorization of F_9, see [Ci90].)

The prime factorization of F_{11} was discovered by Richard Brent in 1989, using a factorization algorithm known as the elliptic curve method (described in detail in [Br89]). There are 617 decimal digits in F_{11}, and $F_{11} = 319,489 \cdot 974,849 \cdot P_{21} \cdot P_{22} \cdot P_{564}$, where P_{21}, P_{22}, and P_{564} are primes with 21, 22, and 564 digits, respectively. It took until 1995 for Brent to completely factor F_{10}. He discovered, using elliptic curve factorization, that $F_{10} = 45,592,577 \cdot 6,487,031,809 \cdot P_{40} \cdot P_{252}$, where P_{40} and P_{252} are primes with 40 and 252 digits, respectively.

Many Fermat numbers are known to be composite because at least one prime factor of these numbers has been found, using results such as Theorem 3.20. It is also known that F_n is composite for $n = 14, 20, 22,$ and 24, but no factors of these numbers have yet been found. The largest n for which it is known that F_n is composite is $n = 2,478,782$. ($F_{382,447}$ was the first Fermat number with more than 100,000 digits shown to be composite; it was shown to be composite in July 1999.) F_{33} is the smallest Fermat number that has not yet been shown to be composite, if it is indeed composite. Because of steady advances in computer software and hardware, we can expect new results on the nature of Fermat numbers and their factorizations to be found at a healthy rate.

The factorization of Fermat numbers is part of the *Cunningham project,* sponsored by the American Mathematical Society. Devoted to building tables of all the known factors of integers of the form $b^n \pm 1$, where $b = 2, 3, 5, 6, 7, 10, 11,$ and 12, the project's name refers to A. J. Cunningham, a colonel in the British army, who compiled a table of factors of integers of this sort in the early years of the twentieth century. The factor tables as of 1988 are contained in [Br88]; the current state of affairs is available over the Internet. Numbers of the form $b^n \pm 1$ are of special interest because of their importance in generating pseudorandom numbers (see Chapter 10), their importance in abstract algebra, and their significance in number theory.

In conjunction with the Cunningham project, a list of the "ten most wanted" integers to be factored is kept by Samuel Wagstaff of Purdue University. For example, until it was factored in 1990, F_9 was on this list. With advances in factoring techniques and computer power, increasingly larger numbers are included on the list. In the early 1980s, the largest had between 50 and 70 decimal digits, in the early 1990s between 90 and 130 decimal digits, and today they have between 190 and 200 decimal digits.

Using the Fermat Numbers to Prove the Infinitude of Primes It is possible to prove that there are infinitely many primes using Fermat numbers. We begin by showing that any two distinct Fermat numbers are relatively prime. The following lemma will be used.

Lemma 3.10. Let $F_k = 2^{2^k} + 1$ denote the kth Fermat number, where k is a nonnegative integer. Then for all positive integers n, we have

$$F_0 F_1 F_2 \cdots F_{n-1} = F_n - 2.$$

Proof. We will prove the lemma using mathematical induction. For $n = 1$, the identity reads

$$F_0 = F_1 - 2.$$

This is obviously true, because $F_0 = 3$ and $F_1 = 5$. Now, let us assume that the identity holds for the positive integer n, so that

$$F_0 F_1 F_2 \cdots F_{n-1} = F_n - 2.$$

With this assumption, we can easily show that the identity holds for the integer $n + 1$, because

$$F_0 F_1 F_2 \cdots F_{n-1} F_n = (F_0 F_1 F_2 \cdots F_{n-1}) F_n$$
$$= (F_n - 2) F_n = (2^{2^n} - 1)(2^{2^n} + 1)$$
$$= (2^{2^n})^2 - 1 = 2^{2^{n+1}} - 1 = F_{n+1} - 2. \qquad \blacksquare$$

This leads to the following theorem.

Theorem 3.21. Let m and n be distinct nonnegative integers. Then the Fermat numbers F_m and F_n are relatively prime.

Proof. Let us assume that $m < n$. By Lemma 3.10, we know that

$$F_0 F_1 F_2 \cdots F_m \cdots F_{n-1} = F_n - 2.$$

Assume that d is a common divisor of F_m and F_n. Then, Theorem 1.8 tells us that

$$d \mid (F_n - F_0 F_1 F_2 \cdots F_m \cdots F_{n-1}) = 2.$$

Hence, either $d = 1$ or $d = 2$. However, since F_m and F_n are odd, d cannot be 2. Consequently, $d = 1$ and $(F_m, F_n) = 1$. $\qquad \blacksquare$

Using Fermat numbers, we now give another proof that there are infinitely many primes. First, we note that by Lemma 3.1 in Section 3.1, every Fermat number F_n has a prime divisor p_n. Because $(F_m, F_n) = 1$, we know that $p_m \neq p_n$ whenever $m \neq n$. Hence, we can conclude that there are infinitely many primes.

The Fermat Primes and Geometry   The Fermat primes are important in geometry. The proof of the following famous theorem of Gauss may be found in [Or88].

Theorem 3.22.   A regular polygon of n sides can be constructed using a ruler and compass if and only if n is the product of a nonnegative power of 2 and a nonnegative number of distinct Fermat primes.

3.6 Exercises

1. Find the prime factorization of each of the following positive integers.

   a) 33,776,925   b) 210,733,237   c) 1,359,170,111

2. Find the prime factorization of each of the following positive integers.

   a) 33,108,075   b) 7,300,977,607   c) 4,165,073,376,607

3. Using the Fermat factorization method, factor each of the following positive integers.

   a) 143   c) 43
   b) 2279   d) 11,413

4. Using the Fermat factorization method, factor each of the following positive integers.

   a) 8051   d) 11,021
   b) 73   e) 3,200,399
   c) 46,009   f) 24,681,023

5. Show that the last two decimal digits of a perfect square must be one of the following pairs: 00, e1, e4, 25, o6, e9, where e stands for any even digit and o stands for any odd digit. (*Hint:* Show that n^2, $(50 + n)^2$, and $(50 - n)^2$ all have the same final decimal digits, and then consider those integers n with $0 \le n \le 25$.)

6. Explain how the result of Exercise 5 can be used to speed up Fermat's factorization method.

7. Show that if the smallest prime factor of n is p, then $x^2 - n$ will not be a perfect square for $x > (n + p^2)/(2p)$, with the single exception $x = (n + 1)/2$.

Exercises 8–10 involve the method of *Draim factorization*. To use this technique to search for a factor of the positive integer $n = n_1$, we start by using the division algorithm, to obtain

$$n_1 = 3q_1 + r_1, \quad 0 \le r_1 < 3.$$

Setting $m_1 = n_1$, we let

$$m_2 = m_1 - 2q_1, \quad n_2 = m_2 + r_1.$$

We use the division algorithm again, to obtain

$$n_2 = 5q_2 + r_2, \quad 0 \le r_2 < 5,$$

and we let

$$m_3 = m_2 - 2q_2, \quad n_3 = m_3 + r_2.$$

We proceed recursively, using the division algorithm, to write

$$n_k = (2k + 1)q_k + r_k, \quad 0 \le r_k < 2k + 1,$$

and we define

$$m_k = m_{k-1} - 2q_{k-1}, \quad n_k = m_k + r_{k-1}.$$

We stop when we obtain a remainder $r_k = 0$.

8. Show that $n_k = kn_1 - (2k + 1)(q_1 + q_2 + \cdots + q_{k-1})$ and that $m_k = n_1 - 2 \cdot (q_1 + q_2 + \cdots + q_{k-1})$.

9. Show that if $(2k + 1) \mid n$, then $(2k + 1) \mid n_k$ and $n = (2k + 1)m_{k+1}$.

10. Factor 5899 using Draim factorization.

In Exercises 11–13, we develop a factorization technique known as *Euler's method*. It is applicable when the integer being factored is odd and can be written as the sum of two squares in two different ways. Let n be odd and let $n = a^2 + b^2 = c^2 + d^2$, where a and c are odd positive integers, and b and d are even positive integers.

11. Let $u = (a - c, b - d)$. Show that u is even, and that if $r = (a - c)/u$ and $s = (d - b)/u$, then $(r, s) = 1$, $r(a + c) = s(d + b)$, and $s \mid (a + c)$.

12. Let $sv = a + c$. Show that $rv = d + b$, $v = (a + c, d + b)$, and v is even.

13. Conclude that n may be factored as $n = [(u/2)^2 + (v/2)^2](r^2 + s^2)$.

14. Use Euler's method to factor each of the following integers.
   a) $221 = 10^2 + 11^2 = 5^2 + 14^2$
   b) $2501 = 50^2 + 1^2 = 49^2 + 10^2$
   c) $1{,}000{,}009 = 1000^2 + 3^2 = 972^2 + 235^2$

15. Show that any number of the form $2^{4n+2} + 1$ can be factored easily by the use of the identity $4x^4 + 1 = (2x^2 + 2x + 1)(2x^2 - 2x + 1)$. Factor $2^{18} + 1$ using this identity.

16. Show that if a is a positive integer and $a^m + 1$ is an odd prime, then $m = 2^n$ for some positive integer n. (*Hint:* Recall the identity $a^m + 1 = (a^k + 1)(a^{k(l-1)} - a^{k(l-2)} + \cdots - a^k + 1)$, where $m = kl$ and l is odd).

17. Show that the last digit in the decimal expansion of $F_n = 2^{2^n} + 1$ is 7 if $n \ge 2$. (*Hint:* Using mathematical induction, show that the last decimal digit of 2^{2^n} is 6.)

18. Use the fact that every prime divisor of $F_4 = 2^{2^4} + 1 = 65{,}537$ is of the form $2^6k + 1 = 64k + 1$ to verify that F_4 is prime. (You should need only one trial division.)

19. Use the fact that every prime divisor of $F_5 = 2^{2^5} + 1$ is of the form $2^7k + 1 = 128k + 1$ to demonstrate that the prime factorization of F_5 is $F_5 = 641 \cdot 6{,}700{,}417$.

20. Find all primes of the form $2^{2^n} + 5$, where n is a nonnegative integer.

21. Estimate the number of decimal digits in the Fermat number F_n.

* 22. What is the greatest common divisor of n and F_n, where n is a positive integer? Prove that your answer is correct.

23. Show that the only integer of the form $2^m + 1$, where m is a positive integer, that is a power of a positive integer (i.e., is of the form n^k, where n and k are positive integers with $k \ge 2$) occurs when $m = 3$.

24. Factoring kn by the Fermat factorization method, where k is a small positive integer, is sometimes easier than factoring n by this method. Show that to factor 901 by the Fermat factorization method, it is easier to factor $3 \cdot 901 = 2703$ than to factor 901.

3.6 Computational and Programming Exercises

Computations and Explorations

Using a computation program such as Maple or *Mathematica,* or programs you have written, carry out the following computations and explorations.

1. Using trial division, find the prime factorization of several integers of your choice exceeding 10,000.

2. Factor several integers of your choice exceeding 10,000, using Fermat factorization.

3. Factor the Fermat numbers F_6 and F_7 using Theorem 3.20.

Programming Projects

Write programs using Maple, *Mathematica,* or a language of your choice to do the following.

1. Given a positive integer n, find the prime factorization of n.

2. Given a positive integer n, perform the Fermat factorization method on n.

3. Given a positive integer n, perform Draim factorization on n (see the preamble to Exercise 8).

4. Check the Fermat number F_n, where n is a positive integer, for prime factors, using Theorem 3.20.

3.7 Linear Diophantine Equations

Consider the following problem: A man wishes to purchase \$510 of travelers' checks. The checks are available only in denominations of \$20 and \$50. How many of each denomination should he buy? If we let x denote the number of \$20 checks and y the number of \$50 checks that he should buy, then the equation $20x + 50y = 510$ must be satisfied. To solve this problem, we need to find all solutions of this equation, where both x and y are nonnegative integers.

A related problem arises when a woman wishes to mail a package. The postal clerk determines the cost of postage to be 83 cents, but only 6-cent and 15-cent stamps are available. Can some combination of these stamps be used to mail the package? To answer this, we first let x denote the number of 6-cent stamps and y the number of 15-cent stamps to be used. Then we must have $6x + 15y = 83$, where both x and y are nonnegative integers.

When we require that solutions of a particular equation come from the set of integers, we have a *diophantine equation.* These equations get their name from the ancient Greek mathematician *Diophantus,* who wrote on equations where solutions are restricted to rational numbers. The equation $ax + by = c$, where a, b, and c are integers, is called a *linear diophantine equation in two variables.*

Note that the pair of integers (x, y) is a solution of the linear diophantine equation $ax + by = c$ if and only if the (x, y) is a lattice point in the plane that lies on the line $ax + by = c$. We illustrate this in Figure 3.2 for the linear diophantine equation $2x + 3y = 5$.

The first person to describe a general solution of linear diophantine equations was the Indian mathematician *Brahmagupta*, who included it in a book he wrote in the seventh century. We now develop the theory for solving such equations. The following theorem tells us when such an equation has solutions, and when there are solutions, explicitly describes them.

Theorem 3.23. Let a and b be integers with $d = (a, b)$. The equation $ax + by = c$ has no integral solutions if $d \nmid c$. If $d \mid c$, then there are infinitely many integral solutions. Moreover, if $x = x_0$, $y = y_0$ is a particular solution of the equation, then all solutions are given by

$$x = x_0 + (b/d)n, \quad y = y_0 - (a/d)n,$$

where n is an integer.

Proof. Assume that x and y are integers such that $ax + by = c$. Then, because $d \mid a$ and $d \mid b$, by Theorem 1.9, $d \mid c$ as well. Hence, if $d \nmid c$, there are no integral solutions of the equation.

Now assume that $d \mid c$. By Theorem 3.8, there are integers s and t with

(3.3) $$d = as + bt.$$

Since $d \mid c$, there is an integer e with $de = c$. Multiplying both sides of (3.3) by e, we have

$$c = de = (as + bt)e = a(se) + b(te).$$

Hence, one solution of the equation is given by $x = x_0$ and $y = y_0$, where $x_0 = se$ and $y_0 = te$.

DIOPHANTUS (c. 250) wrote the *Arithmetica,* which is the earliest known book on algebra; it contains the first systematic use of mathematical notation to represent unknowns in equations and powers of these unknowns. Almost nothing is known about Diophantus, other than that he lived in Alexandria around 250 C.E. The only source of details about his life comes from an epigram found in a collection called the *Greek Anthology:* "Diophantus passed one sixth of his life in childhood, one twelfth in youth, and one seventh as a bachelor. Five years after his marriage was born a son who died four years before his father, at half his father's age." From this the reader can infer that Diophantus lived to the age of 84.

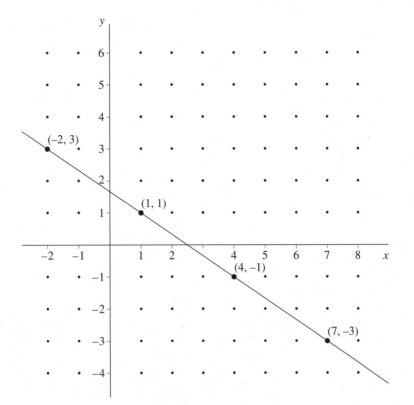

Figure 3.2 *Solutions of* $2x + 3y = 5$ *in integers x and y correspond to the lattice points on the line* $2x + 3y = 5.$

To show that there are infinitely many solutions, let $x = x_0 + (b/d)n$ and $y = y_0 - (a/d)n$, where n is an integer. We will first show that any pair (x, y), with $x = x_0 + (b/d)n$, $y = y_0 - (a/d)n$, where n is an integer, is a solution; then we will show that every solution must have this form. We see that this pair (x, y) is a solution, because

$$ax + by = ax_0 + a(b/d)n + by_0 - b(a/d)n = ax_0 + by_0 = c.$$

BRAHMAGUPTA (598–670), thought to have been born in Ujjain, India, became the head of the astronomical observatory there; this observatory was the center of Indian mathematical studies at that time. Brahmagupta wrote two important books on mathematics and astronomy, *Brahma-sphuta-siddhanta* ("The Opening of the Universe") and *Khandakhadyaka,* written in 628 and 665, respectively. He developed many interesting formulas and theorems in planar geometry, and studied arithmetic progressions and quadratic equations. Brahmagupta developed new algebraic notation, and his understanding of the number system was advanced for his time. He is considered to be the first person to describe a general solution of linear diophantine equations. In astronomy, he studied eclipses, positions of the planets, and the length of the year.

We now show that every solution of the equation $ax + by = c$ must be of the form described in the theorem. Suppose that x and y are integers with $ax + by = c$. Because

$$ax_0 + by_0 = c,$$

by subtraction we find that

$$(ax + by) - (ax_0 + by_0) = 0,$$

which implies that

$$a(x - x_0) + b(y - y_0) = 0.$$

Hence,

$$a(x - x_0) = b(y_0 - y).$$

Dividing both sides of this last equation by d, we see that

$$(a/d)(x - x_0) = (b/d)(y_0 - y).$$

By Theorem 3.6, we know that $(a/d, b/d) = 1$. Using Lemma 3.4, it follows that $(a/d) \mid (y_0 - y)$. Hence, there is an integer n with $(a/d)n = y_0 - y$; this means that $y = y_0 - (a/d)n$. Now, putting this value of y into the equation $a(x - x_0) = b(y_0 - y)$, we find that $a(x - x_0) = b(a/d)n$, which implies that $x = x_0 + (b/d)n$. ∎

The following examples illustrate the use of Theorem 3.23.

Example 3.27.   By Theorem 3.23, there are no integral solutions of the diophantine equation $15x + 6y = 7$, because $(15, 6) = 3$ but $3 \nmid 7$.   ◀

Example 3.28.   By Theorem 3.23, there are infinitely many solutions of the diophantine equation $21x + 14y = 70$, because $(21, 14) = 7$ and $7 \mid 70$. To find these solutions, note that by the Euclidean algorithm, $1 \cdot 21 + (-1) \cdot 14 = 7$, so that $10 \cdot 21 + (-10) \cdot 14 = 70$. Hence, $x_0 = 10$, $y_0 = -10$ is a particular solution. All solutions are given by $x = 10 + 2n$, $y = -10 - 3n$, where n is an integer.   ◀

We will now use Theorem 3.23 to solve the two problems described at the beginning of the section.

Example 3.29.   Consider the problem of forming 83 cents in postage using only 6- and 15-cent stamps. If x denotes the number of 6-cent stamps and y denotes the number of 15-cent stamps, we have $6x + 15y = 83$. Since $(6, 15) = 3$ does not divide 83, by Theorem 3.23 we know that there are no integral solutions. Hence, no combination of 6- and 15-cent stamps gives the correct postage.   ◀

Example 3.30.   Consider the problem of purchasing $510 of travelers' checks, using only $20 and $50 checks. How many of each type of check should be used?

Let x be the number of \$20 checks and let y be the number of \$50 checks. We have the equation $20x + 50y = 510$. Note that the greatest common divisor of 20 and 50 is $(20, 50) = 10$. Because $10 \mid 510$, there are infinitely many integral solutions of this linear diophantine equation. Using the Euclidean algorithm, we find that $20(-2) + 50 = 10$. Multiplying both sides by 51, we obtain $20(-102) + 50(51) = 510$. Hence, a particular solution is given by $x_0 = -102$ and $y_0 = 51$. Theorem 3.23 tells us that all integral solutions are of the form $x = -102 + 5n$ and $y = 51 - 2n$. Because we want both x and y to be nonnegative, we must have $-102 + 5n \geq 0$ and $51 - 2n \geq 0$; thus, $n \geq 20\ 2/5$ and $n \leq 25\ 1/2$. Because n is an integer, it follows that $n = 21, 22, 23, 24$, or 25. Hence, we have the following five solutions: $(x, y) = (3, 9), (8, 7), (13, 5), (18, 3)$, and $(23, 1)$. So the teller can give the customer 3 \$20 checks and 9 \$50 checks, 8 \$20 checks and 7 \$50 checks, 13 \$20 checks and 5 \$50 checks, 18 \$20 checks and 3 \$50 checks, or 23 \$20 checks and 1 \$50 check. ◀

We can extend Theorem 3.23 to cover linear diophantine equations with more than two variables as the following theorem demonstrates.

Theorem 3.24. If $a_1, a_2, \ldots, a_n$ are nonzero positive integers, then the equation $a_1x_1 + a_2x_2 + \cdots + a_nx_n = c$ has an integral solution if and only if $d = (a_1, a_2, \ldots, a_n)$ divides c. Furthermore, when there is a solution, there are infinitely many solutions.

Proof. If there are integers $x_1, x_2, \ldots, x_n$ such that $a_1x_1 + a_2x_2 + \cdots + a_nx_n = c$, then because d divides a_i for $i = 1, 2, \ldots, n$, by Theorem 1.9, d also divides c. Hence, if $d \nmid c$ there are no integral solutions of the equation.

We will use mathematical induction to prove that there are infinitely many integral solutions when $d \mid c$. Note that by Theorem 3.23 this is true when $n = 2$.

Now, suppose that there are infinitely many solutions for all equations in n variables satisfying the hypotheses. By Theorem 3.9, the set of linear combinations $a_nx_n + a_{n+1}x_{n+1}$ is the same as the set of multiples of (a_n, a_{n+1}). Hence, for every integer y there are infinitely many solutions of the linear diophantine equation $a_nx_n + a_{n+1}x_{n+1} = (a_n, a_{n+1})y$. It follows that the original equation in $n + 1$ variables can be reduced to a linear diophantine equation in n variables:

$$a_1x_1 + a_2x_2 + \cdots + a_{n-1}x_{n-1} + (a_n, a_{n+1})y = c.$$

Note that c is divisible by $(a_1, a_2, \ldots, a_{n-1}, (a_n, a_{n+1}))$ because, by Lemma 3.2, this greatest common divisor equals $(a_1, a_2, \ldots, a_n, a_{n+1})$. By the inductive hypothesis, this equation has infinitely many integer solutions, as it is a linear diophantine equation in n variables where the greatest common divisor of the coefficients divides the constant c. It follows that there are infinitely many solutions to the original equation. ∎

A method for solving linear diophantine equations in more than two variables can be found using the reduction in the proof of Theorem 3.24. We leave an application of Theorem 3.24 to the exercises.

3.7 Exercises

1. For each of the following linear diophantine equations, either find all solutions, or show that there are no integral solutions.

   a) $2x + 5y = 11$

   b) $17x + 13y = 100$

   c) $21x + 14y = 147$

   d) $60x + 18y = 97$

   e) $1402x + 1969y = 1$

2. For each of the following linear diophantine equations, either find all solutions, or show that there are no integral solutions.

   a) $3x + 4y = 7$

   b) $12x + 18y = 50$

   c) $30x + 47y = -11$

   d) $25x + 95y = 970$

   e) $102x + 1001y = 1$

3. A Japanese businessman returning home from a trip to North America exchanges his U.S. and Canadian dollars for yen. If he receives 15,286 yen, and received 122 yen for each U.S. and 112 yen for each Canadian dollar, how many of each type of currency did he exchange?

4. A student returning from Europe changes his euros and Swiss francs into U.S. money. If she receives $46.26, and received $1.11 for each euro and 83¢ for each Swiss franc, how much of each type of currency did she exchange?

5. A professor returning home from conferences in Paris and London changes his euros and pounds into U.S. money. If he receives $117.98, and received $1.11 for each euro and $1.69 for each pound, how much of each type of currency did he exchange?

6. The Indian astronomer and mathematician Mahavira, who lived in the ninth century, posed this puzzle: A band of 23 weary travelers entered a lush forest where they found 63 piles each containing the same number of plantains and a remaining pile containing seven plantains. They divided the plantains equally. How many plantains were in each of the 63 piles? Solve this puzzle.

7. A grocer orders apples and oranges at a total cost of $8.39. If apples cost him 25¢ each and oranges cost him 18¢ each, how many of each type of fruit did he order?

8. A shopper spends a total of $5.49 for oranges, which cost 18¢ each, and grapefruit, which cost 33¢ each. What is the minimum number of pieces of fruit the shopper could have bought?

9. A postal clerk has only 14- and 21-cent stamps to sell. What combinations of these may be used to mail a package requiring postage of exactly each of the following amounts?

   a) $3.50   b) $4.00   c) $7.77

10. At a clambake, the total cost of a lobster dinner is $11 and of a chicken dinner is $8. What can you conclude if the total bill is each of the following amounts?

   a) $777   b) $96   c) $69

∗ 11. Find all integer solutions of each of the following linear diophantine equations.

  a) $2x + 3y + 4z = 5$
  b) $7x + 21y + 35z = 8$
  c) $101x + 102y + 103z = 1$

∗ 12. Find all integer solutions of each of the following linear diophantine equations.

  a) $2x_1 + 5x_2 + 4x_3 + 3x_4 = 5$
  b) $12x_1 + 21x_2 + 9x_3 + 15x_4 = 9$
  c) $15x_1 + 6x_2 + 10x_3 + 21x_4 + 35x_5 = 1$

13. Which combinations of pennies, dimes, and quarters have a total value of 99¢?

14. How many ways can change be made for one dollar, using each of the following coins?

  a) dimes and quarters
  b) nickels, dimes, and quarters
  c) pennies, nickels, dimes, and quarters

In Exercises 15–17, we consider simultaneous linear diophantine equations. To solve these, first eliminate all but two variables and then solve the resulting equation in two variables.

15. Find all integer solutions of the following systems of linear diophantine equations.

  a) $x + y + z = 100$
  $x + 8y + 50z = 156$

  b) $x + y + z = 100$
  $x + 6y + 21z = 121$

  c) $x + y + z + w = 100$
  $x + 2y + 3z + 4w = 300$
  $x + 4y + 9z + 16w = 1000$

16. A piggy bank contains 24 coins, all of which are nickels, dimes, or quarters. If the total value of the coins is two dollars, what combinations of coins are possible?

17. Nadir Airways offers three types of tickets on their Boston–New York flights. First-class tickets are $140, second-class tickets are $110, and standby tickets are $78. If 69 passengers pay a total of $6548 for their tickets on a particular flight, how many of each type of ticket were sold?

18. Is it possible to have 50 coins, all of which are pennies, dimes, or quarters, with a total worth $3?

Let a and b be relatively prime positive integers, and let n be a positive integer. A solution (x, y) of the linear diophantine equation $ax + by = n$ is *nonnegative* when both x and y are nonnegative.

∗ 19. Show that whenever $n \geq (a - 1)(b - 1)$, there is a nonnegative solution of $ax + by = n$.

∗ 20. Show that if $n = ab - a - b$, then there are no nonnegative solutions of $ax + by = n$.

∗ 21. Show that there are exactly $(a - 1)(b - 1)/2$ nonnegative integers $n < ab - a - b$ such that the equation has a nonnegative solution.

22. The post office in a small Maine town is left with stamps of only two values. They discover that there are exactly 33 postage amounts that cannot be made up using these stamps, including 46¢. What are the values of the remaining stamps?

* 23. A Chinese puzzle found in the sixth-century work of mathematician Chang Ch'iu-chien, called the "hundred fowls" problem, asks: If a cock is worth five coins, a hen three coins, and three chickens together are worth one coin, how many cocks, hens, and chickens, totaling 100, can be bought for 100 coins? Solve this problem.

* 24. Find all solutions where x and y are integers to the diophantine equation

$$\frac{1}{x} + \frac{1}{y} = \frac{1}{14}.$$

3.7 Computational and Programming Exercises

Computations and Explorations

Using a computation program such as Maple or *Mathematica,* or programs you have written, carry out the following computations and explorations.

1. Determine which positive integers are of the form $ax + by$, where x and y are nonnegative integers and a and b are relatively prime positive integers of your choice. Use your evidence to confirm the results of Exercises 19–21.

Programming Projects

Write programs using Maple, *Mathematica,* or a language of your choice to do the following.

1. Find the solutions of a linear diophantine equation in two variables.

2. Find the positive solutions of a linear diophantine equation in two variables.

3. Find the solutions of a linear diophantine equation in three variables.

* 4. Find all positive integers n for which the linear diophantine equation $ax + by = n$ has no positive solutions (see the preamble to Exercise 19).

4

Congruences

Introduction

The language of congruences was invented by the great German mathematician Gauss. It allows us to work with divisibility relationships in much the same way as we work with equalities. We will develop the basic properties of congruences in this chapter, describe how to do arithmetic with congruences, and study congruences involving unknowns, such as linear congruences. An example leading to a linear congruence is the problem of finding all integers x such that when $7x$ is divided by 11, the remainder is 3. We will also study systems of linear congruences that arise from such problems as the ancient Chinese puzzle that asks for a number that leaves a remainder of 2, 3, and 2, when divided by 3, 5, and 7, respectively. We will learn how to solve systems of linear congruences in one unknown, such as the system that results from this puzzle, using a famous method known as the Chinese remainder theorem. We will also learn how to solve polynomial congruences. Finally, we will introduce a factoring method, known as the Pollard rho method, which we use congruences to specify.

4.1 Introduction to Congruences

The special language of congruences that we introduce in this chapter, which is extremely useful in number theory, was developed at the beginning of the nineteenth century by *Karl Friedrich Gauss*, one of the most famous mathematicians in history.

The language of congruences makes it possible to work with divisibility relationships much as we work with equalities. Prior to the introduction of congruences, the notation used for divisibility relationships was awkward and difficult to work with. The introduction of a convenient notation helped accelerate the development of number theory.

141

Definition.   Let m be a positive integer. If a and b are integers, we say that a is *congruent to b modulo m* if $m \mid (a - b)$.

If a is congruent to b modulo m, we write $a \equiv b \pmod{m}$. If $m \nmid (a - b)$, we write $a \not\equiv b \pmod{m}$, and say that a and b are *incongruent modulo m*. The integer m is called the *modulus* of the congruence. The plural of modulus is *moduli*.

Example 4.1.   We have $22 \equiv 4 \pmod 9$, since $9 \mid (22 - 4) = 18$. Likewise $3 \equiv -6 \pmod 9$ and $200 \equiv 2 \pmod 9$. On the other hand, $13 \not\equiv 5 \pmod 9$ since $9 \nmid (13 - 5) = 8$. ◀

Congruences often arise in everyday life. For instance, clocks work either modulo 12 or 24 for hours and modulo 60 for minutes and seconds; calendars work modulo 7 for days of the week and modulo 12 for months. Utility meters often operate modulo 1000, and odometers usually work modulo 100,000.

In working with congruences, we will sometimes need to translate them into equalities. The following theorem helps us to do this.

Theorem 4.1.   If a and b are integers, then $a \equiv b \pmod m$ if and only if there is an integer k such that $a = b + km$.

Proof.   If $a \equiv b \pmod m$, then $m \mid (a - b)$. This means that there is an integer k with $km = a - b$, so that $a = b + km$.

KARL FRIEDRICH GAUSS (1777–1855) was the son of a bricklayer. It was quickly apparent that he was a prodigy. In fact, at the age of 3, he corrected an error in his father's payroll. In his first arithmetic class, the teacher gave an assignment designed to keep the class busy, namely to find the sum of the first 100 positive integers. Gauss, who was 8 at the time, realized that this sum is $50 \cdot 101 = 5050$, because the terms can be grouped as $1 + 100 = 101$, $2 + 99 = 101, \ldots, 49 + 52 = 101$, and $50 + 51 = 101$. In 1796, Gauss made an important discovery in an area of geometry that had not progressed since ancient times. In particular, he showed that a regular heptadecagon (17-sided polygon) could be drawn using just a ruler and a compass. In 1799, he presented the first rigorous proof of the fundamental theorem of algebra, which states that a polynomial of degree n with real coefficients has exactly n roots. Gauss made fundamental contributions to astronomy, including calculating the orbit of the asteroid Ceres. On the basis of this calculation, Gauss was appointed director of the Göttingen Observatory. He laid the foundations of modern number theory with his book *Disquisitiones Arithmeticae* in 1801. Gauss was called "Princeps Mathematicorum" (the Prince of Mathematicians) by his contemporaries. Although Gauss is noted for his many discoveries in geometry, algebra, analysis, astronomy, and mathematical physics, he had a special interest in number theory. This can be seen from his statement: "Mathematics is the queen of sciences, and the theory of numbers is the queen of mathematics." Gauss made most of his important discoveries early in his life, and spent his later years refining them. Gauss made several fundamental discoveries that he did not reveal. Mathematicians making the same discoveries were often surprised to find that Gauss had described the results years earlier in his unpublished notes.

Conversely, if there is an integer k with $a = b + km$, then $km = a - b$. Hence $m \mid (a - b)$, and consequently, $a \equiv b \pmod{m}$. ∎

Example 4.2.   We have $19 \equiv -2 \pmod 7$ and $19 = -2 + 3 \cdot 7$. ◄

The following proposition establishes some important properties of congruences.

Theorem 4.2.   Let m be a positive integer. Congruences modulo m satisfy the following properties:

(i)   *Reflexive property.* If a is an integer, then $a \equiv a \pmod{m}$.

(ii)   *Symmetric property.* If a and b are integers such that $a \equiv b \pmod{m}$, then $b \equiv a \pmod{m}$.

(iii)   *Transitive property.* If a, b, and c are integers with $a \equiv b \pmod{m}$ and $b \equiv c \pmod{m}$, then $a \equiv c \pmod{m}$.

Proof.

(i)   We see that $a \equiv a \pmod{m}$, since $m \mid (a - a) = 0$.

(ii)   If $a \equiv b \pmod{m}$, then $m \mid (a - b)$. Hence, there is an integer k such that $km = a - b$. This shows that $(-k)m = b - a$, so that $m \mid (b - a)$. Consequently, $b \equiv a \pmod{m}$.

(iii)   If $a \equiv b \pmod{m}$ and $b \equiv c \pmod{m}$, then $m \mid (a - b)$ and $m \mid (b - c)$. Hence, there are integers k and l such that $km = a - b$ and $lm = b - c$. Therefore, $a - c = (a - b) + (b - c) = km + lm = (k + l)m$. It follows that $m \mid (a - c)$ and $a \equiv c \pmod{m}$. ∎

By Theorem 4.2, we see that the set of integers is divided into m different sets called *congruence classes modulo m,* each containing integers that are mutually congruent modulo m.

Example 4.3.   The four congruence classes modulo 4 are given by

$$\ldots \equiv -8 \equiv -4 \equiv 0 \equiv 4 \equiv 8 \equiv \ldots \pmod 4$$
$$\ldots \equiv -7 \equiv -3 \equiv 1 \equiv 5 \equiv 9 \equiv \ldots \pmod 4$$
$$\ldots \equiv -6 \equiv -2 \equiv 2 \equiv 6 \equiv 10 \equiv \ldots \pmod 4$$
$$\ldots \equiv -5 \equiv -1 \equiv 3 \equiv 7 \equiv 11 \equiv \ldots \pmod 4.$$

◄

Suppose that m is a positive integer. Given an integer a, by the division algorithm we have $a = bm + r$, where $0 \leq r \leq m - 1$. We call r the *least nonnegative residue* of a modulo m. We say that r is the result of *reducing a modulo m.* Similarly, when we know that a is not divisible by m, we call r the *least positive residue* of a modulo m.

Another commonly used notation, especially in computer science applications, is $a \bmod m = r$, which denotes that r is the remainder obtained when a is divided by m. For example, $17 \bmod 5 = 2$ and $-8 \bmod 7 = 6$. Although we do not use such notation in this book, it is commonly used in other contexts.

Now note that from the equation $a = bm + r$, it follows that $a \equiv r \pmod{m}$. Hence, every integer is congruent modulo m to one of the integers of the set $0, 1, \ldots, m - 1$, namely the remainder when it is divided by m. Since no two of the integers $0, 1, \ldots, m - 1$ are congruent modulo m, we have m integers such that every integer is congruent to exactly one of these m integers.

Definition.   A *complete system of residues modulo m* is a set of integers such that every integer is congruent modulo m to exactly one integer of the set.

Example 4.4.   The division algorithm shows that the set of integers $0, 1, 2, \ldots, m - 1$ is a complete system of residues modulo m. This is called the set of *least nonnegative residues modulo m.*   ◀

Example 4.5.   Let m be an odd positive integer. Then the set of integers

$$-\frac{m - 1}{2}, -\frac{m - 3}{2}, \ldots, -1, 0, 1, \ldots, \frac{m - 3}{2}, \frac{m - 1}{2},$$

the set of *absolute least residues modulo m,* is a complete system of residues.   ◀

We will often do arithmetic with congruences, which is called *modular arithmetic.* Congruences have many of the same properties that equalities do. First, we show that an addition, subtraction, or multiplication to both sides of a congruence preserves the congruence.

Theorem 4.3.   If $a, b, c,$ and m are integers, with $m > 0$, such that $a \equiv b \pmod{m}$, then

(i) $a + c \equiv b + c \pmod{m}$,

(ii) $a - c \equiv b - c \pmod{m}$,

(iii) $ac \equiv bc \pmod{m}$.

Proof.   Because $a \equiv b \pmod{m}$, we know that $m \mid (a - b)$. From the identity $(a + c) - (b + c) = a - b$, we see that $m \mid ((a + c) - (b + c))$, so that (i) follows. Likewise, (ii) follows from the fact that $(a - c) - (b - c) = a - b$. To show that (iii) holds, note that $ac - bc = c(a - b)$. Because $m \mid (a - b)$, it follows that $m \mid c(a - b)$, and hence, $ac \equiv bc \pmod{m}$.   ∎

Example 4.6.   Because $19 \equiv 3 \pmod{8}$, it follows from Theorem 4.3 that $26 = 19 + 7 \equiv 3 + 7 = 10 \pmod{8}$, $15 = 19 - 4 \equiv 3 - 4 = -1 \pmod{8}$, and $38 = 19 \cdot 2 \equiv 3 \cdot 2 = 6 \pmod{8}$.   ◀

What happens when both sides of a congruence are divided by an integer? Consider the following example.

Example 4.7.   We have $14 = 7 \cdot 2 \equiv 4 \cdot 2 = 8 \pmod{6}$. But we cannot cancel the common factor of 2, because $7 \not\equiv 4 \pmod{6}$.   ◀

This example shows that it is not necessarily true that we preserve a congruence when we divide both sides by an integer. However, the following theorem gives a valid congruence when both sides of a congruence are divided by the same integer.

Theorem 4.4. If a, b, c, and m are integers such that $m > 0, d = (c, m)$, and $ac \equiv bc \pmod{m}$, then $a \equiv b \pmod{m/d}$.

Proof. If $ac \equiv bc \pmod{m}$, we know that $m \mid (ac - bc) = c(a - b)$. Hence, there is an integer k with $c(a - b) = km$. By dividing both sides by d, we have $(c/d)(a - b) = k(m/d)$. Because $(m/d, c/d) = 1$, by Lemma 3.4 it follows that $m/d \mid (a - b)$. Hence, $a \equiv b \pmod{m/d}$. ∎

Example 4.8. Because $50 \equiv 20 \pmod{15}$ and $(10, 15) = 5$, we see that $50/10 \equiv 20/10 \pmod{15/5}$, or $5 \equiv 2 \pmod{3}$. ◀

The following corollary, which is a special case of Theorem 4.4, is used often; it allows us to cancel numbers that are relatively prime to the modulus m in congruences modulo m.

Corollary 4.4.1. If a, b, c, and m are integers such that $m > 0$, $(c, m) = 1$, and $ac \equiv bc \pmod{m}$, then $a \equiv b \pmod{m}$.

Example 4.9. Since $42 \equiv 7 \pmod{5}$ and $(5, 7) = 1$, we can conclude that $42/7 \equiv 7/7 \pmod{5}$, or that $6 \equiv 1 \pmod{5}$. ◀

The following theorem, which is more general than Theorem 4.3, is also useful. Its proof is similar to the proof of Theorem 4.3.

Theorem 4.5. If a, b, c, d, and m are integers such that $m > 0$, $a \equiv b \pmod{m}$, and $c \equiv d \pmod{m}$, then

(i) $a + c \equiv b + d \pmod{m}$,

(ii) $a - c \equiv b - d \pmod{m}$,

(iii) $ac \equiv bd \pmod{m}$.

Proof. Because $a \equiv b \pmod{m}$ and $c \equiv d \pmod{m}$, we know that $m \mid (a - b)$ and $m \mid (c - d)$. Hence, there are integers k and l with $km = a - b$ and $lm = c - d$.

To prove (i), note that $(a + c) - (b + d) = (a - b) + (c - d) = km + lm = (k + l)m$. Hence, $m \mid [(a + c) - (b + d)]$. Therefore, $a + c \equiv b + d \pmod{m}$.

To prove (ii), note that $(a - c) - (b - d) = (a - b) - (c - d) = km - lm = (k - l)m$. Hence, $m \mid [(a - c) - (b - d)]$, so that $a - c \equiv b - d \pmod{m}$.

To prove (iii), note that $ac - bd = ac - bc + bc - bd = c(a - b) + b(c - d) = ckm + blm = m(ck + bl)$. Hence, $m \mid (ac - bd)$. Therefore, $ac \equiv bd \pmod{m}$. ∎

Example 4.10. Because $13 \equiv 3 \pmod 5$ and $7 \equiv 2 \pmod 5$, using Theorem 3.5 we see that $20 = 13 + 7 \equiv 3 + 2 = 5 \pmod 5$, $6 = 13 - 7 \equiv 3 - 2 = 1 \pmod 5$, and $91 = 13 \cdot 7 \equiv 3 \cdot 2 = 6 \pmod 5$. ◄

The following lemma helps us to determine whether a set of m numbers forms a complete set of residues modulo m.

Lemma 4.1. A set of m incongruent integers modulo m forms a complete set of residues modulo m.

Proof. Suppose that a set of m incongruent integers modulo m does not form a complete set of residues modulo m. This implies that at least one integer a is not congruent to any of the integers in the set. Hence, there is no integer in the set congruent modulo m to the remainder of a when it is divided by m. Hence, there can be at most $m - 1$ different remainders of the integers when they are divided by m. It follows (by the pigeonhole principle, which says that if more than n objects are distributed into n boxes, at least two objects are in the same box) that at least two integers in the set have the same remainder modulo m. This is impossible, because these integers are incongruent modulo m. Hence, any m incongruent integers modulo m form a complete system of residues modulo m.
∎

Theorem 4.6. If $r_1, r_2, \ldots, r_m$ is a complete system of residues modulo m, and if a is a positive integer with $(a, m) = 1$, then

$$ar_1 + b, ar_2 + b, \ldots, ar_m + b$$

is a complete system of residues modulo m for any integer b.

Proof. First, we show that no two of the integers

$$ar_1 + b, ar_2 + b, \ldots, ar_m + b$$

are congruent modulo m. To see this, note that if

$$ar_j + b \equiv ar_k + b \pmod m,$$

then, by (ii) of Theorem 4.3, we know that

$$ar_j \equiv ar_k \pmod m.$$

Because $(a, m) = 1$, Corollary 4.4.1 shows that

$$r_j \equiv r_k \pmod m.$$

Given that $r_j \not\equiv r_k \pmod m$ if $j \neq k$, we conclude that $j = k$.

By Lemma 4.1, because the set of integers in question consists of m incongruent integers modulo m, these integers form a complete system of residues modulo m. ∎

The following theorem shows that a congruence is preserved when both sides are raised to the same positive integral power.

Theorem 4.7. If $a, b, k,$ and m are integers such that $k > 0, m > 0,$ and $a \equiv b \pmod{m}$, then $a^k \equiv b^k \pmod{m}$.

Proof. Because $a \equiv b \pmod{m}$, we have $m \mid (a - b)$, and because

$$a^k - b^k = (a - b)(a^{k-1} + a^{k-2}b + \cdots + ab^{k-2} + b^{k-1}),$$

we see that $(a - b) \mid (a^k - b^k)$. Therefore, by Theorem 1.8 it follows that $m \mid (a^k - b^k)$. Hence, $a^k \equiv b^k \pmod{m}$. ∎

Example 4.11. Since $7 \equiv 2 \pmod 5$, Theorem 4.7 tells us that $343 = 7^3 \equiv 2^3 = 8 \pmod 5$. ◄

The following result shows how to combine congruences of two numbers to different moduli.

Theorem 4.8. If $a \equiv b \pmod{m_1}, a \equiv b \pmod{m_2}, \ldots, a \equiv b \pmod{m_k}$, where $a, b, m_1, m_2, \ldots, m_k$ are integers with $m_1, m_2, \ldots, m_k$ positive, then

$$a \equiv b \pmod{[m_1, m_2, \ldots, m_k]},$$

where $[m_1, m_2, \ldots, m_k]$ is the least common multiple of $m_1, m_2, \ldots, m_k$.

Proof. Because $a \equiv b \pmod{m_1}, a \equiv b \pmod{m_2}, \ldots, a \equiv b \pmod{m_k}$, we know that $m_1 \mid (a - b), m_2 \mid (a - b), \ldots, m_k \mid (a - b)$. By Exercise 39 of Section 3.5 we see that

$$[m_1, m_2, \ldots, m_k] \mid (a - b).$$

Consequently,

$$a \equiv b \pmod{[m_1, m_2, \ldots, m_k]}. \quad ∎$$

The following result is an immediate and useful consequence of this theorem.

Corollary 4.8.1. If $a \equiv b \pmod{m_1}, a \equiv b \pmod{m_2}, \ldots, a \equiv b \pmod{m_k}$, where a and b are integers and $m_1, m_2, \ldots, m_k$ are pairwise relatively prime positive integers, then

$$a \equiv b \pmod{m_1 m_2 \cdots m_k}.$$

Proof. Since $m_1, m_2, \ldots, m_k$ are pairwise relatively prime, Exercise 68 of Section 3.5 tells us that

$$[m_1, m_2, \ldots, m_k] = m_1 m_2 \cdots m_k.$$

Hence, by Theorem 4.8, we know that

$$a \equiv b \pmod{m_1 m_2 \cdots m_k}. \quad ∎$$

Modular Exponentiation

In our subsequent studies, we will be working with congruences involving large powers of integers. For example, we will want to find the least positive residue of 2^{644}

modulo 645. If we attempt to find this least positive residue by first computing 2^{644}, we would have an integer with 194 decimal digits, a most undesirable thought. Instead, to find 2^{644} modulo 645 we first express the exponent 644 in binary notation:

$$(644)_{10} = (1010000100)_2.$$

Next, we compute the least positive residues of $2, 2^2, 2^4, 2^8, \ldots, 2^{512}$ by successively squaring and reducing modulo 645. This gives us the congruences

$$
\begin{aligned}
2 &\equiv 2 \pmod{645}, \\
2^2 &\equiv 4 \pmod{645}, \\
2^4 &\equiv 16 \pmod{645}, \\
2^8 &\equiv 256 \pmod{645}, \\
2^{16} &\equiv 391 \pmod{645}, \\
2^{32} &\equiv 16 \pmod{645}, \\
2^{64} &\equiv 256 \pmod{645}, \\
2^{128} &\equiv 391 \pmod{645}, \\
2^{256} &\equiv 16 \pmod{645}, \\
2^{512} &\equiv 256 \pmod{645}.
\end{aligned}
$$

We can now compute 2^{644} modulo 645 by multiplying the least positive residues of the appropriate powers of 2. This gives

$$2^{644} = 2^{512+128+4} = 2^{512}2^{128}2^4 \equiv 256 \cdot 391 \cdot 16 = 1{,}601{,}536 \equiv 1 \pmod{645}.$$

We have just illustrated a general procedure for *modular exponentiation,* that is, for computing b^N modulo m, where $b, m,$ and N are positive integers. We first express the exponent N in binary notation, as $N = (a_k a_{k-1} \ldots a_1 a_0)_2$. We then find the least positive residues of $b, b^2, b^4, \ldots, b^{2^k}$ modulo m, by successively squaring and reducing modulo m. Finally, we multiply the least positive residues modulo m of b^{2^j} for those j with $a_j = 1$, reducing modulo m after each multiplication.

In our subsequent discussions, we will need an estimate for the number of bit operations needed for modular exponentiation. This is provided by the following proposition.

Theorem 4.9. Let $b, m,$ and N be positive integers such that $b < m$. Then the least positive residue of b^N modulo m can be computed using $O((\log_2 m)^2 \log_2 N)$ bit operations.

Proof. To find the least positive residue of b^N modulo m, we can use the algorithm just described. First, we find the least positive residues of $b, b^2, b^4, \ldots, b^{2^k}$ modulo m, where $2^k \leq N < 2^{k+1}$, by successively squaring and reducing modulo m. This requires a total of $O((\log_2 m)^2 \log_2 N)$ bit operations, because we perform $[\log_2 N]$ squarings modulo m, each requiring $O((\log_2 m)^2)$ bit operations. Next, we multiply together the least positive residues of the integers b^{2^j} corresponding to the binary digits of N that are equal to one, and we reduce modulo m after each multiplication. This also requires $O((\log_2 m)^2 \log_2 N)$ bit operations, because there are at most $\log_2 N$ multiplications, each requiring $O((\log_2 m)^2)$ bit operations. Therefore, a total of $O((\log_2 m)^2 \log_2 N)$ bit operations is needed.   ∎

4.1 Exercises

1. Show that each of the following congruences holds.

   a) $13 \equiv 1 \pmod 2$ e) $-2 \equiv 1 \pmod 3$
   b) $22 \equiv 7 \pmod 5$ f) $-3 \equiv 30 \pmod{11}$
   c) $91 \equiv 0 \pmod{13}$ g) $111 \equiv -9 \pmod{40}$
   d) $69 \equiv 62 \pmod 7$ h) $666 \equiv 0 \pmod{37}$

2. Determine whether each of the following pairs of integers is congruent modulo 7.

   a) 1,15   d) $-1,8$
   b) 0,42   e) $-9,5$
   c) 2,99   f) $-1,699$

3. For which positive integers m is each of the following statements true?

   a) $27 \equiv 5 \pmod m$
   b) $1000 \equiv 1 \pmod m$
   c) $1331 \equiv 0 \pmod m$

4. Show that if a is an even integer, then $a^2 \equiv 0 \pmod 4$, and if a is an odd integer, then $a^2 \equiv 1 \pmod 4$.

☞ 5. Show that if a is an odd integer, then $a^2 \equiv 1 \pmod 8$.

6. Find the least nonnegative residue modulo 13 of each of the following integers.

   a) 22   d) -1
   b) 100   e) -100
   c) 1001   f) -1000

7. Find the least positive residue of $1! + 2! + 3! + \cdots + 100!$ modulo each of the following integers.

   a) 2   c) 12
   b) 7   d) 25

8. Show that if $a, b, m,$ and n are integers such that $m > 0, n > 0, n \mid m,$ and $a \equiv b \pmod m$, then $a \equiv b \pmod n$.

9. Show that if $a, b, c,$ and m are integers such that $c > 0, m > 0,$ and $a \equiv b \pmod m$, then $ac \equiv bc \pmod{mc}$.

10. Show that if $a, b,$ and c are integers with $c > 0$ such that $a \equiv b \pmod c$, then $(a, c) = (b, c)$.

11. Show that if $a_j \equiv b_j \pmod m$ for $j = 1, 2, \ldots, n$, where m is a positive integer and a_j, $b_j, j = 1, 2, \ldots, n$, are integers, then

a) $\displaystyle\sum_{j=1}^{n} a_j \equiv \sum_{j=1}^{n} b_j \pmod{m}$.

b) $\displaystyle\prod_{j=1}^{n} a_j \equiv \prod_{j=1}^{n} b_j \pmod{m}$.

In Exercises 12–14, construct tables for arithmetic modulo 6 using the least nonnegative residues modulo 6 to represent the congruence classes.

12. Construct a table for addition modulo 6.

13. Construct a table for subtraction modulo 6.

14. Construct a table for multiplication modulo 6.

15. What time does a clock read

a) 29 hours after it reads 11 o'clock?

b) 100 hours after it reads 2 o'clock?

c) 50 hours before it reads 6 o'clock?

16. Which decimal digits occur as the final digit of a fourth power of an integer?

17. What can you conclude if $a^2 \equiv b^2 \pmod{p}$, where a and b are integers and p is prime?

18. Show that if $a^k \equiv b^k \pmod{m}$ and $a^{k+1} \equiv b^{k+1} \pmod{m}$, where a, b, k, and m are integers with $k > 0$ and $m > 0$ such that $(a, m) = 1$, then $a \equiv b \pmod{m}$. If the condition $(a, m) = 1$ is dropped, is the conclusion that $a \equiv b \pmod{m}$ still valid?

19. Show that if n is an odd positive integer, then
$$1 + 2 + 3 + \cdots + (n - 1) \equiv 0 \pmod{n}.$$
Is this statement true if n is even?

20. Show that if n is an odd positive integer or if n is a positive integer divisible by 4, then
$$1^3 + 2^3 + 3^3 + \cdots + (n - 1)^3 \equiv 0 \pmod{n}.$$
Is this statement true if n is even but not divisible by 4?

21. For which positive integers n is it true that
$$1^2 + 2^2 + 3^2 + \cdots + (n - 1)^2 \equiv 0 \pmod{n}?$$

22. Show by mathematical induction that if n is a positive integer, then $4^n \equiv 1 + 3n \pmod{9}$.

23. Show by mathematical induction that if n is a positive integer, then $5^n \equiv 1 + 4n \pmod{16}$.

24. Give a complete system of residues modulo 13 consisting entirely of odd integers.

25. Show that if $n \equiv 3 \pmod{4}$, then n cannot be the sum of the squares of two integers.

26. Show that if p is prime, then the only solutions of the congruence $x^2 \equiv x \pmod{p}$ are those integers x such that $x \equiv 0$ or 1 $\pmod{p}$.

27. Show that if p is prime and k is a positive integer, then the only solutions of $x^2 \equiv x \pmod{p^k}$ are those integers x such that $x \equiv 0$ or 1 $\pmod{p^k}$.

28. Find the least positive residues modulo 47 of each of the following integers.

a) 2^{32} b) 2^{47} c) 2^{200}

29. Let $m_1, m_2, \ldots, m_k$ be pairwise relatively prime positive integers. Let $M = m_1 m_2 \cdots m_k$ and $M_j = M/m_j$ for $j = 1, 2, \ldots, k$. Show that

$$M_1 a_1 + M_2 a_2 + \cdots + M_k a_k$$

runs through a complete system of residues modulo M when $a_1, a_2, \ldots, a_k$ run through complete systems of residues modulo $m_1, m_2, \ldots, m_k$, respectively.

30. Explain how to find the sum $u + v$ from the least positive residue of $u + v$ modulo m, where u and v are positive integers less than m. (*Hint:* Assume that $u \leq v$, and consider separately the cases where the least positive residue of $u + v$ is less than u, and where it is greater than v.)

31. On a computer with word size w, multiplication modulo n where $n < w/2$ can be performed as outlined. Let $T = [\sqrt{n} + 1/2]$, and $t = T^2 - n$. For each computation, show that all the required computer arithmetic can be done without exceeding the word size. (This method was described by Head [He80]).

    a) Show that $|t| \leq T$.

    b) Show that if x and y are nonnegative integers less than n, then

    $$x = aT + b, \quad y = cT + d,$$

    where a, b, c, and d are integers such that $0 \leq a \leq T$, $0 \leq b < T$, $0 \leq c \leq T$, and $0 \leq d < T$.

    c) Let $z \equiv ad + bc \pmod{n}$, such that $0 \leq z < n$. Show that

    $$xy \equiv act + zT + bd \pmod{n}.$$

    d) Let $ac = eT + f$, where e and f are integers with $0 \leq e \leq T$ and $0 \leq f < T$. Show that

    $$xy \equiv (z + et)T + ft + bd \pmod{n}.$$

    e) Let $v \equiv z + et \pmod{n}$, such that $0 \leq v < n$. Show that we can write

    $$v = gT + h,$$

    where g and h are integers with $0 \leq g \leq T$, $0 \leq h < T$, and such that

    $$xy \equiv hT + (f + g)t + bd \pmod{n}.$$

    f) Show that the right-hand side of the congruence of part (e) can be computed without exceeding the word size, by first finding j such that

    $$j \equiv (f + g)t \pmod{n}$$

    and $0 \leq j < n$, and then finding k such that

    $$k \equiv j + bd \pmod{n}$$

    and $0 \leq k < n$, so that

    $$xy \equiv hT + k \pmod{n}.$$

    This gives the desired result.

32. Develop an algorithm for modular exponentiation from the base 3 expansion of the exponent.

33. Find the least positive residue of each of the following.

    a) 3^{10} modulo 11

    b) 2^{12} modulo 13

    c) 5^{16} modulo 17

    d) 3^{22} modulo 23

    e) Can you propose a theorem from the above congruences?

34. Find the least positive residues of each of the following.

    a) 6! modulo 7

    b) 10! modulo 11

    c) 12! modulo 13

    d) 16! modulo 17

    e) Can you propose a theorem from the above congruences?

∗ 35. Show that for every positive integer m there are infinitely many Fibonacci numbers f_n such that m divides f_n. (*Hint:* Show that the sequence of least positive residues modulo m of the Fibonacci numbers is a repeating sequence.)

36. Prove Theorem 4.7 using mathematical induction.

37. Show that the least nonnegative residue modulo m of the product of two positive integers less than m can be computed using $O(\log^2 m)$ bit operations.

∗ 38. Five men and a monkey are shipwrecked on an island. The men have collected a pile of coconuts which they plan to divide equally among themselves the next morning. Not trusting the other men, one of the group wakes up during the night and divides the coconuts into five equal parts with one left over, which he gives to the monkey. He then hides his portion of the pile. During the night, each of the other four men does exactly the same thing by dividing the pile he finds into five equal parts leaving one coconut for the monkey and hiding his portion. In the morning, the men gather and split the remaining pile of coconuts into five parts and one is left over for the monkey. What is the minimum number of coconuts the men could have collected for their original pile?

∗ 39. Answer the question in Exercise 38, where instead of five men and one monkey, there are n men and k monkeys, and at each stage the monkeys receive one coconut each.

We say that the polynomials $f(x)$ and $g(x)$ are *congruent modulo n as polynomials* if for each power of x the coefficients of that power in $f(x)$ and $g(x)$ are congruent modulo n. For example, $11x^3 + x^2 + 2$ and $x^3 - 4x^2 + 5x + 22$ are congruent as polynomials modulo 5. The notation $f(x) \equiv g(x) \pmod{n}$ is often used to denote that $f(x)$ and $g(x)$ are congruent as polynomials modulo n. In Exercises 40–44 assume that n is a positive integer with $n > 1$ and that all polynomials have integer coefficients.

40. a) Show that if $f(x)$ and $g(x)$ are congruent as polynomials modulo n, then for every integer a, $f(a) \equiv g(a) \pmod{n}$.

    b) Show that it is not necessarily true that $f(x)$ and $g(x)$ are congruent as polynomials modulo n if $f(a) \equiv g(a) \pmod{n}$ for every integer a.

41. Show that if $f_1(x)$ and $g_1(x)$ are congruent as polynomials modulo n and $f_2(x)$ and $g_2(x)$ are congruent as polynomials modulo n, then

    a) $(f_1 + f_2)(x)$ and $(g_1 + g_2)(x)$ are congruent as polynomials modulo n.

    b) $(f_1 f_2)(x)$ and $(g_1 g_2)(x)$ are congruent as polynomials modulo n.

42. Show that if $f(x)$ is a polynomial with integer coefficients and $f(a) \equiv 0 \pmod{n}$, then there is a polynomial $g(x)$ with integer coefficients such that $f(x)$ and $(x - a)g(x)$ are congruent as polynomials modulo n.

43. Suppose that p is prime, $f(x)$ is a polynomial with integer coefficients, $a_1, a_2, \ldots, a_k$ are incongruent integers modulo p, and $f(a_j) \equiv 0 \pmod{p}$ for $j = 1, 2, \ldots, k$. Show that there exists a polynomial $g(x)$ with integer coefficients such that $f(x)$ and $(x - a_1)(x - a_2) \cdots (x - a_k)g(x)$ are congruent as polynomials modulo p.

44. Use Exercise 43 to show that if p is a prime, $f(x)$ is a polynomial with integer coefficients, and x^n is the largest power of x with a coefficient divisible by p, then the congruence $f(x) \equiv 0 \pmod{p}$ has at most p incongruent solutions modulo p.

4.1 Computational and Programming Exercises

Computations and Explorations

Using a computation program such as Maple or *Mathematica,* or programs you have written, carry out the following computations and explorations.

1. Compute the least positive residue modulo 10,403 of 7651^{891}.

2. Compute the least positive residue modulo 10,403 of $7651^{20!}$.

Programming Projects

Write programs using Maple, *Mathematica,* or a language of your choice to do the following.

1. Find the least nonnegative residue of an integer with respect to a fixed modulus.

2. Perform modular addition and subtraction when the modulus is less than half of the word size of the computer.

3. Perform modular multiplication when the modulus is less than half of the word size of the computer, using Exercise 31.

4. Perform modular exponentiation using the algorithm described in the text.

4.2 Linear Congruences

A congruence of the form

$$ax \equiv b \pmod{m},$$

where x is an unknown integer, is called a *linear congruence in one variable*. In this section, we will see that the study of such congruences is similar to the study of linear diophantine equations in two variables.

We first note that if $x = x_0$ is a solution of the congruence $ax \equiv b \pmod{m}$, and if $x_1 \equiv x_0 \pmod{m}$, then $ax_1 \equiv ax_0 \equiv b \pmod{m}$, so that x_1 is also a solution. Hence, if one member of a congruence class modulo m is a solution, then all members of this class are solutions. Therefore, we may ask how many of the m congruence classes modulo m give solutions; this is exactly the same as asking how many incongruent solutions there are modulo m. The following theorem tells us when a linear congruence in one variable has solutions, and if it does, tells exactly how many incongruent solutions there are modulo m.

Theorem 4.10. Let a, b, and m be integers such that $m > 0$ and $(a, m) = d$. If $d \nmid b$, then $ax \equiv b \pmod{m}$ has no solutions. If $d \mid b$, then $ax \equiv b \pmod{m}$ has exactly d incongruent solutions modulo m.

Proof. By Theorem 4.1, the linear congruence $ax \equiv b \pmod{m}$ is equivalent to the linear diophantine equation in two variables $ax - my = b$. The integer x is a solution of $ax \equiv b \pmod{m}$ if and only if there is an integer y such that $ax - my = b$. By Theorem 3.23, we know that if $d \nmid b$, there are no solutions, whereas if $d \mid b$, $ax - my = b$ has infinitely many solutions, given by

$$x = x_0 + (m/d)t, \quad y = y_0 + (a/d)t,$$

where $x = x_0$ and $y = y_0$ is a particular solution of the equation. The values of x given above,

$$x = x_0 + (m/d)t,$$

are the solutions of the linear congruence; there are infinitely many of these.

To determine how many incongruent solutions there are, we find the condition that describes when two of the solutions $x_1 = x_0 + (m/d)t_1$ and $x_2 = x_0 + (m/d)t_2$ are congruent modulo m. If these two solutions are congruent, then

$$x_0 + (m/d)t_1 \equiv x_0 + (m/d)t_2 \pmod{m}.$$

Subtracting x_0 from both sides of this congruence, we find that

$$(m/d)t_1 \equiv (m/d)t_2 \pmod{m}.$$

Now $(m, m/d) = m/d$ since $(m/d) \mid m$, so that by Theorem 4.4, we see that

$$t_1 \equiv t_2 \pmod{d}.$$

This shows that a complete set of incongruent solutions is obtained by taking $x = x_0 + (m/d)t$, where t ranges through a complete system of residues modulo d. One such set is given by $x = x_0 + (m/d)t$, where $t = 0, 1, 2, \ldots, d - 1$.   ∎

A linear congruence where the multiplier a and the modulus m are relatively prime has a unique solution, as Corollary 4.10.1 shows.

Corollary 4.10.1. If a and m are relatively prime integers with $m > 0$ and b is an integer, then the linear congruence $ax \equiv b \pmod{m}$ has a unique solution modulo m.

Proof. Because $(a, m) = 1$, we know that $(a, m) \mid b$. Consequently, by Theorem 4.10, it follows that the congruence $ax \equiv b \pmod{m}$ has exactly $(a, m) = 1$ incongruent solution modulo m.   ∎

We now illustrate the use of Theorem 4.10.

Example 4.12. To find all solutions of $9x \equiv 12 \pmod{15}$, we first note that since $(9, 15) = 3$ and $3 \mid 12$, there are exactly three incongruent solutions. We can find these solutions by first finding a particular solution and then adding the appropriate multiples of $15/3 = 5$.

To find a particular solution, we consider the linear diophantine equation $9x - 15y = 12$. The Euclidean algorithm shows that

$$15 = 9 \cdot 1 + 6$$
$$9 = 6 \cdot 1 + 3$$
$$6 = 3 \cdot 2,$$

so that $3 = 9 - 6 \cdot 1 = 9 - (15 - 9 \cdot 1) = 9 \cdot 2 - 15$. Hence, $9 \cdot 8 - 15 \cdot 4 = 12$, and a particular solution of $9x - 15y = 12$ is given by $x_0 = 8$ and $y_0 = 4$.

From the proof of Theorem 4.10, we see that a complete set of three incongruent solutions is given by $x = x_0 \equiv 8 \pmod{15}$, $x = x_0 + 5 \equiv 13 \pmod{15}$, and $x = x_0 + 5 \cdot 2 \equiv 18 \equiv 3 \pmod{15}$. ◀

Modular Inverses We now consider congruences of the special form $ax \equiv 1 \pmod{m}$. By Theorem 4.10, there is a solution to this congruence if and only if $(a, m) = 1$, and then all solutions are congruent modulo m.

Definition. Given an integer a with $(a, m) = 1$, a solution of $ax \equiv 1 \pmod{m}$ is called an *inverse of a* modulo m.

Example 4.13. Because the solutions of $7x \equiv 1 \pmod{31}$ satisfy $x \equiv 9 \pmod{31}$, 9, and all integers congruent to 9 modulo 31, are inverses of 7 modulo 31. Analogously, since $9 \cdot 7 \equiv 1 \pmod{31}$, 7 is an inverse of 9 modulo 31. ◀

When we have an inverse of a modulo m, we can use it to solve any congruence of the form $ax \equiv b \pmod{m}$. To see this, let $\bar{a}$ be an inverse of a modulo m, so that $a\bar{a} \equiv 1 \pmod{m}$. Then, if $ax \equiv b \pmod{m}$, we can multiply both sides of this congruence by $\bar{a}$ to find that $\bar{a}(ax) \equiv \bar{a}b \pmod{m}$, so that $x \equiv \bar{a}b \pmod{m}$.

Example 4.14. To find the solutions of $7x \equiv 22 \pmod{31}$, we multiply both sides of this congruence by 9, an inverse of 7 modulo 31, to obtain $9 \cdot 7x \equiv 9 \cdot 22 \pmod{31}$. Hence, $x \equiv 198 \equiv 12 \pmod{31}$. ◀

Example 4.15. To find all solutions of $7x \equiv 4 \pmod{12}$, we note that since $(7, 12) = 1$, there is a unique solution modulo 12. To find this, we need only obtain a solution of the linear diophantine equation $7x - 12y = 4$. The Euclidean algorithm gives

$$12 = 7 \cdot 1 + 5$$
$$7 = 5 \cdot 1 + 2$$
$$5 = 2 \cdot 2 + 1$$
$$2 = 1 \cdot 2.$$

Hence, $1 = 5 - 2 \cdot 2 = 5 - (7 - 5 \cdot 1) \cdot 2 = 5 \cdot 3 - 2 \cdot 7 = (12 - 7 \cdot 1) \cdot 3 - 2 \cdot 7 = 12 \cdot 3 - 5 \cdot 7$. Therefore, a particular solution to the linear diophantine equation is $x_0 = -20$ and $y_0 = 12$. Hence, all solutions of the linear congruences are given by $x \equiv -20 \equiv 4 \pmod{12}$. ◄

Later we will want to know which integers are their own inverses modulo p, where p is prime. The following theorem tells us which integers have this property.

Theorem 4.11. Let p be prime. The positive integer a is its own inverse modulo p if and only if $a \equiv 1 \pmod{p}$ or $a \equiv -1 \pmod{p}$.

Proof. If $a \equiv 1 \pmod{p}$ or $a \equiv -1 \pmod{p}$, then $a^2 \equiv 1 \pmod{p}$, so that a is its own inverse modulo p.

Conversely, if a is its own inverse modulo p, then $a^2 = a \cdot a \equiv 1 \pmod{p}$. Hence, $p \mid (a^2 - 1)$. Since $a^2 - 1 = (a - 1)(a + 1)$, either $p \mid (a - 1)$ or $p \mid (a + 1)$. Therefore, either $a \equiv 1 \pmod{p}$ or $a \equiv -1 \pmod{p}$. ■

4.2 Exercises

1. Find all solutions of each of the following linear congruences.

   a) $2x \equiv 5 \pmod 7$ d) $9x \equiv 5 \pmod{25}$

   b) $3x \equiv 6 \pmod 9$ e) $103x \equiv 444 \pmod{999}$

   c) $19x \equiv 30 \pmod{40}$ f) $980x \equiv 1500 \pmod{1600}$

2. Find all solutions of each of the following linear congruences.

   a) $3x \equiv 2 \pmod 7$ d) $15x \equiv 9 \pmod{25}$

   b) $6x \equiv 3 \pmod 9$ e) $128x \equiv 833 \pmod{1001}$

   c) $17x \equiv 14 \pmod{21}$ f) $987x \equiv 610 \pmod{1597}$

3. Find all solutions to the congruence $6{,}789{,}783x \equiv 2{,}474{,}010 \pmod{28{,}927{,}591}$.

4. Suppose that p is prime and that a and b are positive integers with $(p, a) = 1$. The following method can be used to solve the linear congruence $ax \equiv b \pmod p$.

   a) Show that if the integer x is a solution of $ax \equiv b \pmod p$, then x is also a solution of the linear congruence

   $$a_1 x \equiv -b[m/a] \pmod p,$$

   where a_1 is the least positive residue of p modulo a. Note that this congruence is of the same type as the original congruence, with a positive integer smaller than a as the coefficient of x.

   b) When the procedure of part (a) is iterated, one obtains a sequence of linear congruences with coefficients of x equal to $a_0 = a > a_1 > a_2 > \cdots$. Show that there is a positive integer n with $a_n = 1$, so that at the nth stage, one obtains a linear congruence $x \equiv B \pmod p$.

   c) Use the method described in part (b) to solve the linear congruence $6x \equiv 7 \pmod{23}$.

5. An astronomer knows that a satellite orbits the Earth in a period that is an exact multiple of 1 hour that is less than 1 day. If the astronomer notes that the satellite completes 11 orbits in an interval that starts when a 24-hour clock reads 0 hours and ends when the clock reads 17 hours, how long is the orbital period of the satellite?

6. For which integers c, $0 \leq c < 30$, does the congruence $12x \equiv c \pmod{30}$ have solutions? When there are solutions, how many incongruent solutions are there?

7. For which integers c, $0 \leq c < 1001$, does the congruence $154x \equiv c \pmod{1001}$ have solutions? When there are solutions, how many incongruent solutions are there?

8. Find an inverse modulo 13 of each of the following integers.

   a) 2   c) 5
   b) 3   d) 11

9. Find an inverse modulo 17 of each of the following integers.

   a) 4   c) 7
   b) 5   d) 16

10. a) Determine which integers a, where $1 \leq a \leq 14$, have an inverse modulo 14.
    b) Find the inverse of each of the integers from part (a) that have an inverse modulo 14.

11. a) Determine which integers a, where $1 \leq a \leq 30$, have an inverse modulo 30.
    b) Find the inverse of each of the integers from part (a) that have an inverse modulo 30.

12. Show that if $\bar{a}$ is an inverse of a modulo m and $\bar{b}$ is an inverse of b modulo m, then $\bar{a}\,\bar{b}$ is an inverse of ab modulo m.

13. Show that the linear congruence in two variables $ax + by \equiv c \pmod{m}$, where a, b, c, and m are integers, $m > 0$, with $d = (a, b, m)$, has exactly dm incongruent solutions if $d \mid c$, and no solutions otherwise.

14. Find all solutions of each of the following linear congruences in two variables.

    a) $2x + 3y \equiv 1 \pmod 7$ c) $6x + 3y \equiv 0 \pmod 9$
    b) $2x + 4y \equiv 6 \pmod 8$ d) $10x + 5y \equiv 9 \pmod{15}$

15. Let p be an odd prime and k a positive integer. Show that the congruence $x^2 \equiv 1 \pmod{p^k}$ has exactly two incongruent solutions, namely $x \equiv \pm 1 \pmod{p^k}$.

16. Show that the congruence $x^2 \equiv 1 \pmod{2^k}$ has exactly four incongruent solutions, namely $x \equiv \pm 1$ or $\pm(1 + 2^{k-1}) \pmod{2^k}$, when $k > 2$. Show that when $k = 1$ there is one solution and that when $k = 2$ there are two incongruent solutions.

17. Show that if a and m are relatively prime positive integers such that $a < m$, then an inverse of a modulo m can be found using $O(\log^3 m)$ bit operations.

18. Show that if p is an odd prime and a is a positive integer not divisible by p, then the congruence $x^2 \equiv a \pmod p$ has either no solution or exactly two incongruent solutions.

4.2 Computational and Programming Exercises

Computations and Explorations

Using a computation program such as Maple or *Mathematica,* or programs you have written, carry out the following computations and explorations.

1. Find the solutions of $123,456,789x \equiv 9,876,543,210 \pmod{10,000,000,001}$.

2. Find the solutions of $333,333,333x \equiv 87,543,211,376 \pmod{967,454,302,211}$.

3. Find the inverses of 734,342; 499,999; and 1,000,001 modulo 1,533,331.

Programming Projects

Write programs using Maple, *Mathematica,* or a language of your choice to do the following.

1. Solve linear congruences using the method given in the text.

2. Solve linear congruences using the method given in Exercise 4.

3. Find inverses modulo m of integers relatively prime to m, where m is a positive integer.

4. Solve linear congruences using inverses.

5. Solve linear congruences in two variables.

4.3 The Chinese Remainder Theorem

In this and in the following section, we discuss systems of simultaneous congruences. We will study two types of such systems: In the first type, there are two or more linear congruences in one variable, with different moduli. The second type consists of more than one simultaneous congruence in more than one variable, where all congruences have the same modulus.

First, we consider systems of congruences that involve only one unknown, but different moduli. Such systems arose in ancient Chinese puzzles such as the following problem, which appears in *Master Sun's Mathematical Manual,* written late in the third century C.E.. Find a number that leaves a remainder of 1 when divided by 3, a remainder of 2 when divided by 5, and a remainder of 3 when divided by 7. This puzzle leads to the following system of congruences:

$$x \equiv 1 \pmod 3, x \equiv 2 \pmod 5, x \equiv 3 \pmod 7.$$

Problems involving systems of congruences occur in the writings of the Greek mathematician Nicomachus in the first century. They also can be found in the works of Brahmagupta in India in the seventh century. However, it was not until the year 1247 that a general method for solving systems of linear congruences was published by *Ch'in Chiu-Shao* in his *Mathematical Treatise in Nine Sections.* We now present the main theorem concerning the solution of systems of linear congruences in one unknown. This theorem is called the Chinese remainder theorem, most likely because of the contributions of Chinese mathematicians such as Ch'in Chiu-Shao to its solution. (For more information about the history of the Chinese remainder theorem, consult [Ne69], [LiDu87], [Li73], and [Ka98].)

Theorem 4.12. *The Chinese Remainder Theorem.* Let $m_1, m_2, \ldots, m_r$ be pairwise relatively prime positive integers. Then the system of congruences

$$x \equiv a_1 \pmod{m_1},$$
$$x \equiv a_2 \pmod{m_2},$$
$$\vdots$$
$$x \equiv a_r \pmod{m_r},$$

has a unique solution modulo $M = m_1 m_2 \ldots m_r$.

Proof. First, we construct a simultaneous solution to the system of congruences. To do this, let $M_k = M/m_k = m_1 m_2 \cdots m_{k-1} m_{k+1} \cdots m_r$. We know that $(M_k, m_k) = 1$ by Exercise 14 of Section 3.3, because $(m_j, m_k) = 1$ whenever $j \neq k$. Hence, by Theorem 4.10 we can find an inverse y_k of M_k modulo m_k, so that $M_k y_k \equiv 1 \pmod{m_k}$. We now form the sum

$$x = a_1 M_1 y_1 + a_2 M_2 y_2 + \cdots + a_r M_r y_r.$$

The integer x is a simultaneous solution of the r congruences. To demonstrate this, we must show that $x \equiv a_k \pmod{m_k}$ for $k = 1, 2, \ldots, r$. Since $m_k \mid M_j$ whenever $j \neq k$, we have $M_j \equiv 0 \pmod{m_k}$. Therefore, in the sum for x, all terms except the kth term are congruent to 0 $\pmod{m_k}$. Hence, $x \equiv a_k M_k y_k \equiv a_k \pmod{m_k}$, since $M_k y_k \equiv 1 \pmod{m_k}$. We now show that any two solutions are congruent modulo M. Let x_0 and x_1 both be simultaneous solutions to the system of r congruences. Then, for each

CH'IN CHIU-SHAO (1202–1261) was born in the Chinese province of Sichuan. He studied astronomy at Hangzhou, the capital of the Song dynasty. He spent ten years in dangerous and difficult conditions at the frontier, where battles with the Mongols under Genghis Khan were under way. He wrote that he was instructed in mathematics by a "recluse scholar." During his time at the frontier, he investigated mathematical problems. He selected 81 of these, divided them into nine classes, and described them in his book *Mathematical Treatise in Nine Sections*. This book covers systems of linear congruences, the Chinese remainder theorem, algebraic equations, areas of geometrical figures, systems of linear equations, and other topics.

Ch'in Chiu-Shao was considered to be a mathematical genius and was talented in architecture, music, and poetry, as well as in many sports, including archery, fencing, and horsemanship. He held several different positions in government, but was relieved of his duties many times because of corruption. He was considered to be extravagant, boastful, and obsessed with his own advancement. He managed to amass great wealth and through deceit had an immense house constructed at a magnificent site. The back of this house contained a series of rooms for lodging female musicians and singers. Ch'in Chiu-Shao developed a notorious reputation in love affairs.

k, $x_0 \equiv x_1 \equiv a_k \pmod{m_k}$, so that $m_k \mid (x_0 - x_1)$. Using Theorem 4.8, we see that $M \mid (x_0 - x_1)$. Therefore, $x_0 \equiv x_1 \pmod{M}$. This shows that the simultaneous solution of the system of r congruences is unique modulo M. ∎

We illustrate the use of the Chinese remainder theorem by solving the system that arises from the ancient Chinese puzzle.

Example 4.16.   To solve the system

$$x \equiv 1 \pmod{3}$$
$$x \equiv 2 \pmod{5}$$
$$x \equiv 3 \pmod{7},$$

we have $M = 3 \cdot 5 \cdot 7 = 105$, $M_1 = 105/3 = 35$, $M_2 = 105/5 = 21$, and $M_3 = 105/7 = 15$. To determine y_1, we solve $35y_1 \equiv 1 \pmod{3}$, or equivalently, $2y_1 \equiv 1 \pmod{3}$. This yields $y_1 \equiv 2 \pmod{3}$. We find y_2 by solving $21y_2 \equiv 1 \pmod{5}$; this immediately gives $y_2 \equiv 1 \pmod{5}$. Finally, we find y_3 by solving $15y_3 \equiv 1 \pmod{7}$. This gives $y_3 \equiv 1 \pmod{7}$. Hence,

$$x \equiv 1 \cdot 35 \cdot 2 + 2 \cdot 21 \cdot 1 + 3 \cdot 15 \cdot 1$$
$$\equiv 157 \equiv 52 \pmod{105}.$$

We can check that x satisfies this system of congruences whenever $x \equiv 52 \pmod{105}$ by noting that $52 \equiv 1 \pmod{3}$, $52 \equiv 2 \pmod{5}$, and $52 \equiv 3 \pmod{7}$. ◀

There is also an iterative method for solving simultaneous systems of congruences. We illustrate this method with an example.

Example 4.17.   Suppose we wish to solve the system

$$x \equiv 1 \pmod{5}$$
$$x \equiv 2 \pmod{6}$$
$$x \equiv 3 \pmod{7}.$$

We use Theorem 4.1 to rewrite the first congruence as an equality, namely $x = 5t + 1$, where t is an integer. Inserting this expression for x into the second congruence, we find that

$$5t + 1 \equiv 2 \pmod{6},$$

which can easily be solved to show that $t \equiv 5 \pmod{6}$. Using Theorem 4.1 again, we write $t = 6u + 5$, where u is an integer. Hence, $x = 5(6u + 5) + 1 = 30u + 26$. When we insert this expression for x into the third congruence, we obtain

$$30u + 26 \equiv 3 \pmod{7}.$$

When this congruence is solved, we find that $u \equiv 6 \pmod{7}$. Consequently, Theorem 4.1 tells us that $u = 7v + 6$, where v is an integer. Hence,

$$x = 30(7v + 6) + 26 = 210v + 206.$$

Translating this equality into a congruence, we find that

$$x \equiv 206 \pmod{210},$$

and this is the simultaneous solution. ◀

Note that the method we have just illustrated shows that a system of simultaneous questions can be solved by successively solving linear congruences. This can be done even when the moduli of the congruences are not relatively prime as long as congruences are consistent (see Exercises 15–20 at the end of this section).

Computer Arithmetic Using the Chinese Remainder Theorem The Chinese remainder theorem provides a way to perform computer arithmetic with large integers. To store very large integers and do arithmetic with them requires special techniques. The Chinese remainder theorem tells us that given pairwise relatively prime moduli $m_1, m_2, \ldots, m_r$, a positive integer n such that $n < M = m_1 m_2 \cdots m_r$ is uniquely determined by its least positive residues modulo m_j for $j = 1, 2, \ldots, r$. Suppose that the word size of a computer is only 100, but that we wish to do arithmetic with integers as large as 10^6. First, we find pairwise relatively prime integers less than 100 with a product exceeding 10^6; for instance, we can take $m_1 = 99$, $m_2 = 98$, $m_3 = 97$, and $m_4 = 95$. We convert integers less than 10^6 into 4-tuples consisting of their least positive residues modulo m_1, m_2, m_3, and m_4. (To convert integers as large as 10^6 into their list of least positive residues, we need to work with large integers using multiprecision techniques. However, this is done only once for each integer in the input and once for the output.) Then, for instance, to add integers, we simply add their respective least positive residues modulo m_1, m_2, m_3, and m_4, making use of the fact that if $x \equiv x_i \pmod{m_i}$ and $y \equiv y_i \pmod{m_i}$, then $x + y \equiv x_i + y_i \pmod{m_i}$. We then use the Chinese remainder theorem to convert the set of four least positive residues for the sum back to an integer.

The following example illustrates this technique.

Example 4.18. We wish to add $x = 123{,}684$ and $y = 413{,}456$ on a computer of word size 100. We have

$$x \equiv 33 \pmod{99} \quad y \equiv 32 \pmod{99},$$
$$x \equiv \ 8 \pmod{98} \quad y \equiv 92 \pmod{98},$$
$$x \equiv \ 9 \pmod{97} \quad y \equiv 42 \pmod{97},$$
$$x \equiv 89 \pmod{95} \quad y \equiv 16 \pmod{95},$$

so that

$$x + y \equiv 65 \pmod{99},$$
$$x + y \equiv \ 2 \pmod{98},$$
$$x + y \equiv 51 \pmod{97},$$
$$x + y \equiv 10 \pmod{95}.$$

We now use the Chinese remainder theorem to find $x + y$ modulo $99 \cdot 98 \cdot 97 \cdot 95$. We have $M = 99 \cdot 98 \cdot 97 \cdot 95 = 89{,}403{,}930$, $M_1 = M/99 = 903{,}070$, $M_2 = M/98 = 912{,}285$, $M_3 = M/97 = 921{,}690$, and $M_4 = M/95 = 941{,}094$. We need to find the inverse of M_i (mod y_i) for $i = 1, 2, 3, 4$. To do this, we solve the following congruences (using the Euclidean algorithm):

$$903{,}070y_1 \equiv 91y_1 \equiv 1 \pmod{99},$$
$$912{,}285y_2 \equiv 3y_2 \equiv 1 \pmod{98},$$
$$921{,}690y_3 \equiv 93y_3 \equiv 1 \pmod{97},$$
$$941{,}094y_4 \equiv 24y_4 \equiv 1 \pmod{95}.$$

We find that $y_1 \equiv 37 \pmod{99}$, $y_2 \equiv 35 \pmod{98}$, $y_3 \equiv 24 \pmod{97}$, and $y_4 \equiv 4 \pmod{95}$. Hence,

$$\begin{aligned}
x + y &\equiv 65 \cdot 903{,}070 \cdot 37 + 2 \cdot 912{,}285 \cdot 33 + 51 \cdot 921{,}690 \cdot 24 + 10 \cdot 941{,}094 \cdot 4 \\
&= 3{,}397{,}886{,}480 \\
&\equiv 537{,}140 \pmod{89{,}403{,}930}.
\end{aligned}$$

Since $0 < x + y < 89{,}403{,}930$, we conclude that $x + y = 537{,}140$. ◀

On most computers, the word size is a large power of 2, with 2^{35} a common value. Hence, to use modular arithmetic and the Chinese remainder theorem to do computer arithmetic, we need integers less than 2^{35} that are pairwise relatively prime and that multiply together to give a large integer. To find such integers, we use numbers of the form $2^m - 1$, where m is a positive integer. Computer arithmetic with these numbers turns out to be relatively simple (see [Kn97]). To produce a set of pairwise relatively prime numbers of this form, we first prove two lemmas.

Lemma 4.2.   If a and b are positive integers, then the least positive residue of $2^a - 1$ modulo $2^b - 1$ is $2^r - 1$, where r is the least positive residue of a modulo b.

Proof.   From the division algorithm, $a = bq + r$, where r is the least positive residue of a modulo b. We have $2^a - 1 = 2^{bq+r} - 1 = (2^b - 1)(2^{b(q-1)+r} + \cdots + 2^{b+r} + 2^r) + (2^r - 1)$, which shows that the remainder when $2^a - 1$ is divided by $2^b - 1$ is $2^r - 1$; this is the least positive residue of $2^a - 1$ modulo $2^b - 1$. ∎

We use Lemma 4.2 to prove the following result.

Lemma 4.3.   If a and b are positive integers, then the greatest common divisor of $2^a - 1$ and $2^b - 1$ is $2^{(a,b)} - 1$.

Proof.   When we perform the Euclidean algorithm with $a = r_0$ and $b = r_1$, we obtain

$$\begin{aligned}
r_0 &= r_1 q_1 + r_2 & 0 &\leq r_2 < r_1 \\
r_1 &= r_2 q_2 + r_3 & 0 &\leq r_3 < r_2 \\
&\;\;\vdots \\
r_{n-3} &= r_{n-2} q_{n-2} + r_{n-1} & 0 &\leq r_{n-1} < r_{n-2} \\
r_{n-2} &= r_{n-1} q_{n-1},
\end{aligned}$$

where the last remainder, r_{n-1}, is the greatest common divisor of a and b.

Using Lemma 4.2, and the steps of the Euclidean algorithm with $a = r_0$ and $b = r_1$, when we perform the Euclidean algorithm on the pair $2^a - 1 = R_0$ and $2^b - 1 = R_1$, we obtain

$$
\begin{aligned}
R_0 &= R_1 Q_1 + R_2 & R_2 &= 2^{r_2} - 1 \\
R_1 &= R_2 Q_2 + R_3 & R_3 &= 2^{r_3} - 1 \\
&\;\;\vdots \\
R_{n-3} &= R_{n-2} Q_{n-2} + R_{n-1} & R_{n-1} &= 2^{r_{n-1}} - 1 \\
R_{n-2} &= R_{n-1} Q_{n-1}.
\end{aligned}
$$

Here, the last nonzero remainder, $R_{n-1} = 2^{r_{n-1}} - 1 = 2^{(a,b)} - 1$, is the greatest common divisor of R_0 and R_1.   ∎

Using Lemma 4.3, we have the following theorem.

Theorem 4.13.   The positive integers $2^a - 1$ and $2^b - 1$ are relatively prime if and only if a and b are relatively prime.

We can now use Theorem 4.13 to produce a set of pairwise relatively prime integers, each of which is less than 2^{35}, with product greater than a specified integer. Suppose that we wish to do arithmetic with integers as large as 2^{184}. We pick $m_1 = 2^{35} - 1$, $m_2 = 2^{34} - 1$, $m_3 = 2^{33} - 1$, $m_4 = 2^{31} - 1$, $m_5 = 2^{29} - 1$, and $m_6 = 2^{23} - 1$. Since the exponents of 2 in the expressions for the m_j are pairwise relatively prime, by Theorem 4.13, the m_j are pairwise relatively prime. Also, we have $M = m_1 m_2 m_3 m_4 m_5 m_6 > 2^{184}$. We can now use modular arithmetic and the Chinese remainder theorem to perform arithmetic with integers as large as 2^{184}.

Although it is somewhat awkward to do computer operations with large integers using modular arithmetic and the Chinese remainder theorem, there are some definite advantages to this approach. First, on many high-speed computers, operations can be performed simultaneously. So, reducing an operation involving two large integers to a set of operations involving smaller integers, namely the least positive residues of the large integers with respect to the various moduli, leads to simultaneous computations which may be performed more rapidly than one operation with large integers, especially when parallel processing is used. Second, even without taking into account the advantages of simultaneous computations, multiplication of large integers may be done faster using these ideas than with many other multiprecision methods. The interested reader should consult Knuth [Kn97].

4.3   Exercises

1.   Which integers leave a remainder of 1 when divided by both 2 and 3?

2.   Find an integer that leaves a remainder of 1 when divided by either 2 or 5, but that is divisible by 3.

3.   Find an integer that leaves a remainder of 2 when divided by either 3 or 5, but that is divisible by 4.

4. Find all the solutions of each of the following systems of linear congruences.

a) $x \equiv 4 \pmod{11}$
   $x \equiv 3 \pmod{17}$

b) $x \equiv 1 \pmod 2$
   $x \equiv 2 \pmod 3$
   $x \equiv 3 \pmod 5$

c) $x \equiv 0 \pmod 2$
   $x \equiv 0 \pmod 3$
   $x \equiv 1 \pmod 5$
   $x \equiv 6 \pmod 7$

d) $x \equiv 2 \pmod{11}$
   $x \equiv 3 \pmod{12}$
   $x \equiv 4 \pmod{13}$
   $x \equiv 5 \pmod{17}$
   $x \equiv 6 \pmod{19}$

5. Find all the solutions to the system of linear congruences $x \equiv 1 \pmod 2$, $x \equiv 2 \pmod 3$, $x \equiv 3 \pmod 5$, $x \equiv 4 \pmod 7$, and $x \equiv 5 \pmod{11}$.

6. Find all the solutions to the system of linear congruences $x \equiv 1 \pmod{999}$, $x \equiv 2 \pmod{1001}$, $x \equiv 3 \pmod{1003}$, $x \equiv 4 \pmod{1004}$, and $x \equiv 5 \pmod{1007}$.

7. A troop of 17 monkeys store their bananas in 11 piles of equal size, each containing more than 1 banana, with a twelfth pile of 6 left over. When they divide the bananas into 17 equal groups, none remain. What is the smallest number of bananas they can have?

8. As an odometer check, a special counter measures the miles a car travels modulo 7. Explain how this counter can be used to determine whether the car has been driven 49,335; 149,335; or 249,335 miles when the odometer reads 49,335 and works modulo 100,000.

9. Chinese generals counted troops remaining after a battle by lining them up in rows of different lengths, counting the number left over each time, and calculating the total from these remainders. If a general had 1200 troops at the start of a battle and if there were 3 left over when they lined up 5 at a time, 3 left over when they lined up 6 at a time, 1 left over when they lined up 7 at a time, and none left over when they lined up 11 at a time, how many troops remained after the battle?

10. Find an integer that leaves a remainder of 9 when it is divided by either 10 or 11, but that is divisible by 13.

11. Find a multiple of 11 that leaves a remainder of 1 when divided by each of the integers 2, 3, 5, and 7.

12. Solve the following ancient Indian problem: If eggs are removed from a basket 2, 3, 4, 5, and 6 at a time, there remain, respectively, 1, 2, 3, 4, and 5 eggs. But if the eggs are removed 7 at a time, no eggs remain. What is the least number of eggs that could have been in the basket?

13. Show that there are arbitrarily long strings of consecutive integers each divisible by a perfect square greater than 1. (*Hint:* Use the Chinese remainder theorem to show that there is a simultaneous solution to the system of congruences $x \equiv 0 \pmod 4$, $x \equiv -1 \pmod 9$, $x \equiv -2 \pmod{25}$, ..., $x \equiv -k + 1 \pmod{p_k^2}$, where p_k is the kth prime.)

* 14. Show that if a, b, and c are integers such that $(a, b) = 1$, then there is an integer n such that $(an + b, c) = 1$.

In Exercises 15–18, we will consider systems of congruences where the moduli of the congruences are not necessarily relatively prime.

15. Show that the system of congruences

$$x \equiv a_1 \pmod{m_1}$$
$$x \equiv a_2 \pmod{m_2}$$

has a solution if and only if $(m_1, m_2) \mid (a_1 - a_2)$. Show that when there is a solution, it is unique modulo $[m_1, m_2]$. (*Hint:* Write the first congruence as $x = a_1 + km_1$, where k is an integer, and then insert this expression for x into the second congruence.)

16. Using Exercise 15, solve each of the following simultaneous systems of congruences.

a) $x \equiv 4 \pmod 6$ b) $x \equiv 7 \pmod{10}$
   $x \equiv 13 \pmod{15}$ $x \equiv 4 \pmod{15}$

17. Using Exercise 15, solve each of the following simultaneous systems of congruences.

a) $x \equiv 10 \pmod{60}$ b) $x \equiv 2 \pmod{910}$
   $x \equiv 80 \pmod{350}$ $x \equiv 93 \pmod{1001}$

18. Does the system of congruences $x \equiv 1 \pmod 8$, $x \equiv 3 \pmod 9$, and $x \equiv 2 \pmod{12}$ have any simultaneous solutions?

What happens when the moduli in a simultaneous system of more than two congruences in one unknown are not pairwise relatively prime (such as in Exercise 18)? The following exercise provides compatability conditions for there to be a unique solution of such a system, modulo the least common multiple of the moduli.

19. Show that the system of congruences

$$x \equiv a_1 \pmod{m_1}$$
$$x \equiv a_2 \pmod{m_2}$$
$$\vdots$$
$$x \equiv a_r \pmod{m_r}$$

has a solution if and only if $(m_i, m_j) \mid (a_i - a_j)$ for all pairs of integers (i, j), where $1 \leq i < j \leq r$. Show that if a solution exists, then it is unique modulo $[m_1, m_2, \ldots, m_r]$. (*Hint:* Use Exercise 15 and mathematical induction.)

20. Using Exercise 19, solve each of the following systems of congruences.

a) $x \equiv 5 \pmod 6$ d) $x \equiv 2 \pmod 6$
   $x \equiv 3 \pmod{10}$ $x \equiv 4 \pmod 8$
   $x \equiv 8 \pmod{15}$ $x \equiv 2 \pmod{14}$
                                $x \equiv 14 \pmod{15}$

b) $x \equiv 2 \pmod{14}$
   $x \equiv 16 \pmod{21}$ e) $x \equiv 7 \pmod 9$
   $x \equiv 10 \pmod{30}$ $x \equiv 2 \pmod{10}$
                                $x \equiv 3 \pmod{12}$
c) $x \equiv 2 \pmod 9$ $x \equiv 6 \pmod{15}$
   $x \equiv 8 \pmod{15}$
   $x \equiv 10 \pmod{25}$

21. What is the smallest number of lobsters in a tank if 1 lobster is left over when they are removed 2, 3, 5, or 7 at a time, but no lobsters are left over when they are removed 11 at a time?

22. An ancient Chinese problem asks for the least number of gold coins a band of 17 pirates could have stolen. The problem states that when the pirates divided the coins into equal piles, 3 coins were left over. When they fought over who should get the extra coins, one of the pirates was slain. When the remaining pirates divided the coins into equal piles, 10 coins were left over. When the pirates fought again over who should get the extra coins, another pirate was slain. When they divided the coins in equal piles again, no coins were left over. What is the answer to this problem?

23. Solve the following problem originally posed by Ch'in Chiu-Shao (using different weight units). Three farmers equally divide a quantity of rice with a weight that is an integral number of pounds. The farmers each sell their rice, selling as much as possible, at three different markets where the markets use weights of 83 pounds, 110 pounds, and 135 pounds, and only buy rice in multiples of these weights. What is the least amount of rice the farmers could have divided if the farmers return home with 32 pounds, 70 pounds, and 30 pounds, respectively?

24. Using the Chinese remainder theorem, explain how to add and how to multiply 784 and 813 on a computer of word size 100.

A positive integer $x \neq 1$ with n base b digits is called an *automorph to the base b* if the last n base b digits of x^2 are the same as those of x.

* 25. Find the base 10 automorphs with four digits (with initial zeros allowed).

* 26. How many base b automorphs are there with n or fewer base b digits, if b has prime-power factorization $b = p_1^{b_1} p_2^{b_2} \cdots p_k^{b_k}$?

According to the theory of *biorhythms,* there are three cycles in your life that start the day you are born. These are the *physical, emotional,* and *intellectual cycles,* of lengths 23, 28, and 33 days, respectively. Each cycle follows a sine curve with period equal to the length of that cycle, starting with value 0, climbing to value 1 one-quarter of the way through the cycle, dropping back to value 0 one-half of the way through the cycle, dropping further to value -1 three-quarters of the way through the cycle, and climbing back to value 0 at the end of the cycle.

Answer the following questions about biorhythms, measuring time in quarter days (so that the units will be integers).

27. For which days of your life will you be at a triple peak, where all of your three cycles are at maximum values?

28. For which days of your life will you be at a triple nadir, where all three of your cycles have minimum values?

29. When in your life will all three cycles be at a neutral position (value 0)?

A set of congruences to distinct moduli greater than 1 that has the property that every integer satisfies at least one of the congruences is called a *covering set of congruences.*

30. Show that the set of congruences $x \equiv 0 \pmod 2$, $x \equiv 0 \pmod 3$, $x \equiv 1 \pmod 4$, $x \equiv 1 \pmod 6$, and $x \equiv 11 \pmod{12}$ is a covering set of congruences.

31. Show that the set of congruences $x \equiv 0 \pmod 2$, $x \equiv 0 \pmod 3$, $x \equiv 0 \pmod 5$, $x \equiv 0 \pmod 7$, $x \equiv 1 \pmod 6$, $x \equiv 1 \pmod {10}$, $x \equiv 1 \pmod{14}$, $x \equiv 2 \pmod{15}$, $x \equiv 2 \pmod{21}$, $x \equiv 23 \pmod{30}$, $x \equiv 4 \pmod{35}$, $x \equiv 5 \pmod{42}$, $x \equiv 59 \pmod{70}$, and $x \equiv 104 \pmod{105}$ is a covering set of congruences.

* 32. Let m be a positive integer with prime-power factorization $m = 2^{a_0} p_1^{a_1} p_2^{a_2} \cdots p_r^{a_r}$. Show that the congruence $x^2 \equiv 1 \pmod m$ has exactly 2^{r+e} solutions, where $e = 0$ if $a_0 = 0$ or 1, $e = 1$ if $a_0 = 2$, and $e = 2$ if $a_0 > 2$. (*Hint:* Use Exercises 15 and 16 of Section 4.2.)

33. The three children in a family have feet that are 5 inches, 7 inches, and 9 inches long. When they measure the length of the dining room of their house using their feet, they each find that there are 3 inches left over. How long is the dining room?

34. Find all solutions of the congruence $x^2 + 6x - 31 \equiv 0 \pmod{72}$. (*Hint:* First note that $72 = 2^3 3^2$. Find, by trial and error, the solutions of this congruence modulo 8 and modulo 9. Then apply the Chinese remainder theorem.)

35. Find all solutions of the congruence $x^2 + 18x - 823 \equiv 0 \pmod{1800}$. (*Hint:* First note that $1800 = 2^3 3^2 5^2$. Find, by trial and error, the solutions of this congruence modulo 8, modulo 9, and modulo 25. Then apply the Chinese remainder theorem.)

* 36. Give a positive integer R, a prime p that is the only prime between $p - R$ and $p + R$, including the end points, is called *R-reclusive*. Show that for every positive integer R, there are infinitely many R-reclusive primes. (*Hint:* Use the Chinese remainder theorem to find an integer x such that $x - j$ is divisible by p_j and $x + j$ is divisible by p_{R+j}, where p_k is the kth prime. Then invoke Dirichlet's theorem on primes in arithmetic progressions.)

4.3 Computational and Programming Exercises

Computations and Explorations

Using a computation program such as Maple or *Mathematica,* or programs you have written, carry out the following computations and explorations.

1. Solve the simultaneous system of congruences $x \equiv 1 \pmod{12{,}341{,}234{,}567}$, $x \equiv 2 \pmod{750{,}000{,}057}$, and $x \equiv 3 \pmod{1{,}099{,}511{,}627{,}776}$.

2. Solve the simultaneous system of congruences $x \equiv 5269 \pmod{40{,}320}$, $x \equiv 1248 \pmod{11{,}111}$, $x \equiv 16{,}645 \pmod{30{,}003}$, and $x \equiv 2911 \pmod{12{,}321}$.

3. Using Exercise 13 of this section, find a string of 100 consecutive positive integers each divisible by a perfect square. Can you find such a set of smaller integers?

4. Find a covering set of congruences (as described in the preamble to Exercise 30) where the smallest modulus of one of the congruences in the covering set is 3; where the smallest modulus of one of the congruences in the covering set is 6; and where the smallest modulus of one of the congruences in the covering set is 8.

Programming Projects

Write programs using Maple, *Mathematica,* or a language of your choice to do the following.

1. Solve systems of linear congruences of the type found in the Chinese remainder theorem.

2. Solve systems of linear congruences of the type given in Exercises 15–20.

3. Add large integers exceeding the word size of a computer using the Chinese remainder theorem.

4. Multiply large integers exceeding the word size of a computer using the Chinese remainder theorem.

5. Find automorphs to the base b, where b is a positive integer greater than 1 (see the preamble to Exercise 25).

6. Plot biorhythm charts and find triple peaks and triple nadirs (see the preamble to Exercise 27).

4.4 Solving Polynomial Congruences

This section provides a useful tool that can be used to help find solutions of congruences of the form $f(x) \equiv 0 \pmod{m}$, where $f(x)$ is a polynomial of degree greater than 1 with integer coefficients. An example of such a congruence is $2x^3 + 7x - 4 \equiv 0 \pmod{200}$.

We first note that if m has prime-power factorization $m = p_1^{a_1} p_2^{a_2} \cdots p_k^{a_k}$, then solving the congruence $f(x) \equiv 0 \pmod{m}$ is equivalent to finding the simultaneous solutions to the system of congruences

$$f(x) \equiv 0 \pmod{p_i^{a_i}}, \quad i = 1, 2, \ldots, k.$$

Once the solutions of each of the k congruences modulo $p_i^{a_i}$ are known, the solutions of the congruence modulo m can be found by the Chinese remainder theorem. This is illustrated in the following example.

Example 4.19. Solving the congruence

$$2x^3 + 7x - 4 \equiv 0 \pmod{200}$$

reduces to finding the solutions of

$$2x^3 + 7x - 4 \equiv 0 \pmod{8}$$

and

$$2x^3 + 7x - 4 \equiv 0 \pmod{25}$$

since $200 = 2^3 5^2$. The solutions of the congruence modulo 8 are all integers $x \equiv 4 \pmod{8}$ (for x to be a solution x must be even; the cases where x is odd can be quickly checked). In Example 4.20, we will see that the solutions modulo 25 are all integers $x \equiv 16 \pmod{25}$. When we use the Chinese remainder theorem to solve the simultaneous congruences $x \equiv 4 \pmod{8}$ and $x \equiv 16 \pmod{25}$, we find that the solutions are all $x \equiv 116 \pmod{200}$ (as the reader should verify). These are solutions of $2x^3 + 7x - 4 \equiv 0 \pmod{200}$. ◀

We will see that there is a relatively simple way to solve polynomial congruences modulo p^k, once all solutions modulo p are known. We will show that solutions modulo p can be used to find solutions modulo p^2, solutions modulo p^2 can be used to find solutions modulo p^3, and so on. Before introducing the general method, we present an example illustrating the basic idea used to find solutions of a polynomial congruence modulo p^2 from those modulo p.

Example 4.20. The solutions of

$$2x^3 + 7x - 4 \equiv 0 \ (\text{mod } 5)$$

are the integers with $x \equiv 1 \ (\text{mod } 5)$, as can be seen by testing $x = 0, 1, 2, 3,$ and 4. How can we find the solutions modulo 25? We could check all 25 different values $x = 0, 1, 2, \ldots, 24$. However, there is a more systematic method. Since any solution of

$$2x^3 + 7x - 4 \equiv 0 \ (\text{mod } 25)$$

is also a solution modulo 5 and all solutions modulo 5 satisfy $x \equiv 1 \ (\text{mod } 5)$, it follows that $x = 1 + 5t$, where t is an integer. We can solve for t by substituting $1 + 5t$ for x. We obtain

$$2(1 + 5t)^3 + 7(1 + 5t) - 4 \equiv 0 \ (\text{mod } 25).$$

Simplifying, we obtain a linear congruence for t, namely

$$65t + 5 \equiv 15t + 5 \equiv 0 \ (\text{mod } 25).$$

By Theorem 4.4, we can eliminate a factor of 5, so that

$$3t + 1 \equiv 0 \ (\text{mod } 5).$$

The solutions of this congruence are $t \equiv 3 \ (\text{mod } 5)$. This means that the solutions modulo 25 are those x for which $x \equiv 1 + 5t \equiv 1 + 5 \cdot 3 \equiv 16 \ (\text{mod } 25)$. The reader should verify that these are indeed solutions. ◄

We will now introduce a general method that will help us find the solutions of congruences modulo prime powers. In particular, we will show how the solutions of the congruence $f(x) \equiv 0 \ (\text{mod } p^k)$, where p is prime and k is a positive integer with $k \geq 2$, can be found from those of the congruence $f(x) \equiv 0 \ (\text{mod } p^{k-1})$. The solutions of the congruence modulo p^k are said to be *lifted* from those modulo p^{k-1}. The theorem uses $f'(x)$, the derivative of f. However, we will not need results from calculus. Instead, we can define the derivative of a polynomial directly and describe the properties that we will need.

Definition. Let $f(x) = a_n x^n + a_{n-1} x^{n-1} + \cdots + a_1 x + a_0$, where a_i is a real number for $i = 0, 1, 2, \ldots, n$. The *derivative* of $f(x)$, denoted by $f'(x)$, equals $na_n x^{n-1} + (n-1)a_{n-1} x^{n-2} + \cdots + a_1$.

Starting with a polynomial, we can find its derivative and then find the derivative of its derivative, and so on. We can define the kth derivative of a polynomial $f(x)$, denoted by $f^{(k)}(x)$, as the derivative of the $(k-1)$st derivative, that is, $f^{(k)}(x) = (f^{(k-1)})'(x)$.

We will find the following two lemmas helpful. We leave their proofs to the reader.

Lemma 4.4. If $f(x)$ and $g(x)$ are polynomials, then $(f+g)'(x) = f'(x) + g'(x)$ and $(cf)'(x) = c(f'(x))$, where c is a constant. Furthermore, if k is a positive integer, then $(f+g)^{(k)}(x) = f^{(k)}(x) + g^{(k)}(x)$ and $(cf)^{(k)}(x) = c(f^{(k)}(x))$, where c is a constant.

Lemma 4.5. If m and k are positive integers and $f(x) = x^m$, then $f^{(k)}(x) = m(m-1)\cdots(m-k+1)x^{m-k}$.

 We can now state the result that can be used to lift solutions of polynomial congruences. It is called *Hensel's lemma* after the German mathematician *Kurt Hensel*, who discovered it in work leading to the invention of the field of mathematics known as *p*-adic analysis.

Theorem 4.14. *Hensel's Lemma.* Suppose that $f(x)$ is a polynomial with integer coefficients and that k is an integer with $k \geq 2$. Suppose further that r is a solution of the congruence $f(x) \equiv 0 \pmod{p^{k-1}}$. Then,

(i)   if $f'(r) \not\equiv 0 \pmod{p}$, then there is a unique integer t, $0 \leq t < p$, such that $f(r + tp^{k-1}) \equiv 0 \pmod{p^k}$, given by

$$t \equiv -\overline{f'(r)}(f(r)/p^{k-1}) \pmod{p},$$

where $\overline{f'(r)}$ is an inverse of $f'(r)$ modulo p;

(ii) if $f'(r) \equiv 0 \pmod{p}$ and $f(r) \equiv 0 \pmod{p^k}$, then $f(r + tp^{k-1}) \equiv 0 \pmod{p^k}$ for all integers t;

(iii) if $f'(r) \equiv 0 \pmod{p}$ and $f(r) \not\equiv 0 \pmod{p^k}$, then $f(x) \equiv 0 \pmod{p^k}$ has no solutions with $x \equiv r \pmod{p^{k-1}}$.

In case (i), we see that a solution to $f(x) \equiv 0 \pmod{p^{k-1}}$ lifts to a unique solution of $f(x) \equiv 0 \pmod{p^k}$, and in case (ii), such a solution either lifts to p incongruent solutions modulo p^k or to none at all.   ■

KURT HENSEL (1861–1941) was born in Königsberg, Prussia (now Kaliningrad, Russia). He studied mathematics in Berlin, and later in Bonn, under many leading mathematicians, including Kronecker and Weierstrass. Much of his work involved the development of arithmetic in algebraic number fields. Hensel is best known for inventing the *p*-adic numbers in 1902, in work on representations of algebraic numbers in terms of power series. The *p*-adic numbers can be thought of as a completion of the set of rational numbers that is different from the usual completion that produces the set of real numbers. Hensel was able to use the *p*-adic numbers to prove many results in number theory, and these numbers have had a major impact on the development of algebraic number theory. Hensel served as a professor at the University of Marburg until 1930. He was the editor for many years of the famous mathematical journal known as *Crelle's Journal*, whose official name is *Journal für die reine und angewandte Mathematik*.

We will need the following lemma about Taylor expansions for the proof of Hensel's lemma.

Lemma 4.6. If $f(x)$ is a polynomial of degree n with integer coefficients, then

$$f(a+b) = f(a) + f'(a)b + f''(a)b^2/2! + \cdots + f^{(n)}(a)b^n/n!,$$

where the coefficients (namely 1, $f'(a)$, $f''(a)/2! \ldots, f^{(n)}(a)/n!$) are polynomials in a with integer coefficients.

Proof. Every polynomial f of degree n is the sum of multiples of the functions x^m, where $m \leq n$. Furthermore, by Lemma 4.4, we need only establish Lemma 4.6 for the polynomials $f_m(x) = x^m$, where m is a positive integer.

By the binomial theorem, we have

$$(a+b)^m = \sum_{j=0}^{m} \binom{m}{j} a^{m-j} b^j.$$

By Lemma 4.5, we know that $f_m^{(j)}(a) = m(m-1) \cdots (m-j+1) a^{m-j}$. Hence,

$$f_m^{(j)}(a)/j! = \binom{m}{j} a^{m-j}.$$

Because $\binom{m}{j}$ is an integer for all integers m and j such that $0 \leq j \leq m$, the coefficients $f_m(j)/j!$ are integers. This completes the proof. ∎

We now have all the ingredients needed to prove Hensel's lemma.

Proof. If r is a solution of $f(r) \equiv 0 \pmod{p^k}$, then it is also a solution of $f(r) \equiv 0 \pmod{p^{k-1}}$. Hence, it equals $r + tp^{k-1}$ for some integer t. The proof follows once we have determined the conditions on t.

By Lemma 4.6, it follows that

$$f(r + tp^{k-1}) = f(r) + f'(r)tp^{k-1} + \frac{f''(r)}{2!}(tp^{k-1})^2 + \cdots + \frac{f^{(n)}(r)}{n!}(tp^{k-1})^n,$$

where $f^{(k)}(r)/k!$ is an integer for $k = 1, 2, \ldots, n$. Given that $k \geq 2$, it follows that $k \leq m(k-1)$ and $p^k \mid p^{m(k-1)}$ for $2 \leq m \leq n$. Hence,

$$f(r + tp^{k-1}) \equiv f(r) + f'(r)tp^{k-1} \pmod{p^k}.$$

Because $r + tp^{k-1}$ is a solution of $f(r + tp^{k-1}) \equiv 0 \pmod{p^k}$, it follows that $f'(r)tp^{k-1} \equiv -f(r) \pmod{p^k}$.

Furthermore, we can divide this congruence by p^{k-1}, because $f(r) \equiv 0 \pmod{p^{k-1}}$. When we do so and rearrange terms, we obtain a linear congruence in t, namely

$$f'(r)t \equiv -f(r)/p^{k-1} \pmod{p}.$$

By examining its solutions modulo p we can prove the three cases of the theorem.

Suppose that $f'(r) \not\equiv 0 \pmod{p}$. It follows that $(f'(r), p) = 1$. Applying Theorem 4.10, we see that the congruence for t has a unique solution,

$$t \equiv (-f(r)/p^{k-1})\overline{f'(r)} \pmod{p},$$

where $\overline{f'(r)}$ is an inverse of $f'(r)$ modulo p. This establishes case (i).

When $f'(r) \equiv 0 \pmod{p}$, we have $(f'(r), p) = p$. By Theorem 4.10, if $p \mid (f(r)/p^{k-1})$, which holds if and only if $f(r) \equiv 0 \pmod{p^k}$, then all values t are solutions. This means that $x = r + tp^{k-1}$ is a solution for $t = 0, 1, \ldots, p-1$. This establishes case (ii).

Finally, consider the case when $f'(r) \equiv 0 \pmod{p}$, but $p \nmid (f(r)/p^{k-1})$. We have $(f'(r), p) = p$ and $f(r) \not\equiv 0 \pmod{p^k}$; so, by Theorem 4.10, no values of t are solutions. This completes case (iii). ∎

The following corollary shows that we can repeatedly lift solutions, starting with a solution modulo p, when case (i) of Hensel's lemma applies.

Corollary 4.14.1. Suppose that r is a solution of the polynomial congruence $f(x) \equiv 0 \pmod{p}$, where p is a prime. If $f'(r) \not\equiv 0 \pmod{p}$, then there is a unique solution r_k modulo p^k, $k = 2, 3, \ldots$, such that

$$r_k = r_{k-1} - f(r_{k-1})\overline{f'(r)},$$

where $\overline{f'(r)}$ is an inverse of $f'(r)$ modulo p.

Proof. Using the hypotheses, we see by Hensel's lemma that r lifts to a unique solution r_2 modulo p^2 with $r_2 = r + tp$, where $t = -\overline{f'(r)}(f(r)/p)$. Hence,

$$r_2 = r - f(r)\overline{f'(r)}.$$

Because $r_2 \equiv r \pmod{p}$, it follows that $f'(r_2) \equiv f'(r) \not\equiv 0 \pmod{p}$. Using Hensel's lemma again, we see that there is a unique solution r_3 modulo p^3, which can be shown to be $r_3 = r_2 - f(r_2)\overline{f'(r)}$. If we continue in this way, we find that the corollary follows for all integers $k \geq 2$. ∎

The following examples illustrate how Hensel's lemma is applied.

Example 4.21. Find the solutions of

$$x^3 + x^2 + 29 \equiv 0 \pmod{25}.$$

Let $f(x) = x^3 + x^2 + 29$. We see (by inspection) that solutions of $f(x) \equiv 0 \pmod 5$ have $x \equiv 3 \pmod 5$. Because $f'(x) = 3x^2 + 2x$ and $f'(3) = 33 \equiv 3 \not\equiv 0 \pmod 5$, Hensel's lemma tells us that there is a unique solution modulo 25 of the form $3 + 5t$, where

$$t \equiv -\overline{f'(3)}(f(3)/5) \pmod 5.$$

Note that $\overline{f'(3)} = \overline{3} = 2$, because 2 is inverse to 3 modulo 5. Also note that $f(3)/5 = 65/5 = 13$. It follows that $t \equiv -2 \cdot 13 = 4 \pmod 5$. We conclude that $x \equiv 3 + 5 \cdot 4 = 23$ is the unique solution of $f(x) \equiv 0 \pmod{25}$. ◀

Example 4.22. Find the solutions of

$$x^2 + x + 7 \equiv 0 \text{ (mod 27)}.$$

Let $f(x) = x^2 + x + 7$. We find (by inspection) that the solutions of $f(x) \equiv 0 \text{ (mod 3)}$ are the integers with $x \equiv 1 \text{ (mod 3)}$. Because $f'(x) = 2x + 1$, we see that $f'(1) = 3 \equiv 0 \text{ (mod 3)}$. Furthermore, because $f(1) = 9 \equiv 0 \text{ (mod 9)}$, we can apply case (ii) of Hensel's lemma to conclude that $1 + 3t$ is a solution modulo 9 for all integers t. This means that the solutions modulo 9 are $x \equiv 1, 4,$ or 7 (mod 9).

Now, by case (iii) of Hensel's lemma, because $f(1) = 9 \not\equiv 0 \text{ (mod 27)}$, there are no solutions of $f(x) \equiv 0 \text{ (mod 27)}$ with $x \equiv 1 \text{ (mod 9)}$. Because $f(4) = 27 \equiv 0 \text{ (mod 27)}$, by case (ii), $4 + 9t$ is a solution modulo 27 for all integers t. This shows that all $x \equiv 4, 13,$ or 22 (mod 27) are solutions. Finally, by case (iii), because $f(7) = 63 \not\equiv 0 \text{ (mod 27)}$, there are no solutions of $f(x) \equiv 0 \text{ (mod 27)}$ with $x \equiv 7 \text{ (mod 9)}$.

Putting everything together, we see that all solutions of $f(x) \equiv 0 \text{ (mod 27)}$ are those $x \equiv 4, 13,$ or 22 (mod 27). ◄

Example 4.23. What are the solutions of $f(x) = x^3 + x^2 + 2x + 26 \equiv 0 \text{ (mod 343)}$? By inspection, we see that the solutions of $x^3 + x^2 + 2x + 26 \equiv 0 \text{ (mod 7)}$ are the integers $x \equiv 2 \text{ (mod 7)}$. Because $f'(x) = 3x^2 + 2x + 2$, it follows that $f'(2) = 18 \not\equiv 0 \text{ (mod 7)}$. We can use Corollary 4.14.1 to find solutions modulo 7^k for $k = 2, 3, \ldots$. Noting that $\overline{f'(2)} = \overline{4} = 2$, we find that $r_2 = 2 - f(2)\overline{f'(2)} = 2 - 42 \cdot 2 = -82 \equiv 16 \text{ (mod 49)}$, and $r_3 = 16 - f(16)\overline{f'(2)} = 16 - 4410 \cdot 2 = -8804 \equiv 114 \text{ (mod 343)}$. It follows that the solutions modulo 343 are the integers $x \equiv 114 \text{ (mod 343)}$. ◄

4.4 Exercises

1. Find all the solutions of each of the following congruences.
   a) $x^2 + 4x + 2 \equiv 0 \text{ (mod 7)}$
   b) $x^2 + 4x + 2 \equiv 0 \text{ (mod 49)}$
   c) $x^2 + 4x + 2 \equiv 0 \text{ (mod 343)}$

2. Find all the solutions of each of the following congruences.
   a) $x^3 + 8x^2 - x - 1 \equiv 0 \text{ (mod 11)}$
   b) $x^3 + 8x^2 - x - 1 \equiv 0 \text{ (mod 121)}$
   c) $x^3 + 8x^2 - x - 1 \equiv 0 \text{ (mod 1331)}$

3. Find the solutions of the congruence $x^2 + x + 47 \equiv 0 \text{ (mod 2401)}$. (Note that $2401 = 7^4$.)

4. Find the solutions of $x^2 + x + 34 \equiv 0 \text{ (mod 81)}$.

5. Find all solutions of $13x^7 - 42x - 649 \equiv 0 \text{ (mod 1323)}$.

6. Find all solutions of $x^8 - x^4 + 1001 \equiv 0 \text{ (mod 539)}$.

7. Find all solutions of $x^4 + 2x + 36 \equiv 0 \text{ (mod 4375)}$.

8. Find all solutions of $x^6 - 2x^5 - 35 \equiv 0 \text{ (mod 6125)}$.

9. How many incongruent solutions are there to the congruence $5x^3 + x^2 + x + 1 \equiv 0 \pmod{64}$?

10. How many incongruent solutions are there to the congruence $x^5 + x - 6 \equiv 0 \pmod{144}$?

11. Let a be an integer and p a prime such that $(a, p) = 1$. Use Hensel's lemma to solve the congruence $ax \equiv 1 \pmod{p^k}$, for all positive integers k.

* 12. a) Let $f(x)$ be a polynomial with integer coefficients. Let p be a prime, k a positive integer, and j an integer such that $k \geq 2j + 1$. Let a be a solution of $f(a) \equiv 0 \pmod{p^k}$, with p^j exactly dividing $f'(a)$. Show that if $b \equiv a \pmod{p^{k-j}}$, then $f(b) \equiv f(a) \pmod{p^k}$, p^j exactly divides $f'(b)$, and there is a unique t modulo p such that $f(a + tp^{k-j}) \equiv 0 \pmod{p^{k+1}}$. (*Hint:* Using a Taylor expansion, first show that $f(a + tp^{k-j}) \equiv f(a) + tp^{k-j} f'(a) \pmod{p^{2k-2j}}$.)

   b) Show that when the hypotheses of part (a) hold, the solutions of $f(x) \equiv 0 \pmod{p^k}$ may be lifted to solutions of arbitrarily high powers of p.

* 13. How many solutions are there to $x^2 + x + 223 \equiv 0 \pmod{3^j}$, where j is a positive integer? (*Hint:* First find the solutions modulo 3^5 and then apply Exercise 12.)

4.4 Computational and Programming Exercises

Computations and Explorations

Using a computation program such as Maple or *Mathematica,* or programs you have written, carry out the following computations and explorations.

1. Find all solutions of $x^4 - 13x^3 + 11x - 3 \equiv 0 \pmod{7^8}$.

2. Find all solutions of $x^9 + 13x^3 - x + 100{,}336 \equiv 0 \pmod{17^9}$.

Programming Projects

Write programs using Maple, *Mathematica,* or a language of your choice to do the following.

1. Use Hensel's lemma to solve congruences of the form $f(x) \equiv 0 \pmod{p^n}$, where $f(x)$ is a polynomial, p is prime, and n is a positive integer.

4.5 Systems of Linear Congruences

We will consider systems of more than one congruence that involve the same number of unknowns as congruences, where all congruences have the same modulus. We begin our study with an example.

Suppose that we wish to find all integers x and y such that both of the congruences

$$3x + 4y \equiv 5 \pmod{13}$$
$$2x + 5y \equiv 7 \pmod{13}$$

are satisfied. To attempt to find the unknowns x and y, we multiply the first congruence by 5 and the second by 4, to obtain

$$15x + 20y \equiv 25 \ (\text{mod } 13)$$
$$8x + 20y \equiv 28 \ (\text{mod } 13).$$

We subtract the second congruence from the first, to find that

$$7x \equiv -3 \ (\text{mod } 13).$$

Since 2 is an inverse of 7 (mod 13), we multiply both sides of the above congruence by 2. This gives

$$2 \cdot 7x \equiv -2 \cdot 3 \ (\text{mod } 13),$$

which tells us that

$$x \equiv 7 \ (\text{mod } 13).$$

Likewise, we can multiply the first congruence by 2 and the second by 3 (of the original system), to see that

$$6x + 8y \equiv 10 \ (\text{mod } 13)$$
$$6x + 15y \equiv 21 \ (\text{mod } 13).$$

When we subtract the first congruence from the second, we obtain

$$7y \equiv 11 \ (\text{mod } 13).$$

To solve for y, we multiply both sides of this congruence by 2, an inverse of 7 modulo 13. We get

$$2 \cdot 7y \equiv 2 \cdot 11 \ (\text{mod } 13),$$

so that

$$y \equiv 9 \ (\text{mod } 13).$$

What we have shown is that any solution (x, y) must satisfy

$$x \equiv 7 \ (\text{mod } 13), \quad y \equiv 9 \ (\text{mod } 13).$$

When we insert these congruences for x and y into the original system, we see that these pairs actually are solutions:

$$3x + 4y \equiv 3 \cdot 7 + 4 \cdot 9 = 57 \equiv 5 \ (\text{mod } 13)$$
$$2x + 5y \equiv 2 \cdot 7 + 5 \cdot 9 = 59 \equiv 7 \ (\text{mod } 13).$$

Hence, the solutions of this system of congruences are all pairs (x, y) such that $x \equiv 7 \ (\text{mod } 13)$ and $y \equiv 9 \ (\text{mod } 13)$.

We now give a general result concerning certain systems of two congruences in two unknowns. (This result resembles Cramer's rule for solving systems of linear equations.)

Theorem 4.15. Let a, b, c, d, e, f, and m be integers, $m > 0$, such that $(\Delta, m) = 1$, where $\Delta = ad - bc$. Then the system of congruences

$$ax + by \equiv e \ (\text{mod } m)$$
$$cx + dy \equiv f \ (\text{mod } m)$$

has a unique solution modulo m, given by

$$x \equiv \bar{\Delta}(de - bf)(\bmod m)$$
$$y \equiv \bar{\Delta}(af - ce) \ (\bmod m),$$

where $\bar{\Delta}$ is an inverse of Δ modulo m.

Proof. We multiply the first congruence of the system by d and the second by b, to obtain

$$adx + bdy \equiv de(\bmod m)$$
$$bcx + bdy \equiv bf \ (\bmod m).$$

Then we subtract the second congruence from the first, to find that

$$(ad - bc)x \equiv de - bf \ (\bmod m),$$

or, since $\Delta = ad - bc$,

$$\Delta x \equiv de - bf \ (\bmod m).$$

Next, we multiply both sides of this congruence by $\bar{\Delta}$, an inverse of Δ modulo m, to conclude that

$$x \equiv \bar{\Delta}(de - bf) \ (\bmod m).$$

In a similar way, we multiply the first congruence by c and the second by a, to obtain

$$acx + bcy \equiv ce \ (\bmod m)$$
$$acx + ady \equiv af \ (\bmod m).$$

We subtract the first congruence from the second, to find that

$$(ad - bc)y \equiv af - ce \ (\bmod m)$$

or

$$\Delta y \equiv af - ce \ (\bmod m).$$

Finally, we multiply both sides of this congruence by $\bar{\Delta}$ to see that

$$y \equiv \bar{\Delta}(af - ce) \ (\bmod m).$$

We have shown that if (x, y) is a solution of the system of congruences, then

$$x \equiv \bar{\Delta}(de - bf) \ (\bmod m), \quad y \equiv \bar{\Delta}(af - ce) \ (\bmod m).$$

We can easily check that any such pair (x, y) is a solution. When $x \equiv \bar{\Delta}(de - bf) \ (\bmod m)$ and $y \equiv \bar{\Delta}(af - ce) \ (\bmod m)$, we have

$$ax + by \equiv a\bar{\Delta}(de - bf) + b\bar{\Delta}(af - ce)$$
$$\equiv \bar{\Delta}(ade - abf - abf - bce)$$
$$\equiv \bar{\Delta}(ad - bc)e$$
$$\equiv \bar{\Delta}\Delta e$$
$$\equiv e \pmod{m},$$

and

$$cx + dy \equiv c\bar{\Delta}(de - bf) + d\bar{\Delta}(af - ce)$$
$$\equiv \bar{\Delta}(cde - bcf + adf - cde)$$
$$\equiv \bar{\Delta}(ad - bc)f$$
$$\equiv \bar{\Delta}\Delta f$$
$$\equiv f \pmod{m}.$$

This establishes the theorem.   ∎

By similar methods, we may solve systems of n congruences involving n unknowns. However, we will develop the theory of solving such systems, as well as larger systems, by methods taken from linear algebra. Readers unfamiliar with linear algebra may wish to skip the remainder of this section.

Systems of n linear congruences involving n unknowns will arise in our subsequent cryptographic studies. To study such systems when n is large, it is helpful to use the language of matrices. We will use some of the basic notions of matrix arithmetic, which are discussed in most linear algebra texts.

Before we proceed, we need to define congruences of matrices.

Definition.   Let $\mathbf{A}$ and $\mathbf{B}$ be $n \times k$ matrices with integer entries, with (i, j)th entries a_{ij} and b_{ij}, respectively. We say that $\mathbf{A}$ is *congruent to* $\mathbf{B}$ *modulo* m if $a_{ij} \equiv b_{ij} \pmod{m}$ for all pairs (i, j) with $1 \le i \le n$ and $1 \le j \le k$. We write $\mathbf{A} \equiv \mathbf{B} \pmod{m}$ if $\mathbf{A}$ is congruent to $\mathbf{B}$ modulo m.

The matrix congruence $\mathbf{A} \equiv \mathbf{B} \pmod{m}$ provides a succinct way of expressing the nk congruences $a_{ij} \equiv b_{ij} \pmod{m}$ for $1 \le i \le n$ and $1 \le j \le k$.

Example 4.24.   We easily see that

$$\begin{pmatrix} 15 & 3 \\ 8 & 12 \end{pmatrix} \equiv \begin{pmatrix} 4 & 3 \\ -3 & 1 \end{pmatrix} \pmod{11}.$$ ◀

The following proposition will be needed.

Theorem 4.16.   If $\mathbf{A}$ and $\mathbf{B}$ are $n \times k$ matrices with $\mathbf{A} \equiv \mathbf{B} \pmod{m}$, $\mathbf{C}$ is a $k \times p$ matrix, and $\mathbf{D}$ is a $p \times n$ matrix, all with integer entries, then $\mathbf{AC} \equiv \mathbf{BC} \pmod{m}$ and $\mathbf{DA} \equiv \mathbf{DB} \pmod{m}$.

Proof. Let the entries of **A** and **B** be a_{ij} and b_{ij}, respectively, for $1 \le i \le n$ and $1 \le j \le k$, and let the entries of **C** be c_{ij} for $1 \le i \le k$ and $1 \le j \le p$. The (i, j)th entries of **AC** and **BC** are $\sum_{t=1}^{k} a_{it}c_{tj}$ and $\sum_{t=1}^{k} b_{it}c_{tj}$, respectively, for $1 \le i \le n$ and $1 \le j \le p$. Because $\mathbf{A} \equiv \mathbf{B} \pmod{m}$, we know that $a_{it} \equiv b_{it} \pmod{m}$ for all i and k. Hence, by Theorem 4.3, we see that $\sum_{t=1}^{k} a_{it}c_{tj} \equiv \sum_{t=1}^{k} b_{it}c_{tj} \pmod{m}$. Consequently, $\mathbf{AC} \equiv \mathbf{BC} \pmod{m}$.

The proof that $\mathbf{DA} \equiv \mathbf{DB} \pmod{m}$ is similar and is omitted. ∎

Now let us consider the system of congruences

$$a_{11}x_1 + a_{12}x_2 + \cdots + a_{1n}x_n \equiv b_1 \pmod{m}$$
$$a_{21}x_1 + a_{22}x_2 + \cdots + a_{2n}x_n \equiv b_2 \pmod{m}$$
$$\vdots$$
$$a_{n1}x_1 + a_{n2}x_2 + \cdots + a_{nn}x_n \equiv b_n \pmod{m}.$$

Using matrix notation, we see that this system of n congruences is equivalent to the matrix congruence $\mathbf{AX} \equiv \mathbf{B} \pmod{m}$, where

$$\mathbf{A} = \begin{pmatrix} a_{11} & a_{12} & \cdots & a_{1n} \\ a_{21} & a_{22} & \cdots & a_{2n} \\ & & \ddots & \\ a_{n1} & a_{n2} & \cdots & a_{nn} \end{pmatrix}, \quad \mathbf{X} = \begin{pmatrix} x_1 \\ x_2 \\ \vdots \\ x_n \end{pmatrix}, \quad \text{and} \quad \mathbf{B} = \begin{pmatrix} b_1 \\ b_2 \\ \vdots \\ b_n \end{pmatrix}.$$

Example 4.25.   The system

$$3x + 4y \equiv 5 \pmod{13}$$
$$2x + 5y \equiv 7 \pmod{13}$$

can be written as

$$\begin{pmatrix} 3 & 4 \\ 2 & 5 \end{pmatrix} \begin{pmatrix} x \\ y \end{pmatrix} \equiv \begin{pmatrix} 5 \\ 7 \end{pmatrix} \pmod{13}. \quad \blacktriangleleft$$

We now develop a method for solving congruences of the form $\mathbf{AX} \equiv \mathbf{B} \pmod{m}$. This method is based on finding a matrix $\bar{\mathbf{A}}$ such that $\bar{\mathbf{A}}\mathbf{A} \equiv \mathbf{I} \pmod{m}$, where **I** is the identity matrix.

Definition.   If **A** and $\bar{\mathbf{A}}$ are $n \times n$ matrices of integers and $\bar{\mathbf{A}}\mathbf{A} \equiv \mathbf{A}\bar{\mathbf{A}} \equiv \mathbf{I} \pmod{m}$, where $\mathbf{I} = \begin{pmatrix} 1 & 0 & \cdots & 0 \\ 0 & 1 & \cdots & 0 \\ \vdots & & \ddots & \\ 0 & 0 & \cdots & 1 \end{pmatrix}$ is the identity matrix of order n, then $\bar{\mathbf{A}}$ is said to be an *inverse of* **A** *modulo m*.

If $\bar{\mathbf{A}}$ is an inverse of **A** and $\mathbf{B} \equiv \bar{\mathbf{A}} \pmod{m}$, then **B** is also an inverse of **A**. This follows from Theorem 4.16, because $\mathbf{BA} \equiv \bar{\mathbf{A}}\mathbf{A} \equiv \mathbf{I} \pmod{m}$. Conversely, if $\mathbf{B}_1$ and $\mathbf{B}_2$ are both inverses of **A**, then $\mathbf{B}_1 \equiv \mathbf{B}_2 \pmod{m}$. To see this, using Theorem 4.16 and the congruence $\mathbf{B}_1\mathbf{A} \equiv \mathbf{B}_2\mathbf{A} \equiv \mathbf{I} \pmod{m}$, we have $\mathbf{B}_1\mathbf{A}\mathbf{B}_1 \equiv \mathbf{B}_2\mathbf{A}\mathbf{B}_1 \pmod{m}$. Because $\mathbf{AB}_1 \equiv \mathbf{I} \pmod{m}$, we conclude that $\mathbf{B}_1 \equiv \mathbf{B}_2 \pmod{m}$.

Example 4.26.   Given that

$$\begin{pmatrix} 1 & 3 \\ 2 & 4 \end{pmatrix} \begin{pmatrix} 3 & 4 \\ 1 & 2 \end{pmatrix} = \begin{pmatrix} 6 & 10 \\ 10 & 16 \end{pmatrix} \equiv \begin{pmatrix} 1 & 0 \\ 0 & 1 \end{pmatrix} \quad (\text{mod } 5)$$

and

$$\begin{pmatrix} 3 & 4 \\ 1 & 2 \end{pmatrix} \begin{pmatrix} 1 & 3 \\ 2 & 4 \end{pmatrix} = \begin{pmatrix} 11 & 25 \\ 5 & 11 \end{pmatrix} \equiv \begin{pmatrix} 1 & 0 \\ 0 & 1 \end{pmatrix} \quad (\text{mod } 5),$$

we see that the matrix $\begin{pmatrix} 3 & 4 \\ 1 & 2 \end{pmatrix}$ is an inverse of $\begin{pmatrix} 1 & 3 \\ 2 & 4 \end{pmatrix}$ modulo 5.   ◀

The following proposition gives an easy method for finding inverses for 2×2 matrices.

Theorem 4.17.   Let $\mathbf{A} = \begin{pmatrix} a & b \\ c & d \end{pmatrix}$ be a matrix of integers, such that $\Delta = \det \mathbf{A} = ad - bc$ is relatively prime to the positive integer m. Then, the matrix

$$\bar{\mathbf{A}} = \bar{\Delta} \begin{pmatrix} d & -b \\ -c & a \end{pmatrix},$$

where $\bar{\Delta}$ is the inverse of Δ modulo m, is an inverse of $\mathbf{A}$ modulo m.

Proof.   To verify that the matrix $\bar{\mathbf{A}}$ is an inverse of $\mathbf{A}$ modulo m, we need only verify that $\mathbf{A}\bar{\mathbf{A}} \equiv \bar{\mathbf{A}}\mathbf{A} \equiv \mathbf{I} \; (\text{mod } m)$.

To see this, note that

$$\mathbf{A}\bar{\mathbf{A}} \equiv \begin{pmatrix} a & b \\ c & d \end{pmatrix} \bar{\Delta} \begin{pmatrix} d & -b \\ -c & a \end{pmatrix} \equiv \bar{\Delta} \begin{pmatrix} ad - bc & 0 \\ 0 & -bc + ad \end{pmatrix}$$

$$\equiv \bar{\Delta} \begin{pmatrix} \Delta & 0 \\ 0 & \Delta \end{pmatrix} = \begin{pmatrix} \bar{\Delta}\Delta & 0 \\ 0 & \bar{\Delta}\Delta \end{pmatrix} \equiv \begin{pmatrix} 1 & 0 \\ 0 & 1 \end{pmatrix} = \mathbf{I} \; (\text{mod } m)$$

and

$$\bar{\mathbf{A}}\mathbf{A} \equiv \bar{\Delta} \begin{pmatrix} d & -b \\ -c & a \end{pmatrix} \begin{pmatrix} a & b \\ c & d \end{pmatrix} \equiv \bar{\Delta} \begin{pmatrix} ad - bc & 0 \\ 0 & -bc + ad \end{pmatrix}$$

$$\equiv \bar{\Delta} \begin{pmatrix} \Delta & 0 \\ 0 & \Delta \end{pmatrix} = \begin{pmatrix} \bar{\Delta}\Delta & 0 \\ 0 & \bar{\Delta}\Delta \end{pmatrix} \equiv \begin{pmatrix} 1 & 0 \\ 0 & 1 \end{pmatrix} = \mathbf{I} \; (\text{mod } m),$$

where $\bar{\Delta}$ is an inverse of $\Delta \; (\text{mod } m)$, which exists because $(\Delta, m) = 1$.   ■

Example 4.27.   Let $\mathbf{A} = \begin{pmatrix} 3 & 4 \\ 2 & 5 \end{pmatrix}$. Because 2 is an inverse of $\det \mathbf{A} = 7$ modulo 13, we have

$$\bar{\mathbf{A}} \equiv 2 \begin{pmatrix} 5 & -4 \\ -2 & 3 \end{pmatrix} \equiv \begin{pmatrix} 10 & -8 \\ -4 & 6 \end{pmatrix} \equiv \begin{pmatrix} 10 & 5 \\ 9 & 6 \end{pmatrix} \quad (\text{mod } 13).$$ ◀

To provide a formula for an inverse of an $n \times n$ matrix, where n is a positive integer greater than 2, we need a result from linear algebra. It involves the notion of the adjoint of a matrix, which is defined as follows.

Definition.   The *adjoint* of an $n \times n$ matrix $\mathbf{A}$ is the $n \times n$ matrix with (i, j)th entry C_{ji}, where C_{ij} is $(-1)^{i+j}$ times the determinant of the matrix obtained by deleting the ith row and jth column from $\mathbf{A}$. The adjoint of $\mathbf{A}$ is denoted by adj $(\mathbf{A})$, or simply adj $\mathbf{A}$.

Theorem 4.18.   If $\mathbf{A}$ is an $n \times n$ matrix with det $\mathbf{A} \neq 0$, then $\mathbf{A}$ (adj $\mathbf{A}$) $=$ (det $\mathbf{A}$)$\mathbf{I}$, where adj $\mathbf{A}$ is the adjoint of $\mathbf{A}$.

Using this theorem, the following theorem follows readily.

Theorem 4.19.   If $\mathbf{A}$ is an $n \times n$ matrix with integer entries and m is a positive integer such that (det $\mathbf{A}, m$) $= 1$, then the matrix $\bar{\mathbf{A}} = \bar{\Delta}$ (adj $\mathbf{A}$) is an inverse of $\mathbf{A}$ modulo m, where $\bar{\Delta}$ is an inverse of $\Delta =$ det $\mathbf{A}$ modulo m.

Proof.   If (det $\mathbf{A}, m$) $= 1$, then we know that det $\mathbf{A} \neq 0$. Hence, by Theorem 4.18, we have

$$\mathbf{A} \text{ (adj } \mathbf{A}) = (\text{det } \mathbf{A})\mathbf{I} = \Delta \mathbf{I}.$$

Since (det $\mathbf{A}, m$) $= 1$, there is an inverse $\bar{\Delta}$ of $\Delta =$ det $\mathbf{A}$ modulo m. Hence,

$$\mathbf{A}(\bar{\Delta} \text{ adj } \mathbf{A}) \equiv \mathbf{A} \cdot \text{ (adj } \mathbf{A})\bar{\Delta} \equiv \Delta\bar{\Delta}\mathbf{I} \equiv \mathbf{I} \pmod{m},$$

and

$$\bar{\Delta} \text{ (adj } \mathbf{A})\mathbf{A} \equiv \bar{\Delta} \text{ ((adj } \mathbf{A})\mathbf{A}) \equiv \bar{\Delta}\Delta\mathbf{I} \equiv \mathbf{I} \pmod{m}.$$

This shows that $\bar{\mathbf{A}} = \bar{\Delta}$ (adj $\mathbf{A}$) is an inverse of $\mathbf{A}$ modulo m.   ∎

Example 4.28.   Let $\mathbf{A} = \begin{pmatrix} 2 & 5 & 6 \\ 2 & 0 & 1 \\ 1 & 2 & 3 \end{pmatrix}$. Then det $\mathbf{A} = -5$. Furthermore, we have (det $\mathbf{A}, 7$) $= 1$, and we see that 4 is an inverse of det $\mathbf{A} = -5 \pmod 7$. Consequently, we find that

$$\bar{\mathbf{A}} = 4(\text{adj } \mathbf{A}) = 4 \begin{pmatrix} -2 & -3 & 5 \\ -5 & 0 & 10 \\ 4 & 1 & -10 \end{pmatrix} = \begin{pmatrix} -8 & -12 & 20 \\ -20 & 0 & 40 \\ 16 & 4 & -40 \end{pmatrix} \equiv \begin{pmatrix} 6 & 2 & 6 \\ 1 & 0 & 5 \\ 2 & 4 & 2 \end{pmatrix} \pmod 7.$$

◀

We can use an inverse of $\mathbf{A}$ modulo m to solve the system

$$\mathbf{AX} \equiv \mathbf{B} \pmod{m},$$

where (det $\mathbf{A}, m$) $= 1$. By Theorem 4.16, when we multiply both sides of this congruence by an inverse $\bar{\mathbf{A}}$ of $\mathbf{A}$, we obtain

$$\bar{\mathbf{A}}(\mathbf{A}\mathbf{X}) \equiv \bar{\mathbf{A}}\mathbf{B} \pmod{m}$$

$$(\bar{\mathbf{A}}\mathbf{A})\mathbf{X} \equiv \bar{\mathbf{A}}\mathbf{B} \pmod{m}$$

$$\mathbf{X} \equiv \bar{\mathbf{A}}\mathbf{B} \pmod{m}.$$

Hence, we find the solution $\mathbf{X}$ by forming $\bar{\mathbf{A}}\mathbf{B} \pmod{m}$.

Note that this method provides another proof of Theorem 4.15. To see this, let $\mathbf{A}\mathbf{X} = \mathbf{B}$, where $\mathbf{A} = \begin{pmatrix} a & b \\ c & d \end{pmatrix}$, $\mathbf{X} = \begin{pmatrix} x \\ y \end{pmatrix}$, and $\mathbf{B} = \begin{pmatrix} e \\ f \end{pmatrix}$. If $\Delta = \det \mathbf{A} = ad - bc$ is relatively prime to m, then

$$\begin{pmatrix} x \\ y \end{pmatrix} = \mathbf{X} \equiv \bar{\mathbf{A}}\mathbf{B} \equiv \bar{\Delta} \begin{pmatrix} d & -b \\ -c & a \end{pmatrix} \begin{pmatrix} e \\ f \end{pmatrix} = \bar{\Delta} \begin{pmatrix} de & -bf \\ af & -ce \end{pmatrix} \pmod{m}.$$

This demonstrates that (x, y) is a solution if and only if

$$x \equiv \bar{\Delta}(de - bf) \pmod{m}, \quad y \equiv \bar{\Delta}(af - ce) \pmod{m}.$$

Next, we give an example of the solution of a system of three congruences in three unknowns using matrices.

Example 4.29. We consider the system of three congruences

$$2x_1 + 5x_2 + 6x_3 \equiv 3 \pmod{7}$$

$$2x_1 + x_3 \equiv 4 \pmod{7}$$

$$x_1 + 2x_2 + 3x_3 \equiv 1 \pmod{7}.$$

This is equivalent to the matrix congruence

$$\begin{pmatrix} 2 & 5 & 6 \\ 2 & 0 & 1 \\ 1 & 2 & 3 \end{pmatrix} \begin{pmatrix} x_1 \\ x_2 \\ x_3 \end{pmatrix} \equiv \begin{pmatrix} 3 \\ 4 \\ 1 \end{pmatrix} \pmod{7}.$$

We have previously shown that the matrix $\begin{pmatrix} 6 & 2 & 6 \\ 1 & 0 & 5 \\ 2 & 4 & 2 \end{pmatrix}$ is an inverse of $\begin{pmatrix} 2 & 5 & 6 \\ 2 & 0 & 1 \\ 1 & 2 & 3 \end{pmatrix}$ (mod 7). Hence, we have

$$\begin{pmatrix} x_1 \\ x_2 \\ x_3 \end{pmatrix} = \begin{pmatrix} 6 & 2 & 6 \\ 1 & 0 & 5 \\ 2 & 4 & 2 \end{pmatrix} \begin{pmatrix} 3 \\ 4 \\ 1 \end{pmatrix} = \begin{pmatrix} 32 \\ 8 \\ 24 \end{pmatrix} \equiv \begin{pmatrix} 4 \\ 1 \\ 3 \end{pmatrix} \pmod{7}. \quad \blacktriangleleft$$

Before leaving this subject, we should mention that many methods for solving systems of linear equations may be adapted to solve systems of congruences. For instance, Gaussian elimination may be adapted to solve systems of congruences, where division is always replaced by multiplication by inverses modulo m. Also, there is a method for solving systems of congruences analogous to Cramer's rule. We leave the development of these methods as exercises for those readers familiar with linear algebra.

4.5 Exercises

1. Find the solutions of each of the following systems of linear congruences.

   a) $x + 2y \equiv 1 \pmod 5$
      $2x + y \equiv 1 \pmod 5$

   b) $x + 3y \equiv 1 \pmod 5$
      $3x + 4y \equiv 2 \pmod 5$

   c) $4x + y \equiv 2 \pmod 5$
      $2x + 3y \equiv 1 \pmod 5$

2. Find the solutions of each of the following systems of linear congruences.

   a) $2x + 3y \equiv 5 \pmod 7$
      $x + 5y \equiv 6 \pmod 7$

   b) $4x + y \equiv 5 \pmod 7$
      $x + 2y \equiv 4 \pmod 7$

* 3. What are the possibilities for the number of incongruent solutions of the system of linear congruences

$$ax + by \equiv c \pmod p$$
$$dx + ey \equiv f \pmod p,$$

   where p is a prime and a, b, c, d, e, and f are positive integers?

4. Find the matrix $\mathbf{C}$ such that

$$\mathbf{C} \equiv \begin{pmatrix} 2 & 1 \\ 4 & 3 \end{pmatrix} \begin{pmatrix} 4 & 0 \\ 2 & 1 \end{pmatrix} \pmod 5$$

   and all entries of $\mathbf{C}$ are nonnegative integers less than 5.

5. Use mathematical induction to prove that if $\mathbf{A}$ and $\mathbf{B}$ are $n \times n$ matrices with integer entries such that $\mathbf{A} \equiv \mathbf{B} \pmod m$, then $\mathbf{A}^k \equiv \mathbf{B}^k \pmod m$ for all positive integers k.

A matrix $\mathbf{A} \neq \mathbf{I}$ is called *involutory modulo m* if $\mathbf{A}^2 \equiv \mathbf{I} \pmod m$.

6. Show that $\begin{pmatrix} 4 & 11 \\ 1 & 22 \end{pmatrix}$ is involutory modulo 26.

7. Prove or disprove that if $\mathbf{A}$ is a 2×2 involutory matrix modulo m, then $\det \mathbf{A} \equiv \pm 1 \pmod m$.

8. Find an inverse modulo 5 of each of the following matrices.

   a) $\begin{pmatrix} 0 & 1 \\ 1 & 0 \end{pmatrix}$ b) $\begin{pmatrix} 1 & 2 \\ 3 & 4 \end{pmatrix}$ c) $\begin{pmatrix} 2 & 2 \\ 1 & 2 \end{pmatrix}$

9. Find an inverse modulo 7 of each of the following matrices.

   a) $\begin{pmatrix} 1 & 1 & 0 \\ 1 & 0 & 1 \\ 0 & 1 & 1 \end{pmatrix}$ b) $\begin{pmatrix} 1 & 2 & 3 \\ 1 & 2 & 5 \\ 1 & 4 & 6 \end{pmatrix}$ c) $\begin{pmatrix} 1 & 1 & 1 & 0 \\ 1 & 1 & 0 & 1 \\ 1 & 0 & 1 & 1 \\ 0 & 1 & 1 & 1 \end{pmatrix}$

10. Using Exercise 9, find all the solutions of each of the following systems.

a) $x + y \equiv 1 \pmod{7}$
   $x + z \equiv 2 \pmod{7}$
   $y + z \equiv 3 \pmod{7}$

b) $x + 2y + 3z \equiv 1 \pmod{7}$
   $x + 2y + 5z \equiv 1 \pmod{7}$
   $x + 4y + 6z \equiv 1 \pmod{7}$

c) $x + y + z \equiv 1 \pmod{7}$
   $x + y + w \equiv 1 \pmod{7}$
   $x + z + w \equiv 1 \pmod{7}$
   $y + z + w \equiv 1 \pmod{7}$

11. How many incongruent solutions does each of the following systems of congruences have?

a) $x + \ y + \ z \equiv 1 \pmod{5}$
   $2x + 4y + 3z \equiv 1 \pmod{5}$

b) $2x + 3y + \ z \equiv 3 \pmod{5}$
   $x + 2y + 3z \equiv 1 \pmod{5}$
   $2x + \ z \equiv 1 \pmod{5}$

c) $3x + \ y + 3z \equiv 1 \pmod{5}$
   $x + 2y + 4z \equiv 2 \pmod{5}$
   $4x + 3y + 2z \equiv 3 \pmod{5}$

d) $2x + \ y + \ z \equiv 1 \pmod{5}$
   $x + 2y + \ z \equiv 1 \pmod{5}$
   $x + \ y + 2z \equiv 1 \pmod{5}$

* 12. Develop an analogue of Cramer's rule for solving systems of n linear congruences in n unknowns.

* 13. Develop an analogue of Gaussian elimination to solve systems of n linear congruences in m unknowns (where m and n may differ).

A *magic square* is a square array of integers with the property that the sum of the integers in a row or in a column is always the same. In this exercise, we present a method for producing magic squares.

* 14. Show that the n^2 integers $0, 1, \ldots, n^2 - 1$ are put into the n^2 positions of an $n \times n$ square, without putting two integers in the same position, if the integer k is placed in the ith row and jth column, where

$$i \equiv a + ck + e[k/n] \pmod{n},$$
$$j \equiv b + dk + f[k/n] \pmod{n},$$

$1 \le i \le n, 1 \le j \le n$, and a, b, c, d, e, and f are integers with $(cf - de, n) = 1$.

* 15. Show that a magic square is produced in Exercise 14 if $(c, n) = (d, n) = (e, n) = (f, n) = 1$.

* 16. The *positive* and *negative diagonals* of an $n \times n$ square consist of the integers in positions (i, j), where $i + j \equiv k \pmod{n}$ and $i - j \equiv k \pmod{n}$, respectively, where k is a given integer. A square is called *diabolic* if the sum of the integers in a positive or negative diagonal is always the same. Show that a diabolic square is produced using the procedure given in Exercise 14 if $(c + d, n) = (c - d, n) = (e + f, n) = (e - f, n) = 1$.

4.5 Computational and Programming Exercises

Computations and Explorations

Using a computation program such as Maple or *Mathematica,* or programs you have written, carry out the following computations and explorations.

1. Produce 4×4, 5×5, and 6×6 magic squares.

Programming Projects

Write programs using Maple, *Mathematica,* or a language of your choice to do the following.

1. Find the solutions of a system of two linear congruences in two unknowns using Theorem 4.15.

2. Find inverses of 2×2 matrices using Theorem 4.17.

3. Find inverses of $n \times n$ matrices using Theorem 4.19.

4. Solve systems of n linear congruences in n unknowns using inverses of matrices.

5. Solve systems of n linear congruences in n unknowns using an analogue of Cramer's rule (see Exercise 12).

6. Solve systems of n linear congruences in m unknowns using an analogue of Gaussian elimination (see Exercise 13).

7. Given a positive integer, produce an $n \times n$ magic square by the method given in Exercise 14.

4.6 Factoring Using the Pollard Rho Method

In this section, we will describe a factorization method based on congruences that was developed in 1974 by J. M. Pollard. Pollard called this technique the *Monte Carlo method* because it relies on generating integers that behave as though they were randomly chosen; it is now commonly known as the *Pollard rho method,* for reasons which will be explained.

Suppose that n is a large composite integer and that p is its smallest prime divisor. Our goal is to choose integers $x_0, x_1, \ldots, x_s$ so that these integers have distinct least nonnegative residues modulo n, but where their least nonnegative residues modulo p are not all distinct. As can be seen using probabilistic arguments (see [Ri94]), this is likely to be the case when s is large compared to $\sqrt{p}$ but small when compared to $\sqrt{n}$, and the numbers are chosen randomly.

Once we have found integers x_i and x_j, $0 \le i < j \le s$, such that $x_i \equiv x_j \pmod{p}$ but $x_i \not\equiv x_j \pmod{n}$, it follows that $(x_i - x_j, n)$ is a nontrivial divisor of n, as p divides $x_i - x_j$, but n does not. The number $(x_i - x_j, n)$ can be found quickly using the Euclidean algorithm. However, to find $(x_i - x_j, n)$ for each pair (i, j) with $0 \le i < j \le s$ requires that we find $O(s^2)$ greatest common divisors. We will show how to reduce the number of times we must use the Euclidean algorithm.

To find such integers x_i and x_j, we use the following procedure: We start with a seed value x_0 that is chosen randomly and a polynomial function $f(x)$ with integer coefficients of degree greater than 1. We compute the terms x_k, $k = 1, 2, 3, \ldots$, using the recursive definition

$$x_{k+1} \equiv f(x_k) \ (\text{mod } n), \quad 0 \le x_{k+1} < n.$$

The polynomial $f(x)$ should be chosen so that the probability is high that a suitably large number of integers x_i are generated before they repeat. Empirical evidence indicates that the polynomial $f(x) = x^2 + 1$ performs well for this test. The following example illustrates how this sequence is generated.

Example 4.30. Let $n = 8051$, and suppose that $x_0 = 2$ and $f(x) = x^2 + 1$. We find that $x_1 = 5$, $x_2 = 26$, $x_3 = 677$, $x_4 = 7474$, $x_5 = 2839$, $x_6 = 871$, and so on. ◀

Now, note that by the recursive definition of x_k, it follows that if

$$x_i \equiv x_j \ (\text{mod } d),$$

where d is a positive integer, then

$$x_{i+1} \equiv f(x_i) \equiv f(x_j) \equiv x_{j+1} \ (\text{mod } d).$$

It follows that if $x_i \equiv x_j \ (\text{mod } d)$, then the sequence x_k becomes periodic modulo d with a period dividing $j - i$. That is, $x_q \equiv x_r \ (\text{mod } d)$ whenever $q \equiv r \ (\text{mod } j - i)$, and $q \ge i$ and $r \ge i$. It follows that if s is the smallest multiple of $j - i$ that is at least as large as i, then $x_s \equiv x_{2s} \ (\text{mod } d)$.

It follows further that to look for a factor of n, we find the greatest common divisor of $x_{2k} - x_k$ and n for $k = 1, 2, 3, \ldots$. We have found a factor of n when we have found a value k for which $1 < (x_{2k} - x_k, n) < n$. From our observations, we see that it is likely that we will find such an integer k with k close to $\sqrt{p}$.

In practice, when the Pollard rho method is used, the polynomial $f(x) = x^2 + 1$ is often chosen to generate the sequence of integers $x_0, x_1, x_2, \ldots, x_k, \ldots$. Furthermore, the seed $x_0 = 2$ is often used. This choice of polynomial and seed produces a sequence that behaves much like a random sequence for the purposes of this factorization method.

Example 4.31. We use the Pollard rho method with seed $x_0 = 2$ and generator polynomial $f(x) = x^2 + 1$ to find a nontrivial factor of $n = 8051$. We find that $x_1 = 5$, $x_2 = 26$, $x_3 = 677$, $x_4 = 7474$, $x_5 = 2839$, $x_6 = 871$. Using the Euclidean algorithm, it follows that $(x_2 - x_1, 8051) = (26 - 5, 8051) = (21, 8051) = 1$ and $(x_4 - x_2, 8051) = (7474 - 26, 8051) = (7448, 8051) = 1$. However, we find a nontrivial factor of 8051 at the next step, as $(x_6 - x_3, 8051) = (871 - 677, 8051) = (194, 8051) = 97$. We see that 97 is a factor of 8051. ◀

To see why this method is called the Pollard rho method, look at Figure 4.1 on the next page. This figure shows the periodic behavior of the sequence x_i, where $x_0 = 2$

and $x_i + 1 = x_i^2 + 1 \pmod{97}$, $i \geq 1$. The part of this sequence that occurs before the periodicity is the tail of the rho, and the loop is the periodic part.

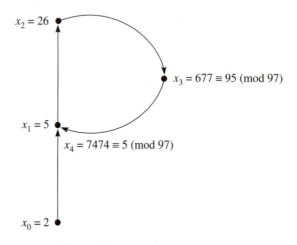

$x_2 = 26$

$x_3 = 677 \equiv 95 \pmod{97}$

$x_1 = 5$

$x_4 = 7474 \equiv 5 \pmod{97}$

$x_0 = 2$

Figure 4.1 *The Pollard rho method.*

The Pollard rho method has proved to be practical for the factorization of integers with moderately large prime factors. In practice, the first attempt to factor a large integer is to do trial division by small primes, say by all primes less than 10,000. Next, the Pollard rho method is used to look for prime factors of intermediate size (up to 10^{15}, for instance). Only after trial division by small primes and the Pollard rho method have failed are the really big guns brought in, such as the quadratic sieve or the elliptic curve method.

4.6 Exercises

1. Use the Pollard rho method with $x_0 = 2$ and $f(x) = x^2 + 1$ to find the prime factorization of each of the following integers.

   a) 133   c) 1927   e) 36,287
   b) 1189   d) 8131   f) 48,227

2. Use the Pollard rho method to factor the integer 1387, with the following seeds and generating polynomials.
   a) $x_0 = 2$, $f(x) = x^2 + 1$
   b) $x_0 = 3$, $f(x) = x^2 + 1$
   c) $x_0 = 2$, $f(x) = x^2 - 1$
   d) $x_0 = 2$, $f(x) = x^3 + x + 1$

* 3. Explain why the choice of $f(x)$ as a linear polynomial, that is, a function of the form $f(x) = ax + b$, where a and b are integers, is a poor choice.

4.6 Computational and Programming Exercises

Computations and Explorations

Using a computation program such as Maple or *Mathematica,* or programs you have written, carry out the following computations and explorations.

1. Use the Pollard rho method to factor ten different integers that have between 15 and 20 decimal digits.

2. Use the Pollard rho method to factor a large number of integers that are close to 100,000, keeping track of the number of steps required. Can you make any conjectures based on your data?

3. Factor $2^{58} + 1$ using the Pollard rho method.

Programming Projects

Write programs using Maple, *Mathematica,* or a language of your choice to do the following.

1. Given a positive integer n, find a prime factor of this integer using the Pollard rho method.

5

Applications of Congruences

Introduction

Congruences have diverse applications. We have already seen some examples of this, such as in Section 4.3, where we saw how large integers can be multiplied on a computer using congruences. This chapter covers a wide variety of interesting applications of congruences. First, we will show how congruences can be used to develop divisibility tests, such as the simple tests you may already know for checking whether an integer is divisible by 3 or by 9. Next, we will develop a congruence that determines the day of the week for any date in history. Then, we will show how congruences can be used to schedule round-robin tournaments. We will discuss some applications of congruences in computer science; for example, we will show how congruences are used in hashing functions, which themselves have many applications, such as determining computer memory locations where data is stored. Finally, we will show how congruences can be used to construct check digits, which are used to determine whether an identification number has been copied in error.

In subsequent chapters, we will discuss additional applications of congruences. For example, in Chapter 8, we will show how congruences can be used in different ways to make messages secret, and in Chapter 10, we will show how congruences can be used to generate pseudorandom numbers.

5.1 Divisibility Tests

You may have learned in primary school that to check whether an integer is divisible by 3, you need only check whether the sum of its digits is divisible by 3. This is an example of a divisibility test that uses the digits of an integer to check whether it is divisible by a particular divisor, without actually dividing the integer by that possible divisor. In this section, we will develop the theory behind such tests. In particular, we will use congruences to develop divisibility tests for integers based on their base b expansions, where b is a positive integer. Taking $b = 10$ will give us the well-known tests for checking integers for divisibility by 2, 3, 4, 5, 7, 9, 11, and 13. Although you may have learned these divisibility tests a long time ago, you will learn why they work here.

Divisibility by Powers of 2   First, we develop tests for divisibility by powers of 2. Let $n = 32,688,048$. It is easy to see that n is divisible by 2 since its last digit is even. Consider the following questions. Does $2^2 = 4$ divide n? Does $2^3 = 8$ divide n? Does $2^4 = 16$ divide n? What is the highest power of 2 that divides n? We will develop a test that does not require that we actually divide n by 4, 8, and successive powers of 2, which answers these questions.

In the following discussion let $n = (a_k a_{k-1} \ldots a_1 a_0)_{10}$. Then $n = a_k 10^k + a_{k-1} 10^{k-1} + \cdots + a_1 10 + a_0$, with $0 \le a_j \le 9$ for $j = 0, 1, 2, \ldots, k$.

Because $10 \equiv 0 \pmod{2}$, it follows that $10^j \equiv 0 \pmod{2^j}$ for all positive integers j. Hence,

$$n \equiv (a_0)_{10} \pmod{2},$$

$$n \equiv (a_1 a_0)_{10} \pmod{2^2},$$

$$n \equiv (a_2 a_1 a_0)_{10} \pmod{2^3},$$

$$\vdots$$

$$n \equiv (a_{k-1} a_{k-2} \ldots a_2 a_1 a_0)_{10} \pmod{2^k}.$$

These congruences tell us that to determine whether an integer n is divisible by 2, we only need to examine its last digit for divisibility by 2. Similarly, to determine whether n is divisible by 4, we only need to check the integer made up of the last two digits of n for divisibility by 4. In general, to test n for divisibility by 2^j, we only need to check the integer made up of the last j digits of n for divisibility by 2^j.

Example 5.1.   Let $n = 32,688,048$. We see that $2 \mid n$ because $2 \mid 8$, $4 \mid n$ because $4 \mid 48$, $8 \mid n$ because $8 \mid 48$, $16 \mid n$ because $16 \mid 8048$, but $32 \nmid n$ since $32 \nmid 88,048$.   ◄

Divisibility by Powers of 5   Next, we develop divisibility tests for powers of 5.

To develop tests for divisibility by powers of 5, first note that because $10 \equiv 0 \pmod{5}$, we have $10^j \equiv 0 \pmod{5^j}$. Hence, divisibility tests for powers of 5 are analogous to those for powers of 2. We only need to check the integer made up of the last j digits of n to determine whether n is divisible by 5^j.

Example 5.2.   Let $n = 15,535,375$. Because $5 \mid 5$, $5 \mid n$, because $25 \mid 75$, $25 \mid n$, because $125 \mid 375$, $125 \mid n$, but because $625 \nmid 5375$, $625 \nmid n$.   ◄

Divisibility by 3 and 9   Next, we develop tests for divisibility by 3 and by 9.

Note that both the congruences $10 \equiv 1 \pmod 3$ and $10 \equiv 1 \pmod 9$ hold. Hence, $10^k \equiv 1 \pmod 3$ and $10^k \equiv 1 \pmod 9$. This gives us the useful congruences

$$(a_k a_{k-1} \cdots a_1 a_0)_{10} = a_k 10^k + a_{k-1} 10^{k-1} + \cdots + a_1 10 + a_0$$
$$\equiv a_k + a_{k-1} + \cdots + a_1 + a_0 \pmod 3 \text{ and } \pmod 9.$$

Hence, we only need to check whether the sum of the digits of n is divisible by 3, or by 9, to see whether n is divisible by 3, or by 9, respectively.

Example 5.3.   Let $n = 4{,}127{,}835$. Then, the sum of the digits of n is $4 + 1 + 2 + 7 + 8 + 3 + 5 = 30$. Because $3 \mid 30$ but $9 \nmid 30$, $3 \mid n$ but $9 \nmid n$.   ◀

Divisibility by 11   A rather simple test can be found for divisibility by 11.

Because $10 \equiv -1 \pmod{11}$, we have

$$(a_k a_{k-1} \ldots a_1 a_0)_{10} = a_k 10^k + a_{k-1} 10^{k-1} + \cdots + a_1 10 + a_0$$
$$\equiv a_k (-1)^k + a_{k-1}(-1)^{k-1} + \cdots - a_1 + a_0 \pmod{11}.$$

This shows that $(a_k a_{k-1} \ldots a_1 a_0)_{10}$ is divisible by 11 if and only if $a_0 - a_1 + a_2 - \cdots + (-1)^k a_k$, the integer formed by alternately adding and subtracting the digits, is divisible by 11.

Example 5.4.   We see that $723{,}160{,}823$ is divisible by 11, because alternately adding and subtracting its digits yields $3 - 2 + 8 - 0 + 6 - 1 + 3 - 2 + 7 = 22$, which is divisible by 11. On the other hand, $33{,}678{,}924$ is not divisible by 11, because $4 - 2 + 9 - 8 + 7 - 6 + 3 - 3 = 4$ is not divisible by 11.   ◀

Divisibility by 7, 11, and 13   Next, we develop a test to simultaneously check for divisibility by the primes 7, 11, and 13.

Note that $7 \cdot 11 \cdot 13 = 1001$ and $10^3 = 1000 \equiv -1 \pmod{1001}$. Hence,

$$(a_k a_{k-1} \ldots a_0)_{10} = a_k 10^k + a_{k-1} 10^{k-1} + \cdots + a_1 10 + a_0$$
$$\equiv (a_0 + 10 a_1 + 100 a_2) + 1000(a_3 + 10 a_4 + 100 a_5)$$
$$+ (1000)^2 (a_6 + 10 a_7 + 100 a_8) + \cdots$$
$$\equiv (100 a_2 + 10 a_1 + a_0) - (100 a_5 + 10 a_4 + a_3)$$
$$+ (100 a_8 + 10 a_7 + a_6) - \cdots$$
$$= (a_2 a_1 a_0)_{10} - (a_5 a_4 a_3)_{10} + (a_8 a_7 a_6)_{10} - \cdots \pmod{1001}.$$

This congruence tells us that an integer is congruent modulo 1001 to the integer formed by successively adding and subtracting the three-digit integers with decimal expansions formed from successive blocks of three decimal digits of the original number, where digits are grouped starting with the rightmost digit. As a consequence, because 7, 11, and 13 are divisors of 1001, to determine whether an integer is divisible by 7, 11, or 13, we only need to check whether this alternating sum and difference of blocks of three digits is divisible by 7, 11, or 13.

Example 5.5. Let $n = 59{,}358{,}208$. Because the alternating sum and difference of the integers formed from blocks of three digits, $208 - 358 + 59 = -91$, is divisible by 7 and 13, but not by 11, we see that n is divisible by 7 and 13, but not by 11. ◄

Divisibility Tests Using Base b Representations   All of the divisibility tests we have developed thus far are based on decimal representations. We now develop divisibility tests using base b representations, where b is a positive integer.

Theorem 5.1. If $d \mid b$ and j and k are positive integers with $j < k$, then $(a_k \cdots a_1 a_0)_b$ is divisible by d^j if and only if $(a_{j-1} \cdots a_1 a_0)_b$ is divisible by d^j.

Proof. Because $b \equiv 0 \pmod{d}$, it follows that $b^j \equiv 0 \pmod{d^j}$. Hence,

$$(a_k a_{k-1} \cdots a_1 a_0)_b = a_k b^k + \cdots + a_j b^j + a_{j-1} b^{j-1} + \cdots + a_1 b + a_0$$

$$\equiv a_{j-1} b^{j-1} + \cdots + a_1 b + a_0$$

$$= (a_{j-1} \cdots a_1 a_0)_b \pmod{d^j}.$$

Consequently, $d^j \mid (a_k a_{k-1} \cdots a_1 a_0)_b$ if and only if $d^j \mid (a_{j-1} \cdots a_1 a_0)_b$. ∎

Theorem 5.1 extends to other bases the divisibility tests of integers expressed in decimal notation by powers of 2 and by powers of 5.

Theorem 5.2. If $d \mid (b - 1)$, then $n = (a_k \ldots a_1 a_0)_b$ is divisible by d if and only if the sum of digits $a_k + \cdots + a_1 + a_0$ is divisible by d.

Proof. Because $d \mid (b - 1)$, we have $b \equiv 1 \pmod{d}$, so that by Theorem 4.7 we have $b^j \equiv 1 \pmod{d}$ for all positive integers j. Hence, $n = (a_k \ldots a_1 a_0)_b = a_k b^k + \cdots + a_1 b + a_0 \equiv a_k + \cdots + a_1 + a_0 \pmod{d}$. This shows that $d \mid n$ if and only if $d \mid (a_k + \cdots + a_1 + a_0)$. ∎

Theorem 5.2 extends to other bases the tests for divisibility of integers expressed in decimal notation by 3 and by 9.

Theorem 5.3. If $d \mid (b + 1)$, then $n = (a_k \ldots a_1 a_0)_b$ is divisible by d if and only if the alternating sum of digits $(-1)^k a_k + \cdots - a_1 + a_0$ is divisible by d.

Proof. Because $d \mid (b + 1)$, we have $b \equiv -1 \pmod{d}$. Hence, $b^j \equiv (-1)^j \pmod{d}$, and consequently, $n = (a_k \ldots a_1 a_0)_b \equiv (-1)^k a_k + \cdots - a_1 + a_0 \pmod{d}$. Hence, $d \mid n$ if and only if $d \mid ((-1)^k a_k + \cdots - a_1 + a_0)$. ∎

Theorem 5.3 extends to other bases the test for divisibility by 11 of integers expressed in decimal notation.

Example 5.6. Let $n = (\text{7F28A6})_{16}$ (in hex notation). Then, because $2 \mid 16$, from Theorem 5.1 we know that $2 \mid n$, because $2 \mid 6$. Likewise, because $4 \mid 16$, we see that $4 \nmid n$, because $4 \nmid 6$. By Theorem 5.2, because $3 \mid (16 - 1), 5 \mid (16 - 1)$, and $15 \mid (16 - 1)$, and $7 + \text{F} + 2 + 8 + \text{A} + 6 = (30)_{16}$, we know that $3 \mid n$, since $3 \mid (30)_{16}$, whereas $5 \nmid n$ and $15 \nmid n$, because $5 \nmid (30)_{16}$ and $15 \nmid (30)_{16}$. Furthermore, by Theorem 5.3, because $17 \mid (16 + 1)$ and $n \equiv 6 - \text{A} + 8 - 2 + \text{F} - 7 = (\text{A})_{16} (\text{mod } 17)$, we conclude that $17 \nmid n$, because $17 \nmid (\text{A})_{16}$. ◀

Example 5.7. Let $n = (1001001111)_2$. Then, using Theorem 5.3 we see that $3 \mid n$, because $n \equiv 1 - 1 + 1 - 1 + 0 - 0 + 1 - 0 + 0 - 1 \equiv 0 \ (\text{mod } 3)$ and $3 \mid (2 + 1)$. ◀

5.1 Exercises

1. Determine the highest power of 2 that divides each of the following positive integers.

   a) 201,984   c) 89,375,744
   b) 1,423,408   d) 41,578,912,246

2. Determine the highest power of 5 that divides each of the following positive integers.

   a) 112,250   c) 235,555,790
   b) 4,860,625   d) 48,126,953,125

3. Which of the following integers are divisible by 3? Of those that are, which are divisible by 9?

   a) 18,381   c) 987,654,321
   b) 65,412,351   d) 78,918,239,735

4. Which of the following integers are divisible by 11?

   a) 10,763,732   c) 674,310,976,375
   b) 1,086,320,015   d) 8,924,310,064,537

5. Find the highest power of 2 that divides each of the following integers.

   a) $(101111110)_2$ c) $(111000000)_2$
   b) $(1010000011)_2$ d) $(1011011101)_2$

6. Determine which of the integers in Exercise 5 are divisible by 3.

7. Which of the following integers are divisible by 2?

   a) $(1210122)_3$ c) $(1112201112)_3$
   b) $(211102101)_3$ d) $(10122222011101)_3$

8. Which of the integers in Exercise 7 are divisible by 4?

9. Which of the following integers are divisible by 3, and which are divisible by 5?

    a) $(3EA235)_{16}$ c) $(F117921173)_{16}$

    b) $(ABCDEF)_{16}$ d) $(10AB987301F)_{16}$

10. Which of the integers in Exercise 9 are divisible by 17?

A *repunit* is an integer with decimal expansion containing all 1s.

11. Determine which repunits are divisible by 3, and which are divisible by 9.

12. Determine which repunits are divisible by 11.

13. Determine which repunits are divisible by 1001. Which are divisible by 7? by 13?

14. Determine which repunits with fewer than 10 digits are prime.

A *base b repunit* is an integer with base b expansion containing all 1s.

15. Determine which base b repunits are divisible by factors of $b - 1$.

16. Determine which base b repunits are divisible by factors of $b + 1$.

A *base b palindromic integer* is an integer whose base b representation reads the same forward and backward.

17. Show that every decimal palindromic integer with an even number of digits is divisible by 11.

18. Show that every base 7 palindromic integer with an even number of digits is divisible by 8.

19. Develop a test for divisibility by 37, based on the fact that $10^3 \equiv 1 \pmod{37}$. Use this to check 443,692 and 11,092,785 for divisibility by 37.

20. Devise a test for integers represented in base b notation to check for divisibility by n, where n is a divisor of $b^2 + 1$. (*Hint:* Split the digits of the base b representation of the integer into blocks of two, starting on the right.)

21. Use the test that you developed in Exercise 20 to decide whether

    a) $(101110110)_2$ is divisible by 5.

    b) $(12100122)_3$ is divisible by 2, and whether it is divisible by 5.

    c) $(364701244)_8$ is divisible by 5, and whether it is divisible by 13.

    d) $(5837041320219)_{10}$ is divisible by 101.

22. An old receipt has faded. It reads 88 chickens at a total of $x4.2y$, where x and y are unreadable digits. How much did each chicken cost?

23. Use a congruence modulo 9 to find the missing digit, indicated by a question mark: $89,878 \cdot 58,965 = 5299\ ?\ 56270$.

We can check a multiplication $c = ab$ by determining whether the congruence $c \equiv ab$ (mod m) is valid, where m is any modulus. If we find that c is not congruent to ab modulo m, then we know that an error has been made. When we take $m = 9$ and use the fact that an integer in decimal notation is congruent modulo 9 to the sum of its digits, this check is called *casting out nines*.

24. Check each of the following multiplications by casting out nines.

a) $875,961 \cdot 2753 = 2,410,520,633$

b) $14,789 \cdot 23,567 = 348,532,367$

c) $24,789 \cdot 43,717 = 1,092,700,713$

25. Is a check of a multiplication by casting out nines foolproof?

26. What combinations of digits of a decimal expansion of an integer are congruent to this integer modulo 99? Use your answer to devise a check for multiplication based on *casting out ninety-nines*. Then use the test to check the multiplications in Exercise 24.

5.1 Computational and Programming Exercises

Computations and Explorations

Using a computation program such as Maple or *Mathematica*, or programs you have written, carry out the following computations and explorations.

1. Determine whether the repunit with n digits is prime, where n is a positive integer not exceeding 30. Can you go further?

Programming Projects

Write programs using Maple, *Mathematica,* or a language of your choice to do the following.

1. Given a positive integer n, determine the highest powers of 2 and of 5 that divide n.

2. Given a positive integer n, test n for divisibility by 3, 7, 9, 11, and 13. (Use congruences modulo 1001 for divisibility by 7 and 13.)

3. Given a positive integer n, determine the highest power of each factor of b that divides an integer from the base b expansion of n.

4. Test a positive integer n, from its base b expansion, for divisibility by factors of $b - 1$ and of $b + 1$.

5.2 The Perpetual Calendar

In this section, we derive a formula that gives us the day of the week of any day of any year. Because the days of the week form a cycle of length seven, we use a congruence modulo 7. We denote each day of the week by a number in the set 0, 1, 2, 3, 4, 5, 6, setting

- *Sunday* $= 0$,
- *Monday* $= 1$,
- *Tuesday* $= 2$,
- *Wednesday* $= 3$,
- *Thursday* $= 4$,
- *Friday* $= 5$,
- *Saturday* $= 6$.

Julius Caesar changed the Egyptian calendar, which was based on a year of exactly 365 days, to a new calendar, called the *Julian calendar,* with a year of average length 365 ¼ days, with leap years every fourth year, to better reflect the true length of the year. However, more recent calculations have shown that the true length of the year is approximately 365.2422 days. As the centuries passed, the discrepancies of 0.0078 days per year added up, so that by the year 1582 approximately 10 extra days had been added unnecessarily in leap years. To remedy this, in 1582 Pope Gregory set up a new calendar. First, 10 days were added to the date, so that October 5, 1582, became October 15, 1582 (and the 6th through the 14th of October were skipped). It was decided that leap years would be precisely the years divisible by 4, except that those exactly divisible by 100, the years that mark centuries, would be leap years only when divisible by 400. As an example, the years 1700, 1800, 1900, and 2100 are not leap years but 1600 and 2000 are. With this arrangement, the average length of a calendar year became 365.2425 days, rather close to the true year of 365.2422 days. An error of 0.0003 days per year remains, which is 3 days per 10,000 years. In the future, this discrepancy will have to be accounted for, and various possibilities have been suggested to correct for this error.

In dealing with calendar dates for various parts of the world, we must also take into account the fact that the Gregorian calendar was not adopted everywhere in 1582. In Britain and what is now the United States, the Gregorian calendar was adopted only in 1752, and by then it was necessary to add 11 days. In these places September 3, 1752, in the Julian calendar became September 14, 1752, in the Gregorian calendar. Japan changed over in 1873, Russia and nearby countries in 1917, while Greece held out until 1923.

We now set up our procedure for finding the day of the week for a given date in the Gregorian calendar. We first must make some adjustments, because the extra day in a leap year comes at the end of February. We take care of this by renumbering the months, starting each year in March, and considering the months of January and February part of the preceding year. For instance, February 2000 is considered the twelfth month of 1999, and May 2000 is considered the third month of 2000. With this convention, for the day of interest, let

- k = day of the month,
- m = month,
  with

January = 11	*July* = 5
February = 12	*August* = 6
March = 1	*September* = 7
April = 2	*October* = 8
May = 3	*November* = 9
June = 4	*December* = 10

- N = year,

  where N is the current year unless the month is January or February in which case N is the previous year, and where $N = 100C + Y$, where

- C = century,
- Y = particular year of the century.

Example 5.8. For the date April 3, 1951, we have $k = 3$, $m = 2$, $N = 1951$, $C = 19$, and $Y = 51$. But note that for February 28, 1951, we have $k = 28$, $m = 12$, $N = 1950$, $C = 19$, and $Y = 50$, because, for our calculations, we consider February to be the twelfth month of the previous year.   ◀

We use March 1 of each year as our basis. Let d_N represent the day of the week of March 1 in year N. We start with the year 1600, and compute the day of the week March 1 falls on in any given year. Note that between March 1 of year $N - 1$ and March 1 of year N, if year N is not a leap year, 365 days have passed; and because $365 \equiv 1$ (mod 7), we see that $d_N \equiv d_{N-1} + 1$ (mod 7), whereas if year N is a leap year, because there is an extra day between the consecutive firsts of March, we see that

$$d_N \equiv d_{N-1} + 2 \text{ (mod 7)}.$$

Hence, to find d_N from d_{1600}, we must first find out how many leap years have occurred between the year 1600 and the year N (not including 1600, but including N); let us call this number x. To compute x, first note that by the division algorithm there are $[(N - 1600)/4]$ years divisible by 4 between 1600 and N, there are $[(N - 1600)/100]$ years divisible by 100 between 1600 and N, and there are $[(N - 1600)/400]$ years divisible by 400 between 1600 and N. Hence,

$$x = [(N - 1600)/4] - [(N - 1600)/100] + [(N - 1600)/400]$$
$$= [N/4] - 400 - [N/100] + 16 + [N/400] - 4$$
$$= [N/4] - [N/100] + [N/400] - 388.$$

(We have used the identity from Example 1.34 to simplify this expression.) Putting this in terms of C and Y, we see that

$$x = [25C + (Y/4)] - [C + (Y/100)] + [(C/4) + (Y/400)] - 388$$
$$= 25C + [Y/4] - C + [C/4] - 388$$
$$\equiv 3C + [C/4] + [Y/4] - 3 \text{ (mod 7)}.$$

Here we have again used the identity from Example 1.4, the inequality $Y/100 < 1$, and the equation $[(C/4) + (Y/400)] = [C/4]$ (which follows from Exercise 19 of Section 1.5, because $Y/400 < 1/4$).

We can now compute d_N from d_{1600} by shifting d_{1600} by one day for every year that has passed, plus an extra day for each leap year between 1600 and N. This gives the following formula:

$$d_N \equiv d_{1600} + N - 1600 + x$$
$$= d_{1600} + 100C + Y - 1600 + 3C + [C/4] + [Y/4] - 3 \text{ (mod 7)}.$$

Simplifying, we have

$$d_N \equiv d_{1600} - 2C + Y + [C/4] + [Y/4] \text{ (mod 7)}.$$

Now that we have a formula relating the day of the week for March 1 of any year to the day of the week of March 1, 1600, we can use the fact that March 1, 1982, is a Monday to find the day of the week of March 1, 1600. For 1982, because $N = 1982$, we have $C = 19$, and $Y = 82$, and since $d_{1982} = 1$, it follows that

$$1 \equiv d_{1600} - 38 + 82 + [19/4] + [82/4] \equiv d_{1600} - 2 \ (\text{mod } 7).$$

Hence, $d_{1600} = 3$, so that March 1, 1600, was a Wednesday. When we insert the value of d_{1600}, the formula for d_N becomes

$$d_N \equiv 3 - 2C + Y + [C/4] + [Y/4] \ (\text{mod } 7).$$

We now use this formula to compute the day of the week of the first day of each month of year N. To do this, we have to use the number of days of the week that the first of the month of a particular month is shifted from the first of the month of the preceding month. The months with 30 days shift the first of the following month up 2 days, because $30 \equiv 2 \ (\text{mod } 7)$, and those with 31 days shift the first of the following month up 3 days, because $31 \equiv 3 \ (\text{mod } 7)$. Therefore, we must add the following amounts:

from March 1 to April 1:	3 days
from April 1 to May 1:	2 days
from May 1 to June 1:	3 days
from June 1 to July 1:	2 days
from July 1 to August 1:	3 days
from August 1 to September 1:	3 days
from September 1 to October 1:	2 days
from October 1 to November 1:	3 days
from November 1 to December 1:	2 days
from December 1 to January 1:	3 days
from January 1 to February 1:	3 days.

We need a formula that gives us the same increments. Notice that we have 11 increments totaling 29 days, so that each increment averages 2.6 days. By inspection, we find that the function $[2.6m - 0.2] - 2$ has exactly the same increments as m goes from 2 to 12, and is zero when $m = 1$. (This formula was originally found by Christian Zeller;[1] he apparently found it by trial and error.) Hence, the day of the week of the first day of month m of year N is given by the least nonnegative residue of $d_N + [2.6m - 0.2] - 2$ modulo 7.

To find W, the day of the week of day k of month m of year N, we simply add $k - 1$ to the formula we have devised for the day of the week of the first day of the same month.

---

[1] Christian Julius Johannes Zeller (1849–1899) was born in Muhlhausen on the Neckar in Germany. He became a priest at Schokingen after completing his theological studies. He served as the principal of a women's college at Markgroningen from 1847 until 1898. He published his formula for the day of the week of a date in 1882.

We obtain the formula:

$$W \equiv k + [2.6m - 0.2] - 2C + Y + [Y/4] + [C/4] \pmod 7.$$

We can use this formula to find the day of the week of any date of any year in the Gregorian calendar.

Example 5.9.   To find the day of the week of January 1, 1900, we have $C = 18$, $Y = 99$, $m = 11$, and $k = 1$ (because we consider January as the eleventh month of the preceding year). Hence, we have $W \equiv 1 + 28 - 36 + 99 + 24 + 4 \equiv 1 \pmod 7$, so that January 1, 1900, was a Monday.   ◀

5.2 Exercises

1. Find the day of the week of the day you were born, and of your birthday this year.

2. Find the day of the week of the following important dates in U. S. history (use the Julian calendar before September 3, 1752, and the Gregorian calendar from September 14, 1752, to the present)

    * a) October 12, 1492   (Columbus sights land in the Caribbean)
    * b) May 6, 1692   (Peter Minuit buys Manhattan from the natives)
    * c) June 15, 1752   (Benjamin Franklin invents the lightning rod)
     d) July 4, 1776   (U. S. Declaration of Independence)
     e) March 30, 1867   (U. S. buys Alaska from Russia)
     f) March 17, 1888   (Great blizzard in the Eastern U. S.)
     g) February 15, 1898   (U. S. Battleship *Maine* blown up in Havana Harbor)
     h) July 2, 1925   (Scopes convicted of teaching evolution)
     i) July 16, 1945   (First atomic bomb exploded)
     j) July 20, 1969   (First man on the moon)
     k) August 9, 1974   (President Nixon resigns)
     l) March 28, 1979   (Three Mile Island nuclear accident)
    m) January 1, 1984   ("Ma Bell" breakup)
     n) December 25, 1991   (Demise of the U.S.S.R.)
     o) June 5, 2027   (First man on Mars)

3. How many times will the 13th of the month fall on a Friday in the year 2020?

4. How many leap years will there be from the year 1 until the year 10,000, inclusive?

5. To correct the small discrepancy between the number of days in a year of the Gregorian calendar and an actual year, it has been suggested that the years exactly divisible by 4000 should not be leap years. Adjust the formula for the day of the week of a given date to take this correction into account.

6. Show that days with the same calendar date in two different years of the same century, 28, 56, or 84 years apart, fall on the identical day of the week.

7. Which of your birthdays, until your one hundredth, fall on the same day of the week as the day you were born?

8. What is the next term in the sequence 1995, 1997, 1998, 1999, 2001, 2002, 2003?

9. What is the next term in the sequence 1700, 1800, 1900, 2100, 2200, 2300?

10. Show that the number of leap years that occur in any 400 consecutive years is always the same and find this number of years.

11. Show the 13th day of each of two consecutive months is a Friday if and only if these months are the February and March of a year for which January 1 falls on a Thursday.

* 12. A new calendar called the *International Fixed Calendar* has been proposed. In this calendar, there are 13 months, including all of our present months, plus a new month, called *Sol,* which is placed between June and July. Each month has 28 days, except for the June of leap years, which has an extra day (leap years are determined the same way as in the Gregorian calendar). There is an extra day, *Year End Day,* which is not in any month, which we may consider as December 29. Devise a perpetual calendar for the International Fixed Calendar to give the day of the week for any calendar date.

13. Show that every year in the Gregorian calendar includes at least one Friday the 13th.

14. Show that for every year of the Gregorian calendar and for every integer k with $1 \le k \le 30$, as the 12 months of the year pass, the kth day of the month falls on all seven days of the week.

15. Given a year in the Gregorian calendar, determine on how many different days of the week the 31st of a month falls.

16. Determine the largest possible number of years in a century during which the month of February has 5 Sundays.

5.2 Computational and Programming Exercises

Computations and Explorations

Using a computation program such as Maple or *Mathematica,* or programs you have written, carry out the following computations and explorations.

1. Find the number of times that the thirteenth of a month falls on a Friday for all years between 1800 and 2300. Can you make and prove a conjecture based on your evidence?

Programming Projects

Write programs using Maple, *Mathematica,* or a language of your choice to do the following.

1. Give the day of the week of any date.

2. Print out a calendar of any year.

3. Print out a calendar for the International Fixed Calendar (see Exercise 12).

5.3 Round-Robin Tournaments

Congruences can be used to schedule round-robin tournaments. In this section, we show how to schedule a tournament for N different teams, so that each team plays every other team exactly once. The method we describe was developed by Freund [Fr56].

First, note that if N is odd, not all teams can be scheduled in each round, because when teams are paired, the total number of teams playing is even. So, if N is odd, we add a dummy team, and if a team is paired with the dummy team during a particular round, it draws a bye in that round and does not play. Hence, we can assume that we always have an even number of teams, with the addition of a dummy team if necessary.

We label the N teams with the integers $1, 2, 3, \ldots, N - 1, N$. We construct a schedule, pairing teams in the following way. We have team i, with $i \neq N$, play team j, with $j \neq N$ and $j \neq i$, in the kth round if $i + j \equiv k \pmod{N - 1}$. This schedules games for all teams in round k, except for team N and the one team i for which $2i \equiv k \pmod{N - 1}$. There is one such team because Theorem 4.10 tells us that the congruence $2x \equiv k \pmod{N - 1}$ has exactly one solution with $1 \leq x \leq N - 1$, because $(2, N - 1) = 1$. We match this team i with team N in the kth round.

We must now show that each team plays every other team exactly once. We consider the first $N - 1$ teams. Note that team i, where $1 \leq i \leq N - 1$, plays team N in round k, where $2i \equiv k \pmod{N - 1}$, and this happens exactly once. In the other rounds, team i does not play the same team twice, for if team i played team j in both rounds k and k', then $i + j \equiv k \pmod{N - 1}$, and $i + j \equiv k' \pmod{N - 1}$, which is an obvious contradiction because $k \not\equiv k' \pmod{N - 1}$. Hence, because each of the first $N - 1$ teams plays $N - 1$ games, and does not play any team more than once, it plays every team exactly once. Also, team N plays $N - 1$ games, and since every other team plays team N exactly once, team N plays every other team exactly once.

Example 5.10. To schedule a round-robin tournament with five teams, labeled 1, 2, 3, 4, and 5, we include a dummy team labeled 6. In round one, team 1 plays team j, where $1 + j \equiv 1 \pmod 5$. This is the team $j = 5$ so that team 1 plays team 5. Team 2 is scheduled in round one with team 4, since the solution of $2 + j \equiv 1 \pmod 5$ is $j = 4$. Because $i = 3$ is the solution of the congruence $2i \equiv 1 \pmod 5$, team 3 is paired with the dummy team 6, and hence, draws a bye in the first round. If we continue this procedure and finish scheduling the other rounds, we end up with the pairings shown in Table 5.1, where the opponent of team i in round k is given in the kth row and ith column. ◀

	Team				
Round	**1**	**2**	**3**	**4**	**5**
1	5	4	bye	2	1
2	bye	5	4	3	2
3	2	1	5	bye	3
4	3	bye	1	5	4
5	4	3	2	1	bye

Table 5.1 *Round-robin schedule for five teams.*

5.3 Exercises

1. Set up a round-robin tournament schedule for the following.

   a) 7 teams   b) 8 teams   c) 9 teams   d) 10 teams

2. In round-robin tournament scheduling, we wish to assign a *home team* and an *away team* for each game so that each of n teams, where n is odd, plays an equal number of home games and away games. Show that if, when $i + j$ is odd, we assign the smaller of i and j as the home team, whereas if $i + j$ is even, we assign the larger of i and j as the home team, then each team plays an equal number of home and away games.

3. In a round-robin tournament scheduling, use Exercise 2 to determine the home team for each game for the following numbers of teams.

   a) 5 teams   b) 7 teams   c) 9 teams

5.3 Computational and Programming Exercises

Computations and Explorations

Using a computation program such as Maple or *Mathematica,* or programs you have written, carry out the following computations and explorations.

1. Construct a round-robin schedule for a tournament with 13 teams, specifying a home team for each game.

Programming Projects

Write programs using Maple, *Mathematica,* or a language of your choice to do the following.

1. Schedule round-robin tournaments for n teams, where n is a positive integer.

2. Using Exercise 2, schedule round-robin tournaments for n teams, where n is an odd positive integer, specifying the home team for each game.

5.4 Hashing Functions

A university wishes to store a file in its computer for each of its students. The identifying number or *key* for each file is the social security number of the student. The social security number is a nine-digit integer, so it is extremely infeasible to reserve a memory location for each possible social security number. Instead, a systematic way to arrange the files in memory, using a reasonable number of memory locations, should be used so that each file can be easily accessed. Systematic methods of arranging files have been developed based on *hashing functions.* A hashing function assigns to the key of each file a particular memory location. Various types of hashing functions have been suggested, but the type most commonly used involves modular arithmetic. We discuss this type of hashing function here; for a general discussion of hashing functions, see Knuth [Kn97] or [CoLeRi01].

Let k be the key of the file to be stored; in our example, k is the social security number of a student. Let m be a positive integer. We define the hashing function $h(k)$ by

$$h(k) \equiv k \ (\text{mod} \ m),$$

where $0 \le h(k) < m$, so that $h(k)$ is the least positive residue of k modulo m. We wish to pick m intelligently, so that the files are distributed in a reasonable way throughout the m different memory locations $0, 1, 2, \ldots, m - 1$.

The first thing to keep in mind is that m should not be a power of the base b that is used to represent the keys. For instance, when using social security numbers as keys, m should not be a power of 10, such as 10^3, because the value of the hashing function would simply be the last several digits of the key; this may not distribute the keys uniformly throughout the memory locations. For instance, the last three digits of early issued social security numbers may often be between 000 and 099, but seldom between 900 and 999. Likewise, it is unwise to use a number dividing $b^k \pm a$, where k and a are small integers for the modulus m. In such a case, $h(k)$ would depend too strongly on the particular digits of the key, and different keys with similar, but rearranged, digits may be sent to the same memory location. For instance, if $m = 111$, then, since $111 \mid (10^3 - 1) = 999$, we have $10^3 \equiv 1 \ (\text{mod} \ 111)$, so that the social security numbers 064 212 848 and 064 848 212 are sent to the same memory location, because

$$h(064 \ 212 \ 848) \equiv 064 \ 212 \ 848 \equiv 064 + 212 + 848 \equiv 1124 \equiv 14 \ (\text{mod} \ 111)$$

and

$$h(064 \ 848 \ 212) \equiv 064 \ 848 \ 212 \equiv 064 + 848 + 212 \equiv 1124 \equiv 14 \ (\text{mod} \ 111).$$

To avoid such difficulties, m should be a prime that approximates the number of available memory locations devoted to file storage. For instance, if there are 5000 memory locations available for storage of 2000 student files, we could pick m to be equal to the prime 4969.

If the hashing function assigns the same memory location to two different files, we say that there is a *collision*. We need a method to resolve collisions, so that files are assigned to unique memory locations. There are two kinds of collision resolution policies. In the first kind, when a collision occurs, extra memory locations are linked together to the first memory location. When one wishes to access a file where this collision resolution policy has been used, it is necessary to first evaluate the hashing function for the particular key involved. Then the list linked to this memory location is searched.

The second kind of collision resolution policy is to look for an open memory location when an occupied location is assigned to a file. Various suggestions have been made for accomplishing this, such as the following techniques.

Starting with our original hashing function $h_0(k) = h(k)$, we define a sequence of memory locations $h_1(k), h_2(k), \ldots$. We first attempt to place the file with key k at location $h_0(k)$. If this location is occupied, we move to location $h_1(k)$. If this is occupied, we move to location $h_2(k)$, and so on.

We can choose the sequence of functions $h_j(k)$ in various ways. The simplest way is to let

$$h_j(k) \equiv h(k) + j \ (\text{mod } m), \quad 0 \le h_j(k) < m.$$

This places the file with key k as near as possible past location $h(k)$. Note that with this choice of $h_j(k)$, all memory locations are checked, so if there is an open location, it will be found. Unfortunately, this simple choice of $h_j(k)$ leads to difficulties; files tend to *cluster*. We see that if $k_1 \ne k_2$ and $h_i(k_1) = h_j(k_2)$ for nonnegative integers i and j, then $h_{i+k}(k_1) = h_{j+k}(k_2)$ for $k = 1, 2, 3, \ldots$, so that exactly the same sequence of locations is traced out once there is a collision. This lowers the efficiency of the search for files in the table. We would like to avoid this problem of clustering, so we choose the function $h_j(k)$ in a different way.

To avoid clustering, we use a technique called *double hashing*. We choose, as before,

$$h(k) \equiv k \ (\text{mod } m),$$

with $0 \le h(k) < m$, where m is prime, as the hashing function. We take a second hashing function

$$g(k) \equiv k + 1 \ (\text{mod } m - 2),$$

where $0 < g(k) \le m - 1$, so that $(g(k), m) = 1$. We take as a *probing sequence*

$$h_j(k) \equiv h(k) + j \cdot g(k) \ (\text{mod } m),$$

where $0 \le h_j(k) < m$. Because $(g(k), m) = 1$, as j runs through the integers $0, 1, 2, \ldots, m - 1$, all memory locations are traced out. The ideal situation would be for $m - 2$ also to be prime, so that the values $g(k)$ are distributed in a reasonable way. Hence, we would like $m - 2$ and m to be twin primes.

Example 5.11.   In our example using social security numbers, both $m = 4969$ and $m - 2 = 4967$ are prime. Our probing sequence is

$$h_j(k) \equiv h(k) + j \cdot g(k) \ (\text{mod } 4969),$$

where $0 \le h_j(k) < 4969$, $h(k) \equiv k \ (\text{mod } 4969)$, and $g(k) \equiv k + 1 \ (\text{mod } 4967)$.

Suppose that we wish to assign memory locations to files for students with the following social security numbers:

$$k_1 = 344\ 401\ 659 \qquad k_6 = 372\ 500\ 191$$
$$k_2 = 325\ 510\ 778 \qquad k_7 = 034\ 367\ 980$$
$$k_3 = 212\ 228\ 844 \qquad k_8 = 546\ 332\ 190$$
$$k_4 = 329\ 938\ 157 \qquad k_9 = 509\ 496\ 993$$
$$k_5 = 047\ 900\ 151 \qquad k_{10} = 132\ 489\ 973.$$

Because $k_1 \equiv 269$, $k_2 \equiv 1526$, and $k_3 \equiv 2854 \ (\text{mod } 4969)$, we assign the first three files to locations 269, 1526, and 2854, respectively.

Because $k_4 \equiv 1526 \pmod{4969}$, but memory location 1526 is taken, we compute $h_1(k_4) \equiv h(k_4) + g(k_4) = 1526 + 216 = 1742 \pmod{4969}$; this follows because $g(k_4) \equiv 1 + k_4 \equiv 216 \pmod{4967}$.

Because location 1742 is free, we assign the fourth file to this location. The fifth, six, seventh, and eighth files go into the available locations 3960, 4075, 2376, and 578, respectively, because $k_5 \equiv 3960$, $k_6 \equiv 4075$, $k_7 \equiv 2376$, and $k_8 \equiv 578 \pmod{4969}$.

We find that $k_9 \equiv 578 \pmod{4969}$; because location 578 is occupied, we compute $h_1(k_9) \equiv h(k_9) + g(k_9) = 578 + 2002 = 2580 \pmod{4969}$, where $g(k_9) \equiv 1 + k_9 \equiv 2002 \pmod{4967}$. Hence, we assign the ninth file to the free location 2580.

Finally, we find that $k_{10} \equiv 1526 \pmod{4969}$, but location 1526 is taken. We compute $h_1(k_{10}) \equiv h(k_{10}) + g(k_{10}) = 1526 + 216 = 1742 \pmod{4969}$, because $g(k_{10}) \equiv 1 + k_{10} \equiv 216 \pmod{4967}$, but location 1742 is taken. Hence, we continue by finding $h_2(k_{10}) \equiv h(k_{10}) + 2g(k_{10}) \equiv 1958 \pmod{4969}$ and in this available location we place the tenth file.

Table 5.2 lists the assignments for the files of students by their social security numbers. In the table, the file locations are shown in boldface. ◄

We wish to find conditions in which double hashing leads to clustering. Hence, we find conditions when

(5.1) $$h_i(k_1) = h_j(k_2)$$

and

(5.2) $$h_{i+1}(k_1) = h_{j+1}(k_2),$$

Social Security Number	$h(k)$	$h_1(k)$	$h_2(k)$
344 401 659	**269**		
325 510 778	**1526**		
212 228 844	**2854**		
329 938 157	1526	**1742**	
047 900 151	**3960**		
372 500 191	**4075**		
034 367 980	**2376**		
546 332 190	**578**		
509 496 993	578	**2580**	
132 489 973	1526	1742	**1958**

Table 5.2 *Hashing function for student files.*

so that the two consecutive terms of two probe sequences agree. If both (5.1) and (5.2) occur, then

(5.3) $$h(k_1) + ig(k_1) \equiv h(k_2) + jg(k_2) \pmod{m}$$

and

(5.4) $$h(k_1) + (i+1)g(k_1) \equiv h(k_2) + (j+1)g(k_2) \pmod{m}.$$

Subtracting congruence (5.3) from (5.4), we obtain

$$g(k_1) \equiv g(k_2) \pmod{m}.$$

Because $0 < g(k) \le m - 1$, the congruence $g(k_1) \equiv g(k_2) \pmod{m}$ implies that $g(k_1) = g(k_2)$. Consequently,

$$k_1 + 1 \equiv k_2 + 1 \pmod{m - 2},$$

which tells us that

$$k_1 \equiv k_2 \pmod{m - 2}.$$

Because $g(k_1) = g(k_2)$, we can simplify congruence (5.3) to obtain

$$h(k_1) \equiv h(k_2) \pmod{m},$$

which shows that

$$k_1 \equiv k_2 \pmod{m}.$$

Consequently, because $(m - 2, m) = 1$, Theorem 4.8 tells us that

$$k_1 \equiv k_2 \pmod{m(m - 2)}.$$

Therefore, the only way that two probing sequences can agree for two consecutive terms is if the two keys involved, k_1 and k_2, are congruent modulo $m(m - 2)$. Hence, clustering is extremely rare. Indeed, if $m(m - 2) > k$ for all keys k, clustering will never occur.

5.4 Exercises

1. A parking lot has 101 parking places. A total of 500 parking stickers are sold and only 50–75 vehicles are expected to be parked at any time. Set up a hashing function and collision resolution policy for assigning parking places based on license plates displaying six-digit numbers.

2. Assign memory locations for students in your class, using as keys the day of the month of birthdays of students, with hashing function $h(K) \equiv K \pmod{19}$, and
   a) with probing sequence $h_j(K) \equiv h(K) + j \pmod{19}$.
   b) with probing sequence $h_j(K) \equiv h(K) + j \cdot g(K), 0 \le j \le 16$, where $g(K) \equiv 1 + K \pmod{17}$.

* **3.** Let a hashing function be $h(K) \equiv K \pmod{m}$, with $0 \le h(K) < m$, and let the probing sequence for collision resolution be $h_j(K) \equiv h(K) + jq \pmod{m}, 0 \le h_j(K) < m$, for $j = 1, 2, \ldots, m - 1$. Show that all memory locations are probed

a) if m is prime and $1 \le q \le m - 1$.

b) if $m = 2^r$ and q is odd.

* **4.** A probing sequence for resolving collisions where the hashing function is $h(K) \equiv K \pmod{m}, 0 \le h(K) < m$, is given by $h_j(K) \equiv h(K) + j(2h(K) + 1) \pmod{m}, 0 \le h_j(K) < m$.

  a) Show that if m is prime, then all memory sequences are probed.

  b) Determine conditions for clustering to occur; that is, when $h_j(K_1) = h_j(K_2)$ and $h_{j+r}(K_1) = h_{j+r}(K_2)$ for $r = 1, 2, \ldots$.

5. Using the hashing function and probing sequence of the example in the text, find open memory locations for the files of additional students with social security numbers $k_{11} = 137\ 612\ 044$, $k_{12} = 505\ 576\ 452$, $k_{13} = 157\ 170\ 996$, $k_{14} = 131\ 220\ 418$. (Add these to the ten files already stored.)

5.4 Computational and Programming Exercises

Computations and Explorations

Using a computation program such as Maple or *Mathematica,* or programs you have written, carry out the following computations and explorations.

1. Assign memory locations to the files of all the students in your class, using the hashing function and probing function from Example 5.11. After doing so, assign memory locations to other files with social security numbers that you make up.

Programming Projects

Write programs using Maple, *Mathematica,* or a language of your choice to assign memory locations to student files, using the hashing function $h(k) \equiv k \pmod{1021}, 0 \le h(k) < 1021$, where the keys are the social security numbers of students,

1. linking files together when collisions occur.

2. using $h_j(k) \equiv h(k) + j \pmod{1021}, j = 0, 1, 2, \ldots$ as the probing sequence.

3. using $h_j(k) \equiv h(k) + j \cdot g(k), j = 0, 1, 2, \ldots$, where $g(k) \equiv 1 + k \pmod{1019}$, as the probing sequence.

5.5 Check Digits

Congruences can be used to check for errors in strings of digits. In this section, we will discuss error detection for bit strings, which are used to represent computer data. Then we will describe how congruences are used to detect errors in strings of decimal digits, which are used to identify passports, checks, books, and other types of objects.

Manipulating or transmitting bit strings can introduce errors. A simple error detection method is to append the bit string $x_1 x_2 \ldots x_n$ with a *parity check bit* x_{n+1} defined by

$$x_{n+1} \equiv x_1 + x_2 + \cdots + x_n \pmod{2},$$

so that $x_{n+1} = 0$ if an even number of the first n bits in the string are 1, whereas $x_{n+1} = 1$ if an odd number of these bits are 1. The appended string $x_1 x_2 \ldots x_n x_{n+1}$ satisfies the congruence

$$(5.5) \qquad\qquad x_1 + x_2 + \cdots + x_n + x_{n+1} \equiv 0 \ (\text{mod } 2).$$

We use this congruence to look for errors.

Suppose that we send $x_1 x_2 \ldots x_n x_{n+1}$, and the string $y_1 y_2 \ldots y_n y_{n+1}$ is received. These two strings are equal, that is, $y_i = x_i$ for $i = 1, 2, \ldots, n + 1$, when there are no errors. But if an error was made, they differ in one or more positions. We check whether

$$(5.6) \qquad\qquad y_1 + y_2 + \cdots + y_n + y_{n+1} \equiv 0 \ (\text{mod } 2)$$

holds. If this congruence fails, at least one error is present, but if it holds, errors may still be present. However, when errors are rare and random, the most common type of error is a single error, which is always detected. In general, we can detect an odd number of errors, but not an even number of errors (see Exercise 4).

Example 5.12. Suppose that we receive 1101111 and 11001000, where the last bit in each string is a parity check bit. For the first string, note that $1 + 1 + 0 + 1 + 1 + 1 + 1 \equiv 0 \ (\text{mod } 2)$, so that either the received string is what was transmitted or it contains an even number of errors. For the second string, note that $1 + 1 + 0 + 0 + 1 + 0 + 0 + 0 \equiv 1 \ (\text{mod } 2)$, so that the received string was not the string sent; we ask for retransmission. ◀

Strings of decimal digits are used for identification numbers in many different contexts. Check digits, computed using a variety of schemes, are used to find errors in these strings. For instance, check digits are used to detect errors in passport numbers. In a scheme used by several European countries, if $x_1 x_2 x_3 x_4 x_5 x_6$ is the identification number of a passport, the check digit x_7 is chosen so that

$$x_7 \equiv 7x_1 + 3x_2 + x_3 + 7x_4 + 3x_5 + x_6 \ (\text{mod } 10).$$

Example 5.13. Suppose that the identification number of a passport is 211894. To find the check digit x_7, we compute

$$x_7 \equiv 7 \cdot 2 + 3 \cdot 1 + 1 \cdot 1 + 7 \cdot 8 + 3 \cdot 9 + 1 \cdot 4 \equiv 5 \ (\text{mod } 10),$$

so that the check digit is 5, and the seven-digit number 2118945 is printed on the passport. ◀

We can always detect a single error in a passport identification number appended with a check digit computed in this way. To see this, suppose that we make an error of a in a digit; that is, $y_j = x_j + a \ (\text{mod } 10)$, where x_j is the correct jth digit and y_j is the incorrect digit that replaces it. From the definition of the check digit, it follows that we change x_7 by either $7a$, $3a$, or $a \ (\text{mod } 10)$, each of which changes x_7. However, errors caused by transposing two digits will be detected if and only if the difference between these two digits is not 5 or -5, that is, if they are not digits x_i and x_j with $|x_i - x_j| = 5$

(see Exercise 7). This scheme also detects a large number of possible errors involving the scrambling of three digits.

ISBNs

We now turn our attention to the use of check digits in publishing. Almost all recent books are identified by their *International Standard Book Number (ISBN)*, which is a ten-digit code assigned by the publisher. For instance, the ISBN for the first edition of this text is 0-201-06561-4. Here the first block of digits, 0, represents the language of the book (English), the second block of digits, 201, represents the publishing company (Addison-Wesley), the third block of digits, 06561, is the number assigned by the publishing company to this book, and the final digit, in this case 4, is the check digit. (The sizes of the blocks differ for different languages and publishers). The check digit in an ISBN can be used to detect the errors most commonly made when ISBNs are copied, namely single errors and errors made when two digits are transposed.

We will describe how this check digit is determined and then show that it can be used to detect the commonly occurring types of errors. Suppose that the ISBN of a book is $x_1 x_2 \ldots x_{10}$, where x_{10} is the check digit. (We ignore the hyphens in the ISBN, because the grouping of digits does not affect how the check digit is computed.) The first nine digits are decimal digits, that is, belong to the set $\{0, 1, 2, 3, 4, 5, 6, 7, 8, 9\}$, whereas the check digit x_{10} is a base 11 digit, belonging to the set $\{0, 1, 2, 3, 4, 5, 6, 7, 8, 9, X\}$, where X is the base 11 digit representing the integer 10 (in decimal notation). The check digit is selected so that the congruence

$$\sum_{i=1}^{10} i x_i \equiv 0 \ (\text{mod } 11)$$

holds. As is easily seen (see Exercise 10), the check digit x_{10} can be computed from the congruence $x_{10} \equiv \sum_{i=1}^{9} i x_i \ (\text{mod } 11)$; that is, the check digit is the remainder upon division by 11 of a weighted sum of the first nine digits.

Example 5.14. We find the check digit for the ISBN of the first edition of this text, which begins with 0-201-06561, by computing

$$x_{10} \equiv 1 \cdot 0 + 2 \cdot 2 + 3 \cdot 0 + 4 \cdot 1 + 5 \cdot 0 + 6 \cdot 6 + 7 \cdot 5 + 8 \cdot 6 + 9 \cdot 1 \equiv 4 (\text{mod } 11).$$

Hence, the ISBN is 0-201-06561-4, as previously stated. Similarly, if the ISBN number of a book begins with 3-540-19102, we find the check digit using the congruence

$$x_{10} \equiv 1 \cdot 3 + 2 \cdot 5 + 3 \cdot 4 + 4 \cdot 0 + 5 \cdot 1 + 6 \cdot 9 + 7 \cdot 1 + 8 \cdot 0 + 9 \cdot 2 \equiv 10 \ (\text{mod } 11).$$

This means that the check digit is X, the base 11 digit for the decimal number 10. Hence, the ISBN number is 3-540-19102-X. ◀

We will show that a single error, or a transposition of two digits, can be detected using the check digit of an ISBN. First, suppose that $x_1 x_2 \ldots x_{10}$ is a valid ISBN, but that this number has been printed as $y_1 y_2 \ldots y_{10}$. We know that $\sum_{i=1}^{10} i x_i \equiv 0 \ (\text{mod } 11)$, because $x_1 x_2 \ldots x_{10}$ is a valid ISBN.

Suppose that exactly one error has been made in printing the ISBN. Then, for some integer j, we have $y_i = x_i$ for $i \neq j$ and $y_j = x_j + a$, where $-10 \leq a \leq 10$ and $a \neq 0$. Here $a = y_j - x_j$ is the error in the jth place. Note that

$$\sum_{i=1}^{10} iy_i = \sum_{i=1}^{10} ix_i + ja \equiv ja \not\equiv 0 \pmod{11}$$

because $\sum_{i=1}^{10} ix_i \equiv 0 \pmod{11}$ and, by Lemma 3.5, it follows that $11 \nmid ja$ because $11 \nmid j$ and $11 \nmid a$. We conclude that $y_1 y_2 \ldots y_{10}$ is not a valid ISBN so that we can investigate the error.

Now suppose that two unequal digits have been transposed; then there are distinct integers j and k such that $y_j = x_k$ and $y_k = x_j$, and $y_i = x_i$ if $i \neq j$ and $i \neq k$. It follows that

$$\sum_{i=1}^{10} iy_i = \sum_{i=1}^{10} ix_i + (jx_k - jx_j) + (kx_j - kx_k) \equiv (j - k)(x_k - x_j) \not\equiv 0 \pmod{11}$$

because $\sum_{i=1}^{10} ix_i \equiv 0 \pmod{11}$, and $11 \nmid (j - k)$ and $11 \nmid (x_k - x_j)$. We see that $y_1 y_2 \ldots y_{10}$ is not a valid ISBN, so that we can detect the interchange of two unequal digits.

We have discussed how a single check digit can be used to detect errors in strings of digits. However, using a single check digit, we cannot detect an error and then correct it, that is, replace the digit in error with the valid one. It is possible to detect and correct an error using additional digits satisfying certain congruences (see Exercises 20 and 22, for example). The reader is referred to any text on coding theory for more information on error detection and correction. Coding theory uses many results from different parts of mathematics, including number theory, abstract algebra, combinatorics, and even geometry. To find good sources of information, consult Chapter 14 of [Ro99a]. We also refer the reader to the excellent articles by J. Gallian on check digits, [Ga92], [Ga91], and [Ga96], [GaWi88], for related information, including how check digits for drivers license numbers are found, and the book [Ki01] entirely devoted to check digits and identification numbers.

5.5 Exercises

1. What is the parity check bit that should be added to each of the following bit strings?

a) 111111   c) 101010   e) 11111111

b) 000000   d) 100000   f) 11001011

2. Suppose that you receive the following bit strings, where the last bit is a parity check bit. Which strings do you know are incorrect?

a) 111111111   b) 0101010101010   c) 1111010101010101

3. Assume that each of the following strings, ending with a parity check bit, was received correctly except for a missing bit indicated with a question mark. What is the missing bit?

   a) 1?11111   b) 000?10101   c) ?0101010100

4. Show that a parity check bit can detect an odd number of errors, but not an even number of errors.

5. Using the check digit scheme described in the text, find the check digit that should be added to the following passport identification numbers.

   a) 132999   b) 805237   c) 645153

6. Are the following passport identification numbers valid, where the seventh digit is the check digit computed as described in the text?

   a) 3300118   b) 4501824   c) 1873336

7. Show that the passport check digit scheme described in the text detects transposition of the digits x_i and x_j if and only if $| x_i - x_j | \neq 5$.

8. The bank identification number printed on a check consists of eight digits, $x_1x_2 \ldots x_8$, followed by a ninth check digit x_9, where $x_9 \equiv 7x_1 + 3x_2 + 9x_3 + 7x_4 + 3x_5 + 9x_6 + 7x_7 + 3x_8 \pmod{10}$.
   a) What is the check digit following the eight-digit identification number 00185403?
   b) Which single errors in bank identification numbers does a check digit computed in this way detect?
   c) Which transpositions of two digits does this scheme detect?

9. What should the check digit be to complete each of the following ISBNs?

   a) 2-113-54001   c) 1-2123-9940
   b) 0-19-081082   d) 0-07-038133

10. Show that the check digit x_{10} in an ISBN $x_1x_2 \ldots x_{10}$ can be computed from the congruence $x_{10} \equiv \sum_{i=1}^{9} ix_i \pmod{11}$.

11. Determine whether each of the following ISBNs is valid.

    a) 0-394-38049-5   c) 0-8218-0123-6   e) 90-6191-705-2
    b) 1-09-231221-3   d) 0-404-50874-X

12. Suppose that one digit, indicated with a question mark, in each of the following ISBNs has been smudged and cannot be read. What should this missing digit be?

    a) 0-19-8?3804-9   b) 91-554-212?-6   c) ?-261-05073-X

13. While copying the ISBN for a book, a clerk accidentally transposed two digits. If the clerk copied the ISBN as 0-07-289095-0 and did not make any other mistakes, what is the correct ISBN for this book?

Retail products are often identified by *Universal Product Codes (UPCs),* the most common of which consists of 12 decimal digits. The first digit identifies a product category, the next five the manufacturer, the following five the particular product, and the last digit is a check digit. The check digit is determined by the following three steps that use the first 11 digits of the UPC. First, digits in odd-numbered positions, starting from the left, are added, and the resulting sum is tripled. Second, the sum of digits in even-numbered positions is added to the result of the first step. Third, the check is found by determining which decimal digit, when added to the overall result of the second step, produces an integer divisible by 10.

14. Give a formula using a congruence that produces the check digit for a UPC from the 11 digits representing the product category, manufacturer, and particular product.

15. Determine whether each of the following 12-digit strings can be the UPC of a product.

    a) 0 47000 00183 6   c) 0 58000 00127 5

    b) 3 11000 01038 9   d) 2 26500 01179 4

16. What is the check digit for the 12-digit UPC code that begins with each of the following 11-digit strings?

    a) 3 81370 02918   c) 0 33003 31439

    b) 5 01175 00557   d) 4 11000 01028

17. Determine whether the 12-digit UPC code can always detect an error in exactly one digit.

18. Determine whether the 12-digit UPC code can always detect the transposition of two digits.

19. Suppose we specify that the valid 10-digit decimal code words $x_1 x_2 \ldots x_{10}$ are those satisfying the congruence $\sum_{i=1}^{10} x_i \equiv 0 \pmod{11}$.
    a) Can we detect all single errors in a code word?
    b) Can we detect transposition of two digits in a code word?

* **20.** Suppose that the only valid 10-digit code words $x_1 x_2 \ldots x_{10}$ are those satisfying the congruences $\sum_{i=1}^{10} x_i \equiv \sum_{i=1}^{10} i x_i \equiv 0 \pmod{11}$.
    a) Show that the valid code words, where the first digits are decimal digits, that is, in the set $\{0, 1, 2, 3, 4, 5, 6, 7, 8, 9\}$, are those where the last two digits satisfy the congruences $x_9 \equiv \sum_{j=1}^{8} (i + 1) x_i \pmod{11}$ and $x_{10} \equiv \sum_{j=1}^{8} (9 - i) x_i \pmod{11}$.
    b) Find the number of valid decimal code words.
    c) Show that any single error in a code word can be detected and corrected, because the location and value of the error can be determined.
    d) Show that we can detect any error caused by transposing two digits in a code word.

21. The government of Norway assigns an 11-digit decimal registration number $x_1 x_2 \ldots x_{11}$ to each of its citizens using a scheme designed by Norwegian number theorist E. Selmer. The digits $x_1 x_2 \ldots x_6$ represent the date of birth, the digits $x_7 x_8 x_9$ identify the particular person born that day, and x_{10} and x_{11} are check digits that are computed using the congruences $x_{10} \equiv 8x_1 + 4x_2 + 5x_3 + 10x_4 + 3x_5 + 2x_6 + 7x_7 + 6x_8 + 9x_9 \pmod{11}$, and $x_{11} \equiv 6x_1 + 7x_2 + 8x_3 + 9x_4 + 4x_5 + 5x_6 + 6x_7 + 7x_8 + 8x_9 + 9x_{10} \pmod{11}$.
    a) Determine the check digits that follow the first nine digits 110491238.
    b) Show that this scheme detects all single errors in a registration number.
    * c) Which double errors are detected?

* **22.** Suppose that we specify that the valid 10-digit code words $x_1 x_2 \ldots x_{10}$, where each digit is a decimal digit, are those satisfying the congruences $\sum_{i=1}^{10} x_i \equiv \sum_{i=1}^{10} i x_i \equiv \sum_{i=1}^{10} i^2 x_i \equiv \sum_{i=1}^{10} i^3 x_i \equiv 0 \pmod{11}$.

  a) How many valid 10-digit code words are there?

  b) Show how any two errors in a code word can be corrected.

  c) Suppose a code word has been received as 0204906710. If two errors have been made, what is the correct code word?

Airline tickets carry 15-digit identification numbers $a_1 a_2 \ldots a_{14} a_{15}$, where a_{15} is a check digit which equals the least nonnegative residue of the integer $a_1 a_2 \ldots a_{14}$ modulo 7.

23. Find the check digit a_{15} when the first 14 digits of the identification of an airplane ticket are

  a) 00032781811224   b) 10238544122339   c) 00611133123278

24. Determine whether these are valid airline ticket identification numbers.

  a) 102284711033122   b) 004113711331240   c) 100261413001533

25. Determine which errors in a single digit can be detected and which cannot be detected using the check digit for airline tickets.

26. Determine which errors involving the transposition of two adjacent digits in the identification number of an airline ticket can be detected and which cannot be detected using the check digit for airline tickets.

The *International Standard Serial Number (ISSN)* used to identify a periodical consists of two blocks of four digits, where the last digit in the second block is a base 11 check digit. As in an ISBN, the character X represents 10 (in decimal notation). The check digit d_8 is determined by the congruence $d_8 \equiv 3d_1 + 4d_2 + 5d_3 + 6d_4 + 7d_5 + 8d_6 + 9d_7 \pmod{11}$.

27. For each of the following initial seven digits of an ISSN, determine the correct check digit.

  a) 0317-847   c) 1063-669
  b) 0423-555   d) 1363-837

28. Is it always possible to detect a single error in an ISSN? That is, is it always possible to detect that an error was made when one digit of an ISSN has been copied incorrectly? Justify your answer.

29. Is it always possible to detect when two consecutive digits in an ISSN have been accidentally transposed? Justify your answer.

5.5 Computational and Programming Exercises

Computations and Explorations

Using a computation program such as Maple or *Mathematica,* or programs you have written, carry out the following computations and explorations.

  1. Check the ISBN numbers of a selection of books to see whether the check digit was computed correctly.

Programming Projects

Write programs using Maple, *Mathematica,* or a language of your choice to do the following.

1. Determine whether a bit string, ending with a parity check bit, has either an odd or an even number of errors.

2. Determine the check digit for an ISBN, given the first nine digits.

3. Determine whether a 10-digit string, where the first nine digits are decimal digits and the last is a decimal digit or an X, is a valid ISBN.

4. Determine whether a 12-digit decimal string is a valid UPC.

6

Some Special Congruences

Introduction

In this chapter, we discuss three congruences that have both theoretical and practical significance: Wilson's theorem shows that when p is prime, the remainder when $(p-1)!$ is divided by p is -1. Fermat's little theorem provides a congruence for the pth powers of integers modulo p. In particular, it shows that if p is prime, then a^p and a have the same remainder when divided by p whenever a is an integer. Euler's theorem provides a generalization of Fermat's little theorem for moduli that are not prime.

These three congruences have many applications. For example, we will explain how Fermat's little theorem can be used as the basis for primality tests and factoring algorithms. We will also discuss composite integers, called pseudoprimes, that masquerade as primes by satisfying the same congruence that primes do in Fermat's little theorem. We will use the fact that pseudoprimes are relatively rare to develop some tests that can provide overwhelming evidence that an integer is prime.

6.1 Wilson's Theorem and Fermat's Little Theorem

In a book published in 1770, English mathematician Edward Waring stated that one of his students, John Wilson, had discovered that $(p-1)!+1$ is divisible by p whenever p is prime. Furthermore, he stated that neither he nor Wilson knew how to prove it. Most likely, Wilson made this conjecture based on numerical evidence. For example, we can easily see that 2 divides $1!+1=2$, 3 divides $2!+1=3$, 5 divides $4!+1=25$, 7 divides $6!+1=721$, and so on. Although Waring thought it would be difficult to find a proof, *Joseph Lagrange* proved this result in 1771. Nevertheless, the fact that p divides $(p-1)!+1$ is known as *Wilson's theorem*. We now state this theorem in the form of a congruence.

Theorem 6.1. *Wilson's Theorem.*   If p is prime, then $(p - 1)! \equiv -1 \pmod{p}$.

Before proving Wilson's theorem, we use an example to illustrate the idea behind the proof.

Example 6.1.   Let $p = 7$. We have $(7 - 1)! = 6! = 1 \cdot 2 \cdot 3 \cdot 4 \cdot 5 \cdot 6$. We will rearrange the factors in the product, grouping together pairs of inverses modulo 7. We note that $2 \cdot 4 \equiv 1 \pmod{7}$ and $3 \cdot 5 \equiv 1 \pmod{7}$. Hence, $6! \equiv 1 \cdot (2 \cdot 4) \cdot (3 \cdot 5) \cdot 6 \equiv 1 \cdot 6 \equiv -1 \pmod{7}$. Thus, we have verified a special case of Wilson's theorem.   ◀

We now use the technique illustrated in the example to prove Wilson's theorem.

Proof.   When $p = 2$, we have $(p - 1)! \equiv 1 \equiv -1 \pmod{2}$. Hence, the theorem is true for $p = 2$. Now let p be a prime greater than 2. Using Theorem 4.10, for each integer a with $1 \leq a \leq p - 1$, there is an inverse $\bar{a}$, $1 \leq \bar{a} \leq p - 1$, with $a\bar{a} \equiv 1 \pmod{p}$. By Theorem 4.11 the only positive integers less than p that are their own inverses are 1 and $p - 1$. Therefore, we can group the integers from 2 to $p - 2$ into $(p - 3)/2$ pairs of integers, with the product of each pair congruent to 1 modulo p. Hence, we have

$$2 \cdot 3 \cdots (p - 3) \cdot (p - 2) \equiv 1 \pmod{p}.$$

We multiply both sides of the this congruence by 1 and $p - 1$ to obtain

$$(p - 1)! = 1 \cdot 2 \cdot 3 \cdots (p - 3)(p - 2)(p - 1) \equiv 1 \cdot (p - 1) \equiv -1 \pmod{p}.$$

This completes the proof.   ■

An interesting observation is that the converse of Wilson's theorem is also true, as the following theorem shows.

Theorem 6.2.   If n is a positive integer with $n \geq 2$ such that $(n - 1)! \equiv -1 \pmod{n}$, then n is prime.

JOSEPH LOUIS LAGRANGE (1736–1813) was born in Italy and studied physics and mathematics at the University of Turin. Although he originally planned to pursue a career in physics, Lagrange's growing interest in mathematics led him to change course. At the age of 19, he was appointed as a mathematics professor at the Royal Artillery School in Turin. In 1766, he filled the post Euler vacated at the Royal Academy of Berlin when Frederick the Great sought him out. Lagrange directed the mathematics section of the Royal Academy for 20 years. In 1787, when his patron Frederick the Great died, Lagrange moved to France at the invitation of Louis XVI, to join the French Academy. In France he had a distinguished career in teaching and writing. He was a favorite of Marie Antoinette, but managed to win the favor of the new regime that came into power after the French Revolution. Lagrange's contributions to mathematics include unifying the mathematical theory of mechanics. He made fundamental discoveries in group theory and helped put calculus on a rigorous foundation. His contributions to number theory include the first proof of Wilson's theorem, and the result that every positive integer can be written as the sum of four squares.

Proof. Assume that n is a composite integer and that $(n-1)! \equiv -1 \pmod{n}$. Because n is composite, we have $n = ab$, where $1 < a < n$ and $1 < b < n$. Because $a < n$, we know that $a \mid (n-1)!$, because a is one of the $n-1$ numbers multiplied together to form $(n-1)!$. Because $(n-1)! \equiv -1 \pmod{n}$ it follows that $n \mid ((n-1)!+1)$. This means, by Theorem 1.8, that a also divides $(n-1)!+1$. By Theorem 1.9, because $a \mid (n-1)!$ and $a \mid ((n-1)!+1)$, we conclude that $a \mid ((n-1)!+1) - (n-1)! = 1$. This is a contradiction, because $a > 1$. ∎

Wilson's theorem can be used to demonstrate that a composite integer is not prime, as Example 6.2 shows.

Example 6.2. Because $(6-1)! = 5! = 120 \equiv 0 \pmod{6}$, Theorem 6.1 verifies the obvious fact that 6 is not prime. ◀

As we can see, Wilson's theorem and its converse give us a primality test. To decide whether an integer n is prime, we determine whether $(n-1)! \equiv -1 \pmod{n}$. Unfortunately, this is an impractical test because $n-2$ multiplications modulo n are needed to find $(n-1)!$, requiring $O(n(\log_2 n)^2)$ bit operations.

Fermat made many important discoveries in number theory, including the fact that p divides $a^{p-1} - 1$ whenever p is prime and a is an integer not divisible by p. He stated this result in a letter to one of his mathematical correspondents, Frènicle de Bessy, in 1640. Fermat did not bother to enclose a proof with his letter, stating that he feared that a proof would be too long. Unlike Fermat's notorious last theorem, discussed in Chapter 13, there is little doubt that Fermat really knew how to prove this theorem (which is called "Fermat's little theorem" to distinguish it from his "last theorem"). Leonhard Euler is responsible for the first published proof, in 1736. Euler also generalized Fermat's little theorem; we will explain how in Section 6.3.

Theorem 6.3. *Fermat's Little Theorem.* If p is prime and a is a positive integer with $p \nmid a$, then $a^{p-1} \equiv 1 \pmod{p}$.

Proof. Consider the $p-1$ integers $a, 2a, \ldots, (p-1)a$. None of these integers are divisible by p, for if $p \mid ja$, then by Lemma 3.4, $p \mid j$, because $p \nmid a$. This is impossible, because $1 \le j \le p-1$. Furthermore, no two of the integers $a, 2a, \ldots, (p-1)a$ are congruent modulo p. To see this, assume that $ja \equiv ka \pmod{p}$, where $1 \le j < k \le p-1$. Then, by Corollary 4.4.1, because $(a, p) = 1$, we have $j \equiv k \pmod{p}$. This is impossible, because j and k are positive integers less than $p-1$.

Because the integers $a, 2a, \ldots, (p-1)a$ are a set of $p-1$ integers all incongruent to 0, and no two are congruent modulo p, we know that the least positive residues of $a, 2a, \ldots, (p-1)a$, taken in some order, must be the integers $1, 2, \ldots, p-1$. As a consequence, the product of the integers $a, 2a, \ldots, (p-1)a$ is congruent modulo p to the product of the first $p-1$ positive integers. Hence,

$$a \cdot 2a \cdots (p-1)a \equiv 1 \cdot 2 \cdots (p-1) \pmod{p}.$$

Therefore,

$$a^{p-1}(p-1)! \equiv (p-1)! \,(\text{mod } p).$$

Because $((p-1)!, p) = 1$, using Corollary 4.4.1, we cancel $(p-1)!$ to obtain

$$a^{p-1} \equiv 1 \,(\text{mod } p). \qquad \blacksquare$$

We illustrate the ideas of the proof with an example.

Example 6.3.   Let $p = 7$ and $a = 3$. Then, $1 \cdot 3 \equiv 3 \,(\text{mod } 7)$, $2 \cdot 3 \equiv 6 \,(\text{mod } 7)$, $3 \cdot 3 \equiv 2 \,(\text{mod } 7)$, $4 \cdot 3 \equiv 5 \,(\text{mod } 7)$, $5 \cdot 3 \equiv 1 \,(\text{mod } 7)$, and $6 \cdot 3 \equiv 4 \,(\text{mod } 7)$. Consequently,

$$(1 \cdot 3) \cdot (2 \cdot 3) \cdot (3 \cdot 3) \cdot (4 \cdot 3) \cdot (5 \cdot 3) \cdot (6 \cdot 3) \equiv 3 \cdot 6 \cdot 2 \cdot 5 \cdot 1 \cdot 4 \,(\text{mod } 7),$$

so that $3^6 \cdot 1 \cdot 2 \cdot 3 \cdot 4 \cdot 5 \cdot 6 \equiv 3 \cdot 6 \cdot 2 \cdot 5 \cdot 1 \cdot 4 \,(\text{mod } 7)$. Hence, $3^6 \cdot 6! \equiv 6! \,(\text{mod } 7)$, and therefore $3^6 \equiv 1 \,(\text{mod } 7)$. ◀

Theorem 6.4.   If p is prime and a is a positive integer, then $a^p \equiv a \,(\text{mod } p)$.

Proof.   If $p \nmid a$, by Fermat's little theorem we know that $a^{p-1} \equiv 1 \,(\text{mod } p)$. Multiplying both sides of this congruence by a, we find that $a^p \equiv a \,(\text{mod } p)$. If $p \mid a$, then $p \mid a^p$ as well, so that $a^p \equiv a \equiv 0 \,(\text{mod } p)$. This finishes the proof, because $a^p \equiv a \,(\text{mod } p)$ if $p \nmid a$ and if $p \mid a$. ∎

Finding the least positive residue of powers of integers is often required in number theory and its applications—especially cryptography, as we will see in Chapter 8. Fermat's little theorem is a useful tool in such computations, as the following example shows.

Example 6.4.   We can find the least positive residue of 3^{201} modulo 11 with the help of Fermat's little theorem. We know that $3^{10} \equiv 1 \,(\text{mod } 11)$. Hence, $3^{201} = (3^{10})^{20} \cdot 3 \equiv 3 \,(\text{mod } 11)$. ◀

A useful application of Fermat's little theorem is provided by the following result.

Theorem 6.5.   If p is prime and a is an integer such that $p \nmid a$, then a^{p-2} is an inverse of a modulo p.

Proof.   If $p \nmid a$, by Fermat's little theorem we have $a \cdot a^{p-2} = a^{p-1} \equiv 1 \,(\text{mod } p)$. Hence, a^{p-2} is an inverse of a modulo p. ∎

Example 6.5.   By Theorem 6.5, we know that $2^9 = 512 \equiv 6 \,(\text{mod } 11)$ is an inverse of 2 modulo 11. ◀

Theorem 6.5 gives us another way to solve linear congruences with respect to prime moduli.

Corollary 6.5.1. If a and b are positive integers and p is prime with $p \nmid a$, then the solutions of the linear congruence $ax \equiv b \pmod{p}$ are the integers x such that $x \equiv a^{p-2}b \pmod{p}$.

Proof. Suppose that $ax \equiv b \pmod{p}$. Because $p \nmid a$, we know from Theorem 6.5 that a^{p-2} is an inverse of $a \pmod{p}$. Multiplying both sides of the original congruence by a^{p-2}, we have

$$a^{p-2}ax \equiv a^{p-2}b \pmod{p}.$$

Hence,

$$x \equiv a^{p-2}b \pmod{p}. \qquad \blacksquare$$

The Pollard $p-1$ Factorization Method

Fermat's little theorem is the basis of a factorization method invented by J. M. Pollard in 1974. This method, known as the *Pollard $p-1$ method,* can find a nontrivial factor of an integer n when n has a prime factor p such that the primes dividing $p-1$ are relatively small.

To see how this method works, suppose that we want to find a factor of the positive integer n. Furthermore, suppose that n has a prime factor p such that $p-1$ divides $k!$, where k is a positive integer. We want $p-1$ to have only small prime factors, so that there is such an integer k that is not too large. For example, if $p = 2269$, then $p-1 = 2268 = 2^2 3^4 7$, so that $p-1$ divides $9!$, but no smaller value of the factorial function.

The reason we want $p-1$ to divide $k!$ is so that we can apply Fermat's little theorem. By Fermat's little theorem we know that $2^{p-1} \equiv 1 \pmod{p}$. Now, since $p-1$ divides $k!$, $k! = (p-1)q$ for some integer q. Hence

$$2^{k!} = 2^{(p-1)q} = (2^{p-1})^q \equiv 1^q = 1 \pmod{p},$$

which implies that p divides $2^{k!} - 1$. Now, let M be the least positive residue of $2^{k!} - 1$ modulo n, so that $M = (2^{k!} - 1) - nt$ for some integer t. We see that p divides M because it divides both $2^{k!} - 1$ and n.

Now, to find a divisor of n, we need only compute the greatest common divisor of M and n, $d = (M, n)$. This can be done rapidly using the Euclidean algorithm. For this divisor d to be a nontrivial divisor, it is necessary that M not be 0. This is the case when n does not itself divide $2^{k!} - 1$, which is likely when n has large prime divisors.

To use this method, we must compute $2^{k!}$, where k is a positive integer. This can be done efficiently because modular exponentiation can be done efficiently. To find the least positive remainder of $2^{k!}$ modulo n, we set $r_1 = 2$ and use the following sequence of computations: $r_2 \equiv r_1^2 \pmod{n}, r_3 \equiv r_2^3 \pmod{n}, \ldots, r_k \equiv r_{k-1}^k \pmod{n}$. We illustrate this procedure in the following example.

Example 6.6. To find $2^{9!}$ (mod 5, 157, 437), we perform the following sequence of computations:

$$r_2 \equiv r_1^2 = 2^2 \equiv 4 \ (\text{mod } 5,157,437)$$

$$r_3 \equiv r_2^3 = 4^3 \equiv 64 \ (\text{mod } 5,157,437)$$

$$r_4 \equiv r_3^4 = 64^4 \equiv 1,304,905 \ (\text{mod } 5,157,437)$$

$$r_5 \equiv r_4^5 = 1,304,905^5 \equiv 404,913 \ (\text{mod } 5,157,437)$$

$$r_6 \equiv r_5^6 = 404,913^6 \equiv 2,157,880 \ (\text{mod } 5,157,437)$$

$$r_7 \equiv r_6^7 = 2,157,880^7 \equiv 4,879,227 \ (\text{mod } 5,157,437)$$

$$r_8 \equiv r_7^8 = 4,879,227^8 \equiv 4,379,778 \ (\text{mod } 5,157,437)$$

$$r_9 \equiv r_8^9 = 4,379,778^9 \equiv 4,381,440 \ (\text{mod } 5,157,437).$$

It follows that $2^{9!} \equiv 4,381,440 \ (\text{mod } 5,157,437)$. ◄

The following example illustrates the use of the Pollard $p - 1$ method to find a factor of the integer 5,157,437.

Example 6.7. To factor 5,157,437 using the Pollard $p - 1$ method, we successively find r_k, the least positive residue of $2^{k!}$ modulo 5,157,437, for $k = 1, 2, 3, \ldots$, as was done in Example 6.6. We compute $(r_k - 1, 5,157,437)$ at each step. To find a factor of 5,157,437 requires nine steps, because $(r_k - 1, 5,157,437) = 1$ for $k = 1, 2, 3, 4, 5, 6, 7, 8$ (as the reader can verify), but $(r_9 - 1, 5,157,437) = (4,381,439, 5,157,437) = 2269$. It follows that 2269 is a divisor of 5,157,437. ◄

The Pollard $p - 1$ method does not always work. However, because nothing in the method depends on the choice of 2 as the base, we can extend the method and find a factor for more integers by using integers other than 2 as the base. In practice, the Pollard $p - 1$ method is used after trial divisions by small primes, but before the heavy artillery of such methods as the quadratic sieve and the elliptic curve method.

6.1 Exercises

1. Show that $10! + 1$ is divisible by 11, by grouping together pairs of inverses modulo 11 that occur in 10!.

2. Show that $12! + 1$ is divisible by 13, by grouping together pairs of inverses modulo 13 that occur in 12!.

3. What is the remainder when 16! is divided by 19?

4. What is the remainder when 5!25! is divided by 31?

5. Using Wilson's theorem, find the least positive residue of $8 \cdot 9 \cdot 10 \cdot 11 \cdot 12 \cdot 13$ modulo 7.

6. What is the remainder when $7 \cdot 8 \cdot 9 \cdot 15 \cdot 16 \cdot 17 \cdot 23 \cdot 24 \cdot 25 \cdot 43$ is divided by 11?

7. What is the remainder when 18! is divided by 437?

8. What is the remainder when 40! is divided by 1763?

9. What is the remainder when 5^{100} is divided by 7?

10. What is the remainder when 6^{2000} is divided by 11?

11. Using Fermat's little theorem, find the least positive residue of $3^{999,999,999}$ modulo 7.

12. Using Fermat's little theorem, find the least positive residue of $2^{1000000}$ modulo 17.

13. Show that $3^{10} \equiv 1 \pmod{11^2}$.

14. Using Fermat's little theorem, find the last digit of the base 7 expansion of 3^{100}.

15. Using Fermat's little theorem, find the solutions of the following linear congruences.

    a) $7x \equiv 12 \pmod{17}$ b) $4x \equiv 11 \pmod{19}$

16. Show that if n is a composite integer with $n \neq 4$, then $(n-1)! \equiv 0 \pmod{n}$.

17. Show that if p is an odd prime, then $2(p-3)! \equiv -1 \pmod{p}$.

18. Show that if n is odd and $3 \nmid n$, then $n^2 \equiv 1 \pmod{24}$.

19. Show that $a^{12} - 1$ is divisible by 35 whenever $(a, 35) = 1$.

20. Show that $a^6 - 1$ is divisible by 168 whenever $(a, 42) = 1$.

21. Show that $42 \mid (n^7 - n)$ for all positive integers n.

22. Show that $30 \mid (n^9 - n)$ for all positive integers n.

23. Show that $1^{p-1} + 2^{p-1} + 3^{p-1} + \cdots + (p-1)^{(p-1)} \equiv -1 \pmod{p}$ whenever p is prime. (It has been conjectured that the converse of this is also true.)

24. Show that $1^p + 2^p + 3^p + \cdots + (p-1)^p \equiv 0 \pmod{p}$ when p is an odd prime.

25. Show that if p is prime and a and b are integers not divisible by p, with $a^p \equiv b^p \pmod{p}$, then $a^p \equiv b^p \pmod{p^2}$.

26. Use the Pollard $p-1$ method to find a divisor of 689.

27. Use the Pollard $p-1$ method to find a divisor of 7,331,117. (For this exercise, you will need to use either a calculator or computational software.)

28. Show that if p and q are distinct primes, then $p^{q-1} + q^{p-1} \equiv 1 \pmod{pq}$.

29. Show that if p is prime and a is an integer, then $p \mid (a^p + (p-1)! \, a)$.

30. Show that if p is an odd prime, then $1^2 3^2 \cdots (p-4)^2 (p-2)^2 \equiv (-1)^{(p+1)/2} \pmod{p}$.

31. Show that if p is prime and $p \equiv 3 \pmod{4}$, then $((p-1)/2)! \equiv \pm 1 \pmod{p}$.

32. a) Let p be prime, and suppose that r is a positive integer less than p such that $(-1)^r r! \equiv -1 \pmod{p}$. Show that $(p-r+1)! \equiv -1 \pmod{p}$.

    b) Using part (a), show that $61! \equiv 63! \equiv -1 \pmod{71}$.

33. Using Wilson's theorem, show that if p is a prime and $p \equiv 1 \pmod{4}$, then the congruence $x^2 \equiv -1 \pmod{p}$ has two incongruent solutions given by $x \equiv \pm((p-1)/2)! \pmod{p}$.

34. Show that if p is a prime and $0 < k < p$, then $(p - k)!(k - 1)! \equiv (-1)^k \pmod{p}$.

35. Show that if n is an integer, then

$$\pi(n) = \sum_{j=2}^{n} \left[\frac{(j-1)! + 1}{j} - \left[\frac{(j-1)!}{j} \right] \right].$$

* **36.** For which positive integers n is $n^4 + 4^n$ prime?

37. Show that the pair of positive integers n and $n + 2$ are twin primes if and only if $4((n - 1)! + 1) + n \equiv 0 \pmod{n(n + 2)}$, where $n \neq 1$.

38. Show that if the positive integers n and $n + k$, where $n > k$ and k is an even positive integer, are both prime, then $(k!)^2((n - 1)! + 1) + n(k! - 1)(k - 1)! \equiv 0 \pmod{n(n + k)}$.

39. Show that if p is prime, then $\binom{2p}{p} \equiv 2 \pmod{p}$.

40. Exercise 74 of Section 3.5 shows that if p is prime and k is a positive integer less than p, then the binomial coefficient $\binom{p}{k}$ is divisible by p. Use this fact and the binomial theorem to show that if a and b are integers, then $(a + b)^p \equiv a^p + b^p \pmod{p}$.

41. Prove Fermat's little theorem by mathematical induction. (*Hint:* In the induction step, use Exercise 40 to obtain a congruence for $(a + 1)^p$.)

* **42.** Using Exercise 30 of Section 4.3, prove *Gauss's generalization of Wilson's theorem,* namely that the product of all the positive integers less than m that are relatively prime to m is congruent to $1 \pmod{m}$, unless $m = 4$, p^t, or $2p^t$, where p is an odd prime and t is a positive integer, in which case it is congruent to $-1 \pmod{m}$.

43. A deck of cards is shuffled by cutting the deck into two piles of 26 cards. Then, the new deck is formed by alternating cards from the two piles, starting with the bottom pile.
   a) Show that if a card begins in the cth position in the deck, it will be in the bth position in the new deck, where $b \equiv 2c \pmod{53}$ and $1 \leq b \leq 52$.
   b) Determine the number of shuffles of the type described above that are needed to return the deck of cards to its original order.

44. Let p be prime and let a be a positive integer not divisible by p. We define the *Fermat quotient* $q_p(a)$ by $q_p(a) = (a^{p-1} - 1)/p$. Show that if a and b are positive integers not divisible by the prime p, then $q_p(ab) \equiv q_p(a) + q_p(b) \pmod{p}$.

45. Let p be prime and let $a_1, a_2, \ldots, a_p$ and $b_1, b_2, \ldots, b_p$ be complete systems of residues modulo p. Show that $a_1b_1, a_2b_2, \ldots, a_pb_p$ is not a complete system of residues modulo p.

* **46.** Show that if n is a positive integer with $n \geq 2$, then n does not divide $2^n - 1$.

* **47.** Let p be an odd prime. Show that $(p - 1)!^{p^{n-1}} \equiv -1 \pmod{p^n}$.

48. Show that if p is a prime with $p > 5$, then $(p - 1)! + 1$ has at least two different prime divisors.

49. Show that if a and n are relatively prime integers with $n > 1$, then n is prime if and only if $(x - a)^n$ and $x^n - a$ are congruent modulo n as polynomials. (Recall from the preamble to Exercise 40 in Section 4.1 that two polynomials are congruent modulo n as polynomials if for each power of x the coefficients of that power in the polynomials are congruent modulo n.) (The proof of Agrawal, Kayal, and Saxena [AgKaSa02] that there is a polynomial-time algorithm for determining whether an integer is prime begins with this result.)

6.1 Computational and Programming Exercises

Computations and Explorations

Using a computation program such as Maple or *Mathematica*, or programs you have written, carry out the following computations and explorations.

1. A *Wilson prime* is a prime p for which $(p-1) \equiv -1 \pmod{p^2}$. Find all Wilson primes less than 10,000.

2. Find all primes p less than 10,000 for which $2^{p-1} \equiv 1 \pmod{p^2}$.

3. Find a factor of each of several different odd integers of your choice using the Pollard $p-1$ method.

4. Verify the conjecture that $1^{n-1} + 2^{n-1} + 3^{n-1} + \cdots + (n-1)^{(n-1)} \not\equiv -1 \pmod{n}$ if n is composite, for as many integers n as you can.

Programming Projects

Write programs using Maple, *Mathematica,* or a language of your choice to do the following.

1. Find all Wilson primes less than a given positive integer n.

2. Find the primes p less than a given positive integer n for which $2^{p-1} \equiv 1 \pmod{p^2}$.

3. Solve linear congruences with prime moduli via Fermat's little theorem.

4. Factor a given positive integer n using the Pollard $p-1$ method.

6.2 Pseudoprimes

Fermat's little theorem tells us that if n is prime and b is any integer, then $b^n \equiv b \pmod{n}$. Consequently, if we can find an integer b such that $b^n \not\equiv b \pmod{n}$, then we know that n is composite.

Example 6.8. We can show that 63 is not prime by observing that

$$2^{63} = 2^{60} \cdot 2^3 = (2^6)^{10} \cdot 2^3 = 64^{10} 2^3 \equiv 2^3 \equiv 8 \not\equiv 2 \pmod{63}. \qquad \blacktriangleleft$$

Using Fermat's little theorem, we can show that an integer is composite. It would be even more useful if it also provided a way to show that an integer is prime. It is commonly reported that the ancient Chinese believed that if $2^n \equiv 2 \pmod{n}$, then n must be prime. This statement is true for $1 \leq n \leq 340$. Unfortunately, the converse of Fermat's little theorem is not true, as the following example, which was discovered by Sarrus in 1919, shows.

Example 6.9. Let $n = 341 = 11 \cdot 31$. By Fermat's little theorem, we see that $2^{10} \equiv 1 \pmod{11}$, so that $2^{340} = (2^{10})^{34} \equiv 1 \pmod{11}$. Also, $2^{340} = (2^5)^{68} \equiv (32)^{68} \equiv 1 \pmod{31}$. Hence, by Corollary 4.8.1, we have $2^{340} \equiv 1 \pmod{341}$. By multiplying both sides of this congruence by 2, we have $2^{341} \equiv 2 \pmod{341}$, even though 341 is not prime.   ◀

Examples such as this lead to the following definition.

Definition. Let b be a positive integer. If n is a composite positive integer and $b^n \equiv b \pmod{n}$, then n is called a *pseudoprime to the base b.*

Note that if $(b, n) = 1$, then the congruence $b^n \equiv b \pmod{n}$ is equivalent to the congruence $b^{n-1} \equiv 1 \pmod{n}$. To see this, note that by Corollary 4.4.1 we can divide both sides of the first congruence by b, because $(b, n) = 1$, to obtain the second congruence. By part (iii) of Theorem 4.3, we can multiply both sides of the second congruence by b to obtain the first. We will often use this equivalent condition.

Example 6.10. The integers $341 = 11 \cdot 31$, $561 = 3 \cdot 11 \cdot 17$, and $645 = 3 \cdot 5 \cdot 43$ are pseudoprimes to the base 2, since it is easily verified that $2^{340} \equiv 1 \pmod{341}$, $2^{560} \equiv 1 \pmod{561}$, and $2^{644} \equiv 1 \pmod{645}$.   ◀

If there are relatively few pseudoprimes to the base b, then checking to see whether the congruence $b^n \equiv b \pmod{n}$ holds is a useful test; only a small fraction of composite numbers pass this test. In fact, there are far fewer pseudoprimes to the base b not exceeding a specified bound than prime numbers not exceeding that bound. In particular, there are 455,052,511 primes, but only 14,884 pseudoprimes to the base 2, less than 10^{10}. Although pseudoprimes to any given base are rare, there are, nevertheless, infinitely many pseudoprimes to any given base. We will prove this for the base 2. The following lemma is useful in the proof.

Lemma 6.1. If d and n are positive integers such that d divides n, then $2^d - 1$ divides $2^n - 1$.

An Historical Inaccuracy

Apparently, the story that the ancient Chinese believed that n is prime if $2^n \equiv 2 \pmod{n}$ is due to a mistaken translation and an error by a nineteenth-century Chinese mathematician. In 1897, J. H. Jeans reported that this statement dates "from the time of Confucius," which seems to be the result of an erroneous translation from the book *The Nine Chapters of Mathematical Art.* In 1869, Alexander Wade published an article, "A Chinese theorem," in the journal *Notes and Queries on China,* crediting the mathematician Li Shan-Lan (1811–1882) for this "theorem." Li learned that this result was false, but the error was perpetuated by later authors. These historical details come from a letter from Chinese mathematician Man-Keung Siu to Paulo Ribenboim (see [Ri96] for more information).

Proof. Given that $d \mid n$, there is a positive integer t with $dt = n$. By setting $x = 2^d$ in the identity $x^t - 1 = (x - 1)(x^{t-1} + x^{t-2} + \cdots + 1)$, we find that $2^n - 1 = (2^d - 1)(2^{d(t-1)} + 2^{d(t-2)} + \cdots + 2^d + 1)$. Consequently, we have $(2^d - 1) \mid (2^n - 1)$. ∎

We can now prove that there are infinitely many pseudoprimes to the base 2.

Theorem 6.6. There are infinitely many pseudoprimes to the base 2.

Proof. We will show that if n is an odd pseudoprime to the base 2, then $m = 2^n - 1$ is also an odd pseudoprime to the base 2. Because we have at least one odd pseudoprime to the base 2, namely $n_0 = 341$, we will be able to construct infinitely many odd pseudoprimes to the base 2 by taking $n_0 = 341$ and $n_{k+1} = 2^{n_k} - 1$ for $k = 0, 1, 2, 3, \ldots$. These integers are all different, because $n_0 < n_1 < n_2 < \cdots < n_k < n_{k+1} < \cdots$.

To continue the proof, let n be an odd pseudoprime to the base 2, so that n is composite and $2^{n-1} \equiv 1 \pmod{n}$. Because n is composite, we have $n = dt$, with $1 < d < n$ and $1 < t < n$. We will show that $m = 2^n - 1$ is also pseudoprime, by first showing that it is composite, and then by showing that $2^{m-1} \equiv 1 \pmod{m}$.

To see that m is composite, we use Lemma 6.1 to note that $(2^d - 1) \mid (2^n - 1) = m$. To show that $2^{m-1} \equiv 1 \pmod{m}$, note that because $2^n \equiv 2 \pmod{n}$, there is an integer k with $2^n - 2 = kn$. Hence, $2^{m-1} = 2^{2^n - 2} = 2^{kn}$. By Lemma 6.1, it follows that $m = (2^n - 1) \mid (2^{kn} - 1) = 2^{m-1} - 1$. Hence, $2^{m-1} - 1 \equiv 0 \pmod{m}$, so that $2^{m-1} \equiv 1 \pmod{m}$. We conclude that m is also a pseudoprime to the base 2. ∎

If we want to know whether an integer n is prime, and we find that $2^{n-1} \equiv 1 \pmod{n}$, we know that n is either prime or a pseudoprime to the base 2. One follow-up approach is to test n with other bases. That is, we check to see whether $b^{n-1} \equiv 1 \pmod{n}$ for various positive integers b. If we find any values of b with $(b, n) = 1$ and $b^{n-1} \not\equiv 1 \pmod{n}$, then we know that n is composite.

Example 6.11. We have seen that 341 is a pseudoprime to the base 2. Because

$$7^3 = 343 \equiv 2 \pmod{341}$$

and

$$2^{10} = 1024 \equiv 1 \pmod{341},$$

we have

$$7^{340} = (7^3)^{113}7 \equiv 2^{113}7 = (2^{10})^{11} \cdot 2^3 \cdot 7$$
$$\equiv 8 \cdot 7 \equiv 56 \not\equiv 1 \pmod{341}.$$

Hence, by the contrapositive of Fermat's little theorem, we see that 341 is composite, because $7^{340} \not\equiv 1 \pmod{341}$. ◀

Carmichael Numbers

Unfortunately, there are composite integers n that cannot be shown to be composite using the above approach, because there are integers that are pseudoprimes to every base, that is, there are composite integers n such that $b^{n-1} \equiv 1 \pmod{n}$, for all b with $(b, n) = 1$. This leads to the following definition.

Definition.   A composite integer n that satisfies $b^{n-1} \equiv 1 \pmod{n}$ for all positive integers b with $(b, n) = 1$ is called a *Carmichael number* (after *Robert Carmichael*, who studied them in the early part of the twentieth century) or an *absolute pseudoprime*.

Example 6.12.   The integer $561 = 3 \cdot 11 \cdot 17$ is a Carmichael number. To see this, note that if $(b, 561) = 1$, then $(b, 3) = (b, 11) = (b, 17) = 1$. Hence, from Fermat's little theorem, we have $b^2 \equiv 1 \pmod 3, b^{10} \equiv 1 \pmod{11}$, and $b^{16} \equiv 1 \pmod{17}$. Consequently, $b^{560} = (b^2)^{280} \equiv 1 \pmod 3$, $b^{560} = (b^{10})^{56} \equiv 1 \pmod{11}$, and $b^{560} = (b^{16})^{35} \equiv 1 \pmod{17}$. Therefore, by Corollary 4.8.1, $b^{560} \equiv 1 \pmod{561}$ for all b with $(b, n) = 1$.
◀

In 1912, Carmichael conjectured that there are infinitely many Carmichael numbers. It took 80 years to resolve this conjecture. In 1992, Alford, Granville, and Pomerance showed that Carmichael was correct.[1] Because of the complicated, nonelementary nature of their proof, we will not describe it here. However, we will prove one of the key ingredients, a theorem that can be used to find Carmichael numbers.

Theorem 6.7.   If $n = q_1 q_2 \ldots q_k$, where the q_j are distinct primes that satisfy $(q_j - 1) \mid (n - 1)$ for all j and $k > 2$, then n is a Carmichael number.

Proof.   Let b be a positive integer with $(b, n) = 1$. Then $(b, q_j) = 1$ for $j = 1, 2, \ldots, k$, and hence, by Fermat's little theorem, $b^{q_j - 1} \equiv 1 \pmod{q_j}$ for $j = 1, 2, \ldots, k$. Because $(q_j - 1) \mid (n - 1)$ for each integer $j = 1, 2, \ldots, k$, there are integers t_j with $t_j(q_j - 1) = n - 1$. Hence, for each j, we know that $b^{n-1} = b^{(q_j-1)t_j} \equiv 1 \pmod{q_j}$. Therefore, by Corollary 4.8.1, we see that $b^{n-1} \equiv 1 \pmod{n}$, and we conclude that n is a Carmichael number. ■

---

[1] In particular, they showed that $C(x)$, the number of Carmichael numbers not exceeding x, satisfies the inequality $C(x) > x^{2/7}$ for sufficiently large numbers x.

ROBERT DANIEL CARMICHAEL (1879–1967) was born in Goodwater, Alabama. He received his B.A. from Lineville College in 1898 and his Ph.D. in 1911 from Princeton University. Carmichael taught at Indiana University from 1911 to 1915, and at the University of Illinois from 1915 until 1947. His thesis, written under the direction of G. D. Birkhoff, was considered the first significant American contribution to differential equations. Carmichael worked in a wide range of areas, including real analysis, differential equations, mathematical physics, group theory, and number theory.

Example 6.13. Theorem 6.7 shows that $6601 = 7 \cdot 23 \cdot 41$ is a Carmichael number, because 7, 23, and 41 are all prime, $6 = (7 - 1) \mid 6600$, $22 = (23 - 1) \mid 6600$, and $40 = (41 - 1) \mid 6600$. ◀

The converse of Theorem 6.7 is also true, that is, all Carmichael numbers are of the form $q_1 q_2 \cdots q_k$, where the q_j are distinct primes and $(q_j - 1) \mid (n - 1)$ for all j. We will prove this fact in Chapter 9.

By the way, we can show that although there are only 43 Carmichael numbers not exceeding 10^6, there are 105,212 of them not exceeding 10^{15}.

Miller's Test

Once the congruence $b^{n-1} \equiv 1 \pmod{n}$, where n is an odd integer, has been verified, another possible approach is to consider the least positive residue of $b^{(n-1)/2}$ modulo n. We note that if $x = b^{(n-1)/2}$, then $x^2 = b^{n-1} \equiv 1 \pmod{n}$. If n is prime, by Theorem 4.11 we know that either $x \equiv 1$ or $x \equiv -1 \pmod{n}$. Consequently, once we have found that $b^{n-1} \equiv 1 \pmod{n}$, we can check to see whether $b^{(n-1)/2} \equiv \pm 1 \pmod{n}$. If this congruence does not hold, then we know that n is composite.

Example 6.14. Let $b = 5$ and let $n = 561$, the smallest Carmichael number. We find that $5^{(561-1)/2} = 5^{280} \equiv 67 \pmod{561}$. Hence, 561 is composite. ◀

To continue developing primality tests, we need the following definitions.

Definition. Let n be a positive integer with $n > 2$ and $n - 1 = 2^s t$, where s is a nonnegative integer and t is an odd positive integer. We say that n passes *Miller's test for the base b* if either $b^t \equiv 1 \pmod{n}$ or $b^{2^j t} \equiv -1 \pmod{n}$ for some j with $0 \le j \le s - 1$.

The following example shows that 2047 passes Miller's test for the base 2.

Example 6.15. Let $n = 2047 = 23 \cdot 89$. Then $2^{2046} = (2^{11})^{186} = (2048)^{186} \equiv 1 \pmod{2047}$, so that 2047 is a pseudoprime to the base 2. Because $2^{2046/2} = 2^{1023} = (2^{11})^{93} = (2048)^{93} \equiv 1 \pmod{2047}$, 2047 passes Miller's test for the base 2. ◀

We now show that if n is prime, then n passes Miller's test for all bases b with $n \nmid b$.

Theorem 6.8. If n is prime and b is a positive integer with $n \nmid b$, then n passes Miller's test for the base b.

Proof. Let $n - 1 = 2^s t$, where s is a nonnegative integer and t is an odd positive integer. Let $x_k = b^{(n-1)/2^k} = b^{2^{s-k}t}$, for $k = 0, 1, 2, \ldots, s$. Because n is prime, Fermat's little theorem tells us that $x_0 = b^{n-1} \equiv 1 \pmod{n}$. By Theorem 4.11, because $x_1^2 = (b^{(n-1)/2})^2 = x_0 \equiv 1 \pmod{n}$, either $x_1 \equiv -1 \pmod{n}$ or $x_1 \equiv 1 \pmod{n}$. If $x_1 \equiv 1 \pmod{n}$, because $x_2^2 = x_1 \equiv 1 \pmod{n}$, either $x_2 \equiv -1 \pmod{n}$ or $x_2 \equiv 1 \pmod{n}$. In general, if we have found that $x_0 \equiv x_1 \equiv x_2 \equiv \cdots \equiv x_k \equiv 1 \pmod{n}$, with $k < s$, then, because $x_{k+1}^2 = x_k \equiv 1 \pmod{n}$, we know that either $x_{k+1} \equiv -1 \pmod{n}$ or $x_{k+1} \equiv 1 \pmod{n}$.

Continuing this procedure for $k = 1, 2, \ldots, s$, we find that either $x_s \equiv 1 \pmod{n}$, or $x_k \equiv -1 \pmod{n}$ for some integer k, with $0 \le k \le s$. Hence, n passes Miller's test for the base b. ∎

If the positive integer n passes Miller's test for the base b, then either $b^t \equiv 1 \pmod{n}$ or $b^{2^j t} \equiv -1 \pmod{n}$ for some j with $0 \le j \le s - 1$, where $n - 1 = 2^s t$ and t is odd.

In either case, we have $b^{n-1} \equiv 1 \pmod{n}$, because $b^{n-1} = (b^{2^j t})^{2^{s-j}}$ for $j = 0, 1, 2, \ldots, s$, so that a composite integer n that passes Miller's test for the base b is automatically a pseudoprime to the base b. With this observation, we are led to the following definition.

Definition.   If n is composite and passes Miller's test for the base b, then we say n is a *strong pseudoprime to the base b*.

Example 6.16.   By Example 6.15, we see that 2047 is a strong pseudoprime to the base 2. ◀

Although strong pseudoprimes are exceedingly rare, there are still infinitely many of them. We demonstrate this for the base 2 with the following theorem.

Theorem 6.9.   There are infinitely many strong pseudoprimes to the base 2.

Proof.   We shall show that if n is a pseudoprime to the base 2, then $N = 2^n - 1$ is a *strong* pseudoprime to the base 2.

Let n be an odd integer that is a pseudoprime to the base 2. Hence, n is composite, and $2^{n-1} \equiv 1 \pmod{n}$. From this congruence, we see that $2^{n-1} - 1 = nk$ for some integer k; furthermore, k must be odd. We have

$$N - 1 = 2^n - 2 = 2(2^{n-1} - 1) = 2^1 nk;$$

this is the factorization of $N - 1$ into an odd integer and a power of 2.

We now note that

$$2^{(N-1)/2} = 2^{nk} = (2^n)^k \equiv 1 \pmod{N},$$

because $2^n = (2^n - 1) + 1 = N + 1 \equiv 1 \pmod{N}$. This demonstrates that N passes Miller's test.

In the proof of Lemma 6.1, we showed that if n is composite, then $N = 2^n - 1$ also is composite. Hence, N passes Miller's test and is composite, so that N is a strong pseudoprime to the base 2. Because every pseudoprime n to the base 2 yields a strong pseudoprime $2^n - 1$ to the base 2, and because there are infinitely many pseudoprimes to the base 2, we conclude that there are infinitely many strong pseudoprimes to the base 2. ∎

The following observations are useful in combination with Miller's test for checking the primality of relatively small integers. The smallest odd strong pseudoprime to the base 2 is 2047, so that if $n < 2047$, n is odd, and n passes Miller's test to the base 2, then n

is prime. Likewise, 1,373,653 is the smallest odd strong pseudoprime to both the bases 2 and 3, giving us a primality test for integers less than 1,373,653. The smallest odd strong pseudoprime to the bases 2, 3, and 5 is 25,326,001, and the smallest odd strong pseudoprime to all the bases 2, 3, 5, and 7 is 3,215,031,751. Furthermore, there are no other strong pseudoprimes to all these bases that are less than $25 \cdot 10^9$. (The reader should verify these statements.) This leads us to a primality test for integers less than $25 \cdot 10^9$. An odd integer n is prime if $n < 25 \cdot 10^9$, n passes Miller's test for the bases 2, 3, 5, and 7, and $n \neq 3{,}215{,}031{,}751$.

Computations show that there are only 101 integers less than 10^{12} that are strong pseudoprimes to the bases 2, 3, and 5 simultaneously. Only 9 of these are also strong pseudoprimes to the base 7, and none of these is a strong pseudoprime to the base 11. The smallest strong pseudoprime to the bases 2, 3, 5, 7, and 11 simultaneously is 2,152,302,898,747. Therefore, if an odd integer n is prime and $n < 2{,}152{,}302{,}898{,}747$, then n is prime if it passes Miller's test for the bases 2, 3, 5, 7, and 11. If we want to test even bigger integers for primality in this way, we can use the observation that no positive integer less than 341,550,071,728,321 is a strong pseudoprime to the bases 2, 3, 5, 7, 11, 13, and 17. A positive odd integer not exceeding this number is prime if it passes Miller's test for the seven primes, 2, 3, 5, 7, 11, 13, and 17.

There is no analogue to a Carmichael number for strong pseudoprimes. This is a consequence of the following theorem.

Theorem 6.10.   If n is an odd composite positive integer, then n passes Miller's test for at most $(n-1)/4$ bases b with $1 \le b \le n-1$.

We prove Theorem 6.10 in Chapter 9. Note that Theorem 6.10 tells us that if n passes Miller's tests for more than $(n-1)/4$ bases less than n, then n must be prime. However, this is a rather lengthy way to show that a positive integer n is prime, worse than performing trial divisions. Miller's test does give an interesting and quick way of showing that an integer n is "probably prime." To see this, take at random an integer b with $1 \le b \le n-1$ (we will see how to make this "random" choice in Chapter 10). From Theorem 6.10, we see that if n is composite, the probability that n passes Miller's test for the base b is less than $1/4$. If we pick k different bases less than n and perform Miller's tests for each of these bases, we are led to the following result.

Theorem 6.11. *Rabin's Probabilistic Primality Test.*   Let n be a positive integer. Pick k different positive integers less than n and perform Miller's test on n for each of these bases. If n is composite, the probability that n passes all k tests is less than $(1/4)^k$.

Let n be a composite positive integer. Using Rabin's probabilistic primality test, if we pick 100 different integers at random between 1 and n and perform Miller's test for each of these 100 bases, then the probability that n passes all the tests is less than 10^{-60}, an extremely small number. In fact, it may be more likely that a computer error was made than that a composite integer passes all 100 tests. Using Rabin's primality test does not definitely prove that an integer n that passes some large number of tests is prime, but does give extremely strong, indeed almost overwhelming, evidence that the integer is prime.

There is a famous conjecture in analytic number theory called the *generalized Riemann hypothesis,* which is a statement about the famous Riemann zeta function, named after the German mathematician *Georg Friedrich Bernhard Riemann*, which is discussed in Section 3.2. The following conjecture is a consequence of this hypothesis.

Conjecture 6.1.   For every composite positive integer n, there is a base b, with $b < 2(\log_2 n)^2$, such that n fails Miller's test for the base b. ∎

If this conjecture is true, as many number theorists believe, the following result provides a rapid primality test.

Theorem 6.12.   If the generalized Riemann hypothesis is valid, then there is an algorithm to determine whether a positive integer n is prime using $O((\log_2 n)^5)$ bit operations.

Proof.   Let b be a positive integer less than n. To perform Miller's test for the base b on n takes $O((\log_2 n)^3)$ bit operations, because this test requires that we perform no more than $\log_2 n$ modular exponentiations, each using $O((\log_2 b)^2)$ bit operations. Assume that the generalized Riemann hypothesis is true. If n is composite, then by Conjecture 6.1, there is a base b with $1 < b < 2(\log_2 n)^2$ such that n fails Miller's test for b. To discover this b requires less than $O((\log_2 n)^3) \cdot O((\log_2 n)^2) = O((\log_2 n)^5)$ bit operations. Hence, using $O((\log_2 n)^5)$ bit operations, we can determine whether n is composite or prime. ∎

The important point about Rabin's probabilistic primality test and Theorem 6.12 is that both results indicate that it is possible to check an integer n for primality using only $O((\log_2 n)^k)$ bit operations, where k is a positive integer. (Also, the recent result of Agrawal, Kayal, and Saxena [AgKaSa02] shows that there is a deterministic test using $O((\log_2 n)^k)$ bit operations.) This contrasts strongly with the problem of factoring. The best algorithm known for factoring an integer requires a number of bit operations

GEORG FRIEDRICH BERNHARD RIEMANN (1826–1866), the son of a minister, was born in Breselenz, Germany. His elementary education came from his father. After completing his secondary education, he entered Göttingen University to study theology. However, he also attended lectures on mathematics. After receiving the approval of his father to concentrate on mathematics, Riemann transfered to Berlin University where he studied under several prominent mathematicians, including Dirichlet and Jacobi. He subsequently returned to Göttingen where he obtained his Ph.D.

Riemann was one of the most imaginative and creative mathematicians of all time. He made fundamental contributions to geometry, mathematical physics, and analysis. He wrote only one paper on number theory, which was eight pages long, but this paper has had tremendous impact. Riemann died of tuberculosis at the early age of 39.

exponential in the square root of the logarithm of the number of bits in the integer being factored, whereas primality testing seems to require only a number of bit operations less than a polynomial in the number of bits of the integer tested. We capitalize on this difference by presenting a recently invented cipher system in Chapter 8.

6.2 Exercises

1. Show that 91 is a pseudoprime to the base 3.

2. Show that 45 is a pseudoprime to the bases 17 and 19.

3. Show that the even integer $n = 161{,}038 = 2 \cdot 73 \cdot 1103$ satisfies the congruence $2^n \equiv 2 \pmod{n}$. The integer 161,038 is the smallest even pseudoprime to the base 2.

4. Show that every odd composite integer is a pseudoprime to both the base 1 and the base -1.

5. Show that if n is an odd composite integer and n is a pseudoprime to the base a, then n is a pseudoprime to the base $n - a$.

* 6. Show that if $n = (a^{2p} - 1)/(a^2 - 1)$, where a is an integer, $a > 1$, and p is an odd prime not dividing $a(a^2 - 1)$, then n is a pseudoprime to the base a. Conclude that there are infinitely many pseudoprimes to any base a. (*Hint*: To establish that $a^{n-1} \equiv 1 \pmod{n}$, show that $2p \mid (n - 1)$, and demonstrate that $a^{2p} \equiv 1 \pmod{n}$.)

7. Show that every composite Fermat number $F_m = 2^{2^m} + 1$ is a pseudoprime to the base 2.

8. Show that if p is prime and $2^p - 1$ is composite, then $2^p - 1$ is a pseudoprime to the base 2.

9. Show that if n is a pseudoprime to the bases a and b, then n is also a pseudoprime to the base ab.

10. Suppose that a and n are relatively prime positive integers. Show that if n is a pseudoprime to the base a, then n is a pseudoprime to the base $\bar{a}$, where $\bar{a}$ is an inverse of a modulo n.

11. a) Show that if n is a pseudoprime to the base a, but not a pseudoprime to the base b, where $(a, n) = (b, n) = 1$, then n is not a pseudoprime to the base ab.

    b) Show that if there is an integer b with $(b, n) = 1$ such that n is not a pseudoprime to the base b, then n is a pseudoprime to less than or equal to $\phi(n)$ different bases a with $1 \le a < n$, where $\phi(n)$ is the number of positive integers not exceeding n that are relatively prime to n. (*Hint*: Show that the sets $a_1, a_2, \ldots, a_r$ and $ba_1, ba_2, \ldots, ba_r$ have no common elements, where $a_1, a_2, \ldots, a_r$ are the bases less than n to which n is a pseudoprime.)

12. Show that 25 is a strong pseudoprime to the base 7.

13. Show that 1387 is a pseudoprime, but not a strong pseudoprime, to the base 2.

14. Show that 1,373,653 is a strong pseudoprime to both bases 2 and 3.

15. Show that 25,326,001 is a strong pseudoprime to bases 2, 3, and 5.

16. Show that the following integers are Carmichael numbers.

    a) $2821 = 7 \cdot 13 \cdot 31$

    b) $10,585 = 5 \cdot 29 \cdot 73$

    c) $29,341 = 13 \cdot 37 \cdot 61$

    d) $314,821 = 13 \cdot 61 \cdot 397$

    e) $278,545 = 5 \cdot 17 \cdot 29 \cdot 113$

    f) $172,081 = 7 \cdot 13 \cdot 31 \cdot 61$

    g) $564,651,361 = 43 \cdot 3361 \cdot 3907$

17. Find a Carmichael number of the form $7 \cdot 23 \cdot q$, where q is an odd prime other than $q = 41$, or show that there are no others.

18. a) Show that every integer of the form $(6m + 1)(12m + 1)(18m + 1)$, where m is a positive integer such that $6m + 1$, $12m + 1$, and $18m + 1$ are all primes, is a Carmichael number.

    b) Conclude from part (a) that $1729 = 7 \cdot 13 \cdot 19$; $294,409 = 37 \cdot 73 \cdot 109$; $56,052,361 = 211 \cdot 421 \cdot 631$; $118,901,521 = 271 \cdot 541 \cdot 811$; and $172,947,529 = 307 \cdot 613 \cdot 919$ are Carmichael numbers.

19. The smallest Carmichael number with six prime factors is $5 \cdot 19 \cdot 23 \cdot 29 \cdot 37 \cdot 137 = 321,197,185$. Verify that this number is a Carmichael number.

∗ **20.** Show that if n is a Carmichael number, then n is square-free.

21. Show that if n is a positive integer with $n \equiv 3 \pmod{4}$, then Miller's test takes $O((\log_2 n)^3)$ bit operations.

6.2 Computational and Programming Exercises

Computations and Explorations

Using a computation program such as Maple or *Mathematica,* or programs you have written, carry out the following computations and explorations.

    1. Determine for which positive integers n, $n \leq 100$, the integer $n \cdot 2^n - 1$ is prime.

    2. Find as many Carmichael numbers of the form $(6m + 1)(12m + 1)(18m + 1)$, where $6m + 1$, $12m + 1$, and $18m + 1$ are all prime, as you can.

    3. Find as many even pseudoprimes to the base 2 that are the product of three primes as you can. Do you think that there are infinitely many?

    4. The integers of the form $n \cdot 2^n + 1$, where n is a positive integer greater than 1, are called *Cullen numbers.* Can you find a prime Cullen number?

Programming Projects

Write programs using Maple, *Mathematica,* or a language of your choice to do the following.

    1. Given a positive integer n, determine whether n satisfies the congruence $b^{n-1} \equiv 1 \pmod{n}$, where b is a positive integer less than n; if it does, then n is either a prime or a pseudoprime to the base b.

2. Given a positive integer n, determine whether n passes Miller's test to the base b; if it does, then n is either prime or a strong pseudoprime to the base b.

3. Perform a primality test for integers less than $25 \cdot 10^9$ based on Miller's test for the bases 2, 3, 5, and 7. (Use the remarks that follow Theorem 6.9.)

4. Perform a primality test for integers less than 2,152,302,898,747 based on Miller's test for the bases 2, 3, 5, 7, and 11. (Use the remarks that follow Theorem 6.9.)

5. Perform a primality test for integers less than 341,550,071,728,321 based on Miller's test for the bases 2, 3, 5, 7, 11, 13, and 17. (Use the remarks that follow Theorem 6.9.)

6. Given an odd positive integer n, determine whether n passes Rabin's probabilistic primality test.

7. Given a positive integer n, find all Carmichael numbers less than a given integer n.

6.3 Euler's Theorem

Fermat's little theorem tells us how to work with certain congruences involving exponents when the modulus is a prime. How do we work with the corresponding congruences modulo a composite integer?

For this purpose, we would like to establish a congruence analogous to that provided by Fermat's little theorem for composite integers. As mentioned in Section 6.1, the great Swiss mathematician *Leonhard Euler* published a proof of Fermat's little theorem in 1736. In 1760, Euler managed to find a natural generalization of the congruence in Fermat's little theorem that holds for composite integers. Before introducing this result, we need to define a special counting function (introduced by Euler) used in the theorem.

Definition.   Let n be a positive integer. The *Euler phi-function* $\phi(n)$ is defined to be the number of positive integers not exceeding n that are relatively prime to n.

In Table 6.1, we display the values of $\phi(n)$ for $1 \leq n \leq 12$. The values of $\phi(n)$ for $1 \leq n \leq 100$ are given in Table 2 of Appendix E.

n	1	2	3	4	5	6	7	8	9	10	11	12
$\phi(n)$	1	1	2	2	4	2	6	4	6	4	10	4

Table 6.1 *The values of Euler's phi-function for $1 \leq n \leq 12$.*

In Chapter 7, we study the Euler phi-function further. In this section, we use the phi-function to give an analogue of Fermat's little theorem for composite moduli. To do this, we need to lay some groundwork.

Definition.   A *reduced residue system modulo n* is a set of $\phi(n)$ integers such that each element of the set is relatively prime to n, and no two different elements of the set are congruent modulo n.

Example 6.17. The set $1, 3, 5, 7$ is a reduced residue system modulo 8. The set $-3, -1, 1, 3$ is also such a set. ◄

We will need the following theorem about reduced residue systems.

Theorem 6.13. If $r_1, r_2, \ldots, r_{\phi(n)}$ is a reduced residue system modulo n, and if a is a positive integer with $(a, n) = 1$, then the set $ar_1, ar_2, \ldots, ar_{\phi(n)}$ is also a reduced residue system modulo n.

Proof. To show that each integer ar_j is relatively prime to n, we assume that $(ar_j, n) > 1$. Then, there is a prime divisor p of (ar_j, n). Hence, either $p \mid a$ or $p \mid r_j$. Thus, we have either $p \mid a$ and $p \mid n$, or $p \mid r_j$ and $p \mid n$. However, we cannot have both $p \mid r_j$ and $p \mid n$, because r_j is a member of a reduced residue system modulo n, and both $p \mid a$ and $p \mid n$ cannot hold because $(a, n) = 1$. Hence, we can conclude that ar_j and n are relatively prime for $j = 1, 2, \ldots, \phi(n)$.

To demonstrate that no two ar_j are congruent modulo n, we assume that $ar_j \equiv ar_k \pmod{n}$, where j and k are distinct positive integers with $1 \leq j \leq \phi(n)$ and $1 \leq k \leq \phi(n)$. Because $(a, n) = 1$, by Corollary 4.4.1 we see that $r_j \equiv r_k \pmod{n}$. This is a contradiction, because r_j and r_k come from the original set of reduced residues modulo n, so that $r_j \not\equiv r_k \pmod{n}$. ∎

We illustrate the use of Theorem 6.13 by the following example.

LEONHARD EULER (1707–1783) was the son of a minister from the vicinity of Basel, Switzerland, who, besides theology, had also studied mathematics. At 13, Euler entered the University of Basel with the aim of pursuing a career in theology, as his father wished. At the university, Euler was tutored in mathematics by Johann Bernoulli, of the famous Bernoulli family of mathematicians, and became friends with Johann's sons Nicklaus and Daniel. His interest in mathematics led him to abandon his plans to follow in his father's footsteps. Euler obtained his master's degree in philosophy at the age of 16. In 1727, Peter the Great invited Euler to join the Imperial Academy in St. Petersburg, at the insistence of Nicklaus and Daniel Bernoulli, who had entered the academy in 1725 when it was founded. Euler spent the years 1727–1741 and 1766–1783 at the Imperial Academy. He spent the interval 1741–1766 at the Royal Academy of Berlin. Euler was incredibly prolific; he wrote more than 700 books and papers, and he left so much unpublished work that the Imperial Academy did not finish publication of Euler's work for 47 years after his death. During his life, his papers accumulated so rapidly that he kept a pile of papers to be published for the academy. They published the top papers in the pile first, so that later results were published before results they superseded or depended on. Euler was blind for the last 17 years of his life, but had a fantastic memory, so that his blindness did not deter his mathematical output. He also had 13 children, and was able to continue his research while a child or two bounced on his knees. The publication of the collected works and letters of Euler, the *Opera Omnia*, by the Swiss Academy of Science will require more than 85 large volumes, of which 76 have aleady been published (as of late 1999).

Example 6.18. The set $1, 3, 5, 7$ is a reduced residue system modulo 8. Because $(3, 8) = 1$, from Theorem 6.13, the set $3 \cdot 1 = 3, 3 \cdot 3 = 9, 3 \cdot 5 = 15, 3 \cdot 7 = 21$ is also a reduced residue system modulo 8. ◄

We now state Euler's theorem.

Theorem 6.14. *Euler's Theorem.* If m is a positive integer and a is an integer with $(a, m) = 1$, then $a^{\phi(m)} \equiv 1 \pmod{m}$.

Before we prove Euler's theorem, we illustrate the idea behind the proof with an example.

Example 6.19. We know that both the sets $1, 3, 5, 7$ and $3 \cdot 1, 3 \cdot 3, 3 \cdot 5, 3 \cdot 7$ are reduced residue systems modulo 8. Hence, they have the same least positive residues modulo 8. Therefore,

$$(3 \cdot 1) \cdot (3 \cdot 3) \cdot (3 \cdot 5) \cdot (3 \cdot 7) \equiv 1 \cdot 3 \cdot 5 \cdot 7 \pmod{8},$$

and

$$3^4 \cdot 1 \cdot 3 \cdot 5 \cdot 7 \equiv 1 \cdot 3 \cdot 5 \cdot 7 \pmod{8}.$$

Because $(1 \cdot 3 \cdot 5 \cdot 7, 8) = 1$, we conclude that

$$3^4 = 3^{\phi(8)} \equiv 1 \pmod{8}. \qquad ◄$$

We now use the ideas illustrated by this example to prove Euler's theorem.

Proof. Let $r_1, r_2, \ldots, r_{\phi(m)}$ denote the reduced residue system made up of the positive integers not exceeding m that are relatively prime to m. By Theorem 6.13, because $(a, m) = 1$, the set $ar_1, ar_2, \ldots, ar_{\phi(m)}$ is also a reduced residue system modulo m. Hence, the least positive residues of $ar_1, ar_2, \ldots, ar_{\phi(m)}$ must be the integers $r_1, r_2, \ldots, r_{\phi(m)}$, in some order. Consequently, if we multiply together all terms in each of these reduced residue systems, we obtain

$$ar_1 ar_2 \cdots ar_{\phi(m)} \equiv r_1 r_2 \cdots r_{\phi(m)} \pmod{m}.$$

Thus,

$$a^{\phi(m)} r_1 r_2 \cdots r_{\phi(m)} \equiv r_1 r_2 \cdots r_{\phi(m)} \pmod{m}.$$

Because $(r_1 r_2 \cdots r_{\phi(m)}, m) = 1$, from Corollary 4.4.1, we can conclude that $a^{\phi(m)} \equiv 1 \pmod{m}$. ∎

We can use Euler's theorem to find inverses modulo m. If a and m are relatively prime, we know that

$$a \cdot a^{\phi(m)-1} = a^{\phi(m)} \equiv 1 \pmod{m}.$$

Hence, $a^{\phi(m)-1}$ is an inverse of a modulo m.

Example 6.20. We know that $2^{\phi(9)-1} = 2^{6-1} = 2^5 = 32 \equiv 5 \pmod 9$ is an inverse of 2 modulo 9. ◀

We can solve linear congruences using this observation. To solve $ax \equiv b \pmod m$, where $(a, m) = 1$, we multiply both sides of this congruence by $a^{\phi(m)-1}$ to obtain

$$a^{\phi(m)-1}ax \equiv a^{\phi(m)-1}b \pmod m.$$

Therefore, the solutions are those integers x such that $x \equiv a^{\phi(m)-1}b \pmod m$.

Example 6.21. The solutions of $3x \equiv 7 \pmod{10}$ are given by $x \equiv 3^{\phi(10)-1} \cdot 7 \equiv 3^3 \cdot 7 \equiv 9 \pmod{10}$, because $\phi(10) = 4$. ◀

6.3 Exercises

1. Find a reduced residue system modulo each of the following integers.

   a) 6   b) 9   c) 10   d) 14   e) 16   f) 17

2. Find a reduced residue system modulo 2^m, where m is a positive integer.

3. Show that if $c_1, c_2, \ldots, c_{\phi(m)}$ is a reduced residue system modulo m, where m is a positive integer with $m \neq 2$, then $c_1 + c_2 + \cdots + c_{\phi(m)} \equiv 0 \pmod m$.

4. Show that if a and m are positive integers with $(a, m) = (a - 1, m) = 1$, then $1 + a + a^2 + \ldots + a^{\phi(m)-1} \equiv 0 \pmod m$.

5. Find the last digit of the decimal expansion of 3^{1000}.

6. Find the last digit of the decimal expansion of $7^{999,999}$.

7. Use Euler's theorem to find the least positive residue of $3^{100,000}$ modulo 35.

8. Show that if a is an integer such that a is not divisible by 3 or such that a is divisible by 9, then $a^7 \equiv a \pmod{63}$.

9. Show that if a is an integer relatively prime to 32,760, then $a^{12} \equiv 1 \pmod{32,760}$.

10. Show that $a^{\phi(b)} + b^{\phi(a)} \equiv 1 \pmod{ab}$, if a and b are relatively prime positive integers.

11. Solve each of the following linear congruences using Euler's theorem.

    a) $5x \equiv 3 \pmod{14}$ b) $4x \equiv 7 \pmod{15}$ c) $3x \equiv 5 \pmod{16}$

12. Show that the solutions to the simultaneous system of congruences

$$x \equiv a_1 \pmod{m_1}$$
$$x \equiv a_2 \pmod{m_2}$$
$$\vdots$$
$$x \equiv a_r \pmod{m_r},$$

where the m_j are pairwise relatively prime, are given by

$$x \equiv a_1 M_1^{\phi(m_1)} + a_2 M_2^{\phi(m_2)} + \cdots + a_r M_r^{\phi(m_r)} \pmod M,$$

where $M = m_1 m_2 \cdots m_r$ and $M_j = M/m_j$ for $j = 1, 2, \ldots, r$.

13. Use Exercise 12 to solve each of the systems of congruences in Exercise 4 of Section 4.3.

14. Use Exercise 12 to solve the system of congruences in Exercise 5 of Section 4.3.

15. Use Euler's theorem to find the last digit in the decimal expansion of 7^{1000}.

16. Use Euler's theorem to find the last digit in the hexadecimal expansion of $5^{1,000,000}$.

17. Find $\phi(n)$ for the integers n with $13 \leq n \leq 20$.

18. Show that every positive integer relatively prime to 10 divides infinitely many repunits (see the preamble to Exercise 11 of Section 5.1). (*Hint:* Note that the n-digit repunit $111 \ldots 11 = (10^n - 1)/9$.)

19. Show that every positive integer relatively prime to b divides infinitely many base b repunits (see the preamble to Exercise 15 of Section 5.1).

* 20. Show that if m is a positive integer, $m > 1$, then $a^m \equiv a^{m-\phi(m)} \pmod{m}$ for all positive integers a.

6.3 Computational and Programming Exercises

Computations and Explorations

Using a computation program such as Maple or *Mathematica,* or programs you have written, carry out the following computations and explorations.

1. Find $\phi(n)$ for all integers n less than 1000. What conjectures can you make about the values of $\phi(n)$?

2. Let $\Phi(n) = \sum_{i=1}^{n} \phi(n)$. Investigate the value of $\Phi(n)/n^2$ for increasingly large values of n, such as $n = 100$, $n = 1000$, and $n = 10,000$. Can you make a conjecture about the limit of this ratio as n grows large without bound?

Programming Projects

Write programs using Maple, *Mathematica,* or a language of your choice to do the following.

1. Construct a reduced residue system modulo n for a given positive integer n.

2. Solve linear congruences using Euler's theorem.

3. Find the solutions of a simultaneous system of linear congruences using Euler's theorem and the Chinese remainder theorem (see Exercise 12).

7

Multiplicative Functions

Introduction

In this chapter, we will study a special class of functions on the set of integers called *multiplicative functions*. A multiplicative function has the property that its value at an integer is the product of its values at each of the prime powers in its prime-power factorization. We will show that some important functions are multiplicative, including the number of divisors function, the sum of divisors function, and the Euler phi-function. We will use the fact that each of these functions is multiplicative to obtain a closed formula for the value of these functions at a positive integer n based on the prime-power factorization of n.

Furthermore, we will study a special type of positive integer, called a *perfect number,* which is equal to the sum of its proper divisors. We will show that all even perfect numbers are generated by a special kind of prime, called a Mersenne prime, which is a prime that is 1 less than a power of 2. The quest for new Mersenne primes has been under way since ancient times, accelerated by the invention of powerful computers, and accelerated even more with the advent of the Internet.

We will also show how the summatory function of an arithmetic function can be used to obtain information about the function itself. The summatory function of a function f takes a value at n equal to the sum of the values of f at each of the positive divisors of n. The famous Möbius inversion formula shows how to obtain the values of f from the values of its summatory function.

7.1 The Euler Phi-Function

The Euler phi-function has the property that its value at an integer n is the product of the values of the Euler phi-function at the prime powers that occur in the factorization of n. Functions with this property are called multiplicative; such functions arise throughout number theory. In this section, we will show that the Euler phi-function is multiplicative. From this fact, we will derive a formula for its values based on prime factorizations. Later in this chapter we will study other multiplicative functions, including the number of divisors function and the sum of divisors function.

We first present some definitions.

Definition.   An *arithmetic function* is a function that is defined for all positive integers.

Throughout this chapter, we are interested in arithmetic functions that have a special property.

Definition.   An arithmetic function f is called *multiplicative* if $f(mn) = f(m)f(n)$ whenever m and n are relatively prime positive integers. It is called *completely multiplicative* if $f(mn) = f(m)f(n)$ for all positive integers m and n.

Example 7.1.   The function $f(n) = 1$ for all n is completely multiplicative, and hence also multiplicative, because $f(mn) = 1$, $f(m) = 1$, and $f(n) = 1$, so that $f(mn) = f(m)f(n)$. Similarly, the function $g(n) = n$ is completely multiplicative, and hence multiplicative, since $g(mn) = mn = g(m)g(n)$.   ◀

If f is a multiplicative function, then we can find a simple formula for $f(n)$ given the prime-power factorization of n. This result is particularly useful, because it shows us how to find $f(n)$ from the values of $f(p_i^{a_i})$ for $i = 1, 2, \ldots, s$, where $n = p_1^{a_1} p_2^{a_2} \ldots p_s^{a_s}$ is the prime-power factorization of n.

Theorem 7.1.   If f is a multiplicative function and if $n = p_1^{a_1} p_2^{a_2} \cdots p_s^{a_s}$ is the prime-power factorization of the positive integer n, then $f(n) = f(p_1^{a_1})f(p_2^{a_2}) \cdots f(p_s^{a_s})$.

Proof.   We will prove this theorem using mathematical induction on the number of different primes in the prime factorization of the integer n. If n has one prime in its prime-power factorization, then $n = p_1^{a_1}$ for some prime p_1, and it follows that the result is trivially true.

Suppose that the theorem is true for all integers with k different primes in their prime-power factorization. Now suppose that n has $k + 1$ different primes in its prime-power factorization, say $n = p_1^{a_1} p_2^{a_2} \cdots p_k^{a_k} p_{k+1}^{a_{k+1}}$. Because f is multiplicative and $(p_1^{a_1} p_2^{a_2} \cdots p_k^{a_k}, p_{k+1}^{a_{k+1}}) = 1$, we see that $f(n) = f(p_1^{a_1} p_2^{a_2} \cdots p_k^{a_k})f(p_{k+1}^{a_{k+1}})$. By the inductive hypothesis, we know that $f(p_1^{a_1} p_2^{a_2} p_3^{a_3} \cdots p_k^{a_k}) = f(p_1^{a_1})f(p_2^{a_2})f(p_3^{a_3}) \cdots f(p_k^{a_k})$. It follows that $f(n) = f(p_1^{a_1})f(p_2^{a_2}) \cdots f(p_k^{a_k})f(p_{k+1}^{a_{k+1}})$. This completes the inductive proof.   ∎

We now return to the Euler phi-function. We first consider its values at primes and then at prime powers.

Theorem 7.2.   If p is prime, then $\phi(p) = p - 1$. Conversely, if p is a positive integer with $\phi(p) = p - 1$, then p is prime.

Proof. If p is prime, then every positive integer less than p is relatively prime to p. Because there are $p - 1$ such integers, we have $\phi(p) = p - 1$. Conversely, if p is not prime, then $p = 1$ or p is composite. If $p = 1$, then $\phi(p) \neq p - 1$ because $\phi(1) = 1$. If p is composite, then p has a divisor d with $1 < d < p$, and, of course, p and d are not relatively prime. Because we know that at least one of the $p - 1$ integers $1, 2, \ldots, p - 1$, namely d, is not relatively prime to p, $\phi(p) \leq p - 2$. Hence, if $\phi(p) = p - 1$, then p must be prime. ∎

We now find the values of the phi-function at prime powers.

Theorem 7.3. Let p be a prime and a a positive integer. Then $\phi(p^a) = p^a - p^{a-1}$.

Proof. The positive integers less than p^a that are not relatively prime to p are those integers not exceeding p^a that are divisible by p. These are the integers kp, where $1 \leq k \leq p^{a-1}$. Since there are exactly p^{a-1} such integers, there are $p^a - p^{a-1}$ integers less than p^a that are relatively prime to p^a. Hence, $\phi(p^a) = p^a - p^{a-1}$. ∎

Example 7.2. Using Theorem 7.3, we find that $\phi(5^3) = 5^3 - 5^2 = 100$, $\phi(2^{10}) = 2^{10} - 2^9 = 512$, and $\phi(11^2) = 11^2 - 11 = 110$. ◀

To find a formula for $\phi(n)$, given the prime factorization of n, it suffices to show that ϕ is multiplicative. We illustrate the idea behind the proof with the following example.

Example 7.3. Let $m = 4$ and $n = 9$, so that $mn = 36$. We list the integers from 1 to 36 in a rectangular chart, as shown in Figure 7.1.

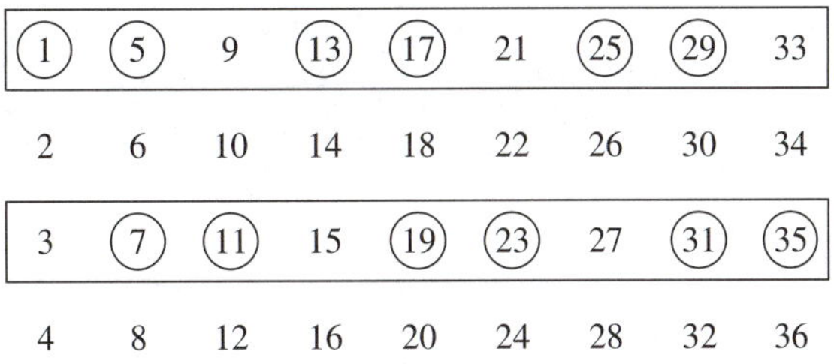

Figure 7.1 *Demonstrating that $\phi(36) = \phi(4)\phi(9)$.*

Neither the second nor the fourth row contains integers relatively prime to 36, since each element in these rows is not relatively prime to 4, and hence not relatively prime to 36. We enclose the other two rows; each element of these rows is relatively prime to 4. Within each of these rows, there are 6 integers relatively prime to 9. We circle these; they are the 12 integers in the list relatively prime to 36. Hence, $\phi(36) = 2 \cdot 6 = \phi(4)\phi(9)$. ◀

We now state and prove the theorem that shows that ϕ is multiplicative.

Theorem 7.4. Let m and n be relatively prime positive integers. Then $\phi(mn) = \phi(m)\phi(n)$.

Proof. We display the positive integers not exceeding mn in the following way.

$$
\begin{array}{ccccc}
1 & m+1 & 2m+1 & \cdots & (n-1)m+1 \\
2 & m+2 & 2m+2 & \cdots & (n-1)m+2 \\
3 & m+3 & 2m+3 & \cdots & (n-1)m+3 \\
\vdots & \vdots & \vdots & & \vdots \\
r & m+r & 2m+r & \cdots & (n-1)m+r \\
\vdots & \vdots & \vdots & & \vdots \\
m & 2m & 3m & \cdots & mn
\end{array}
$$

Now, suppose that r is a positive integer not exceeding m, and suppose that $(m, r) = d > 1$. Then no number in the rth row is relatively prime to mn, because any element of this row is of the form $km + r$, where k is an integer with $1 \le k \le n - 1$, and $d \mid (km + r)$, because $d \mid m$ and $d \mid r$.

Consequently, to find those integers in the display that are relatively prime to mn, we need to look at the rth row only if $(m, r) = 1$. If $(m, r) = 1$ and $1 \le r \le m$, we must determine how many integers in this row are relatively prime to mn. The elements in this row are $r, m + r, 2m + r, \ldots, (n - 1)m + r$. Because $(r, m) = 1$, each of these integers is relatively prime to m. By Theorem 4.6 the n integers in the rth row form a complete system of residues modulo n. Hence, exactly $\phi(n)$ of these integers are relatively prime to n. Because these $\phi(n)$ integers are also relatively prime to m, they are relatively prime to mn.

Because there are $\phi(m)$ rows, each containing $\phi(n)$ integers relatively prime to mn, we can conclude that $\phi(mn) = \phi(m)\phi(n)$. ∎

Combining Theorems 7.3 and 7.4, we derive the following formula for $\phi(n)$.

Theorem 7.5. Let $n = p_1^{a_1} p_2^{a_2} \cdots p_k^{a_k}$ be the prime-power factorization of the positive integer n. Then

$$
\phi(n) = n \left(1 - \frac{1}{p_1} \right) \left(1 - \frac{1}{p_2} \right) \cdots \left(1 - \frac{1}{p_k} \right).
$$

Proof. Because ϕ is multiplicative, Theorem 7.1 tells us that

$$
\phi(n) = \phi(p_1^{a_1})\phi(p_2^{a_2}) \cdots \phi(p_k^{a_k}).
$$

In addition, by Theorem 7.3, we know that

$$
\phi(p_j^{a_j}) = p_j^{a_j} - p_j^{a_j - 1} = p_j^{a_j} \left(1 - \frac{1}{p_j} \right)
$$

for $j = 1, 2, \ldots, k$. Hence,

$$\phi(n) = p_1^{a_1}\left(1 - \frac{1}{p_1}\right) p_2^{a_2}\left(1 - \frac{1}{p_2}\right) \cdots p_k^{a_k}\left(1 - \frac{1}{p_k}\right)$$

$$= p_1^{a_1} p_2^{a_2} \cdots p_k^{a_k}\left(1 - \frac{1}{p_1}\right)\left(1 - \frac{1}{p_2}\right) \cdots \left(1 - \frac{1}{p_k}\right)$$

$$= n\left(1 - \frac{1}{p_1}\right)\left(1 - \frac{1}{p_2}\right) \cdots \left(1 - \frac{1}{p_k}\right).$$

This is the desired formula for $\phi(n)$.   ∎

We illustrate the use of Theorem 7.5 by the following example.

Example 7.4.   Using Theorem 7.5, we note that

$$\phi(100) = \phi(2^2 5^2) = 100\left(1 - \frac{1}{2}\right)\left(1 - \frac{1}{5}\right) = 40$$

and

$$\phi(720) = \phi(2^4 3^2 5) = 720\left(1 - \frac{1}{2}\right)\left(1 - \frac{1}{3}\right)\left(1 - \frac{1}{5}\right) = 192.$$ ◀

Note that $\phi(n)$ is even except when $n = 2$, as the following theorem shows.

Theorem 7.6.   Let n be a positive integer greater than 2. Then $\phi(n)$ is even.

Proof.   Suppose that $n = p_1^{a_1} p_2^{a_2} \cdots p_s^{a_s}$ is the prime-power factorization of n. Because ϕ is multiplicative, it follows that $\phi(n) = \prod_{j=1}^{s} \phi(p_j^{a_j})$. By Theorem 7.3, we know that $\phi(p_j^{a_j}) = p_j^{a_j-1}(p_j - 1)$. We can see that $\phi(p_j^{a_j})$ is even if p_j is an odd prime, because then $p_j - 1$ is even; or if $p_j = 2$ and $a_j > 1$, because then $p_j^{a_j-1}$ is even. Given that $n > 2$, at least one of these two conditions holds, so that $\phi(p_j^{a_j})$ is even for at least one integer j, $1 \le j \le s$. We conclude that $\phi(n)$ is even.   ∎

Let f be an arithmetic function. Then

$$F(n) = \sum_{d \mid n} f(d)$$

represents the sum of the values of f at all the positive divisors of n. The function F is called the *summatory function* of f.

Example 7.5.   If f is an arithmetic function with summatory function F, then

$$F(12) = \sum_{d \mid 12} f(d) = f(1) + f(2) + f(3) + f(4) + f(6) + f(12).$$

For instance, if $f(d) = d^2$ and F is the summatory function of f, then $F(12) = 210$, because

$$\sum_{d \mid 12} d^2 = 1^2 + 2^2 + 3^2 + 4^2 + 6^2 + 12^2$$

$$= 1 + 4 + 9 + 16 + 36 + 144 = 210.$$ ◀

The following result, which states that n is the sum of the values of the phi-function at all the positive divisors of n, will also be useful in the sequel. It says that the summatory function of $\phi(n)$ is the identity function, that is, the function whose value at n is just n.

Theorem 7.7.   Let n be a positive integer. Then

$$\sum_{d\mid n} \phi(d) = n.$$

Proof.   We split the set of integers from 1 to n into classes. Put the integer m into the class C_d if the greatest common divisor of m and n is d. We see that m is in C_d, that is, $(m, n) = d$, if and only if $(m/d, n/d) = 1$. Hence, the number of integers in C_d is the number of positive integers not exceeding n/d that are relatively prime to the integer n/d. From this observation, we see that there are $\phi(n/d)$ integers in C_d. Because we divided the integers 1 to n into disjoint classes and each integer is in exactly one class, n is the sum of the numbers of elements in the different classes. Consequently, we see that

$$n = \sum_{d\mid n} \phi(n/d).$$

As d runs through the positive integers that divide n, n/d also runs through these divisors, so that

$$n = \sum_{d\mid n} \phi(n/d) = \sum_{d\mid n} \phi(d).$$

This proves the theorem.   ∎

Example 7.6.   We illustrate the proof of Theorem 7.7 when $n = 18$. The integers from 1 to 18 can be split into classes C_d, where $d \mid 18$ such that the class C_d contains those integers m with $(m, 18) = d$. We have

$$\begin{aligned}
C_1 &= \{1, 5, 7, 11, 13, 17\} & C_6 &= \{6, 12\} \\
C_2 &= \{2, 4, 8, 10, 14, 16\} & C_9 &= \{9\} \\
C_3 &= \{3, 15\} & C_{18} &= \{18\}.
\end{aligned}$$

We see that the class C_d contains $\phi(18/d)$ integers, as the six classes contain $\phi(18) = 6$, $\phi(9) = 6$, $\phi(6) = 2$, $\phi(3) = 2$, $\phi(2) = 1$, and $\phi(1) = 1$ integers, respectively. We note that $18 = \phi(18) + \phi(9) + \phi(6) + \phi(3) + \phi(2) + \phi(1) = \sum_{d\mid 18} \phi(d)$.   ◀

A useful tool for finding all positive integers n with $\phi(n) = k$, where k is a positive integer, is the equation $\phi(n) = \prod_{i=1}^{k} p_i^{a_i-1}(p_i - 1)$, where the prime-power factorization of n is $n = \prod_{i=1}^{k} p_i^{a_i}$. This is illustrated in the following example.

Example 7.7.   What are the solutions to the equation $\phi(n) = 8$, where n is a positive integer? Suppose that the prime-power factorization of n is $n = p_1^{a_1} p_2^{a_2} \cdots p_k^{a_k}$. Because

$$\phi(n) = \prod_{j=1}^{k} p_j^{a_j-1}(p_j - 1),$$

the equation $\phi(n) = 8$ implies that no prime exceeding 9 divides n (otherwise $\phi(n) >$ $p_j - 1 > 8$). Furthermore, 7 cannot divide n because if it did, $7 - 1 = 6$ would be a factor of $\phi(n)$. It follows that $n = 2^a 3^b 5^c$, where a, b, and c are nonnegative integers. We can also conclude that $b = 0$ or $b = 1$ and that $c = 0$ or $c = 1$; otherwise, 3 or 5 would divide $\phi(n) = 8$.

To find all solutions we need only consider four cases. When $b = c = 0$, we have $n = 2^a$, where $a \geq 1$. This implies that $\phi(n) = 2^{a-1}$, which means that $a = 4$ and $n = 16$. When $b = 0$ and $c = 1$, we have $n = 2^a \cdot 5$, where $a \geq 1$. This implies that $\phi(n) = 2^{a-1} \cdot 4$, so $a = 2$ and $n = 20$. When $b = 1$ and $c = 0$, we have $n = 2^a \cdot 3$, where $a \geq 1$. This implies that $\phi(n) = 2^{a-1} \cdot 2 = 2^a$, so $a = 3$ and $n = 24$. Finally, when $b = 1$ and $c = 1$, we have $n = 2^a \cdot 3 \cdot 5$. We need to consider the case where $a = 0$, as well as the case where $a \geq 1$. When $a = 0$, we have $n = 15$, which is a solution because $\phi(15) = 8$. When $a \geq 1$, we have $\phi(n) = 2^{a-1} \cdot 2 \cdot 4 = 2^{a+2}$. This means that $a = 1$ and $n = 30$. Putting everything together, we see that all the solutions to $\phi(n) = 8$ are $n = 15, 16, 20, 24$ and 30.   ◀

7.1 Exercises

1. Determine whether each of the following arithmetic functions is completely multiplicative. Prove your answers.

   a) $f(n) = 0$ d) $f(n) = \log n$ g) $f(n) = n + 1$
   b) $f(n) = 2$ e) $f(n) = n^2$ h) $f(n) = n^n$
   c) $f(n) = n/2$ f) $f(n) = n!$ i) $f(n) = \sqrt{n}$

2. Find the value of the Euler phi-function at each of the following integers.

   a) 100   d) $2 \cdot 3 \cdot 5 \cdot 7 \cdot 11 \cdot 13$
   b) 256   e) 10!
   c) 1001   f) 20!

3. Show that $\phi(5186) = \phi(5187) = \phi(5188)$.

4. Find all positive integers n such that $\phi(n)$ has each of the following values. Be sure to prove that you have found all solutions.

   a) 1   b) 2   c) 3   d) 4

5. Find all positive integers n such that $\phi(n) = 6$. Be sure to prove that you have found all solutions.

6. Find all positive integers n such that $\phi(n) = 12$. Be sure to prove that you have found all solutions.

7. Find all positive integers n such that $\phi(n) = 24$. Be sure to prove that you have found all solutions.

8. Show that there is no positive integer n such that $\phi(n) = 14$.

9. Can you find a rule involving the Euler phi-function for producing the terms of the sequence $1, 2, 2, 4, 4, 4, 6, 8, 6, \ldots$?

10. Can you find a rule involving the Euler phi-function for producing the terms of the sequence $2, 3, 0, 4, 0, 4, 0, 5, 0, \ldots$?

11. For which positive integers n does $\phi(3n) = 3\phi(n)$?

12. For which positive integers n is $\phi(n)$ divisible by 4?

13. For which positive integers n is $\phi(n)$ equal to $n/2$?

14. For which positive integers n does $\phi(n) \mid n$?

15. Show that if n is a positive integer, then
$$\phi(2n) = \begin{cases} \phi(n) & \text{if } n \text{ is odd;} \\ 2\phi(n) & \text{if } n \text{ is even.} \end{cases}$$

16. Show that if n is a positive integer having k distinct odd prime divisors, then $\phi(n)$ is divisible by 2^k.

17. For which positive integers n is $\phi(n)$ a power of 2?

18. Show that if n is an odd integer, then $\phi(4n) = 2\phi(n)$.

19. Show that if $n = 2\phi(n)$, where n is a positive integer, then $n = 2^j$ for some positive integer j.

20. Let p be prime. Show that $p \nmid n$, where n is a positive integer, if and only if $\phi(np) = (p-1)\phi(n)$.

21. Show that if m and n are positive integers and $(m, n) = p$, where p is prime, then $\phi(mn) = p\phi(m)\phi(n)/(p-1)$.

22. Show that if m and k are positive integers, then $\phi(m^k) = m^{k-1}\phi(m)$.

23. Show that if a and b are positive integers, then
$$\phi(ab) = (a, b)\phi(a)\phi(b)/\phi((a, b)).$$
Conclude that $\phi(ab) > \phi(a)\phi(b)$ when $(a, b) > 1$.

24. Find the least positive integer n such that the following hold.

    a) $\phi(n) \geq 100$ c) $\phi(n) \geq 10{,}000$
    b) $\phi(n) \geq 1000$ d) $\phi(n) \geq 100{,}000$

25. Use the Euler phi-function to show that there are infinitely many primes. (*Hint:* Assume there are only a finite number of primes $p_1, \ldots, p_k$. Consider the value of the Euler phi-function at the product of these primes.)

26. Show that if the equation $\phi(n) = k$, where k is a positive integer, has exactly one solution n, then $36 \mid n$.

27. Show that the equation $\phi(n) = k$, where k is a positive integer, has finitely many solutions in integers n whenever k is a positive integer.

28. Show that if p is prime, $2^a p + 1$ is composite for $a = 1, 2, \ldots, r$ and p is not a Fermat prime, where r is a positive integer, then $\phi(n) = 2^r p$ has no solution.

∗ 29. Show that there are infinitely many positive integers k such that the equation $\phi(n) = k$ has exactly two solutions, where n is a positive integer. (*Hint:* Take $k = 2 \cdot 3^{6j+1}$, where $j = 1, 2, \ldots$.)

30. Show that if n is a positive integer with $n \neq 2$ and $n \neq 6$, then $\phi(n) \geq \sqrt{n}$.

$*$ **31.** Show that if n is a composite positive integer and $\phi(n) \mid n - 1$, then n is square-free and is the product of at least three distinct primes.

32. Show that if m and n are positive integers with $m \mid n$, then $\phi(m) \mid \phi(n)$.

$*$ **33.** Prove Theorem 7.5, using the principle of inclusion-exclusion (see Exercise 16 of Appendix B).

34. Show that a positive integer n is composite if and only if $\phi(n) \leq n - \sqrt{n}$.

35. Let n be a positive integer. Define the sequence of positive integers $n_1, n_2, n_3, \ldots$ recursively by $n_1 = \phi(n)$ and $n_{k+1} = \phi(n_k)$ for $k = 1, 2, 3, \ldots$. Show that there is a positive integer r such that $n_r = 1$.

A multiplicative function is called *strongly multiplicative* if and only if $f(p^k) = f(p)$ for every prime p and every positive integer k.

36. Show that $f(n) = \phi(n)/n$ is a strongly multiplicative function.

Two arithmetic functions f and g may be multiplied using the *Dirichlet product*, which is defined by

$$(f * g)(n) = \sum_{d \mid n} f(d)g(n/d).$$

37. Show that $f * g = g * f$.

38. Show that $(f * g) * h = f * (g * h)$.

We define the ι *function* by

$$\iota(n) = \begin{cases} 1 & \text{if } n = 1; \\ 0 & \text{if } n > 1. \end{cases}$$

39. a) Show that ι is a multiplicative function.

b) Show that $\iota * f = f * \iota = f$ for all arithmetic functions f.

40. The arithmetic function g is said to be the *inverse* of the arithmetic function f if $f * g = g * f = \iota$. Show that the arithmetic function f has an *inverse* if and only if $f(1) \neq 0$. Show that if f has an inverse it is unique. (*Hint:* When $f(1) \neq 0$, find the inverse f^{-1} of f by calculating $f^{-1}(n)$ recursively, using the fact that $\iota(n) = \sum_{d \mid n} f(d) f^{-1}(n/d)$.)

41. Show that if f and g are multiplicative functions, then the Dirichlet product $f * g$ is also multiplicative.

42. Show that if f and g are arithmetic functions, $F = f * g$, and h is the Dirichlet inverse of g, then $f = F * h$.

We define *Liouville's function* $\lambda(n)$, named after French mathematician *Joseph Liouville*, by $\lambda(1) = 1$, and for $n > 1$, $\lambda(n) = (-1)^{a_1 + a_2 + \cdots + a_m}$, where the prime-power factorization of n is $n = p_1^{a_1} p_2^{a_2} \cdots p_m^{a_m}$.

43. Find $\lambda(n)$ for each of the following values of n.

a) 12 c) 210 e) 1001 g) 20!

b) 20 d) 1000 f) 10!

44. Show that $\lambda(n)$ is completely multiplicative.

45. Show that if n is a positive integer, then $\sum_{d|n} \lambda(d)$ equals 0 if n is not a perfect square, and equals 1 if n is a perfect square.

46. Show that if f and g are multiplicative functions, then fg is also multiplicative, where $(fg)(n) = f(n)g(n)$ for every positive integer n.

47. Show that if f and g are completely multiplicative functions, then fg is also completely multiplicative.

48. Show that if f is completely multiplicative, then $f(n) = f(p_1)^{a_1} f(p_2)^{a_2} \cdots f(p_m)^{a_m}$, where the prime-power factorization of n is $n = p_1^{a_1} p_2^{a_2} \cdots p_m^{a_m}$.

A function f that satisfies the equation $f(mn) = f(m) + f(n)$ for all relatively prime positive integers m and n is called *additive*, and if the above equation holds for all positive integers m and n, f is called *completely additive*.

49. Show that the function $f(n) = \log n$ is completely additive.

The function $\omega(n)$ is the function that denotes the number of distinct prime factors of the positive integer n.

50. Find $\omega(n)$ for each of the following integers.

a) 1   b) 2   c) 20   d) 84   e) 128

JOSEPH LIOUVILLE (1809–1882), born in Saint-Omer, France, was the son of a captain in Napoleon's army. He studied mathematics at the Collège St. Louis in Paris, and in 1825 he enrolled in the École Polytechnique; after graduating, he entered the École des Ponts et Chaussées (School of Bridges and Roads). Health problems while working on engineering projects and his interest in theoretical topics convinced him to pursue an academic career. He left the École des Ponts et Chaussées in 1830, but during his time there he wrote papers on electrodynamics, the theory of heat, and partial differential equations.

Liouville's first academic appointment was as an assistant at the École Polytechnique in 1831. He had a teaching load of around 40 hours a week at several different institutions. Some of his less able students complained that he lectured at too high a level. In 1836, Liouville founded the *Journal de Mathématiques Pures et Appliquées,* which played an important role in French mathematics in the nineteenth century. In 1837, he was appointed to lecture at the Collège de France and the following year he was appointed Professor at the École Polytechnique. Besides his academic interests, Liouville was also involved in politics. He was elected to Constituting Assembly in 1848 as a moderate republican, but lost in the election of 1849, embittering him. Liouville was appointed to a chair at the Collège de France in 1851, and the chair of mechanics at the Faculté des Sciences in 1857. Around this time, his heavy teaching load began to take its toll. Liouville was a perfectionist and was unhappy when he could not devote sufficient time to his lectures.

Liouville's work covered many diverse areas of mathematics, including mathematical physics, astronomy, and many areas of pure mathematics. He was the first person to provide an explicit example of a transcendental number. He is also known today for what is now called Sturm-Liouville theory, used in the solution of integral equations, and he made important contributions to differential geometry. His total output exceeds 400 papers in the mathematical sciences, with nearly half of those in number theory alone.

51. Find $\omega(n)$ for each of the following integers.

a) 12   b) 30   c) 32   d) 10!   e) 20!   f) 50!

52. Show that $\omega(n)$ is additive, but not completely additive.

53. Show that if f is an additive function and $g(n) = 2^{f(n)}$, then g is multiplicative.

54. Show that the function n^k is completely multiplicative for every real number k.

7.1 Computational and Programming Exercises

Computations and Explorations

Using a computation program such as Maple or *Mathematica*, or programs you have written, carry out the following computations and explorations.

1. Find $\phi(n)$ when n takes each of the following values.

a) 185,888,434,028   b) 1,111,111,111,111

2. Find the number of iterations of the Euler phi-function required to reach 1, starting with each of the integers in Computation 1.

3. Find the largest integer n such that $\phi(n) \le k$ for each of the following values of k.

a) 1,000,000   b) 10,000,000

4. Find as many positive integers n as you can, such that $\phi(n) = \phi(n+1)$. Can you formulate any conjectures based on the evidence that you have found?

5. Can you find a positive integer n other than 5186 such that $\phi(n) = \phi(n+1) = \phi(n+2)$? Can you find four consecutive positive integers n, $n+1$, $n+2$, $n+3$, such that $\phi(n) = \phi(n+1) = \phi(n+2) = \phi(n+3)$?

6. An open conjecture of D. H. Lehmer asserts that n is prime if $\phi(n)$ divides $n - 1$. Explore the truth of this conjecture.

7. An open conjecture of Carmichael asserts that for every positive integer n there is a positive integer m such that $\phi(m) = \phi(n)$. Gather as much evidence as possible for this conjecture.

Programming Projects

Write programs using Maple, *Mathematica,* or a language of your choice to do the following.

1. Given a positive integer n, find the value of $\phi(n)$.

2. Given a positive integer n, find the number of iterations of the phi-function, starting with n, required to reach 1. (This is the the integer r in Exercise 35.)

3. Given a positive integer k, find the number of solutions of $\phi(n) = k$.

7.2 The Sum and Number of Divisors

As we mentioned in Section 7.1, the number of divisors and the sum of divisors are both multiplicative functions. We will show that these functions are multiplicative, and derive formulas for their values at a positive integer n from the prime factorization of n.

Definition.   The *sum of divisors function,* denoted by σ, is defined by setting $\sigma(n)$ equal to the sum of all the positive divisors of n.

In Table 7.1, we give $\sigma(n)$ for $1 \le n \le 12$. The values of $\sigma(n)$ for $1 \le n \le 100$ are given in Table 2 of Appendix E. (These values can also be computed using Maple or *Mathematica*.)

n	1	2	3	4	5	6	7	8	9	10	11	12
$\sigma(n)$	1	3	4	7	6	12	8	15	13	18	12	28

Table 7.1 *The sum of the divisors for $1 \le n \le 12$.*

Definition.   The *number of divisors function,* denoted by τ, is defined by setting $\tau(n)$ equal to the number of positive divisors of n.

In Table 7.2, we give $\tau(n)$ for $1 \le n \le 12$. The values of $\tau(n)$ for $1 \le n \le 100$ are given in Table 2 of Appendix E. (These values can also be computed using Maple or *Mathematica*.)

n	1	2	3	4	5	6	7	8	9	10	11	12
$\tau(n)$	1	2	2	3	2	4	2	4	3	4	2	6

Table 7.2 *The number of divisors for $1 \le n \le 12$.*

Note that we can express $\sigma(n)$ and $\tau(n)$ in summation notation. It is simple to see that

$$\sigma(n) = \sum_{d \mid n} d$$

and

$$\tau(n) = \sum_{d \mid n} 1.$$

To prove that σ and τ are multiplicative, we use the following theorem.

Theorem 7.8. If f is a multiplicative function, then the summatory function of f, namely $F(n) = \sum_{d \mid n} f(d)$, is also multiplicative.

Before we prove the theorem, we illustrate the idea behind its proof with the following example. Let f be a multiplicative function, and let $F(n) = \sum_{d|n} f(d)$. We will show that $F(60) = F(4)F(15)$. Each of the divisors of 60 may be written as the product of a divisor of 4 and a divisor of 15 in the following way: $1 = 1 \cdot 1$, $2 = 2 \cdot 1$, $3 = 1 \cdot 3$, $4 = 4 \cdot 1$, $5 = 1 \cdot 5$, $6 = 2 \cdot 3$, $10 = 2 \cdot 5$, $12 = 4 \cdot 3$, $15 = 1 \cdot 15$, $20 = 4 \cdot 5$, $30 = 2 \cdot 15$, $60 = 4 \cdot 15$ (in each product, the first factor is the divisor of 4, and the second is the divisor of 15). Hence,

$$
\begin{aligned}
F(60) &= f(1) + f(2) + f(3) + f(4) + f(5) + f(6) + f(10) + f(12) \\
&\quad + f(15) + f(20) + f(30) + f(60) \\
&= f(1 \cdot 1) + f(2 \cdot 1) + f(1 \cdot 3) + f(4 \cdot 1) + f(1 \cdot 5) + f(2 \cdot 3) \\
&\quad + f(2 \cdot 5) + f(4 \cdot 3) + f(1 \cdot 15) + f(4 \cdot 5) + f(2 \cdot 15) + f(4 \cdot 15) \\
&= f(1)f(1) + f(2)f(1) + f(1)f(3) + f(4)f(1) + f(1)f(5) \\
&\quad + f(2)f(3) + f(2)f(5) + f(4)f(3) + f(1)f(15) + f(4)f(5) \\
&\quad + f(2)f(15) + f(4)f(15) \\
&= (f(1) + f(2) + f(4))(f(1) + f(3) + f(5) + f(15)) \\
&= F(4)F(15).
\end{aligned}
$$

We now prove Theorem 7.8 using the idea illustrated by the example.

Proof.   To show that F is a multiplicative function, we must show that if m and n are relatively prime positive integers, then $F(mn) = F(m)F(n)$. So let us assume that $(m, n) = 1$. We have

$$
F(mn) = \sum_{d|mn} f(d).
$$

By Lemma 3.6, because $(m, n) = 1$, each divisor of mn can be written uniquely as the product of relatively prime divisors d_1 of m and d_2 of n, and each pair of divisors d_1 of m and d_2 of n corresponds to a divisor $d = d_1 d_2$ of mn. Hence, we can write

$$
F(mn) = \sum_{\substack{d_1|m \\ d_2|n}} f(d_1 d_2).
$$

Because f is multiplicative, and $(d_1, d_2) = 1$, we see that

$$
\begin{aligned}
F(mn) &= \sum_{\substack{d_1|m \\ d_2|n}} f(d_1) f(d_2) \\
&= \sum_{d_1|m} f(d_1) \sum_{d_2|n} f(d_2) \\
&= F(m)F(n). \qquad \blacksquare
\end{aligned}
$$

We can now use Theorem 7.8 to show that σ and τ are multiplicative.

Corollary 7.8.1.   The sum of divisors function σ and the number of divisors function τ are multiplicative functions.

Proof. Let $f(n) = n$ and $g(n) = 1$. Both f and g are multiplicative. By Theorem 7.8, we see that $\sigma(n) = \sum_{d|n} f(d)$ and $\tau(n) = \sum_{d|n} g(d)$ are multiplicative. ∎

Now that we know that σ and τ are multiplicative, we can derive formulas for their values based on prime factorizations. First, we find formulas for $\sigma(n)$ and $\tau(n)$ when n is the power of a prime.

Lemma 7.1. Let p be prime and a a positive integer. Then

$$\sigma(p^a) = 1 + p + p^2 + \cdots + p^a = \frac{p^{a+1} - 1}{p - 1}$$

and

$$\tau(p^a) = a + 1.$$

Proof. The divisors of p^a are $1, p, p^2, \ldots, p^{a-1}, p^a$. Consequently, p^a has exactly $a + 1$ divisors, so that $\tau(p^a) = a + 1$. Also, we note that $\sigma(p^a) = 1 + p + p^2 + \cdots + p^{a-1} + p^a = \frac{p^{a+1}-1}{p-1}$, using the formula in Example 1.15 for the sum of terms of a geometric progression. ∎

Example 7.8. When we apply Lemma 7.1 with $p = 5$ and $a = 3$, we find that $\sigma(5^3) = 1 + 5 + 5^2 + 5^3 = \frac{5^4 - 1}{5 - 1} = 156$ and $\tau(5^3) = 1 + 3 = 4$. ◄

Lemma 7.1 and Corollary 7.8.1 lead to the following formulas.

Theorem 7.9. Let the positive integer n have prime factorization $n = p_1^{a_1} p_2^{a_2} \cdots p_s^{a_s}$. Then

$$\sigma(n) = \frac{p_1^{a_1+1} - 1}{p_1 - 1} \cdot \frac{p_2^{a_2+1} - 1}{p_2 - 1} \cdot \cdots \cdot \frac{p_s^{a_s+1} - 1}{p_s - 1} = \prod_{j=1}^{s} \frac{p_j^{a_j+1} - 1}{p_j - 1}$$

and

$$\tau(n) = (a_1 + 1)(a_2 + 1) \cdots (a_s + 1) = \prod_{j=1}^{s} (a_j + 1).$$

Proof. Because both σ and τ are multiplicative, we see that $\sigma(n) = \sigma(p_1^{a_1} p_2^{a_2} \cdots p_s^{a_s}) = \sigma(p_1^{a_1})\sigma(p_2^{a_2}) \cdots \sigma(p_s^{a_s})$ and $\tau(n) = \tau(p_1^{a_1} p_2^{a_2} \cdots p_s^{a_s}) = \tau(p_1^{a_1})\tau(p_2^{a_2}) \cdots \tau(p_s^{a_s})$. Inserting the values for $\sigma(p_i^{a_i})$ and $\tau(p_i^{a_i})$ found in Lemma 7.1, we obtain the desired formulas. ∎

We illustrate how to use Theorem 7.9 with the following example.

Example 7.9. Using Theorem 7.9 we find

$$\sigma(200) = \sigma(2^3 5^2) = \frac{2^4 - 1}{2 - 1} \cdot \frac{5^3 - 1}{5 - 1} = 15 \cdot 31 = 465,$$

$$\tau(200) = \tau(2^3 5^2) = (3 + 1)(2 + 1) = 12.$$

Similarly, we have

$$\sigma(720) = \sigma(2^4 \cdot 3^2 \cdot 5) = \frac{2^5 - 1}{2 - 1} \cdot \frac{3^3 - 1}{3 - 1} \cdot \frac{5^2 - 1}{5 - 1} = 31 \cdot 13 \cdot 6 = 2418,$$

$$\tau(2^4 \cdot 3^2 \cdot 5) = (4 + 1)(2 + 1)(1 + 1) = 30. \qquad \blacktriangleleft$$

7.2 Exercises

1. Find the sum of the positive integer divisors of each of the following integers.

   a) 35
   b) 196
   c) 1000
   d) 2^{100}

   e) $2 \cdot 3 \cdot 5 \cdot 7 \cdot 11$
   f) $2^5 3^4 5^3 7^2 11$
   g) 10!
   h) 20!

2. Find the number of positive integer divisors of each of the following integers.

   a) 36
   b) 99
   c) 144

   d) $2 \cdot 3 \cdot 5 \cdot 7 \cdot 11 \cdot 13 \cdot 17 \cdot 19$
   e) $2 \cdot 3^2 \cdot 5^3 \cdot 7^4 \cdot 11^5 \cdot 13^4 \cdot 17^5 \cdot 19^5$
   f) 20!

3. Which positive integers have an odd number of positive divisors?

4. For which positive integers n is the sum of divisors of n odd?

∗ 5. Find all positive integers n with $\sigma(n)$ equal to each of the following integers.

   a) 12
   b) 18
   c) 24

   d) 48
   e) 52
   f) 84

∗ 6. Find the smallest positive integer n with $\tau(n)$ equal to each of the following integers.

   a) 1
   b) 2
   c) 3

   d) 6
   e) 14
   f) 100

7. Show that if $k > 1$ is an integer, then the equation $\tau(n) = k$ has infinitely many solutions.

8. Which positive integers have exactly two positive divisors?

9. Which positive integers have exactly three positive divisors?

10. Which positive integers have exactly four positive divisors?

11. What is the product of the positive divisors of a positive integer n?

12. Show that the equation $\sigma(n) = k$ has at most a finite number of solutions when k is a positive integer.

13. For each of the following sequences, can you find a rule for producing the terms of the sequence that involves the τ and/or the σ function?

    a) $3, 7, 12, 15, 18, 28, 24, 31, \ldots$
    b) $0, 1, 2, 4, 4, 8, 6, 11, \ldots$

c) $1, 2, 4, 6, 16, 12, 64, 24, 36, 48, \ldots$

d) $1, 0, 1, 1, 0, 1, 1, 1, 0, 0, 0, 2, 1, \ldots$

14. For each of the following sequences, can you find a rule for producing the terms of the sequence that involves the τ and/or the σ function?

a) $2, 5, 6, 10, 8, 16, 10, 19, 16, 22, \ldots$

b) $1, 4, 6, 8, 13, 12, 14, 24, 18, \ldots$

c) $6, 8, 10, 14, 15, 21, 22, 26, 27, 33, 34, 35, \ldots$

d) $1, 2, 2, 2, 3, 2, 2, 4, 2, 2, 4, 2, 3, \ldots$

A positive integer n, $n > 1$, is *highly composite,* a concept introduced by the famous Indian mathematician *Srinivasa Ramanujan,* if $\tau(m) < \tau(n)$ for all integers m with $1 \le m < n$.

15. Find the first six highly composite positive integers.

16. Show that if n is a highly composite positive integer and m is a positive integer with $\tau(m) > \tau(n)$, then there exists a highly composite integer k such that $n < k \le m$. Conclude that there are infinitely many highly composite integers.

17. Show that if $n \ge 1$, there exists a highly composite number k such that $n < k \le 2n$. Use this to provide an upper bound on the mth highly composite number, where m is a positive integer.

18. Show that if n is a highly composite positive integer, there exists a positive integer k such that $n = 2^{a_1} 3^{a_2} 5^{a_3} \cdots p_k^{a_k}$, where p_k is the kth prime and $a_1 \ge a_2 \ge \cdots \ge a_k \ge 1$.

* **19.** Find all highly composite numbers of the form $2^a 3^b$, where a and b are nonnegative integers.

Let $\sigma_k(n)$ denote the sum of the kth powers of the divisors of n, so that $\sigma_k(n) = \sum_{d \mid n} d^k$. Note that $\sigma_1(n) = \sigma(n)$.

20. Find $\sigma_3(4)$, $\sigma_3(6)$, and $\sigma_3(12)$.

21. Give a formula for $\sigma_k(p)$, where p is prime.

22. Give a formula for $\sigma_k(p^a)$, where p is prime and a is a positive integer.

23. Show that the function σ_k is multiplicative.

24. Using Exercises 22 and 23, find a formula for $\sigma_k(n)$, where n has prime-power factorization $n = p_1^{a_1} p_2^{a_2} \cdots p_m^{a_m}$.

* **25.** Find all positive integers n such that $\phi(n) + \sigma(n) = 2n$.

* **26.** Show that no two positive integers have the same product of divisors.

27. Show that the number of ordered pairs of positive integers with least common multiple equal to the positive integer n is $\tau(n^2)$.

28. Let n be a positive integer, $n \ge 2$. Define the sequence of integers $n_1, n_2, n_3, \ldots$ by $n_1 = \tau(n)$ and $n_{k+1} = \tau(n_k)$ for $k = 1, 2, 3, \ldots$. Show that there is a positive integer r such that $2 = n_r = n_{r+1} = n_{r+2} = \ldots$.

29. Show that a positive integer n is composite if and only if $\sigma(n) > n + \sqrt{n}$.

30. Let n be a positive integer. Show that $\tau(2^n - 1) \ge \tau(n)$.

* **31.** Show that $\sum_{j=1}^{n} \tau(j) = 2 \sum_{j=1}^{[\sqrt{n}]} [n/j] - [\sqrt{n}]^2$ whenever n is a positive integer. Then use this formula to find $\sum_{j=1}^{100} \tau(j)$.

* **32.** Let a and b be positive integers. Show that $\sigma(a)/a \leq \sigma(ab)/(ab) \leq \sigma(a)\sigma(b)/(ab)$.

* **33.** Show that if a and b are positive integers, then $\sigma(a)\sigma(b) = \sum_{d|(a,b)} d\sigma(ab/d^2)$.

SRINIVASA RAMANUJAN (1887–1920) was born and raised in southern India, near Madras. His father was a clerk in a cloth shop and his mother contributed to the family income by singing at a local temple. Ramanujan studied at a local English language school, displaying a talent in mathematics. At 13 he mastered a textbook used by college students; when he was 15, a university student lent him a copy of *Synopsis of Pure Mathematics,* and Ramanujan decided to work out the more than 6000 results in this book. He graduated from high school in 1904, winning a scholarship to the University of Madras. Enrolling in a fine arts curriculum, he neglected subjects other than mathematics and lost his scholarship. During this time he filled his notebooks with original writings, sometimes rediscovering already published work and at other times making new discoveries.

Lacking a university degree, Ramanujan found it difficult to land a decent job. To survive, he depended on the good will of friends. He tutored students, but his uncoventional ways of thinking and failure to stick to the syllabus caused problems. He was married in 1909 in an arranged marriage to a woman who was 13 years old. Needing to support himself and his wife, he moved to Madras looking for a job. He showed his notebooks to potential employers, but his writings bewildered them. However, a professor at the Presidency College recognized his genius and supported him, and in 1912 he found work as an accounts clerk, which earned him a small salary.

Ramanujan continued his mathematical investigations, publishing his first paper in 1910 in an Indian journal. Realizing that his work was beyond that of Indian mathematicians, he decided to write to leading English mathematicians. Although the first mathematicians turned down his request for help, G. H. Hardy arranged a scholarship for Ramanujan, bringing him to England in 1914. Hardy initially was inclined to turn Ramanujan down, but the mathematical results Ramanujan stated without proof in his letter puzzled Hardy. He examined Ramanujan's writings with the aid of his collaborator, J. E. Littlewood. They decided that Ramanujan was probably a genius, as his statements "could only be written down by a mathematician of the highest class; they must be true, because if they were not true, no one would have the imagination to invent them." Hardy personally tutored Ramanujan and they collaborated for five years, proving significant theorems about the partitions of integers. During this time, Ramanujan made important contributions to number theory, and worked on elliptic functions, infinite series, and continued fractions. Ramanujan had amazing insight involving certain types of functions and series, but his purported theorems on prime numbers were often wrong, illustrating his vague idea of what makes up a correct proof.

Ramanujan was one of the youngest members ever appointed a Fellow of the Royal Society. Unfortunately, in 1917, he became extremely ill. Although it was once thought he contracted turberculosis, it is now thought that he suffered from a vitamin deficiency brought on by his strict vegetarianism and shortages in wartime England. He returned to India in 1919 and continued his mathematical work even while confined to bed. He was highly religious and thought that his mathematical talent came from his family deity, Namaigiri. He said that "an equation for me has no meaning unless it expresses a thought of God." He died in April 1920, leaving several notebooks of unpublished results. Mathematicians have devoted many years of study to the explanation and justification of the results jotted down in Ramanujan's notebooks.

* **34.** Show that if n is a positive integer, then $\left(\sum_{d|n} \tau(d)\right)^2 = \sum_{d|n} \tau(d)^3$.

35. Show that if n is a positive integer, then $\tau(n^2) = \sum_{d|n} 2^{\omega(n)}$, where $\omega(n)$ equals the number of prime divisors of n.

36. Show that $\sum_{d|n} n\sigma(d)/d = \sum_{d|n} d\tau(d)$ whenever n is a positive integer.

* **37.** Find the determinant of the $n \times n$ matrix with (i, j)th entry equal to (i, j).

* **38.** Let n be a positive integer such that $24 \mid (n + 1)$. Show that $\sigma(n)$ is divisible by 24.

39. Show that there are infinitely many pairs of positive integers m, n such that $\phi(m) = \sigma(n)$, if there are infinitely many pairs of twin primes or infinitely many Mersenne primes (that is, primes of the form $2^p - 1$, where p is prime).

40. Prove that $\sum_{d/n} \phi(d) = n$ (Theorem 7.7) as a consequence of Theorem 7.8.

7.2 Computational and Programming Exercises

Computations and Explorations

Using a computation program such as Maple or *Mathematica,* or programs you have written, carry out the following computations and explorations.

1. Find $\tau(n)$, $\sigma(n)$, and $\sigma_2(n)$ (as defined in the preamble to Exercise 20) for each of the following values of n.

   a) 121,110,987,654   b) 11,111,111,111   c) 98,989,898,989

2. Find as many pairs, triples, and quadruples as you can of consecutive integers, each with the same number of positive divisors.

3. Determine the number of iterations required for the sequence $n_1 = \tau(n)$, $n_2 = \tau(n_1), \ldots, n_{k+1} = \tau(n_k), \ldots$ to reach the integer 2, for all positive integers n not exceeding 1000. Formulate some conjectures based on your evidence.

4. Find all the highly composite integers (as defined in the preamble to Exercise 15) not exceeding 10,000.

* **5.** Show that 29,331,862,500 is a highly composite integer.

Programming Projects

Write programs using Maple, *Mathematica,* or a language of your choice to do the following.

1. Given a positive integer n, find $\tau(n)$, the number of positive divisors of n.

2. Given a positive integer n, find $\sigma(n)$, the sum of the positive divisors of n.

3. Given a positive integer n and a positive integer k, find $\sigma_k(n)$, the sum of the kth powers of the positive divisors of n.

4. Given a positive integer n, find the integer r defined in Exercise 28.

5. Given a positive integer n, determine whether n is highly composite.

7.3 Perfect Numbers and Mersenne Primes

Because of certain mystical beliefs, the ancient Greeks were interested in those integers that are equal to the sum of all their proper positive divisors. Such integers are called *perfect numbers.*

Definition.   If n is a positive integer and $\sigma(n) = 2n$, then n is called a *perfect number.*

Example 7.10.   Because $\sigma(6) = 1 + 2 + 3 + 6 = 12$, we see that 6 is perfect. We also note that $\sigma(28) = 1 + 2 + 4 + 7 + 14 + 28 = 56$, so that 28 is another perfect number.

◀

The ancient Greeks knew how to find all even perfect numbers. The following theorem tells us which even positive integers are perfect.

Theorem 7.10.   The positive integer n is an even perfect number if and only if

$$n = 2^{m-1}(2^m - 1),$$

where m is an integer such that $m \geq 2$ and $2^m - 1$ is prime.

Proof.   First, we show that if $n = 2^{m-1}(2^m - 1)$, where $2^m - 1$ is prime, then n is perfect. We note that because $2^m - 1$ is odd, we have $(2^{m-1}, 2^m - 1) = 1$. Because σ is a multiplicative function, we see that

$$\sigma(n) = \sigma(2^{m-1})\sigma(2^m - 1).$$

Lemma 7.1 tells us that $\sigma(2^{m-1}) = 2^m - 1$ and $\sigma(2^m - 1) = 2^m$, because we are assuming that $2^m - 1$ is prime. Consequently,

$$\sigma(n) = (2^m - 1)2^m = 2n,$$

demonstrating that n is a perfect number.

To show that the converse is true, let n be an even perfect number. Write $n = 2^s t$, where s and t are positive integers and t is odd. Because $(2^s, t) = 1$, we see from Lemma 7.1 that

(7.1) $$\sigma(n) = \sigma(2^s t) = \sigma(2^s)\sigma(t) = (2^{s+1} - 1)\sigma(t).$$

Because n is perfect, we have

(7.2) $$\sigma(n) = 2n = 2^{s+1}t.$$

Combining (7.1) and (7.2) shows that

(7.3) $$(2^{s+1} - 1)\sigma(t) = 2^{s+1}t.$$

Because $(2^{s+1}, 2^{s+1} - 1) = 1$, from Lemma 3.4 we see that $2^{s+1} \mid \sigma(t)$. Therefore, there is an integer q such that $\sigma(t) = 2^{s+1}q$. Inserting this expression for $\sigma(t)$ into (7.3) tells us that

$$(2^{s+1} - 1)2^{s+1}q = 2^{s+1}t,$$

and, therefore,

(7.4) $$(2^{s+1} - 1)q = t.$$

Hence, $q \mid t$ and $q \neq t$.

When we add q to both sides of (7.4), we find that

(7.5) $$t + q = (2^{s+1} - 1)q + q = 2^{s+1}q = \sigma(t).$$

We will show that $q = 1$. Note that if $q \neq 1$, then there are at least three distinct positive divisors of t, namely 1, q, and t. This implies that $\sigma(t) \geq t + q + 1$, which contradicts (7.5). Hence, $q = 1$ and, from (7.4), we conclude that $t = 2^{s+1} - 1$. Also, from (7.5), we see that $\sigma(t) = t + 1$, so that t must be prime, because its only positive divisors are 1 and t. Therefore, $n = 2^s(2^{s+1} - 1)$, where $2^{s+1} - 1$ is prime. ∎

By Theorem 7.10, we see that to find even perfect numbers, we must find primes of the form $2^m - 1$. In our search for primes of this form, we first show that the exponent m must be prime.

Theorem 7.11.   If m is a positive integer and $2^m - 1$ is prime, then m must be prime.

Proof.   Assume that m is not prime, so that $m = ab$, where $1 < a < m$ and $1 < b < m$. (Note that $m > 1$, since $2^m - 1$ is prime.) Then

$$2^m - 1 = 2^{ab} - 1 = (2^a - 1)(2^{a(b-1)} + 2^{a(b-2)} + \cdots + 2^a + 1).$$

Because both factors on the right side of the equation are greater than 1, we see that $2^m - 1$ is composite if m is not prime. Therefore, if $2^m - 1$ is prime, then m must also be prime. ∎

By Theorem 7.11, we see that to search for primes of the form $2^m - 1$, we need to consider only integers m that are prime. Integers of the form $2^m - 1$ have been studied in great depth; these integers are named after a French monk of the seventeenth century, *Marin Mersenne*, who studied them.

Definition.   If m is a positive integer, then $M_m = 2^m - 1$ is called the mth *Mersenne number*; if p is prime and $M_p = 2^p - 1$ is also prime, then M_p is called a *Mersenne prime*.

Example 7.11.   The Mersenne number $M_7 = 2^7 - 1$ is prime, whereas the Mersenne number $M_{11} = 2^{11} - 1 = 2047 = 23 \cdot 89$ is composite. ◀

It is possible to prove various theorems that help decide whether Mersenne numbers are prime. One such theorem will now be given. Related results are found in Exercises 37–39 in Section 11.1.

Theorem 7.12.   If p is an odd prime, then any divisor of the Mersenne number $M_p = 2^p - 1$ is of the form $2kp + 1$, where k is a positive integer.

Proof.   Let q be a prime dividing $M_p = 2^p - 1$. By Fermat's little theorem, we know that $q \mid (2^{q-1} - 1)$. Also, from Lemma 3.2, we know that

(7.6)
$$(2^p - 1, 2^{q-1} - 1) = 2^{(p,q-1)} - 1.$$

Because q is a common divisor of $2^p - 1$ and $2^{q-1} - 1$, we know that $(2^p - 1, 2^{q-1} - 1) > 1$. Hence, $(p, q - 1) = p$, because the only other possibility, namely $(p, q - 1) = 1$, would imply from (7.6) that $(2^p - 1, 2^{q-1} - 1) = 1$. Hence $p \mid (q - 1)$ and, therefore, there is a positive integer m such that $q - 1 = mp$. Because q is odd, we see that m must be even, so that $m = 2k$, where k is a positive integer. Hence, $q = mp + 1 = 2kp + 1$. Because any divisor of M_p is a product of prime divisors of M_p, each prime divisor of M_p is of the form $2kp + 1$, and the product of numbers of this form is also of this form, the result follows.   ∎

We can use Theorem 7.12 to help decide whether Mersenne numbers are prime. We illustrate this by the following examples.

Example 7.12.   To decide whether $M_{13} = 2^{13} - 1 = 8191$ is prime, we need only look for a prime factor not exceeding $\sqrt{8191} = 90.504\ldots$. Furthermore, by Theorem 7.12, any such prime divisor must be of the form $26k + 1$. The only candidates for primes dividing M_{13} less than or equal to $\sqrt{M_{13}}$ are 53 and 79. Trial division easily rules out these cases, so that M_{13} is prime.   ◀

Example 7.13.   To decide whether $M_{23} = 2^{23} - 1 = 8,388,607$ is prime, we only need to determine whether M_{23} is divisible by a prime less than or equal to $\sqrt{M_{23}} = 2896.309\ldots$ of the form $46k + 1$. The first prime of this form is 47. A trial division shows that $8,388,607 = 47 \cdot 178,481$, so that M_{23} is composite.   ◀

MARIN MERSENNE (1588–1648) was born in Maine, France, into a family of workers. He attended the College of Mans and the Jesuit College at La Flèche. He continued his education at the Sorbonne, studying theology. He joined the order of the Minims in 1611, a group whose name comes from the word *minimi* indicating that the members considered themselves the least religious order. Besides prayer, members pursued scholarship and study. In 1612, Mersenne became a priest at the Palace Royale in Paris; between 1614 and 1618, he taught philosophy at the Minim Convent in Nevers. He returned to Paris in 1619, where his cell in the Minims de l'Annociade was a meeting place for scientists, philosophers, and mathematicians, including Fermat and Pascal. Mersenne corresponded extensively with scholars throughout Europe, serving as a clearinghouse for new ideas. Mersenne wrote books on mechanics, mathematical physics, mathematics, music, and acoustics. He studied prime numbers and tried unsuccessfully to develop a formula representing all primes. In 1644, he claimed to have the complete list of primes p with $p \leq 257$ for which $2^p - 1$ is prime; this claim was far from accurate. Mersenne is also noted for his defense of two of the most famous men of his time, Descartes and Galileo, from religious critics. He also helped expose alchemists and astrologers as frauds.

Because there are special primality tests for Mersenne numbers, it has been possible to determine whether extremely large Mersenne numbers are prime.

A particularly useful primality test follows, known as the Lucas-Lehmer test after *Edouard Lucas*, who developed the theory the test is based on in the 1870s, and *Derrick H. Lehmer*, who developed a simplified version of the test in 1930. This test has been used to find the largest known Mersenne primes and is being used today in the ongoing search for new Mersenne primes, described later in this section. For most of recent history, the largest known Mersenne prime was the largest known prime as is currently the case. However, from late 1990 until early 1992, the largest known prime was $391{,}581 \cdot 2^{216{,}193} - 1$. Because this number is of the form $k \cdot 2^n - 1$, it was possible to use special tests to show that it is prime.

Theorem 7.13. *The Lucas-Lehmer Test.* Let p be a prime and let $M_p = 2^p - 1$ denote the pth Mersenne number. Define a sequence of integers recursively by setting $r_1 = 4$ and, for $k \geq 2$,

$$r_k \equiv r_{k-1}^2 - 2 \ (\mathrm{mod} \ M_p), 0 \leq r_k < M_p.$$

Then M_p is prime if and only if $r_{p-1} \equiv 0 \ (\mathrm{mod} \ M_p)$.

FRANÇOIS-EDOUARD-ANATOLE LUCAS (1842–1891) was born in Amiens, France, and was educated at the École Normale. After finishing his studies, he worked as an assistant at the Paris Observatory, and during the Franco-Prussian war he served as an artillery officer. After the war he became a teacher at a secondary school. He was considered to be an excellent and entertaining teacher. Lucas was extremely fond of calculating and devised plans for a computer, which unfortunately were never realized. Besides his contributions to number theory, Lucas is also remembered for his work in recreational mathematics. The most famous of his contributions in this area is the well-known tower of Hanoi problem. A freak accident led to Lucas's death. He was gashed in the cheek by a piece of a plate which was accidentally dropped at a banquet. An infection in the resulting wound killed him several days later.

DERRICK H. LEHMER (1905–1991) was born in Berkeley, California. He received his undergraduate degree in 1927 from the University of California and his master's and doctorate degrees from Brown University in 1929 and 1930, respectively. He served on the staffs of the California Institute of Technology, the Institute for Advanced Study, Lehigh University, and Cambridge University before joining the mathematics department at the University of California, Berkeley, in 1940. Lehmer made many contributions to number theory. He invented many special purpose devices for number theoretic computations, some with his father, who was also a mathematician. Lehmer was the thesis advisor of Harold Stark, who in turn was the thesis advisor of the author of this book.

The proof of the Lucas-Lehmer test may be found in [Le80] and [Si64]. We give an example to illustrate how the Lucas-Lehmer test is used.

Example 7.14. Consider the Mersenne number $M_5 = 2^5 - 1 = 31$. Then $r_1 = 4$, $r_2 \equiv 4^2 - 2 = 14 \pmod{31}$, $r_3 \equiv 14^2 - 2 \equiv 8 \pmod{31}$, and $r_4 \equiv 8^2 - 2 \equiv 0 \pmod{31}$. Because $r_4 \equiv 0 \pmod{31}$, we conclude that $M_5 = 31$ is prime. ◀

The Lucas-Lehmer test can be performed quite rapidly, as the following corollary states. It lets us test whether Mersenne numbers are prime without factoring them and makes it possible to determine whether extremely large Mersenne numbers are prime, whereas other numbers of similar size that are not of special form are beyond testing.

Corollary 7.13.2. Let p be prime and let $M_p = 2^p - 1$ denote the pth Mersenne number. It is possible to determine whether M_p is prime using $O(p^3)$ bit operations.

Proof. To determine whether M_p is prime using the Lucas-Lehmer test requires $p - 1$ squarings modulo M_p, each requiring $O((\log M_p)^2) = O(p^2)$ bit operations. Hence, the Lucas-Lehmer test requires $O(p^3)$ bit operations. ∎

It has been conjectured but not proved that there are infinitely many Mersenne primes. However, the search for larger and larger Mersenne primes has been quite successful.

The Search for Mersenne Primes

The history of the search for Mersenne primes can be divided into the eras before and after the advent of computers. In precomputer days, the search was littered with errors and unsubstantiated claims, many turning out to be false. By 1588, Pietro Cataldi had verified that M_{17} and M_{19} were primes, but he also stated, without any justification, that M_p was prime for $p = 23, 29, 31$, and 37 (of these, only M_{31} is prime). In his *Cogitata Physica-Mathematica,* published in 1644, Mersenne claimed (without providing a justification) that M_p is prime for $p = 2, 3, 5, 7, 13, 17, 19, 31, 67, 127$, and 257, and for no other prime p with $p < 257$. In 1772, Euler showed that M_{31} was prime, using trial division by all primes up to 46,337, which is the largest prime not exceeding the square root of M_{31}. In 1811, the English mathematician Peter Barlow wrote in his *Theory of Numbers* that M_{31} would be the greatest Mersenne prime ever found—he thought that no one would ever attempt to find a larger Mersenne prime because they are "merely curious, without being useful." This turned out to be a terrible prediction; not only was Barlow wrong about people finding new Mersenne primes, but he was wrong about their utility, as our subsequent comments will show.

In 1876, Lucas used the test that he had developed to show that M_{67} was composite without finding a factorization; it took an additional 27 years for M_{67} to be factored. The American mathematician Frank Cole devoted 20 years of Sunday-afternoon computations to discover that $M_{67} = 193,707,721 \cdot 761,838,257,287$. When he presented this result at a meeting of the American Mathematical Society in 1903, writing the factorization on a blackboard and not saying a word, the audience gave him a standing ovation, as they understood how much work had been required to find this factorization. The numbers M_{61}, M_{89}, M_{107}, and M_{127} were shown to be prime between 1876 and 1914. But it was not until 1947 that the primality of M_p for all primes p not exceeding 257 was tested, with the help of mechanical calculating machines. When this work was done, it was seen that Mersenne had made exactly five mistakes. He was wrong when he stated that M_{67} and M_{257} are primes, and he failed to include the Mersenne primes M_{61}, M_{89}, and M_{107} in his list.

As we have seen, only 12 Mersenne primes were known before the advent of modern computers, the last of which was discovered in 1914. But since the invention of computers, new Mersenne primes have been found at a fairly steady rate, averaging about one new Mersenne prime every two years since 1950. The first five Mersenne primes found with the help of a computer were the 13th through the 17th Mersenne primes. All five were found in 1952 by Raphael Robinson, using SWAC (the National Bureau of Standards Western Automatic Computer) with the help of D. H. and Emma Lehmer. The 13th and 14th Mersenne primes were found the first day SWAC was used to run the Lucas-Lehmer test, and the other three were found in the following nine months. Compared to computers today, SWAC was primitive. Its total memory was 1152 bytes, and half of this was used for the commands that ran the program. It is interesting to note that Robinson's program to implement the Lucas-Lehmer test was the first program he ever wrote.

Riesel found the 18th Mersenne prime using the Swedish BESK computer, Hurwitz found the 19th and 20th Mersenne primes using the IBM 7090, and Gillies found the 21st, 22nd, and 23rd Mersenne primes using the ILLIAC 2. Tuckerman found the 24th Mersenne prime using the IBM 360.

The 25th and 26th Mersenne primes were found by high school students Laura Nickel and Landon Noll using idle time on the Cyber 174 computer at California State University, Hayward. Nickel and Noll, who were 18 years old at the time, were also studying number theory with D. H. Lehmer and CSU professor Dan Jurca. Their discoveries were announced on the nightly news shows of major networks around the world. Nickel and Noll discovered the 25th Mersenne prime together, while only Noll went on to discover the 26th Mersenne prime by himself.

David Slowinski, working with several different collaborators, discovered the nth Mersenne prime for $n = 27, 28, 30, 31, 32, 33$, and 34 between 1979 and 1996. For example, Slowinski and Gage found the Mersenne prime $M_{1,257,787}$, a number with 378,632 digits, in 1996. The proof that this number is prime took approximately six hours on a Cray supercomputer. The Mersenne prime that Slowinski missed, the 29th, was found by Colquitt and Welsh in 1988 using a NEC SX-2 computer. You may wonder how Slowinski overlooked this prime. The reason is that he did not check whether M_p is prime for consecutive primes, but instead jumped around following hunches about the distribution of Mersenne primes, just as many researchers have done.

The Internet is another factor accelerating the discovery of Mersenne primes. Many people are cooperating to find new Mersenne primes as part of the Great Internet Mersenne Prime Search (GIMPS), founded by George Woltman in 1996. Approximately

15 trillion (10^{12}) floating point operations per second (15 Teraflops) are devoted to GIMPS on PrimeNet, the network linking the distributed computers in GIMPS into one virtual supercomputer. This virtual supercomputer is now the equivalent of more than a dozen of the largest supercomputers in the world, even though most of the individual computers used are Pentium PCs.

The six largest Mersenne primes known, the 35th through the 41st, were all found as part of the GIMPS project, with $M_{1,398,269}$ and $M_{2,976,221}$ discovered to be prime in 1996 and 1997, respectively. The Mersenne prime $M_{2,976,221}$ was shown to be prime using a 100 MHz Pentium computer using about 15 days of CPU time. In January 1998, $M_{3,021,377}$, a number with 909,526 decimal digits, was found to be prime by GIMPS. The lucky person who made this discovery, Roland Clarkson, was a 19-year-old student at California State University, Dominguez Hills, at the time. He used a 200 MHz Pentium computer, taking the equivalent of about a week of full-time CPU processing, to find this prime. The Mersenne $M_{6,972,593}$, a number with 2,098,960 decimal digits, was found in June 1999 by Nayan Hajratwala, a GIMPS participant, using a 350 MHz Pentium computer, using the equivalent of about three weeks of uninterrupted processing.

The 39th Mersenne prime, $M_{13,466,917}$, an integer with 4,053,946 decimal digits, was found in November 2001 by a 20-year-old Canadian university student, Michael Cameron. It took 42 days on an 800 MHz AMD personal computer to show that this number is prime. The 40th Mersenne prime is $M_{20,996,011}$, an integer with 6,320,430 decimal digits, which was shown to be prime in November 2003 by Michael Shafer, a 26-year-old chemical engineering graduate student at Michigan State University. He used a 2.4 GHz Pentium 4 personal computer running for 19 days to make this discovery. The 41st Mersenne prime, and the largest known prime as of June 2004, is $M_{24,036,583}$, an integer with 7,253,733 decimal digits which was shown to be prime in May 2004 by Josh Findley. He used a 2.4 GHz Pentium 4 PC running for 14 days to show this number is prime. The search for new Mersenne primes continues full blast, with more than 60,000 people looking for new ones by running GIMPS software on more than 200,000 personal computers. The next few years will show whether GIMPS can keep up their pace of finding a new Mersenne prime every year or two. (See Table 7.3 for a list of all the currently known Mersenne primes, along with information about their discovery.)

Why do people look for Mersenne primes?   Many people are devoted to the quest for new Mersenne primes. Why do they spend so much time and energy on this task? There are many reasons. The discovery of a new Mersenne prime brings fame and notoriety. Some people may be motivated by the recent cash prizes being offered for finding new Mersenne primes; other people like to contribute to team efforts. By joining GIMPS and PrimeNet, anyone can begin making useful contributions to the search for new Mersenne primes. The quest for new Mersenne primes has sparked the development of new theoretical results, and this has motivated many people; others are interested in the distribution of primes and want evidence to use as the basis for conjectures. Many people have used software for the Lucas-Lehmer test to check out new hardware platforms, as these programs are CPU and computer bus intensive. For example, the Intel Pentium II chip was tested using GIMPS software. Some people would rather have their computer

No.	p	Decimal Digits in M_p	Date of Discovery	Discoverer(s)	Computer Used
1	2	1	ancient times		
2	3	1	ancient times		
3	5	2	ancient times		
4	7	3	ancient times		
5	13	4	1456	anonymous	
6	17	6	1588	Cataldi	
7	19	6	1588	Cataldi	
8	31	10	1772	Euler	
9	61	19	1883	Pervushin	
10	89	27	1911	Powers	
11	107	33	1914	Powers	
12	127	39	1876	Lucas	
13	521	157	1952	Robinson	SWAC
14	607	183	1952	Robinson	SWAC
15	1279	386	1952	Robinson	SWAC
16	2203	664	1952	Robinson	SWAC
17	2281	687	1952	Robinson	SWAC
18	3217	969	1957	Riesel	BESK
19	4253	1281	1961	Hurwitz	IBM 7090
20	4423	1332	1961	Hurwitz	IBM 7090
21	9689	2917	1963	Gillies	ILLIAC 2
22	9941	2993	1963	Gillies	ILLIAC 2
23	11,213	3376	1963	Gillies	ILLIAC 2
24	19,937	6002	1971	Tuckerman	IBM 360/91
25	21,701	6533	1978	Noll, Nickel	Cyber 174
26	23,209	6987	1979	Noll	Cyber 174
27	44,497	13,395	1979	Nelson, Slowinski	Cray 1
28	86,243	25,962	1983	Slowinski	Cray 1
29	110,503	33,265	1988	Colquitt, Welsh	NEC SX-2
30	132,049	39,751	1983	Slowinski	Cray X-MP
31	216,091	65,050	1985	Slowinski	Cray X-MP
32	756,839	227,832	1992	Slowinski, Gage	Cray 2
33	859,433	258,716	1994	Slowinski, Gage	Cray 2
34	1,257,787	378,632	1996	Slowinski, Gage	Cray T94
35	1,398,269	420,921	1996	Armendgaud, Woltman (GIMPS)	90 MHz Pentium
36	2,976,221	895,952	1997	Spence, Woltman (GIMPS)	100 MHz Pentium
37	3,021,377	909,526	1998	Clarkson, Woltman, Kurowski (GIMPS, PrimeNet)	200 MHz Pentium
38	6,972,593	2,098,960	1999	Hajratwala, Woltman, Kurowski (GIMPS, PrimeNet)	350 MHz Pentium
39	13,466,917	4,053,946	2001	Cameron (GIMPS, PrimeNet)	800 MHz AMD
40	20,996,011	6,320,430	2003	Shafer (GIMPS, PrimeNet)	2 GHz Pentium 4
41	24,036,583	7,253,733	2004	Findley (GIMPS, PrimeNet)	2.4 GHz Pentium 4

Table 7.3 *The known Mersenne primes.*

look for Mersenne primes during idle time than run a screen-saver. For these and other reasons, many people look for Mersenne primes.

If you catch the bug and become interested in the search for Mersenne primes, you should investigate the GIMPS Web site, as well as several other relevant Web sites (links for these can be found in Appendix D and on the Web site for this book). At the GIMPS site, you can obtain a program for running the Lucas-Lehmer test, and learn how to join PrimeNet. The GIMPS program for running the Lucas-Lehmer test has been optimized in many ways, so that it runs much more efficiently than a naive implementation of the test. You can reserve a particular range of exponents to check. If history is a guide, it should not be too much longer before the world's record for Mersenne (and all) primes is smashed. If you join GIMPS, you may be the lucky one to break this record!

Odd Perfect Numbers

We have reduced the study of even perfect numbers to the study of Mersenne primes. But are there odd perfect numbers? The answer is still unknown. It is possible to demonstrate that if they exist, odd perfect numbers must have certain properties (see Exercises 32–36, for example). Furthermore, it is known that there are no odd perfect numbers less than 10^{300}; an odd perfect number must have at least eight different prime divisors and at least 37 prime divisors counting multiplicities; and the largest prime factor of the number must be at least 10^{20}. A discussion of odd perfect numbers may be found in [Gu94] or [Ri96], and information about recent results may be found in [BrCote93], [Co87], and [Ha83].

7.3 Exercises

1. Find the six smallest even perfect numbers.

2. Find the seventh and eighth even perfect numbers.

3. Find a factor of each of the following integers.
   a) $2^{15} - 1$
   b) $2^{91} - 1$
   c) $2^{1001} - 1$

A Prime Jackpot
When Nayan Hajratwala found the Mersenne prime $2^{6,972,593} - 1$, he was the first person to find a prime with more than 1 million decimal digits. This made him eligible for a prize of $50,000 from the Electronic Frontier Foundation (EFF), an organization devoted to protecting the health and growth of the Internet. You still have a chance to collect a prize from the EFF by finding large primes. They offer $100,000 for the first person who finds a prime with 10 million digits, a prize that most likely will be claimed within the next few years. Prizes of $150,000 and $250,000 are offered for the first person to find a prime with 100 million and 1 billion decimal digits, respectively. An anonymous donor has funded these prizes to spur cooperative work on scientific problems that involve massive computation.

4. Find a factor of each of the following integers.
   a) $2^{111} - 1$
   b) $2^{289} - 1$
   c) $2^{46,189} - 1$

If n is a positive integer, we say that n is *deficient* if $\sigma(n) < 2n$, and we say that n is *abundant* if $\sigma(n) > 2n$. Every integer is either deficient, perfect, or abundant.

5. Find the six smallest abundant positive integers.

* 6. Find the smallest odd abundant positive integer.

7. Show that every prime power is deficient.

8. Show that any proper divisor of a deficient or perfect number is deficient.

9. Show that any multiple of an abundant or perfect number, other than the perfect number itself, is abundant.

10. Show that if $n = 2^{m-1}(2^m - 1)$, where m is a positive integer such that $2^m - 1$ is composite, then n is abundant.

11. Show that there are infinitely many deficient numbers.

12. Show that there are infinitely many even abundant numbers.

13. Show that there are infinitely many odd abundant numbers.

14. Show that if $n = p^a q^b$, where p and q are distinct odd primes and a and b are positive integers, then n is deficient.

Two positive integers m and n are called an *amicable pair* if $\sigma(m) = \sigma(n) = m + n$.

15. Show that each of the following pairs of integers are amicable pairs.
    a) $220, 284$
    b) $1184, 1210$
    c) $79750, 88730$

16. a) Show that if n is a positive integer with $n \geq 2$, such that $3 \cdot 2^{n-1} - 1$, $3 \cdot 2^n - 1$, and $3^2 \cdot 2^{2n-1} - 1$ are all prime, then $2^n(3 \cdot 2^{n-1} - 1)(3 \cdot 2^n - 1)$ and $2^n(3^2 \cdot 2^{2n-1} - 1)$ form an amicable pair.
    b) Find three amicable pairs using part (a).

An integer n is called *k-perfect* if $\sigma(n) = kn$. Note that a perfect number is 2-perfect.

17. Show that $120 = 2^3 \cdot 3 \cdot 5$ is 3-perfect.

18. Show that $30{,}240 = 2^5 \cdot 3^3 \cdot 5 \cdot 7$ is 4-perfect.

19. Show that $14{,}182{,}439{,}040 = 2^7 \cdot 3^4 \cdot 5 \cdot 7 \cdot 11^2 \cdot 17 \cdot 19$ is 5-perfect.

20. Find all 3-perfect numbers of the form $n = 2^k \cdot 3 \cdot p$, where p is an odd prime.

21. Show that if n is 3-perfect and $3 \nmid n$, then $3n$ is 4-perfect.

An integer n is *k-abundant* if $\sigma(n) > (k + 1)n$.

22. Find a 3-abundant integer.

23. Find a 4-abundant integer.

** 24. Show that for each positive integer k there are an infinite number of k-abundant integers.

A positive integer n is called *superperfect* if $\sigma(\sigma(n)) = 2n$.

25. Show that 16 is superperfect.

26. Show that if $n = 2^q$, where $2^{q+1} - 1$ is prime, then n is superperfect.

27. Show that every even superperfect number is of the form $n = 2^q$, where $2^{q+1} - 1$ is prime.

* **28.** Show that if $n = p^2$, where p is an odd prime, then n is not superperfect.

29. Use Theorem 7.12 to determine whether each of the following Mersenne numbers is prime.

a) M_7 c) M_{17}

b) M_{11} d) M_{29}

30. Use the Lucas-Lehmer test, Theorem 7.13, to determine whether each of the following Mersenne numbers is prime.

a) M_3 c) M_{11}

b) M_7 d) M_{13}

* **31.** Show that if n is a positive integer and $2n + 1$ is prime, then either $(2n + 1) \mid M_n$ or $(2n + 1) \mid (M_n + 2)$. (*Hint:* Use Fermat's little theorem to show that $M_n(M_n + 2) \equiv 0 \pmod{2n + 1}$.)

* **32.** a) Show that if n is an odd perfect number, then $n = p^a m^2$, where p is an odd prime, $p \equiv a \equiv 1 \pmod 4$, and m is an integer.

   b) Use part (a) to show that if n is an odd perfect number, then $n \equiv 1 \pmod 4$.

* **33.** Show that if $n = p^a m^2$ is an odd perfect number, where p is prime, then $n \equiv p \pmod 8$.

* **34.** Show that if n is an odd perfect number, then 3, 5, and 7 are not all divisors of n.

* **35.** Show that if n is an odd perfect number, then n has at least three different prime divisors.

** **36.** Show that if n is an odd perfect number, then n has at least four different prime divisors.

37. Find all positive integers n such that the product of all divisors of n other than n is exactly n^2. (These integers are multiplicative analogues of perfect numbers.)

38. Let n be a positive integer. Define the *aliquot sequence* $n_1, n_2, n_3, \ldots,$ recursively by $n_1 = \sigma(n) - n$ and $n_{k+1} = \sigma(n_k) - n_k$ for $k = 1, 2, 3, \ldots.$ (The word *aliquot* is an adjective that means "contained an exact number of times in something else." Archaically, the *aliquot parts* of an integer were the divisors of this integer.)

   a) Show that if n is perfect, then $n = n_1 = n_2 = n_3 = \cdots$.

   b) Show that if n and m are an amicable pair, then $n_1 = m, n_2 = n, n_3 = m, n_4 = n, \ldots$ and so on; that is, the sequence $n_1, n_2, n_3, \ldots$ is periodic with period 2.

   c) Find the aliquot sequence of integers generated if $n = 12{,}496 = 2^4 \cdot 11 \cdot 71$.

Before computers were used to examine the behavior of aliquot sequences, it was conjectured that for all integers n the aliquot sequence of integers $n_1, n_2, n_3, \ldots$ is bounded. However, evidence obtained from calculations with large integers suggests that some of these sequences are unbounded.

* **39.** Show that if n is a positive integer greater than 1, then the Mersenne number M_n cannot be the power of a positive integer.

7.3 Computational and Programming Exercises

Computations and Explorations

Using a computation program such as Maple or *Mathematica*, or programs you have written, carry out the following computations and explorations.

1. Verify by direct computation that $2^{30}(2^{31} - 1)$ is perfect.

2. Show that the number 154,345,556,085,770,649,600 is a 6-perfect number (as defined in the preamble to Exercise 17).

3. Show that each of the following pairs of integers is an amicable pair (as defined in the preamble to Exercise 15).
   a) 609928, 686072
   b) 643336, 652664
   c) 938304290, 1344480478
   d) 4000783984, 4001351168

4. Find factors of as many Mersenne numbers of the form M_p, where p is prime, as you can, using Theorem 7.12.

5. Verify the primality of as many Mersenne primes as you can, using the Lucas-Lehmer test. (You may want to use GIMPS software to do this.)

6. Join the GIMPS and search for Mersenne primes.

7. Find all amicable pairs, where both integers in the pair are less than 10,000.

8. Show that the aliquot sequence (as defined in Exercise 38) obtained by taking $n = 14,316$ is periodic with period 28.

9. Find as many aliquot sequences as you can that are periodic with period 4.

10. Find the number of terms in the aliquot sequence obtained by taking $n = 138$ before this sequence reaches the integer 1. What is the largest term of the sequence? Can you answer the same question for $n = 276$?

Programming Projects

Write programs using Maple, *Mathematica,* or a language of your choice to do the following.

1. Classify positive integers according to whether they are deficient, perfect, or abundant (see the preamble to Exercise 5).

2. Use Theorem 7.12 to look for factors of Mersenne numbers.

3. Determine whether the Mersenne number $2^p - 1$ is prime, where p is a prime, using the Lucas-Lehmer test.

4. Given a positive integer n, determine if the aliquot sequence defined in Exercise 32 is periodic.

5. Given a positive integer n, find all amicable pairs of integers a, b, where $a \leq n$ and $b \leq n$ (see the preamble to Exercise 15).

7.4 Möbius Inversion

Let f be an arithmetic function. The formula $F(n) = \sum_{d|n} f(d)$ expresses the values of F, the summatory function of f, in terms of the values of f. Can this relationship be inverted? That is, is there a convenient way to express the values of f in terms of those of F? In this section, we will provide a useful formula that does this. We will start with some exploration, to help us see what kind of formula might exist.

Suppose that f is an arithmetic function and F is its summatory function $F(n) = \sum_{d|n} f(d)$. Expanding the definition of $F(n)$ for $n = 1, 2, \ldots, 8$, we see that

$$F(1) = f(1)$$
$$F(2) = f(1) + f(2)$$
$$F(3) = f(1) + f(3)$$
$$F(4) = f(1) + f(2) + f(4)$$
$$F(5) = f(1) + f(5)$$
$$F(6) = f(1) + f(2) + f(3) + f(6)$$
$$F(7) = f(1) + f(7)$$
$$F(8) = f(1) + f(2) + f(4) + f(8),$$

and so on. When we solve these equations successively for $f(n)$, for $n = 1, 2, \ldots, 8$, we find that

$$f(1) = F(1)$$
$$f(2) = F(2) - F(1)$$
$$f(3) = F(3) - F(1)$$
$$f(4) = F(4) - F(2)$$
$$f(5) = F(5) - F(1)$$
$$f(6) = F(6) - F(3) - F(2) + F(1)$$
$$f(7) = F(7) - F(1)$$
$$f(8) = F(8) - F(4).$$

Note that $f(n)$ equals a sum of terms of the form $\pm F(n/d)$, where $d \mid n$. From this evidence, it might be fruitful to look for an identity of the form

$$f(n) = \sum_{d|n} \mu(d) F(n/d),$$

where μ is an arithmetic function. If this identity holds, our computations imply that $\mu(1) = 1$, $\mu(2) = -1$, $\mu(3) = -1$, $\mu(4) = 0$, $\mu(5) = -1$, $\mu(6) = 1$, $\mu(7) = -1$, and $\mu(8) = 0$. Furthermore, $F(p) = f(1) + f(p)$, which implies that $f(p) = F(p) - F(1)$, whenever p is prime. This requires that $\mu(p) = -1$. Moreover, because

$$F(p^2) = f(1) + f(p) + f(p^2),$$

we have

$$f(p^2) = F(p^2) - (F(p) - F(1)) - F(1) = F(p^2) - F(p).$$

This implies that $\mu(p^2) = 0$ for every prime p. Similar reasoning can be used to show that $\mu(p^k) = 0$ for every prime p and integer $k > 1$. If we conjecture that μ is a multiplicative function, the values of μ are determined by those at prime powers. This leads to the following definition.

Definition.   The *Möbius function, $\mu(n)$,* is defined by

$$\mu(n) = \begin{cases} 1 & \text{if } n = 1; \\ (-1)^r & \text{if } n = p_1 p_2 \cdots p_r, \text{ where the } p_i \text{ are distinct primes}; \\ 0 & \text{otherwise.} \end{cases}$$

The Möbius function is named after *August Ferdinand Möbius.*

From the definition, we see that $\mu(n) = 0$ whenever n is divisible by the square of a prime. The only values of n for which $\mu(n) \neq 0$ are those n that are square-free.

Example 7.15.   From the definition of $\mu(n)$, we see that $\mu(1) = 1$, $\mu(2) = -1$, $\mu(3) = -1$, $\mu(4) = \mu(2^2) = 0$, $\mu(5) = -1$, $\mu(6) = \mu(2 \cdot 3) = 1$, $\mu(7) = -1$, $\mu(8) = \mu(2^3) = 0$, $\mu(9) = \mu(3^2) = 0$, and $\mu(10) = \mu(2 \cdot 5) = 1$.   ◀

Example 7.16.   We have $\mu(330) = \mu(2 \cdot 3 \cdot 5 \cdot 11) = (-1)^4 = 1$, $\mu(660) = \mu(2^2 \cdot 3 \cdot 5 \cdot 11) = 0$, and $\mu(4290) = \mu(2 \cdot 3 \cdot 5 \cdot 11 \cdot 13) = (-1)^5 = -1$.   ◀

We now verify that the Möbius function is multiplicative, proceeding directly from its definition.

Theorem 7.14.   The Möbius function $\mu(n)$ is a multiplicative function.

Proof.   Suppose that m and n are relatively prime positive integers. To show that $\mu(n)$ is multiplicative requires that we show that $\mu(mn) = \mu(m)\mu(n)$. To establish this equality, we first consider the case when $m = 1$ or $n = 1$. When $m = 1$, we see that both $\mu(mn)$ and $\mu(m)\mu(n)$ equal $\mu(n)$. The case for $n = 1$ is similar.

AUGUST FERDINAND MÖBIUS (1790–1868) was born in the town of Schulpforta, near Naumburg, Germany. His father was a dancing teacher and his mother was a descendant of Martin Luther. Möbius was taught at home until he was 13, displaying an interest and talent in mathematics at a young age. He received formal training in mathematics from 1803 until 1809, when he entered Leipzig University. He intended to study law, but instead decided to concentrate on subjects more to his interest—mathematics, physics, and astronomy. After pursuing further studies at Göttingen, where he studied astronomy with Gauss, and at Halle, where he studied mathematics with Pfaff, he became professor of astronomy at Leipzig, remaining there until his death. Möbius made contributions to a wide range of subjects, including astronomy, mechanics, projective geometry, optics, statics, and number theory. Today, he is best known for his discovery of a surface with one side, called the *Möbius strip,* which can be formed by taking a strip of paper and connecting two opposite ends after twisting it.

Now suppose that at least one of m and n is divisible by a square of a prime. Then mn is also divisible by the square of a prime. Consequently, $\mu(mn)$ and $\mu(m)\mu(n)$ are both equal to 0. Finally, consider the remaining case when both m and n are square-free integers greater than 1. Suppose that $m = p_1 p_2 \cdots p_s$, where $p_1, p_2, \ldots, p_s$ are distinct primes, and $n = q_1 q_2 \cdots q_t$, where $q_1, q_2, \ldots, q_t$ are distinct primes. Because m and n are relatively prime, no prime occurs in both of the prime factorizations of m and n. Consequently, mn is the product of $s + t$ distinct primes. It follows that $\mu(mn) = (-1)^{s+t} = (-1)^s(-1)^t = \mu(m)\mu(n)$. ∎

We will now show that the summatory function of the Möbius function is a particularly simple function.

Theorem 7.15. The summatory function of the Möbius function at the integer n, $F(n) = \sum_{d|n} \mu(d)$, satisfies

$$\sum_{d|n} \mu(d) = \begin{cases} 1 & \text{if } n = 1; \\ 0 & \text{if } n > 1. \end{cases}$$

Proof. First consider the case when $n = 1$. We have

$$F(1) = \sum_{d|1} \mu(d) = \mu(1) = 1.$$

Next, let $n > 1$. By Theorem 7.8, because μ is a multiplicative function, its summatory function $F(n) = \sum_{d|n} \mu(d)$ is also multiplicative. Now, suppose that p is prime and k is a positive integer. We see that

$$F(p^k) = \sum_{d|p^k} \mu(d) = \mu(1) + \mu(p) + \mu(p^2) + \cdots + \mu(p^k)$$
$$= 1 + (-1) + 0 + \cdots + 0 = 0$$

because $\mu(p^i) = 0$ whenever $i \geq 2$. Finally, suppose that n is a positive integer, $n > 1$, with prime-power factorization $n = p_1^{a_1} p_2^{a_2} \cdots p_t^{a_t}$. Because F is multiplicative, it follows that $F(n) = F(p_1^{a_1}) F(p_2^{a_2}) \cdots F(p_t^{a_t})$. Because each of the factors on the right-hand side of this equation is 0, it follows that $F(n) = 0$. ∎

The Möbius inversion formula provides an answer to the question posed at the beginning of this section. It provides a way to express the values of f in terms of values of its summatory function F. This formula is used extensively in the study of multiplicative functions and can be used to establish new identities involving these functions.

Theorem 7.16. *The Möbius Inversion Formula.* Suppose that f is an arithmetic function and that F is the summatory function of f, so that

$$F(n) = \sum_{d|n} f(d).$$

Then, for all positive integers n,

$$f(n) = \sum_{d|n} \mu(d) F(n/d).$$

Proof. The proof of this formula involves some manipulations of double sums. We proceed as follows, starting with the sum on the right-hand side of the formula, substituting for $F(n/d)$ the expression $\sum_{e|(n/d)} f(e)$, which comes from the definition of the function F as the summatory function of f. We have

$$\sum_{d|n} \mu(d) F(n/d) = \sum_{d|n} \left(\mu(d) \sum_{e|(n/d)} f(e) \right)$$

$$= \sum_{d|n} \left(\sum_{e|(n/d)} \mu(d) f(e) \right).$$

Note that the pairs of integers (d, e) with $d \mid n$ and $e \mid (n/d)$ are the same as those with $e \mid n$ and $d \mid (n/e)$. It follows that

$$\sum_{d|n} \left(\sum_{e|(n/d)} \mu(d) f(e) \right) = \sum_{e|n} \left(\sum_{d|(n/e)} f(e) \mu(d) \right)$$

$$= \sum_{e|n} \left(f(e) \sum_{d|(n/e)} \mu(d) \right).$$

Now we see by Theorem 7.15 that $\sum_{d|(n/e)} \mu(d) = 0$ unless $n/e = 1$. When $n/e = 1$, that is, when $n = e$, this sum equals 1. Consequently,

$$\sum_{e|n} \left(f(e) \sum_{d|(n/e)} \mu(d) \right) = f(n) \cdot 1 = f(n).$$

This completes the proof.   ∎

The Möbius inversion formula can be used to construct many new identities that would be difficult to prove in another manner, as the following example shows.

Example 7.17. The functions $\sigma(n)$ and $\tau(n)$ are the summatory functions of the functions $f(n) = n$ and $f(n) = 1$, respectively, as noted in Section 7.2. That is, $\sigma(n) = \sum_{d|n} d$ and $\tau(n) = \sum_{d|n} 1$. By the Möbius inversion formula, we can conclude that for all integers n,

$$n = \sum_{d|n} \mu(n/d) \sigma(d)$$

and

$$1 = \sum_{d|n} \mu(n/d) \tau(d).$$

Proving these two identities directly would be difficult.   ◄

By Theorem 7.8, we know that if f is a multiplicative function, then so is its summary function, $F(n) = \sum_{d|n} f(d)$. Another useful consequence of the Möbius inversion formula is that we can turn this statement around. That is, if the summatory function F of an arithmetic function f is multiplicative, then so is f.

Theorem 7.17.   Let f be an arithmetic function with summatory $F = \sum_{d|n} f(d)$. Then, if F is multiplicative, f is also multiplicative.

Proof.   Suppose that m and n are relatively prime positive integers. We want to show that $f(mn) = f(m)f(n)$. To show this, first note that by Lemma 3.7, if d is a divisor of mn, then $d = d_1 d_2$ where $d_1 \mid m$, $d_2 \mid n$, and $(d_1, d_2) = 1$. Using the Möbius inversion formula and the fact that μ and F are multiplicative, we see that

$$f(mn) = \sum_{d|mn} \mu(d) F\left(\frac{mn}{d}\right)$$

$$= \sum_{d_1|m,\, d_2|n} \mu(d_1 d_2) F\left(\frac{mn}{d_1 d_2}\right)$$

$$= \sum_{d_1|m,\, d_2|n} \mu(d_1)\mu(d_2) F\left(\frac{m}{d_1}\right) F\left(\frac{n}{d_2}\right)$$

$$= \sum_{d_1|m} \mu(d_1) F\left(\frac{m}{d_1}\right) \cdot \sum_{d_2|n} \mu(d_2) F\left(\frac{n}{d_2}\right)$$

$$= f(m) f(n).\qquad\blacksquare$$

7.4 Exercises

1. Find the following values of the Möbius function.

a) $\mu(12)$ d) $\mu(50)$ g) $\mu(10!)$

b) $\mu(15)$ e) $\mu(1001)$

c) $\mu(30)$ f) $\mu(2 \cdot 3 \cdot 5 \cdot 7 \cdot 11 \cdot 13)$

2. Find the following values of the Möbius function.

a) $\mu(33)$ d) $\mu(740)$ g) $\mu(10!/(5!)^2)$

b) $\mu(105)$ e) $\mu(999)$

c) $\mu(110)$ f) $\mu(3 \cdot 7 \cdot 13 \cdot 19 \cdot 23)$

3. Find the value of $\mu(n)$ for each integer n with $100 \le n \le 110$.

4. Find the value of $\mu(n)$ for each integer n with $1000 \le n \le 1010$.

5. Find all integers n, $1 \le n \le 100$ with $\mu(n) = 1$.

6. Find all composite integers n, $100 \le n \le 200$ with $\mu(n) = -1$.

The *Mertens function* $M(n)$ is defined by $M(n) = \sum_{i=1}^{n} \mu(i)$.

7. Find $M(n)$ for all positive integers not exceeding 10.

8. Find $M(n)$ for $n = 100$.

9. Show that $M(n)$ is the difference between the number of square-free positive integers not exceeding n with an even number of prime divisors and those with an odd number of prime divisors.

10. Show that if n is a positive integer, then $\mu(n)\mu(n+1)\mu(n+2)\mu(n+3) = 0$.

11. Prove or disprove that there are infinitely many positive integers n such that $\mu(n) + \mu(n+1) = 0$.

12. Prove or disprove that there are infinitely many positive integers n such that $\mu(n-1) + \mu(n) + \mu(n+1) = 0$.

13. For how many consecutive integers can the Möbius function $\mu(n)$ take a nonzero value?

14. For how many consecutive integers can the Möbius function $\mu(n)$ take the value 0?

15. Show that if n is a positive integer, then $\phi(n) = n \sum_{d|n} \mu(d)/d$. (*Hint:* Use the Möbius inversion formula.)

16. Use the Möbius inversion formula and the identity $n = \sum_{d|n} \phi(n/d)$, demonstrated in Section 7.1, to show the following.
    a) $\phi(p^t) = p^t - p^{t-1}$, whenever p is prime and t is a positive integer.
    b) $\phi(n)$ is multiplicative.

17. Suppose that f is a multiplicative function with $f(1) = 1$. Show that

$$\sum_{d|n} \mu(d)f(d) = (1 - f(p_1))(1 - f(p_2)) \cdots (1 - f(p_k)),$$

where $n = p_1^{a_1} p_2^{a_2} \ldots p_k^{a_k}$ is the prime-power factorization of n.

18. Use Exercise 17 to find a simple formula for $\sum_{d|n} d\mu(d)$ for all positive integers n.

19. Use Exercise 17 to find a simple formula for $\sum_{d|n} \mu(d)/d$ for all positive integers n.

20. Use Exercise 17 to find a simple formula for $\sum_{d|n} \mu(d)\tau(d)$ for all positive integers n.

21. Use Exercise 17 to find a simple formula for $\sum_{d|n} \mu(d)\sigma(d)$ for all positive integers n.

22. Let n be a positive integer. Show that

$$\prod_{d|n} \mu(d) = \begin{cases} -1 & \text{if } n \text{ is a prime;} \\ 0 & \text{if } n \text{ has a square factor;} \\ 1 & \text{if } n \text{ is square-free and composite.} \end{cases}$$

23. Show that

$$\sum_{d|n} \mu^2(d) = 2^{\omega(n)},$$

where $\omega(n)$ denotes the number of distinct prime factors of n.

24. Use Exercise 23 and the Möbius inversion formula to show that

$$\mu^2(n) = \sum_{d|n} \mu(d)2^{\omega(n/d)}.$$

25. Show that $\sum_{d|n} \mu(d)\lambda(d) = 2^{\omega(n)}$ for all positive integers n, where $\omega(n)$ is the number of distinct prime factors of n. (See the preamble to Exercise 43 in Section 7.1 for a definition of $\lambda(n)$.)

26. Show that $\sum_{d|n} \lambda(n/d)2^{\omega(d)} = 1$ for all positive integers n.

Exercises 27–29 provide a proof of the Möbius inversion formula and Theorem 7.17 using the concepts of the Dirichlet product and the Dirichlet inverse, defined in the exercise set of Section 7.1.

27. Show that the Möbius function $\mu(n)$ is the Dirichlet inverse of the function $\nu(n) = 1$.

28. Use Exercise 38 in Section 7.1 and Exercise 27 to prove the Möbius inversion formula.

29. Prove Theorem 7.17 by noting that if $F = f \star \nu$, where $\nu = 1$ for all positive integers n, then $f = F \star \mu$.

The *Mangoldt function* Λ is defined for all positive integers n by

$$\Lambda(n) = \begin{cases} \log p & \text{if } n = p^k, \text{ where } p \text{ is prime and } k \text{ is a positive integer;} \\ 0 & \text{otherwise.} \end{cases}$$

30. Show that $\sum_{d|n} \Lambda(d) = \log n$ whenever n is a positive integer.

31. Use the Möbius inversion formula and Exercise 30 to show that

$$\Lambda(n) = - \sum_{d|n} \mu(d) \log d.$$

7.4 Computational and Programming Exercises

Computations and Explorations

Using a computation program such as Maple or *Mathematica,* or programs you have written, carry out the following computations and explorations.

1. Find $\mu(n)$ for each of the following values of n.

    a) 421,602,180,943   b) 186,728,732,190   c) 737,842,183,177

2. Find $M(n)$, the value of the Mertens function at n, for each of the following integers. (See the preamble to Exercise 7 for the definition of $M(n)$.)

    a) 1000   b) 10,000   c) 100,000

3. A famous conjecture made in 1897 by F. Mertens, and disproved in 1985 by A. Odlyzko and H. te Riele (in [Odte85]), was that $|M(n)| < \sqrt{n}$ for all positive integers n, where $M(n)$ is the Mertens function. Show that this conjecture, called Mertens' conjecture, is true for all integers n for as large a range as you can. Do not expect to find a counterexample, because the smallest n for which the conjecture is false is fantastically large. What is known is that there is a counterexample less than $3.21 \cdot 10^{64}$. Before the conjecture was shown to be false, it had been checked by computer for all integers n up to 10^{10}. This shows that even a tremendous amount of evidence can be misleading, because the smallest counterexample to a conjecture can nevertheless be titanically large.

Programming Projects

Write programs using Maple, *Mathematica,* or a language of your choice to do the following.

1. Given a positive integer n, find the value of $\mu(n)$.

2. Given a positive integer n, find the value of $M(n)$.

3. Given a positive integer n, check whether Mertens' conjecture holds for n, that is, whether $|M(n)| = |\sum_{i=1}^{n} \mu(i)| \leq \sqrt{n}$.

8

Cryptology

Introduction

How can you make a message secret, so that only the intended recipient of the message can recover it? This problem has interested people since ancient times, especially in diplomacy, military affairs, and commerce. In the modern world, making messages secret has become even more important, especially with the advent of electronic messaging and the Internet. This chapter is devoted to cryptology, the discipline devoted to secrecy systems. We will introduce some of the classical methods for making messages secret, starting with methods used in the Roman Empire, 2000 years ago. We will describe variations and modifications of these classical methods developed in the past two centuries, all based on modular arithmetic, and introduce the basic terminology and concepts of cryptology through our study of these methods. In all these classical systems, two people who wish to communicate privately must share a common secret key.

Since the 1970s, the notion of public key cryptography has been introduced and developed. In public key cryptography, two people who wish to communicate need not share a common key; instead, each person has both a private key that only this person knows and a public key that everyone knows. Using a public key system, you can send someone a message using their public key so that only that person can recover the message, using the corresponding private key. We will introduce the RSA cryptosystem, the most commonly used public key cryptosystem, whose security is based on the difficulty of factoring integers. We will also study a proposed public key cryptosystem, based on the knapsack problem, which (although promising) turned out not to be suitable.

Finally, we will discuss some cryptographic protocols. These are algorithms used to create agreements among two or more parties to achieve some common goal. We will show how cryptographic techniques that we have developed can be used to allow people to share common encryption keys, to sign electronic messages, to play poker electronically, and to share a secret.

8.1 Character Ciphers

Some Terminology

Before discussing specific secrecy systems, we present the basic terminology of secrecy systems. The discipline devoted to secrecy systems is called *cryptology. Cryptography* is the part of cryptology that deals with the design and implementation of secrecy systems, while *cryptanalysis* is aimed at "breaking" (defeating) these systems. A message that is to be altered into a secret form is called *plaintext.* A *cipher,* or *encryption, method* is a procedure method for altering a plaintext message into *ciphertext* by changing the letters of the plaintext using a transformation. The *key* determines a particular transformation from a set of possible transformations. The process of changing plaintext into ciphertext is called *encryption,* or *enciphering,* while the reverse process of changing the ciphertext back to the plaintext by the intended receiver, who possesses knowledge of the method for doing so, is called *decryption,* or *deciphering.* This, of course, is different from the process that someone other than the intended receiver uses to make the message intelligible, through cryptanalysis.

By a *cryptosystem* we mean the collection made up of a set of allowable plaintext messages, a set of possible ciphertext messages, a set of keys where each key specifies a particular encryption function, and the corresponding encryption functions and decryption functions. Formally, a cryptosystem is a system that consists of a finite set $\mathcal{P}$ of possible plaintext messages, a finite set $\mathcal{C}$ of possible ciphertext messages, a *keyspace* $\mathcal{K}$ of possible keys, and for each key k in the keyspace $\mathcal{K}$, an encryption function E_k and a corresponding decryption function D_k, such that $D_k(E_k(x)) = x$ for every plaintext message x.

The Caesar Cipher

In this chapter, we present secrecy systems based on modular arithmetic. The first of these had its origin with Julius Caesar; the newest systems that we will discuss were invented in the late 1970s. In all these systems, we start by translating letters into numbers. We take as our standard alphabet the letters of English and translate them into the integers from 0 to 25, as shown in Table 8.1.

Of course, if we were sending messages in Russian, Greek, Hebrew, or any other language, we would use the appropriate alphabet and range of integers. Also, we may want to include all ASCII characters, including punctuation marks, a symbol to indicate blanks, and the digits for representing numbers as part of the message. However, for

Letter	A	B	C	D	E	F	G	H	I	J	K	L	M	N	O	P	Q	R	S	T	U	V	W	X	Y	Z
Numerical Equivalent	0	1	2	3	4	5	6	7	8	9	10	11	12	13	14	15	16	17	18	19	20	21	22	23	24	25

Table 8.1 *The numerical equivalents of letters.*

the sake of simplicity, we restrict ourselves to the letters of the English alphabet. The transformation of letters to numbered equivalents can be done in many other ways (including translation to bit strings). Here we have chosen a simple and easily understood transformation for simplicity.

First, we discuss secrecy systems based on transforming each letter of the plaintext message into a different letter (or possibly the same) to produce the ciphertext. The encryption methods in these cryptosystems are called *character,* or *monographic, ciphers,* because each character is changed individually to another letter by a *substitution.* Altogether, there are 26! possible ways to produce a monographic transformation. We will discuss some particular monographic transformations based on modular arithmetic.

Julius Caesar used a cipher based on the substitution in which each letter is replaced by the letter three further down the alphabet, with the last three letters shifted to the first three letters of the alphabet. To describe this cipher using modular arithmetic, let P be the numerical equivalent of a letter in the plaintext and C be the numerical equivalent of the corresponding ciphertext letter. Then

$$C \equiv P + 3 \;(\mathrm{mod}\; 26), \quad 0 \le C \le 25.$$

The correspondence between plaintext and ciphertext is given in Table 8.2.

	A	B	C	D	E	F	G	H	I	J	K	L	M	N	O	P	Q	R	S	T	U	V	W	X	Y	Z
Plaintext	0	1	2	3	4	5	6	7	8	9	10	11	12	13	14	15	16	17	18	19	20	21	22	23	24	25
	3	4	5	6	7	8	9	10	11	12	13	14	15	16	17	18	19	20	21	22	23	24	25	0	1	2
Ciphertext	D	E	F	G	H	I	J	K	L	M	N	O	P	Q	R	S	T	U	V	W	X	Y	Z	A	B	C

Table 8.2 *The correspondence of letters for the Caesar cipher.*

To encrypt a message using this transformation, we first change it to its numerical equivalent, grouping letters in blocks of five. Then we transform each number. The grouping of letters into blocks helps to prevent successful cryptanalysis based on recognizing particular words. We illustrate this procedure in Example 8.1

Example 8.1. To encrypt the message

THIS MESSAGE IS TOP SECRET,

we break it into groups of five letters. The message becomes

THISM ESSAG EISTO PSECR ET.

Converting the letters into their numerical equivalents, we obtain

$$\begin{array}{ccccc} 19 & 7 & 8 & 18 & 12 \\ 15 & 18 & 4 & 2 & 17 \end{array} \quad \begin{array}{cc} 4 & 18 \\ 4 & 19. \end{array} \quad \begin{array}{ccc} 18 & 0 & 6 \end{array} \quad \begin{array}{cccc} 4 & 8 & 18 & 19 & 14 \end{array}$$

Using the Caesar transformation $C \equiv P + 3 \pmod{26}$, this becomes

22 10 11 21 15   7 21   21 3 9   7 11 21 22 17
18 21   7   5 20   7 22.

Translating back to letters, we have

WKLVP HVVDJ HLVWR SVHFU HW.

This is the encrypted message.   ◄

The receiver decrypts a message in the following manner. First, the letters are converted to numbers. Then, the relationship $P \equiv C - 3 \pmod{26}$, $0 \leq P \leq 25$, is used to change the ciphertext back to the numerical version of the plaintext, and finally the message is converted to letters.

We illustrate the deciphering procedure in the following example.

Example 8.2.   To decrypt the message

WKLVL VKRZZ HGHFL SKHU

encrypted by the Caesar cipher, we first change these letters into their numerical equivalents, to obtain

22 10 11 21 11   21 10 17 25 25   7 6 7 5 11   18 10 7 20.

Next, we perform the transformation $P \equiv C - 3 \pmod{26}$ to change this to plaintext, and we obtain

19 7 8 18 8   18 7 14 22 22   4 3 4 2 8   15 7 4 17.

We translate this back to letters and recover the plaintext message.

THISI SHOWW EDECI PHER

By combining the appropriate letters into words, we find that the message reads

THIS IS HOW WE DECIPHER   ◄

Affine Transformation

The Caesar cipher is one of a family of similar ciphers described by a *shift transformation*.

$$C \equiv P + k \pmod{26}, \quad 0 \leq C \leq 25,$$

where k is the key representing the size of the shift of letters in the alphabet. There are 26 different transformations of this type, including the case of $k \equiv 0 \pmod{26}$, where letters are not altered, because in this case $C \equiv P \pmod{26}$.

More generally, we will consider transformations of the type

(8.1) $$C \equiv aP + b \pmod{26}, \quad 0 \leq C \leq 25,$$

where a and b are integers with $(a, 26) = 1$. These are called *affine transformations*. Shift transformations are affine transformations with $a = 1$. We require that $(a, 26) = 1$, so that as P runs through a complete system of residues modulo 26, C also does. There are $\phi(26) = 12$ choices for a, and 26 choices for b, giving a total of $12 \cdot 26 = 312$ transformations of this type (one of these is $C \equiv P \pmod{26}$ obtained when $a = 1$ and $b = 0$). If the relationship between plaintext and ciphertext is described by (8.1), then the inverse relationship is given by

$$P \equiv \bar{a}(C - b) \pmod{26}, \quad 0 \le P \le 25,$$

where $\bar{a}$ is an inverse of $a \pmod{26}$, which can be found using the congruence $\bar{a} \equiv a^{\phi(26)-1} = a^{11} \pmod{26}$.

We illustrate how affine transformations work in Example 8.3.

Example 8.3. Let $a = 7$ and $b = 10$ in an affine cipher with $C \equiv aP + b \pmod{26}$, so that $C \equiv 7P + 10 \pmod{26}$. Note that $P \equiv 15(C - 10) \equiv 15C + 6 \pmod{26}$, because 15 is an inverse of 7 modulo 26. The correspondence between letters is given in Table 8.3.

	A	B	C	D	E	F	G	H	I	J	K	L	M	N	O	P	Q	R	S	T	U	V	W	X	Y	Z
Plaintext	0	1	2	3	4	5	6	7	8	9	10	11	12	13	14	15	16	17	18	19	20	21	22	23	24	25
	10	17	24	5	12	19	0	7	14	21	2	9	16	23	4	11	18	25	6	13	20	1	8	15	22	3
Ciphertext	K	R	Y	F	M	T	A	H	O	V	C	J	Q	X	E	L	S	Z	G	N	U	B	I	P	W	D

Table 8.3 *The correspondence of letters for the cipher with $C \equiv 7P + 10$ (mod 26).*

To illustrate how we obtained this correspondence, note that the plaintext letter L with numerical equivalent 11 corresponds to the ciphertext letter J, because $7 \cdot 11 + 10 = 87 \equiv 9 \pmod{26}$ and 9 is the numerical equivalent of J.

To illustrate how to encrypt, note that

PLEASE SEND MONEY

is transformed to

LJMKG MGMXF QEXMW.

Also note that the ciphertext

FEXEN ZMBMK JNHMG MYZMN

corresponds to the plaintext

DONOT REVEA LTHES ECRET,

or, combining the appropriate letters,

DO NOT REVEAL THE SECRET. ◄

We now discuss some of the techniques directed at the cryptanalysis of ciphers based on affine transformations. In attempting to break a monographic cipher, the frequency of letters in the ciphertext is compared with the frequency of letters in ordinary text. This gives information concerning the correspondence between letters. In various frequency counts of English text, one finds the percentages listed in Table 8.4 for the occurrence of the 26 letters of the alphabet. Counts of letter frequencies in other languages may be found in [Fr78] and [Ku76].

Letter	A	B	C	D	E	F	G	H	I	J	K	L	M	N	O	P	Q	R	S	T	U	V	W	X	Y	Z
Frequency (in %)	7	1	3	4	13	3	2	3	8	<1	<1	4	3	8	7	3	<1	8	6	9	3	1	1	<1	2	<1

Table 8.4 *The frequencies of occurrence of the letters of the alphabet.*

From this information, we see that the most frequently occurring letters in typical English text are E, T, N, R, I, O, and A, with E occurring substantially more than the other letters, 13% of the time, and T, N, R, I, O, and A each occurring between 7% and 9% of the time. We can use this information to determine which cipher based on an affine transformation has been used to encrypt a message. We illustrate how this cryptanalysis is done in the following example.

Example 8.4.   Suppose that we know in advance that a shift cipher has been employed to encrypt a message; each letter of the message has been transformed by a correspondence $C \equiv P + k \pmod{26}$, $0 \leq C \leq 25$. To cryptanalyze the ciphertext

> Y F X M P   C E S P Z   C J T D F   D P Q F W   Q Z C P Y
> N T A S P   C T Y R X   P D D L R   P D,

we first count the number of occurrences of each letter in the ciphertext. This is displayed in Table 8.5.

Letter	A	B	C	D	E	F	G	H	I	J	K	L	M	N	O	P	Q	R	S	T	U	V	W	X	Y	Z
Number of Occurrences	1	0	4	5	1	3	0	0	0	1	0	1	1	1	0	7	2	2	2	3	0	0	1	2	3	2

Table 8.5 *The number of occurrences of letters in a ciphertext.*

We notice that the most frequently occurring letter in the ciphertext is P, with the letters C, D, F, T, and Y occurring with relatively high frequency. Our initial guess would be that P represents E, since E is the most frequently occurring letter in English text. If this is so, then $15 \equiv 4 + k \pmod{26}$, so that $k \equiv 11 \pmod{26}$. Consequently, we would have $C \equiv P + 11 \pmod{26}$ and $P \equiv C - 11 \pmod{26}$. This correspondence is given in Table 8.6.

	A	B	C	D	E	F	G	H	I	J	K	L	M	N	O	P	Q	R	S	T	U	V	W	X	Y	Z
Ciphertext	0	1	2	3	4	5	6	7	8	9	10	11	12	13	14	15	16	17	18	19	20	21	22	23	24	25
	15	16	17	18	19	20	21	22	23	24	25	0	1	2	3	4	5	6	7	8	9	10	11	12	13	14
Plaintext	P	Q	R	S	T	U	V	W	X	Y	Z	A	B	C	D	E	F	G	H	I	J	K	L	M	N	O

Table 8.6 *Correspondence of letters for the sample ciphertext.*

Using this correspondence, we attempt to decrypt the message. We obtain

NUMBE   RTHEO   RYISU   SEFUL   FOREN
CIPHE   RINGM   ESSAG   ES.

This can easily be read as

NUMBER THEORY IS USEFUL FOR
ENCIPHERING MESSAGES.

Consequently, we made the correct guess. If we had tried this transformation, and instead of plaintext, it produced garbled text, we would have tried another likely transformation based on the frequency count of letters in the ciphertext. ◄

Example 8.5. Suppose we know that an affine transformation of the form $C \equiv aP + b$ (mod 26), $0 \le C \le 25$, has been used for encryption. For instance, suppose that we wish to cryptanalyze the encrypted message

USLEL   JUTCC   YRTPS   URKLT   YGGFV
ELYUS   LRYXD   JURTU   ULVCU   URJRK
QLLQL   YXSRV   LBRYZ   CYREK   LVEXB
RYZDG   HRGUS   LJLLM   LYPDJ   LJTJU
FALGU   PTGVT   JULYU   SLDAL   TJRWU
SLJFE   OLPU.

The first thing to do is to count the occurrences of each letter; this count is displayed in Table 8.7.

Letter	A	B	C	D	E	F	G	H	I	J	K	L	M	N	O	P	Q	R	S	T	U	V	W	X	Y	Z
Number of Occurrences	2	2	4	4	5	3	6	1	0	10	3	22	1	0	1	4	2	12	7	8	16	5	1	3	10	2

Table 8.7 *The number of occurrences of letters in a ciphertext.*

With this information, we guess that the letter L, which is the most frequently occurring letter in the ciphertext, corresponds to E, while the letter U, which occurs with the second-highest frequency, corresponds to T. This implies, if the transformation is of the form $C \equiv aP + b$ (mod 26), the pair of congruences

$$4a + b \equiv 11 \pmod{26}$$
$$19a + b \equiv 20 \pmod{26}.$$

By Theorem 4.15 we see that the solution of this system is $a \equiv 11 \pmod{26}$ and $b \equiv 19 \pmod{26}$.

If this is the correct enciphering transformation, then using the fact that 19 is an inverse of 11 modulo 26, the deciphering transformation is

$$P \equiv 19(C - 19) \equiv 19C - 361 \equiv 19C + 3 \pmod{26}, \ 0 \le P \le 25$$

This gives the correspondence found in Table 8.8.

	A	B	C	D	E	F	G	H	I	J	K	L	M	N	O	P	Q	R	S	T	U	V	W	X	Y	Z
Ciphertext	0	1	2	3	4	5	6	7	8	9	10	11	12	13	14	15	16	17	18	19	20	21	22	23	24	25
	3	22	15	8	1	20	13	6	25	18	11	4	23	16	9	2	21	14	7	0	19	12	5	24	17	10
Plaintext	D	W	P	I	B	U	N	G	Z	S	L	E	X	Q	J	C	V	O	H	A	T	M	F	Y	R	K

Table 8.8 *The correspondence of letters for the sample ciphertext.*

With this correspondence, we try to read the ciphertext, which becomes

```
T H E B E    S T A P P    R O A C H    T O L E A    R N N U M
B E R T H    E O R Y I    S T O A T    T E M P T    T O S O L
V E E V E    R Y H O M    E W O R K    P R O B L    E M B Y W
O R K I N    G O N T H    E S E E X    E R C I S    E S A S T
U D E N T    C A N M A    S T E R T    H E I D E    A S O F T
H E S U B    J E C T
```

We leave it to the reader to combine the appropriate letters into words to see that the message is intelligible.   ◄

The methods described in this section can be extended to construct cryptosystems more difficult to break than character ciphers. For example, plaintext letters can be shifted by different amounts, as is done in Vigenère ciphers, described in Section 8.2. Additional methods based on enciphering blocks of letters, rather than individual characters will also be described in Section 8.2 and in subsequent sections of this chapter, as will ciphers where the key used to encrypt characters changes from character to character.

8.1 Exercises

1. Using the Caesar cipher, encrypt the message ATTACK AT DAWN.

2. Decrypt the ciphertext message LFDPH LVDZL FRQTX HUHG, which has been encrypted using the Caesar cipher.

3. Encrypt the message SURRENDER IMMEDIATELY using the affine transformation $C \equiv 11P + 18 \pmod{26}$.

4. Encrypt the message THE RIGHT CHOICE using the affine transformation $C \equiv 15P + 14 \pmod{26}$.

5. Decrypt the message YLFQX PCRIT, which was encrypted using the affine transformation $C \equiv 21P + 5 \pmod{26}$.

6. Decrypt the message RTOLK TOIK, which was encrypted using the affine transformation $C \equiv 3P + 24 \pmod{26}$.

7. If the most common letter in a long ciphertext, encrypted by a shift transformation $C \equiv P + k \pmod{26}$ is Q, then what is the most likely value of k?

8. The message KYVMR CLVFW KYVBV PZJJV MVEKV VE was encrypted using a shift transformation $C \equiv P + k \pmod{26}$. Use frequencies of letters to determine the value of k. What is the plaintext message?

9. The message IVQLM IQATQ SMIKP QTLVW VMQAJ MBBMZ BPIVG WCZWE VNZWU KPQVM AMNWZ BCVMK WWSQM was encrypted using a shift transformation $C \equiv P + k \pmod{26}$. Use frequencies of letters to determine the value of k. What is the plaintext message?

10. If the two most common letters in a long ciphertext, encrypted by an affine transformation $C \equiv aP + b \pmod{26}$, are X and Q, respectively, then what are the most likely values for a and b?

11. If the two most common letters in a long ciphertext, encrypted by an affine transformation $C \equiv aP + b \pmod{26}$, are W and B, respectively, then what are the most likely values for a and b?

12. The message MJMZK CXUNM GWIRY VCPUW MPRRW GMIOP MSNYS RYRAZ PXMCD WPRYE YXD was encrypted using an affine transformation $C \equiv aP + b \pmod{26}$. Use frequencies of letters to determine the values of a and b. What is the plaintext message?

13. The message WEZBF TBBNJ THNBT ADZOE TGTYR BZAJN ANOOZ ATWGN ABOVG FNWZV A was encrypted using an affine transformation $C \equiv aP + b \pmod{26}$. the most common letters in the plaintext are A, E, N, and S. What is the plaintext message?

14. The message PJXFJ SWJNX JMRTJ FVSUJ OOJWF OVAJR WHEOF JRWJO DJFFZ BJF was encrypted using an affine transformation $C \equiv aP + b \pmod{26}$. Use frequencies of letters to determine the values of a and b. What is the plaintext message?

Given two ciphers, plaintext may be encrypted by first using one of the ciphers, and then using the other cipher on this result. This procedure produces a *product cipher.*

15. Find the product cipher obtained by using the transformation $C \equiv 5P + 13 \pmod{26}$ followed by the transformation $C \equiv 17P + 3 \pmod{26}$.

16. Find the product cipher obtained by using the transformation $C \equiv aP + b \pmod{26}$ followed by the transformation $C \equiv cP + d \pmod{26}$, where $(a, 26) = (c, 26) = 1$.

8.1 Computational and Programming Exercises

Computations and Explorations

Using a computation program such as Maple or *Mathematica*, or programs you have written, carry out the following computations and explorations.

1. Find the frequency of the letters of the English alphabet in different types of English text, such as in this book, in computer programs, and in a novel.

2. Encrypt some messages using affine transformations, as ciphertexts for your classmates to decipher.

3. Decrypt messages that were enciphered by your classmates using affine transformations, using letter-frequency analysis.

Programming Projects

Write computer programs using Maple, *Mathematica*, or a language of your choice to do the following.

1. Encrypt messages using the Caesar cipher.

2. Encrypt messages using the transformation $C \equiv P + k \pmod{26}$, where k is a given integer.

3. Encrypt messages using the transformation $C \equiv aP + b \pmod{26}$, where a and b are integers with $(a, 26) = 1$.

4. Decrypt messages that have been encrypted using the Caesar cipher.

5. Decrypt messages that have been encrypted using the transformation $C \equiv P + k \pmod{26}$, where k is a given integer.

6. Decrypt messages that have been encrypted using the transformation $C \equiv aP + b \pmod{26}$, where a and b are integers with $(a, 26) = 1$.

∗ 7. Cryptanalyze, using frequency counts, ciphertext that was encrypted using a transformation of the form $C \equiv P + k \pmod{26}$, where k is an unknown integer.

∗ 8. Cryptanalyze, using frequency counts, ciphertext that was encrypted using a transformation of the form $C \equiv aP + b \pmod{26}$, where a and b are unknown integers with $(a, 26) = 1$.

8.2 Block and Stream Ciphers

In Section 8.1, we studied character (or monographic) ciphers based on the substitution of characters. These ciphers are vulnerable to cryptanalysis based on the frequency of letters in the ciphertext. To avoid this weakness, we can use ciphers that substitute for each block of plaintext letters of a specified length a block of ciphertext letters of the same length. Ciphers of this sort are called *block*, or *polygraphic*, ciphers. In this section, we will discuss several varieties of block ciphers, including polygraphic ciphers based on modular arithmetic. We will describe a cipher known since the sixteenth century that employs several different character ciphers determined by a keyword, and a cipher invented by Hill around 1930 (see [Hi31]) that encrypts blocks using modular matrix multiplication. We will also discuss (but not describe in full detail) a more complicated block cipher important in commercial use, the Data Encryption Algorithm. At the end of this section, we will describe another type of cipher, a stream cipher, where the key can change as successive characters (or bits) are encrypted.

Vigenère Ciphers

We begin by describing the *Vigenère cipher*, named for French diplomat and cryptographer *Blaise de Vigenère*. Instead of encrypting each letter of a plaintext message in the same way, we will vary how we encrypt letters. The key of a Vigenère cipher consists of a keyword $\ell_1\ell_2 \ldots \ell_n$. Suppose that the numerical equivalents of the letters $\ell_1, \ell_2, \ldots, \ell_n$ are $k_1, k_2, \ldots, k_n$, respectively. To encrypt a plaintext message, we first split it into blocks of length n. A block consisting of letters with numerical equivalents $p_1, p_2, \ldots, p_n$ is transformed into a ciphertext block of letters with numerical equivalents $c_1, c_2, \ldots, c_n$ using a sequence of shift ciphers with

$$c_i \equiv p_i + k_i \pmod{26}, \quad 0 \le c_i \le 25,$$

for $i = 1, 2, \ldots, n$. The *Vigenère ciphers* are the encryption algorithms for the cryptosystem where blocks of plaintext letters of length n are encrypted to blocks of ciphertext letters of the same length. The keys are n-tuples $(k_1, k_2, \ldots, k_n)$ of letters. (A terminal group of fewer than n dummy letters can be used to fill out a final block.) That is, Vigenère ciphers can be thought of as block ciphers operating on blocks of length n using keys of length n.

Example 8.6. To encrypt the plaintext message MILLENNIUM using the key YT-WOK for a Vigenère cipher, we first translate the message and the key into their numerical

BLAISE DE VIGENÈRE (1523–1596), born in the village of Saint-Pourçain, France, received an excellent education. At 17 he was sent to court, and at 22 to the Diet of Worms as a secretary. He became a secretary for the Duke of Nevers in 1547, and in 1549 he was sent to Rome as a diplomat. While there he read numerous books on cryptography, a subject that he discussed with experts of the papal curia. In 1570, after a long career in diplomacy, interrupted by a period of study, Vigenère retired from court. He married a young wife, turned his annuity over to the poor of Paris, and dedicated himself to writing. He was the author of more than 20 books, the best known being his *Traicté des Chiffres,* written in 1585. In this book, Vigenère provides a comprehensive overview of cryptography. He discusses polyalphabetic ciphers at length and introduces several variations of known polyalphabetic ciphers, including the autokey cipher. Many historians believe that this cipher should have been called the "Vigenère" rather than the simpler one that now bears his name.

Vigenère did not write only about cryptography. His *Traicté des Chiffres* also contains discussions of magic, alchemy, and the mysteries of the universe. His *Traicté des Comètes* helped destroy the myth that God flings comets at Earth to warn people to stop sinning.

equivalents. The letters of the message and the letters of the key translate to

$$p_1p_2p_3p_4p_5p_6p_7p_8p_9p_{10} = 12\ 8\ 11\ 11\ 4\ 13\ 13\ 8\ 20\ 12$$

and

$$k_1k_2k_3k_4k_5 = 24\ 19\ 22\ 14\ 10,$$

respectively. Applying the Vigenère cipher with the specified key, we find that the characters in the encrypted message are:

$$c_1 = p_1 + k_1 = 12 + 24 \equiv 10 \pmod{26}$$
$$c_2 = p_2 + k_2 = 8 + 19 \equiv 1 \pmod{26}$$
$$c_3 = p_3 + k_3 = 11 + 22 \equiv 7 \pmod{26}$$
$$c_4 = p_4 + k_4 = 11 + 14 \equiv 25 \pmod{26}$$
$$c_5 = p_5 + k_5 = 4 + 10 \equiv 14 \pmod{26}$$
$$c_6 = p_6 + k_1 = 13 + 24 \equiv 11 \pmod{26}$$
$$c_7 = p_7 + k_2 = 13 + 19 \equiv 6 \pmod{26}$$
$$c_8 = p_8 + k_3 = 8 + 22 \equiv 4 \pmod{26}$$
$$c_9 = p_9 + k_4 = 20 + 14 \equiv 8 \pmod{26}$$
$$c_{10} = p_{10} + k_5 = 12 + 10 \equiv 22 \pmod{26}.$$

Translating the numerical equivalents of numbers back to letters we see that the encrypted message is KBHZO LGEIW. ◄

Example 8.7. To decrypt the ciphertext message FFFLB CVFX encrypted using a Vigenère cipher with key ZORRO, we first translate the letters of the ciphertext message into their numerical equivalents to obtain $c_1c_2c_3c_4c_5c_6c_7c_8c_9 = 5\ 5\ 5\ 11\ 1\ 2\ 21\ 5\ 23$. The numerical equivalents of the letters in the key are $k_1k_2k_3k_4k_5 = 25\ 14\ 17\ 17\ 14$. To obtain the numerical equivalents of the plaintext letters, we proceed as follows:

$$p_1 \equiv c_1 - k_1 = 5 - 25 \equiv 6 \pmod{26}$$
$$p_2 \equiv c_2 - k_2 = 5 - 14 \equiv 17 \pmod{26}$$
$$p_3 \equiv c_3 - k_3 = 5 - 17 \equiv 14 \pmod{26}$$
$$p_4 \equiv c_4 - k_4 = 11 - 17 \equiv 20 \pmod{26}$$
$$p_5 \equiv c_5 - k_5 = 1 - 14 \equiv 13 \pmod{26}$$
$$p_6 \equiv c_6 - k_1 = 2 - 25 \equiv 3 \pmod{26}$$
$$p_7 \equiv c_7 - k_2 = 21 - 14 \equiv 7 \pmod{26}$$
$$p_8 \equiv c_8 - k_3 = 5 - 17 \equiv 14 \pmod{26}$$
$$p_9 \equiv c_9 - k_4 = 23 - 17 \equiv 6 \pmod{26}.$$

Translating the numerical equivalents back to letters, we see that the plaintext message was GROUNDHOG. ◄

Cryptanalysis of Vigenère Ciphers

The Vigenère cipher was considered unbreakable for many years. It was used extensively to encrypt sensitive information transmitted by telegraphy. However, by the mid-nineteenth century, techniques were developed that could successfully break Vigenère ciphers. In 1863, Friedrich Kasiski, a Prussian military officer, described a method, now known as *Kasiski's test,* for determining the key length of a Vigenère cipher. Once the key length is known, frequency analysis of letters in the ciphertext can be used to determine the characters of the key. As with many discoveries named after their presumed first inventor, Kasiski was not the first person to discover this method. We now know that Charles Babbage discovered the same test in 1854. However, the publication of Babbage's discovery was delayed for many years. The reason for this delay was British national security. The British military used Babbage's test to break secret messages sent by their adversaries and did not want this to become known.

Kasiski's method is based on finding identical strings in ciphertext. When a message is encrypted using a Vigenère cipher with key length n, identical strings of plaintext separated by a multiple of n are encrypted to the same string (see Exercise 5). Kasiski's test is based on locating identical strings in the ciphertext, generally of length three or more, which likely correspond to identical strings in the plaintext. For each pair of identical ciphertext strings, we determine the difference between the positions of their initial characters. Suppose ther are k such pairs of identical strings in the ciphertext and $d_1, d_2, d_3, \ldots, d_k$ are the differences in the positions of their initial characters. If these pairs of identical ciphertext strings really do correspond to identical plaintext strings, the key length n must divide each of the integers $d_i, i = 1, 2, \ldots, k$. It would then follow that n divides the greatest common divisor of these integers, $(d_1, d_2, \ldots, d_k)$.

Because different strings of plaintext may be encrypted to the same ciphertext by different parts of the encyption key, some differences in starting positions of identical strings of ciphertext are extraneous and should be discarded. To overcome this problem, we can compute the greatest common divisor of some, but not all, these differences.

We can run a second test to help us assess whether we have found the correct key length. This test, developed by the famous American cryptographer William Friedman in 1920, estimates the key length of a Vigenère cipher by studying the variation in frequencies of ciphertext letters. Friedman observed that there is considerable variation in the frequencies of the letters in English text, but as the length of the key used in a Vigenère cipher increases, this variation becomes smaller and smaller.

Friedman introduced a measure called the *index of coincidence*. Given a string of n characters $x_1, x_2, \ldots, x_n$, its index of coincidence, denoted by IC, is the probability that two randomly chosen elements of this string are the same. We now assume that we are working with strings of English letters and that the letters $A, B, \ldots, Y$, and Z occur $f_0, f_1, \ldots, f_{24}$, and f_{25} times, respectively, in a string.

Because the ith letter occurs f_i times, there are

$$\binom{f_i}{2} = \frac{f_i(f_i - 1)}{2}$$

ways to choose two of its elements so that both are the ith character. Because there are $\binom{n}{2} = n(n-1)/2$ ways to choose two characters in the string, we can conclude that the index of coincidence for this string is

$$IC = \frac{\sum_{i=0}^{25} f_i(f_i - 1)}{n(n-1)}.$$

Now consider a string of English plaintext. If the plaintext is sufficiently long, we expect the frequencies of letters to approximate their frequencies in typical English (shown in Table 8.4). Suppose that $p_0, p_1, \ldots, p_{25}$ are the expected probabilities of $A, B, \ldots, Y$, and Z, respectively. It follows that the probability two randomly chosen letters are both A is p_0^2, the probability both are B is p_1^2, and so on. Consequently, we would expect the index of coincidence of this plaintext to be approximately

$$\sum_{i=0}^{25} p_i^2 \approx 0.065.$$

(The values $p_i, i = 0, 1, \ldots, 25$ used in this computation can be found in [St02].) Moreover, this reasoning applies for ciphertext produced by character ciphers. For a character cipher the probability of occurrence of a character in ciphertext equals the probability of occurrence of the corresponding plaintext character. Consequently, for ciphertext encrypted with a character cipher, the terms of the sum $\sum_{i=0}^{25} p_i^2$ are permuted, but the sum is not changed.

To use indices of coincidence to determine whether we have guessed correctly that the key has length k, we break the ciphertext message into k different parts. The first part contains characters in positions $1, k+1, 2k+1, \ldots$; the second part contains the characters in positions $2, k+2, 2k+2, \ldots$; and so on. We compute the index of coincidence for each of these different parts separately. If our guess was correct, each of these indices of coincidence should be approximately 0.065. However, if we guessed wrong, these values will most likely be less than 0.065. They probably will be considerably closer to the index of coincidence of a random string of English characters, namely $1/26 \approx 0.038$. (This index of coincidence can be computed using the probabilities of occurrence of letters in typical English text.)

For each part of the ciphertext, we attempt to find the letter of the key that was used to encrypt letters in this part by examining letter frequencies. We determine the most likely possibilities for the letters of the key by determining the letters that are most frequent in the ciphertext and presuming they correspond with the most common letters of English. To determine whether we have guessed correctly, we can compare the frequencies we expect when letters are encrypted by shifting them using this letter of the key with the observed frequencies for this part of the ciphertext.

Once we have made our best guess for each letter of the key, we attempt to decrypt the message using the key we have computed. If we recover a meaningful plaintext message, we presume we have recovered the correct plaintext. On the other hand, if we end up with nonsense, we go back to the drawing board and check out other possibilities.

We illustrate the cryptanalysis of ciphertext encrypted using a Vigenère cipher in the following example.

Example 8.8. Suppose that the ciphertext produced by encrypting plaintext using a Vigenère cipher is

Q W H I D	D N Z E M	W T L M T	B K T I T	E M W L Z
W V C V E	H L T B S	T U D L G	W N U J E	W J E U L
E X W Q O	S L N Z A	N L H Y Q	A L W E H	V O Q W D
V Q T B W	I L U R Y	S T I J W	C L H W W	R N S I H
M N U D I	Y F A V D	E L A G B	L S N Z A	N S M I F
G N Z E M	W A L W L	C X E F A	B Y J T S	S N X L H
Y H U L K	U C L O Z	Z A J H I	H W S M	

We describe the steps we use to break this message. We first use the Kasiski test, looking for repeated triples of letters in the ciphertext. We list our finding in a table.

Triple	Starting positions	Differences in starting positions
EMW	9, 21, 129	12, 108, 120
ZEM	8, 128	120
ZAN	59, 119	60
NZE	7, 127	120
NZA	58, 118	60
LHY	62, 149	87
ALW	66, 132	66

The differences between identical ciphertext blocks of length three are 12, 60, 66, 87, 108, and 120. Because $(12, 60, 66, 87, 108, 120) = 3$, we guess that the key length equals 3.

Assuming that this guess is correct, we split the ciphertext into three separate parts. The first contains the letters in positions $1, 4, 7, \ldots, 169$; the second contains the letters in positions $2, 5, 8, \ldots, 167$; and the third contains the letters in positions, $3, 6, 9, \ldots, 168$. To confirm that our guess is correct, we compute the indices of coincidence for each of these three parts of the ciphertext, obtaining 0.071, 0.109, and 0.091, respectively. (We leave the details of these computations to the reader. See Exercise 12.) One of these numbers is relatively close to the index of coincidence for English text, 0.065, and the other two are even larger. This indicates that 3 might be the correct key length. Because our ciphertext is rather short, we are not too worried that these indices of coincidence are not as close to 0.065 as we might like. Note that if our guess was wrong, we would expect some of these indices of coincidence to be smaller than 0.065, perhaps even near 0.038.

After some work, which we leave to the reader, we find the key used to encrypt the message is USA and the corresponding plaintext is

```
WEHOL    DTHES    ETRUT    HSTOB    ESELF
EVIDE    NTTHA    TALLM    ENARE    CREAT
EDEQU    ALTHA    TTHEY    AREEN    DOWED
BYTHE    IRCRE    ATORW    ITHCE    RTAIN
UNALI    ENABL    ERIGH    TSTHA    TAMON
GTHES    EAREL    IFELI    BERTY    ANDTH
EPURS    UITOF    HAPPI    NESS
```

This plaintext comes from the Declaration of Independence of the United States. It reads: "We hold these truths to be self-evident, that all men are created equal, that they are endowed by their Creator with certain unalienable Rights, that among these are Life, Liberty, and the pursuit of Happiness." For more information on cryptanalysis of Vigenère ciphers, see [St02] and [TrWa02].   ◄

Hill Ciphers

Hill ciphers are block ciphers invented by *Lester Hill* in 1929. To introduce Hill ciphers, we first consider *diagraphic ciphers*; in these ciphers, each block of two letters of plaintext is replaced by a block of two letters of ciphertext. We illustrate this process with an example.

Example 8.9.   To encrypt a message using digraphic Hill ciphers, we first split a message into blocks of two letters (adding a dummy letter, say X, at the end of the message, if necessary, so that the final block has two letters). For instance, the message

THE GOLD IS BURIED IN ORONO

is split up as

TH EG OL DI SB UR IE DI NO RO NO.

Next, these letters are translated into their numerical equivalents (as in previous examples) to obtain

$$19\ 7 \quad 4\ 6 \quad 14\ 11 \quad 3\ 8 \quad 18\ 1 \quad 20\ 17 \quad 8\ 4 \quad 3\ 8$$
$$13\ 14 \quad 17\ 14 \quad 13\ 14.$$

LESTER S. HILL (1891–1961) was born in New York City. He graduated from Columbia College and received his Ph.D. in mathematics from Yale University in 1926. He held positions at the University of Montana, Princeton University, the University of Maine, Yale University, and Hunter College. Hill was interested in applications of mathematics to communications. He developed methods for checking the accuracy of telegraphed code numbers and the encryption method known as the Hill cipher. Hill continued to submit cryptographic papers to the United States Navy mostly dealing with polygraphic ciphers for more than 30 years.

Each block of two plaintext numbers $P_1 P_2$ is converted into a block of two ciphertext numbers $C_1 C_2$ by defining C_1 to be the least nonnegative residue modulo 26 of a linear combination of P_1 and P_2, and defining C_2 to be the least nonnegative residue modulo 26 of a different linear combination of P_1 and P_2. For example, we can let

$$C_1 \equiv 5P_1 + 17P_2 \ (\text{mod } 26), \quad 0 \le C_1 < 26$$
$$C_2 \equiv 4P_1 + 15P_2 \ (\text{mod } 26), \quad 0 \le C_2 < 26,$$

in which case the first block 19 7 is converted to 6 25, because

$$C_1 \equiv 5 \cdot 19 + 17 \cdot 7 \equiv 6 \ (\text{mod } 26)$$
$$C_2 \equiv 4 \cdot 19 + 15 \cdot 7 \equiv 25 \ (\text{mod } 26).$$

After performing this operation on the entire message, the following ciphertext is obtained:

6 25   18 2   23 13   21 2   3 9   25 23   4 14   21 2   17 2   11 18   17 2.

When these blocks are translated into letters, we have the ciphertext message

GZ SC XN VC DJ ZX EO VC RC LS RC.

The decryption procedure for this cryptosystem is obtained by using Theorem 4.15. To find the plaintext block $P_1 P_2$ corresponding to the ciphertext block $C_1 C_2$, we use the relationship

$$P_1 \equiv 17C_1 + 5C_2 \ (\text{mod } 26)$$
$$P_2 \equiv 18C_1 + 23C_2 \ (\text{mod } 26).$$

(The reader should verify that this relationship is implied by Theorem 4.15.) ◀

The digraphic cipher system in Example 8.9 is conveniently described using matrices. For this cryptosystem, we have

$$\begin{pmatrix} C_1 \\ C_2 \end{pmatrix} \equiv \begin{pmatrix} 5 & 17 \\ 4 & 15 \end{pmatrix} \begin{pmatrix} P_1 \\ P_2 \end{pmatrix} \ (\text{mod } 26).$$

By Theorem 4.17, we see that the matrix $\begin{pmatrix} 17 & 5 \\ 18 & 23 \end{pmatrix}$ is an inverse of $\begin{pmatrix} 5 & 17 \\ 4 & 15 \end{pmatrix}$ modulo 26. Hence, Theorem 4.16 tells us that decryption can be done using the relationship

$$\begin{pmatrix} P_1 \\ P_2 \end{pmatrix} \equiv \begin{pmatrix} 17 & 5 \\ 18 & 23 \end{pmatrix} \begin{pmatrix} C_1 \\ C_2 \end{pmatrix} \ (\text{mod } 26).$$

In general, a Hill cryptosystem may be obtained by splitting plaintext into blocks of n letters, translating the letters into their numerical equivalents, and forming ciphertext using the relationship

$$\mathbf{C} \equiv \mathbf{AP} \ (\text{mod } 26),$$

where $\mathbf{A}$ is an $n \times n$ matrix, $(\det \mathbf{A}, 26) = 1$, $\mathbf{C} = \begin{pmatrix} C_1 \\ C_2 \\ \vdots \\ C_n \end{pmatrix}$ and $\mathbf{P} = \begin{pmatrix} P_1 \\ P_2 \\ \vdots \\ P_n \end{pmatrix}$, and

$C_1 C_2 \ldots C_n$ is the ciphertext block that corresponds to the plaintext block $P_1 P_2 \ldots P_n$. Finally, the ciphertext numbers are translated back to letters. For decryption, we use the matrix $\overline{\mathbf{A}}$, an inverse of $\mathbf{A}$ modulo 26, which may be obtained using Theorem 4.19. Because $\overline{\mathbf{A}}\mathbf{A} \equiv \mathbf{I} \pmod{26}$, we have

$$\overline{\mathbf{A}}\mathbf{C} \equiv \overline{\mathbf{A}}(\mathbf{A}\mathbf{P}) \equiv (\overline{\mathbf{A}}\mathbf{A})\mathbf{P} \equiv \mathbf{P} \pmod{26}.$$

Hence, to obtain plaintext from ciphertext, we use the relationship

$$\mathbf{P} \equiv \overline{\mathbf{A}}\mathbf{C} \pmod{26}.$$

Example 8.10.   We illustrate this procedure using $n = 3$ and the encrypting matrix

$$\mathbf{A} = \begin{pmatrix} 11 & 2 & 19 \\ 5 & 23 & 25 \\ 20 & 7 & 1 \end{pmatrix}.$$

Because $\det \mathbf{A} \equiv 5 \pmod{26}$, we have $(\det \mathbf{A}, 26) = 1$. To encrypt a plaintext block of length three, we use the relationship

$$\begin{pmatrix} C_1 \\ C_2 \\ C_3 \end{pmatrix} \equiv \mathbf{A} \begin{pmatrix} P_1 \\ P_2 \\ P_3 \end{pmatrix} \pmod{26}.$$

To encrypt the message STOP PAYMENT, we first split the message into blocks of three letters, adding a final dummy letter X to fill out the last block. We have plaintext blocks.

STO PPA YME NTX.

We translate these letters into their numerical equivalents:

18 19 14   15 15 0   24 12 4   13 19 23.

We obtain the first block of ciphertext in the following way:

$$\begin{pmatrix} C_1 \\ C_2 \\ C_3 \end{pmatrix} \equiv \begin{pmatrix} 11 & 2 & 19 \\ 5 & 23 & 25 \\ 20 & 7 & 1 \end{pmatrix} \begin{pmatrix} 18 \\ 19 \\ 14 \end{pmatrix} \equiv \begin{pmatrix} 8 \\ 19 \\ 13 \end{pmatrix} \pmod{26}.$$

Encrypting the entire plaintext message in the same manner, we obtain the ciphertext message

8 19 13   13 4 15   0 2 22   20 11 0.

Translating this message into letters, we have our ciphertext message

ITN NEP ACW ULA.

The decrypting process for this polygraphic cipher system takes a ciphertext block and obtains a plaintext block using the transformation

$$\begin{pmatrix} P_1 \\ P_2 \\ P_3 \end{pmatrix} \equiv \overline{A} \begin{pmatrix} C_1 \\ C_2 \\ C_3 \end{pmatrix} \pmod{26},$$

where

$$\overline{A} = \begin{pmatrix} 6 & -5 & 11 \\ -5 & -1 & -10 \\ -7 & 3 & 7 \end{pmatrix}$$

is an inverse of A modulo 26, which may be obtained using Theorem 4.19.

Because polygraphic ciphers operate with blocks, rather than with individual letters, they are not vulnerable to cryptanalysis based on letter frequency. However, polygraphic ciphers operating with blocks of size n are vulnerable to cryptanalysis based on frequencies of blocks of size n. For instance, with a digraphic cryptosystem, there are $26^2 = 676$ digraphs, blocks of length two. Studies have been done to compile the relative frequencies of digraphs in typical English text. By comparing the frequencies of digraphs in the ciphertext with the average frequencies of digraphs, it is often possible to successfully attack digraphic ciphers. For example, according to some counts, the most common digraph in English is TH, followed closely by HE. If a Hill digraphic cryptosystem has been employed and the most common digraph is KX, followed by VZ, we may guess that the ciphertext digraphs KX and VZ correspond to TH and HE, respectively. This would mean that the blocks 19 7 and 7 4 are sent to 10 23 and 21 25, respectively. If A is the encrypting matrix, this implies that

$$A \begin{pmatrix} 19 & 7 \\ 7 & 4 \end{pmatrix} \equiv \begin{pmatrix} 10 & 21 \\ 23 & 25 \end{pmatrix} \pmod{26}.$$

Because $\begin{pmatrix} 4 & 19 \\ 19 & 19 \end{pmatrix}$ is an inverse of $\begin{pmatrix} 19 & 7 \\ 7 & 4 \end{pmatrix} \pmod{26}$, we find that

$$A \equiv \begin{pmatrix} 10 & 21 \\ 23 & 25 \end{pmatrix} \begin{pmatrix} 4 & 19 \\ 19 & 19 \end{pmatrix} \equiv \begin{pmatrix} 23 & 17 \\ 21 & 2 \end{pmatrix} \pmod{26},$$

which gives a possible key. After attempting to decrypt the ciphertext using $\overline{A} = \begin{pmatrix} 2 & 9 \\ 5 & 23 \end{pmatrix}$ to transform it, we would know whether our guess was correct.   ◀

In general, if we know n correspondences between plaintext blocks of size n and ciphertext blocks of size n—for instance, if we know that the ciphertext blocks $C_{1j}C_{2j} \ldots C_{nj}, j = 1, 2, \ldots, n$, correspond to the plaintext blocks $P_{1j}P_{2j} \ldots P_{nj}, j = 1, 2, \ldots, n$, respectively—then we have

$$A \begin{pmatrix} P_{1j} \\ \vdots \\ P_{nj} \end{pmatrix} \equiv \begin{pmatrix} C_{1j} \\ \vdots \\ C_{nj} \end{pmatrix} \pmod{26},$$

for $j = 1, 2, \ldots, n$.

These n congruences can be succinctly expressed using the matrix congruence

$$AP \equiv C \pmod{26},$$

where P and C are $n \times n$ matrices with ijth entries P_{ij} and C_{ij}, respectively. If (det P, $26) = 1$, then we can find the encrypting matrix A via

$$A \equiv C\overline{P} \pmod{26},$$

where $\overline{P}$ is an inverse of P modulo 26.

Cryptanalysis using frequencies of polygraphs is only worthwhile for small values of n, where n is the size of the polygraphs. When $n = 10$, for example, there are 26^{10}, which is approximately 1.4×10^{14}, polygraphs of this length. Any analysis of the relative frequencies of these polygraphs is extremely infeasible.

The Data Encryption Standard and Related Ciphers

The most important cipher that has been used for commercial and government applications during the past 20 years is the Data Encryption Algorithm (DEA), which was standardized in 1977 by the federal government as part of the Data Encryption Standard (DES) (Federal Information Processing Standard 46-1). It was developed by IBM and was known as Lucifer before it became a standard. The DEA is a block cipher that encrypts 64-bit blocks using a 64-bit key (where the last 8 bits of the key are parity check bits stripped off before use) transforming them into 64-bit ciphertext blocks.

The encryption procedure used by the DEA is extremely complicated and will not be described in detail here. Basically, a plaintext block of 64 bits is encrypted by first permuting the 64 bits, iterating a function that operates on the left and right halves of a string of 64 bits in a particular way 16 times, and then applying the inverse of the initial permutation. Details of this cipher can be found in [St02] and [MevaVa97]. These details are easily understandable by anyone of the mathematical maturity of students using this text; they are quite lengthy, however.

The DEA is a *symmetric cipher*. Both the sender and the receiver of a message must know the same secret key, which is used for both encryption and decryption. Distributing secure keys for use by the DEA is a difficult problem, which can be addressed using public key cryptography (discussed in Section 8.4).

Although the DEA has not been broken, in the sense that no easy attack on it has been found, it is vulnerable to brute-force analysis. An exhaustive search can now check all 2^{56} possible keys in less than a day. Because of the vulnerability of this algorithm to such attacks, the National Institute of Standards and Technology (NIST) decided not to certify DES for use after 1998.

In November 2000, NIST selected a new algorithm called the *Advanced Encryption Standard (AES)* as the official encryption standard for the U.S. government. This encryption algorithm was developed by two Belgian scientists, Joan Daemen and Vincent Rijmen, and is called *Rijndael* after its creators. The adoption of Rijndael as the Advanced Encryption Standard followed three years of competition among many encryption algorithms submitted as candidates for the standard. The AES algorithm is capable of using 128-, 192-, and 256-bit symmetric keys to encrypt and decrypt 128-bit blocks. The complexity of the AES and the size of the keys that it supports should make it resistant to brute-force attacks for many years. The U.S. government hopes that AES will remain secure for at least 20 years.

Stream Ciphers

The methods discussed so far have the property that the same key is used to determine the particular encryption transformation that is applied to each character (or block). Once a plaintext–ciphertext pair is known, the key can be found. To add additional security, we can change the key used to encrypt successive characters. To discuss this type of encryption, we must first define some terms.

A sequence $k_1, k_2, k_3, \ldots$ of elements from a keyspace $\mathcal{K}$ is called a *keystream*. The encryption function corresponding to the key k_i is denoted by E_{k_i}. A *stream cipher* is a cipher that sends a plaintext string $p_1 p_2 p_3 \ldots$, using a keystream $k_1, k_2, k_3, \ldots$, to a ciphertext string $c_1 c_2 c_3 \ldots$, where $c_i = E_{k_i}(p_i)$. The corresponding decryption function is $D_{d_i}(c_i) = p_i$, where d_i is a decryption key corresponding to the encryption key k_i.

We can generate the keystream for a stream cipher in different ways. For example, we can select the keys at random to construct a keystream, or we can use a *keystream generator*, a function that generates successive keys using an initial sequence of keys (the *seed*), perhaps also using previous plaintext symbols.

The simplest (nontrivial) stream cipher is the *Vernam Cipher*, proposed by *Gilbert Vernam* in 1917 for the automatic encryption and decryption of telegraph messages. In this stream cipher, the keystream is a bit string $k_1 k_2 \ldots k_m$ of the same length as the plaintext message, which is a bit string $p_1 p_2 \ldots p_m$. Plaintext bits are encrypted using the map

$$E_{k_i}(p_i) \equiv k_i + p_i \ (\text{mod } 2).$$

Exactly two different encryption maps are used in a Vernam cipher. When $k_i = 0$, E_{k_i} is the identity map that sends 0 to 0 and 1 to 1. When $k_i = 1$, E_{k_i} is the map that sends 0 to 1 and 1 to 0. The corresponding decryption transformation D_{d_i} is identical to E_{k_i}.

Example 8.11.   When we encrypt the plaintext bit string 0 1111 0111 using a Vernam cipher with keystream 1 1000 1111, we obtain the bit string 1 0111 1000, where each bit is obtained by adding corresponding bits of the plaintext and the keystream. Decrypting this just requires that we repeat the operation.   ◀

Keystreams in the Vernam cipher should be used only once (see Exercise 38). When the keystream of a Vernam cipher is chosen at random and is used to encrypt exactly one plaintext message, it is called a *one-time pad*. It can be shown that a one-time pad is unbreakable, in the sense that someone with a ciphertext string encrypted using a random keystream used only once can do no better than to simply guess at the plaintext string. The problem with the Vernam cipher is that the keystream must be at least as long as the plaintext message, and must be transmitted securely between two parties who want to use a one-time pad. Consequently, the one-time pad is not used except for extremely sensitive communications, mostly of a diplomatic or military nature.

We will describe another stream cipher, the *autokey cipher* invented by Vigenère in the sixteenth century. The autokey cipher uses an initial seed key, which is a single character; subsequent keys are plaintext characters. In particular, the autokey cipher shifts each plaintext character, other than the first character, the numerical equivalent of the previous character modulo 26; it shifts the first character the numerical equivalent of the seed character modulo 26. That is, the autokey cipher encrypts a character p_i according to the transformation

$$c_i \equiv p_i + k_i \pmod{26},$$

where p_i is the numerical equivalent of the ith plaintext character, c_i is the numerical equivalent of the ith ciphertext character, and k_i the numerical equivalent of the ith character of the keystream, is given by $k_1 = s$, where s is the numerical equivalent of the seed character and $k_i = p_{i-1}$ for $i \geq 2$.

To decrypt a message encrypted with the autokey cipher we need to know the seed. We subtract the seed from the first ciphertext character modulo 26 to determine the first plaintext character, and then we subtract the numerical equivalent of each plaintext character modulo 26 from the next ciphertext character to obtain the next plaintext character.

We illustrate how to encrypt and decrypt using the autokey cipher in the following examples.

GILBERT S. VERNAM (1890–1960) was born in Brooklyn, New York. After graduating from Worcester Polytechnic Institute, he took a job at AT&T. He was able to visualize electrical circuits without actually implementing them. He was noted for his cleverness; one story quotes him as asking "What can I invent now?" each evening while stretched out on his couch. At AT&T he developed a method to make transmission via the teletypewriter, the first system that automated cryptology, secure. At AT&T he also developed a technique for encrypted digital images. Vernam also held positions with the International Communications Laboratories and the Postal Telegraph Cable Company. He was granted 65 patents for his inventions in cryptography and in telegraph switching systems.

Example 8.12. To encrypt the plaintext message HERMIT using the autokey cipher with seed X (with numerical equivalent 23), we first translate the letters of HERMIT into their numerical equivalents to obtain 7 4 17 12 8 19. The keystream consists of the numbers 23 7 4 17 12 8. The numerical equivalents of the characters in the ciphertext message are

$$p_1 + k_1 = 7 + 23 \equiv 4 \text{ (mod 26)}$$
$$p_2 + k_2 = 4 + 7 \equiv 11 \text{ (mod 26)}$$
$$p_3 + k_3 = 17 + 4 \equiv 21 \text{ (mod 26)}$$
$$p_4 + k_4 = 12 + 17 \equiv 3 \text{ (mod 26)}$$
$$p_5 + k_5 = 8 + 12 \equiv 20 \text{ (mod 26)}$$
$$p_6 + k_6 = 19 + 8 \equiv 1 \text{ (mod 26)}.$$

Translating back to letters, we see that the ciphertext is ELVDUB.   ◄

Example 8.13. To decrypt the ciphertext message RMNTU encrypted using the autokey cipher with seed F, we first translate the characters of the ciphertext into their numerical equivalents to obtain 17 12 13 19 20. We obtain the numerical equivalent of the first plaintext character by computing

$$p_1 = c_1 - s \equiv 17 - 5 = 12 \text{ (mod 26)}.$$

We obtain the numerical equivalent of successive plaintext characters as follows:

$$p_2 = c_2 - p_1 = 12 - 12 = 0 \text{ (mod 26)}$$
$$p_3 = c_3 - p_2 = 13 - 0 = 13 \text{ (mod 26)}$$
$$p_4 = c_4 - p_3 = 19 - 13 = 6 \text{ (mod 26)}$$
$$p_5 = c_5 - p_4 = 20 - 6 = 14 \text{ (mod 26)}$$

Translating these numerical equivalents back to letters, we find that the plaintext message was MANGO.   ◄

We have only briefly touched the surface of the deep subject of stream ciphers. For more information about them, including descriptions of stream ciphers used in practice, consult [MevaVa97].

8.2 Exercises

1. Use the Vigenère cipher with encrypting key SECRET to encrypt the message

DO NOT OPEN THIS ENVELOPE.

2. Decrypt the following message, which was enciphered using the Vigenère cipher with encrypting key SECRET:

WBRCS LAZGJ MGKMF V.

3. Use the Vigenère cipher with encrypting key TWAIN to encrypt the message

AN ENGLISHMAN IS A PERSON WHO DOES THINGS BECAUSE THEY HAVE BEEN DONE BEFORE. AN AMERICAN IS A PERSON WHO DOES THINGS BECAUSE THEY HAVE NOT BEEN DONE BEFORE.

4. Decrypt the following message, which was enciphered using the Vigenère cipher with encrypting key TWAIN.

P A C W H	E Z U A R	N L T E B	X P E Z A	B P I M F
B J L M N	K J I V T	T H L B U	T P I A G	H X E T R
T N N M Q	T X O C G	H Q R W J	G S O Z Y	W W N L G
A A T P B	N O A V Q	L K F V N	M E O V F	M D A B U
T R E I E	B O E V N	G Z F T B	N N I A U	X Z A V Q
O W N Q F	A A D N E	H I I B Z	T P H M Z	T P I K F
T H O V R	P K U T Q	H Y C C C	R I E M V	Z D T U V
E H I W A	R A A Z F			

5. Suppose a plaintext message is encrypted using a Vigenère cipher. Show that identical strings of characters separated by a multiple of the key length are encrypted to the same string of ciphertext characters.

In Exercises 6–11, use the procedure described in the text to cryptanalyze the given ciphertext, which was encrypted using a Vigenère cipher.

6.
U C Y F C	O O C Q U	C Y F H E	B H F T H	E F E R F
G Q J C K	X V B U V	B S H F T	B L C Z B	S W K U V
B N K W E	H L T I C	G S O U V	B T Z F O	U P B B A
B F O P K	P P T L V	H O B U B	P I P G C	O U I K F

7.
K M K R E	C C W S P	I S N E J	R S X Z I	A L K Z S
Q S L E H	N V W A M	S R I Q M	Y J K M K	R E C C W
X M V O F	E L R L W	W E J C T	J C G A M	Y K J M X
C P W Q W	G L W L F	E L A E F	M R D W F	W J I S P
R W B X Z	C L S P H	O Y C M L	P W Q W A	R M K Y J
S R E D K	M K R E C	C A Z G G	Z Y X D C	E K R S L
F I J Q G	S L P W Y	V F D V G	K	

8.
S I I W Z	F D I B N	H U D E U	W Q J H P	J K R N K
R L A C T	W X B I M	M H M P J	O F U F P	W V E O G
P Q P E L	V P Z Y D	A X I A G	P I T M A	X F S S S
G W P B W	I W O F O	T F W V F	J S X P L	B J O T P
S U D I J	J X F N R	F P A F G	R P S X I	W X J O R
P P X S Q	I			

9.
J W E F F	P R G B A	G D S Z F	Z B T Z J	I B L S P
V D B T P	F X M L V	U G W I D	N W D H O	B N K J T
V L X I J	K P M Z Q	H Q E D W	Q C O B O	V J B Z U
H O I E G	J N V O U	B Y D U Q	N D T U F	U F L Z V
U Q E J V	Q J K F L	S B U P R	W D Q I F	V U J W B
V T H U P	R W J A Y	R V T U K	B D V E F	M E E Z I
E B F X R	X M M K L	D W L O E	P R Y F E	F U O

10.

P D J V J	L F C J W	Z Q L G R	E V M U V	Z O W I D
A J Z P Z	D W E M U	Q L G G I	Q Z Z M E	N Z P J M
Y X S M W	I H Q Q P	D B W I E	K M S F B	G I Q W W
I J W Z E	Y M A I C	T J R R B	M I Y Q S	K P D J V
L A H I Y	L N R R M	A I C Q R	T C W A M	Y O U E E
P D S F S	S S H G T	Y H Q Q P	Y M A I C	O J X E W
Y L P M S	H Z N Y L	P R T Y C	V J C M C	Y X S Q X
W Z N F V	Q Z T Q O	Q X G Z C	W E R Q S	K Z V Q C
L L I W E	W Y L P R	T C L V I	K W W W C	Z N Y L P
K Q M X J				

11.

T U Z T U	W F G C G	L H G T F	G M K G R	F I A S R
K W K R R	D A A G U	W D G T Q	G E Y N B	L I S P Y
Q T N A G	S L R W U	G A X E Y	S U M H R	V A Z A E
W G K N V	M S K S G	Z E E L N	M G N E Q	S T I O Y
M M H U F	L H K Y Y	S U M H R	V A Z F H	D T U N G
Z E E L N	M G N E Q	S T Z H R	O R O G U	L B X O G
Z E X S O	M T Z H R	Q A R S B	D A A G U	W D G T O
G Z U T U	W C R O J	F		

12. Show how we find that the correct key in Example 8.8 is USA once we know the key has length three.

13. Using the digraphic cipher that sends the plaintext block $P_1 P_2$ to the ciphertext block $C_1 C_2$, with

$$C_1 \equiv 3P_1 + 10P_2 \pmod{26}$$
$$C_2 \equiv 9P_1 + 7P_2 \pmod{26}$$

encrypt the message BEWARE OF THE MESSENGER.

14. Using the digraphic cipher that sends the plaintext block $P_1 P_2$ to the ciphertext block $C_1 C_2$, with

$$C_1 \equiv 8P_1 + 9P_2 \pmod{26}$$
$$C_2 \equiv 3P_1 + 11P_2 \pmod{26}$$

encrypt the message DO NOT SHOOT THE MESSENGER.

15. Decrypt the ciphertext message RD SR QO VU QB CZ AN QW RD DS AK OB, which was encrypted using the digraphic cipher that sends the plaintext block $P_1 P_2$ into the ciphertext block $C_1 C_2$, with

$$C_1 \equiv 13P_1 + 4P_2 \pmod{26}$$
$$C_2 \equiv 9P_1 + P_2 \pmod{26}.$$

16. Decrypt the ciphertext message UW DM NK QB EK, which was encrypted using the digraphic cipher that sends the plaintext block $P_1 P_2$ into the ciphertext block $C_1 C_2$, with

$$C_1 \equiv 23P_1 + 3P_2 \pmod{26}$$
$$C_2 \equiv 10P_1 + 25P_2 \pmod{26}.$$

17. A cryptanalyst has determined that the two most common digraphs in a ciphertext message are RH and NI, and guesses that these ciphertext digraphs correspond to the two most common diagraphs in English text, TH and HE. If the plaintext was encrypted using a Hill digraphic cipher described by

$$C_1 \equiv a P_1 + b P_2 \pmod{26}$$
$$C_2 \equiv c P_1 + d P_2 \pmod{26},$$

what are a, b, c, and d?

18. How many pairs of letters remain unchanged when encryption is performed using each of the following digraphic ciphers?

a) $C_1 \equiv 4P_1 + 5P_2 \pmod{26}$
   $C_2 \equiv 3P_1 + P_2 \pmod{26}$

b) $C_1 \equiv 7P_1 + 17P_2 \pmod{26}$
   $C_2 \equiv P_1 + 6P_2 \pmod{26}$

c) $C_1 \equiv 3P_1 + 5P_2 \pmod{26}$
   $C_2 \equiv 6P_1 + 3P_2 \pmod{26}$

19. Show that if the encrypting matrix $\mathbf{A}$ in the Hill cipher system is involutory modulo 26, that is, $\mathbf{A}^2 \equiv \mathbf{I} \pmod{26}$, then $\mathbf{A}$ also serves as a decrypting matrix for this cipher system.

20. A cryptanalyst has determined that the three most common trigraphs (blocks of length three) in a ciphertext are LME, WRI, and ZYC, and guesses that these ciphertext trigraphs correspond to the three most common trigraphs in English text, THE, AND, and THA. If the plaintext was encrypted using a Hill trigraphic cipher described by $\mathbf{C} \equiv \mathbf{AP} \pmod{26}$, what are the entries of the 3×3 encrypting matrix $\mathbf{A}$?

21. Find the product cipher obtained by using the digraphic Hill cipher with encrypting matrix $\begin{pmatrix} 2 & 3 \\ 1 & 17 \end{pmatrix}$ followed by using on the result the digraphic Hill cipher with encrypting matrix $\begin{pmatrix} 5 & 1 \\ 25 & 4 \end{pmatrix}$.

22. Show that the product cipher obtained from two digraphic Hill ciphers is again a digraphic Hill cipher.

23. Show that the product cipher obtained by encrypting first using a Hill cipher with blocks of size m and then using a Hill cipher with blocks of size n is again a Hill cipher that uses blocks of size $[m, n]$.

24. Find the 6×6 encrypting matrix corresponding to the product cipher obtained by first using the Hill cipher with encrypting matrix $\begin{pmatrix} 3 & 1 \\ 2 & 1 \end{pmatrix}$, followed by using the Hill cipher with encrypting matrix $\begin{pmatrix} 1 & 1 & 0 \\ 1 & 0 & 1 \\ 0 & 1 & 1 \end{pmatrix}$.

* 25. In *transposition cipher*, blocks of a specified size are encrypted by permuting their characters in a specified manner. For instance, plaintext blocks of length five, $P_1 P_2 P_3 P_4 P_5$, may be sent to ciphertext blocks $C_1 C_2 C_3 C_4 C_5 = P_4 P_5 P_2 P_1 P_3$. Show that every such transposition cipher is a Hill cipher with an encrypting matrix that contains only 0s and 1s as entries, with the property that each row and each column contains exactly one 1.

Hill ciphers are special cases of block ciphers based on *affine transformations*. To form such a transformation, let $\mathbf{A}$ be an $n \times n$ matrix with integer entries and $(\det \mathbf{A}, 26) = 1$, and let $\mathbf{B}$ be an $n \times 1$ matrix with integer entries. To encrypt a message, we split it into blocks of length n and put the numerical equivalents of the letters in each block into an $n \times 1$ matrix $\mathbf{P}$ (padding the last block with dummy letters, if necessary). We find the corresponding ciphertext block by computing $\mathbf{C} \equiv (\mathbf{AP} + \mathbf{B})$ (mod 26) and translating the entries in $\mathbf{C}$ back into letters.

26. Using the affine transformation $\mathbf{C} \equiv \begin{pmatrix} 3 & 2 \\ 7 & 11 \end{pmatrix} \mathbf{P} + \begin{pmatrix} 8 \\ 19 \end{pmatrix}$ (mod 26) on blocks of two successive letters, encrypt the message HAVE A NICE DAY.

27. What is the decrypting transformation associated with the affine transformation in Exercise 26?

28. What is the decrypting transformation associated with the encrypting transformation $\mathbf{C} \equiv (\mathbf{AP} + \mathbf{B})$ (mod 26), where $\mathbf{A}$ is an $n \times n$ matrix with integer entries and $(\det \mathbf{A}, 26) = 1$, and $\mathbf{B}$ is an $n \times 1$ matrix with integer entries?

29. Decipher the message HG PM QR YN NM that was encrypted using the affine transformation $\mathbf{C} \equiv \begin{pmatrix} 5 & 2 \\ 11 & 15 \end{pmatrix} \mathbf{P} + \begin{pmatrix} 14 \\ 3 \end{pmatrix}$ (mod 26).

30. Explain how you would go about decrypting a message that was encrypted in blocks of length two using an affine transformation $\mathbf{C} \equiv \mathbf{AP} + \mathbf{B}$ (mod 26), where $\mathbf{A}$ is a 2×2 matrix with integer entries and $(\det \mathbf{A}, 26) = 1$, and $\mathbf{B}$ is a 2×1 matrix with integer entries.

31. Explain how you would go about decrypting a message that was encrypted in blocks of length three using an affine transformation $\mathbf{C} \equiv \mathbf{AP} + \mathbf{B}$ (mod 26), where $\mathbf{A}$ is a 3×3 matrix with integer entries and $(\det \mathbf{A}, 26) = 1$, and $\mathbf{B}$ is a 3×1 matrix, with integer entries.

32. Is the product cipher composed of two digraphic block ciphers based on affine transformations also a digraphic block cipher based on an affine transformation?

* **33.** Is the product cipher composed of two block ciphers based on affine transformations, encrypting blocks of length m and blocks of length n, respectively, also a block cipher based on an affine transformation?

34. Encrypt the bit string 11 1010 0011 using the Vernam cipher with keystream 10 0111 1001.

35. Decrypt the bit string 11 1010 0011, assuming that it was encrypted using the Vernam cipher with keystream 10 0111 1001.

36. Encrypt the plaintext message MIDDLETOWN using the autokey cipher with seed Z.

37. Decrypt the ciphertext message ZVRQH DUJIM, assuming that it was encrypted using the autokey cipher with seed I.

38. Show that the Vernam cipher is vulnerable to a known-plaintext attack if a keystream is used repeatedly. In particular, show that if someone can encrypt a bit string and have access to the resulting ciphertext string, the keystring can be found.

39. Show that if a keystream is used to encrypt two different messages using a Vernam cipher, then the bit string obtained by adding corresponding bits of the two messages modulo 2 could be found by someone with the corresponding ciphertext messages. Why might this permit cryptanalysis?

8.2 Computational and Programming Exercises

Computations and Explorations

Using a computation program such as Maple or *Mathematica,* or programs you have written, carry out the following computations and explorations.

1. Encrypt some messages using Vigenère ciphers for your classmates to decrypt.

* 2. Decrypt messages encrypted by your classmates using Vigenère ciphers.

3. Run the Kasiski test on some ciphertexts encrypted using Vigenère ciphers.

4. Find the index of coincidence for some character strings

5. Cryptanalyze some ciphertexts encrypted using Vigenère ciphers.

6. Find the frequencies of digraphs in various types of English texts, such as this text, computer programs, and a novel.

7. Find the frequencies of trigraphs in various types of English texts, such as this text, computer programs, and a novel.

8. Encrypt some messages using Hill ciphers for your classmates to decrypt.

9. Decrypt messages encrypted by your classmates using Hill ciphers.

10. Encrypt and decrypt some long messages using a Vigenère cipher one-time pad, sending these messages to a particular classmate.

11. Encrypt some messages using an autokey cipher for your classmates to decrypt.

12. Decrypt some messages that were encrypted using an autokey cipher by your classmates.

Programming Projects

Write computer programs using Maple, *Mathematica,* or a language of your choice to do the following.

1. Encrypt messages using Vigenère ciphers.

2. Decrypt messages that have been encrypted using Vigenère ciphers.

* 3. Given ciphertext encrypted using a Vigenère cipher, run the Kasiski test to determine the key length of the cipher.

4. Given a string of English characters, find the index of coincidence of this string.

** 5. Cryptanalyze ciphertext encrypted using a Vigenère cipher using the Kasiski test, followed by the Friedman test, which uses the index of coincidence to verify the key length, followed by frequency analysis to find each character of the key. Then use the resulting key to recover the original plaintext.

6. Encrypt messages using a Hill cipher.

7. Decrypt messages that were encrypted using a Hill cipher.

* 8. Cryptanalyze messages that were encrypted using a digraphic Hill cipher, by analyzing the frequency of digraphs in the ciphertext.

9. Encrypt messages using a cipher based on an affine transformation. (See the preamble to Exercise 26.)

10. Decrypt messages that were encrypted using an affine transformation.

11. By analyzing the frequency of digraphs in ciphertext, cryptanalyze messages encrypted using a digraphic block cipher based on an affine transformation.

12. Encrypt messages using the autokey cipher.

13. Decrypt messages that were encrypted using the autokey cipher.

8.3 Exponentiation Ciphers

In this section, we discuss a cipher based on modular exponentiation, which was invented in 1978 by Pohlig and Hellman [PoHe78]. We will see that ciphers produced by this system are resistant to cryptanalysis. (This cipher is of more theoretical than practical significance.)

Let p be an odd prime and let e, the enciphering key, be a positive integer with $(e, p - 1) = 1$. To encrypt a message, we first translate the letters of the message into numerical equivalents (retaining initial zeros in the two-digit numerical equivalents of letters). We use the same relationship we have used before, as shown in Table 8.9

Letter	A	B	C	D	E	F	G	H	I	J	K	L	M	N	O	P	Q	R	S	T	U	V	W	X	Y	Z
Numerical Equivalent	00	01	02	03	04	05	06	07	08	09	10	11	12	13	14	15	16	17	18	19	20	21	22	23	24	25

Table 8.9 *Two-digit numerical equivalents of letters.*

Next, we group the resulting numbers into blocks of $2m$ decimal digits, where $2m$ is the largest positive even integer such that all blocks of numerical equivalents corresponding to m letters (viewed as a single integer with $2m$ decimal digits) are less than p, e.g., if $2525 < p < 252{,}525$, then $m = 2$.

For each plaintext block P, which is an integer with $2m$ decimal digits, we form a ciphertext block C using the relationship

$$C \equiv P^e \pmod{p}, \quad 0 \le C < p.$$

The ciphertext message consists of these ciphertext blocks, which are integers less than p. Notice that different values of e determine different ciphers, hence e is aptly called the enciphering key. We illustrate the encryption technique with the following example.

Example 8.14. Let the prime to be used as the modulus in the encryption procedure be $p = 2633$, and let the encryption key to be used as the exponent in the modular exponentiation be $e = 29$, so that $(e, p - 1) = (29, 2632) = 1$. To encrypt the plaintext message

THIS IS AN EXAMPLE OF AN EXPONENTIATION CIPHER,

we first convert the letters of the message into their numerical equivalents, and then form blocks of length four from these digits, to obtain

$$1907 \quad 0818 \quad 0818 \quad 0013 \quad 0423$$
$$0012 \quad 1511 \quad 0414 \quad 0500 \quad 1304$$
$$2315 \quad 1413 \quad 0413 \quad 1908 \quad 0019$$
$$0814 \quad 1302 \quad 0815 \quad 0704 \quad 1723.$$

Note that we have added the two digits 23, corresponding to the letter X, at the end of the message to fill out the final block of four digits.

We next translate each plaintext block P into a ciphertext block C using the relationship

$$C \equiv P^{29} \ (\text{mod } 2633), \quad 0 \leq C < 2633.$$

For instance, to encrypt the first plaintext block, we compute

$$C \equiv 1907^{29} \equiv 2199 \ (\text{mod } 2633).$$

To efficiently carry out the modular exponentiation, we use the algorithm given in Section 4.1. When we encrypt the blocks, we obtain the ciphertext:

$$2199 \quad 1745 \quad 1745 \quad 1206 \quad 2437$$
$$2425 \quad 1729 \quad 1619 \quad 0935 \quad 0960$$
$$1072 \quad 1541 \quad 1701 \quad 1553 \quad 0735$$
$$2064 \quad 1351 \quad 1704 \quad 1841 \quad 1459.$$ ◄

To decrypt a ciphertext block C, we need to know a decryption key, namely an integer d such that $de \equiv 1 \ (\text{mod } p - 1)$, so that d is an inverse of $e \ (\text{mod } p - 1)$, which exists because $(e, p - 1) = 1$. If we raise the ciphertext block C to the dth power modulo p, we recover your plaintext block P, because

$$C^d \equiv (P^e)^d = P^{ed} \equiv P^{k(p-1)+1} \equiv (P^{p-1})^k P \equiv P \ (\text{mod } p),$$

where $de = k(p - 1) + 1$, for some integer k, because $de \equiv 1 \ (\text{mod } p - 1)$. (Note that we have used Fermat's little theorem to see that $P^{p-1} \equiv 1 \ (\text{mod } p)$.)

Example 8.15.   To decrypt the ciphertext blocks generated using the prime modulus $p = 2633$ and the encryption key $e = 29$, we need an inverse of e modulo $p - 1 = 2632$. An easy computation, as done in Section 4.2, shows that $d = 2269$ is such an inverse. To decrypt the ciphertext block C to define the corresponding plaintext block P, we use the relationship

$$P \equiv C^{2269} \ (\text{mod } 2633).$$

For instance, to decrypt the ciphertext block 2199, we have

$$P \equiv 2199^{2269} \equiv 1907 \ (\text{mod } 2633).$$

Again, the modular exponentiation is carried out using the algorithm given in Section 4.1.   ◄

For each plaintext block P that we encrypt by computing P^e (mod p), we use only $O((\log_2 p)^3)$ bit operations, as Theorem 4.9 demonstrates. Before we decrypt, we need to find an inverse d of e modulo $p - 1$. This can be done using $O(\log^3 p)$ bit operations (see Exercise 15 of Section 4.2), and this must be done only once. Then to recover the plaintext block P from a ciphertext block C, we simply need to compute the least positive residue of C^d modulo p; we can do this using $O((\log_2 p)^3)$ bit operations. Consequently, the process of encryption and decryption using modular exponentiation can be carried out rapidly.

On the other hand, cyptanalysis of messages encrypted using modular exponentiation generally cannot be accomplished rapidly. To see this, suppose that we know the prime p used as the modulus and, moreover, suppose that we know the plaintext block P corresponding to a ciphertext block C, so that

(8.2)
$$C \equiv P^e \;(\text{mod } p).$$

For successful cryptanalysis, we need to find the enciphering key e. This is the discrete logarithm problem, a computationally difficult problem that will be discussed in Chapter 9. Note that when p has more than 200 decimal digits, it is not feasible to solve this problem using a computer.

8.3 Exercises

1. Using the prime $p = 101$ and encryption key $e = 3$, encrypt the message GOOD MORN-ING using modular exponentiation.

2. Using the prime $p = 2621$ and encryption key $e = 7$, encrypt the message SWEET DREAMS using modular exponentiation.

3. What is the plaintext message that corresponds to the ciphertext 01 09 00 12 12 09 24 10 that is produced using modular exponentiation with modulus $p = 29$ and encryption exponent $e = 5$?

4. What is the plaintext message that corresponds to the ciphertext 1213 0902 0539 1208 1234 1103 1374 that is produced using modular exponentiation with modulus $p = 2591$ and encryption key $e = 13$?

5. Show that the encryption and decryption procedures are identical when encryption is done using modular exponentiation with modulus $p = 31$ and enciphering key $e = 11$.

6. With modulus $p = 29$ and unknown encryption key e, modular exponentiation produces the ciphertext 04 19 19 11 04 24 09 15 15. Cryptanalyze the above cipher, if it is also known that the ciphertext block 24 corresponds to the plaintext letter U (with numerical equivalent 20). (*Hint:* First find the logarithm of 24 to the base 20 modulo 29, using some guesswork.)

8.3 Computational and Programming Exercises

Computations and Explorations

Using a computation program such as Maple or *Mathematica,* or programs you have written, carry out the following computations and explorations.

1. Encrypt some messages for your classmates to decrypt using exponentiation ciphers.

2. Decrypt messages encrypted by your classmates using exponentiation ciphers, given the encryption key and prime modulus.

Programming Projects

Write computer programs using Maple, *Mathematica,* or a language of your choice to do the following.

1. Encrypt some messages for your classmates to decrypt using exponentiation ciphers.

2. Decrypt messages encrypted by your classmates using exponentiation ciphers, given the encrypting key and prime modules.

8.4 Public Key Cryptography

The cryptosystems we have discussed so far are all examples of *private key,* or *symmetric* cryptosystems, where the encryption and decryption keys are either the same or can be easily found from each other. For example, in a shift cipher, the encrypting key is an integer k and the corresponding decrypting key is the integer $-k$. In an affine cipher, the encrypting key is a pair (a, b) and the corresponding decrypting key is the pair $(\overline{a}, -\overline{a}b)$, where $\overline{a}$ is an inverse of a modulo 26. In a Hill cipher, the encrypting key is an $n \times n$ matrix $\mathbf{A}$ and the corresponding decrypting key is the $n \times n$ matrix $\overline{\mathbf{A}}$, where $\overline{\mathbf{A}}$ is an inverse of the matrix $\mathbf{A}$ modulo 26. In the Pohlig-Hellman exponentiation cipher, the encrypting key is (e, p), where p is a prime, and the corresponding decrypting key is (d, p), where d is an inverse of e modulo $p - 1$. For the DEA, the encrypting and decrypting keys are exactly the same.

For that reason, if one of the cryptosystems discussed so far is used to establish secure communications within a network, then each pair of communicants must employ an encryption key that is kept secret from the other individuals in the network, because once the encryption key in such a cryptosystem is known, the decryption key can be found using a small amount of computer time. Consequently, to maintain secrecy, the encryption keys must themselves be transmitted over a channel of secure communications.

To avoid assigning a key to each pair of individuals, which must be kept secret from the rest of the network, a new type of cryptosystem, called a *public key* cryptosystem, was invented in the 1970s. In this type of cryptosystem, encrypting keys can be made public, because an unrealistically large amount of computer time is required to find a decrypting transformation from an encrypting transformation. To use a public key cryptosystem to establish secret communications in a network of n individuals, each individual produces a key of the type specified by the cryptosystem, retaining certain private information that went into the construction of the encrypting transformation $E(k)$, obtained from the key

k according to a specified rule. Then a directory of the n keys $k_1, k_2, \ldots, k_n$ is published. When individual i wishes to send a message to individual j, the letters of the message are translated into their numerical equivalents and combined into blocks of specified size. Then, for each plaintext block P a corresponding ciphertext block $C = E_{k_j}(P)$ is computed using the encrypting transformation E_{k_j}. To decrypt the message, individual j applies the decrypting transformation D_{k_j} to each ciphertext block C to find P; that is,

$$D_{k_j}(C) = D_{k_j}(E_{k_j}(P)) = P.$$

Because the decrypting transformation D_{k_j} cannot be found in a realistic amount of time by anyone other than individual j, no unauthorized individuals can decrypt the message, even though they know the key k_j. Furthermore, cryptanalysis of the ciphertext message, even with knowledge of k_j, is extremely infeasible due to the large amount of computer time needed.

Many cryptosystems have been proposed as public key cryptosystems. All but a few have been shown to be unsuitable, by demonstrating that ciphertext messages can be decrypted using a feasible amount of computer time. In this section, we will introduce the most widely used public key cryptosystem, the RSA cryptosystem. In addition, we will introduce several other public key cryptosystems, including the Rabin public key cryptosystem, which we will discuss at the end of this section, and the ElGamal public key cryptosystem, which we will discuss in Chapter 10. The security of these systems rests on the difficulty of two computationally intensive mathematical problems, factoring integers (discussed in Chapter 3) and finding discrete logarithms (to be discussed in Chapter 9). In Section 8.5, we will describe a proposed public key cryptosystem, the knapsack cryptosystem, that turned out not to be suitable as a basis for a public key cryptosystem. (See [MevaVa97] for a comprehensive look at most of the important public key cryptosystems.)

Although public key cryptosystems have many advantages, they are not extensively used for general-purpose encryption. The reason is that encrypting and decrypting in these cryptosystems require too much time and memory on most computers, generally several orders of magnitude more than required for symmetric cryptosystems currently in use. However, public key cryptosystems are used extensively to encrypt keys for symmetric cryptosystems such as DES, so that these keys can be transmitted securely. They are also used in a wide variety of cryptographic protocols, such as in digital signatures (discussed in Section 8.6). They are also particularly useful for applications involving smart cards and electronic commerce.

Also note that in modern cryptography, the cryptosystem used to encrypt messages is publicly known. Consequently, the secrecy of encrypted messages does not depend on the secrecy of the encryption algorithm in use. For symmetric key cryptosystems, the secrecy of messages depends on the secrecy of the encryption key in use and the computational difficulty of finding this key from other information (such as plaintext–ciphertext pairs). For public key cryptosystems, secrecy rests on the secrecy of the decryption key and the computational difficulty of finding this key from the encryption key and other public information (such as plaintext–ciphertext pairs).

The RSA Cryptosystem

The *RSA cryptosystem,* invented by *Ronald Rivest, Adi Shamir, and Leonard Adleman* [RiShAd78] in the 1970s (and patented by them [RiShAd83] in 1983) is a public key cryptosystem based on modular exponentiation, where the keys are pairs (e, n) consisting of an exponent e and a modulus n that is the product of two large primes; that is, $n = pq$, where p and q are large primes, so that $(e, \phi(n)) = 1$. To encrypt a message, we first translate the letters into their numerical equivalents and then form blocks of the largest possible size (with an even number of digits). To encrypt a plaintext block P, we form a ciphertext block C by

$$E(P) = C \equiv P^e \pmod{n}, \quad 0 \le C < n.$$

The decrypting procedure requires knowledge of an inverse d of e modulo $\phi(n)$, which exists because $(e, \phi(n)) = 1$. To decrypt the ciphertext block C, we find

RONALD RIVEST (b. 1948) received his B.A. from Yale University in 1969 and his Ph.D. in computer science from Stanford University in 1974. He is a professor of computer science at M.I.T., and a cofounder of RSA Data Security, Inc. (now a subsidiary of Security Dynamics), the company that holds the patents on the RSA cryptosystem. Rivest has worked in the areas of machine learning, computer algorithms, and VLSI design. He is one of the authors of a popular textbook on algorithms ([ColeRi01]).

ADI SHAMIR (b. 1952) was born in Tel Aviv, Israel. He received his undergraduate degree from Tel Aviv University in 1972, and his Ph.D. in computer science from the Weizmann Institute of Science in 1977. He held a research assistantship at the University of Warwick for one year, and in 1978 he became an assistant professor at M.I.T. He is now a professor in the Applied Mathematics Department at the Weizmann Institute in Israel, where he formed a group to study computer security. Shamir has made many contributions to cryptography besides coinventing the RSA cryptosystem, including cracking the knapsack cryptosystem proposed as a public cryptosystem by Merkle and Hellman, developing numerous cryptographic protocols, and creative cryptanalysis of DES.

LEONARD ADLEMAN (b. 1945) was born in San Francisco, California. He received his B.S. in mathematics and his Ph.D. in computer science from the University of California, Berkeley, in 1968 and 1976, respectively. He was a member of the mathematics faculty at M.I.T. from 1976 until 1980; during his stay at M.I.T., he helped invent the RSA cryptosystem. In 1980 he was appointed to a position in the computer science department of the University of Southern California, and to a chaired professorship in 1985. Adleman has worked in the areas of computational complexity, computer security, immunology, and molecular biology, in addition to his work in cryptography. He coined the term "computer virus." His recent work on computing using DNA has attracted great interest. Adleman served as the technical adviser for the movie *Sneakers,* in which computer security figured prominently.

$$D(C) \equiv C^d = (P^e)^d = P^{ed} = P^{k\phi(n)+1}$$
$$\equiv (P^{\phi(n)})^k P \equiv P \pmod{n},$$

where $ed = k\phi(n) + 1$ for some integer k, because $ed \equiv 1 \pmod{\phi(n)}$, and by Euler's theorem, we have $P^{\phi(n)} \equiv 1 \pmod{n}$, when $(P, n) = 1$ the (probability that P and n are not relatively prime is extremely small; see Exercise 4 at the end of this section). The pair (d, n) is a decrypting key.

Example 8.16. To illustrate how the RSA cryptosystem works, suppose that the encrypting modulus is the product of the two primes 43 and 59 (which are smaller than the large primes that would actually be used); thus, we have $n = 43 \cdot 59 = 2537$ as the modulus. We take $e = 13$ as the exponent; note that we have $(e, \phi(n)) = (13, 42 \cdot 58) = 1$. To encrypt the message

PUBLIC KEY CRYPTOGRAPHY,

we first translate the letters into their numerical equivalents, and then group these numbers together into blocks of four. We obtain

$$
\begin{array}{cccc}
1520 & 0111 & 0802 & 1004 \\
2402 & 1724 & 1519 & 1406 \\
1700 & 1507 & 2423,
\end{array}
$$

where we have added the dummy letter $X = 23$ at the end of the passage to fill out the final block.

We encrypt each plaintext block into a ciphertext block, using the relationship

$$C \equiv P^{13} \pmod{2537}.$$

For instance, when we encrypt the first plaintext block 1520, we obtain the ciphertext block

$$C \equiv (1520)^{13} \equiv 95 \pmod{2537}.$$

Encrypting all the plaintext blocks, we obtain the ciphertext message

$$
\begin{array}{cccc}
0095 & 1648 & 1410 & 1299 \\
0811 & 2333 & 2132 & 0370 \\
1185 & 1957 & 1084.
\end{array}
$$

To decrypt messages that have been encrypted using this RSA cipher, we must find an inverse of $e = 13$ modulo $\phi(2537) = \phi(43 \cdot 59) = 42 \cdot 58 = 2436$. A short computation using the Euclidean algorithm, as done in Section 4.2, shows that $d = 937$ is an inverse of 13 modulo 2436. Consequently, to decrypt the ciphertext block C, we use the relationship

$$P \equiv C^{937} \pmod{2537}, \quad 0 \le P < 2537,$$

which is valid because

$$C^{937} \equiv (P^{13})^{937} \equiv (P^{2436})^5 P \equiv P \pmod{2537}.$$

Note that we have used Euler's theorem to see that

$$P^{\phi(2537)} = P^{2436} \equiv 1 \ (\text{mod } 2537),$$

when $(P, 2537) = 1$ (which is true for all of the plaintext blocks in this example).   ◄

The Security of the RSA Cryptosystem   To understand how the RSA cryptosystem fulfills the requirements of a public key cryptosystem, first note that each individual can find two large primes p and q, each with 100 decimal digits, in just a few minutes of computer time. These primes can be found by picking odd integers with 100 digits at random; by the prime number theorem, the probability that such an integer is prime is approximately $2/\log 10^{100}$. Hence, we expect to find a prime after examining an average of $1/(2/\log 10^{100})$, or approximately 115, such integers. To test these randomly chosen odd integers for primality, we use Rabin's probabilistic primality test (discussed in Section 6.2). For each of these 100-digit odd integers we perform Miller's test for 100 bases less than the integer; the probability that a composite integer passes all these tests is less than 10^{-60}. The procedure we have just outlined requires only a few minutes of computer time to find a 100-digit prime, and each individual need do so only twice.

Once the primes p and q have been found, an encrypting exponent e must be chosen such that $(e, \phi(pq)) = 1$. One suggestion for choosing e is to take any prime greater than both p and q. No matter how e is found, it should be true that $2^e > n = pq$, so that it is impossible to recover the plaintext block P, $P \neq 0$ or 1, just by taking the eth root of the integer C with $C \equiv P^e \ (\text{mod } n)$, $0 \leq C < n$. As long as $2^e > n$, every message, other than $P = 0$ and 1, is encrypted by exponentiation followed by a reduction modulo n.

We note that the modular exponentiation needed for encrypting messages using the RSA cryptosystem can be done using only a few seconds of computer time when the modulus, exponent, and base in the modular exponentiation have as many as 200 decimal digits. Also, using the Euclidean algorithm, we can rapidly find an inverse d of the encryption exponent e modulo $\phi(n)$ when the primes p and q are known, so that $\phi(n) = \phi(pq) = (p-1)(q-1)$ is known.

To see why knowledge of the encrypting key (e, n) does not easily lead to the decrypting key (d, n), note that to find d, an inverse of e modulo $\phi(n)$, requires that we first find $\phi(n) = \phi(pq) = (p-1)(q-1)$. Note that finding $\phi(n)$ is not easier than factoring the integer n. To see why, note that $p + q = n - \phi(n) + 1$ and $p - q = \sqrt{(p+q)^2 - 4pq} = \sqrt{(p+q)^2 - 4n}$ and that $p = \frac{1}{2}[(p+q) + (p-q)]$ and $q = \frac{1}{2}[(p+q) - (p-q)]$. Consequently, p and q can easily be found when $n = pq$ and $\phi(n) = (p-1)(q-1)$ are known. Note that when p and q both have approximately 100 decimal digits, $n = pq$ has approximately 200 decimal digits. Using the fastest factorization algorithm known, millions of years of computer time are required to factor an integer of this size. Also, if the integer d is known, but $\phi(n)$ is not, then n may also be factored easily, since $ed - 1$ is a multiple of $\phi(n)$ and there are special algorithms for factoring an integer n using any multiple of $\phi(n)$ (see [Mi76]).

It has not been proven that it is impossible to decrypt messages encrypted using the RSA cryptosystem without factoring n, but so far no such method has been discovered.

As yet, all decrypting methods that work in general are equivalent to factoring n and, as we have remarked, factoring large integers seems to be an intractable problem, requiring tremendous amounts of computer time. If no method of decrypting RSA messages without factoring the modulus n is found, the security of the RSA system can be maintained as factoring methods and computational power improve, by increasing the size of the modulus. Unfortunately, messages encrypted using the RSA will become vulnerable to attack when factoring the modulus n becomes feasible. This means that extra care should be taken—for example, by using primes p and q each with several hundred digits—to protect the secrecy of messages that must be kept secret for tens, or hundreds, of years.

Note that a few extra precautions should be taken in choosing the primes p and q to be used in the RSA cryptosystem, to prevent the use of special rapid techniques to factor $n = pq$. For example, both $p - 1$ and $q - 1$ should have large prime factors, $(p - 1, q - 1)$ should be small, and p and q should have decimal expansions differing in length by a few digits.

As we have remarked, the security of the RSA cryptosystem depends on the difficulty of factoring large integers. In particular, for the RSA cryptosystem, once the modulus n has been factored it is easy to find the decrypting transformation from the encrypting transformation. Note, however, that it may be possible to somehow find the decrypting transformation from the encrypting transformation without factoring n, although this seems unlikely at present.

Attacks on Implementations of the RSA Cryptosystem

After 20 years of scrutiny, a variety of attacks on particular implementations of the RSA cryptosystem have been devised. These attacks show that care must be taken when implementing RSA to avoid particular vulnerabilities. Note that no fundamental vulnerability has been found that would make RSA unsuitable for use as a public key cryptosystem. We will describe a variety of these attacks. The interested reader should consult [Bo99].

Encrypting the same plaintext message with different keys can lead to a successful *Hastad broadcast attack*. For example, when the encryption exponent 3 is used by three different people with different encryption moduli to encrypt the same plaintext message, someone who has the three ciphertext messages produced can recover the original plaintext. In general, it is possible to recover a plaintext message from ciphertext produced by encrypting the message using different RSA encryption keys when sufficiently many copies of the message have been encrypted. This type of attach can even succeed if the original message is altered for each recipient in a way that produces linearly related plaintext. To avoid this vulnerability, different random paddings of the message should be encrypted.

We now describe a vulnerability of RSA found by M. Wiener [Wi90]. He showed that the decrypting exponent d of an RSA cryptosystem with encrypting key (e, n) can be efficiently determined if $n = pq$, p and q are primes with $q < p < 2q$, and the decrypting exponent d is less than $n^{1/4}/3$. (In Chapter 12 we will use the theory of continued fractions to develop this attack.) This result shows that primes p and q that are not too close together should be used to produce the encrypting modulus and a decrypting exponent d that is relatively large should be used. Although it is customary to first select the encryption key in an RSA cipher, we can make the decrypting exponent large by selecting it first, and then using it to compute the encrypting exponent e.

Disclosing partial information about one of the primes that make up the encrypting modulus n leads to another weakness of the RSA cryptosystem. Suppose that $n = pq$ has m digits. Then knowing the initial $m/4$ or the final $m/4$ digits of p allows n to be efficiently factored. For example, when both p and q have 100 decimal digits, if we know the first 50 or the last 50 digits of p, we will be able to factor n. Details of this partial key disclosure attack can be found in [Co97]. A similar result shows that if we know the last $m/4$ digits of the decrypting exponent d, then we can efficiently find d using $O(e \log e)$ operations. This shows that if the encryption exponent e is small, the decryption exponent d can be found if we know the last $1/4$ of its digits.

The final type of attack we mention was discovered by Paul Kocher in 1995 when he was an undergraduate at Stanford University. He demonstrated that the decryption exponent in the RSA cryptosystem can be determined by carefully measuring the time required for the system to perform a series of decryptions. This provides information that can be used to determine the decryption key d. Fortunately, it is easy to devise methods to thwart this attack. For a description of this attack, see [TrWa02] and the article by Kocher [Ko96a].

The widespread acceptance and use of the RSA cryptosystem makes in an inviting target for attack. That only minor vulnerabilities have been found has given people confidence in the practical use of this cryptosystem. This fuels the search for vulnerabilities in this popular cryptosystem.

The Rabin Cryptosystem

Michael Rabin [Ra79] discovered a variant of the RSA cryptosystem for which factorization of the modulus n has almost the same computational complexity as obtaining the decrypting transformation from the encrypting transformation. To describe Rabin's cryptosystem let $n = pq$, where p and q are odd primes, and let b be an integer with $0 \le b < n$. To encrypt the plaintext message P, we form

$$C \equiv P(P + b) \pmod{n}.$$

We will not discuss the decrypting procedure for Rabin ciphers here, because it relies on some concepts that we have not yet developed (see Exercise 49 in Section 11.1). However, we remark that there are four possible values of P for each ciphertext C such that $C \equiv P(P + b) \pmod{n}$, an ambiguity that complicates the decrypting process. When p and q are known, the decrypting procedure for a Rabin cipher can be carried out rapidly because $O(\log n)$ bit operations are needed.

Rabin has shown that if there is an algorithm for decrypting in this cryptosystem, without knowledge of the primes p and q, that requires $f(n)$ bit operations, then there is an algorithm for the factorization of n requiring only $2(f(n) + \log n)$ bit operations. Hence, the process of decrypting messages encrypted with a Rabin cipher without knowledge of p and q is a problem of computational complexity similar to that of factorization. For more information about the Rabin public key cryptosystem, see [MevaVa97].

8.4 Exercises

1. Find the primes p and q if $n = pq = 14{,}647$ and $\phi(n) = 14{,}400$.

2. Find the primes p and q if $n = pq = 4{,}386{,}607$ and $\phi(n) = 4{,}382{,}136$.

3. Suppose a cryptanalyst discovers a message P that is not relatively prime to the enciphering modulus $n = pq$ used in an RSA cipher. Show that the cryptanalyst can factor n.

4. Show that it is extremely unlikely that a message such as that described in Exercise 3 can be discovered. Do this by demonstrating that the probability that a message P is not relatively prime to n is $\frac{1}{p} + \frac{1}{q} - \frac{1}{pq}$, and if p and q are both larger than 10^{100}, this probability is less than 10^{-99}.

5. What is the ciphertext that is produced when RSA encryption with key $(e, n) = (3, 2669)$ is used to encrypt the message BEST WISHES?

6. What is the ciphertext that is produced when RSA encryption with key $(e, n) = (7, 2627)$ is used to encrypt the message LIFE IS A DREAM?

7. If the ciphertext message produced by RSA encryption with the key $(e, n) = (13, 2747)$ is 2206 0755 0436 1165 1737, what is the plaintext message?

8. If the ciphertext message produced by RSA encryption with the key $(e, n) = (5, 2881)$ is 0504 1874 0347 0515 2088 2356 0736 0468, what is the plaintext message?

9. Encrypt the message SELL NOW using the Rabin cipher $C \equiv P(P + 5) \pmod{2573}$.

10. Encrypt the message LEAVE TOWN using the Rabin cipher $C \equiv P(P + 11) \pmod{3901}$.

11. Suppose that Bob, extremely concerned with security, selects an encrypting modulus n, $n = pq$, where p and q are large primes, and two encrypting exponents e_1 and e_2. He asks Alice to double encrypt messages set to him by first encrypting plaintext using the RSA cipher with encryption key (e_1, n) and then encrypting the resulting ciphertext again using the RSA cipher with encryption key (e_2, n). Does Bob gain any extra security by this double encryption? Justify your answer.

12. Suppose that a plaintext message P is not relatively prime to $n = pq$, where p and q are large primes. Is it possible to successfully decrypt the ciphertext produced by encrypting p using RSA encryption with key (e, n)?

13. Suppose that two parties share a common modulus n in the RSA cryptosystem, but have different encrypting exponents. Show that the plaintext of a message sent to each of these two parties encrypted using each of their RSA keys can be recovered from the ciphertext messages.

14. Show that if the encryption exponent 3 is used for the RSA cryptosystem by three different people with different moduli, a plaintext message P encrypted using each of their keys can be recovered from these resulting three ciphertext messages. (*Hint:* Suppose that the moduli in these three keys are n_1, n_2, and n_3. First find a common

solution to the congruences $x_i \equiv P^3 \pmod{n_i}, i = 1, 2, 3.$) (This is an example of a Hastad broadcast attack.)

15. Describe how an RSA cryptosystem works if the encrypting modulus n is the product of three primes, rather than two primes.

16. Suppose that two people have RSA encrypting keys with encrypting moduli n_1 and n_2, respectively, when $n_1 \neq n_2$. Show how you could break the system if $(n_1, n_2) > 1$.

8.4 Computational and Programming Exercises

Computations and Explorations

Using a computation program such as Maple or *Mathematica*, or programs you have written, carry out the following computations and explorations.

1. Construct a key for the RSA cipher for inclusion in a directory of encryption keys for the members of your class.

2. For each member of your class, encrypt a message using the RSA cipher with the public keys published in the directory.

3. Decipher the messages sent to you by your classmates that were encrypted using your RSA encryption key.

Programming Projects

Write computer programs using Maple, *Mathematica*, or a language of your choice to do the following.

1. Encrypt messages with an RSA cipher.

2. Decrypt messages that were encrypted using an RSA cipher.

8.5 Knapsack Ciphers

In this section, we discuss cryptosystems based on the knapsack problem. Given a set of positive integers $a_1, a_2, \ldots, a_n$ and an integer S, the *knapsack problem* asks which of these integers, if any, add together to give S. Another way to phrase the knapsack problem is to ask for values of $x_1, x_2, \ldots, x_n$, each either 0 or 1, such that

(8.3)
$$S = a_1 x_1 + a_2 x_2 + \cdots + a_n x_n.$$

We use an example to illustrate the knapsack problem.

Example 8.17. Let $(a_1, a_2, a_3, a_4, a_5) = (2, 7, 8, 11, 12)$ and $S = 21$. By inspection, we see that there are two subsets of these five integers that add together to give 21, namely $21 = 2 + 8 + 11 = 2 + 7 + 12$. Equivalently, there are exactly two solutions to the equation $2x_1 + 7x_2 + 8x_3 + 11x_4 + 12x_5 = 21$, with $x_i = 0$ or 1 for $i = 1, 2, 3, 4, 5$. These solutions are $x_1 = x_3 = x_4 = 1, x_2 = x_5 = 0,$ and $x_1 = x_2 = x_5 = 1, x_3 = x_4 = 0.$ ◀

To verify that equation (8.3) holds, where each x_i is either 0 or 1, requires that we perform at most n additions. On the other hand, to search by trial and error for solutions of

(8.3) may require that we check all 2^n possibilities for $(x_1, x_2, \ldots, x_n)$. The best method known for finding a solution of the knapsack problem requires $O(2^{n/2})$ bit operations, which makes a computer solution of a general knapsack problem extremely infeasible even when $n = 100$.

Certain values of the integers $a_1, a_2, \ldots, a_n$ make the solution of the knapsack problem much easier than the solution in the general case. For instance, if $a_j = 2^{j-1}$, to solve $S = a_1 x_1 + a_2 x_2 + \cdots + a_n x_n$, where $x_i = 0$ or 1 for $i = 1, 2, \ldots, n$, simply requires that we find the binary expansion of S. We can also produce easy knapsack problems by choosing the integers $a_1, a_2, \ldots, a_n$ so that the sum of the first $j - 1$ of these integers is always less than the jth integer, that is, so that

$$\sum_{i=1}^{j-1} a_i < a_j, \quad j = 2, 3, \ldots, n.$$

If a sequence of integers $a_1, a_2, \ldots, a_n$ satisfies this inequality, we call the sequence *super-increasing*.

Example 8.18. The sequence 2, 3, 7, 14, 27 is super-increasing because $3 > 2, 7 > 3 + 2, 14 > 7 + 3 + 2$, and $27 > 14 + 7 + 3 + 2$. ◀

To see that knapsack problems involving super-increasing sequences are easy to solve, we first consider an example.

Example 8.19. Let us find the integers from the set 2, 3, 7, 14, 27 that have 37 as their sum. First, we note that since $2 + 3 + 7 + 14 < 27$, a sum of integers from this set can only be greater than 27 if the sum contains the integer 27. Hence, if $2x_1 + 3x_2 + 7x_3 + 14x_4 + 27x_5 = 37$ with each $x_i = 0$ or 1, we must have $x_5 = 1$ and $2x_1 + 3x_2 + 7x_3 + 14x_4 = 10$. Because $14 > 10$, x_4 must be 0 and we have $2x_1 + 3x_2 + 7x_3 = 10$. Because $2 + 3 < 7$, we must have $x_3 = 1$ and therefore $2x_1 + 3x_2 = 3$. Obviously, we have $x_2 = 1$ and $x_1 = 0$. The solution is $37 = 3 + 7 + 27$. ◀

In general, to solve knapsack problems for a super-increasing sequence $a_1, a_2, \ldots, a_n$, that is, to find the values of $x_1, x_2, \ldots, x_n$ with $S = a_1 x_1 + a_2 x_2 + \cdots + a_n x_n$ and $x_i = 0$ or 1 for $i = 1, 2, \ldots, n$ when S is given, we use the following algorithm. First, we find x_n by noting that

$$x_n = \begin{cases} 1 & \text{if } S \geq a_n; \\ 0 & \text{if } S < a_n. \end{cases}$$

Then, we find $x_{n-1}, x_{n-2}, \ldots, x_1$, in succession, using the equations

$$x_j = \begin{cases} 1 & \text{if } S - \sum_{i=j+1}^{n} x_i a_i \geq a_j; \\ 0 & \text{if } S - \sum_{i=j+1}^{n} x_i a_i < a_j, \end{cases}$$

for $j = n - 1, n - 2, \ldots, 1$.

To see that this algorithm works, first note that if $x_n = 0$ when $S \geq a_n$, then $\sum_{i=1}^{n} a_i x_i \leq \sum_{i=1}^{n-1} a_i < a_n \leq S$, contradicting the condition $\sum_{j=1}^{n} a_j x_j = S$. Similarly, if $x_j = 0$ when $S - \sum_{i=j+1}^{n} x_i a_i \geq a_j$, then $\sum_{i=1}^{n} a_i x_i \leq \sum_{i=1}^{j-1} a_i + \sum_{i=j+1}^{n} x_i a_i < a_j + \sum_{i=j+1}^{n} x_i a_i \leq S$, which is again a contradiction.

Using this algorithm, knapsack problems based on super-increasing sequences can be solved extremely quickly. We now discuss a cryptosystem based on this observation, invented by Merkle and Hellman [MeHe78], that was initially considered a good choice for a public key cryptosystem. (We will comment more about this later in this section.)

The ciphers that we describe here are based on transformed super-increasing sequences. To be specific, let $a_1, a_2, \ldots, a_n$ be super-increasing and let m be a positive integer with $m > 2a_n$. Let w be an integer relatively prime to m with inverse $\overline{w}$ modulo m. We form the sequence $b_1, b_2, \ldots, b_n$, where $b_j \equiv wa_j \pmod{m}$ and $0 \leq b_j < m$. We cannot use this special technique to solve a knapsack problem of the type $S = \sum_{i=1}^{n} b_i x_i$, where S is a positive integer, because the sequence $b_1, b_2, \ldots, b_n$ is not super-increasing. However, when $\overline{w}$ is known, we can find

$$(8.4) \qquad \overline{w}S = \sum_{i=1}^{n} \overline{w} b_i x_i \equiv \sum_{i=1}^{n} a_i x_i \pmod{m},$$

because $\overline{w} b_j \equiv a_j \pmod{m}$. From (8.4), we see that

$$S_0 = \sum_{i=1}^{n} a_i x_i,$$

where S_0 is the least positive residue of $\overline{w} S$ modulo m. We can easily solve the equation

$$S_0 = \sum_{i=1}^{n} a_i x_i,$$

because $a_1, a_2, \ldots, a_n$ is super-increasing. This solves the knapsack problem

$$S = \sum_{i=1}^{n} b_i x_i,$$

because $b_j \equiv wa_j \pmod{m}$ and $0 \leq b_j < m$. We illustrate this procedure with an example.

Example 8.20. The super-increasing sequence $(a_1, a_2, a_3, a_4, a_5) = (3, 5, 9, 20, 44)$ can be transformed into the sequence $(b_1, b_2, b_3, b_4, b_5) = (23, 68, 69, 5, 11)$ by taking $b_j \equiv 67a_j \pmod{89}$, for $j = 1, 2, 3, 4, 5$. To solve the knapsack problem $23x_1 + 68x_2 + 69x_3 + 5x_4 + 11x_5 = 84$, we can multiply both sides of this equation by 4, an inverse of 67 modulo 89, and then reduce modulo 89, to obtain the congruence $3x_1 + 5x_2 + 9x_3 + 20x_4 + 44x_5 \equiv 336 \equiv 69 \pmod{89}$. Because $89 > 3 + 5 + 9 + 20 + 44$, we can conclude that $3x_1 + 5x_2 + 9x_3 + 20x_4 + 44x_5 = 69$. The solution of this easy knapsack problem is $x_5 = x_4 = x_2 = 1$ and $x_3 = x_1 = 0$. Hence, the original knapsack problem has as its solution $68 + 5 + 11 = 84$. ◄

The cryptosystem based on the knapsack problem invented by Merkle and Hellman works as follows. Each individual chooses a super-increasing sequence of positive

Letter	Binary Equivalent	Letter	Binary Equivalent
A	00000	N	01101
B	00001	O	01110
C	00010	P	01111
D	00011	Q	10000
E	00100	R	10001
F	00101	S	10010
G	00110	T	10011
H	00111	U	10100
I	01000	V	10101
J	01001	W	10110
K	01010	X	10111
L	01011	Y	11000
M	01100	Z	11001

Table 8.10 *The binary equivalents of letters.*

integers of a specified length, say N (for example, $a_1, a_2, \ldots, a_N$), as well as a modulus m with $m > 2a_N$ and a multiplier w with $(m, w) = 1$. The transformed sequence $b_1, b_2, \ldots, b_n$ is made public. When someone wishes to send a message P to this individual, the message is first translated into a string of zeros and ones using the binary equivalents of letters, as shown in Table 8.10. This string of zeros and ones is next split into segments of length N (for simplicity, we suppose that the length of the string is divisible by N; if not, we can simply fill out the last block with all ones). For each block, a sum is computed using the sequence $b_1, b_2, \ldots, b_N$: for instance, the block $x_1 x_2 \ldots x_N$ gives $S = b_1 x_1 + b_2 x_2 + \cdots + b_N x_N$. Finally, the sums generated by each block form the ciphertext message.

We note that to decipher ciphertext generated by the knapsack cipher, without knowledge of m and w, requires that a group of hard knapsack problems of the form

$$(8.5) \qquad S = b_1 x_1 + b_2 x_2 + \cdots + b_N x_N$$

be solved. On the other hand, when m and w are known, the knapsack problem (8.5) can be transformed into an easy knapsack problem, because

$$\overline{w} S = \overline{w} b_1 x_1 + \overline{w} b_2 x_2 + \cdots + \overline{w} b_N x_N$$
$$\equiv a_1 x_1 + a_2 x_2 + \cdots + a_N x_N \pmod{m},$$

in which $\overline{w} b_j \equiv a_j \pmod{m}$, where $\overline{w}$ is an inverse of w modulo m, so that

$$(8.6) \qquad S_0 = a_1 x_1 + a_2 x_2 + \cdots + a_N x_N,$$

where S_0 is the least positive residue of $\overline{w}S$ modulo m. We have equality in (8.6), because both sides of the equation are positive integers less than m that are congruent modulo m.

We illustrate the encrypting and decrypting procedures of the knapsack cipher with an example. We start with the super-increasing sequence $(a_1, a_2, a_3, a_4, a_5, a_6, a_7, a_8, a_9, a_{10}) = (2, 11, 14, 29, 58, 119, 241, 480, 959, 1917)$. We take $m = 3837$ as the encrypting modulus, so that $m > 2a_{10}$, and $w = 1001$ as the multiplier, so that $(m, w) = 1$, to transform the super-increasing sequence into the sequence $(2002, 3337, 2503, 2170, 503, 172, 3347, 855, 709, 417)$.

To encrypt the message

REPLY IMMEDIATELY,

we first translate the letters of the message into their five-digit binary equivalents, as shown in Table 8.10, and then group these digits into blocks of ten, to obtain

1000100100 0111101011   1100001000

0110001100 0010000011   0100000000

1001100100 0101111000.

For each block of ten binary digits, we form a sum by adding together the appropriate terms of the sequence $(2002, 3337, 2503, 2170, 503, 172, 3347, 855, 709, 417)$ in the slots corresponding to positions of the block containing a digit equal to 1. This gives us

3360 12986 8686 10042 3629 3337 5530 9529.

For instance, we compute the first sum, 3360, by adding 2002, 503, and 855.

To decrypt, we find the least positive residue modulo 3837 of 23 times each sum, because 23 is an inverse of 1001 modulo 3837, and then we solve the corresponding easy knapsack problem with respect to the original super-increasing sequence $(2, 11, 14, 29, 58, 119, 241, 480, 959, 1917)$. For example, to decrypt the first block, we find that $3360 \cdot 23 \equiv 540 \pmod{3837}$, and then note that $540 = 480 + 58 + 2$. This tells us that the first block of plaintext binary digits is 1000100100.

Knapsack ciphers originally seemed to be excellent candidates for use in public key cryptosystems. However, in 1982 Shamir [Sh84] has shown that they are not satisfactory for public key cryptography. The reason is that there is an efficient algorithm for solving knapsack problems involving sequences $b_1, b_2, \ldots, b_n$ with $b_j \equiv wa_j \pmod{m}$, where w and m are relatively prime positive integers and $a_1, a_2, \ldots, a_n$ is a super-increasing sequence. The algorithm found by Shamir can solve these knapsack problems using only $O(P(n))$ bit operations, where P is a polynomial, instead of requiring exponential time, as is required for known algorithms for general knapsack problems involving sequences of a general nature. Although we will not go into the details of the algorithm found by Shamir here, the reader can find these details by consulting [Od90].

There are several possibilities for altering this cryptosystem to avoid the weakness found by Shamir. One such possibility is to choose a sequence of pairs of relatively prime integers $(w_1, m_1), (w_2, m_2), \ldots, (w_r, m_r)$, and then form the series of sequences

$$b_j^{(1)} \equiv w_1 a_j \pmod{m_1}$$

$$b_j^{(2)} \equiv w_2 b_j^{(1)} \pmod{m_2}$$

$$\vdots$$

$$b_j^{(r)} \equiv w_r b_j^{(r-1)} \pmod{m_r},$$

for $j = 1, 2, \ldots, n$. We then use the final sequence $b_1^{(r)}, b_2^{(r)}, \ldots, b_n^{(r)}$ as the encrypting sequence. Unfortunately, efficient algorithms have been found for solving knapsack problems involving sequences obtained by iterating modular multiplications with different moduli.

A comprehensive discussion of knapsack ciphers can be found in [Od90]. This article describes knapsack ciphers and their generalizations, and goes on to explain the attacks that have been found for breaking them.

8.5 Exercises

1. Decide whether each of the following sequences is super-increasing.

   a) $(3, 5, 9, 19, 40)$ c) $(3, 7, 17, 30, 59)$
   b) $(2, 6, 10, 15, 36)$ d) $(11, 21, 41, 81, 151)$

2. Show that if $a_1, a_2, \ldots, a_n$ is a super-increasing sequence, then $a_j \geq 2^{j-1}$ for $j = 1, 2, \ldots, n$.

3. Show that the sequence $a_1, a_2, \ldots, a_n$ is super-increasing if $a_{j+1} > 2a_j$ for $j = 1, 2, \ldots, n - 1$.

4. Find all subsets of the integers 2, 3, 4, 7, 11, 13, 16 that have 18 as their sum.

5. Find the sequence obtained from the super-increasing sequence $(1, 3, 5, 10, 20, 41, 81)$ when modular multiplication is applied with multiplier $w = 17$ and modulus $m = 163$.

6. Encrypt the message BUY NOW using the knapsack cipher based on the sequence obtained from the super-increasing sequence $(17, 19, 37, 81, 160)$, by performing modular multiplication with multiplier $w = 29$ and modulus $m = 331$.

7. Decrypt the ciphertext 402 75 120 325 that was encrypted by the knapsack cipher based on the sequence $(306, 374, 233, 19, 259)$. This sequence is obtained by using modular multiplication with multiplier $w = 17$ and modulus $m = 464$, to transform the super-increasing sequence $(18, 22, 41, 83, 179)$.

8. Find the sequence obtained by applying successively the modular multiplications with multipliers and moduli $(7,92)$, $(11,95)$, and $(6,101)$, respectively, on the super-increasing sequence $(3, 4, 8, 17, 33, 67)$.

9. What process can be employed to decrypt messages that have been encrypted using knapsack ciphers that involve sequences arising from iterating modular multiplications with different moduli?

A *multiplicative knapsack problem* is a problem of the following type: Given positive integers $a_1, a_2, \ldots, a_n$ and a positive integer P, find the subset, or subsets, of these integers with product P, or equivalently, find all solutions of

$$P = a_1^{x_i} a_2^{x_2} \cdots a_n^{x_n},$$

where $x_j = 0$ or 1 for $j = 1, 2, \ldots, n$.

10. Find all products of subsets of the integers 2, 3, 5, 6, 10 equal to 60.

11. Find all products of subsets of the integers 8, 13, 17, 21, 95, 121 equal to 15,960.

12. Show that if the integers $a_1, a_2, \ldots, a_n$ are pairwise relatively prime, then the multiplicative knapsack problem $P = a_1^{x_1} a_2^{x_2} \cdots a_n^{x_n}$, $x_j = 0$ or 1 for $j = 1, 2, \ldots, n$ is easily solved from the prime factorizations of the integers $P, a_1, a_2, \ldots, a_n$, and show that if there is a solution, then it is unique.

13. Show that by taking logarithms to the base b modulo m, where $(b, m) = 1$ and $0 < b < m$, the multiplicative knapsack problem

$$P = a_1^{x_1} a_2^{x_2} \cdots a_n^{x_n}$$

is converted into an additive knapsack problem

$$S = \alpha_1 x_1 + \alpha_2 x_2 + \cdots + \alpha_n x_n,$$

where $S, \alpha_1, \alpha_2, \ldots, \alpha_n$ are the logarithms of $P, a_1, a_2, \ldots, a_n$ to the base b modulo m, respectively.

14. Explain how Exercises 12 and 13 can be used to produce ciphers where messages are easily decrypted when the mutually relatively prime integers $a_1, a_2, \ldots, a_n$ are known, but cannot be decrypted quickly when the integers $\alpha_1, \alpha_2, \ldots, \alpha_n$ are known.

8.5 Computational and Programming Exercises

Computations and Explorations

Using a computation program such as Maple or *Mathematica,* or programs you have written, carry out the following computations and explorations.

1. Starting with a super-increasing sequence that you have constructed, perform modular multiplication with modulus m and multiplier w to find a sequence to serve as your public key for the knapsack cipher.

2. For each of your classmates, encrypt a message using their public key for the knapsack cipher.

3. Decrypt the messages that were sent to you by classmates.

** 4. Using algorithms described in [Od90], solve knapsack problems based on a sequence obtained by modular multiplication of a super-increasing sequence.

Programming Projects

Write computer programs using Maple, *Mathematica,* or a language of your choice to do the following.

1. Solve knapsack problems by trial and error.

2. Solve knapsack problems involving super-increasing sequences.

3. Encrypt messages using knapsack ciphers.

4. Decrypt messages that were encrypted using knapsack ciphers.

5. Encrypt and decrypt messages using knapsack ciphers involving sequences arising from iterating modular multiplications with different moduli.

6. Solve multiplicative knapsack problems involving sequences of mutually relatively prime integers (see Exercise 14).

8.6 Cryptographic Protocols and Applications

In this section, we describe how cryptosystems can be used in protocols, which are algorithms carried out by two or more parties to achieve a specific goal, and in other cryptographic applications. In particular, we will show how two or more people can exchange encryption keys. We will also explain how messages can be signed using the RSA cryptosystem, and how cryptography can be used to allow people to play poker fairly over a network. Finally, we will show how people can share a secret, so that no one person knows the secret, but a large enough group of people can recover the secret by cooperating. These are only a few of the many examples of protocols and applications that we could discuss; the interested reader should consult [MevaVa97] to learn about additional protocols and applications based on the ideas we have covered in this chapter.

Diffie-Hellman Key Exchange

We will now discuss a protocol that allows two parties to exchange a secret key over an insecure communications link without having shared any information in the past. Exchanging keys is a problem of fundamental importance in cryptography. The method that we will describe was invented by Diffie and Hellman in 1976 (see [DiHe76]) and is called the *Diffie-Hellman key agreement protocol*. The common secret key generated by this protocol can be used as a shared key for a symmetric cryptosystem to be used during a particular communication session by parties who have never met or shared any prior information. It has the property that unauthorized parties cannot discover it in a feasible amount of computer time.

To implement this protocol, we need a large prime p and an integer r such that the least positive residue of r^k runs inclusively through all integers from 1 to $n - 1$. (This means that r is a primitive root of p, a concept that we will study in Chapter 9.) Both the large prime p and the integer r are public information.

In this protocol, two parties who want to share a common key each pick a random private value from the set of positive integers between 1 and $p - 2$, inclusive. If the two parties select k_1 and k_2, respectively, the first party sends the second party the integer y_1, where

$$y_1 \equiv r^{k_1} \ (\text{mod } p), \quad 0 < y_1 < p,$$

and the second party finds the common key K by computing

$$K \equiv y_1^{k_2} \equiv r^{k_1 k_2} \ (\text{mod } p), \quad 0 < K < p.$$

Similarly, the second party sends the first party the integer y_2, where

$$y_2 \equiv r^{k_2} \pmod{p}, \quad 0 < y_2 < p,$$

and the first party finds the common key K by computing

$$K \equiv y_2^{k_1} \equiv r^{k_1 k_2} \pmod{p}, \quad 0 < K < p.$$

The security of this key agreement protocol depends on the security of determining the secret key K, given the least positive residues of r^{k_1} and r^{k_2} modulo p; that is, it depends on the difficulty of computing what are known as discrete logarithms modulo p (to be discussed in Chapter 9), which is thought to be a computationally difficult problem. It has been shown (see [Ma94]) that breaking this protocol is equivalent to computing discrete logarithms, when certain conditions hold.

In a similar manner, a common key can be shared by any group of n individuals. If these individuals have keys $k_1, k_2, \ldots, k_n$, they can share the common key

$$K = r^{k_1 k_2 \cdots k_n} \pmod{p}.$$

We leave an explicit description of a method used to produce this common key as a problem for the reader.

The topic of key establishment protocols extends far beyond what we have described here. Many different protocols for establishing shared keys have been developed, including protocols that make use of trusted servers for distributing keys. To learn more about this topic, consult Chapter 12 of [MevaVa97].

Digital Signatures

When we receive an electronic message, how do we know that it has come from the supposed sender? We need a *digital signature* that can tell us that the message must have originated with the party who supposedly sent it. We will show that a public key cryptosystem, such as the RSA cryptosystem, can be used to send "signed" messages. When signatures are used, the recipient of a message is sure that the message came from the sender, and can convince an impartial judge that only the sender could be the source of the message. This authentication is needed for electronic mail, electronic banking, and electronic stock market transactions. To see how the RSA cryptosystem can be used to send signed messages, suppose that individual i wishes to send a signed message to individual j. The first thing that individual i does to a plaintext block P is to compute

$$S = D_{k_i}(P) \equiv P^{d_i} \pmod{n_i},$$

where (d_i, n_i) is the decrypting key for individual i, which only individual i knows. Then, if $n_j > n_i$, where (e_j, n_j) is the encryption key for individual j, individual i encrypts S by forming

$$C = E_{k_j}(S) \equiv S^{e_j} \pmod{n_j}, \quad 0 \leq C < n_j.$$

When $n_j < n_i$, individual i splits S into blocks of size less than n_j and encrypts each block using the encrypting transformation E_{k_j}.

For decrypting, individual j first uses the private decrypting transformation D_{k_j} to recover S, because

$$D_{k_j}(C) = D_{k_j}(E_{k_j}(S)) = S.$$

To find the plaintext message P, supposedly sent by individual i, individual j next uses the public encrypting transformation E_{k_i}, because

$$E_{k_i}(S) = E_{k_i}(D_{k_i}(P)) = P.$$

Here, we have used the identity $E_{k_i}(D_{k_i}(P)) = P$, which follows from the fact that

$$E_{k_i}(D_{k_i}(P)) \equiv (P^{d_i})^{e_i} \equiv P^{d_i e_i} \equiv P \pmod{n_i},$$

because

$$d_i e_i \equiv 1 \pmod{\phi(n_i)}.$$

The combination of the plaintext block P and the signed version S convinces individual j that the message actually came from individual i. Also, individual i cannot deny sending the message, because no one other than individual i could have produced the signed message S from the original message P.

Electronic Poker

An amusing application of exponentiation ciphers has been described by Shamir, Rivest, and Adleman [ShRiAd81]. They show that by using exponentiation ciphers, a fair game of poker may be played by two players, communicating via computers. Suppose that Alex and Betty wish to play poker. First, they jointly choose a large prime p. Next, they individually choose secret keys e_1 and e_2, to be used as exponents in modular exponentiation. Let E_{e_1} and E_{e_2} represent the corresponding encrypting transformations, so that

$$E_{e_1}(M) \equiv M^{e_1} \pmod{p}$$

$$E_{e_2}(M) \equiv M^{e_2} \pmod{p},$$

where M is a plaintext message. Let d_1 and d_2 be the respective inverses of e_1 and e_2 modulo p, and let D_{e_1} and D_{e_2} be the corresponding decrypting transformations, so that

$$D_{e_1}(C) \equiv C^{d_1} \pmod{p}$$

$$D_{e_2}(C) \equiv C^{d_2} \pmod{p},$$

where C is a ciphertext message.

Note that encrypting transformations commute, that is,

$$E_{e_1}(E_{e_2}(M)) = E_{e_2}(E_{e_1}(M)),$$

because $(M^{e_2})^{e_1} \equiv (M^{e_1})^{e_2} \pmod{p}$.

To play electronic poker, the deck of cards is represented by the 52 messages

$$M_1 = \text{"TWO OF CLUBS"}$$
$$M_2 = \text{"THREE OF CLUBS"}$$

$$\vdots$$

$$M_{52} = \text{"ACE OF SPADES"}.$$

When Alex and Betty wish to play poker electronically, they use the following sequence of steps. We suppose that Betty is the dealer.

1. Betty uses her encrypting transformation to encipher the 52 messages for the cards. She obtains $E_{e_2}(M_1), E_{e_2}(M_2), \ldots, E_{e_2}(M_{52})$. Betty shuffles the deck, by randomly reordering the encrypted messages. Then she sends the 52 shuffled encrypted messages to Alex.

2. Alex selects, at random, five of the encrypted messages that Betty has sent him. He returns these five messages to Betty and she decrypts them to find her hand, using her decrypted transformation D_{e_2} because $D_{e_2}(E_{e_2}(M)) = M$ for all messages M. Alex cannot determine which cards Betty has, because he cannot decrypt the encrypted messages $E_{e_2}(M_j), j = 1, 2, \ldots, 52$.

3. Alex selects five other encrypted messages at random. Let these messages be C_1, C_2, C_3, C_4, and C_5, where

$$C_j = E_{e_2}(M_{i_j}),$$

$j = 1, 2, 3, 4, 5$. Alex sends these five previously encrypted messages using his encrypted transformation. He obtains the five messages

$$C_j^* = E_{e_i}((C_j)) = E_{e_i}(E_{e_2}(M_{i_j})),$$

$j = 1, 2, 3, 4, 5$. Alex sends these five messages that have been encrypted twice (first by Betty and afterward by Alex) to Betty.

4. Betty uses her decrypted transformation D_{e_2} to find

$$D_{e_2}(C_j^*) = D_{e_2}(E_{e_1}(E_{e_2}(M_{i_j})))$$
$$= D_{e_2}(E_{e_2}(E_{e_1}(M_{i_j})))$$
$$= E_{e_1}(M_{i_j}),$$

because $E_{e_1}(E_{e_2}(M)) = E_{e_2}(E_{e_1}(M))$ and $D_{e_2}(E_{e_2}(M)) = M$ for all messages M. Betty sends the five messages $E_{e_1}(M_{i_j})$ back to Alex.

5. Alex uses his decrypting transformation D_{e_1} to obtain his hand, because

$$D_{e_1}(E_{e_1}(M_{i_j})) = M_{i_j}.$$

When a game is played where it is necessary to deal additional cards, such as draw poker, the same steps are followed to deal additional cards from the remaining deck. Note that using the procedure we have described, neither player knows the cards in the hand of the other player, and all hands are equally likely for each player. To guarantee that no cheating has occurred, at the end of the game both players reveal their keys, so that each player can verify that the other player was actually dealt the cards claimed.

A description of a possible weakness in this scheme, and how it may be overcome, may be found in the exercise set of Section 11.1.

Secret Sharing

We now discuss another application of cryptography, namely a method for sharing secrets. Suppose that in a communications network there is some vital, but extremely sensitive, information. If this information is distributed to several individuals, it becomes much more vulnerable to exposure; on the other hand, if this information is lost, there are serious consequences. An example of such information is the *master key K* used for access to the password file in a computer system.

To protect this master key K from both loss and exposure, we construct *shadows* $k_1, k_2, \ldots, k_r$, which are given to r different individuals. We will show that the key K can be produced easily from any s of these shadows, where s is a positive integer less than r, whereas the knowledge of less than s of these shadows does not permit the key K to be found. Because at least s different individuals are needed to find K, the key is not vulnerable to exposure. In addition, the key K is not vulnerable to loss, since any s individuals from the r individuals with shadows can produce K. Schemes with properties we have just described are called (s, r)-*threshold schemes*.

To develop a system that can be used to generate shadows with these properties, we use the Chinese remainder theorem. We choose a prime p greater than the key K and a sequence of pairwise relatively prime integers $m_1, m_2, \ldots, m_r$ that are not divisible by p, such that

$$m_1 < m_2 < \cdots < m_r,$$

and

(8.7) $$m_1 m_2 \cdots m_s > p m_r m_{r-1} \cdots m_{r-s+2}.$$

Note that the inequality (8.7) states that the product of the s smallest of the integers m_j is greater than the product of p and the $s - 1$ largest of the integers m_j. From (8.7), we see that if $M = m_1 m_2 \cdots m_s$, then M/p is greater than the product of any set of $s - 1$ of the integers m_j.

Now, let t be a nonnegative integer less than M/p that is chosen at random. Let

$$K_0 = K + tp,$$

so that $0 \leq K_0 \leq M - 1$ (because $0 \leq K_0 = K + tp < p + tp = (t + 1)p \leq (M/p)p = M$).

To produce the shadows $k_1, k_2, \ldots, k_r$, we let k_j be the integer such that

$$k_j \equiv K_0 \pmod{m_j}, \quad 0 \leq k_j < m_j,$$

for $j = 1, 2, \ldots, r$. To see that the master key K can be found by any s individuals from the total of r individuals with shadows, suppose that the s shadows $k_{j_1}, k_{j_2}, \ldots, k_{j_s}$ are available. Using the Chinese remainder theorem, we can easily find the least positive residue of K_0 modulo M_j, where $M_j = m_{j_1} m_{j_2} \cdots m_{j_s}$. Because we know that $0 \leq K_0 < M \leq M_j$, we can determine K_0, and then find $K = K_0 - tp$.

On the other hand, suppose that we know only the $s - 1$ shadows $k_{i_1}, k_{i_2}, \ldots, k_{i_{s-1}}$. By the Chinese remainder theorem, we can determine the least positive residue a of K_0 modulo M_i, where $M_i = m_{i_1} m_{i_2} \cdots m_{i_{s-1}}$. With these shadows, the only information we have about K_0 is that a is the least positive residue of K_0 modulo M_i and $\leq K_0 < M$. Consequently, we only know that

$$K_0 = a + x M_i,$$

where $0 \leq x < M/M_i$. From (8.7), we can conclude that $M/M_i > p$, so that as x ranges through the positive integers less than M/M_i, x takes every value in a full set of residues modulo p. Because $(m_j, p) = 1$ for $j = 1, 2, \ldots, s$, we know that $(M_i, p) = 1$ and, consequently, $a + x M_i$ runs through a full set of residues modulo p as x does. Hence, we see that the knowledge of $s - 1$ shadows is insufficient to determine K_0, as K_0 could be in any of the p congruence classes modulo p.

We use an example to illustrate this threshold scheme.

Example 8.21. Let $K = 4$ be the master key. We will use a (2,3)-threshold scheme of the kind just described, with $p = 7$, $m_1 = 11$, $m_2 = 12$, and $m_3 = 17$, so that $M = m_1 m_2 = 132 > p m_3 = 119$. We pick $t = 14$ randomly from among the positive integers less than $M/p = 132/7$. This gives us

$$K_0 = K + tp = 4 + 14 \cdot 7 = 102.$$

The three shadows k_1, k_2, and k_3 are the least positive residues of K_0 modulo m_1, m_2, and m_3; that is,

$$k_1 \equiv 102 \equiv 3 \pmod{11}$$
$$k_2 \equiv 102 \equiv 6 \pmod{12}$$
$$k_3 \equiv 102 \equiv 0 \pmod{17},$$

so that the three shadows are $k_1 = 3$, $k_2 = 6$, and $k_3 = 0$.

We can recover the master key K from any two of the three shadows. Suppose we know that $k_1 = 3$ and $k_3 = 0$. Using the Chinese remainder theorem, we can determine K_0 modulo $m_1 m_3 = 11 \cdot 17 = 187$; in other words, because $K_0 \equiv 3 \pmod{11}$ and $K_0 \equiv 0 \pmod{17}$, we have $K_0 \equiv 102 \pmod{187}$. Because $0 \leq K_0 < M = 132 < 187$, we know that $K_0 = 102$, and consequently the master key is $K = K_0 - tp = 102 - 14 \cdot 7 = 4$. ◀

For more details on secret sharing schemes, see [MevaVa97].

8.6 Exercises

1. Using the Diffie-Hellman key agreement protocol, find the common key that can be used by two parties with keys $k_1 = 27$ and $k_2 = 31$, when the modulus is $p = 103$ and the base $r = 5$.

2. Using the Diffie-Hellman key agreement protocol, find the common key that can be used by two parties with keys $k_1 = 7$ and $k_2 = 8$, when the modulus is $p = 53$ and the base is $r = 2$.

3. What is the group key K that can be shared by three parties with keys $k_1 = 3, k_2 = 10$, and $k_3 = 5$, using the modulus $p = 601$ and base $r = 7$?

4. What is the group key K that can be shared by four parties with keys $k_1 = 11, k_2 = 12, k_3 = 17$, and $k_4 = 19$, using the modulus $p = 1009$ and base $r = 3$?

* 5. Describe the steps of a protocol that allows n parties to share a common key, as described in the text.

6. Romeo and Juliet have as their RSA keys $(5, 19 \cdot 67)$ and $(3, 11 \cdot 71)$, respectively.

   a) Using the method in the text, what is the signed ciphertext message sent by Romeo to Juliet, when the plaintext message is GOODBYE SWEET LOVE?

   b) Using the method in the text, what is the signed ciphertext message sent by Juliet to Romeo, when the plaintext message is ADIEU FOREVER?

7. Harold and Audrey have as their RSA keys $(3, 23 \cdot 47)$ and $(7, 31 \cdot 59)$, respectively.

   a) Using the method in the text, what is the signed ciphertext sent by Harold to Audrey, when the plaintext message is CHEERS HAROLD?

   b) Using the method in the text, what is the signed ciphertext sent by Audrey to Harold, when the plaintext message is SINCERELY AUDREY?

In Exercises 8 and 9, we present two methods for sending signed messages using the RSA cipher system, avoiding possible changes in block sizes.

* 8. Let H be a fixed integer. Let each individual have two pairs of encrypting keys: $k = (e, n)$ and $k^* = (e, n^*)$ with $n < H < n^*$, where n and n^* are each the product of two primes. Using the RSA cryptosystem, individual i can send a signed message P to individual j by sending $E_{k_j^*}(D_{k_i}(P))$.

   a) Show that it is not necessary to change block sizes when the transformation $E_{k_j^*}$ is applied after D_{k_i} has been applied.

   b) Explain how individual j can recover the plaintext message P, and why no one other than individual i could have sent the message.

   c) Let individual i have encrypting keys $(3, 11 \cdot 71)$ and $(3, 29 \cdot 41)$, so that $781 = 11 \cdot 71 < 1000 < 1189 = 29 \cdot 41$, and let individual j have enciphering keys $(7, 19 \cdot 47)$ and $(7, 31 \cdot 37)$, so that $893 = 19 \cdot 47 < 1000 < 1147 = 31 \cdot 37$. What ciphertext message does individual i send to individual j using the method given at the beginning of this exercise, when the signed plaintext message is HELLO ADAM? What ciphertext message does individual j send to individual i when the signed plaintext message is GOODBYE ALICE?

* 9. a) Show that if individuals i and j have encrypting keys $k_i = (e_i, n_i)$ and $k_j = (e_j, n_j)$, respectively, where both n_i and n_j are products of two distinct primes, then individual

i can send a signed message P to individual j without needing to change the size of blocks, by sending

$$E_{k_j}(D_{k_i}(P)) \text{ if } n_i < n_j$$

$$D_{k_j}(E_{k_i}(P)) \text{ if } n_i < n_j.$$

b) How can individual j recover P?

c) How can individual j guarantee that a message came from individual i?

d) Let $k_i = (11, 47 \cdot 61)$ and $k_j = (13, 43 \cdot 59)$. Using the method described in part (a), what does individual i send to individual j if the message is REGARDS FRED, and what does individual j send to individual i if the message is REGARDS ZELDA?

10. Decompose the master key $K = 5$ into three shadows using a (2,3)-threshold scheme of the type described in the text, with $p = 7$, $m_1 = 11$, $m_2 = 12$, $m_3 = 17$, and $t = 14$, as in Example 8.21.

11. Decompose the master key $K = 3$ into three shadows using a (2,3)-threshold scheme of the type described in the text, with $p = 5$, $m_1 = 8$, $m_2 = 9$, $m_3 = 11$, and $t = 13$.

12. Show how to recover the master key K from each of the three pairs of shadows found in Exercise 10.

13. Show how to recover the master key K from each of the three pairs of shadows found in Exercise 11.

14. Construct a (3,5)-threshold scheme of the type described in the text. Use the scheme to decompose the master key $K = 22$ into five shadows, and show how the master key can be found using one set of three shadows so produced.

8.6 Computational and Programming Exercises

Computations and Explorations

Using a computation program such as Maple or *Mathematica,* or programs you have written, carry out the following computations and explorations.

1. Produce a set of common keys using a prime p with more than 100 digits.

2. Produce some signed messages using the RSA cryptosystem and verify that these messages came from the supposed sender.

3. Construct a (4,6)-threshold scheme that decomposes a master key into six shadows. Distribute these shadows to six members of your class, and then select three different groups of four of these six people, reconstructing the key from the four shadows of the people in each group.

Programming Projects

Write computer programs using Maple, *Mathematica,* or a language of your choice to do the following.

1. Produce common keys for individuals in a network.

2. Send signed messages using an RSA cipher and the method described in the text.

3. Send signed messages using an RSA cipher and the method in Exercise 8.

4. Send signed messages using an RSA cipher and the method in Exercise 9.

∗ **5.** Play electronic poker using encryption via modular exponentiation.

6. Find the shadows in a threshold scheme of the type described in the text.

7. Recover the master key from a set of shadows.

9

Primitive Roots

Introduction

In this chapter, we will investigate the multiplicative structure of the set of integers modulo n, where n is a positive integer. First, we will introduce the concept of the order of an integer modulo n, which is the least power of the integer that leaves a remainder of 1 when it is divided by n. We will study the basic properties of the order of integers modulo n. A positive integer x, such that the powers of x run through all the integers modulo n, where n is a positive integer, is called a primitive root modulo n. We will determine for which integers n there is a primitive root modulo n.

Primitive roots have many uses. For example, when an integer n has a primitive root, discrete logarithms (also called indices) of integers can be defined. These discrete logarithms enjoy many properties analogous to those of logarithms of positive real numbers. Discrete logarithms can be used to simplify computations modulo n.

We will show how the results of this chapter can be used to develop primality tests that are partial converses of Fermat's little theorem. These tests, such as Proth's test, are used extensively to show that numbers of special forms are prime. We will also establish procedures that can be used to certify that an integer is prime.

Finally, we will introduce the concept of the minimal universal exponent modulo n. This is the least exponent U for which $x^U = 1 \pmod{n}$ for all integers x. We will develop a formula for the minimal universal exponent of n, and use this formula to prove some useful results about Carmichael numbers.

9.1 The Order of an Integer and Primitive Roots

In this section, we begin our study of the least positive residues modulo n of powers of an integer a relatively prime to n, where n is a positive integer greater than 1. We will start by studying the *order* of a modulo n, the exponent of the least power of a congruent to 1 modulo n. Then we will study integers a such that the least positive residues of these powers run through all positive integers less than n that are relatively prime to n. Such integers, when they exist, are called *primitive roots* of n. One of our major goals in this chapter will be to determine which positive integers have primitive roots.

The Order of an Integer

By Euler's theorem, if n is a positive integer and if a is an integer relatively prime to n, then $a^{\phi(n)} \equiv 1 \pmod{n}$. Therefore, at least one positive integer x satisfies the congruence $a^x \equiv 1 \pmod{n}$. Consequently, by the well-ordering property, there is a least positive integer x satisfying this congruence.

Definition.   Let a and n be relatively prime positive integers. Then, the least positive integer x such that $a^x \equiv 1 \pmod{n}$ is called the *order of a modulo n*.

We denote the order of a modulo n by $\mathrm{ord}_n a$. This notation was introduced by Gauss in his *Disquisitiones Arithmeticae* in 1801.

Example 9.1.   To find the order of 2 modulo 7, we compute the least positive residues modulo 7 of powers of 2. We find that

$$2^1 \equiv 2 \pmod{7}, \ 2^2 \equiv 4 \pmod{7}, \ 2^3 \equiv 1 \pmod{7}.$$

Therefore, $\mathrm{ord}_7 2 = 3$.

Similarly, to find the order of 3 modulo 7 we compute

$$3^1 \equiv 3 \pmod{7}, \ 3^2 \equiv 2 \pmod{7}, \ 3^3 \equiv 6 \pmod{7},$$
$$3^4 \equiv 4 \pmod{7}, \ 3^5 \equiv 5 \pmod{7}, \ 3^6 \equiv 1 \pmod{7}.$$

We see that $\mathrm{ord}_7 3 = 6$.   ◀

To find all solutions of the congruence $a^x \equiv 1 \pmod{n}$, we need the following theorem.

Theorem 9.1.   If a and n are relatively prime integers with $n > 0$, then the positive integer x is a solution of the congruence $a^x \equiv 1 \pmod{n}$ if and only if $\mathrm{ord}_n a \mid x$.

Proof.   If $\mathrm{ord}_n a \mid x$, then $x = k \cdot \mathrm{ord}_n a$, where k is a positive integer. Hence,

$$a^x = a^{k \cdot \mathrm{ord}_n a} = (a^{\mathrm{ord}_n a})^k \equiv 1 \pmod{n}.$$

Conversely, if $a^x \equiv 1 \pmod{n}$, we first use the division algorithm to write

$$x = q \cdot \mathrm{ord}_n a + r, \quad 0 \leq r < \mathrm{ord}_n a.$$

From this equation, we see that

$$a^x = a^{q \cdot \text{ord}_n a + r} = (a^{\text{ord}_n a})^q a^r \equiv a^r \pmod{n}.$$

Because $a^x \equiv 1 \pmod{n}$, we know that $a^r \equiv 1 \pmod{n}$. From the inequality $0 \leq r < \text{ord}_n a$, we conclude that $r = 0$ because, by definition, $y = \text{ord}_n a$ is the least positive integer such that $a^y \equiv 1 \pmod{n}$. Because $r = 0$, we have $x = q \cdot \text{ord}_n a$. Therefore, $\text{ord}_n a \mid x$. ∎

Example 9.2. We can use Theorem 9.1 and Example 9.1 to determine whether $x = 10$ and $x = 15$ are solutions of $2^x \equiv 1 \pmod{7}$. By Example 9.1, we know that $\text{ord}_7 2 = 3$. Because 3 does not divide 10, but 3 divides 15, by Theorem 9.1 we see that $x = 10$ is not a solution of $2^x \equiv 1 \pmod{7}$, but $x = 15$ is a solution of this congruence. ◀

Theorem 9.1 leads to the following corollary.

Corollary 9.1.1. If a and n are relatively prime integers with $n > 0$, then $\text{ord}_n a \mid \phi(n)$.

Proof. Because $(a, n) = 1$, Euler's theorem tells us that

$$a^{\phi(n)} \equiv 1 \pmod{n}.$$

Using Theorem 9.1, we conclude that $\text{ord}_n a \mid \phi(n)$. ∎

We can use Corollary 9.1.1 as a shortcut when we compute orders. The following example illustrates the procedure.

Example 9.3. To find the order of 7 modulo 9, we first note that $\phi(9) = 6$. Because the only positive divisors of 6 are 1, 2, 3, and 6, by Corollary 9.1.1 these are the only possible values of $\text{ord}_9 7$. Because

$$7^1 \equiv 7 \pmod{9}, 7^2 \equiv 4 \pmod{9}, 7^3 \equiv 1 \pmod{9},$$

it follows that $\text{ord}_9 7 = 3$. ◀

Example 9.4. To find the order of 5 modulo 17, we first note that $\phi(17) = 16$. Because the only positive divisors of 16 are 1, 2, 4, 8, and 16, by Corollary 9.1.1 these are the only possible values of $\text{ord}_{17} 5$. Because

$$5^1 \equiv 5 \pmod{17}, 5^2 \equiv 8 \pmod{17}, 5^4 \equiv 13 \pmod{17},$$

$$5^8 \equiv 16 \pmod{17}, 5^{16} \equiv 1 \pmod{17},$$

we conclude that $\text{ord}_{17} 5 = 16$. ◀

The following theorem will be useful in our subsequent discussions.

Theorem 9.2. If a and n are relatively prime integers with $n > 0$, then $a^i \equiv a^j \pmod{n}$, where i and j are nonnegative integers, if and only if $i \equiv j \pmod{\text{ord}_n a}$.

Proof.   Suppose that $i \equiv j \pmod{\text{ord}_n a}$ and $0 \leq j \leq i$. Then we have $i = j + k \cdot \text{ord}_n a$, where k is a positive integer. Hence

$$a^i = a^{j+k\cdot\text{ord}_n a} = a^j (a^{\text{ord}_n a})^k \equiv a^j \pmod{n},$$

because $a^{\text{ord}_n a} \equiv 1 \pmod{n}$.

Conversely, assume that $a^i \equiv a^j \pmod{n}$ with $i \geq j$. Because $(a, n) = 1$, we know that $(a^j, n) = 1$. Hence, using Corollary 4.4.1 the congruence

$$a^i \equiv a^j a^{i-j} \equiv a^j \pmod{n}$$

implies, by cancellation of a^j, that

$$a^{i-j} \equiv 1 \pmod{n}.$$

By Theorem 9.1, it follows that $\text{ord}_n a$ divides $i - j$, or equivalently, $i \equiv j \pmod{\text{ord}_n a}$.   ∎

The next example illustrates the use of Theorem 9.2.

Example 9.5.   Let $a = 3$ and $n = 14$. By Theorem 9.2, we see that $3^5 \equiv 3^{11} \pmod{14}$, but $3^9 \not\equiv 3^{20} \pmod{14}$, because $\phi(14) = 6$ and $5 \equiv 11 \pmod 6$ but $9 \not\equiv 20 \pmod 6$.   ◀

Primitive Roots

Given an integer n, we are interested in integers a with order modulo n equal to $\phi(n)$, the largest possible order modulo n. As we will show, when such an integer exists, the least positive residues of its powers run through all positive integers relatively prime to n and less than n.

Definition.   If r and n are relatively prime integers with $n > 0$ and if $\text{ord}_n r = \phi(n)$, then r is called a *primitive root modulo n*.

Example 9.6.   We have previously shown that $\text{ord}_7 3 = 6 = \phi(7)$. Consequently, 3 is a primitive root modulo 7. Likewise, because $\text{ord}_7 5 = 6$, as can easily be verified, 5 is also a primitive root modulo 7.   ◀

Euler coined the term *primitive root* in 1773. His purported proof that every prime has a primitive root was incorrect, however. In Section 9.2, we will prove that every prime has a primitive root using the first correct proof of this result by Lagrange in 1769. Gauss also studied primitive roots extensively and provided several additional proofs that every prime has a primitive root.

Not all integers have primitive roots. For instance, there are no primitive roots modulo 8. To see this, note that the only integers less than 8 and relatively prime to 8 are 1, 3, 5, and 7, and $\text{ord}_8 1 = 1$, while $\text{ord}_8 3 = \text{ord}_8 5 = \text{ord}_8 7 = 2$. Because $\phi(8) = 4$, there are no primitive roots modulo 8.

Among the first 30 positive integers, 2, 3, 4, 5, 6, 7, 9, 10, 11, 13, 14, 17, 18, 19, 22, 23, 25, 26, 27, and 29 have primitive roots whereas 8, 12, 15, 16, 20, 21, 24, 28, and 30 do not. (The reader can verify this information; see Exercises 3–6 at the end of this section, for example.) What can we conjecture based on this evidence? In this range, every prime has a primitive root (as Lagrange showed), as does every power of an odd prime (since $9 = 3^2$, $25 = 5^2$, and $27 = 3^3$ have primitive roots), but the only power of 2 that has a primitive root is 4. The other integers in this range with a primitive root are 6, 10, 14, 18, 22, and 26. What do these integers have in common? Each is 2 times an odd prime or power of an odd prime. Using this evidence, we conjecture that a positive integer has a primitive root if it equals $2, 4, p^t$, or $2p^t$, where p is an odd prime and t is a positive integer. Sections 9.2 and 9.3 are devoted to verifying this conjecture.

To indicate one way in which primitive roots are useful, we give the following theorem.

Theorem 9.3. If r and n are relatively prime positive integers with $n > 0$ and if r is a primitive root modulo n, then the integers

$$r^1, r^2, \ldots, r^{\phi(n)}$$

form a reduced residue set modulo n.

Proof. To demonstrate that the first $\phi(n)$ powers of the primitive root r form a reduced residue set modulo n, we need only show that they are all relatively prime to n and that no two are congruent modulo n.

Because $(r, n) = 1$, it follows from Exercise 14 of Section 3.3 that $(r^k, n) = 1$ for any positive integer k. Hence, these powers are all relatively prime to n. To show that no two of these powers are congruent modulo n, assume that

$$r^i \equiv r^j \pmod{n}.$$

By Theorem 9.2, we see that $i \equiv j \pmod{\phi(n)}$. However, for $1 \leq i \leq \phi(n)$ and $1 \leq j \leq \phi(n)$, the congruence $i \equiv j \pmod{\phi(n)}$ implies that $i = j$. Hence, no two of these powers are congruent modulo n. This shows that we do have a reduced residue system modulo n. ∎

Example 9.7. We see that 2 is a primitive root modulo 9, because $2^2 \equiv 4$, $2^3 \equiv 8$, and $2^6 \equiv 1 \pmod 9$. By Theorem 9.3, the first $\phi(9) = 6$ powers of 2 form a reduced residue system modulo 9. These are $2^1 \equiv 2 \pmod 9$, $2^2 \equiv 4 \pmod 9$, $2^3 \equiv 8 \pmod 9$, $2^4 \equiv 7 \pmod 9$, $2^5 \equiv 5 \pmod 9$, and $2^6 \equiv 1 \pmod 9$. ◀

When an integer possesses a primitive root, it usually has many primitive roots. To demonstrate this, we first prove the following theorem.

Theorem 9.4. If $\text{ord}_n a = t$ and if u is a positive integer, then

$$\text{ord}_n(a^u) = t/(t, u).$$

Proof. Let $s = \text{ord}_n(a^u)$, $v = (t, u)$, $t = t_1 v$, and $u = u_1 v$. By Theorem 3.6, we know that $(t_1, u_1) = 1$.

Because $t_1 = t/(t, u)$, we want to show that $\text{ord}_n(a^u) = t_1$. To do this, we will show that $(a^u)^{t_1} \equiv 1 \pmod{n}$ and that if $(a^u)^s \equiv 1 \pmod{n}$, then $t_1 \mid s$. First, note that

$$(a^u)^{t_1} = (a^{u_1 v})^{(t/v)} = (a^t)^{u_1} \equiv 1 \pmod{n},$$

because $\text{ord}_n a = t$. Hence, Theorem 9.1 tells us that $s \mid t_1$.

On the other hand, because

$$(a^u)^s = a^{us} \equiv 1 \pmod{n},$$

we know that $t \mid us$. Hence, $t_1 v \mid u_1 vs$ and, consequently, $t_1 \mid u_1 s$. Because $(t_1, u_1) = 1$, using Lemma 3.4, we see that $t_1 \mid s$.

Now, because $s \mid t_1$ and $t_1 \mid s$, we conclude that $s = t_1 = t/v = t/(t, u)$. This proves the result. ∎

Example 9.8. By Theorem 9.4, we see that $\text{ord}_7 3^4 = 6/(6, 4) = 6/2 = 3$, because we showed in Example 9.1 that $\text{ord}_7 3 = 6$. ◀

The following corollary of Theorem 9.4 tells us which powers of a primitive root are also primitive roots.

Corollary 9.4.1. Let r be a primitive root modulo n, where n is an integer, $n > 1$. Then r^u is a primitive root modulo n if and only if $(u, \phi(n)) = 1$.

Proof. By Theorem 9.4, we know that

$$\text{ord}_n r^u = \text{ord}_n r/(u, \text{ord}_n r)$$
$$= \phi(n)/(u, \phi(n)).$$

Consequently, $\text{ord}_n r^u = \phi(n)$, and r^u is a primitive root modulo n, if and only if $(u, \phi(n)) = 1$. ∎

This leads immediately to the following theorem.

Theorem 9.5. If the positive integer n has a primitive root, then it has a total of $\phi(\phi(n))$ incongruent primitive roots.

Proof. Let r be a primitive root modulo n. Then Theorem 9.3 tells us that the integers $r, r^2, \ldots, r^{\phi(n)}$ form a reduced residue system modulo n. By Corollary 9.4.1, we know that r^u is a primitive root modulo n if and only if $(u, \phi(n)) = 1$. Because there are exactly $\phi(\phi(n))$ such integers u, there are exactly $\phi(\phi(n))$ primitive roots modulo n. ∎

Example 9.9. Let $n = 11$. Note that 2 is a primitive root modulo 11 (see Exercise 3 at the end of this section). Because 11 has a primitive root, by Theorem 9.5 we know that 11 has $\phi(\phi(11)) = 4$ incongruent primitive roots. Because $\phi(11) = 10$, by the proof of Theorem 9.5 we see that we can find these primitive roots by taking the least nonnegative residues of $2^1, 2^3, 2^7$, and 2^9, which are 2, 8, 7, and 6, respectively. In other words, the integers 2, 6, 7, 8 form a complete set of incongruent primitive roots modulo 11.   ◀

9.1 Exercises

1. Determine the following orders.

a) $\text{ord}_5 2$ c) $\text{ord}_{13} 10$
b) $\text{ord}_{10} 3$ d) $\text{ord}_{10} 7$

2. Determine the following orders.

a) $\text{ord}_{11} 3$ c) $\text{ord}_{21} 10$
b) $\text{ord}_{17} 2$ d) $\text{ord}_{25} 9$

3. a) Show that 5 is a primitive root of 6.
b) Show that 2 is a primitive root of 11.

4. Find a primitive root modulo each of the following integers.

a) 4   d) 13
b) 5   e) 14
c) 10   f) 18

5. Show that the integer 12 has no primitive roots.

6. Show that the integer 20 has no primitive roots.

7. How many incongruent primitive roots does 14 have? Find a set of this many incongruent primitive roots modulo 14.

8. How many incongruent primitive roots does 13 have? Find a set of this many incongruent primitive roots modulo 13.

9. Show that if $\bar{a}$ is an inverse of a modulo n, then $\text{ord}_n a = \text{ord}_n \bar{a}$.

10. Show that if n is a positive integer, and a and b are integers relatively prime to n such that $(\text{ord}_n a, \text{ord}_n b) = 1$, then $\text{ord}_n(ab) = \text{ord}_n a \cdot \text{ord}_n b$.

11. What can be said about $\text{ord}_n(ab)$ if a and b are integers relatively prime to n such that $\text{ord}_n a$ and $\text{ord}_n b$ are not necessarily relatively prime?

12. Decide whether it is true that if n is a positive integer and d is a divisor of $\phi(n)$, then there is an integer a with $\text{ord}_n a = d$. Give reasons for your answer.

13. Show that if a is an integer relatively prime to the positive integer m and $\text{ord}_m a = st$, then $\text{ord}_m a^t = s$.

14. Show if m is a positive integer and a is an integer relatively prime to m such that $\text{ord}_m a = m - 1$, then m is prime.

15. Show that r is a primitive root modulo the odd prime p if and only if r is an integer with $(r, p) = 1$ such that

$$r^{(p-1)/q} \not\equiv 1 \pmod{p}$$

for all prime divisors q of $p - 1$.

16. Show that if r is a primitive root modulo the positive integer m, then $\bar{r}$ is also a primitive root modulo m, if $\bar{r}$ is an inverse of r modulo m.

17. Show that $\text{ord}_{F_n} 2 \leq 2^{n+1}$, where $F_n = 2^{2^n} + 1$, is the nth Fermat number.

∗ 18. Let p be a prime divisor of the Fermat number $F_n = 2^{2^n} + 1$.
   a) Show that $\text{ord}_p 2 = 2^{n+1}$.
   b) From part (a), conclude that $2^{n+1} \mid (p - 1)$, so that p must be of the form $2^{n+1}k + 1$.

19. Let $m = a^n - 1$, where a and n are positive integers. Show that $\text{ord}_m a = n$, and conclude that $n \mid \phi(m)$.

∗ 20. a) Show that if p and q are distinct odd primes, then pq is a pseudoprime to the base 2 if and only if $\text{ord}_q 2 \mid (p - 1)$ and $\text{ord}_p 2 \mid (q - 1)$.
   b) Use part (a) to decide which of the following integers are pseudoprimes to the base 2: $13 \cdot 67$, $19 \cdot 73$, $23 \cdot 89$, $29 \cdot 97$.

∗ 21. Show that if p and q are distinct odd primes, then pq is a pseudoprime to the base 2 if and only if $M_p M_q = (2^p - 1)(2^q - 1)$ is a pseudoprime to the base 2.

There is an iterative method known as the *cycling attack* for decrypting messages that were encrypted by an RSA cipher, without knowledge of the decrypting key. Suppose that the public key (e, n) used for encrypting is known, but the decrypting key (d, n) is not. To decrypt a ciphertext block C, we form a sequence $C_1, C_2, C_3, \ldots$, setting $C_1 \equiv C^e \pmod{n}$, $0 < C_1 < n$, and $C_{j+1} \equiv C_j^e \pmod{n}$, $0 < C_{j+1} < n$ for $j = 1, 2, 3, \ldots$.

22. Show that $C_j \equiv C^{e^j} \pmod{n}$, $0 < C_j < n$.

23. Show that there is an index j such that $C_j = C$ and $C_{j-1} = P$, where P is the original plaintext message. Show that this index j is a divisor of $\text{ord}_{\phi(n)} e$.

24. Let $n = 47 \cdot 59$ and $e = 17$. Using iteration, find the plaintext corresponding to the ciphertext 1504.

(*Note:* This iterative method for attacking RSA ciphers is seldom successful in a reasonable amount of time. Moreover, the primes p and q may be chosen so that this attack is almost always futile. See Exercise 19 of Section 9.2.)

9.1 Computational and Programming Exercises

Computations and Explorations

Using a computation program such as Maple or *Mathematica*, or programs you have written, carry out the following computations and explorations.

1. Find $\text{ord}_{52,579} 2$, $\text{ord}_{52,579} 3$, and $\text{ord}_{52,579} 1001$.

2. Find as many integers as you can for which 2 is a primitive root. Do you think that there are infinitely many such integers?

Programming Projects

Write projects using Maple, *Mathematica*, or a language of your choice to do the following.

1. Find the order of a modulo m, when a and m are relatively prime positive integers.

2. Find primitive roots when they exist.

3. Attempt to decrypt RSA ciphers by iteration (see the preamble to Exercise 22).

9.2 Primitive Roots for Primes

In this and the following section, our objective is to determine which integers have primitive roots. In this section, we show that every prime has a primitive root. To do this, we first need to study polynomial congruences.

Let $f(x)$ be a polynomial with integer coefficients. We say that an integer c is a *root of $f(x)$ modulo m* if $f(c) \equiv 0 \pmod{m}$. It is easy to see that if c is a root of $f(x)$ modulo m, then every integer congruent to c modulo m is also a root.

Example 9.10. The polynomial $f(x) = x^2 + x + 1$ has exactly two incongruent roots modulo 7, namely $x \equiv 2 \pmod{7}$ and $x \equiv 4 \pmod{7}$. ◀

Example 9.11. The polynomial $g(x) = x^2 + 2$ has no roots modulo 5. ◀

Example 9.12. Fermat's little theorem tells us that if p is prime, then the polynomial $h(x) = x^{p-1} - 1$ has exactly $p - 1$ incongruent roots modulo p, namely $x \equiv 1, 2, 3, \ldots, p - 1 \pmod{p}$. ◀

We will need the following important theorem concerning roots of polynomials modulo p where p is a prime.

Theorem 9.6. *Lagrange's Theorem.* Let $f(x) = a_n x^n + a_{n-1} x^{n-1} + \cdots + a_1 x + a_0$ be a polynomial of degree $n, n \geq 1$, with integer coefficients and with leading coefficient a_n not divisible by p. Then $f(x)$ has at most n incongruent roots modulo p.

Proof. We use mathematical induction to prove the theorem. When $n = 1$, we have $f(x) = a_1 x + a_0$ with $p \nmid a_1$. A root of $f(x)$ modulo p is a solution of the linear congruence $a_1 x \equiv -a_0 \pmod{p}$. By Theorem 4.10, because $(a_1, p) = 1$, this linear congruence has exactly one solution, so that there is exactly one root modulo p of $f(x)$. Clearly, the theorem is true for $n = 1$.

Now, suppose that the theorem is true for polynomials of degree $n - 1$, and let $f(x)$ be a polynomial of degree n with leading coefficient not divisible by p. Assume that the polynomial $f(x)$ has $n + 1$ incongruent roots modulo p, say $c_0, c_1, \ldots, c_n$, so that $f(c_k) \equiv 0 \pmod{p}$ for $k = 0, 1, \ldots, n$. We have

$$f(x) - f(c_0) = a_n(x^n - c_0^n) + a_{n-1}(x^{n-1} - c_0^{n-1}) + \cdots + a_1(x - c_0)$$

$$= a_n(x - c_0)(x^{n-1} + x^{n-2}c_0 + \cdots + xc_0^{n-2} + c_0^{n-1})$$

$$\quad + a_{n-1}(x - c_0)(x^{n-2} + x^{n-3}c_0 + \cdots + xc_0^{n-3} + c_0^{n-2})$$

$$\quad + \cdots + a_1(x - c_0)$$

$$= (x - c_0)g(x),$$

where $g(x)$ is a polynomial of degree $n - 1$ with leading coefficient a_n. We now show that $c_1, c_2, \ldots, c_n$ are all roots of $g(x)$ modulo p. Let k be an integer, $1 \leq k \leq n$. Because $f(c_k) \equiv f(c_0) \equiv 0 \pmod{p}$, we have

$$f(c_k) - f(c_0) = (c_k - c_0)g(c_k) \equiv 0 \pmod{p}.$$

It follows that $g(c_k) \equiv 0 \pmod{p}$, because $c_k - c_0 \not\equiv 0 \pmod{p}$. Hence, c_k is a root of $g(x)$ modulo p. This shows that the polynomial $g(x)$, which is of degree $n - 1$ and has a leading coefficient not divisible by p, has n incongruent roots modulo p. This contradicts the induction hypothesis. Hence, $f(x)$ must have no more than n incongruent roots modulo p. The induction argument is complete. ∎

We use Lagrange's theorem to prove the following result.

Theorem 9.7. Let p be prime and let d be a divisor of $p - 1$. Then the polynomial $x^d - 1$ has exactly d incongruent roots modulo p.

Proof. Let $p - 1 = de$. Then

$$x^{p-1} - 1 = (x^d - 1)(x^{d(e-1)} + x^{d(e-2)} + \cdots + x^d + 1)$$
$$= (x^d - 1)g(x).$$

From Fermat's little theorem, we see that $x^{p-1} - 1$ has $p - 1$ incongruent roots modulo p. Furthermore, any root of $x^{p-1} - 1$ modulo p is either a root of $x^d - 1$ modulo p or a root of $g(x)$ modulo p.

Lagrange's theorem tells us that $g(x)$ has at most $d(e - 1) = p - d - 1$ roots modulo p. Because every root of $x^{p-1} - 1$ modulo p that is not a root of $g(x)$ modulo p must be a root of $x^d - 1$ modulo p, we know that the polynomial $x^d - 1$ has at least $(p - 1) - (p - d - 1) = d$ incongruent roots modulo p. On the other hand, Lagrange's theorem tells us that it has at most d incongruent roots modulo p. Consequently, $x^d - 1$ has precisely d incongruent roots modulo p. ∎

Theorem 9.7 can be used to prove a useful result that tells us how many incongruent integers have a given order modulo p. Before proving this result, we present a lemma needed for its proof.

Lemma 9.1. Let p be a prime and let d be a positive divisor of $p - 1$. Then the number of positive integers less than p of order d modulo p does not exceed $\phi(d)$.

Proof. For each positive integer d dividing $p - 1$, let $F(d)$ denote the number of positive integers of order d modulo p that are less than p.

If $F(d) = 0$, it is clear that $F(d) \leq \phi(d)$. Otherwise, there is an integer a of order d modulo p. Because $\text{ord}_p a = d$, the integers

$$a, a^2, \ldots, a^d$$

are incongruent modulo p. Furthermore, each of these powers of a is a root of $x^d - 1$ modulo p, because $(a^k)^d = (a^d)^k \equiv 1 \pmod{p}$ for all positive integers k. By Theorem

9.7, we know that $x^d - 1$ has exactly d incongruent roots modulo p, so every root modulo p is congruent to one of these powers of a. However, by Theorem 9.4, we know that the powers of a with order d are those of the form a^k with $(k, d) = 1$. There are exactly $\phi(d)$ such integers k with $1 \leq k \leq d$, and consequently, if there is one element of order d modulo p, there must be exactly $\phi(d)$ such positive integers less than d. Hence, $F(d) \leq \phi(d)$. ∎

We now can determine how many incongruent integers can have a given order modulo p.

Theorem 9.8. Let p be a prime and let d be a positive divisor of $p - 1$. Then the number of incongruent integers of order d modulo p is equal to $\phi(d)$.

Proof. For each positive integer d dividing $p - 1$, let $F(d)$ denote the number of positive integers of order d modulo p that are less than p. Because the order modulo p of an integer not divisible by p divides $p - 1$, it follows that

$$p - 1 = \sum_{d \mid p-1} F(d).$$

By Theorem 7.7, we know that

$$p - 1 = \sum_{d \mid p-1} \phi(d).$$

By Lemma 9.1, $F(d) \leq \phi(d)$ when $d \mid (p - 1)$. This inequality, together with the equality

$$\sum_{d \mid p-1} F(d) = \sum_{d \mid p-1} \phi(d),$$

implies that $F(d) = \phi(d)$ for each positive divisor d of $p - 1$.

Therefore, we can conclude that $F(d) = \phi(d)$, which tells us that there are precisely $\phi(d)$ incongruent integers of order d modulo p. ∎

The following corollary is derived immediately from Theorem 9.8.

Corollary 9.8.1. Every prime has a primitive root.

Proof. Let p be a prime. By Theorem 9.7, we know that there are $\phi(p - 1)$ incongruent integers of order $p - 1$ modulo p. Because each of these is, by definition, a primitive root, p has $\phi(p - 1)$ primitive roots. ∎

The smallest positive primitive root of each prime less than 1000 is given in Table 3 of Appendix E; looking at the table, we see that 2 is the least primitive root of many primes p. Is 2 a primitive root for infinitely many primes? The answer to this question is not known, and it is also unknown when we replace 2 by an integer other than ± 1 or a perfect square. Evidence suggests the truth of the following conjecture made by *Emil Artin*.

Artin's conjecture.   The integer a is a primitive root of infinitely many primes if $a \neq \pm 1$ and a is not a perfect square.

Although Artin's conjecture has not been settled, there are some interesting partial results. For example, one consequence of work by Roger Heath-Brown is that there are at most two primes and three positive square-free integers a such that a is a primitive root of only finitely many primes. One implication of this work is that at least one of the integers 2, 3, and 5 is a primitive root for infinitely many primes.

Many mathematicians have studied the problem of determining bounds on g_p, the smallest primitive root for a prime p. Among the results that have been proved are that

$$g_p > C \log p$$

for some constant C and infinitely many primes p. This result, proved by Fridlender (in 1949), and independently by Salié (in 1950), shows that there are infinitely many primes where the least primitive root is larger than any particular positive integer. However, g_p does not grow very quickly. Grosswald showed (in 1981) that if p is a prime with $p > e^{e^{24}}$, then $g_p < p^{0.499}$. Another interesting result, proved in the problems section of the *American Mathematical Monthly* in 1984, is that for every positive integer M, there are infinitely many primes p such that $M < g_p < p - M$.

9.2 Exercises

1. Find the number of incongruent roots modulo 11 of each of the following polynomials.

a) $x^2 + 2$ c) $x^3 + x^2 + 2x + 2$

b) $x^2 + 10$ d) $x^4 + x^2 + 1$

EMIL ARTIN (1898–1962) was born in Vienna, Austria. He served in the Austrian army during World War I. In 1921, he received a Ph.D. from the University of Leipzig, which he attended both as an undergraduate and as a graduate student. He attended the University of Göttingen from 1922 until 1923. In 1923, he was appointed to a position at the University of Hamburg. Artin was forced to leave Germany in 1937 as a result of Nazi regulations because his wife was Jewish, although he was not. He emigrated to the United States, where he taught at Notre Dame University (1937–1938), Indiana University (1938–1946), and Princeton University (1946–1958). He returned to Germany, taking a position at the University of Hamburg, in 1958.

Artin made major contributions to several areas of abstract algebra, including ring theory and group theory. He also invented the concept of braids structures, defined using the concept of strings woven to form braids, now studied by topologists and algebraists. Artin made major contributions to both analytic and algebraic number theory, beginning with his research involving quadratic fields.

Artin excelled as a teacher and advisor of students. He was also a talented musician who played the harpsichord, clavichord, and flute and was a devotee of old music.

2. Find the number of incongruent roots modulo 13 of each of the following polynomials.

a) $x^2 + 1$ c) $x^3 + 12$

b) $x^2 + 3x + 2$ d) $x^4 + x^2 + x + 1$

3. Find the number of primitive roots of each of the following primes.

a) 7   d) 19

b) 13   e) 29

c) 17   f) 47

4. Find a complete set of incongruent primitive roots of 7.

5. Find a complete set of incongruent primitive roots of 13.

6. Find a complete set of incongruent primitive roots of 17.

7. Find a complete set of incongruent primitive roots of 19.

8. Let r be a primitive root of the prime p with $p \equiv 1 \pmod 4$. Show that $-r$ is also a primitive root.

9. Show that if p is a prime and $p \equiv 1 \pmod 4$, there is an integer x such that $x^2 \equiv -1$ $\pmod p$. (*Hint:* Use Theorem 9.8 to show that there is an integer x of order 4 modulo p.)

10. a) Find the number of incongruent roots modulo 6 of the polynomial $x^2 - x$.

b) Explain why the answer to part (a) does not contradict Lagrange's theorem.

11. a) Use Lagrange's theorem to show that if p is a prime and $f(x)$ is a polynomial of degree n with integer coefficients and more than n roots modulo p, then p divides every coefficient of $f(x)$.

b) Let p be prime. Using part (a), show that every coefficient of the polynomial $f(x) = (x - 1)(x - 2) \cdots (x - p + 1) - x^{p-1} + 1$ is divisible by p.

c) Using part (b), give a proof of Wilson's theorem (Theorem 6.1). (*Hint:* Consider the constant term of $f(x)$.)

12. Find the least positive residue of the product of a set of $\phi(p - 1)$ incongruent primitive roots modulo a prime p.

∗ 13. A systematic method for constructing a primitive root modulo a prime p is outlined in this problem. Let the prime factorization of $\phi(p) = p - 1$ be $p - 1 = q_1^{t_1} q_2^{t_2} \cdots q_r^{t_r}$, where $q_1, q_2, \ldots, q_r$ are prime.

a) Use Theorem 9.8 to show that there are integers $a_1, a_2, \ldots, a_r$ such that $\mathrm{ord}_p a_1 = q_1^{t_1}$, $\mathrm{ord}_p a_2 = q_2^{t_2}, \ldots, \mathrm{ord}_p a_r = q_r^{t_r}$.

b) Use Exercise 10 of Section 9.1 to show that $a = a_1 a_2 \cdots a_r$ is a primitive root modulo p.

c) Follow the procedure outlined in parts (a) and (b) to find a primitive root modulo 29.

∗ 14. Suppose that the composite positive integer n has prime-power factorization $n = p_1^{a_1} p_2^{a_2} \cdots p_r^{a_r}$. Show that the number of incongruent bases modulo n for which n is a pseudoprime to that base is $\prod_{j=1}^{r}(n - 1, p_j - 1)$.

15. Use Exercise 14 to show that every odd composite integer that is not a power of 3 is a pseudoprime to at least two bases other than ± 1.

16. Show that if p is prime and $p = 2q + 1$, where q is an odd prime and a is a positive integer with $1 < a < p - 1$, then $p - a^2$ is a primitive root modulo p.

* 17. a) Suppose that $f(x)$ is a polynomial with integer coefficients of degree $n - 1$. Let $x_1, x_2, \ldots, x_n$ be n incongruent integers modulo p. Show that for all integers x, the congruence

$$f(x) \equiv \sum_{j=1}^{n} f(x_j) \prod_{\substack{i=1 \\ i \neq j}}^{n} (x - x_i)\overline{(x_j - x_i)} \pmod{p}$$

holds, where $\overline{x_j - x_i}$ is an inverse of $x_j - x_i$ modulo p. This technique for finding $f(x)$ modulo p is called *Lagrange interpolation.*

b) Find the least positive residue of $f(5)$ modulo 11 if $f(x)$ is a polynomial of degree 3 with $f(1) \equiv 8$, $f(2) \equiv 2$, and $f(3) \equiv 4 \pmod{11}$.

18. In this exercise, we develop a threshold scheme for protection of master keys in a computer system, different from the scheme discussed in Section 8.6. Let $f(x)$ be a randomly chosen polynomial of degree $r - 1$, with the condition that K, the master key, is the constant term of the polynomial. Let p be a prime, such that $p > K$ and $p > s$. The s shadows $k_1, k_2, \ldots, k_s$ are computed by finding the least positive residue of $f(x_j)$ modulo p for $j = 1, 2, \ldots, s$, where $x_1, x_2, \ldots, x_s$ are randomly chosen integers incongruent modulo p; that is,

$$k_j \equiv f(x_j) \pmod{p}, \quad 0 \leq k_j < p,$$

for $j = 1, 2, \ldots, s$.

a) Use Lagrange interpolation, described in Exercise 17, to show that the master key K can be determined from any r shadows.

b) Show that the master key K cannot be determined from fewer than r shadows.

c) Let $K = 33$, $p = 47$, $r = 4$, and $s = 7$. Let $f(x) = 4x^3 + x^2 + 31x + 33$. Find the seven shadows corresponding to the values of $f(x)$ at 1, 2, 3, 4, 5, 6, 7.

d) Show how to find the master key from the four shadows $f(1)$, $f(2)$, $f(3)$, and $f(4)$.

19. Show that an RSA cipher with encrypting modulus $n = pq$ is resistant to the cycling attack (see the preamble to Exercise 22 of Section 9.1) if $p - 1$ and $q - 1$ have large prime factors p' and q', respectively, and $p' - 1$ and $q' - 1$ have large prime factors p'' and q'', respectively.

9.2 Computational and Programming Exercises

Computations and Explorations

Using a computation program such as Maple or *Mathematica,* or programs you have written, carry out the following computations and explorations.

1. Find the least primitive root for each of the primes 10,007, 10,009, and 10,037.

2. Erdős has asked whether for each sufficiently large prime p there is a prime q for which q is a primitive root of p. What evidence can you find for this conjecture? For which small primes p is the statement in the conjecture false?

Programming Projects

Write programs using Maple, *Mathematica,* or a language of your choice to do the following.

1. Given a prime p, use Exercise 13 to find a primitive root of p.

2. Implement the threshold scheme given in Exercise 18.

9.3   The Existence of Primitive Roots

In the previous section, we showed that every prime has a primitive root. In this section, we will find all positive integers having primitive roots. First, we will show that every power of an odd prime possesses a primitive root.

Primitive Roots Modulo p^2, p Prime   The first step in showing that every power of an odd prime has a primitive root is to show that every square of an odd prime has a primitive root.

Theorem 9.9.   If p is an odd prime with primitive root r, then either r or $r + p$ is a primitive root modulo p^2.

Proof.   Because r is a primitive root modulo p, we know that

$$\text{ord}_p r = \phi(p) = p - 1.$$

Let $n = \text{ord}_{p^2} r$, so that

$$r^n \equiv 1 \pmod{p^2}.$$

Because a congruence modulo p^2 obviously holds modulo p, we have

$$r^n \equiv 1 \pmod{p}.$$

By Theorem 9.1, because $p - 1 = \text{ord}_p r$, it follows that

$$p - 1 \mid n.$$

On the other hand, Corollary 9.1.1 tells us that

$$n \mid \phi(p^2).$$

Because $\phi(p^2) = p(p - 1)$, this implies that $n \mid p(p - 1)$. Because $n \mid p(p - 1)$ and $p - 1 \mid n$, either $n = p - 1$ or $n = p(p - 1)$. If $n = p(p - 1)$, then r is a primitive root modulo p^2, because $\text{ord}_{p^2} r = \phi(p^2)$. Otherwise, we have $n = p - 1$, so that

(9.1) $$r^{p-1} \equiv 1 \pmod{p^2}.$$

Let $s = r + p$. Then, because $s \equiv r \pmod{p}$, s is also a primitive root modulo p. Hence, $\text{ord}_{p^2} s$ equals either $p - 1$ or $p(p - 1)$. We will show that $\text{ord}_{p^2} s = p(p - 1)$ by eliminating the possibility that $\text{ord}_{p^2} s = p - 1$.

To show that $\text{ord}_{p^2} s \neq p - 1$, first note that by the binomial theorem we have

$$s^{p-1} = (r + p)^{p-1} = r^{p-1} + (p - 1)r^{p-2}p + \binom{p-1}{2}r^{p-3}p^2 + \cdots + p^{p-1}$$

$$\equiv r^{p-1} + (p - 1)p \cdot r^{p-2} \pmod{p^2}.$$

Hence, using (9.1), we see that

$$s^{p-1} \equiv 1 + (p - 1)p \cdot r^{p-2} \equiv 1 - pr^{p-2} \pmod{p^2}.$$

From this last congruence, we can show that

$$s^{p-1} \not\equiv 1 \pmod{p^2}.$$

To see this, note that if $s^{p-1} \equiv 1 \pmod{p^2}$, then $pr^{p-2} \equiv 0 \pmod{p^2}$. This last congruence implies that $r^{p-2} \equiv 0 \pmod{p}$, which is impossible because $p \nmid r$ (remember that r is a primitive root of p).

Because $\text{ord}_{p^2} s \neq p - 1$, we can conclude that $\text{ord}_{p^2} s = p(p - 1) = \phi(p^2)$. Consequently, $s = r + p$ is a primitive root of p^2.   ∎

Example 9.13.   The prime $p = 7$ has $r = 3$ as a primitive root. Using observations made in the proof of Theorem 9.9, either $\text{ord}_{49} 3 = 6$ or $\text{ord}_{49} 3 = 42$. However,

$$r^{p-1} = 3^6 \not\equiv 1 \pmod{49}.$$

It follows that $\text{ord}_{49} 3 = 42$. Hence 3 is also a primitive root of $p^2 = 49$.   ◄

We note that it is extremely rare for the congruence

$$r^{p-1} \equiv 1 \pmod{p^2}$$

to hold when r is a primitive root modulo the prime p. Consequently, it is very seldom that a primitive root r modulo the prime p is not also a primitive root modulo p^2. When this occurs, Theorem 9.9 tell us that $r + p$ is a primitive root modulo p^2. The following example illustrates this.

Example 9.14.   Let $p = 487$. For the primitive root 10 modulo 487, we have

$$10^{486} \equiv 1 \pmod{487^2}.$$

Hence, 10 is not a primitive root modulo 487^2 but, by Theorem 9.9, we know that $497 = 10 + 487$ is a primitive root modulo 487^2.   ◄

Primitive Roots Modulo p^k, p Prime and k a Positive Integer   Next, we show that artibrary powers of odd primes have primitive roots.

Theorem 9.10.   Let p be an odd prime. Then p^k has a primitive root for all positive integers k. Moreover, if r is a primitive root modulo p^2, then r is a primitive root modulo p^k, for all positive integers k.

Proof. By Theorem 9.9, we know that p has a primitive root r that is also a primitive root modulo p^2, so that

(9.2) $$r^{p-1} \not\equiv 1 \,(\text{mod } p^2).$$

Using mathematical induction, we will prove that for this primitive root r,

(9.3) $$r^{p^{k-2}(p-1)} \not\equiv 1 \,(\text{mod } p^k)$$

for all positive integers $k, k \geq 2$.

Once we have established congruence, we can show that r is also a primitive root modulo p^k by the following reasoning. Let

$$n = \text{ord}_{p^k} r.$$

By Theorem 8.1, we know that $n \mid \phi(p^k)$. By Theorem 7.3, we have $\phi(p^k) = p^{k-1}(p-1)$. Hence, $r \mid p^k(p-1)$. On the other hand, because

$$r^n \equiv 1 \,(\text{mod } p^k),$$

we also know that

$$r^n \equiv 1 \,(\text{mod } p).$$

By Theorem 9.1, since $\phi(p) = p-1$ we see that $p-1 \mid n$. Because $p-1 \mid n$, and $n \mid p^{k-1}(p-1)$, we know that $n = p^t(p-1)$, where t is an integer such that $0 \leq t \leq k-1$. If $n = p^t(p-1)$ with $t \leq k-2$, then

$$r^{p^{k-2}(p-1)} = (r^{p^t(p-1)})^{p^{k-2-t}} \equiv 1 \,(\text{mod } p^k),$$

which would contradict (9.3). Hence, $\text{ord}_{p^k} r = p^{k-1}(p-1) = \phi(p^k)$. Consequently, r is also a primitive root modulo p^k.

All that remains is to prove (9.3) using mathematical induction. The case of $k = 2$ follows from (9.2). Let us assume that the assertion is true for the positive integer $k \geq 2$. Then

$$r^{p^{k-2}(p-1)} \not\equiv 1 \,(\text{mod } p^k).$$

Because $(r, p) = 1$, we know that $(r, p^{k-1}) = 1$. Consequently, from Euler's theorem, we know that

$$r^{p^{k-2}(p-1)} = r^{\phi(p^{k-1})} \equiv 1 \,(\text{mod } p^{k-1}).$$

Therefore, there is an integer d such that

$$r^{p^{k-2}(p-1)} = 1 + dp^{k-1},$$

where $p \nmid d$, because by hypothesis $r^{p^{k-2}(p-1)} \not\equiv 1 \,(\text{mod } p^k)$. We take the pth power of both sides of the above equation to obtain, via the binomial theorem and using the hypothesis that p is odd,

$$r^{p^{k-1}(p-1)} = (1 + dp^{k-1})^p$$

$$= 1 + p(dp^{k-1}) + \binom{p}{2}(dp^{k-1})^2 + \cdots + (dp^{k-1})^p$$

$$\equiv 1 + dp^k \pmod{p^{k+1}}.$$

Because $p \nmid d$, we can conclude that

$$r^{p^{k-1}(p-1)} \not\equiv 1 \pmod{p^{k+1}}.$$

This completes the proof by induction.   ∎

Example 9.15.   By Example 9.13, we know that $r = 3$ is a primitive root modulo 7 and 7^2. Hence, Theorem 9.10 tells us that $r = 3$ is also a primitive root modulo 7^k for all positive integers k.   ◄

Primitive Roots and Powers of 2   It is now time to discuss whether there are primitive roots modulo powers of 2. We first note that both 2 and $2^2 = 4$ have primitive roots, namely 1 and 3, respectively. For higher powers of 2, the situation is different, as the following theorem shows; there are no primitive roots modulo these powers of 2.

Theorem 9.11.   If a is an odd integer, and if k is an integer, $k \geq 3$, then

$$a^{\phi(2^k)/2} = a^{2^{k-2}} \equiv 1 \pmod{2^k}.$$

Proof.   We prove this result using mathematical induction. If a is an odd integer, then $a = 2b + 1$, where b is an integer. Hence,

$$a^2 = (2b + 1)^2 = 4b^2 + 4b + 1 = 4b(b + 1) + 1.$$

Because either b or $b + 1$ is even, we see that $8 \mid 4b(b + 1)$. By Exercise 5 of Section 4.1, it follows that

$$a^2 \equiv 1 \pmod 8.$$

This is the congruence of interest when $k = 3$.

Now, to complete the induction argument, let us assume that

$$a^{2^{k-2}} \equiv 1 \pmod{2^k}.$$

Then there is an integer d such that

$$a^{2^{k-2}} = 1 + d \cdot 2^k.$$

Squaring both sides of the above equality, we obtain

$$a^{2^{k-1}} = 1 + d2^{k+1} + d^2 2^{2k}.$$

This yields

$$a^{2^{k-1}} \equiv 1 \pmod{2^{k+1}},$$

which completes the induction argument.   ∎

Theorem 9.11 tells us that no power of 2, other than 2 and 4, has a primitive root, because when a is an odd integer, $\text{ord}_{2^k}a \neq \phi(2^k)$, because $a^{\phi(2^k)/2} \equiv 1 \pmod{2^k}$.

Even though there are no primitive roots modulo 2^k for $k \geq 3$, there always is an element of largest possible order, namely $\phi(2^k)/2$, as the following theorem shows.

Theorem 9.12. Let $k \geq 3$ be an integer. Then

$$\text{ord}_{2^k} 5 = \phi(2^k)/2 = 2^{k-2}.$$

Proof. Theorem 9.11 tells us that

$$5^{2^{k-2}} \equiv 1 \pmod{2^k},$$

for $k \geq 3$. By Theorem 9.1, we see that $\text{ord}_{2^k} 5 \mid 2^{k-2}$. Therefore, if we show that $\text{ord}_{2^k} 5 \nmid 2^{k-3}$, we can conclude that

$$\text{ord}_{2^k} 5 = 2^{k-2}.$$

To show that $\text{ord}_{2^k} 5 \nmid 2^{k-3}$, we will prove by mathematical induction that, for $k \geq 3$,

$$5^{2^{k-3}} \equiv 1 + 2^{k-1} \not\equiv 1 \pmod{2^k}.$$

For $k = 3$, we have

$$5 \equiv 1 + 4 \pmod 8.$$

Now, we assume that

$$5^{2^{k-3}} \equiv 1 + 2^{k-1} \pmod{2^k}.$$

This means that there is a positive integer d such that

$$5^{2^{k-3}} = (1 + 2^{k-1}) + d2^k.$$

Squaring both sides, we find that

$$5^{2^{k-2}} = (1 + 2^{k-1})^2 + 2(1 + 2^{k-1})d2^k + (d2^k)^2,$$

so that

$$5^{2^{k-2}} \equiv (1 + 2^{k-1})^2 = 1 + 2^k + 2^{2k-2} \equiv 1 + 2^k \pmod{2^{k+1}}.$$

This completes the induction argument and shows that

$$\text{ord}_{2^k} 5 = \phi(2^k)/2. \qquad \blacksquare$$

Primitive Roots Modulo Integers Not Prime Powers We have now demonstrated that all powers of odd primes possess primitive roots, while the only powers of 2 having primitive roots are 2 and 4. Next, we determine which integers not powers of primes—that is, those integers divisible by two or more primes—have primitive roots. We will demonstrate that the only positive integers not powers of primes that possess primitive roots are twice powers of odd primes.

We first narrow the set of positive integers that we must consider with the following result.

Theorem 9.13. If n is a positive integer that is not a prime power or twice a prime power, then n does not have a primitive root.

Proof. Let n be a positive integer with prime-power factorization

$$n = p_1^{t_1} p_2^{t_1} \cdots p_m^{t_m}.$$

Let us assume that the integer n has a primitive root r. This means that $(r, n) = 1$ and $\mathrm{ord}_n r = \phi(n)$. Because $(r, n) = 1$, we know that $(r, p^t) = 1$, whenever p^t is one of the prime powers occurring in the factorization of n. By Euler's theorem, we know that

$$r^{\phi(p^t)} \equiv 1 \ (\mathrm{mod}\ p^t).$$

Now, let U be the least common multiple of $\phi(p_1^{t_1}), \phi(p_2^{t_2}), \ldots, \phi(p_m^{t_m})$, that is,

$$U = [\phi(p_1^{t_1}), \phi(p_2^{t_2}), \ldots, \phi(p_m^{t_m})].$$

Because $\phi(p_i^{t_i}) \mid U$, we know that

$$r^U \equiv 1 \ (\mathrm{mod}\ p_i^{t_i})$$

for $i = 1, 2, \ldots, m$. Using the Chinese remainder theorem, it now follows that

$$r^U \equiv 1 \ (\mathrm{mod}\ n),$$

which implies that

$$\mathrm{ord}_n r = \phi(n) \leq U.$$

By Theorem 7.4, because ϕ is multiplicative, we have

$$\phi(n) = \phi(p_1^{t_1} p_2^{t_2} \cdots p_m^{t_m}) = \phi(p_1^{t_1})\phi(p_2^{t_2}) \cdots \phi(p_m^{t_m}).$$

This formula for $\phi(n)$ and the inequality $\phi(n) \leq U$ imply that

$$\phi(p_1^{t_1})\phi(p_2^{t_2}) \cdots \phi(p_m^{t_m}) \leq [\phi(p_1^{t_1}), \phi(p_2^{t_2}), \ldots, \phi(p_m^{t_m})].$$

Because the product of a set of integers is less than or equal to their least common multiple only if the integers are pairwise relatively prime (and then the "less than or equal to" relation is really just an equality), the integers $\phi(p_1^{t_1}), \phi(p_2^{t_2}), \ldots, \phi(p_m^{t_m})$ must be pairwise relatively prime.

We note that $\phi(p^t) = p^{t-1}(p - 1)$, so that $\phi(p^t)$ is even if p is odd, or if $p = 2$ and $t \geq 2$. Hence, the numbers $\phi(p_1^{t_1}), \phi(p_2^{t_2}), \ldots, \phi(p_m^{t_m})$ are not pairwise relatively prime unless $m = 1$ and n is a prime power, or $m = 2$ and $n = 2p^t$, where p is an odd prime and t is a positive integer. ∎

We have now limited our consideration to integers of the form $n = 2p^t$, where p is an odd prime and t is a positive integer. We now show that all such integers have primitive roots.

Theorem 9.14. If p is an odd prime and t is a positive integer, then $2p^t$ possesses a primitive root. In fact, if r is a primitive root modulo p^t, then if r is odd, it is also a primitive root modulo $2p^t$; whereas if r is even, $r + p^t$ is a primitive root modulo $2p^t$.

Proof. If r is a primitive root modulo p^t, then

$$r^{\phi(p^t)} \equiv 1 \;(\text{mod } p^t),$$

and no positive exponent smaller than $\phi(p^t)$ has this property. By Theorem 7.4, we note that $\phi(2p^t) = \phi(2)\phi(p^t) = \phi(p^t)$, so that $r^{\phi(2p^t)} \equiv 1 \;(\text{mod } p^t)$.

If r is odd, then

$$r^{\phi(2p^t)} \equiv 1 \;(\text{mod } 2).$$

Thus, by Corollary 4.8.1, we see that $r^{\phi(2p^t)} \equiv 1 \;(\text{mod } 2p^t)$. No smaller power of r is congruent to 1 modulo $2p^t$. Such power would also be congruent to 1 modulo p^t, contradicting the assumption that r is a primitive root of p^t. It follows that r is a primitive root modulo $2p^t$.

On the other hand, if r is even, then $r + p^t$ is odd. Hence,

$$(r + p^t)^{\phi(2p^t)} \equiv 1 \;(\text{mod } 2).$$

Because $r + p^t \equiv r \;(\text{mod } p^t)$, we see that

$$(r + p^t)^{\phi(2p^t)} \equiv 1 \;(\text{mod } p^t).$$

Therefore, $(r + p^t)^{\phi(2p^t)} \equiv 1 \;(\text{mod } 2p^t)$, and as no smaller power of $r + p^t$ is congruent to 1 modulo $2p^t$, we see that $r + p^t$ is a primitive root modulo $2p^t$. ∎

Example 9.16. Earlier in this section we showed that 3 is a primitive root modulo 7^t for all positive integers t. Hence, because 3 is odd, Theorem 9.14 tells us that 3 is also a primitive root modulo $2 \cdot 7^t$ for all positive integers t. For instance, 3 is a primitive root modulo 14.

Similarly, we know that 2 is a primitive root modulo 5^t for all positive integers t. Because $2 + 5^t$ is odd, Theorem 9.14 tells us that $2 + 5^t$ is a primitive root modulo $2 \cdot 5^t$ for all positive integers t. For example, 27 is a primitive root modulo 50. ◀

Putting Everything Together   Combining Corollary 9.8.1 and Theorems 9.10, 9.13, and 9.14, we can now describe which positive integers have a primitive root.

Theorem 9.15. The positive integer n, $n > 1$, possesses a primitive root if and only if

$$n = 2, 4, p^t, \text{ or } 2p^t,$$

where p is an odd prime and t is a positive integer.

9.3 Exercises

1. Which of the integers 4, 10, 16, 22, and 28 have a primitive root?

2. Which of the integers 8, 9, 12, 26, 27, 31, and 33 have a primitive root?

3. Find a primitive root modulo each of the following moduli.

a) 3^2 c) 23^2

b) 5^2 d) 29^2

4. Find a primitive root modulo each of the following moduli.

a) 11^2 c) 17^2

b) 13^2 d) 19^2

5. Find a primitive root for all positive integers k modulo each of the following moduli.

a) 3^k c) 13^k

b) 11^k d) 17^k

6. Find a primitive root for all positive integers k modulo each of the following moduli.

a) 23^k c) 31^k

b) 29^k d) 37^k

7. Find a primitive root modulo each of the following moduli.

a) 10   c) 38

b) 34   d) 50

8. Find a primitive root modulo each of the following moduli.

a) 6   c) 26

b) 18   d) 338

9. Find all the primitive roots modulo 22.

10. Find all the primitive roots modulo 25.

11. Find all the primitive roots modulo 38.

12. Show that there are the same number of primitive roots modulo $2p^t$ as there are modulo p^t, where p is an odd prime and t is a positive integer.

☞ **13.** Show that the integer m has a primitive root if and only if the only solutions of the congruence $x^2 \equiv 1 \pmod{m}$ are $x \equiv \pm 1 \pmod{m}$.

* **14.** Let n be a positive integer possessing a primitive root. Using this primitive root, prove that the product of all positive integers less than n and relatively prime to n is congruent to -1 modulo n. (When n is prime, this result is Wilson's theorem (Theorem 6.1).)

* **15.** Show that although there are no primitive roots modulo 2^k, where k is an integer, $k \geq 3$, every odd integer is congruent modulo 2^n to exactly one of the integers $(-1)^\alpha 5^\beta$, where $\alpha = 0$ or 1 and β is an integer satisfying $0 \leq \beta \leq 2^{k-2} - 1$.

16. Find the smallest odd prime p that has a primitive root r that is not also a primitive root modulo p^2.

9.3 Computational and Programming Exercises

Computations and Explorations

Using a computation program such as Maple or *Mathematica,* or programs you have written, carry out the following computations and explorations.

1. Find as many examples as you can where r is a primitive root of the prime p, but r is not a primitive root of p^2. Can you make any conjectures about how often this occurs?

Programming Projects

Write computer programs using Maple, *Mathematica,* or a language of your choice to do the following.

1. Find primitive roots modulo powers of odd primes.

2. Find primitive roots modulo twice powers of odd primes.

9.4 Index Arithmetic

In this section, we demonstrate how primitive roots may be used to do modular arithmetic. Let r be a primitive root modulo the positive integer m (so that m is of the form described in Theorem 9.15). By Theorem 9.3, we know that the integers

$$r, r^2, r^3, \ldots, r^{\phi(m)}$$

form a reduced system of residues modulo m. From this fact, we see that if a is an integer relatively prime to m, then there is a unique integer x with $1 \leq x \leq \phi(m)$ such that

$$r^x \equiv a \pmod{m}.$$

This leads to the following definition.

Definition.   Let m be a positive integer with primitive root r. If a is a positive integer with $(a, m) = 1$, then the unique integer x with $1 \leq x \leq \phi(m)$ and $r^x \equiv a \pmod{m}$ is called the *index* (or *discrete logarithm*) of a to the base r modulo m. With this definition, we have $r^{\text{ind}_r a} \equiv a \pmod{m}$.

If x is the index of a to the base r modulo m, then we write $x = \text{ind}_r a$, where we do not indicate the modulus m in the notation, as it is assumed to be fixed. From the definition, we know that if a and b are integers relatively prime to m and $a \equiv b \pmod{m}$, then $\text{ind}_r a = \text{ind}_r b$. Indices share many properties of logarithms, but with equalities replaced with congruences modulo $\phi(m)$ (that is why they are called discrete logarithms).

Example 9.17.   Let $m = 7$. We have seen that 3 is a primitive root modulo 7 and that $3^1 \equiv 3 \pmod 7$, $3^2 \equiv 2 \pmod 7$, $3^3 \equiv 6 \pmod 7$, $3^4 \equiv 4 \pmod 7$, $3^5 \equiv 5 \pmod 5$, and $3^6 \equiv 1 \pmod 7$.

Hence, modulo 7, we have

$$\text{ind}_3 1 = 6, \text{ind}_3 2 = 2, \text{ind}_3 3 = 1,$$
$$\text{ind}_3 4 = 4, \text{ind}_3 5 = 5, \text{ind}_3 6 = 3.$$

With a different primitive root modulo 7, we obtain a different set of indices. For instance, calculations show that with respect to the primitive root 5,

$$\text{ind}_5 1 = 6, \text{ind}_5 2 = 4, \text{ind}_5 3 = 5,$$
$$\text{ind}_5 4 = 2, \text{ind}_5 5 = 1, \text{ind}_5 6 = 3. \qquad \blacktriangleleft$$

Properties of Indices   We now develop properties of indices, modulo m similar to those of logarithms, but instead of equalities, we have congruences modulo $\phi(m)$.

Theorem 9.16.   Let m be a positive integer with primitive root r, and let a and b be integers relatively prime to m. Then

(i) $\text{ind}_r 1 \equiv 0 \pmod{\phi(m)}$,

(ii) $\text{ind}_r(ab) \equiv \text{ind}_r a + \text{ind}_r b \pmod{\phi(m)}$,

(iii) $\text{ind}_r a^k \equiv k \cdot \text{ind}_r a \pmod{\phi(m)}$ if k is a positive integer.

Proof of (i).   From Euler's theorem, we know that $r^{\phi(m)} \equiv 1 \pmod{m}$. Because r is a primitive root modulo m, no smaller positive power of r is congruent to 1 modulo m. Hence, $\text{ind}_r 1 = \phi(m) \equiv 0 \pmod{\phi(m)}$.

Proof of (ii).   To prove this congruence, note that from the definition of indices,

$$r^{\text{ind}_r(ab)} \equiv ab \pmod{m}$$

and

$$r^{\text{ind}_r a + \text{ind}_r b} \equiv r^{\text{ind}_r a} \cdot r^{\text{ind}_r b} \equiv ab \pmod{m}.$$

Hence,

$$r^{\text{ind}_r(ab)} \equiv r^{\text{ind}_r a + \text{ind}_r b} \pmod{m}.$$

Using Theorem 9.2, we conclude that

$$\text{ind}_r(ab) \equiv \text{ind}_r a + \text{ind}_r b \pmod{\phi(m)}.$$

Proof of (iii).   To prove the congruence of interest, first note that by definition, we have

$$r^{\text{ind}_r a^k} \equiv a^k \pmod{m}$$

and

$$r^{k \cdot \text{ind}_r a} \equiv (r^{\text{ind}_r a})^k \pmod{m}.$$

Hence,

$$r^{\text{ind}_r a^k} \equiv r^{k \cdot \text{ind}_r a} \pmod{m}.$$

Using Theorem 9.2, this leads us immediately to the congruence we want, namely

$$\text{ind}_r a^k \equiv k \cdot \text{ind}_r a \;(\text{mod } \phi(m)). \qquad \blacksquare$$

Example 9.18. From the previous examples, we see that, modulo 7, $\text{ind}_5 2 = 4$ and $\text{ind}_5 3 = 5$. Because $\phi(7) = 6$, part (ii) of Theorem 9.16 tells us that

$$\text{ind}_5 6 = \text{ind}_5(2 \cdot 3) = \text{ind}_5 2 + \text{ind}_5 3 = 4 + 5 = 9 \equiv 3 \;(\text{mod } 6).$$

Note that this agrees with the value previously found for $\text{ind}_5 6$.

From part (iii) of Theorem 9.16, we see that

$$\text{ind}_5 3^4 \equiv 4 \cdot \text{ind}_5 3 \equiv 4 \cdot 5 = 20 \equiv 2 \;(\text{mod } 6).$$

Note that direct computation gives the same result, because

$$\text{ind}_5 3^4 = \text{ind}_5 81 = \text{ind}_5 4 = 2. \qquad \blacktriangleleft$$

Indices are helpful in the solution of certain types of congruences. Consider the following examples.

Example 9.19. We will use indices to solve the congruence $6x^{12} \equiv 11 \;(\text{mod } 17)$. We find that 3 is a primitive root of 17 (because $3^8 \equiv -1 \;(\text{mod } 17)$). The indices of integers to the base 3 modulo 17 are given in Table 9.1.

a	1	2	3	4	5	6	7	8	9	10	11	12	13	14	15	16
$\text{ind}_3 a$	16	14	1	12	5	15	11	10	2	3	7	13	4	9	6	8

Table 9.1 *Indices to the base* 3 *modulo* 17.

Taking the index of each side of the congruence to the base 3 modulo 17, we obtain a congruence modulo $\phi(17) = 16$, namely

$$\text{ind}_3(6x^{12}) \equiv \text{ind}_3 11 = 7 \;(\text{mod } 16).$$

Using parts (ii) and (iii) of Theorem 9.16, we obtain

$$\text{ind}_3(6x^{12}) \equiv \text{ind}_3 6 + \text{ind}_3(x^{12}) \equiv 15 + 12 \cdot \text{ind}_3 x \;(\text{mod } 16).$$

Hence,

$$15 + 12 \cdot \text{ind}_3 x \equiv 7 \;(\text{mod } 16)$$

or

$$12 \cdot \text{ind}_3 x \equiv 8 \;(\text{mod } 16).$$

From this congruence it follows (as the reader should show) that

$$\text{ind}_3 x \equiv 2 \;(\text{mod } 4).$$

Hence,

$$\text{ind}_3 x \equiv 2, 6, 10, \text{ or } 14 \pmod{16}.$$

Consequently, from the definition of indices, we find that

$$x \equiv 3^2, 3^6, 3^{10}, \text{ or } 3^{14} \pmod{17}.$$

(Note that this congruence holds modulo 17). Because $3^2 \equiv 9$, $3^6 \equiv 15$, $3^{10} \equiv 8$, and $3^{14} \equiv 2 \pmod{17}$, we conclude that

$$x \equiv 9, 15, 8, \text{ or } 2 \pmod{17}.$$

Because each step in the computations is reversible, there are four incongruent solutions of the original congruence modulo 17. ◄

Example 9.20.   We wish to find all solutions of the congruence $7^x \equiv 6 \pmod{17}$. When we take indices to the base 3 modulo 17 of both sides of this congruence, we find that

$$\text{ind}_3(7^x) \equiv \text{ind}_3 6 = 15 \pmod{16}.$$

By part (iii) of Theorem 9.16, we obtain

$$\text{ind}_3(7^x) \equiv x \cdot \text{ind}_3 7 \equiv 11x \pmod{16}.$$

Hence,

$$11x \equiv 15 \pmod{16}.$$

Because 3 is an inverse of 11 modulo 16, we multiply both sides of the linear congruence above by 3, to find that

$$x \equiv 3 \cdot 15 = 45 \equiv 13 \pmod{16}.$$

All steps in this computation are reversible. Therefore, the solutions of

$$7^x \equiv 6 \pmod{17}$$

are given by

$$x \equiv 13 \pmod{16}. \quad ◄$$

The Difficulty of Finding Discrete Logarithms

Given a prime p and a primitive root r, the problem of finding the index (discrete logarithm) of an integer a to the base r modulo m is called the *discrete logarithm problem*. This problem is believed to be as computationally difficult as that of factoring integers. For this reason, it has been used as the basis for several public key cryptosystems, such as the ElGamal cryptosystem discussed in Section 10.2, and protocols, such as the Diffie-Hellman key agreement scheme discussed in Section 8.3. With the growing importance of the discrete logarithm problem in cryptography, a great deal of research has been devoted to constructing efficient algorithms for computing discrete logarithms. The most efficient algorithm known for computing discrete logarithms is the number-field sieve method, which requires approximately the same number of bit operations to find discrete logarithms modulo a prime p as it would to factor a composite number of about the same size as p. To determine how long it takes to solve the discrete logarithm problem modulo a prime p, consult Table 3.2, which shows how long it takes to factor an integer n of the same number of decimal digits as p. For more information about the discrete logarithm problem, and algorithms for solving it, consult [MevaVa97] and the many references cited there.

Power Residues

Indices are also helpful for studying congruences of the form $x^k \equiv a \pmod{m}$, where m is a positive integer with a primitive root and $(a, m) = 1$. Before we study such congruences, we present a definition.

Definition.   If m and k are positive integers and a is an integer relatively prime to m, then we say that a is a *kth power residue of m* if the congruence $x^k \equiv a \pmod{m}$ has a solution.

When m is an integer possessing a primitive root, the following theorem gives a useful criterion for an integer a relatively prime to m to be a kth power residue of m.

Theorem 9.17.   Let m be a positive integer with a primitive root. If k is a positive integer and a is an integer relatively prime to m, then the congruence $x^k \equiv a \pmod{m}$ has a solution if and only if

$$a^{\phi(m)/d} \equiv 1 \pmod{m},$$

where $d = (k, \phi(m))$. Furthermore, if there are solutions of $x^k \equiv a \pmod{m}$, then there are exactly d incongruent solutions modulo m.

Proof.   Let r be a primitive root modulo the positive integer m. We note that the congruence

$$x^k \equiv a \pmod{m}$$

holds if and only if

(9.4) $$k \cdot \mathrm{ind}_r x \equiv \mathrm{ind}_r a \pmod{\phi(m)}.$$

Now let $d = (k, \phi(m))$ and $y = \mathrm{ind}_r x$, so that $x \equiv r^y \pmod{m}$. By Theorem 4.10, we note that if $d \nmid \mathrm{ind}_r a$, then the linear congruence

(9.5) $$ky \equiv \mathrm{ind}_r a \pmod{\phi(m)}$$

has no solutions and, hence, there are no integers x satisfying (9.4). If $d \mid \mathrm{ind}_r a$, then there are exactly d integers y incongruent modulo $\phi(m)$ such that (9.5) holds and, hence, exactly d integers x incongruent modulo m such that (9.4) holds. Because $d \mid \mathrm{ind}_r a$ if and only if

$$(\phi(m)/d)\mathrm{ind}_r a \equiv 0 \pmod{\phi(m)},$$

and this congruence holds if and only if

$$a^{\phi(m)/d} \equiv 1 \pmod{m},$$

the theorem is true. ∎

We note that Theorem 9.17 tells us that if p is a prime, k is a positive integer, and a is an integer relatively prime to p, then a is a kth power residue of p if and only if

$$a^{(p-1)/d} \equiv 1 \pmod{p},$$

where $d = (k, p - 1)$. We illustrate this observation with an example.

Example 9.21. To determine whether 5 is a sixth power residue of 17, that is, whether the congruence

$$x^6 \equiv 5 \pmod{17}$$

has a solution, we determine that

$$5^{16/(6,16)} = 5^8 \equiv -1 \pmod{17}.$$

Hence, 5 is not a sixth power residue of 17. ◄

A table of indices with respect to the least primitive root modulo each prime less than 100 is given in Table 4 of Appendix E.

Proving Theorem 6.10 This proof of Theorem 6.10 is quite long and complicated, but is based only on results already established. We present this proof to give the reader an indication that even elementary proofs can be difficult to create and hard to follow. As you read this proof, follow each part carefully and check each separate case. We restate Theorem 6.10 for convenience.

Theorem 6.10. If n is an odd composite positive integer, then n passes Miller's test for at most $(n - 1)/4$ bases b with $1 \leq b < n - 1$.

We need the following lemma in the proof.

Lemma 9.2. Let p be an odd prime and let e and q be positive integers. Then the number of incongruent solutions of the congruence $x^q \equiv 1 \pmod{p^e}$ is $(q, p^{e-1}(p - 1))$.

Proof. Let r be a primitive root of p^e. By taking indices with respect to r, we see that $x^q \equiv 1 \pmod{p^e}$ if and only if $qy \equiv 0 \pmod{\phi(p^e)}$, where $y = \text{ind}_r x$. Using Theorem 4.10, we see that there are exactly $(q, \phi(p^e))$ incongruent solutions of $qy \equiv 0 \pmod{\phi(p^e)}$. Consequently, there are $(q, \phi(p^e)) = (q, p^{e-1}(p - 1))$ incongruent solutions of $x^q \equiv 1 \pmod{p^e}$. ∎

We now proceed with a proof of Theorem 6.10.

Proof. Let $n - 1 = 2^s t$, where s is a positive integer and t is an odd positive integer. For n of Theorem 6.10 to be a strong pseudoprime to the base b, either

$$b^t \equiv 1 \pmod{n}$$

or

$$b^{2^j t} \equiv -1 \pmod{n}$$

for some integer j with $0 \leq j \leq s - 1$. In either case, we have

$$b^{n-1} \equiv 1 \pmod{n}.$$

Let the prime-power factorization of n be $n = p_1^{e_1} p_2^{e_2} \cdots p_r^{e_r}$. By Lemma 9.2, we know that there are $(n - 1, p_j^{e_j}(p_j - 1)) = (n - 1, p_j - 1)$ incongruent solutions of $x^{n-1} \equiv 1 \pmod{p_j^{e_j}}$, $j = 1, 2, \ldots, r$. Consequently, the Chinese remainder theorem tells us that there are exactly $\prod_{j=1}^{r}(n - 1, p_j - 1)$ incongruent solutions of $x^{n-1} \equiv 1 \pmod{n}$. ∎

We consider two cases.

Case (i). We first consider the case where the prime-power factorization of n contains a prime power $p_k^{e_k}$ with exponent $e_k \geq 2$. Because

$$(p_k - 1)/p_k^{e_k} = \left(1/p_k^{e_k - 1}\right) - \left(1/p_k^{e_k}\right) \leq 2/9$$

(the largest possible value occurs when $p_j = 3$ and $e_j = 2$), we see that

$$\prod_{j=1}^{r}(n - 1, p_j - 1) \leq \prod_{j=1}^{r}(p_j - 1)$$

$$\leq \left(\prod_{\substack{j=1 \\ j \neq k}}^{r} p_j\right)\left(\frac{2}{9} p_k^{e_k}\right)$$

$$\leq \frac{2}{9} n.$$

Because $\frac{2}{9} n \leq \frac{1}{4}(n - 1)$ for $n \geq 9$, it follows that

$$\prod_{j=1}^{r}(n - 1, p_j - 1) \leq (n - 1)/4.$$

Consequently, there are at most $(n - 1)/4$ integers b, $1 \leq b \leq n$, for which n is a strong pseudoprime to the base b.

Case (ii). Now, we consider the case where $n = p_1 p_2 \cdots p_r$, where $p_1, p_2, \ldots, p_r$ are distinct odd primes. Let

$$p_i - 1 = 2^{s_i} t_i, \quad i = 1, 2, \ldots, r,$$

where s_i is a positive integer and t_i is an odd positive integer. We reorder the primes $p_1, p_2, \ldots, p_r$ (if necessary) so that $s_1 \le s_2 \le \cdots \le s_r$. We note that

$$(n - 1, p_i - 1) = 2^{\min(s, s_i)}(t, t_i).$$

The number of incongruent solutions of $x^t \equiv 1 \pmod{p_i}$ is $T_i = (t, t_i)$. From Exercise 22 at the end of this section, there are $2^j T_i$ incongruent solutions of $x^{2^j t} \equiv -1 \pmod{p_i}$ when $0 \le s_i - 1$, and no solutions otherwise. Hence, using the Chinese remainder theorem, there are $T_1 T_2 \cdots T_r$ incongruent solutions of $x^t \equiv 1 \pmod{n}$, and $2^{jr} T_1 T_2 \cdots T_r$ incongruent solutions of $x^{2^j t} \equiv -1 \pmod{n}$ when $0 \le j \le s_1 - 1$. Therefore, there are a total of

$$T_1 T_2 \cdots T_r \left(1 + \sum_{j=0}^{s_1-1} 2^{jr} \right) = T_1 T_2 \cdots T_r \left(1 + \frac{2^{r s_1} - 1}{2^r - 1} \right)$$

integers b, with $1 \le b \le n - 1$, for which n is a strong pseudoprime to the base b.

Now, we note that

$$\phi(n) = (p_1 - 1)(p_2 - 1) \cdots (p_r - 1) = t_1 t_2 \cdots t_r 2^{s_1 + s_2 + \cdots + s_r}.$$

We will show that

$$T_1 T_2 \cdots T_r \left(1 + \frac{2^{r s_1} - 1}{2^r - 1} \right) \le \phi(n)/4,$$

which proves the desired result. Because $T_1 T_2 \cdots T_r \le t_1 t_2 \cdots t_r$, we can achieve our goal by showing that

$$(9.6) \qquad \left(1 + \frac{2^{r s_j} - 1}{2^r - 1} \right) / 2^{s_1 + s_2 + \cdots + s_r} \le \frac{1}{4}.$$

Because $s_1 \le \cdots \le s_r$, we see that

$$\left(1 + \frac{2^{r s_j} - 1}{2^r - 1} \right) / 2^{s_1 + s_2 + \cdots + s_r} \le \left(1 + \frac{2^{r s_j} - 1}{2^r - 1} \right) / 2^{r s_1}$$

$$= \frac{1}{2^{r s_1}} + \frac{2^{r s_1} - 1}{2^{r s_1}(2^r - 1)}$$

$$= \frac{1}{2^{r s_1}} + \frac{1}{2^r - 1} - \frac{1}{2^{r s_1}(2^r - 1)}$$

$$= \frac{1}{2^r - 1} + \frac{2^r - 2}{2^{r s_1}(2^r - 1)}$$

$$\le \frac{1}{2^{r-1}}.$$

From this inequality, we conclude that (9.6) is valid when $r \ge 3$.

When $r = 2$, we have $n = p_1 p_2$, with $p_1 - 1 = 2^{s_1} t_1$ and $p_2 - 1 = 2^{s_2} t_2$, with $s_1 \le s_2$. If $s_1 < s_2$, then (9.6) is again valid, because

$$\left(1 + \frac{2^{2s_1} - 1}{3}\right) / 2^{s_1 + s_2} = \left(1 + \frac{2^{2s_1} - 1}{3}\right) / \left(2^{2s_1} \cdot 2^{s_2 - s_1}\right)$$

$$= \left(\frac{1}{3} + \frac{1}{3 \cdot 2^{s_1 - 1}}\right) / 2^{s_2 - s_1}$$

$$\leq \frac{1}{4}.$$

When $s_1 = s_2$, we have $(n - 1, p_1 - 1) = 2^s T_1$ and $(n - 1, p_2 - 1) = 2^s T_2$. Let us assume that $p_1 > p_2$. Note that $T_1 \neq t_1$, for if $T_1 = t_1$, then $(p_1 - 1) \mid (n - 1)$, so that

$$n = p_1 p_2 \equiv p_2 \equiv 1 \pmod{p_1 - 1},$$

which implies that $p_2 > p_1$, a contradiction. Because $T_1 \neq t_1$, we know that $T_1 \leq t_1/3$. Similarly, if $p_1 < p_2$, then $T_2 \neq t_2$, so that $T_2 \leq t_2/3$. Hence, $T_1 T_2 \leq t_1 t_2/3$, and because $\left(1 + \frac{2^{2s_1} - 1}{3}\right) / 2^{2s_1} \leq \frac{1}{2}$, we have

$$T_1 T_2 \left(1 + \frac{2^{2s_1} - 1}{3}\right) \leq t_1 t_2 2^{2s_1}/6 = \phi(n)/6,$$

proving the theorem for this final case, since $\phi(n)/6 \leq (n - 1)/6 < (n - 1)/4$.

By analyzing the inequalities in the proof of Theorem 6.10, we can see that the probability that n is a strong pseudoprime to the randomly chosen base b, $1 \leq b \leq n - 1$, is close to 1/4 only for integers n with prime factorizations of the form $n = p_1 p_2$, with $p_1 = 1 + 2q_1$ and $p_2 = 1 + 4q_2$, where q_1 and q_2 are odd primes, or $n = q_1 q_2 q_3$, with $p_1 = 1 + 2q_1$, $p_2 = 1 + 2q_2$, and $p_3 = 1 + 2q_3$, where q_1, q_2, and q_3 are distinct odd primes (see Exercise 23).

9.4 Exercises

1. Write out a table of indices modulo 23 with respect to the primitive root 5.

2. Find all the solutions of the following congruences.

   a) $3x^5 \equiv 1 \pmod{23}$ b) $3x^{14} \equiv 2 \pmod{23}$

3. Find all the solutions of the following congruences.

   a) $3^x \equiv 2 \pmod{23}$ b) $13^x \equiv 5 \pmod{23}$

4. For which positive integers a is the congruence $ax^4 \equiv 2 \pmod{13}$ solvable?

5. For which positive integers b is the congruence $8x^7 \equiv b \pmod{29}$ solvable?

6. Find the solutions of $2^x \equiv x \pmod{13}$, using indices to the base 2 modulo 13.

7. Find all the solutions of $x^x \equiv x \pmod{23}$.

8. Show that if p is an odd prime and r is a primitive root of p, then $\text{ind}_r(p - 1) = (p - 1)/2$.

9. Let p be an odd prime. Show that the congruence $x^4 \equiv -1 \pmod{p}$ has a solution if and only if p is of the form $8k + 1$.

10. Prove that there are infinitely many primes of the form $8k + 1$. (*Hint:* Assume that $p_1, p_2, \ldots, p_n$ are the only primes of this form. Let $Q = (2p_1, p_2 \cdots p_n)^k + 1$. Show that Q must have an odd prime factor different than $p_1, p_2, \ldots, p_n$ and, by Exercise 9, necessarily of the form $8k + 1$.)

By Exercise 15 of Section 9.3, we know that if a is an odd positive integer, then there are unique integers α and β with $\alpha = 0$ or 1 and $0 \le \beta \le 2^{k-2} - 1$ such that $a \equiv (-1)^\alpha 5^\beta$ (mod 2^k). Define the *index system of a modulo 2^k* to be equal to the pair (α, β).

11. Find the index system of 7 and 9 modulo 16.

12. Develop rules for the index systems modulo 2^k of products and powers, analogous to the rules for indices.

13. Use the index system modulo 32 to find all solutions of $7x^9 \equiv 11$ (mod 32) and $3^x \equiv 17$ (mod 32).

Let $n = 2^{t_0} p_1^{t_1} p_2^{t_2} \cdots p_m^{t_m}$ be the prime-power factorization of n. Let a be an integer relatively prime to n. Let $r_1, r_2, \ldots, r_m$ be primitive roots of $p_1^{t_1}, p_2^{t_2}, \ldots, p_m^{t_m}$, respectively, and let $\gamma_1 = \text{ind}_{r_1} a \pmod{\phi(p_1^{t_1})}$, $\gamma_2 = \text{ind}_{r_2} a \pmod{\phi(p_2^{t_2})}, \ldots, \gamma_m = \text{ind}_{r_m} a \pmod{\phi(p_m^{t_m})}$. If $t_0 \le 2$, let r_0 be a primitive root of 2^{t_0}, and let $\gamma_0 = \text{ind}_{r_0} a \pmod{\phi(2^{t_0})}$. If $t_0 \ge 3$, let (α, β) be the index system of a modulo 2^k, so that $a \equiv (-1)^\alpha 5^\beta \pmod{2^k}$. Define the *index system of a modulo n* to be $(\gamma_0, \gamma_1, \gamma_2, \ldots, \gamma_m)$ if $t_0 \le 2$ and $(\alpha, \beta, \gamma_1, \gamma_2, \ldots, \gamma_m)$ if $t_0 \ge 3$.

14. Show that if n is a positive integer, then every integer has a unique index system modulo n.

15. Find the index systems of 17 and 41 (mod 120) (in your computations, use 2 as a primitive root of the prime factor 5 of 120).

16. Develop rules for the index systems modulo n of products and powers, analogous to those for indices.

17. Use an index system modulo 60 to find the solutions of $11x^7 \equiv 43$ (mod 60).

18. Let p be a prime, $p > 3$. Show that if $p \equiv 2$ (mod 3), then every integer not divisible by 3 is a third-power, or *cubic*, residue of p, whereas if $p \equiv 1$ (mod 3), an integer a is a cubic residue of p if and only if $a^{(p-1)/3} \equiv 1$ (mod p).

19. Let e be a positive integer with $e \ge 2$. Show that if k is an odd positive integer, then every odd integer a is a kth power residue of 2^e.

* **20.** Let e be a positive integer with $e \ge 2$. Show that if k is even, then an integer a is a kth power residue of 2^e if and only if $a \equiv 1$ (mod $(4k, 2^e)$).

* **21.** Let e be a positive integer with $e \ge 2$. Show that if k is a positive integer, then the number of incongruent kth power residues of 2^e is

$$\frac{2^{e-1}}{(k, 2)(k, 2^{e-2})}.$$

☞ **22.** Let $N = 2^j u$ be a positive integer, with j a nonnegative integer and u an odd positive integer, and let $p - 1 = 2^s t$, where s and t are positive integers with t odd. Show that there are $2^j (t, u)$ incongruent solutions of $x^N \equiv -1 \pmod{p}$ if $0 \le j \le s - 1$, and no solutions otherwise.

* **23.** a) Show that the probability that n is a strong pseudoprime for a base b randomly chosen with $1 \le b \le n - 1$ is near $1/4$ only when n has a prime factorization of the form $n = p_1 p_2$, where $p_1 = 1 + 2q_1$ and $p_2 = 1 + 4q_2$, with q_1 and q_2 prime, or $n = p_1 p_2 p_3$, where $p_1 = 1 + 2q_1$, $p_2 = 1 + 2q_2$, and $p_3 = 1 + 2q_3$, with q_1, q_2, q_3 distinct odd primes.

   b) Find the probability that $n = 49,939 \cdot 99,877$ is a strong pseudoprime to the base b randomly chosen with $1 \le b \le n - 1$.

9.4 Computational and Programming Exercises

Computations and Explorations

Using a computation program such as Maple or *Mathematica,* or programs you have written, carry out the following computations and explorations.

1. Find integers n for which the probability that n is a strong pseudoprime to the randomly chosen base b, $1 \le b \le n - 1$, is close to $1/4$.

Programming Projects

Write programs using Maple, *Mathematica,* or a language of your choice to do the following.

1. Construct a table of indices modulo a particular primitive root of an integer.

2. Using indices solve congruences of the form $ax^b \equiv c \pmod{m}$, where $a, b, c,$ and m are integers with $c > 0, m > 0$, and where m has a primitive root.

3. Find kth power residues of a positive integer m having a primitive root, where k is a positive integer.

4. Find index systems modulo powers of 2 (see the preamble to Exercise 11).

5. Find index systems modulo arbitrary positive integers (see the preamble to Exercise 14).

9.5 Primality Tests Using Orders of Integers and Primitive Roots

In Chapter 6, we saw that the converse of Fermat's little theorem is not true. Fermat's little theorem tells us that if p is prime and a is an integer with $(a, p) = 1$, then $a^{p-1} \equiv 1 \pmod{p}$. Even if $a^{n-1} \equiv 1 \pmod{n}$, where a is a positive integer, n may still be composite. Although the converse of Fermat's little theorem is not true, can we establish partial converses? That is, can we add hypotheses to the converse to make it true?

In this section, we will use the concepts developed in this chapter to prove some partial converses of Fermat's little theorem. We begin with a result known as *Lucas's converse of Fermat's little theorem.* This result was proved by French mathematician Edouard Lucas in 1876.

Theorem 9.18. *Lucas's Converse of Fermat's Little Theorem.*   If n is a positive integer and if an integer x exists such that

$$x^{n-1} \equiv 1 \pmod{n}$$

and

$$x^{(n-1)/q} \not\equiv 1 \pmod{n}$$

for all prime divisors q of $n - 1$, then n is prime.

Proof.   Because $x^{n-1} \equiv 1 \pmod{n}$, Theorem 9.1 tells us that $\mathrm{ord}_n x \mid (n - 1)$. We will show that $\mathrm{ord}_n x = n - 1$. Suppose that $\mathrm{ord}_n x \neq n - 1$. Because $\mathrm{ord}_n x \mid (n - 1)$, there is an integer k with $n - 1 = k \cdot \mathrm{ord}_n x$, and because $\mathrm{ord}_n x \neq n - 1$, we know that $k > 1$. Let q be a prime divisor of k. Then

$$x^{(n-1)/q} = x^{k/(\mathrm{ord}_n x \cdot q)} = (x^{\mathrm{ord}_n x})^{(k/q)} \equiv 1 \pmod{n}.$$

However, this contradicts the hypotheses of the theorem, so we must have $\mathrm{ord}_n x = n - 1$. Now because $\mathrm{ord}_n x \leq \phi(n)$ and $\phi(n) \leq n - 1$, it follows that $\phi(n) = n - 1$. By Theorem 7.2, we know that n must be prime.   ∎

Note that Theorem 9.18 is equivalent to the fact that if there is an integer with order modulo n equal to $n - 1$, then n must be prime. We illustrate the use of Theorem 9.18 with an example.

Example 9.22.   Let $n = 1009$. Then $11^{1008} \equiv 1 \pmod{1009}$. The prime divisors of 1008 are 2, 3, and 7. We see that $11^{1008/2} = 11^{504} \equiv -1 \pmod{1009}$, $11^{1008/3} = 11^{336} \equiv 374 \pmod{1009}$, and $11^{1008/7} = 11^{144} \equiv 935 \pmod{1009}$. Hence, by Theorem 9.18, we know that 1009 is prime.   ◄

The following corollary of Theorem 9.18 gives a slightly more efficient primality test.

Corollary 9.18.1.   If n is an odd positive integer and if x is a positive integer such that

$$x^{(n-1)/2} \equiv -1 \pmod{n}$$

and

$$x^{(n-1)/q} \not\equiv 1 \pmod{n}$$

for all odd prime divisors q of $n - 1$, then n is prime.

Proof.   Because $x^{(n-1)/2} \equiv -1 \pmod{n}$, we see that

$$x^{n-1} = (x^{(n-1)/2})^2 \equiv (-1)^2 \equiv 1 \pmod{n}.$$

Because the hypotheses of Theorem 9.18 are met, we know that n is prime.   ∎

Example 9.23.   Let $n = 2003$. The odd prime divisors of $n - 1 = 2002$ are 7, 11, and 13. Because $5^{2002/2} = 5^{1001} \equiv -1 \pmod{2003}$, $5^{2002/7} = 5^{286} \equiv 874 \pmod{2003}$,

$5^{2002/11} = 5^{183} \equiv 886 \pmod{2003}$, and $5^{2002/13} = 5^{154} \equiv 633 \pmod{2003}$, we see from Corollary 9.18.1 that 2003 is prime. ◀

To determine whether an integer n is prime using either Theorem 9.18 or Corollary 9.18.1, it is necessary to know the prime factorization of $n - 1$. As we have remarked before, finding the prime factorization of an integer is a time-consuming process. Only when we have some a priori information about the factorization of $n - 1$ are the primality tests given by these results practical. Indeed, with such information these tests can be useful. Such a situation occurs with the Fermat numbers; in Chapter 11 we give a primality test for these numbers based on the ideas of this section.

In Chapter 3, we discussed the recent discovery of an algorithm that can prove that an integer n is prime in polynomial time (in the number of digits in the prime). We can prove a weaker result using Corollary 9.18.1, which shows that we can prove that an integer is prime in polynomial time once particular information is known.

Theorem 9.19. If n is prime, this can be proved when sufficient information is available using $O((\log_2 n)^4)$ bit operations.

Proof. We use the second principle of mathematical induction. The induction hypothesis is an estimate for $f(n)$, where $f(n)$ is the total number of multiplications and modular exponentiations needed to verify that the integer n is prime.

We demonstrate that

$$f(n) \leq 3(\log n / \log 2) - 2.$$

First, we note that $f(2) = 1$. We assume that for all primes q, with $q < n$, the inequality

$$f(q) \leq 3(\log n / \log 2) - 2$$

holds.

To prove that n is prime, we use Corollary 9.18.1. Once we have the numbers $2^a, q_1, \ldots, q_t$, and x that supposedly satisfy

(i) $n - 1 = 2^a q_1 q_2 \cdots q_t$,

(ii) q_i is prime for $i = 1, 2, \ldots, t$,

(iii) $x^{(n-1)/2} \equiv -1 \pmod{n}$,

and

(iv) $x^{(n-1)/q_j} \equiv 1 \pmod{n}$, for $i = 1, 2, \ldots, t$,

we need to do t multiplications to check (i), $t + 1$ modular exponentiations to check (iii) and (iv), and $f(q_i)$ multiplications and modular exponentiations to check (ii), that q_i is prime for $i = 1, 2, \ldots, t$. Hence,

$$f(n) = t + (t+1) + \sum_{i=1}^{t} f(q_i)$$

$$\leq 2t + 1 + \sum_{i=1}^{t} ((3 \log q_i / \log 2) - 2).$$

Now, each multiplication requires $O((\log_2 n)^2)$ bit operations and each modular exponentiation requires $O((\log_2 n)^3)$ bit operations. Because the total number of multiplications and modular exponentiations needed is $f(n) = O(\log_2 n)$, the total number of bit operations needed is $O((\log_2 n)(\log_2 n)^3) = O((\log_2 n)^4)$. ∎

Another limited converse of Fermat's little theorem was established by Henry Pocklington in 1914. He showed that the primality of n can be established using a partial factorization of $n - 1$. We use the usual notation $n - 1 = FR$, where F represents the part of $n - 1$ *factored* into primes and R the *remaining* part not factored into primes.

Theorem 9.20. *Pocklington's Primality Test.* Suppose that n is a positive integer with $n - 1 = FR$, where $(F, R) = 1$ and $F > R$. The integer n is prime if there exists an integer a such that $(a^{(n-1)/q} - 1, n) = 1$ whenever q is a prime with $q \mid F$ and $a^{n-1} \equiv 1 \pmod{n}$.

Proof. Suppose that p is a prime divisor of n with $p \leq \sqrt{n}$. Because $a^{n-1} \equiv 1 \pmod{n}$ (where a is the integer assumed to have the properties specified in the hypotheses), if $p \mid n$, we see that $a^{n-1} \equiv 1 \pmod{p}$. It follows that $\mathrm{ord}_p\, a \mid n - 1$. Consequently, there exists an integer t such that $n - 1 = t \cdot \mathrm{ord}_p\, a$.

Now, suppose that q is a prime with $q \mid F$ and that q^e is the power of q appearing in the prime-power factorization of F. We will show that $q \nmid t$. To see this, note that if $q \mid t$, then

$$a^{(n-1)/q} = a^{\mathrm{ord}_p\, a \cdot (t/q)} \equiv 1 \pmod{p}.$$

This implies that $p \mid (a^{(n-1)/q} - 1, n)$ because $p \mid a^{(n-1)/q} - 1$ and $p \mid n$. This contradicts the hypothesis that $(a^{(n-1)/q} - 1, n) = 1$. Consequently, $q \nmid t$. It follows that $q^e \mid \mathrm{ord}_p\, a$. Because for every prime dividing F the power of this prime in the prime-power factorization of F divides $\mathrm{ord}_p\, a$, it follows that $F \mid \mathrm{ord}_p\, a$. Because $\mathrm{ord}_p\, a \mid p - 1$, it follows that $F \mid p - 1$, implying that $F < p$.

Because $F > R$ and $n - 1 = FR$, it follows that $n - 1 < F^2$. Because both $n - 1$ and F^2 are integers, we have $n \leq F^2$, so $p > F \geq \sqrt{n}$. We can conclude that n is prime. ∎

The following example illustrates the use of Pocklington's primality test, where only a partial factorization of $n - 1$ is used to show that n is prime.

Example 9.24. We will use Pocklington's primality test to show that 23801 is prime. With $n = 23801$, we can use the partial factorization of $n - 1 = 23800 = FR$, where $F = 200 = 2^3 5^2$ and $R = 119$, so that $F > R$. Taking $a = 3$, we find (with the help of

computation software) that

$$3^{23800} \equiv 1 \ (\text{mod } 23801)$$

$$3^{23800/2} \equiv -1 \ (\text{mod } 23801)$$

$$3^{23800/5} \equiv 19672 \ (\text{mod } 23801).$$

From this we find (using the Euclidean algorithm) that $(3^{23800/2} - 1, 23801) = (-2, 23801) = 1$ and $(3^{23800/5} - 1, 23801) = (19671, 23801) = 1$. This shows that $n = 23801$ is prime, even though we did not use the complete factorization of $n - 1 = 23800$ (namely, $23800 = 2^3 \cdot 5^2 \cdot 7 \cdot 17$). ◀

We can use Pocklington's primality test to develop another test, which is useful for testing the primality of numbers of special form. This test (which actually predates Pocklington's) was proved by E. Proth in 1878.

Theorem 9.21. *Proth's Primality Test.* Let n be a positive integer with $n = k2^m + 1$, where k is an odd integer and m is an integer with $k < 2^m$. If there is an integer a such that

$$a^{(n-1)/2} \equiv -1 \ (\text{mod } n),$$

then n is prime.

Proof. Let $s = 2^m$ and $t = k$, so that $s > t$ by the hypotheses. If

(9.7) $$a^{(n-1)/2} \equiv -1 \ (\text{mod } n),$$

we can easily show that $(a^{(n-1)/2} - 1, n) = 1$. To see this, note that if $d \mid (a^{(n-1)/2} - 1)$ and $d \mid n$, then by (9.7), $d \mid (a^{(n-1)/2} + 1)$. It follows that d divides $(a^{(n-1)/2} - 1) + (a^{(n-1)/2} + 1) = 2$. Because n is odd, it follows that $d = 1$. Consequently, all the hypotheses of Pocklington's primality test are satisfied, so n is prime. ∎

Example 9.25. We will use Proth's primality test to show that $n = 13 \cdot 2^8 + 1 = 3329$ is prime. First, note that $13 < 2^8 = 256$. Take $a = 3$. We find (with the help of computation software) that

$$3^{(n-1)/2} = 3^{3328/2} = 3^{1664} \equiv -1 \ (\text{mod } 3329).$$

It follows by Proth's primality test that 3329 is prime. ◀

    Proth's primality test has been used extensively to prove the primality of many large numbers of the form $k2^m + 1$. Three of the ten largest primes currently known have been found using Proth's primality test; the rest are Mersenne primes. For a few years, the largest known prime was not a Mersenne prime, but one of the form $k2^m + 1$. You can download PC-based software from the Web for running Proth's primality test, and look for new primes of the form $k2^m + 1$ yourself! If you find one you will receive some small amount of fame, but it will not make you as famous as if you found a new Mersenne prime.

9.5 Exercises

1. Show that 101 is prime using Lucas's converse of Fermat's little theorem with $x = 2$.

2. Show that 211 is prime using Lucas's converse of Fermat's little theorem with $x = 2$.

3. Show that 233 is prime using Corollary 9.18.1 with $x = 3$.

4. Show that 257 is prime using Corollary 9.18.1 with $x = 3$.

5. Show that if an integer x exists such that

$$x^{2^{2^n}} \equiv 1 \pmod{F_n}$$

and

$$x^{2^{(2^n-1)}} \not\equiv 1 \pmod{F_n},$$

then the Fermat number $F_n = 2^{2^n} + 1$ is prime.

* 6. Let n be a positive integer. Show that if the prime-power factorization of $n - 1$ is $n - 1 = p_1^{a_1} p_2^{a_2} \cdots p_t^{a_t}$, and for $j = 1, 2, \ldots, t$, there exists an integer x_j such that

$$x_j^{(n-1)/p_j} \not\equiv 1 \pmod{n}$$

and

$$x_j^{n-1} \equiv 1 \pmod{n},$$

then n is prime.

* 7. Let n be a positive integer such that

$$n - 1 = m \prod_{j=1}^{r} q_j^{a_j},$$

where m is a positive integer, $a_1, a_2, \ldots, a_r$ are positive integers, and $q_1, q_2, \ldots, q_r$ are relatively prime integers greater than one. Furthermore, let $b_1, b_2, \ldots, b_r$ be positive integers such that there exist integers $x_1, x_2, \ldots, x_r$ with

$$x_j^{n-1} \equiv 1 \pmod{n}$$

and

$$(x_j^{(n-1)/q_j} - 1, n) = 1$$

for $j = 1, 2, \ldots, r$, where every prime factor of q_j is greater than or equal to b_j for $j = 1, 2, \ldots, r$, and

$$n < \left(1 + \prod_{j=1}^{r} b_j^{a_j}\right)^2.$$

Show that n is prime.

8. Use Pocklington's primality test to show that 7057 is prime. (*Hint:* Take $F = 2^4 \cdot 3^2 = 144$ and $R = 49$ in $7057 - 1 = 7056 = FR$.)

9. Use Pocklington's primality test to show that 9929 is prime. (*Hint:* Take $F = 136 = 2^3 \cdot 17$ and $R = 73$ in $9929 - 1 = 9928 = FR$.)

10. Use Proth's primality test to show that 449 is prime.

11. Use Proth's primality test to show that 3329 is prime.

* **12.** Show that the integer n is prime if $n - 1 = FR$, where $(F, R) = 1$, B is an integer with $FB > \sqrt{n}$, and R has no prime factors less than B; for each prime q dividing F, there exists an integer a such that $a^{n-1} \equiv 1 \pmod{n}$ and $(a^{(n-1)/q} - 1, n) = 1$; and there exists an integer b greater than 1 such that $b^{n-1} \equiv 1 \pmod{n}$ and $(b^F - 1, n) = 1$.

* **13.** Suppose that $n = hq^k + 1$, where q is prime and $q^k > h$. Show that n is prime if there exists an integer a such that $a^{n-1} \equiv 1 \pmod{n}$ and $(a^{(n-1)/q} - 1, n) = 1$.

* **14.** A *Sierpinski number* is a positive odd integer k for which the integers $k2^n + 1$, where n is an integer with $n > 1$, are all composite. Show 78557 is a Sierpinski number.

9.5 Computational and Programming Exercises

Computations and Explorations

Using a computation program such as Maple or *Mathematica*, or programs you have written, carry out the following computations and explorations.

1. Use Pocklington's primality test to show that 10,998,989 is prime, with $n - 1 = FR$, where $s = 4004$, $t = 2747$, and $a = 3$.

2. Use Pocklington's primality test to show that 111,649,121 is prime.

3. Use Proth's primality test to find as many primes of the form $3 \cdot 2^n + 1$ as you can.

4. Use Proth's primality test to find as many primes of the form $5 \cdot 2^n + 1$ as possible.

5. It has been conjectured that 78557 is the smallest Sierpinski number (see Exercise 14). (Sierpinski showed in 1960 that there are infinitely many Sierpinski numbers.) Can you help verify this conjecture (if it is true) by eliminating any of the integers 4847, 5359, 10223, 19249, 21811, 22699, 24737, 27653, 28433, 33661, 55459, and 67607 from contention? To do so, you will have to find an integer n such that $k2^n + 1$ is prime, where k is an integer on this list. (You can monitor progress on this conjecture at www.seventeenorbust.com.)

6. Give a succinct certification of primality of $F_4 = 2^{2^4} + 1 = 65537$.

Programming Projects

Write programs using Maple, *Mathematica,* or a language of your choice to show that a positive integer n is prime using the following.

1. Lucas's converse of Fermat's little theorem

2. Corollary 9.18.1

3. Pocklington's primality test

4. Proth's primality test

9.6 Universal Exponents

Let n be a positive integer with prime-power factorization

$$n = p_1^{t_1} p_2^{t_2} \cdots p_m^{t_m}.$$

If a is an integer relatively prime to n, then Euler's theorem tells us that

$$a^{\phi(p^t)} \equiv 1 \pmod{p^t},$$

whenever p^t is one of the prime powers occurring in the factorization of n. As in the proof of Theorem 9.13, let

$$U = [\phi(p_1^{t_1}), \phi(p_2^{t_2}), \ldots, \phi(p_m^{t_m})],$$

the least common multiple of the integers $\phi(p_i^{t_i})$, $i = 1, 2, \ldots, m$. Because

$$\phi(p_i^{t_i}) \mid U,$$

for $i = 1, 2, \ldots, m$, using Theorem 9.1 we see that

$$a^U \equiv 1 \pmod{p_i^{t_i}},$$

for $i = 1, 2, \ldots, m$. Hence, by Corollary 4.8.1, it follows that

$$a^U \equiv 1 \pmod{n}.$$

This leads to the following definition.

Definition.   A *universal exponent* of the positive integer n is a positive integer U such that

$$a^U \equiv 1 \pmod{n},$$

for all integers a relatively prime to n.

Example 9.26.   Because the prime-power factorization of 600 is $2^3 \cdot 3 \cdot 5^2$, it follows that $U = [\phi(2^3), \phi(3), \phi(5^2)] = [4, 2, 20] = 20$ is a universal exponent of 600.   ◄

From Euler's theorem, we know that $\phi(n)$ is a universal exponent. As we have already demonstrated, the integer $U = [\phi(p_1^{t_1}), \phi(p_2^{t_2}), \ldots, \phi(p_m^{t_m})]$ is also a universal exponent of $n = p_1^{t_1} p_2^{t_2} \cdots p_m^{t_m}$. We are interested in finding the *smallest* positive universal exponent of n.

Definition.   The least universal exponent of the positive integer n is called the *minimal universal exponent of n*, and is denoted by $\lambda(n)$.

We now find a formula for the minimal universal exponent $\lambda(n)$, based on the prime-power factorization of n.

First, note that if n has a primitive root, then $\lambda(n) = \phi(n)$. Because powers of odd primes possess primitive roots, we know that

$$\lambda(p^t) = \phi(p^t),$$

whenever p is an odd prime and t is a positive integer. Similarly, we have $\lambda(2) = \phi(2) = 1$ and $\lambda(4) = \phi(4) = 2$, because both 2 and 4 have primitive roots. On the other hand, if $t \geq 3$, then we know by Theorem 9.11 that

$$a^{2^{t-2}} \equiv 1 \ (\text{mod } 2^t)$$

and $\text{ord}_{2^t} 5 = 2^{t-2}$, so that we can conclude that $\lambda(2^t) = 2^{t-2}$ if $t \geq 3$.

We have found $\lambda(n)$ when n is a power of a prime. Next, we turn our attention to arbitrary positive integers n.

Theorem 9.22. Let n be a positive integer with prime-power factorization

$$n = 2^{t_0} p_1^{t_1} p_2^{t_2} \cdots p_m^{t_m}.$$

Then $\lambda(n)$, the minimal universal exponent of n, is given by

$$\lambda(n) = [\lambda(2^{t_0}), \phi(p_1^{t_1}), \ldots, \phi(p_m^{t_m})].$$

Moreover, there exists an integer a such that $\text{ord}_n a = \lambda(n)$, the largest possible order of an integer modulo n.

Proof. Let a be an integer with $(a, n) = 1$. For convenience, let

$$M = [\lambda(2_0^2), \phi(p_1^{t_1}), \phi(p_2^{t_2}), \ldots, \phi(p_m^{t_m})].$$

Because M is divisible by all of the integers $\lambda(2^{t_0}), \phi(p_1^{t_1}) = \lambda(p_1^{t_1}), \phi(p_2^{t_2}) = \lambda(p_2^{t_2}), \ldots,$ $\phi(p_m^{t_m}) = \lambda(p_m^{t_m})$, and because $a^{\lambda(p^t)} \equiv 1 \ (\text{mod } p^t)$ for all prime powers in the factorization of n, we see that

$$a^M \equiv 1 \ (\text{mod } p^t),$$

whenever p^t is a prime power occurring in the factorization of n.

Consequently, by Corollary 4.8.1 we can conclude that

$$a^M \equiv 1 \ (\text{mod } n).$$

The last congruence established the fact that M is a universal exponent. We must now show that M is the *least* universal exponent. To do this, we find an integer a such that no positive power smaller than the Mth power of a is congruent to 1 modulo n. With this in mind, let r_i be a primitive root of $p_i^{t_j}$.

We consider the system of simultaneous congruences

$$x \equiv 5 \pmod{2^{t_0}}$$

$$x \equiv r_1 \pmod{p_1^{t_1}}$$

$$x \equiv r_2 \pmod{p_2^{t_2}}$$

$$\vdots$$

$$x \equiv r_m \pmod{p_m^{t_m}}.$$

By the Chinese remainder theorem, there is a simultaneous solution a of this system that is unique modulo $n = 2^{t_0} p_1^{t_1} p_2^{t_2} \cdots p_m^{t_m}$; we will show that $\text{ord}_n a = M$. To prove this claim, assume that N is a positive integer such that

$$a^N \equiv 1 \pmod{n}.$$

Then, if p^t is a prime-power divisor of n, we have

$$a^N \equiv 1 \pmod{p^t},$$

so that

$$\text{ord}_{p^t} a \mid N.$$

But, because a satisfies each of the $m + 1$ congruences of the system, we have

$$\text{ord}_{p^t} a = \lambda(p^t),$$

for each prime power in the factorization. Hence, by Theorem 9.1, we have

$$\lambda(p^t) \mid N,$$

for all prime powers p^t in the factorization of n. Therefore, by Corollary 4.8.1, we know that $M = [\lambda(2^{t_0}), \lambda(p_1^{t_1}), \lambda(p_2^{t_2}), \ldots, \lambda(p_m^{t_m})] \mid N$.

Because $a^M \equiv 1 \pmod{n}$ and $M \mid N$ whenever $a^N \equiv 1 \pmod{n}$, we can conclude that

$$\text{ord}_n a = M.$$

This shows that $M = \lambda(n)$ and simultaneously produces a positive integer a with $\text{ord}_n a = \lambda(n)$. ∎

Example 9.27.   Because the prime-power factorization of 180 is $2^2 \cdot 3^2 \cdot 5$, from Theorem 9.22 it follows that

$$\lambda(180) = [\phi(2^2), \phi(3^2), \phi(5)] = 12.$$

To find an integer a with $\text{ord}_{180} a = 12$, first we find primitive roots modulo 3^2 and 5. For instance, we take 2 and 3 as primitive roots modulo 3^2 and 5, respectively. Then, using the Chinese remainder theorem, we find a solution of the system of congruences

$$a \equiv 3 \pmod{4}$$
$$a \equiv 2 \pmod{9}$$
$$a \equiv 3 \pmod{5},$$

obtaining $a \equiv 83 \pmod{180}$. From the proof of Theorem 9.22, we see that $\text{ord}_{180}83 = 12$.
◄

Example 9.28.   Let $n = 2^6 \cdot 3^2 \cdot 5 \cdot 7 \cdot 13 \cdot 17 \cdot 19 \cdot 37 \cdot 73$. Then, we have

$$\begin{aligned}
\lambda(n) &= [\lambda(2^6), \phi(3^2), \phi(5), \phi(17), \phi(13), \phi(17), \phi(19), \phi(37), \phi(73)] \\
&= [2^4, 2 \cdot 3, 2^2, 2 \cdot 3, 2^2 \cdot 3, 2^4, 2 \cdot 3^2, 2^2 3^2, 2^3 3^2] \\
&= 2^4 \cdot 3^2 \\
&= 144.
\end{aligned}$$

Hence, whenever a is a positive integer relatively prime to $2^6 \cdot 3^2 \cdot 5 \cdot 7 \cdot 13 \cdot 17 \cdot 17 \cdot 19 \cdot 37 \cdot 73$, we know that $a^{144} \equiv 1 \pmod{2^6 \cdot 3^2 \cdot 5 \cdot 17 \cdot 19 \cdot 37 \cdot 37 \cdot 73}$.   ◄

Results About Carmichael Numbers   We now return to the Carmichael numbers, which we discussed in Section 6.2. Recall that a Carmichael number is a composite integer that satisfies $b^{n-1} \equiv 1 \pmod{n}$ for all positive integers b with $(b, n) = 1$. We proved that if $n = q_1 q_2 \cdots q_k$, where $q_1 q_2, \ldots, q_k$ are distinct primes satisfying $(q_j - 1) \mid (n - 1)$ for $j = 1, 2, \ldots, k$, then n is a Carmichael number. Here, we prove the converse of this result.

Theorem 9.23.   If $n > 2$ is a Carmichael number, then $n = q_1 q_2 \cdots q_k$, where the q_j are distinct primes such that $(q_j - 1) \mid (n - 1)$ for $j = 1, 2, \ldots, k$.

Proof.   If n is a Carmichael number, then

$$b^{n-1} \equiv 1 \pmod{n},$$

for all positive integers b with $(b, n) = 1$. Theorem 9.22 tells us that there is an integer a with $\text{ord}_n a = \lambda(n)$, where $\lambda(n)$ is the minimal universal exponent; and because $a^{n-1} \equiv 1 \pmod{n}$, Theorem 9.1 tells us that

$$\lambda(n) \mid (n - 1).$$

Now n must be odd, for if n were even, then $n - 1$ would be odd, but $\lambda(n)$ is even (because $n > 2$), contradicting the fact that $\lambda(n) \mid (n - 1)$.

We now show that n must be the product of distinct primes. Suppose that n has a prime-power factor p^t with $t \geq 2$. Then

$$\lambda(p^t) = \phi(p^t) = p^{t-1}(p - 1) \mid \lambda(n) = n - 1.$$

This implies that $p \mid (n - 1)$, which is impossible because $p \mid n$. Consequently, n must be the product of distinct odd primes, say

$$n = q_1 q_2 \cdots q_k.$$

We conclude the proof by noting that

$$\lambda(q_i) = \phi(q_i) = (q_j - 1) \mid \lambda(n) = n - 1.$$ ∎

We can easily prove more about the prime factorizations of Carmichael numbers.

Theorem 9.24. A Carmichael number must have at least three different odd prime factors.

Proof. Let n be a Carmichael number. Then n cannot have just one prime factor, because it is composite, and is the product of distinct primes. So assume that $n = pq$, where p and q are odd primes with $p > q$. Then

$$n - 1 = pq - 1 = (p - 1)q + (q - 1) \equiv q - 1 \not\equiv 0 \ (\text{mod } p - 1),$$

which shows that $(p - 1) \nmid (n - 1)$. Hence, n cannot be a Carmichael number if it has just two different prime factors. ∎

9.6 Exercises

1. Find $\lambda(n)$, the minimal universal exponent of n, for the following values of n.

   a) 100
   b) 144
   c) 222
   d) 884
   e) $2^4 \cdot 3^3 \cdot 5^2 \cdot 7$
   f) $2^5 \cdot 3^2 \cdot 5^2 \cdot 7^3 \cdot 11^2 \cdot 13 \cdot 17 \cdot 19$
   g) 10!
   h) 20!

2. Find all positive integers n such that $\lambda(n)$ is equal to each of the following integers.

   a) 1
   b) 2
   c) 3
   d) 4
   e) 5
   f) 6

3. Find the largest integer n with $\lambda(n) = 12$.

4. Find an integer with the largest possible order for the following moduli.

   a) 12
   b) 15
   c) 20
   d) 36
   e) 40
   f) 63

5. Show that if m is a positive integer, then $\lambda(m)$ divides $\phi(m)$.

6. Show that if m and n are relatively prime positive integers, then $\lambda(mn) = [\lambda(m), \lambda(n)]$.

7. Let n be the largest positive integer satisfying the equation $\lambda(n) = a$, where a is a fixed positive integer. Show that if m is another solution of $\lambda(m) = a$, then m divides n.

8. Suppose that n is a positive integer. How many incongruent integers are there with maximal order modulo n?

9. Show that if a and m are relatively prime integers, then the solutions of the congruence $ax \equiv b \ (\text{mod } m)$ are the integers x such that $x \equiv a^{\lambda(m)-1}b \ (\text{mod } m)$.

10. Show that if c is a positive integer greater than 1, then the integers $1^c, 2^c, \ldots, (m-1)^c$ form a complete system of residues modulo m if and only if m is square-free and $(c, \lambda(m)) = 1$.

* 11. a) Show that if c and m are positive integers and m is odd, then the congruence $x^c \equiv x \pmod{m}$ has exactly

$$\prod_{j=1}^{r} (1 + (c - 1, \phi(p_j^{a_j})))$$

incongruent solutions, where m has prime-power factorization $m = p_1^{a_1} p_2^{a_2} \cdots p_r^{a_r}$.

b) Show that $x^c \equiv x \pmod{m}$ has exactly r^r solutions if $(c - 1, \phi(m)) = 2$.

12. Use Exercise 11 to show that there are always at least nine plaintext messages that are not changed when encrypted using an RSA cipher.

* 13. Show that 561 is the only Carmichael number of the form $3pq$, where p and q are primes.

* 14. Find all Carmichael numbers of the form $5pq$, where pq are primes.

* 15. Show that there are only a finite number of Carmichael numbers of the form $n = pqr$, where p is a fixed prime, and q and r are also primes.

16. Show that the decrypting exponent d for an RSA cipher with encrypting key (e, n) can be taken to be an inverse of e modulo $\lambda(n)$.

Let n be a positive integer. When $(a, n) = 1$, we define the *generalized Fermat quotient* $q_n(a)$ by $q_n(a) \equiv (a^{\lambda(n)} - 1)/n \pmod{n}$ and $0 \leq q_n(a) < n$.

17. Show that if $(a, n) = (b, n) = 1$, then $q_n(ab) \equiv q_n(a) + q_n(b) \pmod{n}$.

18. Show that if $(a, n) = 1$, then $q_n(a + nc) \equiv q_n(a)_\lambda(n)c\bar{a} \pmod{n}$, where $\bar{a}$ is the inverse of a modulo n.

9.6 Computational and Programming Exercises

Computations and Explorations

Using a computation program such as Maple or *Mathematica,* or programs you have written, carry out the following computations and explorations.

1. Find the universal exponent of all integers less than 1000.

2. Find Carmichael numbers with at least four different prime factors.

Programming Projects

Write programs using Maple, *Mathematica,* or a language of your choice to do the following.

1. Find the minimal universal exponent of a positive integer.

2. Find an integer with the minimal universal exponent of n as its order modulo n.

3. Given a positive integer M, find all positive integers n with minimal universal exponent equal to M.

4. Solve linear congruences using the method of Exercise 9.

10

Applications of Primitive Roots and the Order of an Integer

Introduction

In this chapter, we will introduce applications that rely on the concepts of orders and primitive roots. First, we consider the problem of generating random numbers. Computers can produce random numbers using data generated by hardware or software, but they cannot create long sequences of random numbers this way. To meet the need for long sequences of random numbers in computer programs, procedures have been developed to generate numbers that pass many statistical tests that numbers selected truly at random pass. The numbers that such procedures generate are called pseudorandom numbers. We will introduce several techniques to generate pseudorandom numbers based on modular arithmetic and the concepts of the order of integers and primitive roots.

We will also introduce a public key cryptosystem, known as the ElGamal cryptosystem, defined using the concept of a primitive root of a prime. The security of this cryptosystem is based on the difficulty of the problem of finding discrete logarithms modulo a prime. We will explain how to encrypt and decrypt messages using ElGamal encryption, and how to sign messages in this cryptosystem.

Finally, we will discuss an application of the concepts of the order of an integer and of primitive roots to the splicing of telephone cables.

10.1 Pseudorandom Numbers

Numbers chosen at random are useful in many applications. Random numbers are needed for computer simulations used to study phenomena in areas such as nuclear physics, operations research, and data networking. They can be used to construct random samples so that the behavior of a system can be studied when it is impossible to test all possible cases. Random numbers are used to test the performance of computer algorithms, and to run randomized algorithms that make random choices during their execution. Random numbers are also extensively used in numerical analysis. For instance, random numbers can be used to estimate integrals using Riemann sums, a topic studied in calculus. In number theory, random numbers are used in probabilistic primality tests. In cryptography, random number have many applications, such as in generation of cryptokeys and in the execution of cryptographic protocols.

When we talk about *random numbers,* we mean the terms of a sequence of numbers in which each term is selected by chance without any dependence on the other terms of the sequence, and with a specified probability of lying in a particular interval. (It really makes no sense to say that a particular number, such as 47, is random, although it can be a term of a sequence of random numbers.) Before 1940, scientists requiring random numbers produced them by rolling dice, spinning roulette wheels, picking balls out of an urn, dealing cards, or taking random digits from tabulated data, such as census reports. In the 1940s, machines were invented to produce random numbers, and in the 1950s, computers were used to generate random numbers using random noise generators. However, random numbers produced by a mechanical process often became skewed from malfunctions in computer hardware. Another important problem was that random numbers generated using physical phenomena could not be reproduced to check the results of a computer program.

The idea of generating random numbers using computer programs instead of via mechanical method was first proposed in 1946 by *John von Neumann.* The method he suggested, called the *middle-square method,* works as follows. To generate four-digit random numbers, we start with an arbitrary four-digit number, say 6139. We square this number to obtain 37687321, and we take the middle four digits, 6873, as the second random number. We iterate this procedure to obtain a sequence of random numbers, always squaring and removing the middle four digits to obtain a new random number from the preceding one. (The square of a four-digit number has eight or fewer digits. Those with fewer than eight digits are considered eight-digit numbers by adding initial digits of 0.)

Sequences produced by the middle-square method are, in reality, not randomly chosen. When the initial four-digit number is known, the entire sequence is determined. However, the sequence of numbers produced appears to be random, and the numbers

---

JOHN VON NEUMANN (1903–1957) was born in Budapest, Hungary. In 1930, after holding several positions at universities in Germany, he came to the United States. In 1933, von Neumann became, along with Albert Einstein, one of the first members of the famous Institute for Advanced Study in Princeton, New Jersey. Von Neumann was one of the most versatile mathematical talents of the twentieth century. He invented the mathematical discipline known as game theory; using game theory, he made many important discoveries in mathematical economics. Von Neumann made fundamental contributions to the development of the first computers, and participated in the early development of atomic weapons.

produced are useful for computer simulations. The integers in sequences that have been chosen in some methodical manner, but appear to be random, are called *pseudorandom numbers.*

It turns out that the middle-square method has some unfortunate weaknesses. The most undesirable feature of this method is that, for many choices of the initial integer, the method produces the same small set of numbers over and over. For instance, starting with the four-digit integer 4100 and using the middle-square method, we obtain the sequence 8100, 6100, 2100, 4100, 8100, 6100, 2100, . . . , which only gives four different numbers before repeating.

The Linear Congruential Generation

The most commonly used method for generating pseudorandom numbers, called the *linear congruential method*, was introduced by D. H. Lehmer in 1949. It works as follows: Integers m, a, c, and x_0 are chosen so that $2 \leq a < m, 0 \leq c < m$, and $0 \leq x_0 \leq m$. The sequence of pseudorandom numbers is defined recursively by

$$x_{n+1} \equiv ax_n + c \pmod{m}, \quad 0 \leq x_{n+1} < m,$$

for $n = 0, 1, 2, 3, \ldots$. We call m the *modulus,* a the *multiplier,* c the *increment,* and x_0 the *seed* of the pseudorandom numbers generator. The following examples illustrate the linear congruential method.

Example 10.1.   When we take $m = 12, a = 3, c = 4$, and $x_0 = 5$ in the linear congruential generator, we have $x_1 \equiv 3 \cdot 5 + 4 \equiv 7 \pmod{12}$, so that $x_1 = 7$. Similarly, we find that $x_2 = 1$, because $x_2 \equiv 3 \cdot 7 + r \equiv 1 \pmod{12}$, $x_3 = 7$, because $x_3 \equiv 3 \cdot 1 + r \equiv 7 \pmod{12}$, and so on. Hence, the generator produces just three different integers before repeating. The sequence of pseudorandom numbers obtained is $5, 7, 1, 7, 1, 7, 1, \ldots$. ◄

Example 10.2.   When we take $m = 9, a = 7, c = 4$, and $x_0 = 3$ in the linear congruential generator, we obtain the sequence 3,7,8,6,1,2,0,4,5,3, . . . (as should be verified by the reader). This sequence contains nine different numbers before repeating. ◄

Remark.   For computer simulations it is often necessary to generate pseudorandom numbers between 0 and 1. We can obtain such numbers by using a linear congruential generator to produce pseudorandom numbers $x_i, i = 1, 2, 3, \ldots$ between 0 and m, and then dividing each number by m, obtaining the sequence $x_i/m, i = 1, 2, 3, \ldots$.

The following theorem tells us how to find the terms of a sequence of pseudorandom numbers generated by the linear congruential method directly from the multiplier, the increment, and the seed.

Theorem 10.1.   The terms of the sequence generated by the linear congruential method previously described are given by

$$x_k \equiv a^k x_0 + c(a^k - 1)/(a - 1) \pmod{m}, \quad 0 \leq x_k < m.$$

Proof. We prove this result using mathematical induction. For $k = 1$, the formula is obviously true, because $x_1 \equiv ax_0 + c \pmod{m}$, $0 \le x_1 < m$. Assume that the formula is valid for the kth term, so that

$$x_k \equiv a^k x_0 + c(a^k - 1)/(a - 1) \pmod{m}, \quad 0 \le x_k < m.$$

Because

$$x_{k+1} \equiv ax_k + c \pmod{m}, \quad 0 \le x_{k+1} < m,$$

we have

$$
\begin{aligned}
x_{k+1} &\equiv a(a^k x_0 + c(a^k - 1)/(a - 1)) + c \\
&\equiv a^{k+1} x_0 + c(a(a^k - 1)/(a - 1) + 1) \\
&\equiv a^{k+1} x_0 + c(a^{k+1} - 1)/(a - 1) \pmod{m},
\end{aligned}
$$

which is the correct formula for the $(k + 1)$st term. This demonstrates that the formula is correct for all positive integers k. ∎

The *period length* of a linear congruential pseudorandom number generator is the maximum length of the sequence obtained without repetition. We note that the longest possible period length for a linear congruential generator is the modulus m. The following theorem tells us when this maximum length is obtained.

Theorem 10.2. The linear congruential generator produces a sequence of period length m if and only if $(c, m) = 1$, $a \equiv 1 \pmod{p}$ for all primes p dividing m, and $a \equiv 1 \pmod{4}$ if $4 \mid m$.

Because the proof of Theorem 10.2 is complicated and quite lengthy, we omit it. The reader is referred to [Kn97] for a proof.

The Pure Multiplicative Congruential Method

The case of the linear congruential generator with $c = 0$ is of special interest because of its simplicity. In this case, the method is called the *pure multiplicative congruential method*. We specify the modulus m, multiplier a, and seed x_0. The sequence of pseudorandom numbers is defined recursively by

$$x_{n+1} \equiv ax_n \pmod{m}, \quad 0 < x_{n+1} < m.$$

In general, we can express the pseudorandom numbers generated in terms of the multiplier and seed:

$$x_n \equiv a^n x_0 \pmod{m}, \quad 0 < x_{n+1} < m.$$

If l is the period length of the sequence obtained using this pure multiplicative generator, then l is the smallest positive integer such that

$$x_0 \equiv a^l x_0 \pmod{m}.$$

If $(x_0, m) = 1$, using Corollary 4.4.1 we have

$$a^l \equiv 1 \pmod{m}.$$

From this congruence, we know that the largest possible period length is $\lambda(m)$, where $\lambda(m)$ is the minimal universal exponent modulo m.

For many applications, the pure multiplicative generator is used with the modulus m equal to the Mersenne prime $M_{31} = 2^{31} - 1$. When the modulus m is a prime, the maximum period length is $m - 1$, and this is obtained when a is a primitive root of m. To find a primitive root of M_{31} that can be used with good results, we first demonstrate that 7 is a primitive root of M_{31}.

Theorem 10.3.   The integer 7 is a primitive root of $M_{31} = 2^{31} - 1$.

Proof.   To show that 7 is a primitive root of $M_{31} = 2^{31} - 1$, it is sufficient to show that

$$7^{(M_{31}-1)/q} \not\equiv 1 \pmod{M_{31}},$$

for all prime divisors q of $M_{31} - 1$. With this information, we can conclude that $\mathrm{ord}_{M_{31}} 7 = M_{31} - 1$. To find the factorization of $M_{31} - 1$, we note that

$$
\begin{aligned}
M_{31} - 1 &= 2^{31} - 2 = 2(2^{30} - 1) = 2(2^{15} - 1)(2^{15} + 1) \\
&= 2(2^5 - 1)(2^{10} + 2^5 + 1)(2^5 + 1)(2^{10} - 2^5 + 1) \\
&= 2 \cdot 3^2 \cdot 7 \cdot 11 \cdot 31 \cdot 151 \cdot 331.
\end{aligned}
$$

If we show that

$$7^{(M_{31}-1)/q} \not\equiv 1 \pmod{M_{31}},$$

for $q = 2, 3, 7, 11, 31, 151$, and 331, then we know that 7 is a primitive root of $M_{31} = 2{,}147{,}483{,}647$. Because

$$7^{(M_{31}-1)/2} \equiv 2{,}147{,}483{,}646 \not\equiv 1 \pmod{M_{31}}$$

$$7^{(M_{31}-1)/3} \equiv 1{,}513{,}477{,}735 \not\equiv 1 \pmod{M_{31}}$$

$$7^{(M_{31}-1)/7} \equiv 120{,}536{,}285 \not\equiv 1 \pmod{M_{31}}$$

$$7^{(M_{31}-1)/11} \equiv 1{,}969{,}212{,}174 \not\equiv 1 \pmod{M_{31}}$$

$$7^{(M_{31}-1)/31} \equiv 512 \not\equiv 1 \pmod{M_{31}}$$

$$7^{(M_{31}-1)/151} \equiv 535{,}044{,}134 \not\equiv 1 \pmod{M_{31}}$$

$$7^{(M_{31}-1)/331} \equiv 1{,}761{,}885{,}083 \not\equiv 1 \pmod{M_{31}},$$

we see that 7 is a primitive root of M_{31}.   ∎

In practice, we do not want to use the primitive root 7 as the generator, because the first few integers generated are small. Instead, we find a larger primitive root using Corollary 9.4.1. We use 7^k, where $(k, M_{31} - 1) = 1$. For instance, because $(5, M_{31} - 1) = 1$, we know that $7^5 = 16{,}807$ is a primitive root. Because $(13, M_{31} - 1) = 1$, another possibility is to use $7^{13} \equiv 252{,}246{,}292 \pmod{M_{31}}$ as the multiplier.

The Square Pseudorandom Number Generator

Another example of a pseudorandom number generator is the *square pseudorandom number generator*. Given a positive integer n (the *modulus*) and an initial term x_0 (the *seed*), this generator produces a sequence of pseudorandom numbers using the congruence

$$x_{i+1} \equiv x_i^2 \pmod{n}, \quad 0 \leq x_{i+1} < n.$$

From this definition, we can easily see that

$$x_i \equiv x_0^{2^i} \pmod{n}, \quad 0 \leq x_i < n.$$

Example 10.3. Let $n = 209$ be the modulus and $x_0 = 6$ the seed of the square pseudorandom number generator. The sequence produced by this generator is

$$6, 36, 42, 92, 104, 157, 196, 169, 137, 168, 9, 81, 82, 36, 42, \ldots.$$

We see that this sequence has a a period of length 12. The first term is not part of the period. ◀

We can determine the length of the period of a square pseudorandom number generator using the concept of order modulo n, as the following theorem shows.

Theorem 10.4. The length of the period of the square pseudorandom number with seed x_0 and modulus n is $\mathrm{ord}_s 2$, where the integer s is the odd positive integer such that $\mathrm{ord}_n x_0 = 2^t s$, where t is a nonnegative integer.

Proof. We will show that $\mathrm{ord}_s 2$ divides ℓ, the length of the period of this generator. Suppose that $x_j = x_{j+\ell}$ for some integer j. Then

$$x_0^{2^j} \equiv x_0^{2^{j+\ell}} \pmod{n},$$

which implies that

$$x_0^{2^{j+\ell} - 2^j} \equiv 1 \pmod{n}.$$

Using the definition of the order of an integer modulo n, we see that

$$\mathrm{ord}_n x_0 \mid (2^{j+\ell} - 2^j),$$

or, equivalently, that

(10.1) $2^{j+\ell} \equiv 2^j \pmod{2^t s}.$

Because $2^t \mid (2^{j+\ell} - 2^j)$ and $2^{j+\ell} - 2^j = 2^j (2^\ell - 1)$, we see that $j \geq t$. By congruence (10.1) and Theorem 4.4, it follows that

$$2^{j+\ell-t} \equiv 2^{j-t} \pmod{s}.$$

Using Theorem 9.2, we see that $j + \ell - t \equiv j - t \pmod{\mathrm{ord}_2 s}$. Hence, $\ell \equiv 0 \pmod{\mathrm{ord}_2 s}$, which means that $\mathrm{ord}_2 s$ divides ℓ, the period length.

We will now show that the period ℓ divides $\text{ord}_s 2$. To show that $\text{ord}_s 2$ is a multiple of ℓ, we need only show that there are two terms x_j and $x_j = x_k$ such that $j \equiv k \pmod{\text{ord}_s 2}$. To accomplish this, we suppose that $j \equiv k \pmod{\text{ord}_s 2}$ and that $k \geq j \geq t$. By Theorem 9.2, we see that

$$2^j \equiv 2^k \pmod{s}.$$

Furthermore, we have

$$2^k \equiv 2^j \pmod{2^t},$$

because $2^k - 2^j = 2^j(2^{k-j} - 1)$ and $j \geq t$. By Corollary 4.8.1 and the fact that $(2^t, s) = 1$, we can conclude that

$$2^j \equiv 2^k \pmod{2^t s}.$$

Because $\text{ord}_n x_0 = 2^t s$, we know that

$$\text{ord}_n x_0 \mid (2^k - 2^j),$$

which means that

$$x^{2^k - 2^j} \equiv 1 \pmod{n},$$

which in turn tells us that

$$x^{2^k} \equiv x^{2^j} \pmod{n}.$$

This implies that $x_k = x_j$. We conclude that $\text{ord}_s 2$ must be a multiple of ℓ, completing the proof. ∎

Example 10.4. In Example 10.3, we used the modulus $n = 209$ and the seed $x_0 = 6$ in the square pseudorandom generator. We note that $\text{ord}_{209} 6 = 90$ (as the reader should verify). Because $90 = 2 \cdot 45$, Theorem 10.4 tells us that the period length of this generator is $\text{ord}_{45} 2 = 12$ (as the reader should verify). This is the length we observed when we listed the terms generated. ◀

How can we tell whether the terms of a sequence of pseudorandom numbers are useful for computer simulations and other applications? One method is to see whether these numbers pass statistical tests designed to determine whether a sequence has particular characteristics that a truly random sequence would most likely have. A battery of such tests can be used to evaluate pseudorandom number generators. For example, the frequencies of numbers can be tested, as can the frequencies of pairs of numbers. The frequencies of the appearance of subsequences can be checked, as can the frequency of runs of the same number of various lengths. An autocorrelation test that checks whether there are correlations of the sequence and shifted versions of it may also be helpful. These and other tests are discussed in [Kn97] and [MevaVa97].

For cryptographic applications, pseudorandom number generators must not be predictable. For example, a linear congruential pseudorandom number generator cannot be used for cryptographic applications because, in sequences generated this way, knowledge of several consecutive terms can be used to find other terms. Instead, *cryptographically secure* pseudorandom number generators must be used. These produce sequences such that the terms of the sequence are unpredictable to an adversary with limited computational resources. These notions are made more precise in [MevaVa97], and in [La90].

We have only briefly touched upon the subject of pseudorandom numbers. For a thorough discussion of pseudorandom numbers, see [Kn97], and for a survey of the relationships between pseudorandom number generators and cryptography, see the chapter by Lagarias in [Po90].

10.1 Exercises

1. Find the sequence of two-digit pseudorandom numbers generated using the middle-square method, taking 69 as the seed.

2. Find the first ten terms of the sequence of pseudorandom numbers generated by the linear congruential method with $x_0 = 6$ and $x_{n+1} \equiv 5x_n + 2 \pmod{19}$. What is the period length of this generator?

3. Find the period length of the sequence of pseudorandom numbers generated by the linear congruential method with $x_0 = 2$ and $x_{n+1} \equiv 4x_n + 7 \pmod{25}$.

4. Show that if either $a = 0$ or $a = 1$ is used for the multiplier in the linear congruential method, the result would not be a good choice for a sequence of pseudorandom numbers.

5. Using Theorem 10.2, find those integers a that give period length m, where $(c, m) = 1$, for the linear congruential generator $x_{n+1} \equiv ax_n + c \pmod{m}$, for each of the following moduli.

a) $m = 1000$ c) $m = 10^6 - 1$
b) $m = 30030$ d) $m = 2^{25} - 1$

* 6. Show that every linear congruential pseudorandom number generator can be simply expressed in terms of a linear congruential generator with increment $c = 1$ and seed 0, by showing that the terms generated by the linear congruential generator $x_{n+1} \equiv ax_n + c \pmod{m}$, with seed x_0, can be expressed as $x_n \equiv b \cdot y_n + x_0 \pmod{m}$, where $b \equiv (a - 1)x_0 + c \pmod{m}$, $y_0 = 0$, and $y_{n+1} \equiv ay_n + 1 \pmod{m}$.

7. Find the period length of the pure multiplicative pseudorandom number generator $x_n \equiv cx_{n-1} \pmod{2^{31} - 1}$ for each of the following multipliers c.

a) 2   c) 4   e) 13
b) 3   d) 5   f) 17

8. Show that the maximal possible period length for a pure multiplicative generator of the form $x_{n+1} \equiv ax_n \pmod{2^e}$, $e \geq 3$, is 2^{e-2}. Show that this is obtained when $a \equiv \pm 3 \pmod 8$.

9. Find the sequence of numbers generated by the square pseudorandom number generator with modulus 77 and seed 8.

10. Find the sequence of numbers generated by the square pseudorandom number generator with modulus 1001 and seed 5.

11. Use Theorem 10.4 to find the period length of the pseudorandom sequence in Exercise 9.

12. Use Theorem 10.4 to find the period length of the pseudorandom sequence in Exercise 10.

13. Show that longest possible period of any sequence of pseudorandom numbers generated by the square pseudorandom number generator with modulus 77, regardless of the seed chosen, is 4.

14. What is the longest possible period of any sequence of pseudorandom numbers generated by the square pseudorandom number generator with modulus 989, regardless of the seed chosen?

Another way to generate pseudorandom numbers is to use the *Fibonacci generator.* Let m be a positive integer. Two initial integers x_0 and x_1, both less than m, are specified, and the rest of the sequence is generated recursively by the congruence $x_{n+1} \equiv x_n + x_{n-1} \pmod{m}$, $0 \le x_{n+1} < m$.

15. Find the first eight pseudorandom numbers generated by the Fibonacci generator with modulus $m = 31$ and initial values $x_0 = 1$ and $x_1 = 24$.

16. Find a good choice for the multiplier a in the pure multiplicative pseudorandom number generator $x_{n+1} \equiv ax_n \pmod{101}$. (*Hint:* Find a primitive root of 101 that is not too small.)

17. Find a good choice for the multiplier a in the pure multiplicative pseudorandom number generator $x_n \equiv ax_{n-1} \pmod{2^{25} - 1}$. (*Hint:* Find a primitive root of $2^{25} - 1$ and then take an appropriate power of this root.)

18. Find the multiplier a and increment c of the linear congruential pseudorandom number generator $x_{n+1} \equiv ax_n + c \pmod{1003}$, $0 \le x_{n+1} < 1003$, if $x_0 = 1$, $x_2 = 402$, and $x_3 = 361$.

19. Find the multiplier a of the pure multiplicative pseudorandom number generator $x_{n+1} \equiv ax_n \pmod{1000}$, $0 \le x_{n+1} < 1000$, if 313 and 145 are consecutive terms generated.

20. The *discrete exponential generator* takes a positive integer x_0 as its seed and generates pseudorandom numbers $x_1, x_2, x_3, \ldots$ using the recursive definition $x_{n+1} \equiv g^{x_n} \pmod{p}$, $0 < x_{n+1} < p$, for $n = 0, 1, 2, \ldots$, where p is an odd prime and g is a primitive root modulo p.
   a) Find the sequence of pseudorandom numbers generated by the discrete exponential generator with $p = 17$, $g = 3$, and $x_0 = 2$.
   b) Find the sequence of pseudorandom numbers generated by the discrete exponential generator with $p = 47$, $g = 5$, and $x_0 = 3$.
   c) Given a term of a sequence of pseudorandom numbers generated by using a discrete exponential generator, can the previous term be found easily when the prime p and primitive root g are known?

21. Another method of generating pseudorandom numbers is to use the *power generator* with parameters m, d. Here, m is a positive integer and d is a positive integer relatively prime to $\phi(m)$. The generator starts with a positive integer x_0 as its seed and generates pseudorandom numbers $x_1, x_2, x_3, \ldots$ using the recursive definition $x_{n+1} \equiv x_n^d \pmod{m}$, $0 < x_{n+1} < m$.
   a) Find the sequence of pseudorandom numbers generated by a power generator with $m = 15$, $d = 3$, and seed $x_0 = 2$.
   b) Find the sequence of pseudorandom numbers generated by a power generator with $m = 23$, $d = 3$, and seed $x_0 = 3$.

10.1 Computational and Programming Exercises

Computations and Explorations

Using a computation program such as Maple or *Mathematica,* or programs you have written, carry out the following computations and explorations.

1. Examine the behavior of the sequence of five-digit pseudorandom numbers produced by the middle-square method, starting with different choices of the initial term.

2. Find the period length of different linear congruential pseudorandom generators of your choice.

3. How long is the period of the linear congruential pseudorandom number generator with $a = 65,539$, $c = 0$, and $m = 2^{31}$?

4. How long is the period of the linear congruential pseudorandom number generator with $a = 69,069$, $c = 1$, and $m = 2^{32}$?

5. Find a seed that produces the longest possible period length for the square pseudorandom number generator with modulus 2867.

6. Show that the square pseudorandom number generator with modulus 9,992,503 and seed 564 has a period length of 924.

7. Find the period length of different *quadratic congruential* pseudorandom number generators; that is, generators of the form $x_{n+1} \equiv (ax_n^2 + bx_n + c) \pmod{m}$, $0 \le x_{n+1} < m$, where a, b, and c are integers. Can you find conditions that guarantee that the period of this generator is m?

8. Determine the length of the period of the Fibonacci generator described in the preamble to Exercise 15 for various choices of the modulus m. Do you think this is a good generator of pseudorandom numbers?

9. There are a variety of empirical tests to measure the randomness of pseudorandom number generators. Ten such tests are described in Knuth [Kn97]. Look up these tests and apply some of them to different pseudorandom number generators.

Programming Projects

Write programs using Maple, *Mathematica,* or a language of your choice to generate pseudorandom numbers using the following generators.

1. The middle-square generator

2. The linear congruential generator

3. The pure multiplicative generator

4. The square generator

5. The Fibonacci generator (see the preamble to Exercise 15)

6. The discrete exponential generator (see Exercise 20)

7. The power generator (see Exercise 21)

10.2 The ElGamal Cryptosystem

In Chapter 8, we introduced the RSA public key cryptosystem. The security of the RSA cryptosystem is based on the difficulty of factoring integers. In this section, we introduce another public key cryptosystem known as the ElGamal cryptosystem, invented by T. ElGamal in 1985. Its security is based on the difficulty of finding discrete logarithms modulo a large prime. (Recall that if p is a prime and r is a primitive root of p, the discrete logarithm of an integer a is the exponent x for which $r^x \equiv a \pmod{p}$.)

In the ElGamal cryptosystem, each person selects a prime p, a primitive root r of p, and an integer a with $0 \leq a \leq p - 1$. This exponent is the private key, that is, it is the information kept secret by that person. The corresponding public key is (p, r, b), where b is the integer with

$$b \equiv r^a \pmod{p}, \ 0 \leq a \leq p - 1.$$

In the following example, we illustrate how keys for the ElGamal cryptosystem are selected.

Example 10.5. To generate a public and private key for the ElGamal cryptosystem, we first select a prime p. Here we will take $p = 2539$. (This four-digit prime is selected to illustrate how the cryptosystem works; in practice, a prime with several hundred digits should be used.) Next, we need a primitive root of this prime p. We select the primitive root $r = 2$ of 2539 (as the reader should verify). Next, we choose an integer a with $0 \leq a \leq 2538$. We choose $a = 14$. This exponent a is the private key. The corresponding public key is the triple $(p, r, b) = (2539, 2, 1150)$, because $b \equiv 2^{14} \equiv 1150 \pmod{2539}$. ◀

Before we encrypt a message using the ElGamal cryptosystem, we will translate letters into their numerical equivalents and then form blocks of the largest possible size (with an even number of digits), as we did when we encrypted messages in Section 8.4 using the RSA cryptosystem. (This is just one of many ways to translate messages made up of characters into integers.) To encrypt a message to be sent to the person with public key (p, r, b), we first select a random number k with $1 \leq k \leq p - 2$. For each plaintext block P, we compute the integers γ and δ with

$$\gamma \equiv r^k \pmod{p}, \quad 0 \leq \gamma \leq p - 1$$

and

$$\delta \equiv P \cdot b^k \pmod{p}, \quad 0 \leq \delta \leq p - 1.$$

The ciphertext corresponding to the plaintext block P is the ordered pair $E(P) = (\gamma, \delta)$. The plaintext message P has been hidden by multiplying it by b^k to produce δ. This hidden message is transmitted together with γ. Only the person with the secret key a can compute b^k and γ, and use this to recover the original message.

When messages are encrypted using the ElGamal cryptosystem, the ciphertext corresponding to a plaintext block is twice as long as the original plaintext block. We say that this encryption method has a *message expansion factor* of 2. The random number k

is included in the encryption procedure to increase security in several ways that we will describe later in this section.

Decrypting a message encrypted using ElGamal encryption depends on knowledge of a, the private key. The first step of the decryption of a ciphertext pair (γ, δ) is to compute $\overline{\gamma^a}$. This is done by computing γ^{p-1-a} modulo p. Then, the pair $C = (\gamma, \delta)$ is decrypted by computing

$$D(C) = \overline{\gamma^a}\delta.$$

To see that this recovers the plaintext message note that

$$D(C) \equiv \overline{\gamma^a}\delta \pmod{p}$$
$$\equiv \overline{r^{ka}} \cdot Pb^k \pmod{p}$$
$$\equiv \overline{(r^a)}^k Pb^k \pmod{p}$$
$$\equiv \overline{b^k} Pb^k \pmod{p}$$
$$\equiv \overline{b^k}b^k P \pmod{p}$$
$$\equiv P \pmod{p}.$$

Example 10.6 illustrates encryption and decryption using the ElGamal cryptosystem.

Example 10.6.   We will encrypt the message

PUBLIC KEY CRYPTOGRAPHY

using the ElGamal cryptosystem with the public key we constructed in Example 10. In Example 8.16, we encrypted this same message using the RSA cryptosystem. We translated the letters into their numerical equivalents and then grouped numbers into blocks of four decimal digits. We can use this same grouping here because the largest possible block is 2525. The blocks we obtained were

1520 0111 0802 1004
2402 1724 1519 1406
1700 1507 2423,

where the dummy letter X is translated into 23 at the end of the passage to fill out the final block.   ◄

To encrypt these blocks, we first select a random number k with $1 \leq k \leq 2537$ (we will use the same k for each block here; in practice, a different number k is chosen for each block to ensure a higher level of security). Picking $k = 1443$, we encrypt each plaintext block P in a ciphertext block, using the relationship $E(C) = (\gamma, \delta)$, with

$$\gamma \equiv 2^{1443} \equiv 2141 \pmod{2539}$$

and

$$\delta \equiv P \cdot 1150^{1443} \pmod{2539}, \quad 0 \leq \delta \leq 2538.$$

For example, the first block is encrypted to $(2141, 216)$, because

$$\gamma \equiv 2^{1443} \equiv 2141 \ (\text{mod } 2539)$$

and

$$\delta \equiv 1520 \cdot 1150^{1443} \equiv 216 \ (\text{mod } 2539).$$

When we encrypt each block, we obtain the following ciphertext message:

$$(2141, 0216) \quad (2141, 1312) \quad (2141, 1771) \quad (2141, 1185)$$
$$(2141, 2132) \quad (2141, 1177) \quad (2141, 1938) \quad (2141, 2231)$$
$$(2141, 1177) \quad (2141, 1938) \quad (2141, 1694).$$

To decrypt a ciphertext block, we compute

$$D(C) \equiv \overline{\gamma^{14}} \delta \ (\text{mod } 2539).$$

For example, to decrypt the second ciphertext block $(2141, 1312)$, we compute

$$D((2141, 1312)) \equiv \overline{2141^{14}} \cdot 1312$$
$$\equiv \overline{1430} \cdot 1312$$
$$\equiv 2452 \cdot 1312$$
$$\equiv 111 \ (\text{mod } 2539).$$

We have used the fact that 2452 is an inverse of 1430 modulo 2539. This inverse can be found using the extended Euclidean algorithm, as the reader should verify. (We have also used the fact that $2141^{14} \equiv 1430 \ (\text{mod } 2539)$.)

As mentioned, the security of the ElGamal cryptosystem is based on the difficulty of determining the private key a from the public key (p, r, b), an instance of the discrete logarithm problem, a computationally difficult problem described in Section 9.4. Breaking the ElGamal encryption method requires the recovery of a message P given the public key (p, r, b) together with the encrypted message (γ, δ) without knowledge of the private key a. Although there may be another way to do this other than solving a discrete logarithm problem, it is widely thought that this is a computationally difficult problem.

Signing Messages in the ElGamal Cryptosystem

We will describe a procedure invented by T. ElGamal in 1985 for signing messages using the ElGamal cryptosystem. Suppose that a person's public key is (p, r, b) and his private key is a, so that $b \equiv r^a \ (\text{mod } p)$. To sign a message P, the person with private key a does the following: First, he selects an integer k with $(k, p - 1) = 1$. Next, he computes γ, where

$$\gamma \equiv r^k \ (\text{mod } p), \quad 0 \le \gamma \le p - 1$$

and

$$s \equiv (P - a\gamma)\bar{k} \ (\text{mod } p - 1), \quad 0 \le s \le p - 2.$$

The signature on the message P is the pair (γ, s). Note that this signature depends on the value of the random integer k and can only be computed with knowledge of the private key a.

To see that this is a valid signature scheme, note that we know the public key (p, r, b), hence we can verify that the message came from the person who supposedly sent it. To do this, we compute

$$V_1 \equiv \gamma^s b^\gamma \pmod{p}, \quad 0 \le V_1 \le p - 1$$

and

$$V_2 \equiv r^P \pmod{p}, \quad 0 \le V_2 \le p - 1.$$

For this signature to be valid, we must have $V_1 = V_2$. If the signature is valid, then

$$V_1 \equiv \gamma^s b^\gamma \pmod{p}$$
$$\equiv \gamma^{(P-a\gamma)\overline{k}} b^\gamma \pmod{p}$$
$$\equiv (\gamma^{\overline{k}})^{P-a\gamma} b^\gamma \pmod{p}$$
$$\equiv r^{(P-a\gamma)} b^\gamma \pmod{p}$$
$$\equiv r^P \overline{r^{a\gamma}} b^\gamma \pmod{p}$$
$$\equiv r^P \overline{b^\gamma} b^\gamma \pmod{p}$$
$$\equiv r^P \pmod{p}$$
$$= V_2.$$

A different integer k should be chosen to sign each message in the ElGamal signature scheme. If the same integer k is chosen for two signatures, it can be found from these signatures, making it possible to find the private key a (see Exercise 8). Another concern is whether someone could forge a signature on a message P by selecting an integer k and computing $\gamma \equiv r^k \pmod{p}$ using the public key (p, r, b). To complete the signature, this person also would have to compute $s = (P - a\gamma)\overline{k} \pmod{p - 1}$. She cannot easily find a, because computing a from b requires that a discrete logarithm be found, namely the discrete logarithm of b with respect to r modulo p. Not knowing a, a person could select a value of s at random. The probability that this would work is only $1/p$, which is close to zero when p is large.

Example 10.7 illustrates how a message is signed using the ElGamal signature scheme.

Example 10.7. Suppose that a person has a public ElGamal key of $(p, r, b) = (2539, 2, 1150)$ with corresponding private ElGamal key $a = 14$. To sign the plaintext message $P = 111$, they first choose the integer $k = 457$, selected at random with $1 \le k \le 2538$ and $(k, 2538) = 1$. Note that $\overline{457} = 2227 \pmod{2538}$. ◄

The signature of this plaintext message 111 is found by computing

$$\gamma \equiv 2^{457} \equiv 1079 \pmod{2539}$$

and

$$s \equiv (111 - 14 \cdot 1079) \cdot 2227 \equiv 1139 \pmod{2538}.$$

Anyone who has this signature (1079, 1139) and the message 111 can verify that the signature is valid by computing

$$1150^{1079} 1079^{1139} \equiv 1158 \pmod{2539}$$

and

$$2^{111} \equiv 1158 \pmod{2539}.$$

The ElGamal signature scheme has been modified to create another signature scheme that is widely used, known as the *Digital Signature Algorithm (DSA)*. The DSA was incorporated in 1994 as a U.S. government standard, Federal Information Processing Standard (FIPS) 186, commonly known as the *Digital Signature Standard*. To learn how the ElGamal signature scheme was modified to produce the DSA, consult [St95] and [MevaVa97].

10.2 Exercises

1. Encrypt the message HAPPY BIRTHDAY using the ElGamal cryptosystem with the public key $(p, r, b) = (2551, 6, 33)$. Show how the resulting ciphertext can be decrypted using the private key $a = 13$.

2. Encrypt the message DO NOT PASS GO using the ElGamal cryptosystem with the public key $(2591, 7, 591)$. Show how the resulting ciphertext can be decrypted using the private key $a = 99$.

3. Decrypt the message (2161, 660), (2161, 1284), (2161, 1467) encrypted using the ElGamal cryptosystem with public key $(2713, 5, 193)$ corresponding to the private key 17.

4. Decrypt the message (1061, 2185), (1061, 733), (1061, 1096) encrypted using the ElGamal cryptosystem with public key $(2677, 2, 1410)$ corresponding to the private key 133.

5. Find the signature produced by the ElGamal signature scheme for the plaintext message $P = 823$ with public key $(p, r, b) = (2657, 3, 801)$, private key $a = 211$, and where the integer $k = 101$ is selected to construct the signature. Show how this signature is verified.

6. Find the signature produced by the ElGamal signature scheme for the plaintext message $P = 2525$ with public key $(p, r, b) = (2543, 5, 1615)$, private key $a = 99$, and where the integer $k = 257$ is selected to construct the signature. Show how this signature is verified.

7. Show that if the same random number k is used to encrypt two plaintext messages P_1 and P_2 using ElGamal encryption, then P_2 can be found once the plaintext message P_1 is known.

8. Show that if the same integer k is used to sign two different messages using the ElGamal signature scheme, producing signatures (γ_1, s_1) and (γ_2, s_2), the integer k can be found from these signatures as long as $s_1 \not\equiv s_2 \pmod{p-1}$. Show that once k has been found, the private key a is easily found.

10.2 Computational and Programming Exercises

Computations and Explorations

Using a computation program such as Maple or *Mathematica,* or programs you have written, carry out the following computations and explorations.

1. Construct a private key, public key pair for the ElGamal cryptosystem for each member of your class. Put together a directory of the public keys.

2. For each member of your class, encrypt a message using the ElGamal cryptosystem using the public keys published in the directory.

3. Decrypt the messages sent to you by your classmates that were encrypted using your ElGamal public key.

Programming Projects

Write programs using Maple, *Mathematica,* or a language of your choice to do the following things.

1. Encrypt messages using an ElGamal cryptosystem.

2. Decrypt messages that were encrypted using an ElGamal cryptosystem.

3. Sign messages using the ElGamal cryptosystem.

10.3 An Application to the Splicing of Telephone Cables

An interesting application of the preceding material involves the splicing of telephone cables. We base our discussion on the explosion in [Or88], relating the contents of an original article by Lawther [La35], reporting on work done for the Southwestern Bell Telephone Company.

To develop the application, we first make the following definition.

Definition. Let m be a positive integer and let a be an integer relatively prime to m. The ± 1-*exponent of a modulo m* is the smallest positive integer x such that

$$a^x \equiv \pm 1 \pmod{m}.$$

We are interested in determining the largest possible ± 1-exponent of an integer modulo m; we denote this by $\lambda_0(m)$. The following two theorems relate the value of the maximal ± 1-exponent $\lambda_0(m)$ to $\lambda(m)$, the minimal universal exponent modulo m.

First, we consider positive integers that possess primitive roots.

Theorem 10.5. If m is a positive integer, $m > 2$, with a primitive root, then the maximal ± 1-exponent $\lambda_0(m)$ equals $\phi(m)/2 = \lambda(m)/2$.

Proof. We first note that if m has a primitive root, then $\lambda(m) = \phi(m)$. By Theorem 7.6, we know that $\phi(m)$ is even, so that $\phi(m)/2$ is an integer, if $m > 2$. Euler's theorem tells us that

$$a^{\phi(m)} = (a^{\phi(m)/2})^2 \equiv 1 \pmod{m},$$

for all integers a with $(a, m) = 1$. By Exercise 13 of Section 9.3, we know that when m has a primitive root, the only solutions of $x^2 \equiv 1 \pmod{m}$ are $x \equiv \pm 1 \pmod{m}$. Hence,

$$a^{\phi(m)/2} \equiv \pm 1 \pmod{m}.$$

This implies that

$$\lambda_0(m) \leq \phi(m)/2.$$

Now, let r be a primitive root of modulo m with ± 1-exponent e. Then

$$r^e \equiv \pm 1 \pmod{m},$$

so that

$$r^{2e} \equiv 1 \pmod{m}.$$

Because $\text{ord}_m r = \phi(m)$, Theorem 9.1 tells us that $\phi(m) \mid 2e$, or equivalently, that $(\phi(m)/2) \mid e$. Hence, the maximum ± 1-exponent $\lambda_0(m)$ is at least $\phi(m)/2$. However, we know that $\lambda(m) \leq \phi(m)/2$. Consequently, $\lambda_0(m) = \phi(m)/2 = \lambda(m)/2$. ∎

We now will find the maximal ± 1-exponent of integers without primitive roots.

Theorem 10.6. If m is a positive integer without a primitive root, then the maximal ± 1-exponent $\lambda_0(m)$ equals $\lambda(m)$, the minimal universal exponent of m.

Proof. We first show that if a is an integer of order $\lambda(m)$ modulo m with ± 1-exponent e such that

$$a^{\lambda(m)/2} \not\equiv -1 \pmod{m},$$

then $e = \lambda(m)$. Consequently, once we have found such an integer a, we will have shown that $\lambda_0(m) = \lambda(m)$.

Assume that a is an integer of order $\lambda(m)$ modulo m with ± 1-exponent e such that

$$a^{\lambda(m)/2} \not\equiv -1 \pmod{m}.$$

Because $a^e \equiv \pm 1 \pmod{m}$, it follows that $a^{2e} \equiv 1 \pmod{m}$. By Theorem 9.1, we know that $\lambda(m) \mid 2e$. Because $\lambda(m) \mid 2e$ and $e \leq \lambda(m)$, either $e = \lambda(m)/2$ or $e = \lambda(m)$. To see that $e \neq \lambda(m)/2$, note that $a^e \equiv \pm 1 \pmod{m}$, but $a^{\lambda(m)/2} \not\equiv 1 \pmod{m}$, because $\text{ord}_m a = \lambda(m)$, and $a^{\lambda(m)/2} \not\equiv -1 \pmod{m}$, by hypothesis. Therefore, we can conclude tht if $\text{ord}_m a = \lambda(m)$, a has ± 1-exponent e, and $a^e \equiv -1 \pmod{m}$, then $e = \lambda(m)$.

We now find an integer a with the desired properties. Let the prime-power factorization of m be $m = 2^{t_0} p_1^{t_1} p_2^{t_2} \cdots p_s^{t_s}$. We consider several cases.

We first consider those m with at least two different odd prime factors. Among the prime powers $p_i^{t_i}$ dividing m, let $p_j^{t_j}$ be one with the smallest power of 2 dividing $\phi(p_j^{t_j})$. Let r_i be a primitive root of $p_i^{t_i}$ for $i = 1, 2, \ldots, s$. Let a be an integer satisfying the simultaneous congruences

$$a \equiv 3 \pmod{2^{t_0}},$$
$$a \equiv r_i \pmod{p_i^{t_i}} \quad \text{for all } i \text{ with } i \neq j,$$
$$a \equiv r_j^2 \pmod{p_j^{t_j}}.$$

Such an integer a is guaranteed to exist by the Chinese remainder theorem. Note that

$$\operatorname{ord}_m a = [\lambda(2^{t_0}), \phi(p_i^{t_2}), \ldots, \phi(p_j^{t_j})/2, \ldots, \phi(p_s^{t_s})],$$

and, by our choice of $p_j^{t_j}$, we know that this least common multiple equals $\lambda(m)$. Because $a \equiv r_j^2 \pmod{p_j^{t_j}}$, it follows that $a^{\phi(p_j^{t_j})/2} \equiv r_j^{\phi(p_j^{t_j})} \equiv 1 \pmod{p_j^{t_j}}$. Because $\phi(p_j^{t_j})/2 \mid \lambda(m)/2$, we know that

$$a^{\lambda(m)/2} \equiv 1 \pmod{p_j^{t_j}},$$

so that

$$a^{\lambda(m)/2} \not\equiv -1 \pmod{m}.$$

Consequently, the ± 1-exponent of a is $\lambda(m)$.

The next case that we consider deals with integers of the form $m = 2^{t_0} p^{t_1}$, where p is an odd prime, $t_1 \geq 1$ and $t_0 \geq 2$, because m has no primitive roots. When $t_0 = 2$ or 3, we have

$$\lambda(m) = [2, \phi(p_1^{t_1})] = \phi(p_1^{t_1}).$$

Let a be a solution of the simultaneous congruences

$$a \equiv 1 \pmod{4}$$
$$a \equiv r \pmod{p_t^{t_1}},$$

where r is a primitive root of $(p_1^{t_1})$. We see that $\operatorname{ord}_m a = \lambda(m)$. Because

$$a^{\lambda(m)/2} \equiv 1 \pmod{4},$$

we know that

$$a^{\lambda(m)/2} \not\equiv -1 \pmod{m}.$$

Consequently, the ± 1-exponent of a is $\lambda(m)$.

When $t_0 \leq 4$, let a be a solution of the simultaneous congruences

$$a \equiv 3 \ (\mathrm{mod}\ 2^{t_0})$$

$$a \equiv r \ (\mathrm{mod}\ p_t^{t_1});$$

the Chinese remainder theorem tells us that such an integer exists. We see that $\mathrm{ord}_m a = \lambda(m)$. Because $4 \mid \lambda(2^{t_0})$, we know that $4 \mid \lambda(m)$. Hence,

$$a^{\lambda(m)/2} \equiv 3^{\lambda(m)/2} \equiv (3^2)^{\lambda(m)/4} \equiv 1 \ (\mathrm{mod}\ 8).$$

Thus,

$$a^{\lambda(m)/2} \not\equiv -1 \ (\mathrm{mod}\ m),$$

so that the ± 1-exponent of a is $\lambda(m)$.

Finally, when $m = 2^{t_0}$ with $t_0 \geq 3$, we know from Theorem 9.12 that $\mathrm{ord}_m 5 = \lambda(m)$, but

$$5^{\lambda(m)/2} \equiv (5^2)^{\lambda(m)/4} \equiv 1 \ (\mathrm{mod}\ 8).$$

Therefore, we see that

$$5^{\lambda(m)/2} \not\equiv -1 \ (\mathrm{mod}\ m);$$

we conclude that the ± 1-exponent of 5 is $\lambda(m)$.

This finishes the argument, because we have dealt with all cases where m does not have a primitive root. ■

We now develop a system for splicing telephone cables. Telephone cables are made up of concentric layers of insulated copper wire, as illustrated in Figure 10.1, and are produced in sections of specified length.

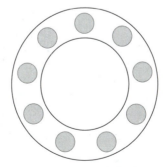

Figure 10.1 *A cross-section of one layer of a telephone cable.*

Telephone lines are constructed by splicing together sections of cable. When two wires are adjacent in the same layer in multiple sections of the cable, there are often problems with interference and crosstalk. Consequently, two wires adjacent in the same layer in one section should not be adjacent in the same layer in any nearby sections. For practical purposes, the splicing system should be simple. We use the following rules to describe the system: Wires in concentric layers are spliced to wires in the corresponding layers of the next section, following the identical splicing direction at each connection. In a layer with m wires, we connect the wire in position j in one section, where $l \leq j \leq m$, to the wire in position $S(j)$ in the next section, where $S(j)$ is the least positive residue of $1 + (j - 1)s$ modulo m. Here, s is called the *spread* of the splicing system. We see that when a wire in one section is spliced to a wire in the next section, the adjacent wire in the first section is spliced to the wire in the next section in the position obtained by counting forward s modulo m from the position of the last wire spliced in this section. To have a one-to-one correspondence between wires of adjacent sections, we require that the spread s be relatively prime to the number of wires m. This shows that if wires in positions j and k are sent to the same wire in the next section, then $S(j) = S(k)$ and

$$1 + (j - 1)s \equiv 1 + (k - 1)s \pmod{m},$$

so that $js \equiv ks \pmod{m}$. Because $(m, s) = 1$, from Corollary 4.4.1 we see that $j \equiv k \pmod{m}$, which is impossible.

Example 10.8.   Let us connect nine wires with a spread of 2. We have the correspondence

$$
\begin{array}{lll}
1 \to 1 & 2 \to 3 & 3 \to 5 \\
4 \to 7 & 5 \to 9 & 6 \to 2 \\
7 \to 4 & 8 \to 6 & 9 \to 8,
\end{array}
$$

as illustrated in Figure 10.2.   ◀

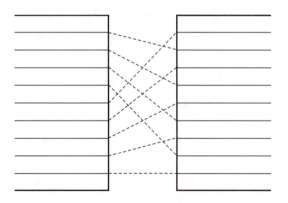

Figure 10.2 *Splicing of nine wires with a spread of 2.*

The following result tells us the correspondence of wires in the first section of cable to the wires in the nth section.

Theorem 10.7.   Let $S_n(j)$ denote the position of the wire in the nth section spliced to the jth wire of the first section. Then

$$S_n(j) \equiv 1 + (j - 1)s^{n-1} \pmod{m}.$$

Proof. For $n = 2$, by the rules for the splicing system, we have

$$S_2(j) \equiv 1 + (j - 1)s \pmod{m},$$

so the proposition is true for $n = 2$. Now assume that

$$S_n(j) \equiv 1 + (j - 1)s^{n-1} \pmod{m}.$$

Then, in the next section, we have the wire in position $S_n(j)$ spliced to the wire in position.

$$\begin{aligned} S_{n+1}(j) &\equiv 1 + (S_n(j) - 1)s \\ &\equiv 1 + ((j - 1)s^{n-1})s \\ &\equiv 1 + (j - 1)s^n \pmod{m}. \end{aligned}$$

This shows that the proposition is true. ∎

In the splicing system, we want to have wires adjacent in one section separated as long as possible in the following sections. Theorem 10.7 tells us that after n splices, the adjacent wires in the jth and $(j + 1)$th positions are connected to wires in positions $S_n(j) \equiv 1 + (j - 1)s^n \pmod{m}$ and $S_n(j + 1) \equiv 1 + js^n \pmod{m}$, respectively. These wires are adjacent in the nth section if, and only if,

$$S_n(j) - S_n(j + 1) \equiv \pm 1 \pmod{m},$$

or, equivalently,

$$(1 + (j - 1)s^n) - (1 + js^n) \equiv \pm 1 \pmod{m},$$

which holds if and only if

$$s^n \equiv \pm 1 \pmod{m}.$$

We can now apply the material at the beginning of this section. To keep wires that are adjacent in the first section separated as long as possible thereafter, we should pick for the spread s an integer with maximal ± 1-exponent $\lambda_0(m)$.

Example 10.9. With 100 wires, we should choose a spread s so that the ± 1-exponent of s is $\lambda_0(100) = \lambda(100) = 20$. The appropriate computations show that $s = 3$ is such a spread. ◀

10.3 Exercises

1. Find the maximal ± 1-exponent of each of the following positive integers.

a) 17   c) 24   e) 99
b) 22   d) 36   f) 100

2. Find an integer with maximal ± 1-exponent modulo each of the following positive integers.

    a) 13   c) 15   e) 36

    b) 14   d) 25   f) 60

3. Devise a splicing scheme for telephone cables containing each of the following number of wires.

    a) 50 wires   b) 76 wires   c) 125 wires

* 4. Show that using any splicing system of telephone cables with m wires arranged in a concentric layer, adjacent wires in one section can be kept separated in at most $[(m-1)/2]$ successive sections of cable. Show that when m is prime, this upper limit is achieved using the system developed in this section.

10.3 Computational and Programming Exercises

Computations and Explorations

Using a computation program such as Maple or *Mathematica,* or programs you have written, carry out the following computations and explorations.

1. Find the maximal ± 1-exponent of each positive integer less than 1000.

Programming Projects

Write programs using Maple, *Mathematica,* or a language of your choice to generate pseudorandom numbers using the following generators.

1. Given an integer m, find the maximal ± 1-exponent of m.

2. Develop a scheme for splicing telephone cables as described in this section.

11

Quadratic Residues

Introduction

When is an integer a a perfect square modulo a prime p? The work of the great number theorists Euler, Legendre, and Gauss on this and related questions led to the development of much of modern number theory. In this chapter, we develop results, both old and new, created in the study of such questions. We first define the concept of a quadratic residue, an integer a that is a square modulo p, and establish basic properties of quadratic residues. We introduce the Legendre symbol, a notation that tells us whether an integer is a quadratic residue of p, and develop its basic properties. We state and prove two important criteria, discovered by Euler and by Gauss, for determining whether a is a quadratic residue modulo p, and use these criteria to determine whether -1 and 2 are quadratic residues of p.

We also show that an integer that is a perfect square modulo pq, where p and q are primes, has exactly four incongruent square roots modulo pq. Modular square roots are used extensively in cryptography, such as in a protocol for fairly choosing a random bit ("flipping a coin electronically"). We will also illustrate (in the last section of the chapter) how modular square roots can be used in an interactive protocol to show that a person has some secret information, without revealing this information.

Suppose that p and q are distinct odd primes. We can ask whether p is a square modulo q and whether q is a square modulo p. Is there any relationship between the answers to these two questions? In this chapter, we will show that these answers are closely related in a way specified by the famous theorem called the law of quadratic reciprocity. This law was observed by Euler and Legendre, and ultimately proved by Gauss at the end of the eighteenth century. We will present one of the many proofs of this famous theorem, selected because it is one of the easiest to understand. The law of quadratic reciprocity has both theoretical and practical implications. We show how it can be used in computations and to prove useful results, such as Pepin's test, which can be used to determine whether Fermat numbers are prime.

The Legendre symbol, which tells us whether an integer is a quadratic residue modulo p, can be generalized to the Jacobi symbol. We will establish the basic properties of Jacobi symbols and show that they satisfy a reciprocity law that is a consequence of the law of quadratic reciprocity. We show how Jacobi symbols can be used to simplify computations of Legendre symbols. We also use Jacobi symbols to introduce a particular type of pseudoprime, known as an Euler pseudoprime, which is an integer that masquerades as a prime by satisfying Euler's criteria for quadratic residues. We will use this concept to develop a probabilistic primality test.

11.1 Quadratic Residues and Nonresidues

Let p be an odd prime and a an integer relatively prime to p. In this chapter, we devote our attention to the question: Is a a perfect square modulo p? We begin with a definition.

Definition.   If m is a positive integer, we say that the integer a is a *quadratic residue of m* if $(a, m) = 1$ and the congruence $x^2 \equiv a \pmod{m}$ has a solution. If the congruence $x^2 \equiv a \pmod{m}$ has no solution, we say that a is a *quadratic nonresidue of m.*

Example 11.1.   To determine which integers are quadratic residues of 11, we compute the squares of the integers $1, 2, 3, \ldots, 10$. We find that $1^2 \equiv 10^2 \equiv 1 \pmod{11}$, $2^2 \equiv 9^2 \equiv 4 \pmod{11}$, $3^2 \equiv 8^2 \equiv 9 \pmod{11}$, $4^2 \equiv 7^2 \equiv 5 \pmod{11}$, and $5^2 \equiv 6^2 \equiv 3 \pmod{11}$. Hence, the quadratic residues of 11 are $1, 3, 4, 5, 9$; the integers $2, 6, 7, 8, 10$ are quadratic nonresidues of 11.   ◄

Note that the quadratic residues of the positive integer m are just the kth power residues of m with $k = 2$, as defined in Section 9.4. We will show that if p is an odd prime, then there are exactly as many quadratic residues as quadratic nonresidues of p among the integers $1, 2, \ldots, p - 1$. To demonstrate this fact, we use the following lemma.

Lemma 11.1.   Let p be an odd prime and a an integer not divisible by p. Then, the congruence

$$x^2 \equiv a \pmod{p}$$

has either no solutions or exactly two incongruent solutions modulo p.

Proof.   If $x^2 \equiv a \pmod{p}$ has a solution, say $x = x_0$, then we can easily demonstrate that $x = -x_0$ is a second incongruent solution. Because $(-x_0)^2 = x_0^2 \equiv a \pmod{p}$, we see that $-x_0$ is a solution. We note that $x_0 \not\equiv -x_0 \pmod{p}$, for if $x_0 \equiv -x_0 \pmod{p}$, then we have $2x_0 \equiv 0 \pmod{p}$. This is impossible because p is odd and $p \nmid x_0$ because $x_0^2 \equiv a \pmod{p}$ and $p \nmid a$.

To show that there are no more than two incongruent solutions, assume that $x = x_0$ and $x = x_1$ are both solutions of $x^2 \equiv a \pmod{p}$. Then we have $x_0^2 \equiv x_1^2 \equiv a \pmod{p}$, so that $x_0^2 - x_1^2 = (x_0 + x_1)(x_0 - x_1) \equiv 0 \pmod{p}$. Hence, $p \mid (x_0 + x_1)$ or $p \mid (x_0 - x_1)$, so that $x_1 \equiv -x_0 \pmod{p}$ or $x_1 \equiv x_0 \pmod{p}$. Therefore, if there is a solution of $x^2 \equiv a \pmod{p}$, there are exactly two incongruent solutions. ∎

This leads us to the following theorem.

Theorem 11.1. If p is an odd prime, then there are exactly $(p-1)/2$ quadratic residues of p and $(p-1)/2$ quadratic nonresidues of p among the integers $1, 2, \ldots, p-1$.

Proof. To find all the quadratic residues of p among the integers $1, 2, \ldots, p-1$, we compute the least positive residues modulo p of the squares of the integers $1, 2, \ldots, p-1$. Because there are $p-1$ squares to consider, and because each congruence $x^2 \equiv a \pmod{p}$ has either zero or two solutions, there must be exactly $(p-1)/2$ quadratic residues of p among the integers $1, 2, \ldots, p-1$. The remaining $p - 1 - (p-1)/2 = (p-1)/2$ positive integers less than $p-1$ are quadratic nonresidues of p. ∎

Primitive roots and indices, studied in Chapter 9, provide an alternative method for proving results about quadratic residues.

Theorem 11.2. Let p be a prime and let r be a primitive root of p. If a is an integer not divisible by p, then a is a quadratic residue of p if $\text{ind}_r a$ is even, and a is a quadratic nonresidue of p if $\text{ind}_r a$ is odd.

Proof. Suppose that $\text{ind}_r a$ is even. Then $(r^{\text{ind}_r a/2})^2 \equiv a \pmod{p}$, which shows that a is a quadratic residue of p. Now suppose that a is a quadratic residue of p. Then there exists an integer x such that $x^2 \equiv a \pmod{p}$. It follows that $\text{ind}_r x^2 = \text{ind}_r a$. By Part (iii) of Theorem 9.16, it follows that $2 \cdot \text{ind}_r x \equiv \text{ind}_r a \pmod{\phi(p)}$, so $\text{ind}_r a$ is even. We have shown that a is a quadratic residue of p if and only if $\text{ind}_r a$ is even. It follows that a is a quadratic nonresidue of p if and only if $\text{ind}_r a$ is odd. ∎

Note that by Theorem 11.2, every primitive root of an odd prime p is a quadratic nonresidue of p.

We illustrate how the relationship between primitive roots and indices and quadratic residues can be used to prove results about quadratic residues by giving an alternative proof of Theorem 11.1.

Proof. Let p be an odd prime with primitive root r. By Theorem 11.2, the quadratic residues of p among the integers $1, 2, \ldots, p-1$ are those with even index to the base r. It follows that the quadratic residues of a in this set are the least positive residues of r^k, where k is an even integer with $1 \le k \le p-1$. The result follows because there are exactly $(p-1)/2$ such integers. ∎

The special notation associated with quadratic residues is described in the following definition.

Definition.   Let p be an odd prime and a be an integer not divisible by p. The *Legendre symbol* $\left(\frac{a}{p}\right)$ is defined by

$$\left(\frac{a}{p}\right) = \begin{cases} 1 & \text{if } a \text{ is a quadratic residue of } p; \\ -1 & \text{if } a \text{ is a quadratic nonresidue of } p. \end{cases}$$

   This symbol is named after the French mathematician *Adrien-Marie Legendre*, who introduced the use of this notation.

Example 11.2.   The previous example shows that the Legendre symbols $\left(\frac{a}{11}\right)$, $a = 1, 2, \ldots, 10$, have the following values:

$$\left(\frac{1}{11}\right) = \left(\frac{3}{11}\right) = \left(\frac{4}{11}\right) = \left(\frac{5}{11}\right) = \left(\frac{9}{11}\right) = 1,$$

$$\left(\frac{2}{11}\right) = \left(\frac{6}{11}\right) = \left(\frac{7}{11}\right) = \left(\frac{8}{11}\right) = \left(\frac{10}{11}\right) = -1. \qquad \blacktriangleleft$$

We now present a criterion for deciding whether an integer is a quadratic residue of a prime. This criterion is useful in demonstrating properties of the Legendre symbol.

Theorem 11.3. *Euler's Criterion.*   Let p be an odd prime and let a be a positive integer not divisible by p. Then

$$\left(\frac{a}{p}\right) \equiv a^{(p-1)/2} \pmod{p}.$$

Proof.   First, assume that $\left(\frac{a}{p}\right) = 1$. Then, the congruence $x^2 \equiv a \pmod{p}$ has a solution, say $x = x_0$. Using Fermat's little theorem, we see that

$$a^{(p-1)/2} = \left(x_0^2\right)^{(p-1/2)} = x_0^{p-1} \equiv 1 \pmod{p}.$$

Hence, if $\left(\frac{a}{p}\right) = 1$, we know that $\left(\frac{a}{p}\right) \equiv a^{(p-1)/2} \pmod{p}$.

ADRIEN-MARIE LEGENDRE (1752–1833) was born into a well-to-do family. He was a professor at the École Militaire in Paris from 1775 to 1780. In 1795, he was appointed professor at the École Normale. His memoir *Recherches d'Analyse Indetermineé*, published in 1785, contains a discussion of the law of quadratic reciprocity, a statement of Dirichlet's theorem on primes in arithmetic progressions, and a discussion of the representation of positive integers as the sum of three squares. He established the $n = 5$ case of Fermat's last theorem. Legendre wrote a textbook on geometry, *Eléments de géométrie*, that was used for more than 100 years, and served as a model for other textbooks. Legendre made fundamental discoveries in mathematical astronomy and geodesy, and gave the first treatment of the law of least squares.

Now consider the case where $\left(\frac{a}{p}\right) = -1$. Then, the congruence $x^2 \equiv a \pmod{p}$ has no solutions. By Theorem 4.10, for each integer i with $(i, p) = 1$ there is an integer j such that $ij \equiv a \pmod{p}$. Furthermore, because the congruence $x^2 \equiv a \pmod{p}$ has no solutions, we know that $i \neq j$. Thus, we can group the integers $1, 2, \ldots, p - 1$ into $(p - 1)/2$ pairs, each with product a. Multiplying these pairs together, we find that

$$(p - 1)! \equiv a^{(p-1)/2} \pmod{p}.$$

Because Wilson's theorem tells us that $(p - 1)! \equiv -1 \pmod{p}$, we see that

$$-1 \equiv a^{(p-1)/2} \pmod{p}.$$

In this case, we also have $\left(\frac{a}{p}\right) \equiv a^{(p-1)/2} \pmod{p}$. ∎

Example 11.3. Let $p = 23$ and $a = 5$. Because $5^{11} \equiv -1 \pmod{23}$, Euler's criterion tells us that $\left(\frac{5}{23}\right) = -1$. Hence, 5 is a quadratic nonresidue of 23. ◄

We now prove some properties of the Legendre symbol.

Theorem 11.4. Let p be an odd prime and a and b be integers not divisible by p. Then

(i) if $a \equiv b \pmod{p}$, then $\left(\frac{a}{p}\right) = \left(\frac{b}{p}\right)$.

(ii) $\left(\frac{a}{p}\right)\left(\frac{b}{p}\right) = \left(\frac{ab}{p}\right)$.

(iii) $\left(\frac{a^2}{p}\right) = 1$.

Proof of (i). If $a \equiv b \pmod{p}$, then $x^2 \equiv a \pmod{p}$ has a solution if and only if $x^2 \equiv b \pmod{p}$ has a solution. Hence $\left(\frac{a}{p}\right) = \left(\frac{b}{p}\right)$.

Proof of (ii). By Euler's criterion, we know that

$$\left(\frac{a}{p}\right) \equiv a^{(p-1)/2} \pmod{p}, \quad \left(\frac{b}{p}\right) \equiv b^{(p-1)/2} \pmod{p},$$

and

$$\left(\frac{ab}{p}\right) \equiv (ab)^{(p-1)/2} \pmod{p}.$$

Hence,

$$\left(\frac{a}{p}\right)\left(\frac{b}{p}\right) \equiv a^{(p-1)/2}b^{(p-1)/2} = (ab)^{(p-1)/2} \equiv \left(\frac{ab}{p}\right) \pmod{p}.$$

Because the only possible values of a Legendre symbol are ± 1, we conclude that

$$\left(\frac{a}{p}\right)\left(\frac{b}{p}\right) = \left(\frac{ab}{p}\right).$$

Proof of (iii).   Because $\left(\frac{a}{p}\right) = \pm 1$, from part (ii) it follows that

$$\left(\frac{a^2}{p}\right) = \left(\frac{a}{p}\right)\left(\frac{a}{p}\right) = 1.$$ ∎

Part (ii) of Theorem 11.4 has the following interesting consequence. The product of two quadratic residues, or of two quadratic nonresidues, of a prime is a quadratic residue of that prime, whereas the product of a quadratic residue and a quadratic nonresidue of a prime is a quadratic nonresidue.

Relatively simple proofs of Theorems 11.3 and 11.4 can be constructed using the concepts of primitive roots and indices, together with Theorem 11.2. (See Exercises 30 and 31 at the end of this section.)

When is −1 a Quadratic Residue of the Prime p?

For which odd primes not exceeding 20 is −1 a quadratic residue? Since $2^2 \equiv -1$ (mod 5), $5^2 \equiv -1$ (mod 13) and $4^2 \equiv -1$ (mod 17), we see that −1 is a quadratic residue of 5, 13, and 17. However it is easy to see (as the reader should verify) that the congruence $x^2 \equiv -1$ (mod p) has no solution when $p = 3, 7, 11$, and 19. This evidence leads to the conjecture that −1 is a quadratic residue of the prime p if and only if $p \equiv 1$ (mod 4).

Using Euler's criterion, we can prove this conjecture.

Theorem 11.5.   If p is an odd prime, then

$$\left(\frac{-1}{p}\right) = \begin{cases} 1 & \text{if } p \equiv 1 \text{ (mod 4)}; \\ -1 & \text{if } p \equiv -1 \text{ (mod 4)}. \end{cases}$$

Proof.   By Euler's criterion, we know that

$$\left(\frac{-1}{p}\right) \equiv (-1)^{(p-1)/2} \text{ (mod } p).$$

If $p \equiv 1$ (mod 4), then $p = 4k + 1$ for some integer k. Thus,

$$(-1)^{(p-1)/2} = (-1)^{2k} = 1,$$

so that $\left(\frac{-1}{p}\right) = 1$. If $p \equiv 3$ (mod 4), then $p = 4k + 3$ for some integer k. Thus,

$$(-1)^{(p-1)/2} = (-1)^{2k+1} = -1,$$

so that $\left(\frac{-1}{p}\right) = -1$.   ∎

Gauss's Lemma

The following elegant result of Gauss provides another criterion to determine whether an integer a relatively prime to the prime p is a quadratic residue of p.

Lemma 11.2. *Gauss's Lemma.* Let p be an odd prime and a an integer with $(a, p) = 1$. If s is the number of least positive residues of the integers $a, 2a, 3a, \ldots, ((p-1)/2)a$ that are greater than $p/2$, then $\left(\frac{a}{p}\right) = (-1)^s$.

Proof. Consider the integers $a, 2a, \ldots, ((p-1)/2)a$. Let $u_1, u_2, \ldots, u_s$ be the least positive residues of those that are greater than $p/2$, and let $v_1, v_2, \ldots, v_t$ be the least positive residues of those integers that are less than $p/2$. Because $(ja, p) = 1$ for all j with $1 \leq j \leq (p-1)/2$, these least positive residues are in the set $1, 2, \ldots, p-1$.

We will show that $p - u_1, p - u_2, \ldots, p - u_s, v_1, v_2, \ldots, v_t$ comprise the set of integers $1, 2, \ldots, (p-1)/2$, in some order. To see this, we need only show that no two of these integers are congruent modulo p, because there are exactly $(p-1)/2$ numbers in the set, and all are positive integers not exceeding $(p-1)/2$.

Clearly, no two of the u_i are congruent modulo p and no two of the v_j are congruent modulo p; if a congruence of either of these two sorts held, we would have $ma \equiv na$ (mod p), where m and n are both positive integers not exceeding $(p-1)/2$. Because $p \nmid a$, this would imply that $m \equiv n$ (mod p), which is impossible.

In addition, one of the integers $p - u_i$ cannot be congruent to a v_j, for if such a congruence held, we would have $ma \equiv p - na$ (mod p), so that $ma \equiv -na$ (mod p). Because $p \nmid a$, this would imply that $m \equiv -n$ (mod p), which is impossible because both m and n are in the set $1, 2, \ldots, (p-1)/2$.

Now that we know that $p - u_1, p - u_2, \ldots, p - u_s, v_1, v_2, \ldots, v_t$ are the integers $1, 2, \ldots, (p-1)/2$, in some order, we conclude that

$$(p - u_1)(p - u_2) \cdots (p - u_s)v_1 v_2 \cdots v_t \equiv \left(\frac{p-1}{2}\right)! \,(\text{mod } p),$$

which implies that

(11.1) $$(-1)^s u_1 u_2 \cdots u_s v_1 v_2 \cdots v_t \equiv \left(\frac{p-1}{2}\right)! \,(\text{mod } p).$$

But, because $u_1, u_2, \ldots, u_s, v_1, v_2, \ldots, v_t$ are the least positive residues of $a, 2a, \ldots, ((p-1)/2)a$ we also know that

(11.2) $$u_1 u_2 \cdots u_s v_1 v_2 \cdots v_t \equiv a \cdot 2a \cdots ((p-1)/2))a$$
$$= a^{\frac{p-1}{2}} ((p-1)/2)! \,(\text{mod } p).$$

Hence, from (11.1) and (11.2), we see that

$$(-1)^s a^{\frac{p-1}{2}} ((p-1)/2)! \equiv ((p-1)/2)! \,(\text{mod } p)$$

Because $(p, ((p-1)/2)!) = 1$, this congruence implies that

$$(-1)^s a^{\frac{p-1}{2}} \equiv 1 \,(\text{mod } p).$$

By multiplying both sides by $(-1)^s$, we obtain

$$a^{\frac{p-1}{2}} \equiv (-1)^s \pmod{p}.$$

Because Euler's criterion tells us that $a^{\frac{p-1}{2}} \equiv \left(\frac{a}{p}\right) \pmod{p}$, it follows that

$$\left(\frac{a}{p}\right) \equiv (-1)^s \pmod{p},$$

establishing Gauss's lemma.   ∎

Example 11.4.   Let $a = 5$ and $p = 11$. To find $\left(\frac{5}{11}\right)$ by Gauss's lemma, we compute the least positive residues of $1 \cdot 5, 2 \cdot 5, 3 \cdot 5, 4 \cdot 5$, and $5 \cdot 5$. These are $5, 10, 4, 9$, and 3, respectively. Because exactly two of these are greater than $11/2$, Gauss's lemma tells us that $\left(\frac{5}{11}\right) = (-1)^2 = 1$.   ◀

When is 2 a Quadratic Residue of a Prime p?

For which odd primes not exceeding 50 is 2 a quadratic residue? Since $3^2 \equiv 2$ (mod 7), $6^2 \equiv 2$ (mod 17), $5^2 \equiv 2$ (mod 23), $8^2 \equiv 2$ (mod 31), $17^2 \equiv 2$ (mod 41), and $7^2 \equiv 2$ (mod 47), we see that 2 is a quadratic residue of $7, 17, 23, 31, 41$, and 47. However (as the reader should verify) $x^2 \equiv 2 \pmod{p}$ has no solution when $p = 3, 5, 11, 13, 19, 29, 37$, and 43. Is there a pattern to the primes p for which 2 is a quadratic residue modulo p? Examining these primes and noting that whether 2 is a quadratic residue of p seems to depend on the congruence of p modulo 8, we conjecture that 2 is a quadratic residue of the odd prime p if and only if $p \equiv \pm 1 \pmod{8}$. Using Gauss's lemma, we can prove this conjecture.

Theorem 11.6.   If p is an odd prime, then

$$\left(\frac{2}{p}\right) = (-1)^{(p^2-1)/8}.$$

Hence, 2 is a quadratic residue of all primes $p \equiv \pm 1 \pmod{8}$ and a quadratic nonresidue of all primes $p \equiv \pm 3 \pmod{8}$.

Proof.   By Gauss's lemma, we know that if s is the number of least positive residues of the integers

$$1 \cdot 2, \ 2 \cdot 2, \ 3 \cdot 2, \ \ldots, \ ((p-1)/2) \cdot 2$$

that are greater than $p/2$, then $\left(\frac{2}{p}\right) = (-1)^s$. Because all of these integers are less than p, we need only count those greater than $p/2$ to find how many have least positive residues greater than $p/2$.

The integer $2j$, where $1 \leq j \leq (p-1)/2$, is less than $p/2$ when $j \leq p/4$. Hence, there are $[p/4]$ integers in the set less than $p/2$. Consequently, there are

$s = (p-1)/2 - [p/4]$ greater than $p/2$. Therefore, by Gauss's lemma, we see that

$$\left(\frac{2}{p}\right) = (-1)^{\frac{p-1}{2} - [p/4]}.$$

To prove the theorem, it is enough to show that for every odd integer p,

(11.3)
$$\frac{p-1}{2} - [p/4] \equiv \frac{p^2 - 1}{8} \pmod{2}.$$

Note that (11.3) holds for a positive integer p if and only if it holds for $p + 8$. This follows because

$$\frac{(p+8)-1}{2} - [(p+8)/4] = \left(\frac{p-1}{2} + 4\right) - ([p/4] + 2) \equiv \frac{p-1}{2} - [p/4] \pmod{2}$$

and

$$\frac{(p+8)^2 - 1}{8} = \frac{p^2 - 1}{8} + 2p + 8 \equiv \frac{p^2 - 1}{8} \pmod{2}.$$

Thus we can conclude that (11.3) holds for every odd integer n if it holds for $p = \pm 1$ and ± 3. We leave it to the reader to verify that (11.3) holds for these four values of p.

It follows that for every prime p we have $\left(\frac{2}{p}\right) = (-1)^{(p^2-1)/8}$.

From the computations of the congruence class of $(p^2 - 1)/8 \pmod 2$, we see that $\left(\frac{2}{p}\right) = 1$ if $p \equiv \pm 1 \pmod 8$, while $\left(\frac{2}{p}\right) = -1$ if $p \equiv \pm 3 \pmod 8$. ∎

Example 11.5. By Theorem 11.6, we see that

$$\left(\frac{2}{7}\right) = \left(\frac{2}{17}\right) = \left(\frac{2}{23}\right) = \left(\frac{2}{31}\right) = 1,$$

whereas

$$\left(\frac{2}{3}\right) = \left(\frac{2}{5}\right) = \left(\frac{2}{11}\right) = \left(\frac{2}{13}\right) = \left(\frac{2}{19}\right) = \left(\frac{2}{29}\right) = -1. \quad ◄$$

We now present an example to show how to evaluate some Legendre symbols.

Example 11.6. To evaluate $\left(\frac{317}{11}\right)$, we use part (i) of Theorem 11.4 to obtain

$$\left(\frac{317}{11}\right) = \left(\frac{9}{11}\right) = \left(\frac{3}{11}\right)^2 = 1,$$

because $317 \equiv 9 \pmod{11}$.

To evaluate $\left(\frac{89}{13}\right)$, because $89 \equiv -2 \pmod{13}$, we have

$$\left(\frac{89}{13}\right) = \left(\frac{-2}{13}\right) = \left(\frac{-1}{13}\right)\left(\frac{2}{13}\right).$$

Because $13 \equiv 1$ (mod 4), Theorem 11.5 tells us that $\left(\frac{-1}{13}\right) = 1$. Because $13 \equiv -3$ (mod 8), we see from Theorem 11.6 that $\left(\frac{2}{13}\right) = -1$. Consequently, $\left(\frac{89}{13}\right) = -1$.   ◀

In the next section we will state and prove one of the most intriguing and challenging results of elementary number theory, the *law of quadratic reciprocity*. This theorem relates the values of $\left(\frac{p}{q}\right)$ and $\left(\frac{q}{p}\right)$, where p and q are odd primes. The law of quadratic reciprocity has many implications, both theoretical and practical, as we will see throughout this chapter. From a computational standpoint, we will see that it can help us evaluate Legendre symbols.

Modular Square Roots

Suppose that $n = pq$, where p and q are distinct odd primes, and suppose that the congruence $x^2 \equiv a$ (mod n), where $0 < a < n$ and $(a, n) = 1$, has a solution $x = x_0$. We will show that there are exactly four incongruent solutions modulo n. In other words, we will show that a has four incongruent *square roots modulo* n. To see this, let $x_0 \equiv x_1$ (mod p), $0 < x_1 < p$, and let $x_0 \equiv x_2$ (mod q), $0 < x_2 < q$. Then the congruence $x^2 \equiv a$ (mod p) has exactly two incongruent solutions modulo p, namely $x \equiv x_1$ (mod p) and $x \equiv p - x_1$ (mod p). Similarly, the congruence $x_2 \equiv a$ (mod q) has exactly two incongruent solutions modulo q, namely $x \equiv x_2$ (mod q) and $x \equiv q - x_2$ (mod q).

From the Chinese remainder theorem, there are exactly four incongruent solutions of the congruence $x^2 \equiv a$ (mod n); these four incongruent solutions are the unique solutions modulo pq of the four sets of simultaneous congruences:

(i) $x \equiv x_1$ (mod p)   (iii) $x \equiv p - x_1$ (mod p)
       $x \equiv x_2$ (mod q), $x \equiv x_2$ (mod q),

(ii) $x \equiv x_1$ (mod p)   (iv) $x \equiv p - x_1$ (mod p)
       $x \equiv q - x_2$ (mod q), $x \equiv q - x_2$ (mod q).

We denote solutions of (i) and (ii) by x and y, respectively. Solutions of (iii) and (iv) are easily seen to be $n - y$ and $n - x$, respectively.

We also note that when $p \equiv q \equiv 3$ (mod 4), the solutions of $x^2 \equiv a$ (mod p) and of $x^2 \equiv a$ (mod q) are $x \equiv \pm a^{(p+1)/4}$ (mod p) and $x \equiv \pm a^{(q+1)/4}$ (mod q), respectively. By Euler's criterion, we know that $a^{(p-1)/2} \equiv \left(\frac{a}{p}\right) = 1$ (mod p) and $a^{(q-1)/2} \equiv \left(\frac{a}{q}\right) = 1$ (mod q) (recall that we are assuming that $x^2 \equiv a$ (mod pq) has a solution, so that a is a quadratic residue of both p and q). Hence,

$$(a^{(p+1)/4})^2 = a^{(p+1)/2} = a^{(p-1)/2} \cdot a \equiv a \text{ (mod } p)$$

and

$$(a^{(q+1)/4})^2 = a^{(q+1)/2} = a^{(q-1)/2} \cdot a \equiv a \text{ (mod } q).$$

Using the Chinese remainder theorem, together with the explicit solutions just constructed, we can easily find the four incongruent solutions of $x^2 \equiv a \pmod{n}$. The following example illustrates this procedure.

Example 11.7. Suppose that we know à priori that the congruence

$$x^2 \equiv 860 \pmod{11{,}021}$$

has a solution. Because $11{,}021 = 103 \cdot 107$, to find the four incongruent solutions we solve the congruences

$$x^2 \equiv 860 \equiv 36 \pmod{103}$$

and

$$x^2 \equiv 860 \equiv 4 \pmod{107}.$$

The solutions of these congruences are

$$x \equiv \pm 36^{(103+1)/4} \equiv \pm 36^{26} \equiv \pm 6 \pmod{103}$$

and

$$x \equiv \pm 4^{(107+1)/4} \equiv \pm 4^{27} \equiv \pm 2 \pmod{107},$$

respectively. Using the Chinese remainder theorem, we obtain $x \equiv \pm 212, \pm 109$ (mod 11,021) as the solutions of the four systems of congruences described by the four possible choices of signs in the system of congruences $x \equiv \pm 6 \pmod{103}$, $x \equiv \pm 2$ (mod 107). ◀

Flipping Coins Electronically

An interesting and useful application of the properties of quadratic residues is a method to "flip coins" electronically, invented by Blum [Bl82]. This method takes advantage of the difference in the length of time needed to find primes and needed to factor integers that are the products of two primes, also the basis of the RSA cipher discussed in Chapter 8.

We now describe a method for electronically flipping coins. Suppose that Bob and Alice are communicating electronically. Alice picks two distinct large primes p and q, with $p \equiv q \equiv 3 \pmod{4}$. Alice sends Bob the integer $n = pq$. Bob picks, at random, a positive integer x less than n and sends to Alice the integer a with $x^2 \equiv a \pmod{n}$, $0 < a < n$. Alice finds the four solutions of $x^2 \equiv a \pmod{n}$, namely $x, y, n - x$, and $n - y$. Alice picks one of these four solutions and sends it to Bob. Note that since $x + y \equiv 2x_1 \not\equiv 0 \pmod{p}$ and $x + y \equiv 0 \pmod{q}$, we have $(x + y, n) = q$, and similarly $(x + (n - y), n) = p$. Thus, if Bob receives either y or $n - y$, he can rapidly factor n by using the Euclidean algorithm to find one of the two prime factors of n. On the other hand, if Bob receives either x or $n - x$, he has no way to factor n in a reasonable length of time.

Consequently, Bob wins the coin flip if he can factor n, whereas Alice wins if Bob cannot factor n. From previous comments, we know that there is an equal chance for

Bob to receive a solution of $x^2 \equiv a \pmod{n}$ that helps him rapidly factor n, or a solution of $x^2 \equiv a \pmod{n}$ that does not help him factor n. Hence, the coin flip is fair.

11.1 Exercises

1. Find all of the quadratic residues of each of the following integers.

    a) 3   b) 5   c) 13   d) 19

2. Find all of the quadratic residues of each of the following integers.

    a) 7   b) 8   c) 15   d) 18

3. Find the value of the Legendre symbols $\left(\frac{j}{5}\right)$ for $j = 1, 2, 3, 4$.

4. Find the value of the Legendre symbols $\left(\frac{j}{7}\right)$ for $j = 1, 2, 3, 4, 5, 6$.

5. Evaluate the Legendre symbol $\left(\frac{7}{11}\right)$

    a) using Euler's criterion.
    b) using Gauss's lemma.

6. Let a and b be integers not divisible by the prime p. Show that either one or all three of the integers a, b, and ab are quadratic residues of p.

7. Show that if p is an odd prime, then

$$\left(\frac{-2}{p}\right) = \begin{cases} 1 & \text{if } p \equiv 1 \text{ or } 3 \pmod{8}; \\ -1 & \text{if } p \equiv -1 \text{ or } -3 \pmod{8}. \end{cases}$$

8. Show that if the prime-power factorization of n is

$$n = p_1^{2t_1+1} p_2^{2t_2+1} \cdots p_k^{2t_k+1} p_{k+1}^{2t_{k+1}} \cdots p_m^{2t_m}$$

    and q is a prime not dividing n, then

$$\left(\frac{n}{q}\right) = \left(\frac{p_1}{q}\right)\left(\frac{p_2}{q}\right) \cdots \left(\frac{p_k}{q}\right).$$

9. Show that if p is prime and $p \equiv 3 \pmod{4}$, then $[(p-1)/2]! \equiv (-1)^t \pmod{p}$, where t is the number of positive integers less than $p/2$ that are nonquadratic residues of p.

10. Show that if b is a positive integer not divisible by the prime p, then

$$\left(\frac{b}{p}\right) + \left(\frac{2b}{p}\right) + \left(\frac{3b}{p}\right) + \cdots + \left(\frac{(p-1)b}{p}\right) = 0.$$

11. Let p be prime and a be a quadratic residue of p. Show that if $p \equiv 1 \pmod{4}$, then $-a$ is also a quadratic residue of p, whereas if $p \equiv 3 \pmod{4}$, then $-a$ is a quadratic nonresidue of p.

12. Consider the quadratic congruence $ax^2 + bx + c \equiv 0 \pmod{p}$, where p is prime and a, b, and c are integers with $p \nmid a$.

    a) Let $p = 2$. Determine which quadratic congruences (mod 2) have solutions.

b) Let p be an odd prime and let $d = b^2 - 4ac$. Show that the congruence $ax^2 + bx + c \equiv 0 \pmod{p}$ is equivalent to the congruence $y^2 \equiv d \pmod{p}$, where $y = 2ax + b$. Conclude that if $d \equiv 0 \pmod{p}$, then there is exactly one solution x modulo p; if d is a quadratic residue of p, then there are two incongruent solutions; and if d is a quadratic nonresidue of p, then there are no solutions.

13. Find all solutions of the following quadratic congruences.
   a) $x^2 + x + 1 \equiv 0 \pmod 7$
   b) $x^2 + 5x + 1 \equiv 0 \pmod 7$
   c) $x^2 + 3x + 1 \equiv 0 \pmod 7$

14. Show that if p is prime and $p \geq 7$, then there are always two consecutive quadratic residues of p. (*Hint:* First show that at least one of 2, 5, and 10 is a quadratic residue of p.)

* 15. Show that if p is prime and $p \geq 7$, then there are always two quadratic residues of p that differ by 2.

16. Show that if p is prime and $p \geq 7$, then there are always two quadratic residues of p that differ by 3.

17. Show that if a is a quadratic residue of the prime p, then the solutions of $x^2 \equiv a \pmod p$ are
   a) $x \equiv \pm a^{n+1} \pmod p$, if $p = 4n + 3$.
   b) $x \equiv \pm a^{n+1}$ or $\pm 2^{2n+1}a^{n+1} \pmod p$, if $p = 8n + 5$.

* 18. Show that if p is a prime and $p = 8n + 1$, and r is a primitive root modulo p, then the solutions of $x^2 \equiv \pm 2 \pmod p$ are given by

$$x \equiv \pm(r^{7n} \pm r^n) \pmod p,$$

where the $\pm$ sign in the first congruence corresponds to the $\pm$ sign inside the parentheses in the second congruence.

19. Find all solutions of the congruence $x^2 \equiv 1 \pmod{15}$.

20. Find all solutions of the congruence $x^2 \equiv 58 \pmod{77}$.

21. Find all solutions of the congruence $x^2 \equiv 207 \pmod{1001}$.

22. Let p be an odd prime, e a positive integer, and a an integer relatively prime to p. Show that the congruence $x^2 \equiv a \pmod{p^e}$ has either no solutions or exactly two incongruent solutions.

* 23. Let p be an odd prime, e a positive integer, and a an integer relatively prime to p. Show that there is a solution to the congruence $x^2 \equiv a \pmod{p^{e+1}}$ if and only if there is a solution to the congruence $x^2 \equiv a \pmod{p^e}$. Use Exercise 22 to conclude that the congruence $x^2 \equiv a \pmod{p^e}$ has no solutions if a is a quadratic nonresidue of p, and exactly two incongruent solutions modulo p if a is a quadratic residue of p.

24. Let n be an odd integer. Find the number of incongruent solutions modulo n of the congruence $x^2 \equiv a \pmod n$, where n has prime-power factorization $n = p_1^{t_1} p_2^{t_2} \cdots p_m^{t_m}$, in terms of the Legendre symbols $\left(\frac{a}{p_1}\right), \ldots, \left(\frac{a}{p_m}\right)$. (*Hint:* Use Exercise 23.)

25. Find the number of incongruent solutions of each of the following congruences.
   a) $x^2 \equiv 31 \pmod{75}$

b) $x^2 \equiv 16 \pmod{105}$

c) $x^2 \equiv 46 \pmod{231}$

d) $x^2 \equiv 1156 \pmod{3^2 5^3 7^5 11^6}$

* **26.** Show that the congruence $x^2 \equiv a \pmod{2^e}$, where e is an integer, $e \geq 3$, has either no solutions or exactly four incongruent solutions. (*Hint:* Use the fact that $(\pm x)^2 \equiv (2^{e-1} \pm x)^2 \pmod{2^e}$.)

27. Show that there are infinitely many primes of the form $4k + 1$. (*Hint:* Assume that $p_1, p_2, \ldots, p_n$ are the only such primes. Form $N = 4(p_1 p_2 \cdots p_n)^2 + 1$, and show, using Theorem 11.5, that N has a prime factor of the form $4k + 1$ that is not one of $p_1, p_2, \ldots, p_n$.)

* **28.** Show that there are infinitely many primes of each of the following forms.

a) $8k + 3$ b) $8k + 5$ c) $8k + 7$

(*Hint:* For each part, assume that there are only finitely many primes $p_1, p_2, \ldots, p_n$ of the particular form. For part (a), look at $(p_1 p_2 \cdots p_n)^2 + 2$; for part (b), look at $(p_1 p_2 \cdots p_n)^2 + 4$; and for part (c), look at $(4 p_1 p_2 \cdots p_n)^2 - 2$. In each part, show that there is a prime factor of this integer of the required form not among the primes $p_1, p_2, \ldots, p_n$. Use Theorems 11.5 and 11.6.)

29. Let p and q be odd primes with $p \equiv q \equiv 3 \pmod 4$ and let a be a quadratic residue of $n = pq$. Show that exactly one of the four incongruent square roots of a modulo pq is a quadratic residue of n.

30. Prove Theorem 11.3 using the concept of primitive roots and indices.

31. Prove Theorem 11.4 using the concept of primitive roots and indices.

32. Let p be an odd prime. Show that there are $(p - 1)/2 - \phi(p - 1)$ quadratic nonresidues of p that are not primitive roots of p.

* **33.** Let p and $q = 2p + 1$ both be odd primes. Show that the $p - 1$ primitive roots of q are the quadratic nonresidues of q, other than the nonresidue $2p$ of q.

* **34.** Show that if p and $q = 4p + 1$ are both primes and if a is a quadratic nonresidue of q with $\mathrm{ord}_q a \neq 4$, then a is a primitive root of q.

* **35.** Show that a prime p is a Fermat prime if and only if every quadratic nonresidue of p is also a primitive root of p.

* **36.** Show that a prime divisor p of the Fermat number $F_n = 2^{2^n} + 1$ must be of the form $2^{n+2}k + 1$. (*Hint:* Show that $\mathrm{ord}_p 2 = 2^{n+1}$. Then show that $2^{(p-1)/2} \equiv 1 \pmod p$ using Theorem 11.6. Conclude that $2^{n+1} \mid (p - 1)/2$.)

* **37.** a) Show that if p is a prime of the form $4k + 3$ and $q = 2p + 1$ is prime, then q divides the Mersenne number $M_p = 2^p - 1$. (*Hint:* Consider the Legendre symbol $\left(\frac{2}{q}\right)$.)

     b) From part (a), show that $23 \mid M_{11}, 47 \mid M_{23}$, and $503 \mid M_{251}$.

* **38.** Show that if n is a positive integer and $2n + 1$ is prime, and if $n \equiv 0$ or $3 \pmod 4$, then $2n + 1$ divides the Mersenne number $M_n = 2^n - 1$, whereas if $n \equiv 1$ or $2 \pmod 4$, then $2n + 1$ divides $M_n + 2 = 2^n + 1$. (*Hint:* Consider the Legendre symbol $\left(\frac{2}{2n+1}\right)$ and use Theorem 11.5.)

39. Show that if p is an odd prime, then every prime divisor q of the Mersenne number M_p must be of the form $q = 8k \pm 1$, where k is a positive integer. (*Hint:* Use Exercise 38.)

40. Show how Exercise 39, together with Theorem 7.12, can be used to help show that M_{17} is prime.

∗ **41.** Show that if p is an odd prime, then

$$\sum_{j=1}^{p-2} \left(\frac{j(j+1)}{p} \right) = -1.$$

(*Hint:* First show that $\left(\frac{j(j+1)}{p} \right) = \left(\frac{\bar{j}+1}{p} \right)$, where $\bar{j}$ is an inverse j of modulo p.)

∗ **42.** Let p be an odd prime. Among pairs of consecutive positive integers less than p, let **(RR)**, **(RN)**, **(NR)**, and **(NN)** denote the number of pairs of two quadratic residues, of a quadratic residue followed by a quadratic nonresidue, of a quadratic nonresidue followed by a quadratic residue, and of two quadratic nonresidues, respectively.

a) Show that

$$\textbf{(RR)} + \textbf{(RN)} = \frac{1}{2}(p - 2 - (-1)^{(p-1)/2})$$

$$\textbf{(NR)} + \textbf{(NN)} = \frac{1}{2}(p - 2 + (-1)^{(p-1)/2})$$

$$\textbf{(RR)} + \textbf{(NR)} = \frac{1}{2}(p - 1) - 1$$

$$\textbf{(RN)} + \textbf{(NN)} = \frac{1}{2}(p - 1).$$

b) Using Exercise 41, show that

$$\sum_{j=1}^{p-2} \left(\frac{j(j+1)}{p} \right) = \textbf{(RR)} + \textbf{(NN)} - \textbf{(RN)} - \textbf{(NR)} = -1.$$

c) From parts (a) and (b), find **(RR)**, **(RN)**, **(NR)**, and **(NN)**.

43. Use Theorem 9.16 to prove Theorem 11.1.

∗ **44.** Let p and q be odd primes. Show that 2 is a primitive root of q, if $q = 4p + 1$.

∗ **45.** Let p and q be odd primes. Show that 2 is a primitive root of q, if p is of the form $4k + 1$ and $q = 2p + 1$.

∗ **46.** Let p and q be odd primes. Show that -2 is a primitive root of q, if p is of the form $4k - 1$ and $q = 2p + 1$.

∗ **47.** Let p and q be odd primes. Show that -4 is a primitive root of q, if $q = 2p + 1$.

48. Find the solutions of $x^2 \equiv 482 \pmod{2773}$ (note that $2773 = 47 \cdot 59$).

∗ **49.** In this exercise, we develop a method for decrypting messages encrypted using a Rabin cipher. Recall that the relationship between a ciphertext block C and the corresponding plaintext block P in a Rabin cipher is $C \equiv P(P + \bar{2}b) \pmod{n}$, where $n = pq$, p and q are distinct odd primes, and b is a positive integer less than n.

    a) Show that $C + a \equiv (P + \overline{2}b)^2 \pmod{n}$, where $a \equiv (\overline{2}b)^2 \pmod{n}$, and $\overline{2}$ is an inverse of 2 modulo n.

    b) Using the algorithm in the text for solving congruences of the type $x^2 \equiv a \pmod{n}$, together with part (a), show how to find a plaintext block P from the corresponding ciphertext block C. Explain why there are four possible plaintext messages. (This ambiguity is a disadvantage of Rabin ciphers.)

    c) Decrypt the ciphertext message 1819 0459 0803 that was encrypted using the Rabin cryptosytem with $b = 3$ and $n = 47 \cdot 59 = 2773$.

50. Let p be an odd prime, and let C be the ciphertext obtained in modular exponentiation, with exponent e and modulus p, from the plaintext P, that is, $C \equiv P^e \pmod{p}$, $0 < C < n$, where $(e, p - 1) = 1$. Show that C is a quadratic residue of p if and only if P is a quadratic residue of p.

∗ 51. a) Show that the second player in a game of electronic poker (see Section 8.6) can obtain an advantage by noting which cards have numerical equivalents that are quadratic residues modulo p. (*Hint:* Use Exercise 50.)

    b) Show that the advantage of the second player noted in part (a) can be eliminated if the numerical equivalents of cards that are quadratic nonresidues are all multiplied by a fixed quadratic nonresidue.

∗ 52. Show that if the probing sequence for resolving collisions in a hashing scheme is $h_j(K) \equiv h(K) + aj + bj^2 \pmod{m}$, where $h(K)$ is a hashing function, m is a positive integer, and a and b are integers with $(b, m) = 1$, then only half the possible file locations are probed. This is called the *quadratic search*.

We say that x and y form a *chain of quadratic residues* modulo p if x, y, and $x + y$ are all quadratic residues modulo p.

53. Find a chain $x, y, x + y$ of quadratic residues modulo 11.

54. Is there a chain of quadratic residues modulo 7?

11.1 Computational and Programming Exercises

Computations and Explorations

Using a computation program such as Maple or *Mathematica,* or programs you have written, carry out the following computations and explorations.

1. Find the value of each of the following Legendre symbols: $\left(\frac{1521}{451{,}879}\right)$, $\left(\frac{222{,}344}{21{,}155{,}500{,}207}\right)$, $\left(\frac{6{,}818{,}811}{15{,}454{,}356{,}666{,}611}\right)$.

2. Show that the prime $p = 30{,}059{,}924{,}764{,}123$ has $\left(\frac{q}{p}\right) = -1$ for all primes q with $2 \leq q \leq 181$.

3. A set of integers $x_1, x_2, \ldots, x_n$, where n is a positive integer, is called *chain of quadratic residues* if all sums of consecutive subsets of these numbers are quadratic residues. Show that the integers 1, 4, 45, 94, 261, 310, 344, 387, 393, 394, and 456 form a chain of quadratic residues modulo 631. (*Note:* There are 66 values to check.)

4. Find the smallest quadratic nonresidue of each prime less than 1000.

5. Find the smallest quadratic nonresidue of 100 randomly selected primes between 100,000 and 1,000,000, and 100 randomly selected primes between 100,000,000 and 1,000,000,000. Can you make any conjectures based on your evidence?

6. Use numerical evidence to determine for which odd primes p there are more quadratic residues a of p with $1 \leq a \leq (p-1)/2$ than there are with $(p+1)/2 \leq a \leq p-1$.

7. Let p be a prime with $p \equiv 3 \pmod 4$. It has been proved that if R is the largest number of consecutive quadratic residues of p and N is the largest number of consecutive quadratic nonresidues of p, then $R = N < \sqrt{p}$. Verify this result for all primes of this type less than 1000.

8. Let p be a prime with $p \equiv 1 \pmod 4$. It has been conjectured that if N is the largest number of consecutive quadratic nonresidues of p, then $N < \sqrt{p}$ when p is sufficiently large. Find evidence for this conjecture. For which small primes does this inequality fail?

9. Find the four modular square roots of $4{,}609{,}126$ modulo $14{,}438{,}821 = 4003 \cdot 3607$.

10. Find the square roots of $11{,}535$ modulo $142{,}661$. Which one is a quadratic residue of $142{,}661$?

Programming Projects

Write computer programs using Maple, *Mathematica,* or a language of your choice to do the following.

1. Evaluate Legendre symbols using Euler's criterion.

2. Evaluate Legendre symbols using Gauss's lemma.

3. Given a positive integer n that is the product of two distinct primes both congruent to 3 modulo 4, find the four square roots of the least positive residue of x^2, where x is an integer relatively prime to n.

* 4. Flip coins electronically using the procedure described in this section.

** 5. Decrypt messages that were encrypted using a Rabin cryptosystem (see Exercise 49).

11.2 The Law of Quadratic Reciprocity

Suppose that p and q are distinct odd primes. Suppose further that we know whether q is a quadratic residue of p. Do we also know whether p is a quadratic residue of q? The answer to this question was found by Euler in the mid-1700s. He found the answer by examining numerical evidence, but he did not prove that his answer was correct. Later, in 1785, Legendre reformulated Euler's answer, in its modern, elegant form, in a theorem known as the *law of quadratic reciprocity.* This theorem tells us whether the congruence $x^2 \equiv q \pmod p$ has solutions, once we know whether there are solutions of $x^2 \equiv p \pmod q$.

Theorem 11.7. *The Law of Quadratic Reciprocity.* Let p and q be distinct odd primes. Then

$$\left(\frac{p}{q}\right)\left(\frac{q}{p}\right) = (-1)^{\frac{p-1}{2} \cdot \frac{q-1}{2}}.$$

Legendre published several proposed proofs of this theorem, but each of his proofs contained a serious gap. The first correct proof was provided by Gauss, who claimed to have rediscovered this result when he was 18 years old. Gauss devoted considerable attention to his search for a proof. In fact, he wrote that "for an entire year this theorem tormented me and absorbed my greatest efforts until at last I obtained a proof."

Once Gauss found his first proof in 1796, he continued searching for different proofs. He found at least six different proofs of the law of quadratic reciprocity. His goal in looking for more proofs was to find an approach that could be generalized to higher powers. In particular, he was interested in cubic and biquadratic residues of primes; that is, he was interested in determining when, given a prime p and an integer a not divisible by p, the congruences $x^3 \equiv a \pmod{p}$ and $x^4 \equiv a \pmod{p}$ are solvable. With his sixth proof, Gauss finally succeeded in his goal, as this proof could be generalized to higher powers. (See [IrRo91], [Go98], and [Le00] for more information about Gauss's proofs and the generalization to higher power residues.)

Finding new and different approaches did not stop with Gauss. Some of the well-known mathematicians who have published original proofs of the law of quadratic reciprocity are Cauchy, Dedekind, Dirichlet, Kronecker, and Eisenstein. One count in 1921 stated that there were 56 different proofs of the law of quadratic reciprocity, and in 1963 an article published by M. Gerstenhaber [Ge63] offered the 152nd proof of the law of quadratic reciprocity. In 2000, Franz Lemmermeyer [Le00] compiled a comprehensive list of 192 proofs of quadratic reciprocity, noting for each proof the year, the prover, and the method of proof. Lemmermeyer maintains a current version of this on the Web; as of early 2004, 207 different proofs were listed. According to his count, Gerstenhaber's proof is number 153 and eight of the proofs were completed since 2000. It will be interesting to see if new proofs continue to be found at the rate of one per year. (See Exercise 17 for an outline of the 207th proof.)

Although many of the different proofs of the law of quadratic reciprocity are similar, they encompass an amazing variety of approaches. The ideas in different approaches can have useful consequences. For example, the ideas behind Gauss's first proof, which is a complicated argument using mathematical induction, were of little interest to mathematicians for more than 175 years, until they were used in the 1970s in computations in an advanced area of algebra known as K-theory.

The version of the law of quadratic reciprocity that we have stated and proved is different from the version originally conjectured by Euler. This version, which we now state, turns out to be equivalent to the version we have stated as Theorem 11.7. Euler formulated this version based on the evidence of many computations of special cases.

Theorem 11.8. Suppose that p is an odd prime and a is an integer not divisible by p. If q is a prime with $p \equiv \pm q \pmod{4a}$, then $\left(\frac{a}{p}\right) = \left(\frac{a}{q}\right)$.

This version of the law of quadratic reciprocity shows that the value of the Legendre symbol $\left(\frac{a}{p}\right)$ depends only on the residue class of p modulo $4a$, and that the value of $\left(\frac{a}{p}\right)$ takes the same value for all primes p with remainder r or $4a - r$ when divided by $4a$.

We leave it to the reader as Exercises 10 and 11 to show that this form of the law of quadratic reciprocity is equivalent to the form given in Theorem 11.7. We also ask the reader to prove, in Exercise 12, this form of quadratic reciprocity directly, using Gauss's lemma.

Before we prove the law of quadratic reciprocity, we will discuss its consequences and how it is used to evaluate Legendre symbols. We first note that the quantity $(p-1)/2$ is even when $p \equiv 1 \pmod 4$ and odd when $p \equiv 3 \pmod 4$. Consequently, we see that $\frac{p-1}{2} \cdot \frac{q-1}{2}$ is even if $p \equiv 1 \pmod 4$ or $q \equiv 1 \pmod 4$, whereas $\frac{p-1}{2} \cdot \frac{q-1}{2}$ is odd if $p \equiv q \equiv 3 \pmod 4$. Hence, we have

$$\left(\frac{p}{q}\right)\left(\frac{q}{p}\right) = \begin{cases} 1 & \text{if } p \equiv 1 \pmod 4 \text{ or } q \equiv 1 \pmod 4 \text{ (or both);} \\ -1 & \text{if } p \equiv q \equiv 3 \pmod 4. \end{cases}$$

Because the only possible values of $\left(\frac{p}{q}\right)$ and $\left(\frac{q}{p}\right)$ are ± 1, we see that

$$\left(\frac{p}{q}\right) = \begin{cases} \left(\frac{q}{p}\right) & \text{if } p \equiv 1 \pmod 4 \text{ or } q \equiv 1 \pmod 4 \text{ (or both);} \\ -\left(\frac{q}{p}\right) & \text{if } p \equiv q \equiv 3 \pmod 4. \end{cases}$$

This means that if p and q are odd primes, then $\left(\frac{p}{q}\right) = \left(\frac{q}{p}\right)$, unless both p and q are congruent to 3 modulo 4, and in that case, $\left(\frac{p}{q}\right) = -\left(\frac{q}{p}\right)$.

Example 11.8. Let $p = 13$ and $q = 17$. Because $p \equiv q \equiv 1 \pmod 4$, the law of quadratic reciprocity tells us that $\left(\frac{13}{17}\right) = \left(\frac{17}{13}\right)$. By part (i) of Theorem 11.4, we know that $\left(\frac{17}{13}\right) = \left(\frac{4}{13}\right)$, and from part (iii) of Theorem 11.4, it follows that $\left(\frac{4}{13}\right) = \left(\frac{2^2}{13}\right) = 1$. Combining these equalities, we conclude that $\left(\frac{13}{17}\right) = 1$.   ◀

Example 11.9. Let $p = 7$ and $q = 19$. Because $p \equiv q \equiv 3 \pmod 4$, by the law of quadratic reciprocity, we know that $\left(\frac{7}{19}\right) = -\left(\frac{19}{7}\right)$. From part (i) of Theorem 11.4, we see that $\left(\frac{19}{7}\right) = \left(\frac{5}{7}\right)$. Again, using the law of quadratic reciprocity, because $5 \equiv 1 \pmod 4$ and $7 \equiv 3 \pmod 4$, we have $\left(\frac{5}{7}\right) = \left(\frac{7}{5}\right)$. By part (i) of Theorem 11.4 and Theorem 11.6, we know that $\left(\frac{7}{5}\right) = \left(\frac{2}{5}\right) = -1$. Hence, $\left(\frac{7}{19}\right) = 1$.   ◀

We can use the law of quadratic reciprocity and Theorems 11.4 and 11.6 to evaluate Legendre symbols. Unfortunately, prime factorizations must be computed to evaluate Legendre symbols in this way.

Example 11.10. We will calculate $\left(\frac{713}{1009}\right)$ (note that 1009 is prime). We factor $713 = 23 \cdot 31$, so that by part (ii) of Theorem 11.4, we have

$$\left(\frac{713}{1009}\right) = \left(\frac{23 \cdot 31}{1009}\right) = \left(\frac{23}{1009}\right)\left(\frac{31}{1009}\right).$$

To evaluate the two Legendre symbols on the right side of this equality, we use the law of quadratic reciprocity. Because $1009 \equiv 1 \pmod 4$, we see that

$$\left(\frac{23}{1009}\right) = \left(\frac{1009}{23}\right), \quad \left(\frac{31}{1009}\right) = \left(\frac{1009}{31}\right).$$

Using Theorem 11.4, part (i), we have

$$\left(\frac{1009}{23}\right) = \left(\frac{20}{23}\right), \quad \left(\frac{1009}{31}\right) = \left(\frac{17}{31}\right).$$

By parts (ii) and (iii) of Theorem 11.4, it follows that

$$\left(\frac{20}{23}\right) = \left(\frac{2^2 \cdot 5}{23}\right) = \left(\frac{2^2}{23}\right)\left(\frac{5}{23}\right) = \left(\frac{5}{23}\right).$$

The law of quadratic reciprocity, part (i) of Theorem 11.4, and Theorem 11.6 tell us that

$$\left(\frac{5}{23}\right) = \left(\frac{23}{5}\right) = \left(\frac{3}{5}\right) = \left(\frac{5}{3}\right) = \left(\frac{2}{3}\right) = -1.$$

Thus, $\left(\frac{23}{1009}\right) = -1$.

Likewise, using the law of quadratic reciprocity, Theorem 11.4, and Theorem 11.6, we find that

$$\left(\frac{17}{31}\right) = \left(\frac{31}{17}\right) = \left(\frac{14}{17}\right) = \left(\frac{2}{17}\right)\left(\frac{7}{17}\right) = \left(\frac{7}{17}\right) = \left(\frac{17}{7}\right) = \left(\frac{3}{7}\right)$$

$$= -\left(\frac{7}{3}\right) = -\left(\frac{4}{3}\right) = -\left(\frac{2^2}{3}\right) = -1.$$

Consequently, $\left(\frac{31}{1009}\right) = -1$.

Therefore, $\left(\frac{713}{1009}\right) = (-1)(-1) = 1.$ ◄

A Proof of the Law of Quadratic Reciprocity

We now present a proof of the law of quadratic reciprocity originally given by *Max Eisenstein*. This proof is a simplification of the third proof given by Gauss. This simplification was made possible by the following lemma of Eisenstein, which will help us reduce the proof of the law of quadratic reciprocity to counting lattice points in triangles.

Lemma 11.3. If p is an odd prime and a is an odd integer not divisible by p, then

$$\left(\frac{a}{p}\right) = (-1)^{T(a,p)},$$

where

$$T(a, p) = \sum_{j=1}^{(p-1)/2} [ja/p].$$

Proof.   Consider the least positive residues of the integers $a, 2a, \ldots, ((p-1)/2)a$; let $u_1, u_2, \ldots, u_s$ be those greater than $p/2$ and let $v_1, v_2, \ldots, v_t$ be those less than $p/2$. The division algorithm tells us that

$$ja = p[ja/p] + \text{ remainder},$$

where the remainder is one of the u_j or v_j. By adding the $(p-1)/2$ equations of this sort, we obtain

(11.4)
$$\sum_{j=1}^{(p-1)/2} ja = \sum_{j=1}^{(p-1)/2} p[ja/p] + \sum_{j=1}^{s} u_j + \sum_{j=1}^{t} v_j.$$

As we showed in the proof of Gauss's lemma, the integers $p - u_1, \ldots, p - u_s, v_1, \ldots, v_t$ are precisely the integers $1, 2, \ldots, (p-1)/2$, in some order. Hence, summing all these integers, we obtain

(11.5)
$$\sum_{j=1}^{(p-1)/2} j = \sum_{j=1}^{s}(p - u_j) + \sum_{j=1}^{t} v_j = ps - \sum_{j=1}^{s} u_j + \sum_{j=1}^{t} v_j.$$

Subtracting (11.5) from (11.4), we find that

$$\sum_{j=1}^{(p-1)/2} ja - \sum_{j=1}^{(p-1)/2} j = \sum_{j=1}^{(p-1)/2} p[ja/p] - ps + 2\sum_{j=1}^{s} u_j$$

FERDINAND GOTTHOLD MAX EISENSTEIN (1823–1852) suffered from poor health his entire life. He moved with his family to England, Ireland, and Wales before returning to Germany. In Ireland, Eisenstein met Sir William Rowan Hamilton, who stimulated his interest in mathematics by giving him a paper that discussed the impossibility of solving quintic equations in radicals. On his return to Germany in 1843, at the age of 20, Eisenstein entered the University of Berlin.

Eisenstein amazed the mathematical community when he quickly began producing new results soon after entering the university. In 1844, Eisenstein met Gauss in Göttingen where they discussed reciprocity for cubic residues. Gauss was extremely impressed by Eisenstein, and tried to obtain financial support for him. Gauss wrote to the explorer and scientist Alexander von Humboldt that the talent Eisenstein had was "that nature bestows upon only a few in each century." Eisenstein was amazingly prolific. In 1844, he published 16 papers in Volume 27 of *Crelle's Journal* alone. In the third semester of his studies, he received an honorary doctorate from the University of Breslau. Eistenstein was appointed to an unsalaried position as a Privatdozent at the University of Berlin; however, after 1847, Eisenstein's health worsened so much that he was mostly confined to bed. Nevertheless, his mathematical output continued unabated. After spending a year in Sicily in a futile attempt to improve his health, he returned to Germany where he died from tuberculosis at the age of 29. His early death was considered a tremendous loss by mathematicians.

or, equivalently, because $T(a, p) = \sum_{j=1}^{(p-1)/2}[ja/p]$,

$$(a - 1)\sum_{j=1}^{(p-1)/2} j = pT(a, p) - ps + 2\sum_{j=1}^{s} u_j.$$

Reducing this last equation modulo 2, because a and p are odd, yields

$$0 \equiv T(a, p) - s \pmod{2}.$$

Hence,

$$T(a, p) \equiv s \pmod{2}.$$

To finish the proof, we note that from Gauss's lemma,

$$\left(\frac{a}{p}\right) = (-1)^s.$$

Consequently, because $(-1)^s = (-1)^{T(a,p)}$, it follows that

$$\left(\frac{a}{p}\right) = (-1)^{T(a,p)}. \qquad \blacksquare$$

Although Lemma 11.3 is used primarily as a tool in the proof of the law of quadratic reciprocity, it can also be used to evaluate Legendre symbols.

Example 11.11. To find $\left(\frac{7}{11}\right)$ using Lemma 11.3, we evaluate the sum

$$\sum_{j=1}^{5}[7j/11] = [7/11] + [14/11] + [21/11] + [28/11] + [35/11]$$

$$= 0 + 1 + 1 + 2 + 3 = 7.$$

Hence, $\left(\frac{7}{11}\right) = (-1)^7 = -1$.

Likewise, to find $\left(\frac{11}{7}\right)$, we note that

$$\sum_{j=1}^{3}[11j/7] = [11/7] + [22/7] + [33/7] = 1 + 3 + 4 = 8,$$

so that $\left(\frac{11}{7}\right) = (-1)^8 = 1.$ ◀

Before we present a proof of the law of quadratic reciprocity, we use an example to illustrate the method of proof.

Let $p = 7$ and $q = 11$. We consider pairs of integers (x, y) with $1 \le x \le (7 - 1)/2 = 3$ and $1 \le y \le (11 - 1)/2 = 5$. There are 15 such pairs. We note that none of these pairs satisfies $11x = 7y$, because the equality $11x = 7y$ implies that $11 \mid 7y$, so that either $11 \mid 7$, which is absurd, or $11 \mid y$, which is impossible because $1 \le y \le 5$.

We divide these 15 pairs into two groups, depending on the relative sizes of $11x$ and $7y$, as shown in Figure 11.1.

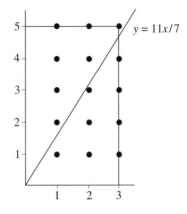

Figure 11.1 *Counting lattice points to determine* $\left(\frac{7}{11}\right)\left(\frac{11}{7}\right)$.

The pairs of integers (x, y) with $1 \le x \le 3$, $1 \le y \le 5$, and $11x > 7y$ are precisely those pairs satisfying $1 \le x \le 3$ and $1 \le y \le 11x/7$. For a fixed integer x with $1 \le x \le 3$, there are $[11x/7]$ allowable values of y. Hence, the total number of pairs satisfying $1 \le x \le 3$, $1 \le y \le 5$, and $11x > 7y$ is

$$\sum_{j=1}^{3} [11j/7] = [11/7] + [22/7] + [33/7] = 1 + 3 + 4 = 8;$$

these eight pairs are $(1, 1)$, $(2, 1)$, $(2, 2)$, $(2, 3)$, $(3, 1)$, $(3, 2)$, $(3, 3)$, and $(3, 4)$.

The pairs of integers (x, y) with $1 \le x \le 3$, $1 \le y \le 5$, and $11x < 7y$ are precisely those pairs satisfying $1 \le y \le 5$ and $1 \le x \le 7y/11$. For a fixed integer y with $1 \le y \le 5$, there are $[7y/11]$ allowable values of x. Hence, the total number of pairs satisfying $1 \le x \le 3$, $1 \le y \le 5$, and $11x < 7y$ is

$$\sum_{j=1}^{5} [7j/11] = [7/11] + [14/11] + [21/11] + [28/11] + [35/11]$$
$$= 0 + 1 + 1 = 2 = 3 = 7.$$

These seven pairs are $(1, 2)$, $(1, 3)$, $(1, 4)$, $(1, 5)$, $(2, 4)$, $(2, 5)$, and $(3, 5)$.

Consequently, we see that

$$\frac{11-1}{2} \cdot \frac{7-1}{2} = 5 \cdot 3 = 15 = \sum_{j=1}^{3} [11j/7] + \sum_{j=1}^{5} [7j/11] = 8 + 7.$$

Hence,

$$(-1)^{\frac{11-1}{2} \cdot \frac{7-1}{2}} = (-1)^{\sum_{j=1}^{3}[11j/7] + \sum_{j=1}^{5}[7j/11]}$$
$$= (-1)^{\sum_{j=1}^{3}[11j/7]} (-1)^{\sum_{j=1}^{5}[7j/11]}.$$

Because Lemma 11.3 tells us that $\left(\frac{11}{7}\right) = (-1)^{\sum_{j=1}^{3}[11j/7]}$ and $\left(\frac{7}{11}\right) = (-1)^{\sum_{j=1}^{5}[7j/11]}$, we see that $\left(\frac{7}{11}\right)\left(\frac{11}{7}\right) = (-1)^{\frac{7-1}{2}\cdot\frac{11-1}{2}}$.

This establishes the special case of the law of quadratic reciprocity when $p = 7$ and $q = 11$.

We now prove the law of quadratic reciprocity, using the idea illustrated in the example.

Proof. We consider pairs of integers (x, y) with $1 \le x \le (p - 1)/2$ and $1 \le y \le (q - 1)/2$. There are $\frac{p-1}{2} \cdot \frac{q-1}{2}$ such pairs. We divide these pairs into two groups, depending on the relative sizes of qx and py, as shown in Figure 11.2

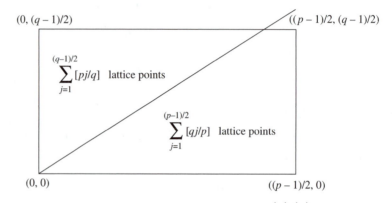

Figure 11.2 *Counting lattice points to determine* $\left(\frac{p}{q}\right)\left(\frac{q}{p}\right)$.

First, we note that $qx \ne py$ for all these pairs. For if $qx = py$, then $q \mid py$, which implies that $q \mid p$ or $q \mid y$. However, because q and p are distinct primes, we know that $q \nmid p$, and because $1 \le y \le (q - 1)/2$, we know that $q \nmid y$.

To enumerate the pairs of integers (x, y) with $1 \le x \le (p - 1)/2$, $1 \le y \le (q - 1)/2$, and $qx > py$, we note that these pairs are precisely those where $1 \le x \le (p - 1)/2$ and $1 \le y \le qx/p$. For each fixed value of the integer x, with $1 \le x \le (p - 1)/2$, there are $[qx/p]$ integers satisfying $1 \le y \le qx/p$. Consequently, the total number of pairs of integers (x, y) with $1 \le x \le (p - 1)/2$, $1 \le y \le (q - 1)/2$, and $qx > py$ is $\sum_{j=1}^{(p-1)/2}[qj/p]$.

We now consider the pairs of integers (x, y) with $1 \le x \le (p - 1)/2$, $1 \le y \le (q - 1)/2$, and $qx < py$. These pairs are precisely the pairs of integers (x, y) with $1 \le y \le (q - 1)/2$ and $1 \le x \le py/q$. Hence, for each fixed value of the integer y, where $1 \le y \le (q - 1)/2$, there are exactly $[py/q]$ integers x satisfying $1 \le x \le py/q$. This shows that the total number of pairs of integers (x, y) with $1 \le x \le (p - 1)/2$, $1 \le y \le (q - 1)/2$, and $qx < py$ is $\sum_{j=1}^{(q-1)/2}[pj/q]$.

Adding the numbers of pairs in these classes, and recalling that the total number of such pairs is $\frac{p-1}{2} \cdot \frac{q-1}{2}$, we see that

$$\sum_{j=1}^{(p-1)/2} [qj/p] + \sum_{j=1}^{(q-1)/2} [pj/q] = \frac{p-1}{2} \cdot \frac{q-1}{2},$$

or, using the notation of Lemma 11.3,

$$T(q, p) + T(p, q) = \frac{p-1}{2} \cdot \frac{q-1}{2}.$$

Hence,

$$(-1)^{T(q,p)+T(p,q)} = (-1)^{T(q,p)}(-1)^{T(p,q)} = (-1)^{\frac{p-1}{2} \cdot \frac{q-1}{2}}.$$

Lemma 11.2 tells us that $(-1)^{T(q,p)} = \left(\frac{q}{p}\right)$ and $(-1)^{T(p,q)} = \left(\frac{p}{q}\right)$. Hence

$$\left(\frac{p}{q}\right)\left(\frac{q}{p}\right) = (-1)^{\frac{p-1}{2} \cdot \frac{q-1}{2}}.$$

This concludes the proof of the law of quadratic reciprocity. ∎

The law of quadratic reciprocity has many applications. One use is to prove the validity of the following primality test for Fermat numbers.

Theorem 11.9. *Pepin's Test.* The Fermat number $F_m = 2^{2^m} + 1$ is prime if and only if

$$3^{(F_m-1)/2} \equiv -1 \pmod{F_m}.$$

Proof. We will first show that F_m is prime if the congruence in the statement of the theorem holds. Assume that

$$3^{(F_m-1)/2} \equiv -1 \pmod{F_m}.$$

Then, by squaring both sides, we obtain

$$3^{F_m-1} \equiv -1 \pmod{F_m}.$$

From this congruence, we see that if p is a prime dividing F_m, then

$$3^{F_m-1} \equiv -1 \pmod{p},$$

and hence,

$$\text{ord}_p 3 \mid (F_m - 1) = 2^{2^m}.$$

Consequently, $\text{ord}_p 3$ must be a power of 2. However,

$$\text{ord}_p 3 \nmid 2^{2^m - 1} = (F_m - 1)/2,$$

because $3^{(F_m-1)/2} \equiv -1 \pmod{F_m}$. Hence, the only possibility is that $\text{ord}_p 3 = 2^{2^m} = F_m - 1$. Because $\text{ord}_p 3 = F_m - 1 \leq p - 1$ and $p \mid F_m$, we see that $p = F_m$ and, consequently, F_m must be prime.

Conversely, if $F_m = 2^{2^m} + 1$ is prime for $m \geq 1$, then the law of quadratic reciprocity tells us that

(11.6)
$$\left(\frac{3}{F_m}\right) = \left(\frac{F_m}{3}\right) = \left(\frac{2}{3}\right) = -1,$$

because $F_m \equiv 1 \pmod 4$ and $F_m \equiv 2 \pmod 3$.

Now, using Euler's criterion, we know that

(11.7)
$$\left(\frac{3}{F_m}\right) \equiv 3^{(F_m-1)/2} \pmod{F_m}.$$

By the two equations involving $\left(\frac{3}{F_m}\right)$, (11.6) and (11.7), we conclude that

$$3^{(F_m-1)/2} \equiv -1 \pmod{F_m}.$$

This finishes the proof. ∎

Example 11.12. Let $m = 2$. Then $F_2 = 2^{2^2} + 1 = 17$ and

$$3^{(F_2-1)/2} = 3^8 \equiv -1 \pmod{17}.$$

By Pepin's test, we see that $F_2 = 17$ is prime.

Let $m = 5$. Then $F_5 = 2^{2^5} + 1 = 2^{32} + 1 = 4{,}294{,}967{,}297$. We note that

$$3^{(F_5-1)/2} = 3^{2^{31}} = 3^{2{,}146{,}483{,}648} \equiv 10{,}324{,}303 \not\equiv -1 \pmod{4{,}294{,}967{,}297}.$$

Hence, by Pepin's test, we see that F_5 is composite. ◀

11.2 Exercises

1. Evaluate each of the following Legendre symbols.

   a) $\left(\frac{3}{53}\right)$ c) $\left(\frac{15}{101}\right)$ e) $\left(\frac{111}{991}\right)$

   b) $\left(\frac{7}{79}\right)$ d) $\left(\frac{31}{641}\right)$ f) $\left(\frac{105}{1009}\right)$

2. Using the law of quadratic reciprocity, show that if p is an odd prime, then

   $$\left(\frac{3}{p}\right) = \begin{cases} 1 & \text{if } p \equiv \pm 1 \pmod{12}; \\ -1 & \text{if } p \equiv \pm 5 \pmod{12}. \end{cases}$$

3. Show that if p is an odd prime, then

   $$\left(\frac{-3}{p}\right) = \begin{cases} 1 & \text{if } p \equiv 1 \pmod 6; \\ -1 & \text{if } p \equiv -1 \pmod 6. \end{cases}$$

4. Find a congruence describing all primes for which 5 is a quadratic residue.

5. Find a congruence describing all primes for which 7 is a quadratic residue.

6. Show that there are infinitely many primes of the form $5k + 4$. (*Hint:* Let n be a positive integer and form $Q = 5(n!)^2 - 1$. Show that Q has a prime divisor of the form $5k + 4$ greater than n. To do this, use the law of quadratic reciprocity to show that if a prime p divides Q, then $\left(\frac{p}{5}\right) = 1$.)

7. Use Pepin's test to show that the following Fermat numbers are primes.

   a) $F_1 = 5$ b) $F_3 = 257$ c) $F_4 = 65{,}537$

* 8. Use Pepin's test to conclude that 3 is a primitive root of every Fermat prime.

* 9. In this exercise, we give another proof of the law of quadratic reciprocity. Let p and q be distinct odd primes. Let $\mathbf{R}$ be the interior of the rectangle with vertices $\mathbf{Q} = (0, 0)$, $\mathbf{A} = (p/2, 0)$, $\mathbf{B} = (q/2, 0)$, and $\mathbf{C} = (p/2, q/2)$, as shown.

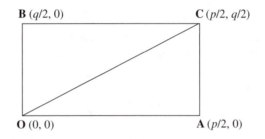

$\mathbf{B}\ (q/2, 0)$ $\mathbf{C}\ (p/2, q/2)$

$\mathbf{O}\ (0, 0)$ $\mathbf{A}\ (p/2, 0)$

   a) Show that the number of lattice points (points with integer coordinates) in $\mathbf{R}$ is $\frac{p-1}{2} \cdot \frac{q-1}{2}$.

   b) Show that there are no lattice points on the diagonal connecting $\mathbf{O}$ and $\mathbf{C}$.

   c) Show that the number of lattice points in the triangle with vertices $\mathbf{O}$, $\mathbf{A}$, and $\mathbf{C}$ is $\sum_{j=1}^{(p-1)/2} [jq/p]$.

   d) Show that the number of lattice points in the triangle with vertices $\mathbf{O}$, $\mathbf{B}$, and $\mathbf{C}$ is $\sum_{j=1}^{(q-1)/2} [jp/q]$.

   e) Conclude from parts (a), (b), (c), and (d) that
   $$\sum_{j=1}^{(p-1)/2} [jq/p] + \sum_{j=1}^{(q-1)/2} [jp/q] = \frac{p-1}{2} \cdot \frac{q-1}{2}.$$
   Derive the law of quadratic reciprocity using this equation and Lemma 11.2.

Exercises 10 and 11 ask that you show that Euler's form of the law of quadratic reciprocity (Theorem 11.8) and the form given in Theorem 11.7 are equivalent.

10. Show that Euler's form of the law of quadratic reciprocity, Theorem 11.8, implies the law of quadratic reciprocity as stated in Theorem 11.7. (*Hint:* Consider separately the cases when $p \equiv q \pmod 4$ and $p \not\equiv q \pmod 4$.)

11. Show that the law of quadratic reciprocity as stated in Theorem 11.7 implies Euler's form of the law of quadratic reciprocity, Theorem 11.8. (*Hint:* First consider the cases when $a = 2$ and when a is an odd prime. Then consider the case when a is composite.)

12. Prove Euler's form of the law of quadratic reciprocity, Theorem 11.8, using Gauss's lemma. (*Hint:* Show that to find $\left(\frac{a}{p}\right)$, we need only find the parity of the number

of integers k satisfying one of the inequalities $(2t-1)(p/2a) \le k \le t(p/a)$ for $t = 1, 2, \ldots, 2u-1$, where $u = a/2$ if a is even and $u = (a-1)/2$ if a is odd. Then, take $p = 4am + r$ with $0 < r < 4a$, and show that finding the parity of the number of integers k satisfying one of the inequalities listed is the same as finding the parity of the number of integers satisfying one of the inequalities $(2t-1)r/2a \le k \le tr/a$ for $t = 1, 2, \ldots, 2u-1$. Show that this number depends only on r. Then, repeat the last step of the argument with r replaced by $4a - r$).

Exercise 13 asks that you fill in the details of a proof of the law of quadratic reciprocity originally developed by Eisenstein. This proof requires familiarity with the complex numbers.

13. A complex number ζ is an *nth root of unity*, where n is a positive integer, if $\zeta^n = 1$. If n is the least integer for which $\zeta^n = 1$, then ζ is called a *primitive nth root of unity*. Recall that $e^{2\pi i} = 1$.

a) Show that $e^{(2\pi i/n)k}$ is a root of unity if k is an integer with $0 \le k \le n-1$, which is primitive if and only if $(k, n) = 1$.

b) Show that if ζ is an nth root of unity and $m \equiv \ell \pmod{n}$, then $\zeta^m = \zeta^\ell$. Furthermore, show that if ζ is a primitive nth root of unity and $\zeta^m = \zeta^\ell$, then $m \equiv \ell \pmod{n}$.

c) Define $f(z) = e^{2\pi iz} - e^{-2\pi iz} = 2i \sin(2\pi z)$. Show that $f(z+1) = f(z)$ and $f(-z) = -f(z)$, and that the only real zeros of $f(z)$ are the numbers $n/2$, where n is an integer.

d) Show that if n is a positive integer, then $x^n - y^n = \prod_{k=0}^{n-1}(\zeta^k x - \zeta^{-k} y)$, where $\zeta = e^{2\pi i/n}$.

e) Show that if n is an odd positive integer and $f(z)$ is as defined in part (c), then

$$\frac{f(nz)}{f(z)} = \prod_{k=1}^{(n-1)/2} f\left(z + \frac{k}{n}\right) f\left(z - \frac{k}{n}\right).$$

f) Show that if p is an odd prime and a is an integer not divisible by p, then

$$\prod_{\ell=1}^{(p-1)/2} f\left(\frac{\ell a}{p}\right) = \left(\frac{a}{p}\right) \prod_{\ell=1}^{(p-1)/2} f\left(\frac{\ell}{p}\right).$$

g) Prove the law of quadratic reciprocity using parts (e) and (f), starting with

$$\prod_{\ell=1}^{(p-1)/2} f\left(\frac{\ell q}{p}\right) = \left(\frac{q}{p}\right) \prod_{\ell=1}^{(p-1)/2} f\left(\frac{\ell}{p}\right).$$

(*Hint:* Use part (e) to obtain a formula for $f\left(\frac{\ell q}{p}\right) / f\left(\frac{\ell}{p}\right)$.)

14. Suppose that p is an odd prime with $\left(\frac{n}{p}\right) = -1$, where $n = k2^m + 1$ with $k < 2^m$ for some integers k and m. Show that n is prime if and only if $p^{(n-1)/2} \equiv -1 \pmod{n}$. (*Hint:* Use Proth's theorem from Section 9.5 for the "only if" part and Euler's criterion and the law of quadratic reciprocity for the "if" part.)

15. The integer $p = 1 + 8 \cdot 3 \cdot 5 \cdot 7 \cdot 11 \cdot 13 \cdot 17 \cdot 19 \cdot 23 = 892{,}371{,}481$ is prime (as the reader can verify using computational software). Show that for all primes q with $q \le 23$, $\left(\frac{q}{p}\right) = 1$. Conclude that there is no quadratic nonresidue of p less than 29 and that p has

no primitive root less than 29. (This fact is a particular case of the result established in the following exercise.)

16. In this exercise, we will show that given any integer M, there exist infinitely many primes p such that $M < r_p < p - M$, where r_p is the least primitive root modulo p.

   a) Let $q_1 = 2, q_2 = 3, q_3 = 5, \ldots, q_n$ be all the primes not exceeding M. Using Dirichlet's theorem on primes in arithmetic progressions, there is a prime $p = 1 + 8q_1 q_2 \cdots q_n r$, where r is a positive integer. Show that $\left(\frac{-1}{p}\right) = 1$, $\left(\frac{2}{p}\right) = 1$, and that $\left(\frac{q_i}{p}\right) = 1$ for $i = 2, 3, \ldots, n$.

   b) Deduce that all integers $t + kp$ with $-M \le t + kp \le M$, where t is an arbitrarily chosen integer, are quadratic residues modulo p and hence not primitive roots modulo p. Show that this implies the result of interest.

* 17. New proofs of the law of quadratic reciprocity are found surprisingly often. In this exercise we fill in the steps of a proof discovered by Kim [Ki04], the 207th proof of quadratic reciprocity according to the count by Lemmermeyer. To set up the proof, let R be the set of integers a such that $1 \le a \le \frac{pq-1}{2}$ and $(a, pq) = 1$, let S be the set of integers a with $1 \le a \le \frac{pq-1}{2}$ and $(a, p) = 1$, and let T be the set of integers a with $q \cdot 1, q \cdot 2, \ldots, q \cdot \frac{p-1}{2}$. Finally, let $A = \prod_{a \in R} a$.

   a) Show that T is a subset of S and that $R = S - T$.

   b) Use part (a) and Euler's criterion to show that $A \equiv (-1)^{\frac{q-1}{2}} \left(\frac{q}{p}\right) \pmod{p}$.

   c) Show that $A \equiv (-1)^{\frac{p-1}{2}} \left(\frac{p}{q}\right) \pmod{q}$ by switching the roles of p and q in parts (a) and (b).

   d) Use parts (b) and (c) to show that $(-1)^{\frac{q-1}{2}} \left(\frac{q}{p}\right) = (-1)^{\frac{p-1}{2}} \left(\frac{p}{q}\right)$ if and only if $A \equiv \pm 1 \pmod{pq}$.

   e) Show that $A \equiv 1$ or $-1 \pmod{pq}$ if and only if $p \equiv q \equiv 1 \pmod 4$.

   (*Hint:* First show that $A \equiv \pm \prod_{a \in U} a \pmod{pq}$, where $U = \{a \in R \mid a^2 \equiv \pm 1 \pmod{pq}\}$ by pairing together elements of R that have either 1 or -1 as their product. Then consider the solutions of each of the congruences $a^2 \equiv 1 \pmod{pq}$ and $a^2 \equiv -1 \pmod{pq}$.)

   f) Conclude from parts (d) and (e) that $(-1)^{\frac{q-1}{2}} \left(\frac{q}{p}\right) = (-1)^{\frac{p-1}{2}} \left(\frac{p}{q}\right)$ if and only if $p \equiv q \equiv 1 \pmod 4$. Deduce the law of quadratic reciprocity from this congruence.

11.2 Computational and Programming Exercises

Computations and Explorations

Using a computation program such as Maple or *Mathematica,* or programs you have written, carry out the following computations and explorations.

1. Use Pepin's test to show that the Fermat numbers F_6, F_7, and F_8 are all composite. Can you go further?

Programming Projects

Write programs using Maple, *Mathematica,* or a language of your choice to do the following.

1. Evaluate Legendre symbols, using the law of quadratic reciprocity.

2. Given a positive integer n, determine whether the nth Fermat number F_n is prime, using Pepin's test.

11.3 The Jacobi Symbol

In this section, we define the Jacobi symbol, named after German mathematician *Carl Jacobi* who introduced it. The Jacobi symbol is a generalization of the Legendre symbol studied in the previous two sections. Jacobi symbols are useful in the evaluation of Legendre symbols and in the definition of a type of pseudoprime.

Definition.   Let n be an odd positive integer with prime factorization $n = p_1^{t_1} p_2^{t_2} \cdots p_m^{t_m}$ and let a be an integer relatively prime to n. Then, the *Jacobi symbol* $\left(\frac{a}{n}\right)$ is defined by

$$\left(\frac{a}{n}\right) = \left(\frac{a}{p_1^{t_1} p_2^{t_2} \cdots p_m^{t_m}}\right) = \left(\frac{a}{p_1}\right)^{t_1} \left(\frac{a}{p_2}\right)^{t_2} \cdots \left(\frac{a}{p_m}\right)^{t_m},$$

where the symbols on the right-hand side of the equality are Legendre symbols.

Example 11.13.   From the definition of the Jacobi symbol, we see that

$$\left(\frac{2}{45}\right) = \left(\frac{2}{3^2 \cdot 5}\right) = \left(\frac{2}{3}\right)^2 \left(\frac{2}{5}\right) = (-1)^2(-1) = -1$$

and

$$\left(\frac{109}{385}\right) = \left(\frac{109}{5 \cdot 7 \cdot 11}\right) = \left(\frac{109}{5}\right)\left(\frac{109}{7}\right)\left(\frac{109}{11}\right) = \left(\frac{4}{5}\right)\left(\frac{4}{7}\right)\left(\frac{10}{11}\right)$$

$$= \left(\frac{2}{5}\right)^2 \left(\frac{2}{7}\right)^2 \left(\frac{-1}{11}\right) = (-1)^2 1^2 (-1) = -1. \quad \blacktriangleleft$$

CARL GUSTAV JACOB JACOBI (1804–1851) was born into a well-to-do German banking family. Jacobi received an excellent early education at home. He studied at the University of Berlin, mastered mathematics through the texts of Euler, and obtained his doctorate in 1825. In 1826, he became a lecturer at the University of Königsberg; he was appointed a professor there in 1831. Besides his work in number theory, Jacobi made important contributions to analysis, geometry, and mechanics. He was also interested in the history of mathematics and was a catalyst in the publication of the collected works of Euler, a job not yet completed although it was begun more than 125 years ago!

When n is prime, the Jacobi symbol is the same as the Legendre symbol. However, when n is composite, the value of the Jacobi symbol $\left(\frac{a}{n}\right)$ does *not* tell us whether the congruence $x^2 \equiv a \pmod{n}$ has solutions. We do know that if the congruence $x^2 \equiv a \pmod{n}$ has solutions, then $\left(\frac{a}{n}\right) = 1$. To see this, note that if p is a prime divisor of n and if $x^2 \equiv a \pmod{n}$ has solutions, then the congruence $x^2 \equiv a \pmod{p}$ also has solutions. Thus, $\left(\frac{a}{p}\right) = 1$. Consequently, $\left(\frac{a}{n}\right) = \prod_{j=1}^{m} \left(\frac{a}{p_j}\right)^{t_j} = 1$, where the prime factorization of n is $n = p_1^{t_1} p_2^{t_2} \cdots p_m^{t_m}$. To see that it is possible that $\left(\frac{a}{n}\right) = 1$ when there are no solutions to $x^2 \equiv a \pmod{n}$, let $a = 2$ and $n = 15$. Note that $\left(\frac{2}{15}\right) = \left(\frac{2}{3}\right)\left(\frac{2}{5}\right) = (-1)(-1) = 1$. However, there are no solutions to $x^2 \equiv 2 \pmod{15}$, because the congruences $x^2 \equiv 2 \pmod{3}$ and $x^2 \equiv 2 \pmod{5}$ have no solutions.

We now show that the Jacobi symbol enjoys some properties similar to those of the Legendre symbol.

Theorem 11.10. Let n be an odd positive integer and let a and b be integers relatively prime to n. Then

  (i)   if $a \equiv b \pmod{n}$, then $\left(\frac{a}{n}\right) = \left(\frac{b}{n}\right)$.

  (ii) $\left(\frac{ab}{n}\right) = \left(\frac{a}{n}\right)\left(\frac{b}{n}\right)$.

  (iii) $\left(\frac{-1}{n}\right) = (-1)^{(n-1)/2}$.

  (iv) $\left(\frac{2}{n}\right) = (-1)^{(n^2-1)/8}$.

Proof. In the proof of this theorem, we use the prime factorization $n = p_1^{t_1} p_2^{t_2} \cdots p_m^{t_m}$.

Proof of (i). We know that if p is a prime dividing n, then $a \equiv b \pmod{p}$. Hence, from Theorem 11.4 (i), we have $\left(\dfrac{a}{p}\right) = \left(\dfrac{b}{p}\right)$. Consequently, we see that

$$\left(\frac{a}{n}\right) = \left(\frac{a}{p_1}\right)^{t_1}\left(\frac{a}{p_2}\right)^{t_2} \cdots \left(\frac{a}{p_m}\right)^{t_m} = \left(\frac{b}{p_1}\right)^{t_1}\left(\frac{b}{p_2}\right)^{t_2} \cdots \left(\frac{b}{p_m}\right)^{t_m} = \left(\frac{b}{n}\right).$$

Proof of (ii). From Theorem 11.4 (ii), we know that $\left(\dfrac{ab}{p_i}\right) = \left(\dfrac{a}{p_i}\right)\left(\dfrac{b}{p_i}\right)$ for $i = 1, 2, 3, \ldots, m$. Hence,

$$\left(\frac{ab}{n}\right) = \left(\frac{ab}{p_1}\right)^{t_1}\left(\frac{ab}{p_2}\right)^{t_2} \cdots \left(\frac{ab}{p_m}\right)^{t_m}$$

$$= \left(\frac{a}{p_1}\right)^{t_1}\left(\frac{b}{p_1}\right)^{t_1}\left(\frac{a}{p_2}\right)^{t_2}\left(\frac{b}{p_2}\right)^{t_2} \cdots \left(\frac{a}{p_m}\right)^{t_m}\left(\frac{b}{p_m}\right)^{t_m}$$

$$= \left(\frac{a}{n}\right)\left(\frac{b}{n}\right).$$

Proof of (iii).   Theorem 11.5 tells us that if p is prime, then $\left(\frac{-1}{p}\right) = (-1)^{(p-1)/2}$. Consequently,

$$\left(\frac{-1}{n}\right) = \left(\frac{-1}{p_1}\right)^{t_1} \left(\frac{-1}{p_2}\right)^{t_2} \cdots \left(\frac{-1}{p_m}\right)^{t_m}$$
$$= (-1)^{t_1(p_1-1)/2 + t_2(p_2-1)/2 + \cdots + t_m(p_m-1)/2}.$$

From the prime factorization of n, we have

$$n = (1 + (p_1 - 1))^{t_1}(1 + (p_2 - 1))^{t_2} \cdots (1 + (p_m - 1))^{t_m}.$$

Because $p_i - 1$ is even, it follows that

$$(1 + (p_i - 1))^{t_i} \equiv 1 + t_i(p_i - 1) \pmod 4$$

and

$$(1 + t_i(p_i - 1))(1 + t_j(p_j - 1)) \equiv 1 + t_i(p_i - 1) + t_j(p_j - 1) \pmod 4.$$

Therefore,

$$n \equiv 1 + t_1(p_1 - 1) + t_2(p_2 - 1) + \cdots + t_m(p_m - 1) \pmod 4,$$

which imples that

$$(n - 1)/2 \equiv t_1(p_1 - 1)/2 + t_2(p_2 - 1)/2 + \cdots + t_m(p_m - 1)/2 \pmod 2.$$

Combining this congruence for $(n - 1)/2$ with the expression for $\left(\frac{-1}{n}\right)$ shows that $\left(\frac{-1}{n}\right) = (-1)^{(n-1)/2}$.

Proof of (iv).   If p is prime, then $\left(\frac{2}{p}\right) = (-1)^{(p^2-1)/8}$. Hence,

$$\left(\frac{2}{n}\right) = \left(\frac{2}{p_1}\right)^{t_1} \left(\frac{2}{p_2}\right)^{t_2} \cdots \left(\frac{2}{p_m}\right)^{t_m} = (-1)^{t_1(p_1^2-1)/8 + t_2(p_2^2-1)/8 + \cdots + t_m(p_m^2-1)/8}.$$

As in the proof of (iii), we note that

$$n^2 = (1 + (p_1^2 - 1))^{t_1}(1 + (p_2^2 - 1))^{t_2} \cdots (1 + (p_m^2 - 1))^{t_m}.$$

Because $p_i^2 - 1 \equiv 0 \pmod 8$ for $i = 1, 2, \ldots, m$, we see that

$$(1 + (p_i^2 - 1))^{t_i} \equiv 1 + t_i(p_i^2 - 1) \pmod{64}$$

and

$$(1 + t_i(p_i^2 - 1))(1 + t_j(p_j^2 - 1)) \equiv 1 + t_i(p_i^2 - 1) + t_j(p_j^2 - 1) \pmod{64}.$$

Hence,

$$n^2 \equiv 1 + t_1(p_1^2 - 1) + t_2(p_2^2 - 1) + \cdots + t_m(p_m^2 - 1) \pmod{64},$$

which implies that

$$(n^2 - 1)/8 = t_1(p_1^2 - 1)/8 + t_2(p_2^2 - 1)/8 + \cdots + t_m(p_m^2 - 1)/8 \pmod{8}.$$

Combining this congruence for $(n^2 - 1)/8$ with the expression for $\left(\frac{2}{n}\right)$ tells us that $\left(\frac{2}{n}\right) = (-1)^{(n^2-1)/8}$. ∎

We now demonstrate that the reciprocity law holds for the Jacobi symbol as well as the Legendre symbol.

Theorem 11.11. *The Reciprocity Law for Jacobi Symbols.* Let n and m be relatively prime odd positive integers. Then

$$\left(\frac{n}{m}\right)\left(\frac{m}{n}\right) = (-1)^{\frac{m-1}{2} \cdot \frac{n-1}{2}}.$$

Proof. Let the prime factorizations of m and n be $m = p_1^{a_1} p_2^{a_2} \cdots p_s^{a_s}$ and $n = q_1^{b_1} q_2^{b_2} \cdots q_r^{b_r}$. We see that

$$\left(\frac{m}{n}\right) = \prod_{i=1}^{r} \left(\frac{m}{q_i}\right)^{b_i} = \prod_{i=1}^{r}\prod_{j=1}^{s} \left(\frac{p_j}{q_j}\right)^{b_i a_j}$$

and

$$\left(\frac{n}{m}\right) = \prod_{j=1}^{s} \left(\frac{n}{p_j}\right)^{a_j} = \prod_{j=1}^{s}\prod_{i=1}^{r} \left(\frac{q_i}{p_j}\right)^{a_j b_i}.$$

Thus,

$$\left(\frac{m}{n}\right)\left(\frac{n}{m}\right) = \prod_{i=1}^{r}\prod_{j=1}^{s} \left[\left(\frac{p_j}{q_i}\right)\left(\frac{q_i}{p_j}\right)\right]^{a_j b_i}.$$

By the law of quadratic reciprocity, we know that

$$\left(\frac{p_j}{q_i}\right)\left(\frac{q_i}{p_j}\right) = (-1)^{\left(\frac{p_j-1}{2}\right)\left(\frac{q_i-1}{2}\right)}.$$

Hence,

$$(11.8) \quad \left(\frac{m}{n}\right)\left(\frac{n}{m}\right) = \prod_{i=1}^{r}\prod_{j=1}^{s} (-1)^{a_j\left(\frac{p_j-1}{2}\right)b_i\left(\frac{q_i-1}{2}\right)} = (-1)^{\sum_{i=1}^{r}\sum_{j=1}^{s} a_j\left(\frac{p_j-1}{2}\right)b_i\left(\frac{q_i-1}{2}\right)}.$$

We note that

$$\sum_{i=1}^{r}\sum_{j=1}^{s} a_j\left(\frac{p_j-1}{2}\right)b_i\left(\frac{q_i-1}{2}\right) = \sum_{j=1}^{s} a_j\left(\frac{p_j-1}{2}\right)\sum_{i=1}^{r} b_i\left(\frac{q_i-1}{2}\right).$$

As we demonstrated in the proof of Theorem 11.10 (iii),

$$\sum_{j=1}^{s} a_j \left(\frac{p_j - 1}{2} \right) \equiv \frac{m-1}{2} \pmod{2}$$

and

$$\sum_{i=1}^{r} b_i \left(\frac{q_i - 1}{2} \right) \equiv \frac{n-1}{2} \pmod{2}.$$

Thus,

(11.9) $$\sum_{i=1}^{r}\sum_{j=1}^{s} a_j \left(\frac{p_j - 1}{2} \right) b_i \left(\frac{q_i - 1}{2} \right) \equiv \frac{m-1}{2} \cdot \frac{n-1}{2} \pmod{2}.$$

Therefore, by equations (11.8) and (11.9), we can conclude that

$$\left(\frac{m}{n} \right) \left(\frac{n}{m} \right) = (-1)^{\frac{m-1}{2} \cdot \frac{n-1}{2}}. \qquad \blacksquare$$

An Algorithm for Computing Jacobi Symbols   We now develop an efficient algorithm for evaluating Jacobi symbols. Let a and b be relatively prime positive integers with $a > b$. Let $R_0 = a$ and $R_1 = b$. Using the division algorithm and factoring out the highest power of two dividing the remainder, we obtain

$$R_0 = R_1 q_1 + 2^{s_1} R_2,$$

where s_1 is a nonnegative integer and R_2 is an odd positive integer less than R_1. When we successively use the division algorithm, and factor out the highest power of two that divides remainders, we obtain

$$R_1 = R_2 q_2 + 2^{s_2} R_3$$
$$R_2 = R_3 q_3 + 2^{s_3} R_4$$

$$\vdots$$

$$R_{n-3} = R_{n-2} q_{n-2} + 2^{s_{n-2}} R_{n-1}$$
$$R_{n-2} = R_{n-1} q_{n-1} + 2^{s_{n-1}} \cdot 1,$$

where s_j is a nonnegative integer and R_j is an odd positive integer less than R_{j-1} for $j = 2, 3, \ldots, n - 1$. Note that the number of divisions required to reach the final equation does not exceed the number of divisions required to find the greatest common divisor of a and b using the Euclidean algorithm.

We illustrate this sequence of equations with the following example.

Example 11.14.   Let $a = 401$ and $b = 111$. Then

$$401 = 111 \cdot 3 + 2^2 \cdot 17$$
$$111 = 17 \cdot 6 + 2^0 \cdot 9$$
$$17 = 9 \cdot 1 + 2^3 \cdot 1. \qquad \blacktriangleleft$$

Using the sequence of equations that we have described, together with the properties of the Jacobi symbol, we prove the following theorem, which gives an algorithm for evaluating Jacobi symbols.

Theorem 11.12. Let a and b be positive integers with $a > b$. Then

$$\left(\frac{a}{b}\right) = (-1)^{s_1 \frac{R_1^2-1}{8} + \cdots + s_{n-1} \frac{R_{n-1}^2-1}{8} + \frac{R_1-1}{2} \cdot \frac{R_2-1}{2} + \cdots + \frac{R_{n-2}-1}{2} \cdot \frac{R_{n-1}-1}{2}},$$

where the integers R_j and s_j, $j = 1, 2, \ldots, n-1$, are as previously described.

Proof. From the first equation with (i), (ii), and (iv) of Theorem 11.10, we have

$$\left(\frac{a}{b}\right) = \left(\frac{R_0}{R_1}\right) = \left(\frac{2^{S_1}R_2}{R_1}\right) = \left(\frac{2}{R_1}\right)^{s_1}\left(\frac{R_2}{R_1}\right) = (-1)^{s_1 \frac{R_1^2-1}{8}}\left(\frac{R_2}{R_1}\right).$$

Using Theorem 11.11, the reciprocity law for Jacobi symbols, we have

$$\left(\frac{R_2}{R_1}\right) = (-1)^{\frac{R_1-1}{2} \cdot \frac{R_2-1}{2}}\left(\frac{R_1}{R_2}\right),$$

so that

$$\left(\frac{a}{b}\right) = (-1)^{\frac{R_1-1}{2} \cdot \frac{R_2-1}{2} + s_1 \frac{R_1^2-1}{8}}\left(\frac{R_1}{R_2}\right).$$

Similarly, using the subsequent divisions, we find that

$$\left(\frac{R_{j-1}}{R_j}\right) = (-1)^{\frac{R_j-1}{2} \cdot \frac{R_{j+1}-1}{2} + s_1 \frac{R_j^2-1}{8}}\left(\frac{R_j}{R_{j+1}}\right)$$

for $j = 2, 3, \ldots, n-1$. When we combine all the equalities, we obtain the desired expression for $\left(\frac{a}{b}\right)$. ∎

The following example illustrates the use of Theorem 11.12.

Example 11.15. To evaluate $\left(\frac{401}{111}\right)$, we use the sequence of divisions in Example 11.14 and Theorem 11.12. This tells us that

$$\left(\frac{401}{111}\right) = (-1)^{2 \cdot \frac{111^2-1}{8} + 0 \cdot \frac{17^2-1}{8} + 3 \cdot \frac{9^2-1}{8} + \frac{111-1}{2} \cdot \frac{17-1}{2} + \frac{17-1}{2} \cdot \frac{9-1}{2}} = 1. \quad \blacktriangleleft$$

The following corollary describes the computational complexity of the algorithm for evaluating Jacobi symbols given in Theorem 11.12.

Corollary 11.12.1. Let a and b be relatively prime positive integers with $a > b$. Then the Jacobi symbol $\left(\frac{a}{b}\right)$ can be evaluated using $O((\log_2 b)^3)$ bit operations.

Proof. To find $\left(\frac{a}{b}\right)$ using Theorem 11.12, we perform a sequence of $O(\log_2 b)$ divisions. To see this, note that the number of divisions does not exceed the number of divisions needed to find (a, b) using the Euclidean algorithm. Thus, by Lamé's theorem, we know that $O(\log_2 b)$ divisions are needed. Each division can be done using $O((\log_2 b)^2)$ bit operations. Each pair of integers R_j and s_j can be found using $O(\log_2 b)$ bit operations once the appropriate division has been carried out.

Consequently, $O((\log_2 b)^3)$ bit operations are required to find the integers R_j, s_j, $j = 1, 2, \ldots, n - 1$ from a and b. Finally, to evaluate the exponent of -1 in the expression for $\left(\frac{a}{b}\right)$ in Theorem 11.12, we use the last three bits in the binary expansions of R_j, $j = 1, 2, \ldots, n - 1$ and the last bit in the binary expansions of s_j, $j = 1, 2, \ldots, n - 1$. Therefore, we use $O(\log_2 b)$ additional bit operations to find $\left(\frac{a}{b}\right)$. Because $O((\log_2 b)^3) + O(\log_2 b) = O((\log_2 b)^3)$, the corollary holds. ∎

We can improve this corollary if we use more care when estimating the number of bit operations used by divisions. In particular, we can show that $O((\log_2 b)^2)$ bit operations suffice for evaluating $\left(\frac{a}{b}\right)$. We leave this as an exercise.

11.3 Exercises

1. Evaluate each of the following Jacobi symbols.

   a) $\left(\frac{5}{21}\right)$ c) $\left(\frac{111}{1001}\right)$ e) $\left(\frac{2663}{3299}\right)$

   b) $\left(\frac{27}{101}\right)$ d) $\left(\frac{1009}{2307}\right)$ f) $\left(\frac{10001}{20003}\right)$

2. For which positive integers n that are relatively prime to 15 does the Jacobi symbol $\left(\frac{15}{n}\right)$ equal 1?

3. For which positive integers n that are relatively prime to 30 does the Jacobi symbol $\left(\frac{30}{n}\right)$ equal 1?

Suppose that $n = pq$, where p and q are primes. We say that the integer a is a *pseudo-square* modulo n if a is a quadratic nonresidue of n, but $\left(\frac{a}{n}\right) = 1$.

4. Show that if a is a pseudo-square modulo n, then $\left(\frac{a}{p}\right) = \left(\frac{a}{q}\right) = -1$.

5. Find all the pseudo-squares modulo 21.

6. Find all the pseudo-squares modulo 35.

7. Find all the pseudo-squares modulo 143.

8. Let a and b be relatively prime integers such that b is odd and positive and $a = (-1)^s 2^t q$, where q is odd. Show that

$$\left(\frac{a}{b}\right) = (-1)^{\frac{b-1}{2} \cdot s + \frac{b^2-1}{8} \cdot t} \left(\frac{q}{b}\right).$$

9. Let n be an odd square-free positive integer. Show that there is an integer a such that $(a, n) = 1$ and $\left(\frac{a}{n}\right) = -1$.

10. Let n be an odd square-free positive integer.

    a) Show that $\sum \left(\frac{k}{n}\right) = 0$, where the sum is taken over all k in a reduced set of residues modulo n. (*Hint:* Use Exercise 9.)

b) From part (a), show that the number of integers in a reduced set of residues modulo n such that $\left(\frac{k}{n}\right) = 1$ is equal to the number with $\left(\frac{k}{n}\right) = -1$.

* **11.** Let a and $b = r_0$ be relatively prime odd positive integers such that

$$a = r_0 q_1 + \varepsilon_1 r_1$$
$$r_0 = r_1 q_2 + \varepsilon_2 r_2$$

$$\vdots$$

$$r_{n-1} = r_{n-1} q_{n-1} + \varepsilon_n r_n,$$

where q_i is a nonnegative even integer, $\varepsilon_i = \pm 1$, r_i is a positive integer with $r_i < r_{i-1}$, for $i = 1, 2, \ldots, n_j$, and $r_n = 1$. These equations are obtained by successively using the modified division algorithm given in Exercise 18 of Section 1.5.

a) Show that Jacobi symbol $\left(\frac{a}{b}\right)$ is given by

$$\left(\frac{a}{b}\right) = (-1)^{\left(\frac{r_0-1}{2} \cdot \frac{\varepsilon_1 r_1 - 1}{2} + \frac{r_1-1}{2} \cdot \frac{\varepsilon_2 r_2 - 1}{2} + \cdots + \frac{r_{n-1}-1}{2} \cdot \frac{\varepsilon_n r_n - 1}{2}\right)}.$$

b) Show that the Jacobi symbol $\left(\frac{a}{b}\right)$ is given by

$$\left(\frac{a}{b}\right) = (-1)^T,$$

where T is the number of integers i, $1 \le i \le n$, with $r_{i-1} \equiv \varepsilon_i r_i \equiv 3 \pmod 4$.

* **12.** Show that if a and b are odd integers and $(a, b) = 1$, then the following reciprocity law holds for the Jacobi symbol:

$$\left(\frac{a}{|b|}\right)\left(\frac{b}{|a|}\right) = \begin{cases} -(-1)^{\frac{a-1}{2}\frac{b-1}{2}} & \text{if } a < 0 \text{ and } b < 0; \\ (-1)^{\frac{a-1}{2}\frac{b-1}{2}} & \text{otherwise.} \end{cases}$$

In Exercises 13–19, we deal with the *Kronecker symbol* (named after Leopold Kronecker), which is defined as follows. Let a be a positive integer that is not a perfect square such that $a \equiv 0$ or $1 \pmod 4$. We define

$$\left(\frac{a}{2}\right) = \begin{cases} 1 & \text{if } a \equiv 1 \pmod 8; \\ -1 & \text{if } a \equiv 5 \pmod 8. \end{cases}$$

$$\left(\frac{a}{p}\right) = \text{the Legendre symbol } \left(\frac{a}{p}\right) \text{ if } p \text{ is an odd prime such that } p \nmid a.$$

$$\left(\frac{a}{n}\right) = \prod_{j=1}^{r} \left(\frac{a}{p_j}\right)^{t_j} \text{ if } (a, n) = 1 \text{ and } n = \prod_{j=1}^{r} p_j^{t_j} \text{ is the prime factorization of } n.$$

13. Evaluate each of the following Kronecker symbols,

a) $\left(\frac{5}{12}\right)$ b) $\left(\frac{13}{20}\right)$ c) $\left(\frac{101}{200}\right)$

For Exercises 14–19, let a be a positive integer that is not a perfect square such that $a \equiv 0$ or $1 \pmod 4$.

14. Show that $\left(\frac{a}{2}\right) = \left(\frac{2}{|a|}\right)$ if $2 \nmid a$, where the symbol on the right is a Jacobi symbol.

15. Show that if n_1 and n_2 are positive integers and if $(a_1, n_1, n_2) = 1$, then $\left(\frac{a}{n_1 n_2}\right) = \left(\frac{a}{n_1}\right) \cdot \left(\frac{a}{n_2}\right)$.

$*$ **16.** Show that if n is a positive integer relatively prime to a and if a is odd, then $\left(\frac{a}{n}\right) = \left(\frac{n}{|a|}\right)$, whereas if a is even and $a = 2^s t$, where t is odd, then

$$\left(\frac{a}{n}\right) = \left(\frac{2}{n}\right)^s (-1)^{\frac{t-1}{2} \cdot \frac{n-1}{2}} \left(\frac{n}{|t|}\right).$$

$*$ **17.** Show that if n_1 and n_2 are positive integers relatively prime to a and $n_1 \equiv n_2 \pmod{|a|}$, then $\left(\frac{a}{n_1}\right) = \left(\frac{a}{n_2}\right)$.

LEOPOLD KRONECKER (1823–1891) was born in Liegnitz, Prussia, to prosperous Jewish parents. His father was a successful businessman and his mother came from a wealthy family. As a child, Kronecker was taught by private tutors. He later entered the Liegnitz Gymnasium where he was taught mathematics by the number theorist Kummer. Kronecker's mathematical talents were quickly recognized by Kummer, who encouraged Kronecker to engage in mathematics research. In 1841, Kronecker entered Berlin University where he studied mathematics, astronomy, meteorology, chemistry, and philosophy. In 1845, Kronecker wrote his doctoral thesis on algebraic number theory; his supervisor was Dirichlet.

Kronecker could have begun a promising academic career, but instead he returned to Liegnitz to help manage the banking business of an uncle. In 1848, Kronecker married a daughter of this uncle. During his time back in Liegnitz, Kronecker continued his research for his own enjoyment. In 1855, when his family obligations eased, Kronecker returned to Berlin. He was eager to participate in the mathematical life of the university. Not holding a university post, he did not teach any classes. However, he was extremely active in research and he published extensively in number theory, elliptic functions and algebra, and their interconnections. In 1860, Kronecker was elected to the Berlin Academy, giving him the right to lecture at Berlin University. He took advantage of this opportunity and lectured on number theory and other mathematical topics. Kronecker's lectures were considered very demanding but were also considered to be stimulating. Unfortunately, he was not a popular teacher with average students; most of these dropped out of his courses by the end of the semester.

Kronecker was a strong believer in constructive mathematics, thinking that mathematics should be concerned only with finite numbers and with a finite number of operations. He doubted the validity of nonconstructive existence proofs and was opposed to objects defined nonconstructively, such as irrational numbers. He did not believe that transcendental numbers could exist. He is famous for his statement: "God created the integers, all else is the work of man." Kronecker's belief in constructive mathematics was not shared by most of his colleagues, although he was not the only prominent mathematician to hold such beliefs. Many mathematicians found it difficult to get along with Kronecker, especially because he was prone to fallings out over mathematical disagreements. Also, Kronecker was self-conscious about his short height, reacting badly even to good-natured references to his short stature.

* **18.** Show that if $a \neq 0$, then there exists a positive integer n such that $\left(\frac{a}{n}\right) = -1$.

* **19.** Show that if $a \neq 0$, then $\left(\frac{a}{|a|-1}\right) = \begin{cases} 1 & \text{if } a > 0; \\ -1 & \text{if } a < 0. \end{cases}$

20. Show that if a and b are relatively prime integers with $a < b$, then Jacobi symbol $\left(\frac{a}{b}\right)$ can be evaluated using $O((\log_2 b)^2)$ bit operations.

11.3 Computational and Programming Exercises

Computations and Explorations

Using a computation program such as Maple or *Mathematica,* or programs you have written, carry out the following computations and explorations.

1. Find the value of each of the Legendre symbol $\left(\frac{1,656,169}{2,355,151}\right)$.

2. Find the value of the following Jacobi symbols: $\left(\frac{9343}{65,518,791}\right)$, $\left(\frac{54371}{5,400,207,333}\right)$, $\left(\frac{320001}{11,111,111,111,111}\right)$.

Programming Projects

Write computer programs using Maple, *Mathematica,* or a language of your choice to do the following.

1. Evaluate Jacobi symbols using the method of Theorem 11.12.

2. Evaluate Jacobi symbols using Exercises 8 and 11.

3. Evaluate Kronecker symbols (as defined in the preamble to Exercise 13).

11.4 Euler Pseudoprimes

Let p be an odd prime number and let b be an integer not divisible by p. By Euler's criterion, we know that

$$b^{(p-1)/2} \equiv \left(\frac{b}{p}\right) \pmod{p}.$$

Hence, if we wish to test the positive integer n for primality, we can take an integer b, with $(b, n) = 1$, and determine whether

$$b^{(n-1)/2} \equiv \left(\frac{b}{n}\right) \pmod{n},$$

where the symbol on the right-hand side of the congruence is the Jacobi symbol. If we find that this congruence fails, then n is composite.

Example 11.16. Let $n = 341$ and $b = 2$. We calculate that $2^{170} \equiv 1 \pmod{341}$. Because $341 \equiv -3 \pmod{8}$, using Theorem 11.10 (iv), we see that $\left(\frac{2}{341}\right) = -1$. Consequently, $2^{170} \not\equiv \left(\frac{2}{341}\right) \pmod{341}$. This demonstrates that 341 is not prime. ◀

Thus, we can define a type of pseudoprime based on Euler's criterion.

Definition. An odd, composite, positive integer n that satisfies the congruence

$$b^{(n-1)/2} \equiv \left(\frac{b}{n}\right) \pmod{n},$$

where b is a positive integer, is called an *Euler pseudoprime to the base b*.

An Euler pseudoprime to the base b is a composite integer that masquerades as a prime by satisfying the congruence given in the definition.

Example 11.17. Let $n = 1105$ and $b = 2$. We calculate that $2^{552} \equiv 1 \pmod{1105}$. Because $1105 \equiv 1 \pmod{8}$, we see that $\left(\frac{2}{1105}\right) = 1$. Hence, $2^{552} \equiv \left(\frac{2}{1105}\right) \pmod{1105}$. Because 1105 is composite, it is an Euler pseudoprime to the base 2. ◀

The following theorem shows that every Euler pseudoprime to the base b is a pseudoprime to this base.

Theorem 11.13. If n is an Euler pseudoprime to the base b, then n is a pseudoprime to the base b.

Proof. If n is an Euler pseudoprime to the base b, then

$$b^{(n-1)/2} \equiv \left(\frac{b}{n}\right) \pmod{n}.$$

Hence, by squaring both sides of this congruence, we find that

$$(b^{(n-1)/2})^2 \equiv \left(\frac{b}{n}\right)^2 \pmod{n}.$$

Because $\left(\frac{b}{n}\right) = \pm 1$, we see that $b^{n-1} \equiv 1 \pmod{n}$, which means that n is a pseudoprime to the base b. ∎

Not every pseudoprime is an Euler pseudoprime. For example, the integer 341 is not an Euler pseudoprime to the base 2, as we have shown, but is a pseudoprime to this base.

We know that every Euler pseudoprime is a pseudoprime. Next, we show that every strong pseudoprime is an Euler pseudoprime.

Theorem 11.14. If n is a strong pseudoprime to the base b, then n is an Euler pseudoprime to this base.

Proof. Let n be a strong pseudoprime to the base b. Then, if $n - 1 = 2^s t$, where t is odd, either $b^t \equiv 1 \pmod{n}$ or $b^{2^r t} \equiv -1 \pmod{n}$, where $0 \le r \le s - 1$. Let $n = \prod_{i=1}^{m} p_i^{a_i}$ be the prime-power factorization of n.

First, consider the case where $b^t \equiv 1 \pmod{n}$. Let p be a prime divisor of n. Because $b^t \equiv 1 \pmod{p}$, we know that $\operatorname{ord}_p b \mid t$. Because t is odd, we see that $\operatorname{ord}_p b$ is also odd. Hence, $\operatorname{ord}_p b \mid (p - 1)/2$, because $\operatorname{ord}_p b$ is an odd divisor of the even integer $\phi(p) = p - 1$. Therefore,

$$b^{(p-1)/2} \equiv 1 \pmod{p}.$$

Consequently, by Euler's criterion, we have $\left(\dfrac{b}{p}\right) = 1$.

To compute the Jacobi symbol $\left(\dfrac{b}{n}\right)$, we note that $\left(\dfrac{b}{p}\right) = 1$ for all primes p dividing n. Hence,

$$\left(\frac{b}{n}\right) = \left(\frac{b}{\prod_{i=1}^{m} p_i^{a_i}}\right) = \prod_{i=1}^{m} \left(\frac{b}{p_i}\right)^{a_i} = 1.$$

Because $b^t \equiv 1 \pmod{n}$, we know that $b^{(n-1)/2} = (b^t)^{2^{s-1}} \equiv 1 \pmod{n}$. Therefore, we have

$$b^{(n-1)/2} \equiv \left(\frac{b}{n}\right) \equiv 1 \pmod{n}.$$

We conclude that n is an Euler pseudoprime to the base b.

Next, we consider the case where

$$b^{2^r t} \equiv -1 \pmod{n}$$

for some r with $0 \le r \le s - 1$. If p is a prime divisor of n, then

$$b^{2^r t} \equiv -1 \pmod{p}.$$

Squaring both sides of this congruence, we obtain

$$b^{2^{r+1} t} \equiv 1 \pmod{p},$$

which implies that $\operatorname{ord}_p b \mid 2^{r+1} t$, but that $\operatorname{ord}_p b \nmid 2^r t$. Hence,

$$\operatorname{ord}_p b = 2^{r+1} c,$$

where c is an odd integer. Because $\operatorname{ord}_p b \mid (p - 1)$ and $2^{r+1} \mid \operatorname{ord}_p b$, it follows that $2^{r+1} \mid (p - 1)$. Therefore, we have $p = 2^{r+1} d + 1$, where d is an integer. Because

$$b^{(\operatorname{ord}_p b)/2} \equiv -1 \pmod{p},$$

we have

$$\left(\frac{b}{p}\right) \equiv b^{(p-1)/2} = b^{(\operatorname{ord}_p b/2)((p-1)/\operatorname{ord}_p b)}$$

$$\equiv (-1)^{(p-1)/\operatorname{ord}_p b} = (-1)^{(p-1)/2^{r+1} c} \pmod{p}.$$

Because c is odd, we know that $(-1)^c = -1$. Hence,

$$(11.10) \qquad \left(\frac{b}{p}\right) = (-1)^{(p-1)/2^{r+1}} = (-1)^d,$$

recalling that $d = (p-1)/2^{r+1}$. Because each prime p_i dividing n is of the form $p_i = 2^{r+1}d_i + 1$, it follows that

$$n = \prod_{i=1}^{m} p_i^{a_i}$$

$$= \prod_{i=1}^{m} (2^{r+1}d_i + 1)^{a_i}$$

$$\equiv \prod_{i=1}^{m} (1 + 2^{r+1}a_i d_i)$$

$$\equiv 1 + 2^{r+1} \sum_{i=1}^{m} a_i d_i \pmod{2^{2r+2}}.$$

Therefore,

$$t 2^{s-1} = (n-1)/2 \equiv 2^r \sum_{i=1}^{m} a_i d_i \pmod{2^{r+1}}.$$

This congruence implies that

$$t 2^{s-1-r} \equiv \sum_{i=1}^{m} a_i d_i \pmod{2}$$

and

$$(11.11) \qquad b^{(n-1)/2} = (b^{2^r t})^{2^{s-1-r}} \equiv (-1)^{2^{s-1-r}} = (-1)^{\sum_{i=1}^{m} a_i d_i} \pmod{n}.$$

On the other hand, from (11.10), we have

$$\left(\frac{b}{n}\right) = \prod_{i=1}^{m} \left(\frac{b}{p_i}\right)^{a_i} = \prod_{i=1}^{m} ((-1)^{d_i})^{a_i} = \prod_{i=1}^{m} (-1)^{a_i d_i} = (-1)^{\sum_{i=1}^{m} a_i d_i}.$$

Therefore, combining the preceding equation with (11.11), we see that

$$b^{(n-1)/2} \equiv \left(\frac{b}{n}\right) \pmod{n}.$$

Consequently, n is an Euler pseudoprime to the base b. ∎

Although every strong pseudoprime to the base b is an Euler pseudoprime to this base, note that not every Euler pseudoprime to the base b is a strong pseudoprime to the base b, as the following example shows.

Example 11.18. We have previously shown that the integer 1105 is an Euler pseudoprime to the base 2. However, 1105 is not a strong pseudoprime to the base 2, because

$$2^{(1105-1)/2} = 2^{552} \equiv 1 \pmod{1105},$$

whereas

$$2^{(1105-1)/2^2} = 2^{276} \equiv 781 \not\equiv \pm 1 \pmod{1105}. \qquad \blacktriangleleft$$

Although an Euler pseudoprime to the base b is not always a strong pseudoprime to this base, when certain additional conditions are met, an Euler pseudoprime to the base b is, in fact, a strong pseudoprime to this base. The following two theorems give results of this kind.

Theorem 11.15. If $n \equiv 3 \pmod{4}$ and n is an Euler pseudoprime to the base b, then n is a strong pseudoprime to the base b.

Proof. From the congruence $n \equiv 3 \pmod{4}$, we know that $n - 1 = 2 \cdot t$, where $t = (n-1)/2$ is odd. Because n is an Euler pseudoprime to the base b, it follows that

$$b^t = b^{(n-1)/2} \equiv \left(\frac{b}{n} \right) \pmod{n}.$$

Because $\left(\frac{b}{n} \right) = \pm 1$, we know that either $b^t \equiv 1 \pmod{n}$ or $b^t \equiv -1 \pmod{n}$.

Hence, one of the congruences in the definition of a strong pseudoprime to the base b must hold. Consequently, n is a strong pseudoprime to the base b. $\qquad \blacksquare$

Theorem 11.16. If n is an Euler pseudoprime to the base b and $\left(\frac{b}{n} \right) = -1$, then n is a strong pseudoprime to the base b.

Proof. We write $n - 1 = 2^s t$, where t is odd and s is a positive integer. Because n is an Euler pseudoprime to the base b, we have

$$b^{2^{s-1}t} = b^{(n-1)/2} \equiv \left(\frac{b}{n} \right) \pmod{n}.$$

But because $\left(\frac{b}{n} \right) = -1$, we see that

$$b^{t 2^{s-1}} \equiv -1 \pmod{n}.$$

This is one of the congruences in the definition of a strong pseudoprime to the base b. Because n is composite, it is a strong pseudoprime to the base b. $\qquad \blacksquare$

Using the concept of Euler pseudoprimality, we will develop a probabilistic primality test. This test was first suggested by Solovay and Strassen [SoSt 77].

Before presenting the test, we give some helpful lemmas.

Lemma 11.4. If n is an odd positive integer that is not a perfect square, then there is at least one integer b with $1 < b < n$, $(b, n) = 1$, and $\left(\frac{b}{n} \right) = -1$, where $\left(\frac{b}{n} \right)$ is the Jacobi symbol.

Proof.   If n is prime, the existence of such an integer b is guaranteed by Theorem 11.1. If n is composite, because n is not a perfect square, we can write $n = rs$, where $(r, s) = 1$ and $r = p^e$, with p an odd prime and e an odd positive integer.

Now, let t be a quadratic nonresidue of the prime p; such a t exists by Theorem 11.1. We use the Chinese remainder theorem to find an integer b such that $1 < b < n, (b, n) = 1$, and such that b satisfies the two congruences

$$b \equiv t \pmod{r}$$
$$b \equiv 1 \pmod{s}.$$

Then

$$\left(\frac{b}{r}\right) = \left(\frac{b}{p^e}\right) = \left(\frac{b}{p}\right)^e = (-1)^e = -1$$

and $\left(\frac{b}{s}\right) = 1$. Because $\left(\frac{b}{n}\right) = \left(\frac{b}{r}\right)\left(\frac{b}{s}\right)$, it follows that $\left(\frac{b}{n}\right) = -1$.   ∎

Lemma 11.5.   Let n be an odd composite integer. Then there is at least one integer b with $1 < b < n, (b, n) = 1$, and

$$b^{(n-1)/2} \not\equiv \left(\frac{b}{n}\right) \pmod{n}.$$

Proof.   Assume, for all positive integers not exceeding n and relatively prime to n, that

$$(11.12) \qquad\qquad b^{(n-1)/2} \equiv \left(\frac{b}{n}\right) \pmod{n}.$$

Squaring both sides of this congruence tells us that

$$b^{n-1} \equiv \left(\frac{b}{n}\right)^2 \equiv (\pm 1)^2 = 1 \pmod{n},$$

if $(b, n) = 1$. Hence, n must be a Carmichael number. Therefore, by Theorem 9.24, we know that $n = q_1 q_2 \cdots q_r$, where $q_1, q_2, \ldots, q_r$ are distinct odd primes.

We will now show that

$$b^{(n-1)/2} \equiv 1 \pmod{n}$$

for all integers b with $1 \le b \le n$ and $(b, n) = 1$. Suppose that b is an integer such that

$$b^{(n-1)/2} \equiv -1 \pmod{n}.$$

We use the Chinese remainder theorem to find an integer a with $1 < a < n, (a, n) = 1$, and

$$a \equiv b \pmod{q_1}$$
$$a \equiv 1 \pmod{q_2 q_3 \cdots q_r}.$$

Then we observe that

(11.13) $$a^{(n-1)/2} \equiv b^{(n-1)/2} \equiv -1 \ (\text{mod } q_1).$$

whereas

(11.14) $$a^{(n-1)/2} \equiv 1 \ (\text{mod } q_2 q_3 \cdots q_r).$$

From congruences (11.13) and (11.14), we see that

$$a^{(n-1)/2} \not\equiv \pm 1 \ (\text{mod } n),$$

contradicting congruence (11.12). Hence, we must have

$$b^{(n-1)/2} \equiv 1 \ (\text{mod } n),$$

for all b with $1 \le b \le n$ and $(b, n) = 1$. Consequently, from the definition of an Euler pseudoprime, we know that

$$b^{(n-1)/2} \equiv \left(\frac{b}{n} \right) = 1 \ (\text{mod } n),$$

for all b with $1 \le b \le n$ and $(b, n) = 1$. However, Lemma 11.4 tells us that this is impossible. Hence, the original assumption is false. There must be at least one integer b with $1 < b < n$, $(b, n) = 1$, and

$$b^{(n-1)/2} \not\equiv \left(\frac{b}{n} \right) \ (\text{mod } n). \qquad \blacksquare$$

We can now state and prove the theorem that is the basis of the probabilistic primality test.

Theorem 11.17. Let n be an odd composite integer. Then the number of positive integers less than n and relatively prime to n that are bases to which n is an Euler pseudoprime does not exceed $\phi(n)/2$.

Proof. By Lemma 11.5, we know that there is an integer b with $1 < b < n$, $(b, n) = 1$, and

(11.15) $$b^{(n-1)/2} \not\equiv \left(\frac{b}{n} \right) \ (\text{mod } n).$$

Now, let $a_1, a_2, \ldots, a_m$ denote the positive integers less than n satisfying $1 \le a_j \le n$, $(a_j, n) = 1$, and

(11.16) $$a_j^{(n-1)/2} \equiv \left(\frac{a_j}{n} \right) \ (\text{mod } n),$$

for $j = 1, 2, \ldots, m$.

Let $r_1, r_2, \ldots, r_m$ be the least positive residues of the integers $ba_1, ba_2, \ldots, ba_m$ modulo n. We note that the integers r_j are distinct and that $(r_j, n) = 1$ for $j = 1, 2, \ldots, m$. Furthermore,

(11.17) $$r_j^{(n-1)/2} \not\equiv \left(\frac{r_j}{n} \right) \ (\text{mod } n);$$

for, if it were true that

$$r_j^{(n-1)/2} \equiv \left(\frac{r_j}{n}\right) \pmod{n},$$

then we would have

$$(ba_j)^{(n-1)/2} \equiv \left(\frac{ba_j}{n}\right) \pmod{n},$$

which would imply that

$$b^{(n-1)/2}a_j^{(n-1)/2} \equiv \left(\frac{b}{n}\right)\left(\frac{a_j}{n}\right) \pmod{n},$$

and because (11.16) holds, we would have

$$b^{(n-1)/2} \equiv \left(\frac{b}{n}\right),$$

contradicting (11.15).

Because a_j, $j = 1, 2, \ldots, m$, satisfies the congruence (11.16), whereas r_j, $j = 1, 2, \ldots, m$, does not, as (11.17) shows, we know that these two sets of integers share no common elements. Hence, looking at the two sets together, we have a total of $2m$ distinct positive integers less than n and relatively prime to n. Because there are $\phi(n)$ integers less than n that are relatively prime to n, we can conclude that $2m \le \phi(n)$, so that $m \le \phi(n)/2$. This proves the theorem. ∎

By Theorem 11.17, we see that if n is an odd composite integer, when an integer b is selected at random from the integers $1, 2, \ldots, n - 1$, the probability that n is an Euler pseudoprime to the base b is less than 1/2. This leads to the following probabilistic primality test.

Theorem 11.18. *The Solovay-Strassen Probabilistic Primality Test.* Let n be a positive integer. Select, at random, k integers $b_1, b_2, \ldots, b_k$ from the integers $1, 2, \ldots, n - 1$. For each of these integers b_j, $j = 1, 2, \ldots, k$, determine whether

$$b_j^{(n-1)/2} \equiv \left(\frac{b_j}{n}\right) \pmod{n}.$$

If any of these congruences fails, then n is composite. If n is prime, then all these congruences hold. If n is composite, the probability that all k congruences hold is less than $1/2^k$. Therefore, if n passes this test when k is large, then n is "almost certainly prime."

Because every strong pseudoprime to the base b is an Euler pseudoprime to this base, more composite integers pass the Solovay-Strassen probabilistic primality test than the Rabin probabilistic primality test, although both require $O(k(\log_2 n)^3)$ bit operations.

11.4 Exercises

1. Show that the integer 561 is an Euler pseudoprime to the base 2.

2. Show that the integer 15,841 is an Euler pseudoprime to the base 2, a strong pseudoprime to the base 2 and a Carmichael number.

3. Show that if n is an Euler pseudoprime to the bases a and b, then n is an Euler pseudoprime to the base ab.

4. Show that if n is an Euler pseudoprime to the base b, then n is also an Euler pseudoprime to the base $n - b$.

5. Show that if $n \equiv 5 \pmod 8$ and n is an Euler pseudoprime to the base 2, then n is a strong pseudoprime to the base 2.

6. Show that if $n \equiv 5 \pmod{12}$ and n is an Euler pseudoprime to the base 3, then n is a strong pseudoprime to the base 3.

7. Find a congruence condition for an Euler pseudoprime n to the base 5 that guarantees that n is a strong pseudoprime to the base 5.

** 8. Let the composite positive integer n have prime-power factorization $n = p_1^{a_1} p_2^{a_2} \cdots p_m^{a_m}$, where $p_j = 1 + 2^{k_j} q_j$ for $j = 1, 2, \ldots, m$, where $k_1 \le k_2 \le \cdots \le k_m$, and where $n = 1 + 2^k q$. Show that n is an Euler pseudoprime to exactly

$$\delta_n \prod_{j=1}^{m} ((n-1)/2, p_j - 1)$$

different bases b with $1 \le b < n$, where

$$\delta_n = \begin{cases} 2 & \text{if } k_1 = k; \\ 1/2 & \text{if } k_j < k \text{ and } a_j \text{ is odd for some } j; \\ 1 & \text{otherwise.} \end{cases}$$

9. For how many integers b, $1 \le b < 561$, is 561 an Euler pseudoprime to the base b?

10. For how many integers b, $1 \le b < 1729$, is 1729 an Euler pseudoprime to the base b?

11.4 Computational and Programming Exercises

Computations and Explorations

Using a computation program such as Maple or *Mathematica,* or programs you have written, carry out the following computations and explorations.

1. Find all Euler pseudoprimes to the base 2 less than 1,000,000. Do the same thing for the bases 3, 5, 7, and 11. Devise a primality test based on your results.

2. Find 10 integers, each with between 50 and 60 decimal digits, that are "probably prime" because they pass more than 20 iterations of the Solovay-Strassen probabilistic primality test.

Programming Projects

Write computer programs using Maple, *Mathematica,* or a language of your choice to do the following.

1. Given an integer n and a positive integer b greater than 1, determine whether n passes the test for Euler pseudoprimes to the base b.

2. Given an integer n, perform the Solovay-Strassen probabilistic primality test on n.

11.5 Zero-Knowledge Proofs

Suppose that you want to convince another person that you have some important private information, without revealing this information. For example, you may want to convince someone that you know the prime factorization of a 200-digit positive integer without telling them the prime factors. Or you may have a proof of an important theorem and you want to convince the mathematical community that you have such a proof without revealing it. In this section we will discuss methods, commonly known as *zero-knowledge* or *minimum-disclosure proofs,* that can be used to convince someone that you have certain private, verifiable information, without revealing it. Zero-knowledge proofs were invented in the mid-1980s.

In a zero-knowledge proof, there are two parties, the *prover,* the person who has the secret information, and the *verifier,* who wants to be convinced that the prover has this secret information. When a zero-knowledge proof is used, the probability is extremely small that someone who does not have the information can successfully cheat the verifier by masquerading as the prover. Moreover, the verifier learns nothing, or almost nothing, about the information other than that the prover possesses it. In particular, the verifier cannot convince a third party that the verifier knows this information.

Remark: Because zero-knowledge proofs supply the verifier with a small amount of information, zero-knowledge proofs are more properly called *minimum-disclosure proofs.* Nevertheless, we will use the original terminology for such proofs.

We will illustrate the use of zero-knowledge proofs by describing several examples of such proofs, each based on the ease of finding square roots modulo products of two primes compared with the difficulty of finding square roots when the two primes are not known. (See Section 11.1 for a discussion of this topic.)

Our first example presents a proposed scheme for a zero-knowledge proof that turned out to have a flaw making it unsuitable for this use. Nevertheless, we introduce this scheme as our first example because it illustrates the concept of zero-knowledge proofs and is relatively simple. Moreover, understanding why it fails to be a valid scheme for zero-knowledge proofs adds valuable insight (see Exercise 11). In this scheme Paula, the *prover*, attempts to convince Vince, the *verifier*, that she knows the prime factors of n, where n is the product of two large primes p and q, without helping him find these two prime factors.

When this scheme was originally devised, it was thought that someone who does not know p and q would be unable to find the square root of y modulo n in a reasonable amount of time, unlike Paula who knows these primes. This turns out not to be the case, as Exercise 11 illustrates.

The proposed scheme is based on iterating the following procedure.

(i) Vince, who knows n, but not p and q, chooses an integer x at random. He computes y, the least nonnegative residue of x^4 modulo n and sends this to Paula.

(ii) When Paula receives y, she computes its square root modulo n. (We will explain how she can do this after describing the steps of the procedure.) This square root is the least positive residue of x^2 modulo n. She sends this integer to Vince.

(iii) Vince checks Paula's answer by finding the remainder of x^2 when it is divided by n.

To see why Paula can find the least positive residue of x^2 modulo n in step (ii), note that because she knows p and q, she can easily find the four square roots of x^4 modulo n. Next, note that only one of the four square roots of x^4 modulo n is a quadratic residue modulo n (see Exercise 3). So, to find x^2, she can select the correct square root of the four square roots of x^4 modulo n by computing the value of the Legendre symbols of each of these square roots modulo p and modulo q. Note that someone who does not know p and q is unable to find the square root of y modulo n in a reasonable amount of time, unlike Paula, who knows these primes.

We illustrate this procedure in the following example.

Example 11.19. Suppose that Paula's private information is her factorization of $n = 103 \cdot 239 = 24{,}617$. She can use the procedure just described to convince Vince that she knows the primes $p = 103$ and $q = 239$ without revealing them to him. (In practice, primes p and q with hundreds of digits would be used, rather than the small primes used in this example.)

To illustrate the procedure, suppose that in step (i) Vince selects the integer 9134 at random. He computes the least positive residue of 9134^4 modulo 24,617, which equals 20,682. He sends the integer 20,682 to Paula.

In step (ii), Paula determines the integer x^2 using the congruences

$$x^2 \equiv \pm 20{,}682^{(103+1)/4} = \pm 20{,}682^{26} \equiv \pm 59 \pmod{103}$$

$$x^2 \equiv \pm 20{,}682^{(239+1)/4} = \pm 20{,}682^{60} \equiv \pm 75 \pmod{239}.$$

(Note that we have used the fact that when $p \equiv q \equiv 3 \pmod 4$, the solutions of $x^2 \equiv a \pmod p$ and $x^2 \equiv a \pmod q$ are $x^2 \equiv \pm a^{(p+1)/4} \pmod p$ and $x^2 \equiv \pm a^{(q+1)/4} \pmod q$, respectively.)

Because x^2 is a quadratic residue modulo $24{,}627 = 103 \cdot 239$, we know that it also is a quadratic residue modulo 103 and 239. Computing Legendre symbols, we find that $\left(\frac{59}{103}\right) = 1$, $\left(\frac{-59}{103}\right) = -1$, $\left(\frac{75}{239}\right) = 1$, and $\left(\frac{-75}{239}\right) = -1$. Therefore, Paula finds x^2 by solving the system $x^2 \equiv 59 \pmod{103}$ and $x^2 \equiv 75 \pmod{239}$. When she solves this system, she concludes that $x^2 \equiv 2943 \pmod{24{,}617}$.

In step (iii), Vince checks Paula's answer by noting that $x^2 = 9134^2 \equiv 2943$ (mod 24,617).   ◄

We now describe a method to verify the identity of the prover, based on zero-knowledge techniques, invented by Shamir in 1985. We again suppose that $n = pq$, where p and q are two large primes both congruent to 3 modulo 4. Let I be a positive integer that represents some particular information, such as a personal identification number. The prover selects a small positive integer c, which has the property that the integer v obtained by concatenating I with c (the number obtained by writing the digits of I followed by the digits of c) is a quadratic residue modulo n. (The number c can be found by trial and error, with probability close to 1/2.) The prover can easily find u, a square root of v modulo n.

The prover convinces the verifier that she knows the primes p and q using an interactive proof. Each cycle of the proof is based on the following steps.

(i)   The prover, Paula, chooses a random number r, and sends to the verifier a message containing two values: x, where $x \equiv r^2$ (mod n), $0 \leq x < n$, and y, where $y \equiv v\bar{x}$ (mod n), $0 \leq y < n$. Here, as usual, $\bar{x}$ is an inverse of x modulo n.

(ii) The verifier, Vince, checks that $xy \equiv v$ (mod n) and chooses, at random, a bit b, which he sends to the prover.

(iii) If the bit b sent by Vince is 0, Paula sends r to Vince. Otherwise, if the bit b is 1, Paula sends the least positive residue of $u\,\bar{r}$ modulo n, where $\bar{r}$ is an inverse of r modulo n.

(iv) Vince computes the square of what Paula has sent. If Vince sent a 0, he checks that this square is x, that is, that $r^2 \equiv x$ (mod n). If he sent a 1, he checks that this square is y, that is, that $s^2 \equiv y$ (mod n).

This procedure is also based on the fact that the prover can find u, a square root of v modulo n, whereas someone who does not know p and q will not be able to compute a square root modulo n in a reasonable amount of time.

The four steps of this procedure form one cycle. Cycles can be repeated sufficiently often to guarantee a high degree of security, as we will subsequently describe.

We illustrate this type of zero-knowledge proof with the following example.

Example 11.20.   Suppose Paula wants to verify her identity to Vince by convincing him that she knows the prime factors of $n = 31 \cdot 61 = 1891$. Her identification number is $I = 391$. Note that 391 is a quadratic residue of 1891 because, as the reader can verify, it is a quadratic residue of both 31 and 61, so she can take $v = 391$ (that is, in this case, she does not have to concatenate an integer c with I). Paula finds that $u = 239$ is a square root of 391 modulo 1891. She can easily perform this calculation, because she knows the primes 31 and 61. (Note that we have selected small primes p and q in this example to illustrate the procedure. In practice, primes with hundreds of digits should be used.)

We illustrate one cycle of this procedure. In step (i), Paula chooses a random number, say $r = 998$. She sends Vince two numbers, $x \equiv r^2 \equiv 998^2 \equiv 1338 \pmod{1891}$ and $y \equiv v\,\overline{x} \equiv 391 \cdot 1296 \equiv 1839 \pmod{1891}$.

In step (ii), Vince checks that $xy \equiv 1338 \cdot 1839 \equiv 391 \pmod{1891}$ and chooses, at random, a bit b, say $b = 1$, which he sends to Paula.

In step (iii), Paula sends $s \equiv u\,\overline{r} = 239 \cdot 1855 \equiv 851 \pmod{1891}$ to Vince. Finally, in step (iv), Vince checks that $s^2 \equiv 851^2 \equiv 1839 \equiv y \pmod{1891}$. ◀

Note that if the prover sends the verifier both r and s, the verifier will know the private information $u = rs$, which is the secret information held by the prover. By passing the test with sufficiently many cycles, the prover has shown that she can produce either r or s on request. It follows that she must know u because, in each cycle, she knows both r and s. The choice of the random bit by the verifier makes it impossible for someone to fix the procedure by using numbers that have been rigged to pass the test. For example, someone could compute the square of a known number r and send $x = r^2$, instead of choosing a random number. Similarly, someone could select a number x such that $v\overline{x}$ is a known square. However, it is impossible to do precalculations to make both x and y the squares of known numbers without knowing u.

Because the bit chosen by the verifier is chosen at random, the probability that it will be a 0 is 1/2, as is the probability that it will be a 1. If someone does not know u, the square root of v, the probability that they will pass one iteration of this test is almost exactly 1/2. Consequently, the probability that someone masquerading as the prover will pass the test with 30 cycles is approximately $1/2^{30}$, which is less than one in a billion.

A variation of this procedure, known as the Fiat-Shamir method, is the basis for verification procedures used by smart cards, such as for verifying personal identification numbers.

Next, we describe a method that can be used to prove, using a zero-knowledge proof, that someone has certain information. Suppose that the prover, Paula, has information represented by a sequence of numbers $v_1, v_2, \ldots, v_m$, where $1 \le v_j < n$ for $j = 1, 2, \ldots, m$. Here, as before, n is the product of two primes p and q that are both congruent to 3 modulo 4. Paula makes public the sequence of integers $s_1, s_2, \ldots, s_m$, where $s_j \equiv \overline{v}_j^{\,2} \pmod{n}$, $1 \le s_j < n$. Paula wants to convince the verifier, Vince, that she knows the private information $v_1, v_2, \ldots, v_m$, without revealing this information to Vince. What Vince knows is her public moduli n and her public information $s_1, s_2, \ldots, s_m$.

The following procedure can be used to convince Vince she has this information. Each cycle of the procedure has the following steps.

(i)   Paula chooses a random number r and computes $x = r^2$, which she sends to Vince.

(ii) Vince selects a subset S of the set $\{1, 2, \ldots, m\}$ and sends this subset to Paula.

(iii) Paula computes y, the least positive residue modulo n of the product of r and the integers v_j, with j in S, that is, $y \equiv r \prod_{j \in S} v_j \pmod{n}$, $0 \le y < n$.

(iv) Vince verifies that $x \equiv y^2 z \pmod{n}$, where z is the product of the integers c_j, with j in S, that is, $z \equiv \prod_{j \in S} s_j \pmod{n}$, $0 \leq z < n$.

Note that the congruence in step (iv) holds, because

$$y^2 z \equiv r^2 \prod_{j \in S} v_j^2 \prod_{j \in S} s_j$$

$$\equiv r^2 \prod_{j \in S} v_j^2 \bar{v}_j^2$$

$$\equiv r^2 \pmod{n}.$$

The random number r is used so that the verifier cannot determine the value of the integer v_j, part of the secret information, by selecting the set $S = \{j\}$. When this procedure is carried out, the verifier is given no new information that will help him determine the private information $c_1, \ldots, c_m$.

We illustrate one cycle of this interactive zero-knowledge proof in the following example.

Example 11.21. Suppose that Paula wants to convince Vince that she has secret information, which is represented by the integers $v_1 = 1144$, $v_2 = 877$, $v_3 = 2001$, $v_4 = 1221$, $v_5 = 101$. Her secret modulus is $n = 47 \cdot 53 = 2491$. (In practice, primes with hundreds of digits are used rather than the small primes used in this example.)

Her public information consists of the integers s_j, with $s_j \equiv \bar{v}_j^2 \pmod{2491}$, $0 < s_j < 2491$, $j = 1, 2, 3, 4, 5$. It follows, after routine calculation, that her public information consists of the integers $s_1 = 197$, $s_2 = 2453$, $s_3 = 1553$, $s_4 = 941$, and $s_5 = 494$.

Paula can convince Vince that she has the secret information using the procedure described in the text. We describe one cycle of the procedure. In step (i), Paula chooses a random number, say $r = 1253$. Next, she sends $x = 679$, the least positive residue of r^2 modulo 2491, to Vince.

In step (ii), Vince selects a subset of $\{1, 2, 3, 4, 5\}$, say $s = \{1, 3, 4, 5\}$, and informs Paula of this choice.

In step (iii), Paula computes the number y, with $0 \leq y < 2491$ and

$$y \equiv r v_1 v_3 v_4 v_5$$
$$\equiv 1253 \cdot 1144 \cdot 2001 \cdot 1221 \cdot 101$$
$$\equiv 68 \pmod{2491}.$$

Consequently, she sends $y = 68$ to Vince.

Finally, in step (iv), Vince confirms that $x \equiv y^2 s_1 s_3 s_4 s_5 \pmod{2491}$ by verifying that $x = 679 \equiv 68^2 \cdot 197 \cdot 1553 \cdot 941 \cdot 494 \pmod{2491}$.

Vince can ask Paula to run through more cycles of this procedure to verify that she does have the secret information. He stops when he feels that the probability that she is cheating is small enough to satisfy his needs. ◄

How can the prover cheat in this interactive procedure for zero-knowledge proofs of information? That is, how can the prover fool the verifier into thinking that she really knows the private information $c_1, \ldots, c_m$ when she does not? The only obvious way is for the prover to guess the set S before the verifier supplies this; in step (1), to take $x = r^2 \prod_{j \in S} v_j^2$; and in step (iii), to take $y = 4$. Because there are 2^m possible sets S (as there are that many subsets of $\{1, 2, \ldots, m\}$), the probability that someone not knowing the private information fools the verifier using this technique is $1/2^m$. Furthermore, when this cycle is iterated T times, the probability decreases to $1/2^{mT}$. For instance, if $m = 10$ and $T = 3$, the probability of the verifier being fooled is less than one in a billion.

In this section, we have only briefly touched upon zero-knowledge proofs. The reader interested in learning more about this subject should refer to the chapter by Goldwasser in [Po90], as well as to the reference supplied in that chapter.

11.5 Exercises

1. Suppose that $n = 3149 = 47 \cdot 67$ and that $x^4 \equiv 2070 \pmod{3149}$. Find the least nonnegative residue of x^2 modulo 3149.

2. Suppose that $n = 11{,}021 = 103 \cdot 107$ and that $x^4 \equiv 1686 \pmod{11{,}021}$. Find the least nonnegative residue of x^2 modulo 11,021.

3. Suppose that $n = pq$, where p and q are primes both congruent to 3 modulo 4, and that x is an integer relatively prime to n. Show that of the four square roots of x^4 modulo n, only one is the least nonnegative residue of a square of an integer.

4. Suppose that Paula has identification number 1760 and modulus $1961 = 37 \cdot 53$. Show how she verifies her identity to Vince in one cycle of the Shamir procedure, if she selects the random number 1101 and he chooses 1 as his random bit.

5. Suppose that Paula has identification number 7 and modulus $1411 = 17 \cdot 83$. Show how she verifies her identify to Vince in one cycle of the Shamir procedure, if she selects the random number 822 and he chooses 1 as his random bit.

6. Run through the steps used to verify that the prover has the secret information in Example 11.21, when the random number $r = 888$ is selected by the prover in step (i) and the verifier selects the subset $\{2, 3, 5\}$ of $\{1, 2, 3, 4, 5\}$.

7. Run through the steps used to verify that the prover has the secret information in Example 11.21, when the random number $r = 1403$ is selected by the prover in step (i) and the verifier selects the subset $\{1, 5\}$ of $\{1, 2, 3, 4, 5\}$.

8. Let $n = 2491 = 47 \cdot 53$. Suppose that Paula's identification information consists of the sequence of six numbers $v_1 = 881$, $v_2 = 1199$, $v_3 = 2144$, $v_4 = 110$, $v_5 = 557$, and $v_6 = 2200$.

   a) Find Paula's public identification information, $s_1, s_2, s_3, s_4, s_5, s_6$.

   b) Suppose that Paula selects at random the number $r = 1091$, and Vince chooses the subset $S = 2, 3, 5, 6$ and sends this to Paula. Find the number that Paula computes and sends back to Vince.

   c) What computation does Vince make to verify Paula's knowledge of her secret information?

9. Let $n = 3953 = 59 \cdot 67$. Suppose that Paula's identification information consists of the sequence of six numbers $v_1 = 1001$, $v_2 = 21$, $v_3 = 3097$, $v_4 = 989$, $v_5 = 157$, and $v_6 = 1039$.

   a) Find Paula's public identification information $s_1, s_2, s_3, s_4, s_5, s_6$.

   b) Suppose that Paula selects at random the number $r = 403$, and Vince chooses the subset $S = \{1, 2, 4, 6\}$ and sends this to Paula. Find the number that Paula computes and sends back to Vince.

   c) What computation does Vince make to verify Paula's knowledge of her secret information?

10. Suppose that $n = pq$, where p and q are large odd primes and that you are able to efficiently extract square roots modulo n without knowing p and q. Show that you can, with probability close to 1, find the prime factors p and q. (*Hint:* Base your algorithm on the following procedure. Select an integer x. Exract a square root of the least nonnegative residue of x^2 modulo n. You will need to show that there is a 1/2 chance that you found a square root not congruent to $\pm x$ modulo n.)

11. In this exercise, we expose a flaw in the proposed scheme of a zero-knowledge proof presented prior to Example 11.19. Suppose that Vince randomly chooses integers w until he finds a value of w for which the Jacobi symbol $\left(\frac{w}{n}\right)$ equals -1 and that he sends Paula z, the least nonnegative residue of w^2 modulo n. Show that Vince can factor n once Paula sends back the square root of z that she computes.

11.5 Computational and Programming Exercises

Computations and Explorations

Using a computation program such as Maple or *Mathematica,* or programs you have written, carry out the following computations and explorations.

1. Give one of your classmates the integer n, where $n = pq$ and p and q are primes with more than 50 decimal digits, both congruent to 3 modulo 4. Convince your classmate that you know both p and q using a zero-knowledge proof.

2. Convince one of your classmates that you know a secret in the form of a sequence of 10 positive integers each less than 10,000, using the zero-knowledge proof described in the text.

Programming Projects

Write computer programs using Maple, *Mathematica,* or a language of your choice to do the following.

1. Given n, the product of two distinct primes both congruent to 3 modulo 4, and the least positive residue of x^4 modulo n, where x is an integer relatively prime to n, find the least positive residue of x^2 modulo n.

12

Decimal Fractions and Continued Fractions

Introduction

In this chapter, we will discuss the representation of rational and irrational numbers as decimal fractions and continued fractions. We will show that every rational number can be expressed as a terminating or periodic decimal fraction, and provide some results that tell us the length of the period of the decimal fraction of a rational number. We will also construct irrational numbers using decimal fractions, and show how decimal fractions can be used to express a transcendental number and to demonstrate that the set of real numbers is uncountable.

Continued fractions provide a useful way of expressing numbers. We will show that every rational number has a finite continued fraction; that every irrational number has an infinite continued fraction and that continued fractions are the best rational approximations to numbers. We will establish a key result that will tell us that the set of quadratic irrationals can be characterized as the set of numbers with periodic continued fractions. Finally, we will show how continued fractions can be used to help factor integers.

12.1 Decimal Fractions

In this section, we discuss the representation of rational and irrational numbers as decimal fractions. We first consider base b expansions of real numbers, where b is a positive integer, $b > 1$. Let α be a positive real number, and let $a = [\alpha]$ be the integer part of α, so that $\gamma = \alpha - [\alpha]$ is the fractional part of α and $\alpha = a + \gamma$ with $0 \leq \gamma < 1$. By Theorem 2.1, the integer a has a unique base b expansion. We now show that the fractional part γ also has a unique base b expansion.

Theorem 12.1.   Let γ be a real number with $0 \le \gamma < 1$, and let b be a positive integer, $b > 1$. Then γ can be uniquely written as

$$\gamma = \sum_{j=1}^{\infty} c_j/b^j,$$

where the coefficients c_j are integers with $0 \le c_j \le b - 1$ for $j = 1, 2, \ldots$, with the restriction that for every positive integer N there is an integer n with $n \ge N$ and $c_n \ne b - 1$.

In the proof of Theorem 12.1, we deal with infinite series. We will use the following formula for the sum of the terms of an infinite geometric series.

Theorem 12.2.   Let a and r be real numbers with $|r| < 1$. Then

$$\sum_{j=0}^{\infty} ar^j = a/(1 - r).$$

Most books on calculus or mathematical analysis contain a proof of Theorem 12.2 (see [Ru64], for instance).

We can now prove Theorem 12.1.

Proof.   We first let

$$c_1 = [b\gamma],$$

so that $0 \le c_1 \le b - 1$, because $0 \le b\gamma < b$. In addition, let

$$\gamma_1 = b\gamma - c_1 = b\gamma - [b\gamma],$$

so that $0 \le \gamma_1 < 1$ and

$$\gamma = \frac{c_1}{b} + \frac{\gamma_1}{b}.$$

We recursively define c_k and γ_k for $k = 2, 3, \ldots$, by

$$c_k = [b\gamma_{k-1}]$$

and

$$\gamma_k = b\gamma_{k-1} - c_k$$

so that $0 \le c_k \le b - 1$, because $0 \le b\gamma_{k-1} < b$ and $0 \le \gamma_k < 1$. Then, it follows that

$$\gamma = \frac{c_1}{b} + \frac{c_2}{b^2} + \cdots + \frac{c_n}{b^n} + \frac{\gamma_n}{b^n}.$$

Because $0 \le \gamma_n < 1$, we see that $0 \le \gamma_n/b^n < 1/b^n$. Consequently,

$$\lim_{n \to \infty} \gamma_n/b^n = 0.$$

Therefore, we can conclude that

$$\gamma = \lim_{n \to \infty} \left(\frac{c_1}{b} + \frac{c_2}{b^2} + \cdots + \frac{c_n}{b_n} \right)$$

$$= \sum_{j=1}^{\infty} c_j/b^j.$$

To show that this expansion is unique, assume that

$$\gamma = \sum_{j=1}^{\infty} c_j/b^j = \sum_{j=1}^{\infty} d_j/b^j,$$

where $0 \le c_j \le b - 1$ and $0 \le d_j \le b - 1$ and, for every positive integer N, there are integers n and m with $c_n \ne b - 1$ and $d_m \ne b - 1$. Assume that k is the smallest index for which $c_k \ne d_k$, and assume that $c_k > d_k$ (the case $c_k < d_k$ is handled by switching the roles of the two expansions). Then

$$0 = \sum_{j=1}^{\infty} (c_j - d_j)/b^j = (c_k - d_k)/b^k + \sum_{j=k+1}^{\infty} (d_j - c_j)/b^j,$$

so that

(12.1)
$$(c_k - d_k)/b^k = \sum_{j=k+1}^{\infty} (d_j - c_j)/b^j.$$

Because $c_k > d_k$, we have

(12.2)
$$(c_k - d_k)/b^k \ge 1/b^k,$$

whereas

(12.3)
$$\sum_{j=k+1}^{\infty} (d_j - c_j)/b^j \le \sum_{j=k+1}^{\infty} (b-1)/b^j$$

$$= (b-1)\frac{1/b^{k+1}}{1 - 1/b}$$

$$= 1/b^k,$$

where we have used Theorem 12.2 to evaluate the sum on the right-hand side of the inequality. Note that equality holds in (12.3) if and only if $d_j - c_j = b - 1$ for all j with $j \ge k + 1$, and this occurs if and only if $d_j = b - 1$ and $c_j = 0$ for $j \ge k + 1$. However, such an instance is excluded by the hypotheses of the theorem. Hence, the inequality in (12.3) is strict, and therefore (12.2) and (12.3) contradict (12.1). This shows that the base b expansion of α is unique. ∎

The unique expansion of a real number in the form $\sum_{j=1}^{\infty} c_j/b^j$ is called the *base b expansion* of this number and is denoted by $(.c_1 c_2 c_3 \ldots)_b$.

To find the base b expansion $(.c_1c_2c_3 \ldots)_b$ of a real number γ, we can use the recursive formula for the digits given in the proof of Theorem 12.1, namely

$$c_k = [b\gamma_{k-1}], \quad \gamma_k = b\gamma_{k-1} - [b\gamma_{k-1}],$$

where $\gamma_0 = \gamma$, for $k = 1, 2, 3, \ldots$. (Note that there is also an explicit formula for these digits—see Exercise 21.)

Example 12.1.   Let $(.c_1c_2c_3 \ldots)_b$ be the base 8 expansion of 1/6. Then

$$c_1 = \left[8 \cdot \frac{1}{6}\right] = 1, \quad \gamma_1 = 8 \cdot \frac{1}{6} - 1 = \frac{1}{3},$$

$$c_2 = \left[8 \cdot \frac{1}{3}\right] = 2, \quad \gamma_2 = 8 \cdot \frac{1}{3} - 2 = \frac{2}{3},$$

$$c_3 = \left[8 \cdot \frac{2}{3}\right] = 5, \quad \gamma_3 = 8 \cdot \frac{2}{3} - 5 = \frac{1}{3},$$

$$c_4 = \left[8 \cdot \frac{1}{3}\right] = 2, \quad \gamma_4 = 8 \cdot \frac{1}{3} - 2 = \frac{2}{3},$$

$$c_5 = \left[8 \cdot \frac{2}{3}\right] = 5, \quad \gamma_5 = 8 \cdot \frac{2}{3} - 5 = \frac{1}{3},$$

and so on. We see that the expansion repeats; hence,

$$1/6 = (.1252525 \ldots)_8. \qquad \blacktriangleleft$$

We will now discuss base b expansions of rational numbers. We will show that a number is rational if and only if its base b expansion is periodic or terminates.

Definition.   A base b expansion $(.c_1c_2c_3 \ldots)_b$ is said to *terminate* if there is a positive integer n such that $c_n = c_{n+1} = c_{n+2} = \cdots = 0$.

Example 12.2.   The decimal expansion of 1/8, $(.125000 \ldots)_{10} = (.125)_{10}$, terminates. Also, the base 6 expansion of 4/9, $(.24000 \ldots)_6 = (.24)_6$, terminates. $\blacktriangleleft$

To describe those real numbers with terminating base b expansion, we prove the following theorem.

Theorem 12.3.   The real number $\alpha, 0 \leq \alpha < 1$, has a terminating base b expansion if and only if α is rational and can be written as $\alpha = r/s$, where $0 \leq r < s$ and every prime factor of s also divides b.

Proof.   First, suppose that α has a terminating base b expansion,

$$\alpha = (.c_1c_2 \ldots c_n)_b.$$

Then

$$\alpha = \frac{c_1}{b} + \frac{c_2}{b^2} + \cdots + \frac{c_n}{b^n}$$

$$= \frac{c_1 b^{n-1} + c_2 b^{n-2} + \cdots + c_n}{b^n},$$

so that α is rational, and can be written with a denominator divisible only by primes dividing b.

Conversely, suppose that $0 \le \alpha < 1$, and

$$\alpha = r/s,$$

where each prime dividing s also divides b. Hence, there is a power of b, say b^N, that is divisible by s (for instance, take N to be the largest exponent in the prime-power factorization of s). Then

$$b^N \alpha = b^N r/s = ar,$$

where $sa = b^N$, and a is a positive integer because $s | b^N$. Now let $(a_m a_{m-1} \ldots a_1 a_0)_b$ be the base b expansion of ar. Then

$$\alpha = ar/b^N = \frac{a_m b^m + a_{m-1} b^{m-1} + \cdots + a_1 b + a_0}{b^N}$$

$$= a_m b^{m-N} + a_{m-1} b^{m-1-N} + \cdots + a_1 b^{1-N} + a_0 b^{-N}$$

$$= (.00 \ldots a_m a_{m-1} \ldots a_1 a_0)_b.$$

Hence, α has a terminating base b expansion. ∎

Note that every terminating base b expansion can be written as a nonterminating base b expansion with a tail-end consisting entirely of the digit $b - 1$, because $(.c_1 c_2 \ldots c_m)_b = (.c_1 c_2 \ldots c_m - 1 \ b - 1 \ b - 1 \ldots)_b$. For instance, $(.12)_{10} = (.11999 \ldots)_{10}$. This is why we require in Theorem 12.1 that for every integer N there is an integer n such that $n > N$ and $c_n \ne b - 1$; without this restriction, base b expansions would not be unique.

A base b expansion that does not terminate may be *periodic*, for instance,

$$1/3 = (.333 \ldots)_{10},$$
$$1/6 = (.1666 \ldots)_{10},$$

and

$$1/7 = (.142857142857142857 \ldots)_{10}.$$

Definition. A base b expansion $(.c_1 c_2 c_3 \ldots)_b$ is called *periodic* if there are positive integers N and k such that $c_{n+k} = c_n$ for $n \ge N$.

We denote by $(.c_1c_2 \ldots c_{N-1}\overline{c_N \ldots c_{N+k-1}})_b$ the periodic base b expansion $(.c_1c_2 \ldots c_{N-1}c_N \ldots c_{N+k-1}c_N \ldots c_{N+k-1}c_N \ldots)_b$. For instance, we have

$$1/3 = (.\bar{3})_{10},$$
$$1/6 = (.1\bar{6})_{10},$$

and

$$1/7 = (.\overline{142857})_{10}.$$

Note that the periodic parts of the decimal expansions of 1/3 and 1/7 begin immediately, whereas in the decimal expansion of 1/6 the digit 1 precedes the periodic part of the expansion. We call the part of a periodic base b expansion preceding the periodic part the *pre-period,* and the periodic part the *period,* where we take the period to have minimal possible length.

Example 12.3.   The base 3 expansion of 2/45 is $(.00\overline{1012})_3$. The pre-period is $(00)_3$ and the period is $(1012)_3$.

The next theorem tells us that the rational numbers are those real numbers with periodic or terminating base b expansions. Moreover, the theorem gives the lengths of the pre-period and periods of base b expansions of rational numbers.

Theorem 12.4.   Let b be a positive integer. Then a periodic base b expansion represents a rational number. Conversely, the base b expansion of a rational number either terminates or is periodic. Further, if $0 < \alpha < 1$, $\alpha = r/s$, where r and s are relatively prime positive integers, and $s = TU$, where every prime factor of T divides b and $(U, b) = 1$, then the period length of the base b expansion of α is $\text{ord}_U b$, and the pre-period length is N, where N is the smallest positive integer such that $T | b^N$.

Proof.   First, suppose that the base b expansion of α is periodic, so that

$$\alpha = (.c_1c_2 \ldots c_N \overline{c_{N+1} \ldots c_{N+k}})_b$$

$$= \frac{c_1}{b} + \frac{c_2}{b^2} + \cdots + \frac{c_N}{b^N} + \left(\sum_{j=0}^{\infty} \frac{1}{b^{jk}} \right) \left(\frac{c_{N+1}}{b^{N+1}} + \cdots + \frac{c_{N+k}}{b^{N+k}} \right)$$

$$= \frac{c_1}{b} + \frac{c_2}{b^2} + \cdots + \frac{c_N}{b^N} + \left(\frac{b^k}{b^k - 1} \right) \left(\frac{c_{N+1}}{b^{N+1}} + \cdots + \frac{c_{N+K}}{b^{N+k}} \right),$$

where we have used Theorem 12.2 to see that

$$\sum_{j=0}^{\infty} \frac{1}{b^{jk}} = \frac{1}{1 - \frac{1}{b^k}} = \frac{b^k}{b^k - 1}.$$

Because α is the sum of rational numbers, it is rational.

Conversely, suppose that $0 < \alpha < 1$, $\alpha = r/s$, where r and s are relatively prime positive integers, $s = TU$, where every prime factor of T divides b, $(U, b) = 1$, and N is the smallest integer such that $T \mid b^N$.

Because $T \mid b^N$, we have $aT = b^N$, where a is a positive integer. Hence,

$$(12.4) \qquad b^N \alpha = b^N \frac{r}{TU} = \frac{ar}{U}.$$

Furthermore, we can write

$$(12.5) \qquad \frac{ar}{U} = A + \frac{C}{U},$$

where A and C are integers with

$$0 \le A < b^N, \quad 0 < C < U,$$

and $(C, U) = 1$. (The inequality for A follows because $0 < b^N \alpha = \frac{ar}{U} < b^N$, which results from the inequality $0 < \alpha < 1$ when both sides are multiplied by b^N). The fact that $(C, U) = 1$ follows easily from the condition $(r, s) = 1$. By Theorem 12.1, A has a base b expansion $A = (a_n a_{n-1} \ldots a_1 a_0)_b$.

If $U = 1$, then the base b expansion of α terminates as shown. Otherwise let $v = \mathrm{ord}_U b$. Then,

$$(12.6) \qquad b^v \frac{C}{U} = \frac{(tU + 1)C}{U} = tC + \frac{C}{U},$$

where t is an integer, because $b^v \equiv 1 \pmod{U}$. However, we also have

$$(12.7) \qquad b^v \frac{C}{U} = b^v \left(\frac{c_1}{b} + \frac{c_2}{b^2} + \cdots + \frac{c_v}{b^v} + \frac{\gamma_v}{b^v} \right),$$

where $(.c_1 c_2 c_3 \ldots)_b$ is the base b expansion of $\frac{C}{U}$, so that

$$c_k = [b\gamma_{k-1}], \quad \gamma_k = b\gamma_{k-1} - [b\gamma_{k-1}],$$

where $\gamma_0 = \frac{C}{U}$, for $k = 1, 2, 3, \ldots$. From (12.7), we see that

$$(12.8) \qquad b^v \frac{C}{U} = \left(c_1 b^{v-1} + c_2 b^{v-2} + \cdots + c_v \right) + \gamma_v.$$

Equating the fractional parts of (12.6) and (12.8), noting that $0 \le \gamma_v < 1$, we find that

$$\gamma_v = \frac{C}{U}.$$

Consequently, we see that

$$\gamma_v = \gamma_0 = \frac{C}{U},$$

so that from the recursive definition of $c_1, c_2, \ldots$ we can conclude that $c_{k+v} = c_k$ for $k = 1, 2, 3, \ldots$. Hence, $\frac{C}{U}$ has a periodic base b expansion

$$\frac{C}{U} = (.\overline{c_1 c_2 \dots c_v})_b.$$

Combining (12.4) and (10.5), and inserting the base b expansions of A and $\frac{C}{U}$, we have

(12.9) $b^N \alpha = (a_n a_{n-1} \dots a_1 a_0 . \overline{c_1 c_2 \dots c_v})_b.$

Dividing both sides of (12.9) by b^N, we obtain

$$\alpha = (.00 \dots a_n a_{n-1} \dots a_1 a_0 \overline{c_1 c_2 \dots c_v})_b,$$

(where we have shifted the decimal point in the base b expansion of $b^N \alpha$ N spaces to the left to obtain the base b expansion of α). In this base b expansion of α, the pre-period $(.00 \dots a_n a_{n-1} \dots a_1 a_0)_b$ is of length N, beginning with $N - (n+1)$ zeros, and the period length is v.

We have shown that there is a base b expansion of α with a pre-period of length N and a period of length v. To finish the proof, we must show that we cannot regroup the base b expansion of α, so that either the pre-period has length less than N, or the period has length less than v. To do this, suppose that

$$\alpha = (.c_1 c_2 \dots c_M \overline{c_{M+1} \dots c_{M+k}})_b$$

$$= \frac{c_1}{b} + \frac{c_2}{b_2} + \dots + \frac{c_M}{b^M} + \left(\frac{b^k}{b^k - 1}\right)\left(\frac{c_{M+1}}{b^{M+1}} + \dots + \frac{c_{M+k}}{b^{M+k}}\right)$$

$$= \frac{(c_1 b^{M-1} + c_2 b^{M-2} + \dots + c_M)(b^k - 1) + (c_{M+1} b^{k-1} + \dots + c_{M+k})}{b^M (b^k - 1)}.$$

Because $\alpha = r/s$, with $(r, s) = 1$, we see that $s | b^M (b^k - 1)$. Consequently, $T | b^M$ and $U | (b^k - 1)$. Hence, $M \geq N$, and $v | k$ (by Theorem 9.1, because $b^k \equiv 1 \pmod{U}$ and $v = \text{ord}_U b$). Therefore, the pre-period length cannot be less than N and the period length cannot be less than v.   ∎

We can use Theorem 12.4 to determine the lengths of the pre-period and period of decimal expansions. Let $\alpha = r/s$, $0 < \alpha < 1$, and $s = 2^{s_1} 5^{s_2} t$, where $(t, 10) = 1$. Then, by Theorem 12.4, the pre-period has length $\max(s_1, s_2)$ and the period has length $\text{ord}_t 10$.

Example 12.4. Let $\alpha = 5/28$. Because $28 = 2^2 \cdot 7$, Theorem 12.4 tells us that the pre-period has length two and the period has length $\text{ord}_7 10 = 6$. As $5/28 = (.17\overline{857142})$, we see that these lengths are correct.   ◄

Note that the pre-period and period lengths of a rational number r/s, in lowest terms, depend only on the denominator s, and not on the numerator r.

We observe that by Theorem 12.4 a base b expansion that is not terminating and is not periodic represents an irrational number.

Example 12.5. The number with decimal expansion

$$\alpha = .10100100010000 \dots,$$

consisting of a one followed by a zero, a one followed by two zeros, a one followed by three zeroes, and so on, is irrational because this decimal expansion does not terminate and is not periodic.   ◀

The number α in the preceding example is concocted so that its decimal expansion is clearly not periodic. To show that naturally occurring numbers such as e and π are irrational, we cannot use Theorem 12.4, because we do not have explicit formulas for the decimal digits of these numbers. No matter how many decimal digits of their expansions we compute, we still cannot conclude that they are irrational from this evidence, because the period could be longer than the number of digits that we have computed.

Transcendental Numbers

The French mathematician Liouville was the first person to show that a particular number is transcendental. (Recall from Section 1.1 that a transcendental number is one that is not the root of a polynomial with integer coefficients.) The number that Liouville showed is transcendental is the number

$$\alpha = \sum_{i=1}^{\infty} \frac{1}{10^{i!}} = 0.110001000000000000000000100\dots.$$

This number has a 1 in the $n!$th place for each positive integer n and a 0 elsewhere. To show that this number is transcendental, Liouville proved the following theorem, which shows that algebraic numbers cannot be approximated very well by rational numbers. In particular, this theorem provides a lower bound for how well an algebraic number of degree n can be approximated by rational numbers. Note that an *algebraic number of degree n* is a real number that is a root of a polynomial of degree n with integer coefficients which is not a root of any polynomial with integer coefficients of degree less than n.

Theorem 12.5.   If α is an algebraic number of degree n, where n is a positive integer greater than 1, then there exists a positive real number C such that

$$\left| \alpha - \frac{p}{q} \right| > C/q^n$$

for every rational number p/q, where $q > 0$.

Because the proof of Theorem 12.5, although not difficult, relies on calculus, we will not supply it here. We refer the reader to [HaWr79] for a proof. We will be content to use this theorem to show that Liouville's number is transcendental.

Corollary 12.5.1.   The number $\alpha = \sum_{i=1}^{\infty} 1/10^{i!}$ is transcendental.

Proof.   First, note that α is not rational, because its decimal expansion does not terminate and is not periodic. To see that it is not periodic, note that there are increasingly larger numbers of 0s between successive 1s in the expansion.

Let p_k/q_k denote the sum of the first k terms in the sum defining α. Note that $q_k = 10^{k!}$. Because $10^{i!} \geq 10^{(k+1)!i}$ whenever $i \geq k+1$, we have

$$\left| \alpha - \frac{p_k}{q_k} \right| = \sum_{i=k+1}^{\infty} \frac{1}{10^{i!}} < \sum_{i=k+1}^{\infty} \frac{1}{(10^{(k+1)!})^i}.$$

Because

$$\sum_{i=k+1}^{\infty} \frac{1}{10^{(k+1)!i}} = \frac{1}{10^{(k+1)!} - 1} \leq \frac{2}{10^{(k+1)!}},$$

it follows that

$$\left| \alpha - \frac{p_k}{q_k} \right| < \frac{2}{10^{(k+1)!}}.$$

It therefore follows that α cannot be algebraic, for if it were algebraic of degree n, then by Theorem 12.5 there would be a positive real number C such that $|\alpha - p_k/q_k| > C/q_k^n$. This is not the case, because we have seen that $|\alpha - p_k/q_k| < 2/q_k^{k+1}$, and taking k to be sufficiently larger than n produces a contradiction. ∎

The notion of the decimal expansion of real numbers can be used to show that the set of real numbers is not *countable*. A *countable set* is one that can be put into a one-to-one correspondence with the set of positive integers. Equivalently, the elements of a countable set can be listed as the terms of a sequence. The element corresponding to the integer 1 is listed first, the element corresponding to the integer 2 is listed second, and so on. We will give the proof found by German mathematician *Georg Cantor.*

Theorem 12.6.   The set of real numbers is an uncountable set.

Proof.   We assume that the set of real numbers is countable. Then the subset of all real numbers between 0 and 1 would also be countable, as a subset of a countable set is also

GEORG CANTOR (1845–1918) was born in St. Petersburg, Russia, where his father was a successful merchant. When he was 11, his family moved to Germany to escape the harsh weather of Russia. Cantor developed his interest in mathematics while in German high schools. He attended university at Zurich and later at the University of Berlin, studying under the famous mathematicians Kummer, Weierstrass, and Kronecker. He received his doctorate in 1867 for work in number theory. Cantor took a position at the University of Halle in 1869, a position that he held until he retired in 1913.

Cantor is considered the founder of set theory; he is also noted for his contributions to mathematical analysis. Many mathematicians had extremely high regard for Cantor's work, such as Hilbert, who said that it was "the finest product of mathematical genius and one of the supreme achievements of purely intellectual human activity." Besides mathematics, Cantor was interested in philosophy, and wrote papers connecting his theory of sets and metaphysics.

Cantor was married in 1874 and had five children. He had a melancholy temperament that was balanced by his wife's happy disposition. He received a large inheritance from his father, but since he was poorly paid as a professor at Halle, he applied for a better-paying position at the University of Berlin. His appointment there was blocked by Kronecker, who did not agree with Cantor's views on set theory. Unfortunately, Cantor suffered from mental illness throughout the later years of his life; he died of a heart attack in 1918 in a psychiatric clinic.

countable (as the reader should verify). With this assumption, the set of real numbers between 0 and 1 can be listed as terms of a sequence $r_1, r_2, r_3, \ldots$. Suppose that the decimal expansions of these real numbers are

$$r_1 = 0.d_{11}d_{12}d_{13}d_{14}\ldots$$
$$r_2 = 0.d_{21}d_{22}d_{23}d_{24}\ldots$$
$$r_3 = 0.d_{31}d_{32}d_{33}d_{34}\ldots$$
$$r_4 = 0.d_{41}d_{42}d_{43}d_{44}\ldots$$

and so on. Now form a new real number r with the decimal expansion $0.d_1d_2d_3d_4\ldots$, where the decimal digits are determined by $d_i = 4$ if $d_{ii} \neq 4$ and $d_i = 5$ if $d_{ii} = 4$.

Because every real number has a unique decimal expansion (when the possibility that the expansion has a tail end that consists entirely of 9s is excluded), the real number r that we constructed is between 0 and 1 and is not equal to any of the real numbers $r_1, r_2, r_3, \ldots$, because the decimal is a real number r between 0 and 1 not in the list, the assumption that all real numbers between 0 and 1 could be listed is false. It follows that the set of real numbers between 0 and 1, and hence the set of all real numbers, is uncountable. ∎

12.1 Exercises

1. Find the decimal expansion of each of the following numbers.

   a) 2/5   c) 12/13   e) 1/111
   b) 5/12   d) 8/15   f) 1/1001

2. Find the base 8 expansions of each of the following numbers.

   a) 1/3   c) 1/5   e) 1/12
   b) 1/4   d) 1/6   f) 1/22

3. Find the fraction, in lowest terms, represented by each of the following expansions.

   a) .12   b) .1$\overline{2}$ c) .$\overline{12}$

4. Find the fraction, in lowest terms, represented by each of the following expansions.

   a) $(.123)_7$ c) $(.\overline{17})_{11}$
   b) $(.0\overline{13})_6$ d) $(.\overline{ABC})_{16}$

5. For which positive integers b does the base b expansion of 11/210 terminate?

6. Find the pre-period and period lengths of the decimal expansion of each of the following rational numbers.

   a) 7/12   c) 1/75   e) 13/56
   b) 11/30   d) 10/23   f) 1/61

7. Find the pre-period and period lengths of the base 12 expansions of each of the following rational numbers.

   a) 1/4   c) 7/10   e) 17/132
   b) 1/8   d) 5/24   f) 7/360

8. Let b be a positive integer. Show that the period length of the base b expansion of $1/m$ is $m - 1$ if and only if m is prime and b is a primitive root of m.

9. For which primes p does the decimal expansion of $1/p$ have period length equal to each of the following integers?

a) 1   c) 3   e) 5
b) 2   d) 4   f) 6

10. Find the base b expansion of each of the following numbers.

a) $1/(b - 1)$ b) $1/(b + 1)$

11. Let b be an integer with $b > 2$. Show that the base b expansion of $1/(b - 1)^2$ is $(.\overline{0123 \ldots b - 3\, b - 1})_b$.

12. Show that the real number with base b expansion

$$(.0123 \ldots b - 1\, 101112 \ldots)_b,$$

constructed by successively listing the base b expansions of the integers, is irrational.

13. Show that

$$\frac{1}{b} + \frac{1}{b^4} + \frac{1}{b^9} + \frac{1}{b^{16}} + \frac{1}{b^{25}} + \cdots.$$

is irrational, whenever b is a positive integer larger than one.

14. Let $b_1, b_2, b_3, \ldots$ be an infinite sequence of positive integers greater than one. Show that every real number can be represented as

$$c_0 + \frac{c_1}{b_1} + \frac{c_2}{b_1 b_2} + \frac{c_3}{b_1 b_2 b_3} + \cdots,$$

where $c_0, c_1, c_2, c_3, \ldots$ are integers such that $0 \le c_k < k$ for $k = 1, 2, 3, \ldots$.

15. Show that every real number has an expansion

$$c_0 + \frac{c_1}{1!} + \frac{c_2}{2!} + \frac{c_3}{3!} + \cdots,$$

where $c_0, c_1, c_2, c_3, \ldots$ are integers and $0 \le c_k < k$ for $k = 1, 2, 3, \ldots$.

16. Show that every rational number has a terminating expansion of the type described in Exercise 15.

∗ 17. Suppose that p is a prime and the base b expansion of $1/p$ is $(.\overline{c_1 c_2 \ldots c_{p-1}})_b$, so that the period length of the base b expansion of $1/p$ is $p - 1$. Show that if m is a positive integer with $1 \le m < p$, then

$$m/p = (.\overline{c_{k+1} \ldots c_{p-1} c_1 c_2 \ldots c_{k-1} c_k})_b,$$

where k is the least positive residue of $\text{ind}_b m$ modulo p.

∗ 18. Show that if p is prime and $1/p = (.\overline{c_1 c_2 \ldots c_k})_b$ has an even period length, $k = 2t$, then $c_j + c_{j+t} = b - 1$ for $j = 1, 2, \ldots, t$.

19. For which positive integers n is the length of the period of the binary expansion of $1/n$ equal to $n - 1$?

20. For which positive integers n is the length of the period of the decimal expansion of $1/n$ equal to $n - 1$?

21. Suppose that b is a positive integer. Show that the coefficients in the base b expansion of the real number $\gamma = \sum_{j=1}^{\infty} c_j/b^j$ with $0 \leq \gamma < 1$ are given by the formula $c_j = [\gamma b^j] - b[\gamma b^{j-1}]$ for $j = 1, 2, \ldots$. (*Hint:* First, show that $0 \leq [\gamma b^j] - b[\gamma b^{j-1}] \leq b - 1$. Then, show that $\sum_{j=1}^{N}([\gamma b^j] - b[\gamma b^{j-1}])/b^j = \gamma - (\gamma b^N[\gamma b^N]/b^N)$ and let $N \to \infty$.)

22. Use the formula in Exercise 21 to find the base 14 expansion of $1/6$.

23. Show that the number
$$\sum_{i=1}^{\infty}(-1)^{a_i}/10^{i!}$$
is transcendental for all sequences of positive integers $a_1, a_2, \ldots$.

24. Is the set of all real numbers with decimal expansions consisting of only 0s and 1s countable?

* 25. Show that the number e is irrational.

26. Pseudorandom numbers can be generated using the base m expansion of $1/P$, where P is a positive integer relatively prime to m. We set $x_n = c_{j+n}$, where j, the position of the seed, is a positive integer and $1/P = (.c_1c_2c_3 \ldots)_m$. This is called the $1/P$ *generator*. Find the first ten terms of the pseudorandom sequence generator with each of the following parameters.
    a) $m = 7$, $P = 19$, and $j = 6$
    b) $m = 8$, $P = 21$, and $j = 5$

12.1 Computational and Programming Exercises

Computations and Explorations

Using a computation program such as Maple or *Mathematica,* or programs you have written, carry out the following computations and explorations.

1. Find the pre-period and period of the decimal expansions of 212/31597, 1053/4437189, and 81327/16666699.

2. Find as many positive integers n as you can such that the length of the period of the decimal expansion of $1/n$ is $n - 1$.

3. Find the first 10,000 terms of the decimal expansion of π. Can you find any patterns? Make some conjectures about this expansion.

4. Find the first 10,000 terms of the decimal expansion of e. Can you find any patterns? Make some conjectures about this expansion.

Programming Projects

Write computer programs using Maple, *Mathematica,* or a language of your choice to do the following.

1. Find the base b expansion of a rational number, where b is a positive integer.

2. Find the numerator and denominator of a rational number in lowest terms from its base b expansion.

3. Find the pre-period and period lengths of the base b expansion of a rational number, where b is a positive integer.

4. Generate pseudorandom numbers using the $1/P$ generator (introduced in Exercise 26) with modulus m and seed in position j, where P and m are relatively prime positive integers greater than 1 and j is a positive integer.

12.2 Finite Continued Fractions

Using the Euclidean algorithm, we can express rational numbers as *continued fractions*. For instance, the Euclidean algorithm produces the following sequence of equations:

$$62 = 2 \cdot 23 + 16$$
$$23 = 1 \cdot 16 + 7$$
$$16 = 2 \cdot 7 + 2$$
$$7 = 3 \cdot 2 + 1.$$

When we divide both sides of each equation by the divisor of that equation, we obtain

$$\frac{62}{23} = 2 + \frac{16}{23} = 2 + \frac{1}{23/16}$$

$$\frac{23}{16} = 1 + \frac{7}{16} = 1 + \frac{1}{16/7}$$

$$\frac{16}{7} = 2 + \frac{2}{7} = 2 + \frac{1}{7/2}$$

$$\frac{7}{2} = 3 + \frac{1}{2}.$$

By combining these equations, we find that

$$\frac{62}{23} = 2 + \frac{1}{23/16}$$

$$= 2 + \cfrac{1}{1 + \cfrac{1}{16/7}}$$

$$= 2 + \cfrac{1}{1 + \cfrac{1}{2 + \cfrac{1}{7/2}}}$$

$$= 2 + \cfrac{1}{1 + \cfrac{1}{2 + \cfrac{1}{3 + \cfrac{1}{2}}}}.$$

The final expression in this string of equations is a continued fraction expansion of 62/23.

We now define continued fractions.

Definition. A *finite continued fraction* is an expression of the form

$$a_0 + \cfrac{1}{a_1 + \cfrac{1}{a_2 + \cfrac{1}{\ddots + \cfrac{1}{a_{n-1} + \cfrac{1}{a_n}}}}},$$

where $a_0, a_1, a_2, \ldots, a_n$ are real numbers with $a_1, a_2, a_3, \ldots, a_n$ positive. The real numbers $a_1, a_2, \ldots, a_n$ are called the *partial quotients* of the continued fraction. The continued fraction is called *simple* if the real numbers $a_0, a_1, \ldots, a_n$ are all integers.

Because it is cumbersome to fully write out continued fractions, we use the notation $[a_0; a_1, a_2, \ldots, a_n]$ to represent the continued fraction in the definition of a finite continued fraction.

We will now show that every finite simple continued fraction represents a rational number. Later we will demonstrate that every rational number can be expressed as a finite simple continued fraction.

Theorem 12.7. Every finite simple continued fraction represents a rational number.

Proof. We will prove the theorem using mathematical induction. For $n = 1$, we have

$$[a_0; a_1] = a_0 + \frac{1}{a_1} = \frac{a_0 a_1 + 1}{a_0},$$

which is rational. Now, we assume that for the positive integer k the simple continued fraction $[a_0; a_1, a_2, \ldots, a_k]$ is rational whenever $a_0, a_1, \ldots, a_k$ are integers with $a_1, \ldots, a_k$ positive. Let $a_0, a_1, \ldots, a_{k+1}$ be integers with $a_1, \ldots, a_{k+1}$ positive. Note that

$$[a_0; a_1, \ldots, a_{k+1}] = a_0 + \frac{1}{[a_1; a_2, \ldots, a_k, a_{k+1}]}.$$

By the induction hypothesis, $[a_1; a_2, \ldots, a_k, a_{k+1}]$ is rational; hence, there are integers r and s, with $s \neq 0$, such that this continued fraction equals r/s. Then

$$[a_0; a_1, \ldots, a_k, a_{k+1}] = a_0 + \frac{1}{r/s} = \frac{a_0 r + s}{r},$$

which is again a rational number.   ∎

We now show, using the Euclidean algorithm, that every rational number can be written as a finite simple continued fraction.

Theorem 12.8. Every rational number can be expressed by a finite simple continued fraction.

Proof.   Let $x = a/b$, where a and b are integers with $b > 0$. Let $r_0 = a$ and $r_1 = b$. Then, the Euclidean algorithm produces the following sequence of equations:

$$
\begin{aligned}
r_0 &= r_1 q_1 + r_2 & 0 &< r_2 < r_1, \\
r_1 &= r_2 q_2 + r_3 & 0 &< r_3 < r_2, \\
r_2 &= r_3 q_3 + r_4 & 0 &< r_4 < r_3, \\
&\ \ \vdots & & \\
r_{n-3} &= r_{n-2} q_{n-2} + r_{n-1} & 0 &< r_{n-1} < r_{n-2}, \\
r_{n-2} &= r_{n-1} q_{n-1} + r_n & 0 &< r_n < r_{n-1}, \\
r_{n-1} &= r_n q_n.
\end{aligned}
$$

In these equations, $q_2, q_3, \ldots, q_n$ are positive integers. Writing these equations in fractional form, we have

$$
\frac{a}{b} = \frac{r_0}{r_1} = q_1 + \frac{r_2}{r_1} = q_1 + \frac{1}{r_1/r_2}
$$

$$
\frac{r_1}{r_2} = q_2 + \frac{r_3}{r_2} = q_2 + \frac{1}{r_2/r_3}
$$

$$
\frac{r_2}{r_3} = q_3 + \frac{r_4}{r_3} = q_3 + \frac{1}{r_3/r_4}
$$

$$
\vdots
$$

$$
\frac{r_{n-3}}{r_{n-2}} = q_{n-2} + \frac{r_{n-1}}{r_{n-2}} = q_{n-2} + \frac{1}{r_{n-2}/r_{n-1}}
$$

$$
\frac{r_{n-2}}{r_{n-1}} = q_{n-1} + \frac{r_n}{r_{n-1}} = q_{n-1} + \frac{1}{r_{n-1}/r_n}
$$

$$
\frac{r_{n-1}}{r_n} = q_n.
$$

Substituting the value of r_1/r_2 from the second equation into the first equation, we obtain

$$
(12.10) \qquad \frac{a}{b} = q_1 + \cfrac{1}{q_2 + \cfrac{1}{r_2/r_3}}.
$$

Similarly, substituting the value of r_2/r_3 from the third equation into (12.10), we obtain

$$
\frac{c}{b} = q_1 + \cfrac{1}{q_2 + \cfrac{1}{q_3 + \cfrac{1}{r_3/r_4}}}.
$$

Continuing in this manner, we find that

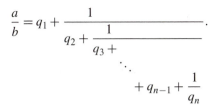

Hence, $\frac{a}{b} = [q_1; q_2, \ldots, q_n]$. This shows that every rational number can be written as a finite simple continued fraction. ∎

We note that continued fractions for rational numbers are not unique. From the identity

$$a_n = (a_n - 1) + \frac{1}{1},$$

we see that

$$[a_0; a_1, a_2, \ldots, a_{n-1}, a_n] = [a_0; a_1, a_2, \ldots, a_{n-1}, a_n - 1, 1]$$

whenever $a_n > 1$.

Example 12.6. We have

$$\frac{7}{11} = [0; 1, 1, 1, 3] = [0; 1, 1, 1, 2, 1]. \qquad \blacktriangleleft$$

In fact, it can be shown that every rational number can be written as a finite simple continued fraction in exactly two ways, one with an odd number of terms, the other with an even number (see Exercise 12 at the end of this section).

Next, we will discuss the numbers obtained from a finite continued fraction by cutting off the expression at various stages.

Definition. The continued fraction $[a_0; a_1, a_2, \ldots, a_k]$, where k is a nonnegative integer less than or equal to n, is called the *kth convergent* of the continued fraction $[a_0; a_1, a_2, \ldots, a_n]$. The kth convergent is denoted by C_k.

In our subsequent work, we will need some properties of the convergents of a continued fraction. We now develop these properties, starting with a formula for the convergents.

Theorem 12.9. Let $a_0, a_1, a_2, \ldots, a_n$ be real numbers, with $a_1, a_2, \ldots, a_n$ positive. Let the sequences $p_0, p_1, \ldots, p_n$ and $q_0, q_1, \ldots, q_n$ be defined recursively by

$$p_0 = a_0 \qquad\qquad q_0 = 1$$
$$p_1 = a_0 a_1 + 1 \qquad q_1 = a_1$$

and

$$p_k = a_k p_{k-1} + p_{k-2} \qquad q_k = a_k q_{k-1} + q_{k-2}$$

for $k = 2, 3, \ldots, n$. Then the kth convergent $C_k = [a_0; a_1, \ldots, a_k]$ is given by

$$C_k = p_k/q_k.$$

Proof. We will prove this theorem using mathematical induction. For $k = 0$, we have

$$C_0 = [a_0] = a_0/1 = p_0/q_0.$$

For $k = 1$, we see that

$$C_1 = [a_0; a_1] = a_0 + \frac{1}{a_1} = \frac{a_0 a_1 + 1}{a_1} = \frac{p_1}{q_1}.$$

Hence, the theorem is valid for $k = 0$ and $k = 1$.

Now assume that the theorem is true for the positive integer k, where $2 \leq k < n$. This means that

$$(12.11) \qquad C_k = [a_0; a_1, \ldots, a_k] = \frac{p_k}{q_k} = \frac{a_k p_{k-1} + p_{k-2}}{a_k q_{k-1} + q_{k-2}}.$$

Because of the way in which the p_j's and q_j's are defined, we see that the real numbers $p_{k-1}, p_{k-2}, q_{k-2}$, depend only on the partial quotients $a_0, a_1, \ldots, a_{k-1}$. Consequently, we can replace the real number a_k by $a_k + 1/a_{k+1}$ in (12.11), to obtain

$$\begin{aligned}
C_{k+1} = \left[a_0; a_1, \ldots, a_k, a_{k+1} \right] &= \left[a_o; a_1, \ldots, a_{k-1}, a_k + \frac{1}{a_{k+1}} \right] \\
&= \frac{\left(a_k + \frac{1}{a_{k+1}} \right) p_{k-1} + p_{k-2}}{\left(a_k + \frac{1}{a_{k+1}} \right) q_{k-1} + q_{k-2}} \\
&= \frac{a_{k+1}(a_k p_{k-1} + p_{k-2}) + p_{k-1}}{a_{k+1}(a_k q_{k-1} + q_{k-2}) + q_{k-1}} \\
&= \frac{a_{k+1} p_k + p_{k-1}}{a_{k+1} q_k + q_{k-1}} \\
&= \frac{p_{k+1}}{q_{k+1}}.
\end{aligned}$$

This finishes the proof by induction. ∎

We will illustrate how to use Theorem 12.9 with the following example.

Example 12.7.   We have $173/55 = [3; 6, 1, 7]$. We compute the sequences p_j and q_j for $j = 0, 1, 2, 3$, by

$$\begin{aligned}
p_0 &= 3 & q_0 &= 1 \\
p_1 &= 3 \cdot 6 + 1 = 19 & q_1 &= 6 \\
p_2 &= 1 \cdot 19 + 3 = 22 & q_2 &= 1 \cdot 6 + 1 = 7 \\
p_3 &= 7 \cdot 22 + 19 = 173 & q_3 &= 7 \cdot 7 + 6 = 55.
\end{aligned}$$

Hence, the convergents of the above continued fraction are

$$C_0 = p_0/q_0 = 3/1 = 3$$
$$C_1 = p_1/q_1 = 19/6$$
$$C_2 = p_2/q_2 = 22/7$$
$$C_3 = p_3/q_3 = 173/55.$$ ◄

We now state and prove another important property of the convergents of a continued fraction.

Theorem 12.10. Let $C_k = p_k/q_k$ be the kth convergent of the continued fraction $[a_0; a_1, \ldots, a_n]$, where k is a positive integer, $1 \le k \le n$. If p_k are as defined in Theorem 12.9, then

$$p_k q_{k-1} - p_{k-1} q_k = (-1)^{k-1}.$$

Proof. We use mathematical induction to prove the theorem. For $k = 1$, we have

$$p_1 q_0 - p_0 q_1 = (a_0 a_1 + 1) \cdot 1 - a_0 a_1 = 1.$$

Assume that the theorem is true for an integer k, where $1 \le k < n$, so that

$$p_k q_{k-1} - p_{k-1} q_k = (-1)^{k-1}.$$

Then we have

$$p_{k+1} q_k - p_k q_{k+1} = (a_{k+1} p_k + p_{k-1}) q_k - p_k (a_{k+1} q_k + q_{k-1})$$
$$= p_{k-1} q_k - p_k q_{k-1} = -(-1)^{k-1} = (-1)^k,$$

so that the theorem is true for $k + 1$. This finishes the proof by induction. ∎

We illustrate this theorem with the example that we used to illustrate Theorem 12.9.

Example 12.8. For the continued fraction $[3; 6, 1, 7]$ we have

$$p_0 q_1 - p_1 q_0 = 3 \cdot 6 - 19 \cdot 1 = -1$$
$$p_1 q_2 - p_2 q_1 = 19 \cdot 7 - 22 \cdot 6 = 1$$
$$p_2 q_3 - p_3 q_2 = 22 \cdot 55 - 173 \cdot 7 = -1.$$

As a consequence of Theorem 12.10, we see that for $k = 1, 2, \ldots$, the convergents p_k/q_k of a simple continued fraction are in lowest terms. Corollary 12.10.1 demonstrates this. ◄

Corollary 12.10.1. Let $C_k = p_k/q_k$ be the kth convergent of the simple continued fraction $[a_0; a_1, \ldots, a_n]$, where the integers p_k and q_k are as defined in Theorem 12.9. Then the integers p_k and q_k are relatively prime.

Proof. Let $d = (p_k, q_k)$. By Theorem 12.10, we know that

$$p_k q_{k-1} - q_k p_{k-1} = (-1)^{k-1}.$$

Hence,

$$d \mid (-1)^{k-1}.$$

Therefore, $d = 1$.   ∎

We also have the following useful corollary of Theorem 12.10.

Corollary 12.10.2.   Let $C_k = p_k/q_k$ be the kth convergent of the simple continued fraction $[a_0; a_1, a_2, \ldots, a_k]$. Then

$$C_k - C_{k-1} = \frac{(-1)^{k-1}}{q_k q_{k-1}}$$

for all integers k with $1 \le k \le n$. Also,

$$C_k - C_{k-2} = \frac{a_k(-1)^k}{q_k q_{k-2}}$$

for all integers k with $2 \le k \le n$.

Proof.   By Theorem 12.10, we know that $p_k q_{k-1} - q_k p_{k-1} = (-1)^{k-1}$.

We obtain the first identity,

$$C_k - C_{k-1} = \frac{p_k}{q_k} - \frac{p_{k-1}}{q_{k-1}} = \frac{(-1)^{k-1}}{q_k q_{k-1}},$$

by dividing both sides by $q_k q_{k-1}$.

To obtain the second identity, note that

$$C_k - C_{k-2} = \frac{p_k}{q_k} - \frac{p_{k-2}}{q_{k-2}} = \frac{p_k q_{k-2} - p_{k-2} q_k}{q_k q_{k-2}}.$$

Because $p_k = a_k p_{k-1} + p_{k-2}$ and $q_k = a_k q_{k-1} + q_{k-2}$, we see that the numerator of the fraction on the right is

$$\begin{aligned}
p_k q_{k-2} - p_{k-2} q_k &= (a_k p_{k-1} + p_{k-2}) q_{k-2} - p_{k-2} (a_k q_{k-1} + q_{k-2}) \\
&= a_k (p_{k-1} q_{k-2} - p_{k-2} q_{k-1}) \\
&= a_k (-1)^{k-2},
\end{aligned}$$

using Theorem 12.10 to see that $p_{k-1} q_{k-2} - p_{k-2} q_{k-1} = (-1)^{k-2}$.

Therefore, we find that

$$C_k - C_{k-2} = \frac{a_k(-1)^k}{q_k q_{k-2}}.$$

This is the second identity of the corollary.   ∎

Using Corollary 12.10.2, we can prove the following theorem, which is useful when developing infinite continued fractions.

Theorem 12.11. Let C_k be the kth convergent of the finite simple continued fraction $[a_0; a_1, a_2, \ldots, a_n]$. Then

$$C_1 > C_3 > C_5 > \cdots,$$
$$C_0 < C_2 < C_4 < \cdots,$$

and every odd-numbered convergent C_{2j+1}, $j = 0, 1, 2, \ldots$, is greater than every even-numbered convergent C_{2j}, $j = 0, 1, 2, \ldots$.

Proof. Because Corollary 12.10.2 tells us that, for $k = 2, 3, \ldots, n$,

$$C_k - C_{k-2} = \frac{a_k(-1)^k}{q_k q_{k-2}}$$

we know that

$$C_k < C_{k-2}$$

when k is odd, and

$$C_k > C_{k-2}$$

when k is even. Hence,

$$C_1 > C_3 > C_5 > \cdots$$

and

$$C_0 < C_2 < C_4 < \cdots.$$

To show that every odd-numbered convergent is greater than every even-numbered convergent, note that from Corollary 12.10.2, we have

$$C_{2m} - C_{2m-1} = \frac{(-1)^{2m-1}}{q_{2m} q_{2m-1}} < 0,$$

so that $C_{2m-1} > C_{2m}$. To compare C_{2k} and C_{2j-1}, we see that

$$C_{2j-1} > C_{2j+2k-1} > C_{2j+2k} > C_{2k}.$$

so that every odd-numbered convergent is greater than every even-numbered convergent.
■

Example 12.9. Consider the finite simple continued fraction $[2; 3, 1, 1, 2, 4]$. Then the convergents are

$$
\begin{aligned}
C_0 &= & 2/1 &= 2 \\
C_1 &= & 7/3 &= 2.3333\ldots \\
C_2 &= & 9/4 &= 2.25 \\
C_3 &= & 16/7 &= 2.2857\ldots \\
C_4 &= & 41/18 &= 2.2777\ldots \\
C_5 &= & 180/79 &= 2.2784\ldots.
\end{aligned}
$$

We see that

$$C_0 = 2 < C_2 = 2.25 < C_4 = 2.2777\ldots$$
$$< C_5 = 2.2784\ldots < C_3 = 2.2857\ldots < C_1 = 2.3333\ldots. \qquad \blacktriangleleft$$

12.2 Exercises

1. Find the rational number, expressed in lowest terms, represented by each of the following simple continued fractions.

a) $[2; 7]$ e) $[1; 1]$

b) $[1; 2, 3]$ f) $[1; 1, 1]$

c) $[0; 5, 6]$ g) $[1; 1, 1, 1]$

d) $[3; 7, 15, 1]$ h) $[1; 1, 1, 1, 1]$

2. Find the rational number, expressed in lowest terms, represented by each of the following simple continued fractions.

a) $[10; 3]$ e) $[2; 1, 2, 1, 1, 4]$

b) $[3; 2, 1]$ f) $[1; 2, 1, 2]$

c) $[0; 1, 2, 3]$ g) $[1; 2, 1, 2, 1]$

d) $[2; 1, 2, 1]$ h) $[1; 2, 1, 2, 1, 2]$

3. Find the simple continued fraction expansion, not terminating with the partial quotient of 1, of each of the following rational numbers.

a) 18/13   c) 19/9   e) $-931/1005$

b) 32/17   d) 310/99   f) 831/8110

4. Find the simple continued fraction expansion, not terminating with the partial quotient of 1, of each of the following rational numbers.

a) 6/5   c) 19/29   e) $-943/1001$

b) 22/7   d) 5/999   f) 873/4867

5. Find the convergents of each of the continued fractions found in Exercise 3.

6. Find the convergents of each of the continued fractions found in Exercise 4.

7. Show that the convergents that you found in Exercise 5 satisfy Theorem 12.11.

8. Let f_k denote the kth Fibonacci number. Find the simple continued fraction, terminating with the partial quotient of 1, of f_{k+1}/f_k, where k is a positive integer.

9. Show that if the simple continued fraction expression of the rational number $\alpha, \alpha > 1$, is $[a_0; a_1, \ldots, a_k]$, then the simple continued fraction expression of $1/\alpha$ is $[0; a_1, \ldots, a_k]$.

☞ **10.** Show that if $a_0 > 0$, then

$$p_k/p_{k-1} = [a_k; a_{k-1}, \ldots, a_1, a_0]$$

and

$$q_k/q_{k-1} = [a_k; a_{k-1}, \ldots, a_2, a_1],$$

where $C_{k-1} = p_{k-1}/q_{k-1}$ and $C_k = p_k/q_k, k \geq 1$, are successive convergents of the continued fraction $[a_0; a_1, \ldots, a_n]$. (*Hint:* Use the relation $p_k = a_k p_{k-1} + p_{k-2}$ to show that $p_k/p_{k-1} = a_k + 1/(p_{k-1}/p_{k-2})$.)

☞ **11.** Show that $q_k \geq f_k$ for $k = 1, 2, \ldots$, where $C_k = p_k/q_k$ is the kth convergent of the simple continued fraction $[a_0; a_1, \ldots, a_n]$ and f_k denotes the kth Fibonacci number.

12. Show that every rational number has exactly two finite simple continued fraction expansions.

* **13.** Let $[a_0; a_1, a_2, \ldots, a_n]$ be the simple continued fraction expansion of r/s, where $(r, s) = 1$ and $r \geq 1$. Show that this continued fraction is symmetric, that is, $a_0 = a_n, a_1 = a_{n-1}, a_2 = a_{n-2}, \ldots$, if and only if $r|(s^2 + 1)$ if n is odd and $r|(s^2 - 1)$ if n is even. (*Hint:* Use Exercise 10 and Theorem 12.10.)

* **14.** Explain how finite continued fractions for rational numbers, with both plus and minus signs allowed, can be generated from the division algorithm given in Exercise 18 of Section 1.5.

15. Let $a_0, a_1, a_2, \ldots, a_k$ be real numbers with $a_1, a_2, \ldots$ positive, and let x be a positive real number. Show that $[a_0; a_1, \ldots, a_k] < [a_0; a_1 \ldots, a_k + x]$ if k is odd and $[a_0; a_1, \ldots, a_k] > [a_0; a_1 \ldots, a_k + x]$ if k is even.

16. Determine whether n can be expressed as the sum of positive integers a and b, where all the partial quotients of the finite simple continued fraction of a/b are either 1 or 2, for each of the following integers n.

a) 13   c) 19   e) 27
b) 17   d) 23   f) 29

12.2 Computational and Programming Exercises

Computations and Explorations

Using a computation program such as Maple or *Mathematica,* or programs you have written, carry out the following computations and explorations.

1. Find the simple continued fractions of 1001/3000, 10,001/30,000, and 100,001/300,000.

2. Find the finite continued fractions of x and $2x$ for 20 different rational numbers. Can you find a rule for finding the finite simple continued fraction of $2x$ from that of x?

3. Determine for each integer $n, n \leq 1000$, whether there are integers a and b with $n = a + b$ such that the partial quotients of the continued fraction of a/b are all either 1 or 2. Can you make any conjectures?

Programming Projects

Write programs using Maple, *Mathematica,* or a language of your choice to do the following.

1. Find the simple continued fraction expansion of a rational number.

2. Find the convergents of a finite simple continued fraction, and find the rational number that this continued fraction represents.

12.3 Infinite Continued Fractions

Suppose that we have an infinite sequence of positive integers $a_0; a_1, a_2, \ldots$. How can we define the infinite continued fraction $[a_0; a_1, a_2, \ldots]$? To make sense of infinite continued fractions, we need a result from mathematical analysis. We state the result, and refer the reader to a mathematical analysis text, such as [Ru64], for a proof.

Theorem 12.12.   Let $x_0, x_1, x_2, \ldots$ be a sequence of real numbers such that $x_0 < x_1 < x_2 < \cdots$ and $x_k < U$ for $k = 0, 1, 2, \ldots$ for some real number U, or $x_0 > x_1 > x_2 > \ldots$ and $x_k > L$ for $k = 0, 1, 2, \ldots$ for some real number L. Then the terms of the sequence $x_0, x_1, x_2, \ldots$ tend to a limit x, that is, there exists a real number x such that

$$\lim_{k \to \infty} x_k = x.$$

Theorem 12.12 tells us that the terms of an infinite sequence tend to a limit in two special situations: when the terms of the sequence are increasing and all are less than an upper bound, and when the terms of the sequence are decreasing and all are greater than a lower bound.

We can now define infinite continued fractions as limits of finite continued fractions, as the following theorem shows.

Theorem 12.13.   Let $a_0, a_1, a_2, \ldots$ be an infinite sequence of integers with $a_1, a_2, \ldots$ positive, and let $C_k = [a_0; a_1, a_2, \ldots, a_k]$. Then the convergents C_k tend to a limit α, that is,

$$\lim_{k \to \infty} C_k = \alpha.$$

Before proving Theorem 12.13, we note that the limit α described in the statement of the theorem is called the value of the *infinite simple continued fraction* $[a_0; a_1, a_2, \ldots]$.

To prove Theorem 12.13, we will show that the infinite sequence of even-numbered convergents is increasing and has an upper bound and that the infinite sequence of odd-numbered convergents is decreasing and has a lower bound. We then show that the limits of these two sequences, guaranteed to exist by Theorem 12.12, are in fact equal.

Proof.   Let m be an even positive integer. By Theorem 12.11, we see that

$$C_1 > C_3 > C_5 > \cdots > C_{m-1},$$
$$C_0 < C_2 < C_4 < \cdots < C_m,$$

and $C_{2j} < C_{2k+1}$ whenever $2j \le m$ and $2k + 1 < m$. By considering all possible values of m, we see that

$$C_1 > C_3 > C_5 > \cdots > C_{2n-1} > C_{2n+1} > \cdots,$$
$$C_0 < C_2 < C_4 < \cdots < C_{2n-2} < C_{2n} < \cdots,$$

and $C_{2j} > C_{2k+1}$ for all positive integers j and k. We see that the hypotheses of Theorem 12.12 are satisfied for each of the two sequences $C_1, C_3, C_2, \ldots$ and $C_0, C_2, C_4, \ldots$.

Hence, the sequence $C_1, C_3, C_5, \ldots$ tends to a limit α_1 and the sequence $C_0, C_2, C_4, \ldots$ tends to a limit α_2, that is,

$$\lim_{n \to \infty} C_{2n+1} = \alpha_1$$

and

$$\lim_{n \to \infty} C_{2n} = \alpha_2.$$

Our goal is to show that these two limits α_1 and α_2 are equal. Using Corollary 12.10.2, we have

$$C_{2n+1} - C_{2n} = \frac{p_{2n+1}}{q_{2n+1}} - \frac{p_{2n}}{q_{2n}} = \frac{(-1)^{(2n+1)-1}}{q_{2n+1}q_{2n}} = \frac{1}{q_{2n+1}q_{2n}}.$$

Because $q_k \geq k$ for all positive integers k (see Exercise 11 of Section 12.2), we know that

$$\frac{1}{q_{2n+1}q_{2n}} < \frac{1}{(2n+1)(2n)},$$

and hence,

$$C_{2n+1} - C_{2n} = \frac{1}{q_{2n+1}q_{2n}}$$

tends to zero, that is,

$$\lim_{n \to \infty} (C_{2n+1} - C_{2n}) = 0.$$

Hence, the sequences $C_1, C_3, C_5, \ldots$ and $C_0, C_2, C_4, \ldots$ have the same limit, because

$$\lim_{n \to \infty} (C_{2n+1} - C_{2n}) = \lim_{n \to \infty} C_{2n+1} - \lim_{n \to \infty} C_{2n} = 0.$$

Therefore, $\alpha_1 = \alpha_2$, and we conclude that all the convergents tend to the limit $\alpha = \alpha_1 = \alpha_2$. This finishes the proof of the theorem. ∎

Previously, we showed that rational numbers have finite simple continued fractions. Next, we will show that the value of any infinite simple continued fraction is irrational.

Theorem 12.14. Let $a_0, a_1, a_2, \ldots$ be integers with $a_1, a_2, \ldots$ positive. Then $[a_0; a_1, a_2, \ldots]$ is irrational.

Proof. Let $\alpha = [a_0; a_1, a_2, \ldots]$ and let

$$C_k = p_k/q_k = [a_0; a_1, a_2, \ldots a_k]$$

denote the kth convergent of α. When n is a positive integer, Theorem 12.13 shows that $C_{2n} < \alpha < C_{2n+1}$, so that

$$0 < \alpha - C_{2n} < C_{2n+1} - C_{2n}.$$

However, by Corollary 12.10.2, we know that

$$C_{2n+1} - C_{2n} = \frac{1}{q_{2n+1}q_{2n}},$$

which means that

$$0 < \alpha - C_{2n} = \alpha - \frac{p_{2n}}{q_{2n}} < \frac{1}{q_{2n+1}q_{2n}},$$

and, therefore, we have

$$0 < \alpha q_{2n} - p_{2n} < \frac{1}{q_{2n+1}}.$$

Assume that α is rational, so that $\alpha = a/b$, where a and b are integers with $b \neq 0$. Then

$$0 < \frac{aq_{2n}}{b} - p_{2n} < \frac{1}{q_{2n+1}},$$

and by multiplying this inequality by b, we see that

$$0 < aq_{2n} - bp_{2n} < \frac{b}{q_{2n+1}}.$$

Note that $aq_{2n} - bp_{2n}$ is an integer for all positive integers n. However, because $q_{2n+1} > 2n + 1$, for each integer n there is an integer n_0 such that $q_{2n_0+1} > b$, so that $b/q_{2n_0+1} < 1$. This is a contradiction, because the integer $aq_{2n_0} - bp_{2n_0}$ cannot be between 0 and 1. We conclude that α is irrational. ∎

We have demonstrated that every infinite simple continued fraction represents an irrational number. We will now show that every irrational number can be uniquely expressed by an infinite simple continued fraction, by first constructing such a continued fraction, and then by showing that it is unique.

Theorem 12.15. Let $\alpha = \alpha_0$ be an irrational number, and define the sequence $a_0, a_1, a_2, \ldots$ recursively by

$$a_k = [\alpha_k] \quad \alpha_{k+1} = 1/(\alpha_k - a_k)$$

for $k = 0, 1, 2, \ldots$. Then α is the value of the infinite simple continued fraction $[a_0; a_1, a_2, \ldots]$.

Proof. From the recursive definition of the integers a_k, we see that a_k is an integer for every k. Furthermore, using mathematical induction, we can show that α_k is irrational for every nonnegative integer k and that, as a consequence, α_{k+1} exists. First, note that $\alpha_0 = \alpha$ is irrational, so that $\alpha_0 \neq a_0 = [\alpha_0]$ and $\alpha_1 = 1/(\alpha_0 - a_0)$ exists.

Next, we assume that α_k is irrational. As a consequence, α_{k+1} exists. We can easily see that α_{k+1} is also irrational, because the relation

$$\alpha_{k+1} = 1/(\alpha_k - a_k)$$

implies that

(12.12)
$$\alpha_k = a_k + \frac{1}{\alpha_{k+1}},$$

and if α_{k+1} were rational, then α_k would also be rational. Now, because α_k is irrational and a_k is an integer, we know that $\alpha_k \neq a_k$, and

$$a_k < \alpha_k < a_k + 1,$$

so that

$$0 < \alpha_k - a_k < 1.$$

Hence,

$$\alpha_{k+1} = 1/(\alpha_k - a_k) > 1$$

and, consequently,

$$a_{k+1} = [\alpha_{k+1}] \geq 1$$

for $k = 0, 1, 2, \ldots$. This means that all the integers $a_1, a_2, \ldots$ are positive.

Note that by repeatedly using (12.12), we see that

$$\alpha = \alpha_0 = a_0 + \frac{1}{\alpha_1} = [a_0; \alpha_1]$$

$$= a_0 + \cfrac{1}{a_1 + \cfrac{1}{\alpha_2}} = [a_0; a_1, a_2]$$

$$\vdots$$

$$= a_0 + \cfrac{1}{a_1 + \cfrac{1}{a_2 + \cfrac{\ddots}{+ a_k + \cfrac{1}{\alpha_{k+1}}}}} = [a_0; a_1, a_2, \ldots, a_k, \alpha_{k+1}].$$

What we must now show is that the value of $[a_0; a_1, a_2, \ldots, a_k, \alpha_{k+1}]$ tends to α as k tends to infinity, that is, as k grows without bound. By Theorem 12.9, we see that

$$\alpha = [a_0; a_1, \ldots, a_k, \alpha_{k+1}] = \frac{\alpha_{k+1} p_k + p_{k-1}}{\alpha_{k+1} q_k + q_{k-1}},$$

where $C_j = p_j/q_j$ is the jth convergent of $[a_0; a_1, a_2, \ldots]$. Hence,

$$\alpha - C_k = \frac{\alpha_{k+1}p_k + p_{k-1}}{\alpha_{k+1}q_k + q_{k-1}} - \frac{p_k}{q_k}$$

$$= \frac{-(p_k q_{k-1} - p_{k-1}q_k)}{(\alpha_{k+1}q_k + q_{k-1})q_k}$$

$$= \frac{-(-1)^{k-1}}{(\alpha_{k+1}q_k + q_{k-1})q_k},$$

where we have used Theorem 12.10 to simplify the numerator on the right-hand side of the second equality. Because

$$\alpha_{k+1}q_k + q_{k-1} > a_{k+1}q_k + q_{k-1} = q_{k+1},$$

we see that

$$|\alpha - C_k| < \frac{1}{q_k q_{k+1}}.$$

Because $q_k > k$ (from Exercise 11 of Section 12.2), we note that $1/(q_k q_{k+1})$ tends to zero as k tends to infinity. Hence, C_k tends to α as k tends to infinity or, phrased differently, the value of the infinite simple continued fraction $[a_0; a_1, a_2, \ldots]$ is α. ∎

To show that the infinte simple continued fraction that represent an irrational number is unique, we prove the following theorem.

Theorem 12.16. If the two infinite simple continued fractions $[a_0; a_1, a_2, \ldots]$ and $[b_0; b_1, b_2, \ldots]$ represent the same irrational number, then $a_k = b_k$ for $k = 0, 1, 2, \ldots$.

Proof. Suppose that $\alpha = [a_0; a_1, a_2, \ldots]$. Then, because $C_0 = a_0$ and $C_1 = a_0 + 1/a_1$, Theorem 12.11 tells us that

$$a_0 < \alpha < a_0 + 1/a_1,$$

so that $a_0 = [\alpha]$. Further, we note that

$$[a_0; a_1, a_2, \ldots] = a_0 + \frac{1}{[a_1; a_2, a_3, \ldots]},$$

because

$$\alpha = [a_0; a_1, a_2, \ldots] = \lim_{k \to \infty} [a_0; a_1, a_2, \ldots, a_k]$$

$$= \lim_{k \to \infty} \left(a_0 + \frac{1}{[a_1; a_2, a_3, \ldots, a_k]} \right)$$

$$= a_0 + \frac{1}{\lim_{k \to \infty} [a_1; a_2, \ldots, a_k]}$$

$$= a_0 + \frac{1}{[a_1; a_2, a_3, \ldots]}.$$

Suppose that

$$[a_0; a_1, a_2, \ldots] = [b_0; b_1, b_2, \ldots].$$

Our remarks show that

$$a_0 = b_0 = [\alpha]$$

and that

$$a_0 + \cfrac{1}{[a_1; a_2, \ldots]} = b_0 + \cfrac{1}{[b_1; b_2, \ldots]},$$

so that

$$[a_1; a_2, \ldots] = [b_1; b_2, \ldots].$$

Now, assume that $a_k = b_k$, and that $[a_{k+1}; a_{k+2}, \ldots] = [b_{k+1}; b_{k+2}, \ldots]$. Using the same argument, we see that $a_{k+1} = b_{k+1}$, and

$$a_{k+1} + \cfrac{1}{[a_{k+2}; a_{k+3}, \ldots]} = b_{k+1} + \cfrac{1}{[b_{k+1}; b_{k+3}, \ldots]},$$

which implies that

$$[a_{k+2}; a_{k+3} \ldots] = [b_{k+2}; b_{k+3}, \ldots].$$

Hence, by mathematical induction, we see that $a_k = b_k$ for $k = 0, 1, 2, \ldots$. ∎

To find the simple continued fraction expansion of a real number, we use the algorithm given in Theorem 12.15. We illustrate this procedure with the following example.

Example 12.10. Let $\alpha = \sqrt{6}$. We find that

$$a_0 = \left[\sqrt{6}\right] = 2, \qquad \alpha_1 = \cfrac{1}{\sqrt{6} - 2} = \cfrac{\sqrt{6} + 2}{2},$$

$$a_1 = \left[\cfrac{\sqrt{6} + 2}{2}\right] = 2, \qquad \alpha_2 = \cfrac{1}{\left(\frac{\sqrt{6}+2}{2}\right) - 2} = \sqrt{6} + 2,$$

$$a_2 = \left[\sqrt{6} + 2\right] = 4, \qquad \alpha_3 = \cfrac{1}{\left(\sqrt{6} + 2\right) - 4} = \cfrac{\sqrt{6} + 2}{2} = \alpha_1.$$

Because $\alpha_3 = \alpha_1$, we see that $a_3 = a_1, a_4 = a_2, \ldots$, and so on. Hence

$$\sqrt{6} = [2; 2, 4, 2, 4, 2, 4, \ldots].$$

The simple continued fraction of $\sqrt{6}$ is periodic. We will discuss periodic simple continued fractions in the next section. ◀

The convergents of the infinite simple continued fraction of an irrational number are good approximations to α. This leads to the following theorem, which we introduced in Exercise 34 of Section 1.1.

Theorem 12.17. *Dirichlet's Theorem on Diophantine Approximation.* If α is an irrational number, then there are infinitely many rational numbers p/q such that

$$|\alpha - p/q| < 1/q^2.$$

Proof. Let p_k/q_k be the kth convergent of the continued fraction of α. Then, by the proof of Theorem 12.15, we know that

$$|\alpha - p_k/q_k| < 1/(q_k q_{k+1}).$$

Because $q_k < q_{k+1}$, it follows that

$$|\alpha - p_k/q_k| < 1/q_k^2.$$

Consequently, the convergents of α, p_k/q_k, $k = 1, 2, \ldots$, are infinitely many rational numbers meeting the conditions of the theorem.   ∎

The next theorem and corollary show that the convergents of the simple continued fraction of α are the *best rational approximations* to α, in the sense that p_k/q_k is closer to α than any other rational number with a denominator less than q_k.

Theorem 12.18.   Let α be an irrational number and let p_j/q_j, $j = 1, 2, \ldots$, be the convergents of the infinite simple continued fraction of α. If r and s are integers with $s > 0$ and if k is a positive integer such that

$$|s\alpha - r| < |q_k\alpha - p_k|,$$

then $s \geq q_{k+1}$.

Proof.   Assume that $|s\alpha - r| < |q_k\alpha - p_k|$, but that $1 \leq s < q_{k+1}$. We consider the simultaneous equations

$$p_k x + p_{k+1} y = r$$
$$q_k x + q_{k+1} y = s.$$

By multiplying the first equation by q_k and the second by p_k, and then subtracting the second from the first, we find that

$$(p_{k+1}q_k - p_k q_{k+1})y = rq_k - sp_k.$$

By Theorem 12.10, we know that $p_{k+1}q_k - p_k q_{k+1} = (-1)^k$, so that

$$y = (-1)^k(rq_k - sp_k).$$

Similarly, multiplying the first equation by q_{k+1} and the second by p_{k+1}, and then subtracting the first from the second, we find that

$$x = (-1)^k(sp_{k+1} - rq_{k+1}).$$

We note that $s \neq 0$ and $y \neq 0$. If $x = 0$, then $sp_{k+1} = rq_{k+1}$. Because $(p_{k+1}, q_{k+1}) = 1$, Lemma 3.4 tells us that $q_{k+1} | s$, which implies that $q_{k+1} \leq s$, contrary to our assumption. If $y = 0$, then $r = p_k x$ and $s = q_k x$, so that

$$|s\alpha - r| = |x| \, |q_k\alpha - p_k| \geq |q_k\alpha - p_k|,$$

because $|x| \geq 1$, contrary to our assumption.

We will now show that x and y have opposite signs. First, suppose that $y < 0$. Because $q_k x = s - q_{k+1} y$, we know that $x > 0$, because $q_k x > 0$ and $q_k > 0$. When $y > 0$, because $q_{k+1} y \geq q_{k+1} > s$, we see that $q_k x = s - q_{k+1} y < 0$, so that $x < 0$.

By Theorem 12.11, we know that either $p_k/q_k < \alpha < p_{k+1}/q_{k+1}$ or that $p_{k+1}/q_{k+1} < \alpha < p_k/q_k$. In either case, we easily see that $q_k \alpha - p_k$ and $q_{k+1} \alpha - p_{k+1}$ have opposite signs.

From the simultaneous equations we started with, we see that

$$|s\alpha - r| = |(q_k x + q_{k+1} y)\alpha - (p_k x + p_{k+1} y)|$$
$$= |x(q_k \alpha - p_k) + y(q_{k+1} \alpha - p_{k+1})|.$$

Combining the conclusions of the previous two paragraphs, we see that $x(q_k \alpha - p_k)$ and $y(q_{k+1} \alpha - p_{k+1})$ have the same sign, so that

$$|s\alpha - r| = |x| \, |q_k \alpha - p_k| + |y| \, |q_{k+1} \alpha - p_{k+1}|$$
$$\geq |x| \, |q_k \alpha - p_k|$$
$$\geq |q_k \alpha - p_k|,$$

because $|x| \geq 1$. This contradicts our assumption.

We have shown that our assumption is false and, consequently, the proof is complete.
∎

Corollary 12.18.1. Let α be an irrational number and let p_j/q_j, $j = 1, 2, \ldots$ be the convergents of the infinite simple continued fraction of α. If r/s is a rational number, where r and s are integers with $s > 0$, and if k is a positive integer such that

$$|\alpha - r/s| < |\alpha - p_k/q_k|,$$

then $s > q_k$.

Proof. Suppose that $s \leq q_k$ and that

$$|\alpha - r/s| < |\alpha - p_k/q_k|.$$

By multiplying these two inequalities, we find that

$$s|\alpha - r/s| < q_k|\alpha - p_k/q_k|,$$

so that

$$|s\alpha - r| < |q_k \alpha - p_k|,$$

violating the conclusion of Theorem 12.18. ∎

Example 12.11. The simple continued fraction of the real number π is $\pi = [3; 7, 15, 1, 292, 1, 1, 1, 2, 1, 3, \ldots]$. Note that there is no discernible pattern in the sequence of partial quotients. The convergents of this continued fraction are the best rational approximations to π. The first five are 3, 22/7, 333/106, 355/113, and 103,993/33,102. We conclude from Corollary 12.18.1 that 22/7 is the best rational approximation of π with denominator less than or equal to 105, and so on. ◀

Finally, we conclude this section with a result that shows that any sufficiently close rational approximation to an irrational number must be a convergent of the infinite simple continued fraction expansion of this number.

Theorem 12.19.   If α is an irrational number and if r/s is a rational number in lowest terms, where r and s are integers with $s > 0$ such that

$$|\alpha - r/s| < 1/(2s^2),$$

then r/s is a convergent of the simple continued fraction expansion of α.

Proof.   Assume that r/s is not a convergent of the simple continued fraction expansion of α. Then, there are successive convergents p_k/q_k and p_{k+1}/q_{k+1} such that $q_k \leq s < q_{k+1}$. By Theorem 12.18, we see that

$$|q_k\alpha - p_k| \leq |s\alpha - r| = s|\alpha - r/s| < 1/(2s).$$

Dividing by q_k, we obtain

$$|\alpha - p_k/q_k| < 1/(2sq_k).$$

Because we know that $|sp_k - rq_k| \geq 1$ (we know that $sp_k - rq_k$ is a nonzero integer because $r/s \neq p_k/q_k$), it follows that

$$
\begin{aligned}
\frac{1}{sq_k} &\leq \frac{|sp_k - rq_k|}{sq_k} \\
&= \left| \frac{p_k}{q_k} - \frac{r}{s} \right| \\
&\leq \left| \alpha - \frac{p_k}{q_k} \right| + \left| \alpha - \frac{r}{s} \right| \\
&< \frac{1}{2sq_k} + \frac{1}{2s^2}
\end{aligned}
$$

(where we have used the triangle inequality to obtain the second inequality). Hence, we see that

$$1/2sq_k < 1/2s^2.$$

Consequently,

$$2sq_k > 2s^2,$$

which implies that $q_k > s$, contradicting the assumption.   ∎

Applying Continued Fractions to Attack the RSA Cryptosystem   We can use a version of Theorem 12.19 for rational numbers to explain why an attack on certain implementations of RSA ciphers works. We leave it as an exercise to prove that this version of Theorem 12.19 is valid.

Theorem 12.20. *Wiener's Low Encryption Exponent Attack on RSA.*   Suppose that $n = pq$, where p and q are odd primes with $q < p < 2q$ and that $d < n^{1/4}/3$. Then, given an RSA encryption key (e, n), the decryption key can be found using $O((\log n)^3)$ bit operations.

Proof. We will base the proof on approximation of a rational number by continued fractions. First note that because $de \equiv 1 \pmod{\phi(n)}$, there is an integer k such that $de - 1 = k\phi(n)$. Dividing both sides of this equation by $d\phi(n)$, we find that

$$\frac{e}{\phi(n)} - \frac{1}{d\phi(n)} = \frac{k}{d},$$

which implies that

$$\frac{e}{\phi(n)} - \frac{k}{d} = \frac{1}{d\phi(n)}.$$

This shows that the fraction k/d is a good approximation of $e/\phi(n)$.

Note also that $q < \sqrt{n}$, because $q < p$ and $n = pq$ by the hypotheses of the theorem. Using the hypothesis that $q < p$, it follows that

$$p + q - 1 \le 2q + q - 1 = 3q - 1 < 3\sqrt{n}.$$

Because $\phi(n) = n - p - q + 1$, we see that $n - \phi(n) = n - (n - p - q + 1) = p + q - 1 < 3\sqrt{n}$.

We can make use of this last inequality to show that k/d is an excellent approximation of e/n. We see that

$$\left| \frac{e}{n} - \frac{k}{d} \right| = \left| \frac{de - kn}{nd} \right|$$

$$= \left| \frac{(de - k\phi(n)) - (kn + k\phi(n))}{nd} \right|$$

$$= \left| \frac{1 - k(n - \phi(n))}{nd} \right| \le \frac{3k\sqrt{n}}{nd} = \frac{3k}{d\sqrt{n}}.$$

Because $e < \phi(n)$, we see that $ke < k\phi(n) = de - 1 < de$. This implies that $k < d$. We now use the hypothesis that $d < n^{1/4}/3$ to see that $k < n^{1/4}/3$.

It follows that

$$\left| \frac{e}{n} - \frac{k}{d} \right| \le \frac{3k\sqrt{n}}{nd} \le \frac{3(n^{1/4}/3)\sqrt{n}}{nd} = \frac{1}{dn^{1/4}} < \frac{1}{2d^2}.$$

We now use the version of Theorem 12.19 for rational numbers. By this theorem, we know that k/d is a convergent of the continued fraction expansion of e/n. Note also that both e and n are public information. Consequently, to find k/d we need only examine the convergents of e/n. Because k/d is a reduced fraction, to check each convergent to see whether it equals k/d, we suppose that its denominator equals k. We then use this value to compute $\phi(n)$, because $\phi(n) = (de - 1)/k$. We use this purported value of $\phi(n)$ and the value of n to factor n (see the discussion in Section 8.4 to see how this is done). Once we have found k/d, we know d because k/d is a reduced fraction and d is its denominator. To see that k/d is reduced, note that $ed - k\phi(n) = 1$, which implies, by

Theorem 3.8, that $(d, k) = 1$. Because computing all convergents of a rational number with denominator n uses $O((\log n)^3)$ bit operations, we see that d can be found using $O((\log n)^3)$ bit operations. ∎

12.3 Exercises

1. Find the simple continued fractions of each of the following real numbers.

   a) $\sqrt{2}$ c) $\sqrt{5}$

   b) $\sqrt{3}$ d) $(1 + \sqrt{5})/2$

2. Find the first five partial quotients of the simple continued fractions of each of the following real numbers.

   a) $\sqrt[3]{2}$ c) $(e - 1)/(e + 1)$

   b) 2π d) $(e^2 - 1)/(e^2 + 1)$

3. Find the best rational approximation to π with a denominator less than or equal to 100,000.

4. The infinite simple continued fraction expansion of the number e is

$$e = [2; 1, 2, 1, 1, 4, 1, 1, 6, 1, 1, 8, \ldots].$$

   a) Find the first eight convergents of the continued fraction of e.

   b) Find the best rational approximation to e having a denominator less than or equal to 536.

* 5. Let α be an irrational number with simple continued fraction expansion $\alpha = [a_0; a_1, a_2, \ldots]$. Show that the simple continued fraction of $-\alpha$ is $[-a_0 - 1; 1, a_1 - 1, a_2, a_3, \ldots]$ if $a_1 > 1$ and $[-a_0 - 1; a_2 + 1, a_3, \ldots]$ if $a_1 = 1$.

* 6. Show that if p_k/q_k and p_{k+1}/q_{k+1} are consecutive convergents of the simple continued fraction of an irrational number α, then

$$|\alpha - p_k/q_k| < 1/(2q_k^2)$$

   or

$$|\alpha - p_{k+1}/q_{k+1}| < 1/(2q_{k+1}^2).$$

   (*Hint:* First show that $|\alpha - p_{k+1}/q_{k+1}| + |\alpha - p_k/q_k| = |p_{k+1}/q_{k+1} - p_k/q_k| = 1/(q_k q_{k+1})$.)

☞ 7. Let α be an irrational number, $\alpha > 1$. Show that the kth convergent of the simple continued fraction of $1/\alpha$ is the reciprocal of the $(k - 1)$th convergent of the simple continued fraction of α.

* 8. Let α be an irrational number, and let p_j/q_j denote the jth convergent of the simple continued fraction expansion of α. Show that at least one of any three consecutive convergents satisfies the inequality

$$|\alpha - p_j/q_j| < 1/(\sqrt{5}q_j^2).$$

Conclude that there are infinitely many rational numbers p/q, where p and q are integers with $q \neq 0$, such that

$$|\alpha - p/q| < 1/(\sqrt{5}q^2).$$

* **9.** Show that if $\alpha = (1 + \sqrt{5})/2$, and $c > \sqrt{5}$, then there are only a finite number of rational numbers p/q, where p and q are integers, $q \neq 0$, such that

$$|\alpha - p/q| < 1/(cq^2).$$

(*Hint:* Consider the convergents of the simple continued fraction expansion of $\sqrt{5}$.)

If α and β are two real numbers, we say that β is *equivalent* to α if there are integers a, b, c, and d such that $ad - bc = \pm 1$ and $\beta = \frac{a\alpha + b}{c\alpha + d}$.

10. Show that a real number α is equivalent to itself.

11. Show that if α and β are real numbers with β equivalent to α, then α is equivalent to β. Hence, we can say that two numbers α and β are equivalent.

12. Show that if α, β, and λ are real numbers such that α and β are equivalent and β and λ are equivalent, then α and λ are equivalent.

13. Show that any two rational numbers are equivalent.

* **14.** Show that two irrational numbers α and β are equivalent if and only if the tails of their simple continued fractions agree, that is, if $\alpha = [a_0; a_1, a_2, \ldots, a_j, c_1, c_2, c_3, \ldots]$, $\beta = [b_0; b_1, b_2, \ldots, b_k, c_1, c_2, c_3, \ldots]$, where $a_i, i = 0, 1, 2, \ldots, j$; $b_i, i = 0, 1, 2, \ldots, k$; and $c_i, i = 1, 2, 3, \ldots$ are integers, all positive except perhaps a_0 and b_0.

Let α be an irrational number, and let the simple continued fraction expansion of α be $\alpha = [a_0; a_1, a_2, \ldots]$. Let p_k/q_k denote, as usual, the kth convergent of this continued fraction. We define the *pseudoconvergents* of this continued fraction to be

$$p_{k,t}/q_{k,t} = (tp_{k-1} + p_{k-2})/(tq_{k-1} + q_{k-2}),$$

where k is a positive integer, $k \geq 2$, and t is an integer with $0 < t < a_k$.

15. Show that each pseudoconvergent is in lowest terms.

* **16.** Show that the sequence of rational numbers $p_{k,2}/q_{k,2}, \ldots, p_{k,a_{k-1}}/q_{k,a_{k-1}}, p_k/q_k$ is increasing if k is even, and decreasing if k is odd.

* **17.** Show that if r and s are integers with $s > 0$ such that

$$|\alpha - r/s| \leq |\alpha - p_{k,t}/q_{k,t}|,$$

where k is a positive integer and $0 < t < a_k$, then $s > q_{k,t}$ or $r/s = p_{k-1}/q_{k-1}$. This shows that the closest rational approximations to a real number are the convergents and pseudoconvergents of its simple continued fraction.

18. Find the pseudoconvergents of the simple continued fraction of π for $k = 2$.

19. Find a rational number r/s that is closer to π than $22/7$ with denominator s less than 106. (*Hint:* Use Exercise 17.)

20. Find the rational number r/s that is closest to e with denominator s less than 100.

21. Show that the version of Theorem 12.19 for rational numbers is valid. That is, show that if a, b, c, and d are all integers with b and d nonzero, $(a, b) = (c, d) = 1$ and

$$\left| \frac{a}{b} - \frac{c}{d} \right| < \frac{1}{2d^2},$$

then c/d is a convergent of the continued fraction expansion of a/b.

22. Show that computing all convergents of a rational number with denominator n can be done using $O((\log n)^3)$ bit operations.

12.3 Computational and Programming Exercises

Computations and Explorations

Using a computation program such as Maple or *Mathematica*, or programs you have written, carry out the following computations and explorations.

1. Compute the first 100 partial quotients of each of the real numbers in Exercise 2.

2. Compute the first 100 partial quotients of the simple continued fraction of e^2. From this, find the rule for the partial quotients of this simple continued fraction.

3. Compute the first 1000 partial quotients of the simple continued fraction of π. What is the largest partial quotient that appears? How often does the integer 1 appear as a partial quotient?

Programming Projects

Write programs in Maple, *Mathematica*, or a language of your choice to do the following.

1. Given a real number x, find the simple continued fraction of x.

2. Given an irrational number x and a positive integer n, find the best rational approximation to x with denominator not exceeding n.

12.4 Periodic Continued Fractions

We call the infinite simple continued fraction $[a_0; a_1, a_2, \ldots]$ *periodic* if there are positive integers N and k such that $a_n = a_{n+k}$ for all positive integers n with $n \geq N$. We use the notation

$$[a_0; a_1, a_2, \ldots, a_{N-1}, \overline{a_N, a_{N+1}, a_{N+k-1}}]$$

to express the periodic infinite simple continued fraction

$$[a_0; a_1, a_2, \ldots, a_{N-1}, a_N, a_{N+1}, \ldots, a_{N+k-1}a_N, a_{N+1}, \ldots].$$

For instance, $[1; 2, \overline{3, 4}]$ denotes the infinite simple continued fraction $[1; 2, 3, 4, 3, 4, 3, 4, \ldots]$.

In Section 12.1, we showed that the base b expansion of a number is periodic if and only if the number is rational. To characterize those irrational numbers with periodic infinite simple continued fractions, we need the following definition.

Definition.   The real number α is said to be a *quadratic irrational* if α is irrational and is a root of a quadratic polynomial with integer coefficients, that is,

$$A\alpha^2 + B\alpha + C = 0,$$

where A, B, and C are integers and $A \neq 0$.

Example 12.12.   Let $\alpha = 2 + \sqrt{3}$. Then α is irrational, for if α were rational, then by Exercise 3 of Section 1.1, $\alpha - 2 = \sqrt{3}$ would be rational, contradicting Theorem 3.18. Next, note that

$$\alpha^2 - 4\alpha + 1 = (7 + 4\sqrt{3}) - 4(2 + \sqrt{3}) + 1 = 0.$$

Hence, α is a quadratic irrational.   ◀

We will show that the infinite simple continued fraction of an irrational number is periodic if and only if this number is a quadratic irrational. Before we do this, we first develop some useful results about quadratic irrationals.

Lemma 12.1.   The real number α is a quadratic irrational if and only if there are integers a, b, and c with $b > 0$ and $c \neq 0$, such that b is not a perfect square and

$$\alpha = (a + \sqrt{b})/c.$$

Proof.   If α is a quadratic irrational, then α is irrational, and there are integers A, B, and C such that $A\alpha^2 + B\alpha + C = 0$. From the quadratic formula, we know that

$$\alpha = \frac{-B \pm \sqrt{B^2 - 4AC}}{2A}.$$

Because α is a real number, we have $B^2 - 4AC > 0$, and because α is irrational, $B^2 - 4AC$ is not a perfect square and $A \neq 0$. By either taking $a = -B, b = B^2 - 4AC$, and $c = 2A$, or $a = B, b = B^2 - 4AC$, and $c = -2A$, we have our desired representation of α.

Conversely, if

$$\alpha = (a + \sqrt{b})/c,$$

where a, b, and c are integers with $b > 0, c \neq 0$, and b not a perfect square, then by Exercise 3 of Section 1.1 and Theorem 3.18, we can easily see that α is irrational. Furthermore, we note that

$$c^2\alpha^2 - 2ac\alpha + (a^2 - b) = 0,$$

so that α is a quadratic irrational.   ∎

The following lemma will be used when we show that periodic simple continued fractions represent quadratic irrationals.

Lemma 12.2.   If α is a quadratic irrational and if r, s, t, and u are integers, then $(r\alpha + s)/(t\alpha + u)$ is either rational or a quadratic irrational.

Proof. From Lemma 12.1, there are integers a, b, and c with $b > 0, c \neq 0$, and b not a perfect square, such that

$$\alpha = (a + \sqrt{b})/c.$$

Thus,

$$\frac{r\alpha + s}{t\alpha + u} = \left[\frac{r(a + \sqrt{b})}{c} + s\right] \bigg/ \left[\frac{t(a + \sqrt{b})}{c} + u\right]$$

$$= \frac{(ar + cs) + r\sqrt{b}}{(at + cu) + t\sqrt{b}}$$

$$= \frac{[(ar + cs) + r\sqrt{b}][(at + cu) - t\sqrt{b}]}{[(at + cu) + t\sqrt{b}][(at + cu) - t\sqrt{b}]}$$

$$= \frac{[(ar + cs)(at + cu) - rtb] + [r(at + cu) - t(ar + cs)]\sqrt{b}}{(at + cu)^2 - t^2 b}.$$

Hence, by Lemma 12.1, $(r\alpha + s)/(t\alpha + u)$ is a quadratic irrational, unless the coefficient of $\sqrt{b}$ is zero, which would imply that this number is rational. ∎

In our subsequent discussions of simple continued fractions of quadratic irrationals, we will use the notion of the conjugate of a quadratic irrational.

Definition.   Let $\alpha = (a + \sqrt{b})/c$ be a quadratic irrational. Then the *conjugate* of α, denoted by α', is defined by $\alpha' = (a - \sqrt{b})/c$.

Lemma 12.3.   If the quadratic irrational α is a root of the polynomial $Ax^2 + Bx + C = 0$, then the other root of this polynomial is α', the conjugate of α.

Proof. From the quadratic formula, we see that the two roots of $Ax^2 + Bx + C = 0$ are

$$\frac{-B \pm \sqrt{B^2 - 4AC}}{2A}.$$

If α is one of these roots, then α' is the other root, because the sign of $\sqrt{B^2 - 4AC}$ is reversed to obtain α' from α. ∎

The following lemma tells us how to find the conjugates of arithmetic expressions involving quadratic irrationals.

Lemma 12.4.   If $\alpha_1 = (a_1 + b_1\sqrt{d})/c_1$ and $\alpha_2 = (a_2 + b_2\sqrt{d})/c_2$ are rational or quadratic irrationals, then

   (i) $(\alpha_1 + \alpha_2)' = \alpha_1' + \alpha_2'$

   (ii) $(\alpha_1 + \alpha_2)' = \alpha_1' - \alpha_2'$

   (iii) $(\alpha_1\alpha_2)' = \alpha_1'\alpha_2'$

   (iv) $(\alpha_1/\alpha_2)' = \alpha_1'/\alpha_2'.$

The proof of (iv) will be given here; the proofs of the other parts are easier, and appear at the end of this section as problems for the reader.

Proof of (iv). Note that

$$\alpha_1/\alpha_2 = \frac{(a_1 + b_1\sqrt{d})/c_1}{(a_2 + b_2\sqrt{d})/c_2}$$

$$= \frac{c_2(a_1 + b_1\sqrt{d})(a_2 - b_2\sqrt{d})}{c_1(a_2 + b_2\sqrt{d})(a_2 - b_2\sqrt{d})}$$

$$= \frac{(c_2a_1a_2 - c_2b_1b_2d) + (c_2a_2b_1 - c_2a_1b_2)\sqrt{d}}{c_1(a_2^2 - b_2^2d)},$$

whereas

$$\alpha_1'/\alpha_2' = \frac{(a_1 - b_1\sqrt{d})/c_1}{(a_2 - b_2\sqrt{d})/c_2}$$

$$= \frac{c_2(a_1 - b_1\sqrt{d})(a_2 + b_2\sqrt{d})}{c_1(a_2 - b_2\sqrt{d})(a_2 + b_2\sqrt{d})}$$

$$= \frac{(c_2a_1a_2 - c_2b_1b_2d) - (c_2a_2b_1 - c_2a_1b_2)\sqrt{d}}{c_1(a_2^2 - b_2^2d)}.$$

Hence, $(\alpha_1/\alpha_2)' = \alpha_1'/\alpha_2'$. ∎

The fundamental result about periodic simple continued fractions is called Lagrange's theorem (although part of the theorem was proved by Euler). (Note that this theorem is different from Lagrange's theorem on polynomial congruences discussed in Chapter 9. In this chapter, we do not refer to that result.) Euler proved in 1737 that a periodic infinite simple continued fraction represents a quadratic irrational. Lagrange showed in 1770 that a quadratic irrationality has a periodic continued fraction.

Theorem 12.21. *Lagrange's Theorem.* The infinite simple continued fraction of an irrational number is periodic if and only if this number is a quadratic irrational.

We first prove that a periodic continued fraction represents a quadratic irrational. The converse, that the simple continued fraction of a quadratic irrational is periodic, will be proved after a special algorithm for obtaining the continued fraction of a quadratic irrational is developed.

Proof. Let the simple continued fraction of α be periodic, so that

$$\alpha = [a_0; a_1, a_2, \ldots a_{N-1}, \overline{a_N, a_{N+1}, \ldots, a_{N+k}}].$$

Now, let

$$\beta = [\overline{a_N; a_{N+1}, \ldots, a_{N+k}}].$$

Then

$$\beta = [a_N; a_{N+1}, \ldots, a_{N+k}, \beta],$$

and by Theorem 12.9, it follows that

(12.13)
$$\beta = \frac{\beta p_k + p_{k-1}}{\beta q_k + q_{k-1}},$$

where p_k/q_k and p_{k-1}/q_{k-1} are convergents of $[a_N; a_{N+1}, \ldots, a_{N+k}]$. Because the simple continued fraction of β is infinite, β is irrational, and by (12.13), we have

$$q_k\beta^2 + (q_{k-1} - p_k)\beta - p_{k-1} = 0,$$

so that β is a quadratic irrational. Now, note that

$$\alpha = [a_0; a_1, a_2, \ldots, a_{N-1}, \beta],$$

so that, from Theorem 12.11, we have

$$\alpha = \frac{\beta p_{N-1} + p_{N-2}}{\beta q_{N-1} + q_{N-2}},$$

where p_{N-1}/q_{N-1} and p_{N-2}/q_{N-2} are convergents of $[a_0; a_1, a_2, \ldots, a_{N-1}]$. Because β is a quadratic irrational, Lemma 12.2 tells us that α is also a quadratic irrational (we know that α is irrational because it has an infinite simple continued fraction expansion). ∎

The following example shows how to use the proof of Theorem 12.21 to find the quadratic irrational represented by a periodic simple continued fraction.

Example 12.13.   Let $x = [3; \overline{1, 2}]$. By Theorem 12.21, we know that x is a quadratic irrational. To find the value of x, we let $x = [3; y]$, where $y = [\overline{1; 2}]$, as in the proof of Theorem 12.21. We have $y = [1; 2, y]$, so that

$$y = 1 + \frac{1}{2 + \frac{1}{y}} = \frac{3y + 1}{2y + 1}.$$

It follows that $2y^2 - 2y - 1 = 0$. Because y is positive, by the quadratic formula, we have $y = \frac{1+\sqrt{3}}{2}$. Because $x = 3 + \frac{1}{y}$, we have

$$x = 3 + \frac{2}{1 + \sqrt{3}} = 3 + \frac{2 - \sqrt{3}}{-2} = \frac{4 + \sqrt{3}}{2}. \qquad \blacktriangleleft$$

To develop an algorithm for finding the simple continued fraction of a quadratic irrational, we need the following lemma.

Lemma 12.5.   If α is a quadratic irrational, then α can be written as

$$\alpha = (P + \sqrt{d})/Q,$$

where P, Q, and d are integers, $Q \neq 0$, $d > 0$, d is not a perfect square, and $Q|(d - P^2)$.

Proof.   Because α is a quadratic irrational, Lemma 12.1 tells us that

$$\alpha = (a + \sqrt{b})/c,$$

where a, b, and c are integers, $b > 0$, and $c \neq 0$. We multiply both the numerator and the denominator of this expression for α by $|c|$ to obtain

$$\alpha = \frac{a|c| + \sqrt{bc^2}}{c|c|}$$

(where we have used the fact that $|c| = \sqrt{c^2}$). Now, let $P = a|c|$, $Q = c|c|$, and $d = bc^2$. Then P, Q, and d are integers, $Q \neq 0$, because $c \neq 0$, $d > 0$ (because $b > 0$). d is not a perfect square because b is not a perfect square and, finally, $Q|(d - P^2)$ because $d - P^2 = bc^2 - a^2c^2 = c^2(b - a^2) = \pm Q(b - a^2)$.   ∎

We now present an algorithm for finding the simple continued fractions of quadratic irrationals.

Theorem 12.22.   Let α be a quadratic irrational, so that by Lemma 12.5 there are integers P_0, Q_0, and d such that

$$\alpha = (P_0 + \sqrt{d})/Q_0,$$

where $Q_0 \neq 0$, $d > 0$, d is not a perfect square, and $Q_0|(d - P_0^2)$. Recursively define

$$\alpha_k = (P_k + \sqrt{d})/Q_k,$$
$$a_k = [\alpha_k],$$
$$P_{k+1} = a_k Q_k - P_k,$$
$$Q_{k+1} = (d - P_{k+1}^2)/Q_k,$$

for $k = 0, 1, 2, \ldots$. Then, $\alpha = [a_0; a_1, a_2, \ldots]$.

Proof.   Using mathematical induction, we will show that P_k and Q_k are integers with $Q_k \neq 0$ and $Q_k|(d - P_k^2)$, for $k = 0, 1, 2, \ldots$. First, note that this assertion is true for $k = 0$ from the hypotheses of the theorem. Next,, assume that P_k and Q_k are integers with $Q_k \neq 0$ and $Q_k|(d - P_k^2)$. Then,

$$P_{k+1} = a_k Q_k - P_k$$

is also an integer. Further,

$$Q_{k+1} = (d - P_{k+1}^2)/Q_k$$
$$= [d - (a_k Q_k - P_k)^2]/Q_k$$
$$= (d - P_k^2)/Q_k + (2a_k P_k - a_k^2 Q_k).$$

Because $Q_k|(d - P_k^2)$, by the induction hypothesis, we see that Q_{k+1} is an integer, and because d is not a perfect square, we see that $d \neq P_k^2$, so that $Q_{k+1} = (d - P_{k+1}^2)/Q_k \neq 0$. Because

$$Q_k = (d - P_{k+1}^2)/Q_{k+1},$$

we can conclude that $Q_{k+1}|(d - P_{k+1}^2)$. This finishes the inductive argument.

To demonstrate that the integers $a_0, a_1, a_2, \ldots$ are the partial quotients of the simple continued fraction of α, we use Theorem 12.15. If we can show that

$$\alpha_{k+1} = 1/(\alpha_k - a_k),$$

for $k = 0, 1, 2, \ldots$, then we know that $\alpha = [a_0; a_1, a_2, \ldots]$. Note that

$$
\begin{aligned}
\alpha_k - a_k &= \frac{P_k + \sqrt{d}}{Q_k} - a_k \\
&= [\sqrt{d} - (a_k Q_k - P_k)]/Q_k \\
&= (\sqrt{d} - P_{k+1})/Q_k \\
&= (\sqrt{d} - P_{k+1})(\sqrt{d} + P_{k+1})/Q_k(\sqrt{d} + P_{k+1}) \\
&= (d - P_{k+1}^2)/(Q_k(\sqrt{d} + P_{k+1})) \\
&= Q_k Q_{k+1}/(Q_k(\sqrt{d} + P_{k+1})) \\
&= Q_{k+1}/(\sqrt{d} + P_{k+1}) \\
&= 1/\alpha_{k+1},
\end{aligned}
$$

where we have used the defining relation for Q_{k+1} to replace $d - P_{k+1}^2$ with $Q_k Q_{k+1}$. Hence, we can conclude that $\alpha = [a_0; a_1, a_2, \ldots]$. ∎

We illustrate the use of the algorithm given in Theorem 12.22 with the following example.

Example 12.14.   Let $\alpha = (3 + \sqrt{7})/2$. Using Lemma 12.5, we write

$$\alpha = (6 + \sqrt{28})/4,$$

where we set $P_0 = 6$, $Q_0 = 4$, and $d = 28$. Hence, $a_0 = [\alpha] = 2$, and

$$P_1 = 2 \cdot 4 - 6 = 2, \qquad \alpha_1 = (2 + \sqrt{28})/6,$$
$$Q_1 = (28 - 2^2)/4 = 6, \qquad a_1 = [(2 + \sqrt{28})/6] = 1,$$

$$P_2 = 1 \cdot 6 - 2 = 4, \qquad \alpha_2 = (4 + \sqrt{28})/2$$
$$Q_2 = (28 - 4^2)/6 = 2, \qquad a_2 = [(4 + \sqrt{28})/2] = 4,$$

$$P_3 = 4 \cdot 2 - 4 = 4, \qquad \alpha_3 = (4 + \sqrt{28})/6,$$
$$Q_3 = (28 - 4^2)/2 = 6 \qquad a_3 = [(4 + \sqrt{28})/6] = 1,$$

$$P_4 = 1 \cdot 6 - 4 = 2, \qquad \alpha_4 = (\sqrt{28})/4,$$
$$Q_4 = (28 - 2^2)/6 = 4, \qquad a_4 = [(2 + \sqrt{28})/4] = 1,$$

$$P_5 = 1 \cdot 4 - 2 = 2, \qquad \alpha_5 = (\sqrt{28})/6,$$
$$Q_5 = (28 - 2^2)/4 = 6, \qquad a_5 = [(2 + \sqrt{28})/6] = 1,$$

and so on, with repetition, because $P_1 = P_5$ and $Q_1 = Q_5$. Hence, we see that

$$(3 + \sqrt{7})/2 = [2; 1, 4, 1, 1, 1, 4, 1, 1, \ldots]$$
$$= [2; \overline{1, 4, 1, 1}]. \qquad \blacktriangleleft$$

We now finish the proof of Lagrange's theorem by showing that the simple continued fraction expansion of a quadratic irrational is periodic.

Proof of Theorem 12.21 (continued).   Let α be a quadratic irrational, so that by Lemma 12.5, we can write α as

$$\alpha = (P_0 + \sqrt{d})/Q_0.$$

Furthermore, by Theorem 12.20, we have $\alpha = [a_0; a_1, a_2, \ldots]$, where

$$\alpha_k = (P_k + \sqrt{d})/Q_k,$$
$$a_k = [\alpha_k],$$
$$P_{k+1} = a_k Q_k - P_k,$$
$$Q_{k+1} = (d - P_{k+1}^2)/Q_k,$$

for $k = 0, 1, 2, \ldots$.

Because $\alpha = [a_0; a_1, a_2, \ldots, \alpha_k]$, Theorem 12.11 tells us that

$$\alpha = (p_{k-1}\alpha_k + p_{k-2})/(q_{k-1}\alpha_k + q_{k-2}).$$

Taking conjugates of both sides of this equation, and using Lemma 12.4, we see that

(12.14) $$\alpha' = (p_{k-1}\alpha_k' + p_{k-2})/(q_{k-1}\alpha_k' + q_{k-2}).$$

When we solve (12.14) for α'_k, we find that

$$\alpha'_k = \frac{-q_{k-2}}{q_{k-1}} \left(\frac{\alpha' - \frac{p_{k-2}}{q_{k-2}}}{\alpha' - \frac{p_{k-1}}{q_{k-1}}} \right).$$

Note that the convergents p_{k-2}/q_{k-2} and p_{k-1}/q_{k-1} tend to α as k tends to infinity, so that

$$\left(\alpha' - \frac{p_{k-2}}{q_{k-2}} \right) \Big/ \left(\alpha' - \frac{p_{k-1}}{q_{k-1}} \right)$$

tends to 1. Hence, there is an integer N such that $\alpha'_k < 0$ for $k \geq N$. Because $\alpha_k > 0$ for $k > 1$, we have

$$\alpha_k - \alpha'_k = \frac{P_k + \sqrt{d}}{Q_k} - \frac{P_k - \sqrt{d}}{Q_k} = \frac{2\sqrt{d}}{Q_k} > 0,$$

so that $Q_k > 0$ for $k \geq N$.

Because $Q_k Q_{k+1} = d - P_{k+1}^2$, we see that for $k \geq N$,

$$Q_k \leq Q_k Q_{k+1} = d - P_{k+1}^2 \leq d.$$

Also for $k \geq N$, we have

$$P_{k+1}^2 \leq d = P_{k+1}^2 - Q_k Q_{k+1},$$

so that

$$-\sqrt{d} < P_{k+1} < \sqrt{d}.$$

From the inequalities $0 \leq Q_k \leq d$ and $-\sqrt{d} < P_{k+1} < \sqrt{d}$, which hold for $k \geq N$, we see that there are only a finite number of possible values for the pair of integers P_k, Q_k for $k > N$. Because there are infinitely many integers k with $k \geq N$, there are two integers i and j such that $P_i = P_j$ and $Q_i = Q_j$ with $i < j$. Hence, from the defining relation for α_k, we see that $\alpha_i = \alpha_j$. Consequently, we can see that $a_i = a_j, a_{i+1} = a_{j+1}, a_{i+2} = a_{j+2}, \ldots$. Hence,

$$\alpha = [a_0; a_1, a_2, \ldots, a_{i-1}, a_i, a_{i+1}, \ldots, a_{j-1}, a_i, a_{i+1}, \ldots, a_{j-1}, \ldots]$$
$$= [a_0; a_1, a_2, \ldots, a_{i-1}, \overline{a_i, a_{i+1}, \ldots, a_{j-1}}].$$

This shows that α has a periodic simple continued fraction.   ∎

Purely Periodic Continued Fractions   Next, we investigate those periodic simple continued fractions that are *purely periodic,* that is, those without a pre-period.

Definition.   The continued fraction $[a_0; a_1, a_2, \ldots]$ is *purely periodic* if there is an integer n such that $a_k = a_{n+k}$, for $k = 0, 1, 2, \ldots$, so that

$$[a_0; a_1, a_2, \ldots] = \overline{[a_0; a_1, a_2, a_3, \ldots, a_{n-1}]}.$$

Example 12.15. The continued fraction $[\overline{2; 3}] = (1 + \sqrt{3})/2$ is purely periodic, whereas $[2; \overline{2, 4}] = \sqrt{6}$ is not.   ◄

The next definition and theorem describe those quadratic irrationals with purely periodic simple continued fractions.

Definition. A quadratic irrational α is called *reduced* if $\alpha > 1$ and $-1 < \alpha' < 0$, where α' is the conjugate of α.

Theorem 12.23. The simple continued fraction of the quadratic irrational α is purely periodic if and only if α is reduced. Further, if α is reduced and $\alpha = [\overline{a_0; a_1, a_2, \ldots, a_n}]$, then the continued fraction of $-1/\alpha'$ is $[\overline{a_n; a_{n-1}, \ldots, a_0}]$.

Proof. First, assume that α is a reduced quadratic irrational. Recall from Theorem 12.18 that the partial fractions of the simple continued fraction of α are given by

$$a_k = [\alpha_k], \quad \alpha_{k+1} = 1/(\alpha_k - a_k),$$

for $k = 0, 1, 2, \ldots$, where $\alpha_0 = \alpha$. We see that

$$1/\alpha_{k+1} = \alpha_k - a_k,$$

and by taking conjugates and using Lemma 12.4, we see that

$$(12.15) \qquad\qquad 1/\alpha'_{k+1} = \alpha'_k - a_k.$$

We can prove, by mathematical induction, that $-1 < \alpha'_k < 0$ for $k = 0, 1, 2, \ldots$. First, note that because $\alpha_0 = \alpha$ is reduced, $-1 < \alpha'_0 < 0$. Now, assume that $-1 < \alpha'_k < 0$. Then, because $a_k \geq 1$ for $k = 0, 1, 2, \ldots$ (note that $a_0 \geq 1$ because $\alpha > 1$), we see from (12.15) that

$$1/\alpha'_{k+1} < -1,$$

so that $-1 < \alpha'_{k+1} < 0$. Hence, $-1 < \alpha'_k < 0$ for $k = 0, 1, 2, \ldots$.

Next, note that from (12.15) we have

$$\alpha'_k = a_k + 1/\alpha'_{k+1},$$

and because $-1 < \alpha'_k < 0$, it follows that

$$-1 < a_k + 1/\alpha'_{k+1} < 0.$$

Consequently,

$$-1 - 1/\alpha'_{k+1} < a_k < -1/\alpha'_{k+1},$$

so that

$$a_k = [-1/\alpha'_{k+1}].$$

Because α is a quadratic irrational, the proof of Lagrange's theorem shows that there are nonnegative integers i and j, $i < j$, such that $\alpha_i = \alpha_j$, and hence with $-1/\alpha'_i = -1/\alpha'_j$.

Because $a_{i-1} = [-1/\alpha_i']$ and $a_{j-1} = [-1/\alpha_j']$, we see that $a_{i-1} = a_{j-1}$. Furthermore, because $\alpha_{i-1} = a_{i-1} + 1/\alpha_i$ and $\alpha_{j-1} = a_{j-1} + 1/\alpha_j$, we also see that $\alpha_{i-1} = \alpha_{j-1}$. Continuing this argument, we see that $\alpha_{i-2} = \alpha_{j-2}, \alpha_{j-3} = \alpha_{j-3}, \ldots$, and, finally, that $\alpha_0 = \alpha_{j-1}$. Because

$$\alpha_0 = \alpha = [a_0; a_1, \ldots, a_{j-i-1}, \alpha_{j-1}]$$
$$= [a_0; a_1, \ldots, a_{j-i-1}, \alpha_0]$$
$$= [\overline{a_0; a_1, \ldots, a_{j-i-1}}],$$

we see that the simple continued fraction of α is purely periodic.

To prove the converse, assume that α is a quadratic irrational with a purely periodic continued fraction $\alpha = [\overline{a_0; a_1, a_2, \ldots, a_k}]$. Because $\alpha = [a_0; a_1, a_2, \ldots, a_k, \alpha]$, Theorem 12.11 tells that

(12.16) $$\alpha = \frac{\alpha p_k + p_{k-1}}{\alpha q_k + q_{k-1}},$$

where p_{k-1}/q_{k-1} and p_k/q_k are the $(k-1)$th and kth convergents of the continued fraction expansion of α. From (12.16), we see that

(12.17) $$q_k \alpha^2 + (q_{k-1} - p_k)\alpha - p_{k-1} = 0.$$

Now, let β be the quadratic irrational such that $\beta = [\overline{a_k; a_{k-1}, \ldots, a_1, a_0}]$, that is, with the period of the simple continued fraction for α reversed. Then $\beta = [a_k; a_{k-1}, \ldots, a_1, a_0, \beta]$, so that by Theorem 12.11, it follows that

(12.18) $$\beta = \frac{\beta p_k' + p_{k-1}'}{\beta q_k' + q_{k-1}'},$$

where p_{k-1}'/q_{k-1}' and p_K'/q_k' are the $(k-1)$th and kth convergents of the continued fraction expansion of β. Note, however, from Exercise 10 of Section 12.2, that

$$p_k/p_{k-1} = [a_k; a_{k-1}, \ldots, a_1, a_0] = p_k'/q_k'$$

and

$$q_k/q_{k-1} = [a_k; a_{k-1}, \ldots, a_2, a_1] = p_{k-1}'/q_{k-1}'.$$

Because p_{k-1}'/q_{k-1}' and p_k'/q_k' are convergents, we know that they are in lowest terms. Also, p_k/p_{k-1} and q_k/q_{k-1} are in lowest terms, because Theorem 12.12 tells us that $p_k q_{k-1} - p_{k-1} q_k = (-1)^{k-1}$. Hence,

$$p_k' = p_k, \quad q_k' = p_{k-1}$$

and

$$p_{k-1}' = q_k, \quad q_{k-1}' = q_{k-1}.$$

Inserting these values into (12.18), we see that

$$\beta = \frac{\beta p_k + q_k}{\beta p_{k-1} + q_{k-1}}.$$

Therefore, we know that

$$p_{k-1}\beta^2 + (q_{k-1} - p_k)\beta - q_k = 0.$$

This imples that

(12.19) $$q_k(-1/\beta)^2 + (q_{k-1} - p_k)(-1/\beta) - p_{k-1} = 0.$$

By (12.17) and (12.19), we see that the two roots of the quadratic equation

$$q_k x^2 + (q_{k-1} - p_k)x - p_{k-1} = 0$$

are α and $-1/\beta$, so that by the quadratic equation, we have $\alpha' = -1/\beta$. Because $\beta = [\overline{a_n; a_{n-1}, \ldots, a_1, a_0}]$, we see that $\beta > 1$, so that $-1 < \alpha' = -1/\beta < 0$. Hence, α is a reduced quadratic irrational.

Furthermore, note that because $\beta = -1/\alpha'$, it follows that

$$-1/\alpha' = [\overline{a_n; a_{n-1}, \ldots, a_1, a_0}]. \qquad \blacksquare$$

We now find the form of the periodic simple continued fraction of $\sqrt{D}$, where D is a positive integer that is not a perfect square. Although $\sqrt{D}$ is not reduced, because its conjugate, $-\sqrt{D}$, is not between -1 and 0, the quadratic irrational $[\sqrt{D}] + \sqrt{D}$ is reduced because its conjugate, $[\sqrt{D}] - \sqrt{D}$, does lie between -1 and 0. Therefore, from Theorem 12.23, we know that the continued fraction of $[\sqrt{D}] + \sqrt{D}$ is purely periodic. Because the initial partial quotient of the simple continued fraction of $[\sqrt{D}] + \sqrt{D}$ is $[[\sqrt{D}] + \sqrt{D}] = 2[\sqrt{D}] = 2a_0$, where $a_0 = [\sqrt{D}]$, we can write

$$[\sqrt{D}] + \sqrt{D} = [\overline{2a_0; a_1, a_2, \ldots, a_n}]$$
$$= [2a_0; a_1, a_2, \ldots, a_n, 2a_0, a_1, \ldots, a_n].$$

Subtracting $a_0 = \sqrt{D}$ from both sides of this equality, we find that

$$\sqrt{D} = [a_0; a_1, a_2, \ldots, 2a_0, a_1, a_2, \ldots 2a_0, \ldots]$$
$$= [a_0; \overline{a_1, a_2, \ldots, a_n, 2a_0}].$$

To obtain even more information about the partial quotients of the continued fraction of $\sqrt{D}$, we note that from Theorem 12.23, the simple continued fraction expansion of $-1/([\sqrt{D}] - \sqrt{D})$ can be obtained from that for $[\sqrt{D}] + \sqrt{D}$ by reversing the period, so that

$$1/(\sqrt{D} - [\sqrt{D}]) = [\overline{a_n; a_{n-1}, \ldots, a_1, 2a_0}].$$

But also note that

$$\sqrt{D} - [\sqrt{D}] = [0; \overline{a_1, a_2, \ldots, a_n, 2a_0}],$$

so that by taking reciprocals, we find that

$$1/(\sqrt{D} - [\sqrt{D}]) = [\overline{a_1; a_2, \ldots, a_n, 2a_0}].$$

Therefore, when we equate these two expressions for the simple continued fraction of $1/(\sqrt{D} - [\sqrt{D}])$, we obtain

$$a_1 = a_n, a_2 = a_{n-1}, \ldots, a_n = a_1,$$

so that the periodic part of the continued fraction for $\sqrt{D}$ is symmetric from the first to the penultimate term.

In conclusion, we see that the simple continued fraction of $\sqrt{D}$ has the form

$$\sqrt{D} = [a_0; \overline{a_1, a_2, \ldots, a_2, a_1, 2a_0}].$$

We illustrate this with some examples.

Example 12.16. Note that

$$\sqrt{23} = [4; \overline{1, 3, 1, 8}],$$
$$\sqrt{31} = [5, \overline{1, 1, 3, 5, 3, 1, 1, 10}],$$
$$\sqrt{46} = [6; \overline{1, 2, 1, 1, 2, 6, 2, 1, 1, 2, 1, 12}],$$
$$\sqrt{76} = [8; \overline{1, 2, 1, 1, 5, 4, 5, 1, 1, 2, 1, 16}],$$

and

$$\sqrt{97} = [9; \overline{1, 5, 1, 1, 1, 1, 1, 1, 5, 1, 18}],$$

where each continued fraction has a pre-period of length 1, and a period ending with twice the first partial quotient, which is symmetric from the first to the next-to-the-last term.   ◀

The simple continued fraction expansions of $\sqrt{d}$ for positive integers d such that d is not a perfect square and $d < 100$ can be found in Table 5 of Appendix D.

12.4 Exercises

1. Find the simple continued fractions of each of the following numbers.

   a) $\sqrt{7}$ c) $\sqrt{23}$ e) $\sqrt{59}$
   b) $\sqrt{11}$ d) $\sqrt{47}$ f) $\sqrt{94}$

2. Find the simple continued fractions of each of the following numbers.

   a) $\sqrt{101}$ c) $\sqrt{107}$ e) $\sqrt{203}$
   b) $\sqrt{103}$ d) $\sqrt{201}$ f) $\sqrt{209}$

3. Find the simple continued fractions of each of the following numbers.

   a) $1 + \sqrt{2}$ b) $\left(2 + \sqrt{5}\right)/3$ c) $\left(5 - \sqrt{7}\right)/4$

4. Find the simple continued fractions of each of the following numbers.

   a) $\left(1 + \sqrt{3}\right)/2$ b) $\left(14 + \sqrt{37}\right)/3$ c) $\left(13 - \sqrt{2}\right)/7$

5. Find the quadratic irrational with each of the following simple continued fraction expansions.

   a) $[2; 1, \overline{5}]$ b) $[2; \overline{1, 5}]$ c) $[\overline{2; 1, 5}]$

6. Find the quadratic irrational with each of the following simple continued fraction expansions.

   a) $[1; 2, \overline{3}]$ b) $[1; \overline{2, 3}]$ c) $[\overline{1; 2, 3}]$

7. Find the quadratic irrational with each of the following simple continued fraction expansions.

   a) $[3; \overline{6}]$ b) $[4; \overline{8}]$ c) $[5; \overline{10}]$ d) $[6; \overline{12}]$

8. a) Let d be a positive integer. Show that the simple continued fraction of $\sqrt{d^2 + 1}$ is $[d; \overline{2d}]$.
   b) Use part (a) to find the simple continued fractions of $\sqrt{101}$, $\sqrt{290}$, and $\sqrt{2210}$.

9. Let d be an integer, $d \geq 2$.
   a) Show that the simple continued fraction of $\sqrt{d^2 - 1}$ is $[d - 1; \overline{1, 2d - 2}]$.
   b) Show that the simple continued fraction of $\sqrt{d^2 - d}$ is $[d - 1; \overline{2, 2d - 2}]$.
   c) Use parts (a) and (b) to find the simple continued fractions of $\sqrt{99}$, $\sqrt{110}$, $\sqrt{272}$, and $\sqrt{600}$.

10. a) Show that if d is an integer, $d \geq 3$, then the simple continued fraction of $\sqrt{d^2 - 2}$ is $[d - 1; \overline{1, d - 2, 1, 2d - 2}]$.
    b) Show that if d is a positive integer, then the simple continued fraction of $\sqrt{d^2 + 2}$ is $[d; \overline{d, 2d}]$.
    c) Find the simple continued fraction expansions of $\sqrt{47}$, $\sqrt{51}$, and $\sqrt{287}$.

11. Let d be an odd positive integer.
    a) Show that the simple continued fraction of $\sqrt{d^2 + 4}$ is $[d; \overline{(d - 1)/2, 1, 1, (d - 1)/2, 2d}]$, if $d > 1$.
    b) Show that the simple continued fraction of $\sqrt{d^2 - 4}$ is $[d - 1; \overline{1, (d - 3)/2, 2, (d - 3)/2, 1, 2d - 2}]$, if $d > 3$.

12. Show that the simple continued fraction of $\sqrt{d}$, where d is a positive integer, has period length one if and only if $d = a^2 + 1$, where a is a nonegative integer.

13. Show that the simple continued fraction of $\sqrt{d}$, where d is a positive integer, has period length two if and only if $d = a^2 + b$, where a and b are integers, $b > 1$, and $b|2a$.

14. Prove that if $\alpha_1 = (a_1 + b_1\sqrt{d})/c_1$ and $\alpha_2 = (a_2 + b_2\sqrt{d})/c_2$ are quadratic irrationals, then the following hold.
    a) $(\alpha_1 + \alpha_2)' = \alpha_1' + \alpha_2'$
    b) $(\alpha_1 - \alpha_2)' = \alpha_1' - \alpha_2'$
    c) $(\alpha_1\alpha_2)' = \alpha_1' \cdot \alpha_2'$

15. Which of the following quadratic irrationals have purely periodic continued fractions?

a) $1 + \sqrt{5}$ c) $4 + \sqrt{17}$ e) $(3 + \sqrt{23})/2$

b) $2 + \sqrt{8}$ d) $(11 - \sqrt{10})/9$ f) $(17 + \sqrt{188})/3$

16. Suppose that $\alpha = (a + \sqrt{b})/c$, where a, b, and c are integers, $b > 0$, and b is not a perfect square. Show that α is a reduced quadratic irrational if and only if $0 < a < \sqrt{b}$ and $\sqrt{b} - a < c < \sqrt{b} + a < 2\sqrt{b}$.

17. Show that if α is a reduced quadratic irrational, then $-1/\alpha'$ is also a reduced quadratic irrational.

* **18.** Let k be a positive integer. Show that there are not infinitely many positive integers D, such that the simple continued fraction expansion of $\sqrt{D}$ has a period of length k. (*Hint:* Let $a_1 = 2, a_2 = 5$, and for $k \geq 3$, let $a_k = 2a_{k-1} + a_{k-2}$. Show that if $D = (ta_k + 1)^2 + 2ta_{k-1} + 1$, where t is a nonnegative integer, then $\sqrt{D}$ has a period of length $k + 1$.)

* **19.** Let k be a positive integer. Let $D_k = (3^k + 1)^2 + 3$. Show that the simple continued fraction of $\sqrt{D_k}$ has a period of length $6k$.

12.4 Computational and Programming Exercises

Computations and Explorations

Using a computation program such as Maple or *Mathematica*, or programs you have written, carry out the following computations and explorations.

1. Find the simple continued fraction of $\sqrt{100{,}007}$, $\sqrt{1{,}000{,}007}$, and $\sqrt{10{,}000{,}007}$.

2. Find the smallest positive integer D such that the length of the period of the simple continued fraction of $\sqrt{D}$ is 10, 100, 1000, and 10,000.

3. Find the length of the largest period of the simple continued fraction of $\sqrt{D}$, where D is a positive integer less than 1003, less than 10,000, and less than 100,000. Can you make any conjectures?

4. Look for patterns in the continued fractions of $\sqrt{D}$ for many different values of D.

Programming Projects

Write programs in Maple, *Mathematica*, or a language of your choice to do the following.

* **1.** Find the quadratic irrational that is the value of a periodic simple continued fraction.

2. Find the periodic simple continued fraction expansion of a quadratic irrational.

12.5 Factoring Using Continued Fractions

We can factor the positive integer n if we can find positive integers x and y such that $x^2 - y^2 = n$ and $x - y \neq 1$. This is the basis of the Fermat factorization method discussed in Section 3.6. However, it is possible to factor n if we can find positive integers x and y that satisfy the weaker condition

(12.20) $x^2 \equiv y^2 \pmod{n}, \quad 0 < y < x < n, \quad \text{and} \quad x + y \neq n.$

To see this, note that if (12.20) holds, then n divides $x^2 - y^2 = (x + y)(x - y)$, and n divides neither $x - y$ nor $x + y$. It follows that $(n, x - y)$ and $(n, x + y)$ are divisors of n that do not equal 1 or n. We can find these divisors rapidly using the Euclidean algorithm.

Example 12.17. Note that $29^2 - 17^2 = 841 - 289 = 552 \equiv 0 \pmod{69}$. Because $29^2 - 17^2 = (29 - 17)(29 + 17) \equiv 0 \pmod{69}$, both $(29 - 17, 69) = (12, 69)$ and $(29 + 17, 69) = (46, 69)$ are divisors of 69 not equal to either 1 or 69; using the Euclidean algorithm, we find that these factors are $(12, 69) = 3$ and $(46, 69) = 23$. ◀

The continued fraction expansion of $\sqrt{n}$ can be used to find solutions of the congruence $x^2 \equiv y^2 \pmod n$. The following theorem is the basis for this.

Theorem 12.24. Let n be a positive integer that is not a perfect square. Define $\alpha_k = (P_k + \sqrt{n})/Q_k$, $a_k = [\alpha_k]$, $P_{k+1} = a_k Q_k - P_k$, and $Q_{k+1} = (n - P_{k+1}^2)/Q_k$, for $k = 0, 1, 2, \ldots$, where $\alpha_0 = \sqrt{n}$. Furthermore, let p_k/q_k denote the kth convergent of the simple continued fraction expansion of $\sqrt{n}$. Then,

$$p_k^2 - n q_k^2 = (-1)^{k-1} Q_{k+1}.$$

The proof of Theorem 12.24 depends on the following useful lemma.

Lemma 12.6. Let $r + s\sqrt{n} = t + u\sqrt{n}$, where r, s, t, and u are rational numbers and n is a positive integer that is not a perfect square. Then, $r = t$ and $s = u$.

Proof. Because $r + s\sqrt{n} = t + u\sqrt{n}$, we see that if $s \neq u$, then

$$\sqrt{n} = \frac{r - t}{u - s}.$$

Because $(r - t)/(u - s)$ is rational and $\sqrt{n}$ is irrational, it follows that $s = u$, and consequently, that $r = t$. ∎

We can now prove Theorem 12.24.

Proof. Because $\sqrt{n} = \alpha_0 = [a_0; a_1, a_2, \ldots, a_k, \alpha_{k+1}]$, Theorem 12.9 tells us that

$$\sqrt{n} = \frac{\alpha_{k+1} p_k + p_{k-1}}{\alpha_{k+1} q_k + q_{k-1}}.$$

Because $\alpha_{k+1} = (P_{k+1} + \sqrt{n})/Q_{k+1}$, we have

$$\sqrt{n} = \frac{(P_{k+1} + \sqrt{n}) p_k + Q_{k+1} p_{k-1}}{(P_{k+1} + \sqrt{n}) q_k + Q_{k+1} q_{k-1}}.$$

Therefore, we see that

$$n q_k + (P_{k+1} q_k + Q_{k+1} q_{k-1})\sqrt{n} = (P_{k+1} p_k + Q_{k+1} p_{k-1}) + p_k \sqrt{n}.$$

By Lemma 12.6, we see that $n q_k = P_{k+1} p_k + Q_{k+1} p_{k-1}$ and $P_{k+1} q_k + Q_{k+1} q_{k-1} = p_k$. When we multiply the first of these two equations by q_k and the second by p_k, subtract the first from the second, and then simplify, we obtain

$$p_k^2 - nq_k^2 = (p_k q_{k-1} - p_{k-1}q_k)Q_{k+1} = (-1)^{k-1}Q_{k+1},$$

where we have used Theorem 12.10 to complete the proof.   ∎

We now outline the technique known as the *continued fraction algorithm* for factoring an integer n, which was proposed by D. H. Lehmer and R. E. Powers in 1931, and further developed by J. Brillhart and M. A. Morrison in 1975 (see [LePo31] and [MoBr75] for details). Suppose that the terms p_k, q_k, Q_k, a_k, and α_k have their usual meanings in the computation of the continued fraction expansion of $\sqrt{n}$. By Theorem 12.24, it follows that for every nonnegative integer k,

$$p_k^2 \equiv (-1)^{k-1}Q_{k+1} \pmod{n},$$

where p_k and Q_{k+1} are as defined in the statement of the theorem. Now, suppose that k is odd and that Q_{k+1} is a square, that is, $Q_{k+1} = s^2$, where s is a positive integer. Then $p_k^2 \equiv s^2 \pmod{n}$, and we may be able to use this congruence of two squares modulo n to find factors of n. Summarizing, to factor n we carry out the algorithm described in Theorem 12.10 to find the continued fraction expansion of $\sqrt{n}$. We look for squares among the terms with even indices in the sequence $\{Q_k\}$. Each such occurrence may lead to a nonproper factor of n (or may just lead to the factorization $n = 1 \cdot n$). We illustrate this technique with several examples.

Example 12.18.   We can factor 1037 using the continued fraction algorithm. Take $\alpha = \sqrt{1037} = (0 + \sqrt{1037})/1$ with $P_0 = 0$ and $Q_0 = 1$, and generate the terms P_k, Q_k, α_k, and a_k. We look for squares among the terms with even indices in the sequence $\{Q_k\}$. We find that $Q_1 = 13$ and $Q_2 = 49$. Because $49 = 7^2$ is a square, and the index of Q_2 is even, we examine the congruence $p_1^2 \equiv (-1)^2 Q_2 \pmod{1037}$. Computing the terms of the sequence $\{p_k\}$, we find that $p_1 = 129$. This gives the congruence $129^2 \equiv 49 \pmod{1037}$. Hence, $129^2 - 7^2 = (129 - 7)(129 + 7) \equiv 0 \pmod{1037}$. This produces the factors $(129 - 7, 1037) = (122, 1037) = 61$ and $(129 + 7, 1037) = (136, 1037) = 17$ of 1037.   ◄

Example 12.19.   We can use the continued fraction algorithm to find factors of 1,000,009 (we follow computations of [Ri85]). We have $Q_1 = 9$, $Q_2 = 445$, $Q_3 = 873$, and $Q_4 = 81$. Because $81 = 9^2$ is a square, we examine the congruence $p_3^2 \equiv (-1)^4 Q_4 \pmod{1,000,009}$. However, $p_3 = 2,000,009 \equiv -9 \pmod{1,000,009}$, so that $p_3 + 9$ is divisible by 1,000,009. It follows that we do not get any proper factors of 1,000,009 from this.

We continue until we reach another square in the sequence $\{Q_k\}$ with k even. This happens when $k = 18$ with $Q_{18} = 16$. Calculating p_{17} gives $p_{17} = 494,881$. From the congruence $p_{17}^2 \equiv (-1)^{18} Q_{18} \pmod{1,000,009}$, we have $494,881^2 \equiv 4^2 \pmod{1,000,009}$. It follows that $(494881 - 4, 1000009) = (494877, 1000009) = 293$ and $(494881 + 4, 1000009) = (494885, 1000009) = 3413$ are factors of 1,000,009.   ◄

More powerful techniques based on continued fraction expansions are known. These are described in [Di84], [Gu75], and [WaSm87]. We describe one such generalization in the exercises.

12.5 Exercises

1. Find factors of 119 using the congruence $19^2 \equiv 2^2 \pmod{119}$.

2. Factor 1537 using the continued fraction algorithm.

3. Factor the integer 13,290,059 using the continued fraction algorithm. (*Hint:* Use a computer program to generate the integers Q_k for the continued fraction for $\sqrt{13{,}290{,}059}$. You will need more than 50 terms.)

4. Let n be a positive integer and let $p_1, p_2, \ldots$, and p_m be primes. Suppose that there exist integers $x_1, x_2, \ldots, x_r$ such that

$$x_1^2 \equiv (-1)^{e_{01}} p_1^{e_{11}} \cdots p_m^{e_{m1}} \pmod{n},$$
$$x_2^2 \equiv (-1)^{e_{02}} p_1^{e_{12}} \cdots p_m^{e_{m2}} \pmod{n},$$
$$\vdots$$
$$x_r^2 \equiv (-1)^{e_{0r}} p_1^{e_{1r}} \cdots p_m^{e_{mr}} \pmod{n},$$

where

$$e_{01} + e_{02} + \cdots + e_{0r} = 2e_0$$
$$e_{11} + e_{12} + \cdots + e_{1r} = 2e_1$$
$$\vdots$$
$$e_{m1} + e_{m2} + \cdots + e_{mr} = 2e_m.$$

Show that $x^2 \equiv y^2 \pmod{n}$, where $x = x_1 x_2 \cdots x_r$ and $y = (-1)^{e_0} p_1^{e_1} \cdots p_r^{e_r}$. Explain how to factor n using this information. Here the primes $p_1, \ldots, p_r$, together with -1, are called the *factor base.*

5. Show that 143 can be factored by setting $x_1 = 17$ and $x_2 = 19$, taking the factor base to be $\{3, 5\}$.

6. Let n be a positive integer and let $p_1, p_2, \ldots, p_r$ be primes. Suppose that $Q_{k_i} = \prod_{j=1}^{r} p_j^{k_{ij}}$ for $i = 1, \ldots, t$, where the integers Q_j have their usual meaning with respect to the continued fraction of $\sqrt{n}$. Explain how n can be factored if $\sum_{i=1}^{t} k_i$ is even and $\sum_{i=1}^{t} k_{ij}$ is even for $j = 1, 2, \ldots, r$.

7. Show that 12,007,001 can be factored using the continued fraction expansions of $\sqrt{12{,}007{,}001}$ with factor base $-1, 2, 31, 71, 97$. (*Hint:* Use the factorizations $Q_1 = 2^3 \cdot 97$, $Q_{12} = 2^4 \cdot 71$, $Q_{28} = 2^{11}$, $Q_{34} = 31 \cdot 97$, and $Q_{41} = 31 \cdot 71$, and show that $P_0 P_{11} P_{27} P_{33} P_{40} = 9{,}815{,}310$.)

8. Factor 197,209 using the continued fraction expansion of $\sqrt{197{,}209}$ and factor base 2, 3, 5.

12.5 Computational and Programming Exercises

Computations and Explorations

Using a computational program such as Maple or *Mathematica*, or programs you have written, carry out the following computations and explorations.

   1. Use the continued fraction algorithm to factor $F_7 = 2^{2^7} + 1$.

* **2.** Use the continued fraction algorithm to find the prime factorization of N_{11}, where N_j is the jth term of the sequence defined by $N_1 = 2$, $N_{j+1} = p_1 p_2 \ldots p_j + 1$, where p_j is the largest prime factor of N_j. (For example, $N_2 = 3$, $N_3 = 7$, $N_4 = 43$, $N_5 = 1807$, and so on.)

Programming Projects

Write programs using Maple or *Mathematica*, or a language of your choice to do the following things.

   * **1.** Factor positive integers using the continued fraction algorithmn.

** **2.** Factor positive integers using factor bases and continued fraction expansions (see Exercise 6).

13

Some Nonlinear Diophantine Equations

Introduction

An equation with the restriction that only integer (or sometimes rational) solutions are sought is called a diophantine equation. We have already studied a simple type of diophantine equation, namely linear diophantine equations (Section 3.6). We learned how all solutions in integers of a linear diophantine equation can be found. But what about nonlinear diophantine equations?

It is a deep theorem (beyond the scope of this text) that there is no general method for solving all nonlinear diophantine equations. However, many results have been established about particular nonlinear diophantine equations, as well as certain families of nonlinear diophantine equations. This chapter addresses several types of nonlinear diophantine equations. First, we will consider the diophantine equation $x^2 + y^2 = z^2$, satisfied by the lengths of the sides of a right triangle. We will be able to provide an explicit formula for all of its solutions in integers.

After studying the diophantine equation $x^2 + y^2 = z^2$, we will consider the famous diophantine equation $x^n + z^n = z^n$, where n is an integer greater than 2. That is, we will be interested in whether the sum of the nth powers of two integers can also be the nth power of an integer, where none of the three integers equals 0. Fermat stated that there are no solutions of this diophantine equation when $n > 2$ (a statement known as Fermat's last theorem), but for more than 350 years no one could find a proof. The first proof of this theorem was discovered by Andrew Wiles in 1995, which ended one of the greatest challenges of mathematics. The proof of Fermat's last theorem is far beyond the scope of this book, but we will be able to provide a proof for the case when $n = 4$.

Next, we will consider the problem of representing integers as the sums of squares. We will determine which integers can be written as the sum of two squares. Furthermore, we will prove that every positive integer is the sum of four squares.

Finally, we will study the diophantine equation $x^2 - dy^2 = 1$, known as Pell's equation. We will show that the solutions of this equation can be found using the simple continued fraction of $\sqrt{d}$, providing another example of the usefulness of continued fractions.

13.1 Pythagorean Triples

The Pythagorean theorem tells us that the sum of the squares of the lengths of the legs of a right triangle equals the square of the length of the hypotenuse. Conversely, any triangle for which the sum of the squares of the lengths of the two shortest sides equals the square of the third side is a right triangle. Consequently, to find all right triangles with integral side lengths, we need to find all triples of positive integers x, y, z satisfying the diophantine equation

(13.1) $$x^2 + y^2 = z^2.$$

Triples of positive integers satisfying this equation are called *Pythagorean triples* after the ancient Greek mathematician *Pythagoras*.

Example 13.1.   The triples 3, 4, 5; 6, 8, 10; and 5, 12, 13 are Pythagorean triples because $3^2 + 4^2 = 5^2, 6^2 + 8^2 = 10^2$, and $5^2 + 12^2 = 13^2$. ◄

Unlike most nonlinear diophantine equations, it is possible to explicitly describe all the integral solutions of (13.1). Before developing the result describing all Pythagorean triples, we need a definition.

Definition.   A Pythagorean triple x, y, z is called *primitive* if $(x, y, z) = 1$.

PYTHAGORAS (c. 572–c. 500 B.C.E.) was born on the Greek island of Samos. After extensive travels and studies, Pythagoras founded his famous school at the Greek port of Crotona, in what is now southern Italy. Besides being an academy devoted to the study of mathematics, philosophy, and science, the school was the site of a brotherhood sharing secret rites. The Pythagoreans, as the members of this brotherhood were called, published nothing and ascribed all their discoveries to Pythagoras himself. However, it is believed that Pythagoras himself discovered what is now called the Pythagorean theorem, namely that $a^2 + b^2 = c^2$, where a, b, and c are the lengths of the two legs and of the hypotenuse of a right triangle, respectively. The Pythagoreans believed that the key to understanding the world lay with natural numbers and form. Their central tenet was "Everything is Number." Because of their fascination with the natural numbers, the Pythagoreans made many discoveries in number theory. In particular, they studied perfect numbers and amicable numbers for the mystical properties they felt these numbers possessed.

Example 13.2. The Pythagorean triples 3, 4, 5 and 5, 12, 13 are primitive, whereas the Pythagorean triple 6, 8, 10 is not. ◀

Let x, y, z be a Pythagorean triple with $(x, y, z) = d$. Then there are integers x_1, y_1, z_1 with $x = dx_1, y = dy_1, z = dz_1$, and $(x_1, y_1, z_1) = 1$. Furthermore, because

$$x^2 + y^2 = z^2$$

we have

$$(x/d)^2 + (y/d)^2 = (z/d)^2,$$

so that

$$x_1^2 + y_1^2 = z_1^2.$$

Hence, x_1, y_1, z_1 is a primitive Pythagorean triple, and the original triple x, y, z is simply an integral multiple of this primitive Pythagorean triple.

Also note that any integral multiple of a primitive (or for that matter any) Pythagorean triple is again a Pythagorean triple. If x_1, y_1, z_1 is a primitive Pythagorean triple, then we have

$$x_1^2 + y_1^2 = z_1^2,$$

and hence,

$$(dx_1)^2 + (dy_1)^2 = (dz_1)^2,$$

so that dx_1, dy_1, dz_1, is a Pythagorean triple.

Consequently, all Pythagorean triples can be found by forming integral multiples of primitive Pythagorean triples. To find all primitive Pythagorean triples, we need some lemmas. The first lemma tells us that any two integers of a primitive Pythagorean triple are relatively prime.

Lemma 13.1. If x, y, z is a primitive Pythagorean triple, then $(x, y) = (x, z) = (y, z) = 1$.

Proof. Suppose that x, y, z is a primitive Pythagorean triple and $(x, y) > 1$. Then, there is a prime p such that $p \mid (x, y)$, so that $p \mid x$ and $p \mid y$. Because $p \mid x$ and $p \mid y$, we know that $p \mid (x^2 + y^2) = z^2$. Because $p \mid z^2$, we can conclude that $p \mid z$. This is a contradiction, because $(x, y, z) = 1$. Therefore, $(x, y) = 1$. In a similar manner we can easily show that $(x, z) = (y, z) = 1$. ∎

Next, we establish a lemma about the parity of the integers of a primitive Pythagorean triple.

Lemma 13.2. If x, y, z is a primitive Pythagorean triple, then x is even and y is odd or x is odd and y is even.

Proof. Let x, y, z be a primitive Pythagorean triple. By Lemma 13.1, we know that $(x, y) = 1$, so that x and y cannot both be even. Also x and y cannot both be odd. If x and y were both odd, then we would have

$$x^2 \equiv y^2 \equiv 1 \ (\text{mod } 4),$$

so that

$$z^2 = x^2 + y^2 \equiv 2 \ (\text{mod } 4).$$

This is impossible. Therefore, x is even and y is odd, or vice versa. ∎

The final lemma that we need is a consequence of the fundamental theorem of arithmetic. It tells us that two relatively prime integers that multiply together to give a square must both be squares.

Lemma 13.3. If $r, s,$ and t are positive integers such that $(r, s) = 1$ and $rs = t^2$, then there are integers m and n such that $r = m^2$ and $s = n^2$.

Proof. If $r = 1$ or $s = 1$, then the lemma is obviously true, so we may suppose that $r > 1$ and $s > 1$. Let the prime-power factorizations of $r, s,$ and t be

$$r = p_1^{a_1} p_2^{a_2} \cdots p_u^{a_u},$$
$$s = p_{u+1}^{a_{u+1}} p_{u+2}^{a_{u+2}} \cdots p_v^{a_v},$$

and

$$t = q_1^{b_1} q_2^{b_2} \cdots q_k^{b_k}.$$

Because $(r, s) = 1$, the primes occurring in the factorizations of r and s are distinct. Because $rs = t^2$, we have

$$p_1^{a_1} p_2^{a_2} \cdots p_u^{a_u} p_{u+1}^{a_{u+1}} p_{u+2}^{a_{u+2}} \cdots p_v^{a_v} = q_1^{2b_1} q_2^{2b_2} \cdots q_k^{2b_k}.$$

From the fundamental theorem of arithmetic, the prime-powers occurring on the two sides of the above equation are the same. Hence, each p_i must be equal to q_j for some j with matching exponents, so that $a_i = 2b_j$. Consequently, every exponent a_i is even, and therefore $a_i/2$ is an integer. We see that $r = m^2$ and $s = n^2$, where m and n are the integers

$$m = p_1^{a_1/2} p_2^{a_2/2} \cdots p_u^{a_u/2}$$

and

$$n = p_{u+1}^{a_{u+1}/2} p_{u+2}^{a_{u+2}/2} \cdots p_v^{a_v/2}.$$ ∎

We can now prove the desired result that describes all primitive Pythagorean triples.

Theorem 13.1. The positive integers x, y, z form a primitive Pythagorean triple, with y even, if and only if there are relatively prime positive integers m and n, $m > n$, with

m odd and n even or m even and n odd, such that

$$x = m^2 - n^2,$$
$$y = 2mn,$$
$$z = m^2 + n^2.$$

Proof. Let x, y, z be a primitive Pythagorean triple. We will show that there are integers m and n as specified in the statement of the theorem. Lemma 13.2 tells us that x is odd and y is even, or vice versa. Because we have assumed that y is even, x and z are both odd. Hence, $z + x$ and $z - x$ are both even, so that there are positive integers r and s with $r = (z + x)/2$ and $s = (z - x)/2$.

Because $x^2 + y^2 = z^2$, we have $y^2 = z^2 - x^2 = (z + x)(z - x)$. Hence,

$$\left(\frac{y}{2}\right)^2 = \left(\frac{z+x}{2}\right)\left(\frac{z-x}{2}\right) = rs.$$

We note that $(r, s) = 1$. To see this, let $(r, s) = d$. Because $d \mid r$ and $d \mid s, d \mid (r + s) = z$ and $d \mid (r - s) = x$. This means that $d \mid (x, z) = 1$, so that $d = 1$.

Using Lemma 13.3, we see that there are positive integers m and n such that $r = m^2$ and $s = n^2$. Writing $x, y,$ and z in terms of m and n, we have

$$x = r - s = m^2 - n^2,$$
$$y = \sqrt{4rs} = \sqrt{4m^2n^2} = 2mn,$$
$$z = r + s = m^2 + n^2.$$

We also see that $(m, n) = 1$, because any common divisor of m and n must also divide $x = m^2 - n^2, y = 2mn,$ and $z = m^2 + n^2$, and we know that $(x, y, z) = 1$. We also note that m and n cannot both be odd, for if they were, then $x, y,$ and z would all be even, contradicting the condition $(x, y, z) = 1$. Because $(m, n) = 1$ and m and n cannot both be odd, we see that m is even and n is odd, or vice versa. This shows that every primitive Pythagorean triple has the appropriate form.

To complete the proof, we must show that every triple

$$x = m^2 - n^2,$$
$$y = 2mn,$$
$$z = m^2 + n^2,$$

where m and n are positive integers $m > n, (m, n) = 1$, and $m \not\equiv n$ (mod 2), forms a primitive Pythagorean triple. First note that $m^2 - n^2, 2mn, m^2 + n^2$ forms a Pythagorean triple since

$$\begin{aligned}
x^2 + y^2 &= (m^2 - n^2)^2 + (2mn)^2 \\
&= (m^4 - 2m^2n^2 + n^4) + 4m^2n^2 \\
&= m^4 + 2m^2n^2 + n^4 \\
&= (m^2 + n^2)^2 \\
&= z^2.
\end{aligned}$$

To see that this triple forms a primitive Pythagorean triple, we must show that these values of x, y, and z are mutually relatively prime. Assume for the sake of contradiction that $(x, y, z) = d > 1$. Then, there is a prime $p \mid (x, y, z)$. We note that $p \neq 2$, because x is odd (because $x = m^2 - n^2$, where m^2 and n^2 have opposite parity). Also, note that because $p \mid x$ and $p \mid z$, $p \mid (z + x) = 2m^2$ and $p \mid (z - x) = 2n^2$. Hence, $p \mid m$ and $p \mid n$, contradicting the fact that $(m, n) = 1$. Therefore, $(x, y, z) = 1$, and x, y, z is a primitive Pythagorean triple, concluding the proof. ∎

The following example illustrates the use of Theorem 13.1 to produce a Pythagorean triple.

Example 13.3. Let $m = 5$ and $n = 2$, so that $(m, n) = 1$, $m \not\equiv n \pmod 2$, and $m > n$. Hence, Theorem 13.1 tells us that

$$x = m^2 - n^2 = 5^2 - 2^2 = 21,$$
$$y = 2mn = 2 \cdot 5 \cdot 2 = 20,$$
$$z = m^2 + n^2 = 5^2 + 2^2 = 29$$

is a primitive Pythagorean triple. ◀

We list the primitive Pythagorean triple generated using Theorem 13.1 with $m \leq 6$ in Table 13.1.

m	n	$x = m^2 - n^2$	$y = 2mn$	$z = m^2 + n^2$
2	1	3	4	5
3	2	5	12	13
4	1	15	8	17
4	3	7	24	25
5	2	21	20	29
5	4	9	40	41
6	1	35	12	37
6	5	11	60	61

Table 13.1 *Some primitive Pythagorean triples.*

13.1 Exercises

1. a) Find all primitive Pythagorean triples x, y, z with $z \leq 40$.
   b) Find all Pythagorean triples x, y, z with $z \leq 40$.

2. Show that if x, y, z is a primitive Pythagorean triple, then either x or y is divisible by 3.

3. Show that if x, y, z is a primitive Pythagorean triple, then exactly one of x, y, and z is divisible by 5.

4. Show that if x, y, z is a primitive Pythagorean triple, then at least one of x, y, and z is divisible by 4.

5. Show that every positive integer greater than 2 is part of at least one Pythagorean triple.

6. Let $x_1 = 3, y_1 = 4, z_1 = 5$, and let x_n, y_n, z_n, for $n = 2, 3, 4, \ldots$, be defined recursively by

$$x_{n+1} = 3x_n + 2z_n + 1,$$
$$y_{n+1} = 3x_n + 2z_n + 2,$$
$$z_{n+1} = 4x_n + 3z_n + 2.$$

Show that x_n, y_n, z_n is a Pythagorean triple.

7. Show that if x, y, z is a Pythagorean triple with $y = x + 1$, then x, y, z is one of the Pythagorean triples given in Exercise 6.

8. Find all solutions in positive integers of the diophantine equation $x^2 + 2y^2 = z^2$.

9. Find all solutions in positive integers of the diophantine equation $x^2 + 3y^2 = z^2$.

∗ 10. Find all solutions in positive integers of the diophantine equation $w^2 + x^2 + y^2 = z^2$.

11. Find all Pythagorean triples containing the integer 12.

12. Find formulas for the integers of all Pythagorean triples x, y, z with $z = y + 1$.

13. Find formulas for the integers of all Pythagorean triples x, y, z with $z = y + 2$.

∗ 14. Show that the number of Pythagorean triples x, y, z (with $x^2 + y^2 = z^2$) with a fixed integer x is $(\tau(x^2) - 1)/2$ if x is odd, and $(\tau(x^2/4) - 1)/2$ if x is even.

∗ 15. Find all solutions in positive integers of the diophantine equation $x^2 + py^2 = z^2$, where p is a prime.

16. Find all solutions in positive integers of the diophantine equation $1/x^2 + 1/y^2 = 1/z^2$.

17. Show that $f_n f_{n+3}$, $2f_{n+1} f_{n+2}$, and $f_{n+1}^2 + f_{n+2}^2$ form a Pythagorean triple, where f_k denotes the kth Fibonacci number.

18. Find the length of the sides of all right triangles, where the sides have integer lengths and the area equals the perimeter.

13.1 Computational and Programming Exercises

Computations and Explorations

Using a computation program such as Maple or *Mathematica,* or programs you have written, carry out the following computations and explorations.

1. Find as many Pythagorean triples x, y, z as you can, where each of x, y, and z is 1 less than the square of an integer. Do you think that there are infinitely many such triples?

Programming Projects

Write programs using Maple, *Mathematica,* or a language of your choice to do the following.

1. Find all Pythagorean triples x, y, z with x, y, and z less than a given bound.

2. Find all Pythagorean triples containing a given integer.

13.2 Fermat's Last Theorem

In the previous section, we showed that the diophantine equation $x^2 + y^2 = z^2$ has infinitely many solutions in nonzero integers x, y, z. What happens when we replace the exponent 2 in this equation with an integer greater than 2? Next to the discussion of the equation $x^2 + y^2 = z^2$ in his copy of the works of Diophantus, Fermat wrote in the margin:

"However, it is impossible to write a cube as the sum of two cubes, a fourth power as the sum of two fourth powers and in general any power as the sum of two similar powers. For this I have discovered a truly wonderful proof, but the margin is too small to contain it."

Fermat did have a proof of this theorem for the special case of $n = 4$. We will present a proof for this case, using his basic methods, later in this section. Although we will never know for certain whether Fermat had a proof of this result for all integers $n > 2$, mathematicians believe it is extremely unlikely that he did. By 1800, all other statements that he made in the margins of his copy of the works of Diophantus were resolved; some were proved and some were shown to be false. Nevertheless, the following theorem is called *Fermat's last theorem*.

Theorem 13.2. *Fermat's Last Theorem.* The diophantine equation

$$x^n + y^n = z^n$$

has no solutions in nonzero integers x, y, and z when n is an integer with $n \geq 3$.

Note that if we could show that the diophantine equation

$$x^p + y^p = z^p$$

has no solution in nonzero integers x, y, and z whenever p is an odd prime, we would know that Fermat's last theorem is true (see Exercise 2 at the end of this section).

The quest for a proof of Fermat's last theorem challenged mathematicians for more than 350 years. Many great mathematicians have worked on this problem without ultimate success. However, a long series of interesting partial results was established, and new areas of number theory were born as mathematicians attempted to solve this problem. The first major development was Euler's proof in 1770 of Fermat's last theorem for the case $n = 3$. (That is, he showed that there are no solutions of the equation $x^3 + y^3 = z^3$ in nonzero integers.) Euler's proof contained an important error, but Legendre managed to fill in the gap soon afterward.

 In 1805, French mathematician *Sophie Germain* proved a general result about Fermat's last theorem, as opposed to a proof for a particular value of the exponent n. She showed that if p and $2p + 1$ are both primes, then $x^p + y^p = z^p$ has no solutions in integers x, y, and z, with $xyz \neq 0$ when $p \nmid xyz$. As a special case, she showed that if $x^5 + y^5 = z^5$, then one of the integers x, y, and z must be divisible by 5. In 1825, both Dirichlet and Legendre, in independent work, completed the proof of the case when $n = 5$, using the method of infinite descent used by Fermat to prove the $n = 4$ case (and which we will demonstrate later in this section). Fourteen years later, the case of $n = 7$ was settled by Lamé, also using a proof by infinite descent.

In the mid-nineteenth century, mathematicians took some new approaches in attempts to prove Fermat's last theorem for all exponents n. The greatest success in this direction was made by the German mathematician *Ernst Kummer*. He realized that a potentially promising approach, based on the assumption that unique factorization into primes held for certain sets of algebraic integers, was doomed to failure. To overcome this

SOPHIE GERMAIN (1776–1831) was born in Paris and educated at home, using her father's extensive library as a resource. She decided as a young teenager to study mathematics when she discovered that Archimedes was murdered by the Romans. She started by reading the works of Euler and Newton. Although Germain did not attend classes, she learned from university course notes that she managed to obtain. After reading the notes from Lagrange's lectures, she sent him a letter under the pseudonym M. Leblanc. Lagrange, impressed with the insights displayed in this letter, decided to meet M. Leblanc; he was surprised to find that its author was a young woman. Germain corresponded under the pseudonym M. LeBlanc with many mathematicians, including Legrende who included many of her discoveries in his book *Theorie des Nombres*. She also made important contributions to the mathematical theories of elasticity and acoustics. Gauss was impressed by her work and recommended that she receive a doctorate from the University of Göttingen. Unfortunately, she died just before she was to receive this degree.

ERNST EDUARD KUMMER (1810–1893) was born in Sorau, Prussia (now Germany). His father, a physician, died in 1813. Kummer received private tutoring before entering the Gymnasium in Sorau in 1819. In 1828, he entered the University of Halle to study theology; his training for philosophy included the study of mathematics. Inspired by his mathematics instructor H. F. Scherk, he switched to mathematics as his major field of study. Kummer was awarded a doctorate from the University of Halle in 1831, and began teaching at the Gymnasium in Sorau, his old school, that same year. The following year he took a similar position teaching at the Gymnasium in Liegnitz (now the Polish city of Legnica), holding the post for ten years. His research on topics in function theory, including extensions of Gauss's work on hypergeometric series, attracted the attention of leading German mathematicians. They worked to find him a university position.

In 1842, Kummer was appointed to a position at the University of Breslau (now Wroclaw, Poland) and began working on number theory. In 1843, in an attempt to prove Fermat's last theorem, he introduced the concept of "ideal numbers." Although this did not lead to a proof of Fermat's last theorem, Kummer's ideas led to the development of new areas of abstract algebra and the new subject of algebraic number theory. In 1855, he moved to the University of Berlin where he remained until his retirement in 1883.

Kummer was a popular instructor. He was noted for the clarity of his lectures as well as his sense of humor and concern for his students. He was married twice. His first wife, the cousin of Dirichlet's wife, died in 1848, eight years after she and Kummer were married.

difficulty, Kummer developed a theory that supported unique factorization into primes. His basic idea was the concept of "ideal numbers." Using this concept, Kummer could prove Fermat's last theorem for a large class of primes called regular primes. Although there are primes, and perhaps infinitely many primes, that are irregular, Kummer's work showed that Fermat's last theorem was true for many values of n. In particular, Kummer's work showed that Fermat's last theorem was true for all prime exponents less than 100 other than 37, 59, and 67, since these are the only primes less than 100 that are irregular. Kummer's introduction of "ideal numbers" gave birth to the subject of algebraic number theory, which blossomed into a major field of study, and to the part of abstract algebra known as ring theory. The exponents Kummer's work did not address—37, 59, 67, and other relatively irregular primes—fell to a variety of more powerful techniques in subsequent years.

In 1986, German mathematician Gerhard Frey made the first connection of Fermat's last theorem to the subject of elliptic curves. His work surprised mathematicians by linking two seemingly unrelated areas. Frey also managed to show (in 1983) that $x^n + y^n = z^n$ can have only a finite number of solutions in nonzero integers. Of course, if this finite number was shown to be zero for $n \geq 3$, Fermat's last theorem would be proved.

Computers were used to run several different numerical tests that could verify that Fermat's last theorem was true for particular values of n. By 1977, Sam Wagstaff used such tests (and several years of computer time) to verify that Fermat's last theorem held for all exponents n with $n \leq 125,000$. By 1993, such tests had been used to verify that Fermat's last theorem was true for all exponents n with $n < 4 \cdot 10^6$. However, at that time, no proof of Fermat's last theorem seemed to be in sight.

Then, in 1993, *Andrew Wiles,* a professor at Princeton University, shocked the mathematical world when he showed that he could prove Fermat's last theorem. He did this in a series of lectures in Cambridge, England. He had given no hint that the subject of his lectures was a proof of this notorious theorem. The proof he outlined was the culmination of seven years of solitary work. It used a vast array of highly sophisticated

ANDREW WILES (b. 1953) became interested in Fermat's last theorem at the age of 10 when, during a visit to his local library, he found a book stating the problem. He was struck that though it looked simple, none of the great mathematicians could solve it, and he knew that he would never let this problem go. In 1971, Wiles entered Merton College, Oxford. He graduated with his B.A. in 1974, and entered Clare College, Cambridge, where he pursued his doctorate, working on the theory of elliptic curves under John Coates. He was a Research Fellow at Clare College and a Benjamin Pierce Assistant Professor at Harvard from 1977 until 1980. In 1981, he held a post at the Institute for Advanced Study in Princeton, and in 1982 he was appointed to a professorship at Princeton University. He was awarded a Guggenheim Fellowship in 1985, and spent a year studying at the Institut des Hautes Études Scientifique and the École Normale Supérieure in Paris. Ironically, he did not realize that during his years of work in the field of elliptic curves he was learning techniques that would someday help him solve the problem that obsessed him.

Wiles's Seven-Year Quest

In 1986, Wiles learned of work by Frey and Ribet that showed that Fermat's last theorem follows from a conjecture in the theory of elliptic curves, known as the Shimura-Taniyama conjecture. Realizing that this led to a possible strategy for proving the theorem, he abandoned his ongoing research and devoted himself entirely to working on Fermat's last theorem.

During the first few years of this work he talked to colleagues about his progress. However, he decided that talking to others generated too much interest and was too distracting. During his seven years of concentrated, solitary work on Fermat's last theorem he decided that he only had time for "his problem" and his family. His best way to relax during time away from his work was to spend time with his young children.

In 1993, Wiles revealed to several colleagues that he was close to a proof of Fermat's last theorem. After filling what he thought were the remaining gaps, he presented an outline of his proof at Cambridge. Although there had been false alarms in the past about promising proofs of Fermat's last theorem, mathematicians generally believed Wiles had a valid proof. However, a subtle but serious error in reasoning was found when he wrote up his results for publication. Wiles worked diligently, with the help of a former student, for more than a year, almost giving up in frustration, before he found a way to fill the gap.

Wiles's success has brought him countless awards and accolades. It has also brought him peace of mind. He has said that "having solved this problem there's certainly a sense of loss, but at the same time there is this tremendous sense of freedom. I was so obsessed by this problem that for eight years I was thinking about it all the time—when I woke up in the morning to when I went to sleep at night. That particular odyssey is now over. My mind is at rest."

The Wolfskehl Prize

There was added incentive besides fame to prove Fermat's last theorem. In 1908, the German industrialist Paul Wolfskehl bequeathed a prize of 100,000 marks to the Göttingen Academy of Sciences, to be awarded to the first person to publish a proof of Fermat's last theorem. Unfortunately, thousands of incorrect proofs were published in a vain attempt to win the prize, with more than 1000 published, usually as privately printed pamphlets, between 1908 and 1912 alone. (Many people, often without serious mathematical training and sometimes without a clear notion of what a correct proof is, attempt to solve famous problems such as this one even if no prize is available.) Even though Wiles's proof was acclaimed to be correct, it took two years for the Göttingen Academy of Sciences to award the Wolfskehl prize to Wiles; they wanted to be certain the proof was really correct.

Contrary to rumors that the prize had been reduced by inflation to almost nothing, maybe even a pfennig (a German penny), Wiles received approximately $50,000. The prize of 100,000 marks, originally worth around $1,500,000, had been reduced to approximately $500,000 after World War I by German hyperinflation, and the introduction of the deutsche mark after World War II further reduced its value. Many people have speculated about why Wolfskehl left such a large prize for a proof of Fermat's last theorem. People with a romantic slant enjoyed the rumor that, suicidal after being jilted by his true love, he had regained his will to live when he found out about Fermat's last theorem. However, more realistic biographical research indicates that he donated the money to spite his wife, Marie, whom he was forced to marry by his family. He did not want his fortune going to her after he died, so instead it went to the first person who could prove Fermat's last theorem.

methods related to the theory of elliptic curves. Knowledgeable mathematicians were impressed with Wiles's arguments. Word began to spread that Fermat's last theorem had finally been proved. However, when Wiles's 200-page manuscript was studied carefully, a serious problem was found. Although it appeared for a time that it might not be possible to fill the gap in the proof, more than a year later, Wiles (with the help of R. Taylor) managed to fill in the remaining portions of the proof. In 1995, Wiles published his revised proof of Fermat's last theorem, now only 125 pages long. This version passed careful review. Wiles's 1995 proof marked the end of the more than 350-year search for a proof of Fermat's last theorem.

Wiles's proof of Fermat's last theorem is one of those rare mathematical discoveries covered by the popular media. An excellent NOVA episode about this discovery was produced by PBS (information on this show can be found at the PBS Web site). Another source of general information about the proof is *Fermat's Enigma: The Epic Quest to Solve the World's Greatest Mathematical Problem* by Simon Singh ([Si97]). A thorough treatment of the proof, including the mathematics of elliptic curves used in it, can be found in [CoSiSt97]. The original proof by Wiles was published in the *Annals of Mathematics* in 1995 ([Wi95]).

Readers interested in learning more about the history of Fermat's last theorem, and how investigations relating to this conjecture led to the genesis of the theory of algebraic numbers, are encouraged to consult [Ed96], [Ri79], and [Va96].

The Proof for *n* = 4

The proof we will give for the case when $n = 4$ uses the *method of infinite descent* devised by Fermat. This method is an offshoot of the well-ordering property, and shows that a diophantine equation has no solutions by showing that for every solution there is a "smaller" solution, contradicting the well-ordering property.

Using the method of infinite descent, we will show that the diophantine equation $x^4 + y^4 = z^2$ has no solutions in nonzero integers x, y, and z. This is stronger than showing Fermat's last theorem is true for $n = 4$, because any $x^4 + y^4 = z^4 = (z^2)^2$ gives a solution of $x^4 + y^4 = z^2$.

Theorem 13.3.   The diophantine equation

$$x^4 + y^4 = z^2$$

has no solutions in nonzero integers x, y, and z.

Proof.   Assume that this equation has a solution in nonzero integers x, y, and z. Because we may replace any number of the variables with their negatives without changing the validity of the equation, we may assume that x, y, and z are positive integers.

We may also suppose that $(x, y) = 1$. To see this, let $(x, y) = d$. Then $x = dx_1$ and $y = dy_1$, with $(x_1, y_1) = 1$, where x_1 and y_1 are positive integers. Because $x^4 + y^4 = z^2$, we have

$$(dx_1)^4 + (dy_1)^4 = z^2,$$

so that

$$d^4(x_1^4 + y_1^4) = z^2.$$

Hence, $d^4 \mid z^2$ and, by Exercise 43 of Section 3.5, we know that $d^2 \mid z$. Therefore, $z = d^2 z_1$, where z_1 is a positive integer. Thus,

$$d^4(x_1^4 + y_1^4) = (d^2 z_1)^2 = d^4 z_1^2,$$

so that

$$x_1^4 + y_1^4 = z_1^4.$$

This gives a solution of $x^4 + y^4 = z^2$ in positive integers $x = x_1, y = y_1$, and $z = z_1$ with $(x_1, y_1) = 1$.

So suppose that $x = x_0, y = y_0$, and $z = z_0$ is a solution of $x^4 + y^4 = z^2$, where x_0, y_0, and z_0 are positive integers with $(x_0, y_0) = 1$. We will show that there is another solution in positive integers $x = x_1, y = y_1$, and $z = z_1$ with $(x_1, y_1) = 1$, such that $z_1 < z_0$.

Because $x_0^4 + y_0^4 = z_0^2$, we have

$$(x_0^2)^2 + (y_0^2)^2 = z_0^2,$$

so that x_0^2, y_0^2, z_0 is a Pythagorean triple. Furthermore, we have $(x_0^2, y_0^2) = 1$, for if p is a prime such that $p \mid x_0^2$ and $p \mid y_0^2$, then $p \mid x_0$ and $p \mid y_0$, contradicting the fact that $(x_0, y_0) = 1$. Hence, x_0^2, y_0^2, z_0 is a primitive Pythagorean triple, and by Theorem 13.1, we know that there are positive integers m and n with $(m, n) = 1$, $m \not\equiv n \pmod 2$, and

$$x_0^2 = m^2 - n^2,$$
$$y_0^2 = 2mn,$$
$$z_0 = m^2 + n^2,$$

where we have interchanged x_0^2 and y_0^2, if necessary, to make y_0^2 the even integer of this part.

From the equation for x_0^2, we see that

$$x_0^2 + n^2 = m^2.$$

Because $(m, n) = 1$, it follows that x_0, n, m is a primitive Pythagorean triple, m is odd, and n is even. Again, using Theorem 13.1, we see that there are positive integers r and s with $(r, s) = 1$, $r \not\equiv s \pmod 2$, and

$$x_0 = r^2 - s^2,$$
$$n = 2rs,$$
$$m = r^2 + s^2.$$

Because m is odd and $(m, n) = 1$, we know that $(m, 2n) = 1$. We note that because $y_0^2 = (2n)m$, Lemma 13.3 tells us that there are positive integers z_1 and w with $m = z_1^2$ and $2n = w^2$. Because w is even, $w = 2v$, where v is a positive integer, so that

$$v^2 = n/2 = rs.$$

Because $(r, s) = 1$, Lemma 13.3 tells us that there are positive integers x_1 and y_1 such that $r = x_1^2$ and $s = y_1^2$. Note that because $(r, s) = 1$, it easily follows that $(x_1, y_1) = 1$. Hence,

$$x_1^4 + y_1^4 = r^2 + s^2 = m = z_1^2,$$

where x_1, y_1, z_1 are positive integers with $(x_1, y_1) = 1$. Moreover, we have $z_1 < z_0$, because

$$z_1 \leq z_1^4 = m^2 < m^2 + n^2 = z_0.$$

To complete the proof, assume that $x^4 + y^4 = z^2$ has at least one integral solution. By the well-ordering property, we know that among the solutions in positive integers, there is a solution with the smallest value z_0 of the variable z. However, we have shown that from this solution we can find another solution with a smaller value of the variable z, leading to a contradiction. This completes the proof by the method of infinite descent. ∎

Conjectures About Some Diophantine Equations

The resolution of a longstanding conjecture in mathematics often leads to new conjectures, and this certainly is the case for Fermat's last theorem. For example, Andrew Beal, a banker and amateur mathematician, conjectured that a generalized version of Fermat's last theorem is true, where the exponents on the three terms in the equation $x^n + y^n = z^n$ are allowed to be different.

Beal's Conjecture   The equation $x^a + y^b = z^c$ has no solutions in positive integers x, y, z, a, b, c, where $a \geq 3$, $b \geq 3$, and $c \geq 3$ and $(x, y) = (y, z) = (x, z) = 1$.

Beal's conjecture has not been solved. To generate interest in his conjecture, Andrew Beal has offered a prize of \$100,000 for a proof or a counterexample.

The proof of Fermat's last theorem in the 1990s settled what was the best-known conjecture related to diophantine equations. Surprisingly, in 2002, another well-known, longstanding conjecture about diophantine equations was also settled. In 1844, the Belgian mathematician Eugene Catalan conjectured that the only consecutive positive integers that are both powers (squares, cubes, or higher powers) of integers are $8 = 2^3$ and $9 = 3^2$. In other words, he made the following conjecture.

The Catalan Conjecture   The diophantine equation

$$x^m - y^n = 1$$

has no solutions in positive integers x, y, m, and n, where $m \geq 2$ and $n \geq 2$, other than $x = 3$, $y = 2$, and $m = 2$, and $n = 3$.

Certain cases of the Catalan conjecture have been settled since the fourteenth century when Levi ben Gerson proved that 8 and 9 were the only consecutive integers that are powers of 2 and 3. That is, he showed that if $3^n - 2^m \neq \pm 1$, where m and n are positive integers with $m \geq 2$ and $n \geq 2$, then $m = 3$ and $n = 2$. In the eighteenth century, Euler used the method of infinite descent to prove that the only consecutive cube and square are 8 and 9. That is, he proved that the only solution of the diophantine equation $x^3 - y^2 = \pm 1$ is $x = 2$ and $y = 3$. Additional progress was made during the nineteenth and early twentieth centuries, and in 1976, R. Tijdeman showed that the Catalan equation had at most a finite number of solutions. It was not until 2002 that the Catalan conjecture was settled, when Preda Mihailescu finally proved that this conjecture is correct.

A new conjecture has been formulated which attempts to unify Fermat's last theorem and Mihailescu's theorem proving the Catalan conjecture.

Fermat-Catalan Conjecture The equation $x^a + y^b = z^c$ has at most finitely many solutions if $(x, y) = (y, z) = (x, z) = 1$ and $\frac{1}{a} + \frac{1}{b} + \frac{1}{c} < 1$.

The Fermat-Catalan conjecture remains open. At the present time, ten solutions of this diophantine equation are known that satisfy the hypotheses. They are:

$$1 + 2^3 = 3^2,$$
$$2^5 + 7^2 = 3^4,$$
$$7^3 + 13^2 = 2^9,$$
$$2^7 + 17^3 = 71^2,$$
$$3^5 + 11^4 = 122^2,$$
$$17^7 + 76271^3 = 21063928^2,$$
$$1414^3 + 2213459^2 = 65^7,$$
$$9262^3 + 15312283^2 = 113^7,$$
$$43^8 + 96222^3 = 30042907^2,$$
$$33^8 + 1549034^2 = 15613^3.$$

The abc Conjecture

In 1985, Joseph Oesterlé and David Masser formulated a conjecture that intrigues many mathematicians. If true, their conjecture could be used to resolve questions about many well-known diophantine equations. Before stating the conjecture we need to introduce some notation.

Definition. If n is a positive integer, then rad(n) is the product of the distinct prime factors of n. Note that rad(n) is also called the *squarefree* part of n because it can be obtained by eliminating all the factors that produce squares from the prime factorization of n.

Example 13.4.   If $n = 2^4 \cdot 3^2 \cdot 5^3 \cdot 7^2 \cdot 11$, then $\text{rad}(n) = 2 \cdot 3 \cdot 5 \cdot 7 \cdot 11 = 2310$.   ◀

We can now state the conjecture.

abc Conjecture   For every real number $\epsilon > 0$ there exists a constant $K(\epsilon)$ such that if a, b, and c are integers such that $a + b = c$ and $(a, b) = 1$, then

$$\max(|a|, |b|, |c|) < K(\epsilon)(\text{rad}(abc))^{1+\epsilon}.$$

Many deep results have been shown to be consequences of this conjecture. It would take us too far afield to develop the background and motivation for the abc conjecture. To learn about the origins of the conjecture and its consequences, see [GrTu02] and [Ma00]. In the following example we will show how the abc conjecture can be used to prove a result related to Fermat's last theorem.

Example 13.5.   We can apply the abc conjecture to obtain a partial solution of Fermat's last theorem. We follow an argument of Granville and Tucker [GrTu02]. Suppose that

$$x^n + y^n = z^n,$$

where x, y, and z are pairwise relatively prime integers. Let $a = x^n$, $b = y^n$, and $c = z^n$. We can estimate $\text{rad}(abc) = \text{rad}(x^n y^n z^n)$ by noting that

$$\text{rad}(x^n y^n z^n) = \text{rad}(xyz) \leq xyz < z^3.$$

The equality $\text{rad}(x^n y^n z^n) = \text{rad}(xyz)$ holds because the primes dividing $x^n y^n z^n$ are the same as the primes dividing xyz. The first inequality follows because $\text{rad}(m) \leq m$ for

---

LEVI BEN GERSON (1288–1344), born at Bagnols in southern France, was a man of many talents. He was a Jewish philosopher and biblical scholar, a mathematician, an astronomer, and a physician. Most likely he made his living by practicing medicine, especially since he never held a rabbinical post. Little is known about the particulars of his life other than that he lived in Orange and later in Avignon. In 1321, Levi wrote *The Book of Numbers* dealing with arithmetical operations, including the extraction of roots. Later in life, he wrote *On Sines, Chords and Arcs*, a book dealing with trigonometry, which gives sine tables that were long noted for their accuracy. In 1343, the bishop of Meaux asked Levi to write a commentary on the first five books of Euclid, which he called *The Harmony of Numbers*. Levi also invented an instrument to measure the angular distance between celestial objects called Jacob's staff. He observed both lunar and solar eclipses and proposed new astronomical models based on the data he collected. His philosophical writings are extensive. They are considered to be major contributions to medieval philosophy.

Levi maintained contacts with prominent Christians, and was noted for the universality of his thinking. Pope Clement VI even translated some of Levi's astronomical writings into Latin, and the astronomer Kepler made use of this translation. Levi was fortunate to live in Provence, where popes provided some protection to Jews, rather than another part of France. However, at times persecution made it difficult for Levi to work, even preventing him from obtaining important volumes of Jewish scholarship.

every positive integer m and the last inequality holds because x and y are positive, so that $x < z$ and $y < z$.

Now applying the abc conjecture and noting that $\max(|a|, |b|, |c|) = z^n$, for every $\epsilon > 0$, there exists a constant $K(\epsilon) > 0$ such that

$$z^n \leq K(\epsilon)(z^3)^{1+\epsilon}.$$

If we can take $\epsilon = 1/6$ and $n \geq 4$, it is easy to see that $n - 3(1 + \epsilon) \geq n/8$. This implies that

$$z^n \leq K(1/6)^8,$$

where $K(1/6)$ is the value of the constant $K(\epsilon)$ for $\epsilon = 1/6$. It follows that $z \leq K(1/6)^{8/n}$. Consequently, in a solution of $x^n + y^n = z^n$ with $n \geq 4$, the numbers x, y, and z are all less than a fixed bound, which implies that there are only finitely many such solutions. ◄

13.2 Exercises

1. Show that if x, y, z is a Pythagorean triple and n is an integer with $n > 2$, then $x^n + y^n \neq z^n$.

2. Show that Fermat's last theorem is a consequence of Theorem 13.3, and of the assertion that $x^p + y^p = z^p$ has no solutions in nonzero integers when p is an odd prime.

3. Using Fermat's little theorem, show that if p is prime, and
    a) if $x^{p-1} + y^{p-1} = z^{p-1}$, then $p \mid xyz$.
    b) if $x^p + y^p = z^p$, then $p \mid (x + y - z)$.

4. Show that the diophantine equation $x^4 - y^4 = z^2$ has no solutions in nonzero integers using the method of infinite descent.

5. Using Exercise 4, show that the area of a right triangle with integer sides is never a perfect square.

EUGÈNE CATALAN (1814–1894) was born in Bruges, Belgium. He graduated from the École Polytechnique in 1835. He then was appointed to a teaching post at Châlons sur Marne. Catalan obtained a lectureship in descriptive geometry at the École Polytechnique in 1838, with the help of his schoolmate Joseph Liouville who was impressed by Catalan's mathematical talents. Unfortunately, Catalan's career was aversely affected by the reaction of the authorities to his political activity in favor of the French Republic. Catalan published extensively on topics in number theory and other areas of mathematics. Perhaps he is best known for his definition of the numbers now known as Catalan numbers, which appear in so many contexts in enumeration problems. He used these numbers to solve the problem of determining the number of regions produced by the dissection of a polygon into triangles by nonintersecting diagonals. It turns out that Catalan was not the first to solve this problem, because it was solved in the eighteenth century by Segner, who presented a less elegant solution than Catalan.

* **6.** Show that the diophantine equation $x^4 + 4y^4 = z^2$ has no solutions in nonzero integers.

* **7.** Show that the diophantine equation $x^4 + 8y^4 = z^2$ has no solutions in nonzero integers.

8. Show that the diophantine equation $x^4 + 3y^4 = z^2$ has infinitely many solutions.

9. Find all solutions in the rational numbers of the diophantine equation $y^2 = x^4 + 1$.

A diophantine equation of the form $y^2 = x^3 + k$, where k is an integer, is called a *Bachet equation* after Claude Bachet, a French mathematician of the early seventeenth century.

10. Show that the Bachet equation $y^2 = x^3 + 7$ has no solutions. (*Hint:* Consider the congruence resulting by first adding 1 to both sides of the equation and reducing modulo 4.)

* **11.** Show that the Bachet equation $y^2 = x^3 + 23$ has no solutions in integers x and y. (*Hint:* Look at the congruence obtained by reducing this equation modulo 4.)

* **12.** Show that the Bachet equation $y^2 = x^3 + 45$ has no solutions in integers x and y. (*Hint:* Look at the congruence obtained by reducing this equation modulo 8.)

13. Show that in a Pythagorean triple there is at most one perfect square.

14. Show that the diophantine equation $x^2 + y^2 = z^3$ has infinitely many integer solutions, by showing that for each positive integer k, the integers $x = 3k^2 - 1$, $y = k(k^2 - 3)$, and $z = k^2 + 1$ form a solution.

15. This exercise asks for a proof of a theorem proved by Sophie Germain in 1805. Suppose that n and p are odd primes, such that $p \mid xyz$ whenever x, y, and z are integers such that $x^n + y^n + z^n \equiv 0 \pmod{p}$. Further suppose that there are no solutions of the congruence $w^n \equiv n \pmod{p}$. Show that if x, y, and z are integers such that $x^n + y^n + z^n = 0$, then $n \mid xyz$.

CLAUDE GASPAR BACHET DE MÉZIRIAC (1581–1638) was born in Bourg-en-Bresse, France. his father was an aristocrat and was the highest judicial officer in the province. His early education took place at a house of the Jesuit order of the Duchy of Savoy. Later, he studied under the Jesuits in Lyon, Padua, and Milan. In 1601, he entered the Jesuit Order in Milan where it is presumed that he taught. Unfortunately, he became ill in 1602 and left the Jesuit Order. He resolved to live a life of leisure on his estate at Bourg-en-Bresse, which produced a considerable annual income for him. Bachet married in 1612. Bachet spent almost all of his life living on his estate, except for 1619–1620 when he lived in Paris. While in Paris, it was suggested that he become tutor to Louis XIII. This led to a hasty departure from the royal court.

Bachet's work in number theory concentrated on diophantine equations. In 1612, he presented a complete discussion on the solution of linear diophantine equations. In 1621, Bachet conjectured that every positive integer can be written as the sum of four squares; he checked his conjecture for all integers up to 325. Also, in 1621, Bachet discussed the diophantine equation that now bears his name. He is best known, however, for his Latin translation from the original Greek of Diophantus' book *Arithmetica*. It was in his copy of this book that Fermat wrote his marginal note about what we now call Fermat's last theorem. Bachet also wrote books on mathematical puzzles. His writings were the basis of most later books on mathematical recreations. Bachet discovered a method of constructing magic squares. He was elected to the French Academy in 1635.

16. Show that the diophantine equation $w^3 + x^3 + y^3 = z^3$ has infinitely many nontrivial solutions. (*Hint:* Take $w = 9zk^4$, $x = z(1 - 9k^3)$, and $y = 3zk(1 - 3k^3)$, where z and k are nonzero integers.)

17. Can you find four consecutive positive integers such that the sum of the cubes of the first three is the cube of the fourth integer?

18. Prove that the diophantine equation $w^4 + x^4 = y^4 + z^4$ has infinitely many nontrivial solutions. (*Hint:* Follow Euler by taking $w = m^7 + m^5n^2 - 2m^3n^4 + 3m^2n^5 + mn^6$, $x = m^6n - 3m^5n^2 - 2m^4n^3 + m^2n^5 + n^7$, $y = m^7 + m^5n^2 - 2m^3n^4 - 3m^2n^5 + mn^6$, and $z = m^6n + 3m^5n^2 - 2m^4n^3 + m^2n^5 + n^7$, where m and n are positive integers.)

19. Show that the only solution of the diophantine equation $3^n - 2^m = -1$ in positive integers m and n is $m = 2$ and $n = 1$.

20. Show that the only solution of the diophantine equation $3^n - 2^m = 1$ in positive integers m and n is $m = 3$ and $n = 2$.

21. The diophantine equation $x^2 + y^2 + z^2 = 3xyz$ is called *Markov's equation*.
    a) Show that if $x = a$, $y = b$, and $z = c$ is a solution of Markov's equation, then $x = a$, $y = b$, and $z = 3ab - c$ is also a solution of Markov's equation.
    * b) Show that every solution in integers of Markov's equation is generated starting with the solution $x = 1$, $y = 1$, and $z = 1$ and successively using part (a).

** 22. Apply the abc conjecture to the Catalan equation $x^m - y^n = 1$, where m and n are integers with $m \geq 2$ and $n \geq 2$ to obtain a partial solution of the Catalan conjecture.

** 23. Apply the abc conjecture to show that there are no solutions to Beal's conjecture when the exponents are sufficiently large.

The positive integer d is called a *congruent number* if there is a right triangle of area d with sides that have rational numbers as their length. (Unfortunately, the terminology for congruent numbers is easily confused with the terminology for the congruence of numbers). The problem determining which positive integers are congruent numbers is more than a millennium old (see [Gu94]).

24. a) Show that d is a congruent number if and only if there are positive rational numbers a, b, and c such that $ab = 2d$ and $a^2 + b^2 = c^2$.
    b) Show that 5, 6, and 7 are congruent numbers by considering right triangles with sides of length 3/2, 20/3, and 41/6; sides of length 3, 4, and 5; and sides of length 35/12, 24/5, and 337/60, respectively. Also, show that 24 and 30 are congruent numbers.

25. a) Show that 1 is a congruent number if and only if there is a right triangle with area equal to a perfect square with sides of integer length.
    b) Use part (a) and Theorem 13.1 to show that if 1 is a congruent number, then there is a solution in positive integers of the diophantine equation $x^2 + y^4 = z^4$. Deduce from this fact and Exercise 4 that 1 is not a congruent number.

In 1983, J. Tunnell characterized congruent numbers using the theory of elliptic curves (see [Ko96] for details). Suppose that d is a squarefree positive integer, $a = 1$ when d is odd and $a = 2$ when d is even, n is the number of triples of integers (x, y, z) such that $x^2 + 2ay^2 + 8z^3 = d/a$, and m is the number of triples of integers (x, y, z) such that $x^2 + 2ay^2 + 32z^2 = d/a$. Tunnell showed that if $n \neq 2m$, then d is not a congruent number.

He also showed that if $n = 2m$ and a well-known conjecture about elliptic curves is true, then d is a congruent number.

26. a) Show that $m = n = 2$, when $d = 1$ or $d = 2$.
    b) Show that $m = n = 4$, when $d = 3$ or $d = 10$.
    c) Show that $n = 12$ and $m = 2$, when $d = 11$.
    d) Show that $n = 8$ and $m = 4$, when $d = 34$.
    e) Show that $n = m = 0$, when d is of the form $8k + j$, where k is a positive integer and $j = 5, 6$, or 7.
    f) Using Tunnell's theorem and parts (a), (b), and (c), show that 1, 2, 3, 10, and 11 are not congruent numbers.
    g) Tunnell's conjecture implies that 34 is a congruent number. Show that 34 is a congruent number by finding a right triangle with sides of rational length with area 34.

13.2 Computational and Programming Exercises

Computations and Explorations

Using a computation program such as Maple or *Mathematica*, or programs you have written, carry out the following computations and explorations.

1. Euler conjectured that no sum of fewer than n nth powers of nonzero integers is equal to the nth power of an integer. Show that this conjecture is false (as was shown in 1966 by Lander and Parkin) by finding four fifth powers of integers whose sum is also the fifth power of an integer. Can you find other counterexamples to Euler's claim?

2. Given a positive integer n, find as many pairs of equal sums of nth powers as you can.

Programming Projects

Write programs using Maple, *Mathematica*, or a language of your choice to do the following.

1. Given a positive integer n, search for solutions of the diophantine equation $x^n + y^n = z^n$.

2. Generate solutions of the diophantine equation $x^2 + y^2 = z^3$ (see Exercise 16).

3. Given a positive integer k, search for solutions in integers of Bachet's equation $y^2 = x^3 + k$.

4. Generate the solutions of Markov's equation, defined in Exercise 21.

13.3 Sums of Squares

Mathematicians throughout history have been interested in problems regarding the representation of integers as sums of squares. Diophantus, Fermat, Euler, and Lagrange are among the mathematicians who made important contributions to the solution of such problems. In this section, we discuss two questions of this kind: Which integers are the sum of two squares? What is the least integer n such that every positive integer is the sum of n squares?

We begin by considering the first question. Not every positive integer is the sum of two squares. In fact, n is not the sum of two squares if it is of the form $4k + 3$. To see this, note that because $a^2 \equiv 0$ or $1 \pmod 4$ for every integer a, $x^2 + y^2 \equiv 0, 1,$ or $2 \pmod 4$.

To conjecture which integers are the sum of two squares, we first examine some small positive integers.

Example 13.6. Among the first 20 positive integers, note that

$1 = 0^2 + 1^2$,	11 is not the sum of two squares,
$2 = 1^2 + 1^2$,	12 is not the sum of two squares,
3 is not the sum of two squares,	$13 = 3^2 + 2^2$,
$4 = 2^2 + 0^2$,	14 is not the sum of two squares,
$5 = 1^2 + 2^2$,	15 is not the sum of two squares,
6 is not the sum of two squares,	$16 = 4^2 + 0^2$,
7 is not the sum of two squares,	$17 = 4^2 + 1^2$,
$8 = 2^2 + 2^2$,	$18 = 3^2 + 3^2$,
$9 = 3^2 + 0^2$,	19 is not the sum of two squares,
$10 = 3^2 + 1^2$,	$20 = 2^2 + 4^2$. ◄

It is not immediately obvious from the evidence in Example 13.6 which integers, in general, are the sum of two squares. (Can you see anything in common among those positive integers not representable as the sum of two squares?)

We now begin a discussion that will show that the prime factorization of an integer determines whether this integer is the sum of two squares. There are two reasons for this. The first is that the product of two integers that are sums of two squares is again the sum of two squares; the second is that a prime is representable as the sum of two squares if and only if it is not of the form $4k + 3$. We will prove both of these results. Then we will state and prove the theorem that specifies which integers are the sum of two squares.

The proof that the product of sums of two squares is again the sum of two squares relies on an important algebraic identity that we will use several times in this section.

Theorem 13.4. If m and n are both sums of two squares, then mn is also the sum of two squares.

Proof. Let $m = a^2 + b^2$ and $n = c^2 + d^2$. Then

(13.2) $$mn = (a^2 + b^2)(c^2 + d^2) = (ac + bd)^2 + (ad - bc)^2.$$

The reader can easily verify this identity by expanding all the terms. ∎

Example 13.7. Because $5 = 2^2 + 1^2$ and $13 = 3^2 + 2^2$, it follows from (13.2) that

$$65 = 5 \cdot 13 = (2^2 + 1^2)(3^2 + 2^2)$$
$$= (2 \cdot 3 + 1 \cdot 2)^2 + (2 \cdot 2 - 1 \cdot 3)^2 = 8^2 + 1^2. \quad ◄$$

One crucial result is that every prime of the form $4k + 1$ is the sum of two squares. To prove this result we will need the following lemma.

Lemma 13.4.   If p is a prime of the form $4m + 1$, where m is an integer, then there exist integers x and y such that $x^2 + y^2 = kp$ for some positive integer k with $k < p$.

Proof.   By Theorem 11.4, we know that -1 is a quadratic residue of p. Hence, there is an integer $a, a < p$, such that $a^2 \equiv -1 \pmod{p}$. It follows that $a^2 + 1 = kp$ for some positive integer k. Hence, $x^2 + y^2 = kp$, where $x = a$ and $y = 1$. From the inequality $kp = x^2 + y^2 \leq (p - 1)^2 + 1 < p^2$, we see that $k < p$.   ∎

We can now prove the following theorem, which tells us that all primes not of the form $4k + 3$ are the sum of two squares.

Theorem 13.5.   If p is a prime, not of the form $4k + 3$, then there are integers x and y such that $x^2 + y^2 = p$.

Proof.   Note that 2 is the sum of two squares, because $1^2 + 1^2 = 2$. Now, suppose that p is a prime of the form $4k + 1$. Let m be the smallest positive integer such that $x^2 + y^2 = mp$ has a solution in integers x and y. By Lemma 13.4, there is such an integer less than p; by the well-ordering property, a least such integer exists. We will show that $m = 1$.

Assume that $m > 1$. Let a and b be defined by

$$a \equiv x \pmod{m}, \quad b \equiv y \pmod{m}$$

and

$$-m/2 < a \leq m/2, \quad -m/2 < b \leq m/2.$$

It follows that $a^2 + b^2 \equiv x^2 + y^2 = mp \equiv 0 \pmod{m}$. Hence, there is an integer k such that

$$a^2 + b^2 = km.$$

We have

$$(a^2 + b^2)(x^2 + y^2) = (km)(mp) = km^2 p.$$

By equation (13.2), we have

$$(a^2 + b^2)(x^2 + y^2) = (ax + by)^2 + (ay - bx)^2.$$

Furthermore, because $a \equiv s \pmod{m}$ and $b \equiv y \pmod{m}$, we have

$$ax + by \equiv x^2 + y^2 \equiv 0 \pmod{m}$$
$$ay - bx \equiv xy - yx \equiv 0 \pmod{m}.$$

Hence, $(ax + by)/m$ and $(ay - bx)/m$ are integers, so that

$$\left(\frac{ax + by}{m}\right)^2 + \left(\frac{ax - by}{m}\right)^2 = km^2 p/m^2 = kp$$

is the sum of two squares. If we show that $0 < k < m$, this will contradict the choice of m as the minimum positive integer such that $x^2 + y^2 = mp$ has a solution in integers. We know that $a^2 + b^2 = km$, $-m/2 < a \le m/2$, and $-m/2 < b \le m/2$. Hence, $a^2 \le m^2/4$ and $b^2 \le m^2/4$. We have

$$0 \le km = a^2 + b^2 \le 2(m^2/4) = m^2/2.$$

Consequently, $0 \le k \le m/2$. It follows that $k < m$. All that remains is to show that $k \ne 0$. If $k = 0$, we have $a^2 + b^2 = 0$. This implies that $a = b = 0$, so that $x \equiv y \equiv 0 \pmod m$, which shows that $m \mid x$ and $m \mid y$. Because $x^2 + y^2 = mp$, this implies that $m^2 \mid mp$, which implies that $m \mid p$. Because m is less than p, this implies that $m = 1$, which is what we wanted to prove. ∎

We can now put all the pieces together, and prove the fundamental result that classifies the positive integers that are representable as the sum of two squares.

Theorem 13.6. The positive integer n is the sum of two squares if and only if each prime factor of n of the form $4k + 3$ occurs to an even power in the prime factorization of n.

Proof. Suppose that in the prime factorization of n there are no primes of the form $4k + 3$ that appear to an odd power. We write $n = t^2u$, where u is the product of primes. No primes of the form $4k + 3$ appear in u. By Theorem 13.5, each prime in u can be written as the sum of two squares. Applying Theorem 13.4 one time fewer than the number of different primes in u shows that u is also the sum of two squares, say

$$u = x^2 + y^2.$$

It then follows that n is also the sum of two squares, namely

$$n = (tx)^2 + (ty)^2.$$

Now, suppose that there is a prime p, $p \equiv 3 \pmod 4$, that occurs in the prime factorization of n to an odd power, say the $(2j + 1)$th power. Furthermore, suppose that n is the sum of two squares, that is,

$$n = x^2 + y^2.$$

Let $(x, y) = d$, $a = x/d$, $b = y/d$, and $m = n/d^2$. It follows that $(a, b) = 1$ and

$$a^2 + b^2 = m.$$

Suppose that p^k is the largest power of p that divides d. Then m is divisible by $p^{2j-2k+1}$, and $2j - 2k + 1$ is at least 1 because it is nonnegative; hence, $p \mid m$. We know that p does not divide a, for if $p \mid a$, then $p \mid b$, because $b^2 = m - a^2$ and $(a, b) = 1$.

Thus, there is an integer z such that $az \equiv b \pmod p$. It follows that

$$a^2 + b^2 \equiv a^2 + (az)^2 = a^2(1 + z^2) \pmod p.$$

Because $a^2 + b^2 = m$ and $p \mid m$, we see that

$$a^2(1 + z^2) \equiv 0 \pmod p.$$

Because $(a, p) = 1$, it follows that $1 + z^2 \equiv 0 \pmod{p}$. This implies that $z^2 \equiv -1 \pmod{p}$, which is impossible because -1 is not a quadratic residue of p, because $p \equiv 3 \pmod 4$. This contradiction shows that n could not have been the sum of two squares. ∎

Because there are positive integers not representable as the sum of two squares, we can ask whether every positive integer is the sum of three squares. The answer is no, as it is impossible to write 7 as the sum of three squares (as the reader should show). Because three squares do not suffice, we ask whether four squares do. The answer to this is yes, as we will show. Fermat wrote that he had a proof of this fact, although he never published it (and most historians of mathematics believe that he actually had such a proof). Euler was unable to find a proof, although he made substantial progress toward a solution. It was in 1770 that Lagrange presented the first published solution.

The proof that every positive integer is the sum of four squares depends on the following theorem, which shows that the product of two integers both representable as the sum of four squares can also be so represented. Just as with the analogous result for two squares, there is an important algebraic identity used in the proof.

Theorem 13.7.   If m and n are positive integers that are each the sum of four squares, then mn is also the sum of four squares.

Proof.   Let $m = a^2 + b^2 + c^2 + d^2$ and $n = e^2 + f^2 + g^2 + h^2$. The fact that mn is also the sum of four squares follows from the following algebraic identity:

$$(13.3) \qquad mn = (a^2 + b^2 + c^2 + d^2)(e^2 + f^2 + g^2 + h^2)$$
$$= (ae + bf + cg + dh)^2 + (af - be + ch - dg)^2$$
$$+ (ag - bh - ce + df)^2 + (ah + bg - cf - de)^2.$$

The reader can easily verify this identity by multiplying all the terms. ∎

We illustrate the use of Theorem 13.7 with an example.

Example 13.8.   Because $7 = 2^2 + 1^2 + 1^2 + 1^2$ and $10 = 3^2 + 1^2 + 0^2 + 0^2$, from (13.3) it follows that

$$70 = 7 \cdot 10 = (2^2 + 1^2 + 1^2 + 1^2)(3^2 + 1^2 + 0^2 + 0^2)$$
$$= (2 \cdot 3 + 1 \cdot 1 + 1 \cdot 0 + 1 \cdot 0)^2 + (2 \cdot 1 - 1 \cdot 3 + 1 \cdot 0 - 1 \cdot 0)^2$$
$$+ (2 \cdot 0 - 1 \cdot 0 - 1 \cdot 3 = 1 \cdot 1)^2 + (2 \cdot 0 + 1 \cdot 0 - 1 \cdot 1 - 1 \cdot 3)^2$$
$$= 7^2 + 1^2 + 2^2 + 4^2. \qquad \blacktriangleleft$$

We will now begin our work to show that every prime is the sum of four squares. We begin with a lemma.

Lemma 13.5.   If p is an odd prime, then there exists an integer $k, k < p$, such that

$$kp = x^2 + y^2 + z^2 + w^2$$

has a solution in integers x, y, z, and w.

Proof. We will first show that there are integers x and y such that

$$x^2 + y^2 + 1 \equiv 0 \pmod{p}$$

with $0 \le x < p/2$ and $0 \le y < p/2$.

Let

$$S = \left\{ 0^2, 1^2, \ldots, \left(\frac{p-1}{2} \right)^2 \right\}$$

and

$$T = \left\{ -1 - 0^2, -1 - 1^2, \ldots, -1 - \left(\frac{p-1}{2} \right)^2 \right\}.$$

No two elements of S are congruent modulo p (because $x^2 \equiv y^2 \pmod{p}$ implies that $x \equiv \pm y \pmod{p}$). Likewise, no two elements of T are congruent modulo p. It is easy to see that the set $S \cup T$ contains $p + 1$ distinct integers. By the pigeonhole principle, there are two integers in this union that are congruent modulo p. It follows that there are integers x and y such that $x^2 \equiv -1 - y^2 \pmod{p}$ with $0 \le x \le (p-1)/2$ and $0 \le y < (p-1)/2$. We have

$$x^2 + y^2 + 1 \equiv 0 \pmod{p};$$

it follows that $x^2 + y^2 + 1 + 0^2 = kp$ for some integer k. Because $x^2 + y^2 + 1 < 2((p-1)/2)^2 + 1 < p^2$, it follows that $k < p$.   ∎

We can now prove that every prime is the sum of four squares.

Theorem 13.8. Let p be a prime. Then the equation $x^2 + y^2 + z^2 + w^2 = p$ has a solution, where x, y, z, and w are integers.

Proof. The result is true when $p = 2$, because $2 = 1^2 + 1^2 + 0^2 + 0^2$. Now, assume that p is an odd prime. Let m be the smallest integer such that $x^2 + y^2 + z^2 + w^2 = mp$ has a solution, where x, y, z, and w are integers. (By Lemma 13.5, such integers exist, and by the well-ordering property, there is a minimal such integer.) The theorem will follow if we can show that $m = 1$. To do this, we assume that $m > 1$ and find a smaller such integer.

If m is even, then either all of x, y, z, and w are odd, all are even, or two are odd and two are even. In all these cases, we can rearrange these integers (if necessary) so that $x \equiv y \pmod{2}$ and $z \equiv w \pmod{2}$. It then follows that $(x-y)/2, (x+y)/2, (z-w)/2$, and $(x+w)/2$ are integers, and

$$\left(\frac{x-y}{2} \right)^2 + \left(\frac{x+y}{2} \right)^2 + \left(\frac{z-w}{2} \right)^2 + \left(\frac{z+w}{2} \right)^2 = (m/2)p.$$

This contradicts the minimality of m.

Now suppose that m is odd and $m > 1$. Let $a, b, c,$ and d be integers such that

$$a \equiv x \pmod{m}, \quad b \equiv y \pmod{m}, \quad c \equiv z \pmod{m}, \quad d \equiv w \pmod{m},$$

and

$$-m/2 < a < m/2, \quad -m/2 < b < m/2, \quad -m/2 < c < m/2, \quad -m/2 < d < m/2.$$

We have

$$a^2 + b^2 + c^2 + d^2 \equiv x^2 + y^2 + z^2 + w^2 \pmod{m};$$

hence,

$$a^2 + b^2 + c^2 + d^2 = km$$

for some integer k, and

$$0 \le a^2 + b^2 + c^2 + d^2 < 4(m/2)^2 = m^2.$$

Consequently, $0 \le k < m$. If $k = 0$, we have $a = b = c = d = 0$, so that $x \equiv y \equiv z \equiv w \equiv 0$ (mod m). From this, it follows that $m^2 \mid mp$, which is impossible because $1 < m < p$. It follows that $k > 0$.

We have

$$(x^2 + y^2 + z^2 + w^2)(a^2 + b^2 + c^2 + d^2) = mp \cdot km = m^2 kp.$$

But by the identity in the proof of Theorem 13.7, we have

$$(ax + by + cz + dw)^2 + (bx - ay + dz - cw)^2$$
$$+ (cx - dy - az + bw)^2 + (dx + cy - bz - aw)^2 = m^2 kp.$$

Each of the four terms being squared is divisible by m, because

$$ax + by + cz + dw \equiv x^2 + y^2 + z^2 + w^2 \equiv 0 \pmod{m},$$
$$bx - ay + dz - cw \equiv yx - xy + wz - zw \equiv 0 \pmod{m},$$
$$cx - dy - az + bw \equiv zx - wy - xz + yw \equiv 0 \pmod{m},$$
$$dx + cy - bz - aw \equiv wx + zy - yz - xw \equiv 0 \pmod{m}.$$

Let $X, Y, Z,$ and W be the integers obtained by dividing these quantities by m, that is,

$$X = (ax + by + cz + dw)/m,$$
$$Y = (bx - ay + dz - cw)/m,$$
$$Z = (cx - dy - az + bw)/m,$$
$$W = (dx + cy - bz - aw)/m.$$

It then follows that

$$X^2 + Y^2 + Z^2 + W^2 = m^2 kp/m^2 = kp.$$

But this contradicts the choice of m; hence, m must be 1. ∎

We now can state and prove the fundamental theorem about representations of integers as sums of four squares.

Theorem 13.9. Every positive integer is the sum of the squares of four integers.

Proof. Suppose that n is a positive integer. Then, by the fundamental theorem of arithmetic, n is the product of primes. By Theorem 13.8, each of these prime factors can be written as the sum of four squares. Applying Theorem 13.7 a sufficient number of times, it follows that n is also the sum of four squares. ∎

We have shown that every positive integer can be written as the sum of four squares. As mentioned, this theorem was originally proved by Lagrange in 1770. Around the same time, the English mathematician Edward Waring generalized this problem. He stated, but did not prove, that every positive integer is the sum of 9 cubes of nonnegative integers, the sum of 19 fourth powers of nonnegative integers, and so on. We can phrase this conjecture in the following way.

EDWARD WARING (1736–1798) was born in Old Heath in Shropshire, England, where his father was a farmer. As a youth, Edward attended Shrewsbury School. He entered Magdalene College, Cambridge, in 1753, winning a scholarship qualifying him for a reduced fee if he also worked as a servant. His mathematical talents quickly impressed his teachers and he was elected a fellow of the college in 1754, graduating in 1757. Noted by many as a prodigy, Waring was nominated for the Lucasian Chair of Mathematics at Cambridge in 1759; after some controversy, he was confirmed as the Lucasian professor in 1760 at the age of 23.

Waring's most important work was *Meditationes Algebraicae*, which covered topics in the theory of equations, number theory, and geometry. In this book he makes one of the first important contributions to the part of abstract algebra now known as Galois theory. It was also in this book that he stated without proof that every integer is equal to the sum of not more than 9 cubes, that every integer is the sum of not more than 19 fourth powers, and so on—the result we now call Waring's theorem. To honor his contributions in the *Meditationes Algebraicae*, Waring was elected a Fellow of the Royal Society in 1763. However, few scholars read the book because of its difficult subject matter and because Waring was a poor communicator who used a notation that made his work hard to understand.

Surprisingly, Waring also studied medicine while holding his chair in mathematics. He graduated with an M.D. in 1767 and for a brief time practiced medicine at several hospitals, before giving up medicine in 1770. His lack of success in medicine has been attributed to his shy manner and poor eyesight. Waring was able to pursue medicine while holding his chair in mathematics because he did not present lectures on mathematics. In fact, Waring was noted as a poor communicator with handwriting almost impossible to read. Regrettably, this is not such a rare trait among mathematics professors!

Waring was married to Mary Oswell in 1776. He and his wife lived in the town of Shrewsbury for a while, but his wife did not like the town. The couple later moved to Waring's country estate.

Waring was considered by his contemporaries to possess an odd combination of vanity and modesty, but with vanity predominating. He is recognized as one of the greatest English mathematicians of his time, although his poor communication skills limited his reputation while he was alive. Moreover, according to one account, near the end of his life he fell into a deep religious melancholy which approached insanity and prevented him from accepting several awards.

Waring's Problem.   If k is a positive integer, is there an integer $g(k)$ such that every positive integer can be written as the sum of $g(k)$ kth powers of nonnegative integers, and no smaller number of kth powers will suffice?

Lagrange's theorem shows that we can take $g(2) = 4$ (because there are integers that are not the sum of three squares). In the nineteenth century, mathematicians showed that such an integer $g(k)$ exists for $3 \leq k \leq 8$ and $k = 10$. But it was not until 1906 that David Hilbert showed that for every positive integer k, there is a constant $g(k)$ such that every positive integer may be expressed as the sum of $g(k)$ kth powers of nonnegative integers. Hilbert's proof is extremely complicated and is not constructive, so that it gives no formula for $g(k)$. It is now known that $g(3) = 9$, $g(4) = 19$, $g(5) = 37$, and

$$g(k) = [(3/2)^k] + 2^k - 2$$

for $6 \leq k \leq 471{,}600{,}000$. Proofs of these formulas rely on nonelementary results from analytical number theory. There are still many unanswered questions about the values of $g(k)$.

Although every positive integer can be written as the sum of 9 cubes, it is known that the only positive integers not representable as the sum of 8 cubes are 23 and 239. It is also known that every sufficiently large integer can be represented as the sum of at most 7 cubes. Observations of this sort lead to the definition of the function $G(k)$, which equals the least positive integer such that all sufficiently large positive integers can be represented as the sum of at most $G(k)$ kth powers. The preceding remarks imply that $G(3) \leq 7$. It is also not hard to see that $G(3) \geq 4$, because no positive integer n with $n \equiv \pm 4 \pmod 9$ can be expressed as the sum of three cubes (see Exercise 22). This implies that $4 \leq G(3) \leq 7$. It may surprise you to learn that it is still not known whether $G(3) = 4, 5, 6,$ or 7. The value of $G(k)$ is extremely difficult to determine; the only known values of $G(k)$ are $G(2) = 4$ and $G(4) = 16$. The best currently known inequalities for $G(k)$, with $k = 5, 6, 7,$ and 8 are $6 \leq G(5) \leq 17, 9 \leq G(6) \leq 24, 8 \leq G(7) \leq 32,$ and $32 \leq G(8) \leq 42$.

The interested reader can learn about recent results regarding Waring's problem by consulting the numerous articles on this problem described in [Le74]. The paper of Wunderlich and Kubina [WuKu90] established the upper limit of the range for which it has been verified that $g(k)$ is given by this formula.

13.3 Exercises

1. Given that $13 = 3^2 + 2^2$, $29 = 5^2 + 2^2$, and $50 = 7^2 + 1^2$, write each of the following integers as the sum of two squares.

   a) $377 = 13 \cdot 29$ c) $1450 = 29 \cdot 50$
   b) $650 = 13 \cdot 50$ d) $18{,}850 = 13 \cdot 29 \cdot 50$

2. Determine whether each of the following integers can be written as the sum of two squares.

a) 19   d) 45   g) 99

b) 25   e) 65   h) 999

c) 29   f) 80   i) 1000

3. Represent each of the following integers as the sum of two squares.

a) 34   c) 101   e) 21,658

b) 90   d) 490   f) 324,608

4. Show that a positive integer is the difference of two squares if and only if it is not of the form $4k + 2$, where k is an integer.

5. Represent each of the following integers as the sum of three squares if possible.

a) 3   c) 11   e) 23

b) 90   d) 18   f) 28

6. Show that the positive integer n is not the sum of three squares of integers if n is of the form $8k + 7$, where k is an integer.

7. Show that the positive integer n is not the sum of three squares of integers if n is of the form $4^m(8k + 7)$, where m and k are nonnegative integers.

8. Prove or disprove that the sum of two integers each representable as the sum of three squares of integers is also thus representable.

9. Given that $7 = 2^2 + 1^2 + 1^2 + 1^2$, $15 = 3^2 + 2^2 + 1^2 + 1^2$, and $34 = 4^2 + 4^2 + 1^2 + 1^2$, write each of the following integers as the sum of four squares.

a) $105 = 7 \cdot 15$ c) $238 = 7 \cdot 34$

b) $510 = 15 \cdot 34$ d) $3570 = 7 \cdot 15 \cdot 34$

10. Write each of the following positive integers as the sum of four squares.

a) 6   c) 21   e) 99

b) 12   d) 89   f) 555

11. Show that every integer $n, n \geq 170$, is the sum of the squares of five positive integers. (*Hint:* Write $m = n - 169$ as the sum of the squares of four integers, and use the fact that $169 = 13^2 = 12^2 + 5^2 = 12^2 + 4^2 + 3^2 = 10^2 + 8^2 + 2^2 + 1^2$.)

12. Show that the only positive integers that are not expressible as the sum of five squares of positive integers are $1, 2, 3, 4, 6, 7, 9, 10, 12, 15, 18, 33$. (*Hint:* Use Exercise 11, show that each of these integers cannot be expressed as stated, and then show all remaining positive integers less than 170 can be expressed as stated.)

* **13.** Show that there are arbitrarily large integers that are not the sums of the squares of four positive integers.

We outline a second proof for Theorem 13.5 in Exercises 14–15.

* **14.** Show that if p is prime and a is an integer not divisible by p, then there exist integers x and y such that $ax \equiv y \pmod{p}$ with $0 < |x| < \sqrt{p}$ and $0 < |y| < \sqrt{p}$. This result is called *Thue's lemma* after Norwegian mathematician *Axel Thue*. (*Hint:* Use the pigeonhole principle to show that there are two integers of the form $au - v$, with $0 \le u \le [\sqrt{p}]$ and $0 \le v \le [\sqrt{p}]$, that are congruent modulo p. Construct x and y from the two values of u and the two values of v, respectively.)

15. Use Exercise 14 to prove Theorem 13.5. (*Hint:* Show that there is an integer a with $a^2 \equiv -1 \pmod{p}$. Then apply Thue's lemma with this value of a.)

16. Show that 23 is the sum of nine cubes of nonnegative integers but not the sum of eight cubes of nonnegative integers.

Exercises 17–21 give an elementary proof that $g(4) \le 50$.

17. Show that

$$\sum_{1 \le i < j \le 4} \left((x_i + x_j)^4 + (x_i - x_j)^4 \right) = 6 \left(\sum_{k=1}^{4} x_k^2 \right)^2.$$

(*Hint:* Start with the identity $(x_i + x_j)^4 + (x_i - x_j)^4 = 2x_i^4 + 12x_i^2 x_j^2 + 2x_j^4$.)

18. Show from Exercise 17 that every integer of the form $6n^2$, where n is a positive integer, is the sum of 12 fourth powers.

19. Use Exercise 18 and the fact that every positive integer is the sum of four squares to show that every positive integer of the form $6m$, where m is a positive integer, can be written as the sum of 48 fourth powers.

20. Show that the integers $0, 1, 2, 81, 16, 17$ form a complete system of residues modulo 6, each of which is the sum of at most two fourth powers. Show from this that every integer n with $n > 81$ can be written as $6m + k$, where m is a positive integer and k comes from this complete system of residues. Conclude from this that every integer n with $n < 81$ is the sum of 50 fourth powers.

21. Show that every positive integer n with $n \le 81$ is the sum of at most 50 fourth powers. (*Hint:* For $51 \le n \le 81$, start by using three terms equal to 2^4.) Conclude from this exercise and Exercise 20 that $g(4) \le 50$.

22. Show that no positive integer n, $n \equiv \pm 4 \pmod 9$, is the sum of three cubes.

AXEL THUE (1863–1922) was born in Tönsberg, Norway. He received his degree from the University of Oslo in 1889. He studied under the German mathematician Lie in Liepzig and in Berlin from 1891 until 1894, and he was professor of applied mechanics at the University of Oslo from 1903 until 1922. Thue was the first person to study the problem of finding an infinite sequence over a finite alphabet that does not contain any occurrences of adjacent identical blocks. His work on the approximations of algebraic numbers was seminal, and was later improved by Siegel and by Roth. Using his results, he managed to prove that certain diophantine equations such as $y^3 - 2x^3 = 1$ have a finite number of solutions. Edmund Landau characterized Thue's theorem on approximation as "the most important discovery in elementary number theory that I know."

23. Show that $G(4) \leq 15$ by showing that if n is a positive integer with $n \equiv 15 \pmod{16}$, then n cannot be represented as the sum of fewer than 15 fourth powers of integers.

24. Use the fact that 31 is not the sum of 15 fourth powers and the method of infinite descent, to show that no positive integer of the form $31 \cdot 16^m$ is the sum of 15 fourth powers. (*Hint:* Suppose that $\sum_{i=1}^{15} x_i^4 = 31 \cdot 16^m$. Show that each x_i must be even, so that $\sum_{i=1}^{15} (x_i/2)^4 = 31 \cdot 16^{m-1}$.)

13.3 Computational and Programming Exercises

Computations and Explorations

Using a computation program such as Maple or *Mathematica,* or programs you have written, carry out the following computations and explorations.

1. Find the number of ways that each integer less than 100 can be written as the sum of two squares. (Count the sum $(\pm x^2) + (\pm y^2)$ four times, once for each choice of signs.)

2. Using numerical evidence, make a conjecture concerning which positive integers can be expressed as the sum of three squares. (Be sure to consult Exercise 7.)

3. Explore which positive integers can be written as the sum of n cubes of nonnegative integers for $n = 2, 3, 4, 5$.

Programming Projects

Write programs using Maple, *Mathematica,* or a language of your choice to do the following.

* **1.** Determine whether a positive integer n can be represented as the sum of two squares and so represent it if possible.

* **2.** Given a positive integer n, represent n as the sum of four squares.

13.4 Pell's Equation

In this section, we study diophantine equations of the form

(13.4) $$x^2 - dy^2 = n,$$

where d and n are fixed integers. When $d < 0$ and $n < 0$, there are no solutions of (13.4). When $d < 0$ and $n > 0$, there can be at most a finite number of solutions, because the equation $x^2 - dy^2 = n$ implies that $|x| \leq \sqrt{n}$ and $|y| \leq \sqrt{n/|d|}$. Also, note that when d is a perfect square, say $d = D^2$, then

$$x^2 - dy^2 = x^2 - D^2 y = (x + Dy)(x - Dy) = n.$$

Hence, any solution of (13.4), when d is a perfect square, corresponds to a simultaneous solution of the equations

$$x + Dy = a,$$
$$x - Dy = b,$$

where a and b are integers such that $n = ab$. In this case, there are only a finite number of solutions, because there is at most one solution in integers of these two equations for each factorization $n = ab$.

For the rest of this section, we are interested in the diophantine equation $x^2 - dy^2 = n$, where d and n are integers and d is a positive integer that is not a perfect square. As the following theorem shows, the simple continued fraction of $\sqrt{d}$ is very useful for the study of this equation.

Theorem 13.10.   Let d and n be integers such that $d > 0$, d is not a perfect square, and $|n| < \sqrt{d}$. If $x^2 - dy^2 = n$, then x/y is a convergent of the simple continued fraction of $\sqrt{d}$.

Proof.   First consider the case where $n > 0$. Because $x^2 - dy^2 = n$, we see that

$$(13.5) \qquad\qquad (x + y\sqrt{d})(x - y\sqrt{d}) = n.$$

From (13.5), we see that $x - y\sqrt{d} > 0$, so that $x > y\sqrt{d}$. Consequently,

$$\frac{x}{y} - \sqrt{d} > 0,$$

and, because $0 < n < \sqrt{d}$, we see that

$$\begin{aligned}
\frac{x}{y} - \sqrt{d} &= \frac{(x - \sqrt{d}\, y)}{y} \\
&= \frac{x^2 - dy^2}{y(x + y\sqrt{d})} \\
&< \frac{n}{y(2y\sqrt{d})} \\
&< \frac{\sqrt{d}}{2y^2\sqrt{d}} \\
&= \frac{1}{2y^2}.
\end{aligned}$$

Because $0 < \frac{x}{y} - \sqrt{d} < \frac{1}{2y^2}$, Theorem 12.19 tells us that x/y must be a convergent of the simple continued fraction of $\sqrt{d}$.

When $n < 0$, we divide both sides of $x^2 - dy^2 = n$ by $-d$, to obtain

$$y^2 - (1/d)x^2 = -n/d.$$

By a similar argument to that given when $n > 0$, we see that y/x is a convergent of the simple continued fraction expansion of $1/\sqrt{d}$. Therefore, from Exercise 7 of Section 12.3, we know that $x/y = 1/(y/x)$ must be a convergent of the simple continued fraction of $\sqrt{d} = 1/(1/\sqrt{d})$. ∎

We have shown that solutions of the diophantine equation $x^2 - dy^2 = n$, where $|n| < \sqrt{d}$, are given by the convergents of the simple continued fraction expansion of $\sqrt{d}$. We will restate Theorem 12.24 here, replacing n by d, because it will help us to use these convergents to find solutions of this diophantine equation.

Theorem 12.24. Let d be a positive integer that is not a perfect square. Define $\alpha_k = (P_k + \sqrt{d})/Q_k$, $a_k = [\alpha_k]$, $P_{k+1} = a_k Q_k - P_k$, and $Q_{k+1} = (d - P_{k+1}^2)/Q_k$, for $k = 0, 1, 2, \ldots$, where $\alpha_0 = \sqrt{d}$. Furthermore, let p_k/q_k denote the kth convergent of the simple continued fraction expansion of $\sqrt{d}$. Then

$$p_k^2 - dq_k^2 = (-1)^{k-1}Q_{k+1}.$$

The special case of the diophantine equation $x^2 - dy^2 = n$ with $n = 1$ is called *Pell's equation*, after *John Pell*. Although Pell played an important role in the mathematical community of his day, he played only a minor part in solving the equation named in his honor. The problem of finding the solutions of this equation has a long history. Special cases of Pell's equations are discussed in ancient works by Archimedes and Diophantus. Moreover, the twelfth-century Indian mathematician *Bhaskara* described a method for finding the solutions of Pell's equation. In more recent times, in a letter written in 1657, Fermat posed to the "mathematicians of Europe" the problem of showing that there are infinitely many integral solutions of the equation $x^2 - dy^2 = 1$, when d is a positive integer greater than 1 that is not a square. Soon afterward, the English mathematicians Wallis and Brouncker developed a method to find these solutions, but did not provide a proof that their method works. Euler provided all the theory needed for a proof in a paper published in 1767, and Lagrange published such a proof in 1768. The methods of Wallis and Brouncker, Euler, and Lagrange all are related to the use of the continued fraction of $\sqrt{d}$. We will show how this continued fraction is used to find the solutions of Pell's equation. In particular, we will use Theorems 13.9 and 12.24 to find all solutions of

JOHN PELL (1611–1683), the son of a clergyman, was born in Sussex, England, and was educated at Trinity College, Cambridge. He became a schoolmaster instead of following his father's wishes that he enter the clergy. After developing a reputation for scholarship in both mathematics and languages, he took a position at the University of Amsterdam. He remained there until, at the request of the Prince of Orange, he joined the faculty of a new college at Breda. Among Pell's writings in mathematics are a book, *Idea of Mathematics*, as well as many pamphlets and articles. He corresponded and discussed mathematics with the leading mathematicians of his day, including Leibniz and Newton, the inventors of calculus. Euler may have called $x^2 - dy^2 = 1$ "Pell's equation" because he was familiar with a book in which Pell augmented the work of other mathematicians on the solutions of the equation $x^2 - 12y^2 = n$.

Pell was involved with diplomacy; he served in Switzerland as an agent of Oliver Cromwell and he joined the English diplomatic service in 1654. He finally decided to join the clergy in 1661, when he took his holy orders and became chaplain to the Bishop of London. Unfortunately, at the time of his death, Pell was living in abject poverty.

Pell's equation and the related equation $x^2 - dy^2 = -1$. More information about Pell's equation can be found in [Ba03], a book entirely devoted to this equation.

Theorem 13.11. Let d be a positive integer that is not a perfect square. Let p_k/q_k denote the kth convergent of the simple continued fraction of $\sqrt{d}, k = 1, 2, 3 \ldots$, and let n be the period length of this continued fraction. Then, when n is even, the positive solutions of the diophantine equation $x^2 - dy^2 = 1$ are $x = p_{jn-1}, y = q_{jn-1}, j = 1, 2, 3 \ldots$, and the diophantine equation $x^2 - dy^2 = -1$ has no solutions. When n is odd, the positive solutions of $x^2 - dy^2 = 1$ are $x = p_{2jn-1}, y = q_{2jn-1}, j = 1, 2, 3, \ldots$, and the solutions of $x^2 - dy^2 = -1$ are $x = p_{(2j-1)n-1}, y = q_{(2j-1)n-1}, j = 1, 2, 3, \ldots$.

Proof. Theorem 13.9 tells us that if x_0, y_0 is a positive solution of $x^2 - dy^2 = \pm 1$. then $x_0 = p_k, y_0 = q_k$, where p_k/q_k is a convergent of the simple continued fraction of $\sqrt{d}$. On the other hand, from Theorem 12.24, we know that

$$p_k^2 - dq_k^2 = (-1)^{k-1}Q_{k+1},$$

where Q_{k+1} is as defined as in the statement of Theorem 12.24.

Because the period of the continued expansion of $\sqrt{d}$ is n, we know that $Q_{jn} = Q_0 = 1$ for $j = 1, 2, 3, \ldots$, because $\sqrt{d} = \frac{P_0 + \sqrt{d}}{Q_0}$. Hence,

$$p_{jn-1}^2 - d\,q_{jn-1}^2 = (-1)^{jn} Q_{nj} = (-1)^{jn}.$$

This equation shows that when n is even, p_{jn-1}, q_{jn-1} is a solution of $x^2 - dy^2 = 1$ for $j = 1, 2, 3 \ldots$, and when n is odd, p_{2jn-1}, q_{2jn-1} is a solution of $x^2 - dy^2 = 1$ and $p_{2(j-1)n-1}, q_{2(j-1)n-1}$ is a solution of $x^2 - dy^2 = -1$ for $j = 1, 2, 3, \ldots$.

To show that the diophantine equations $x^2 - dy^2 = 1$ and $x^2 - dy^2 = -1$ have no solutions other than those already found, we will show that $Q_{k+1} = 1$ implies that $n \mid k$ and that $Q_j \neq -1$ for $j = 1, 2, 3, \ldots$.

We first note that if $Q_{k+1} = 1$, then

BHASKARA (1114–1185) was born in Biddur, in the Indian state of Mysore. Bhaskara was the head of the astronomical observatory at Ujjain, the center of mathematical studies in India for many centuries. He is the best known of all Indian mathematicians of his era. Bhaskara's works on mathematics include *Lilavati* (The Beautiful) and *Bijaganita* (Seed Counting), which are both textbooks that cover parts of algebra, arithmetic, and geometry. Bhaskara studied systems of linear equations in more unknowns than equations, and knew many combinatorial formulas. He investigated the solutions of many different diophantine equations. In particular, he solved the equation $x^2 - dy^2 = 1$ in integers for $d = 8, 11, 32, 61$, and 67, using what he called the "cycle method." One illustration of his keen computational skill is his discovery of the solution of $x^2 - 61y^2 = 1$ with $x = 1,766,319,049$ and $y = 226,153,980$. Bhaskara also wrote several important books on astronomy, including the *Siddhantasiromani*.

$$\alpha_{k+1} = P_{k+1} + \sqrt{d}.$$

Because $\alpha_{k+1} = [a_{k+1}; a_{k+2}, \ldots]$, the continued fraction expansion of α_{k+1} is purely periodic. Hence, Theorem 12.23 tells us that $-1 < \alpha_{k+1} = P_{k+1} - \sqrt{d} < 0$. This implies that $P_{k+1} = [\sqrt{d}]$, so that $\alpha_k - \alpha_0$, and $n \mid k$.

To see that $Q_j \neq -1$ for $j = 1, 2, 3, \ldots$, note that $Q_j = -1$ implies that $\alpha_j = -P_j - \sqrt{d}$. Because α_j has a purely periodic simple continued fraction expansion, we know that

$$-1 < \alpha'_j = -P_j + \sqrt{d} < 0$$

and

$$\alpha_j = -P_j - \sqrt{d} > 1.$$

From the first of these inequalities, we see that $P_j > -\sqrt{d}$, and from the second, we see that $P_j < -1 - \sqrt{d}$. Because these two inequalities for p_j are contradictory, we see that $Q_j \neq -1$.

Because we have found all solutions of $x^2 - dy^2 = 1$ and $x^2 - dy^2 = -1$, where x and y are positive integers, we have completed the proof. ∎

We illustrate the use of Theorem 13.10 with the following examples.

Example 13.9. Because the simple continued fraction of $\sqrt{13}$ is $[3; \overline{1, 1, 1, 1, 6}]$, the positive solutions of the diophantine equation $x^2 - 13y^2 = 1$ are $p_{10j-1}, q_{10j-1}, j = 1, 2, 3, \ldots$, where p_{10j-1}/q_{10j-1} is the $(10j - 1)$th convergent of the simple continued fraction expansion of $\sqrt{13}$. The least positive solution is $p_9 = 649, q_9 = 180$. The positive solutions of the diophantine equation $x^2 - 13y^2 = -1$ are $p_{10j-6}, j = 1, 2, 3, \ldots$; the least positive solution is $p_4 = 18, q + 4 = 5$. ◀

Example 13.10. Because the continued fraction of $\sqrt{14}$ is $[3; \overline{1, 2, 1, 6}]$, the positive solutions of $x^2 - 14y^2 = 1$ are $p_{4j-1}, q_{4j-1}, j = 1, 2, 3, \ldots$, where p_{4j-1}/q_{j-1} is the jth convergent of the simple continued fraction expansion of $\sqrt{14}$. The least positive solution is $p_3 = 15, q_3 = 4$. The diophantine equation $x^2 - 14y^2 = -1$ has no solutions, because the period length of the simple continued fraction expansion of $\sqrt{14}$ is even. ◀

We conclude this section with the following theorem, which shows how to find all the positive solutions of Pell's equation, $x^2 - dy^2 = 1$, from the least positive solution, without finding subsequent convergents of the continued fraction expansion of $\sqrt{d}$.

Theorem 13.12. Let x_1, y_1 be the least positive solution of the diophantine equation $x^2 - dy^2 = 1$, where d is a positive integer that is not a perfect square. Then all positive solutions x_k, y_k are given by

$$x_k + y_k\sqrt{d} = (x_1 + y_1\sqrt{d})^k$$

for $k = 1, 2, 3, \ldots$. (Note that x_k and y_k are determined by the use of Lemma 13.4.)

Proof. We must show that x_k, y_k is a solution for $k = 1, 2, 3, \ldots$, and that every solution is of this form.

To show that x_k, y_k is a solution, first note that by taking conjugates, it follows that $x_k - y_k\sqrt{d} = (x_1 - y_1\sqrt{d})^k$ because, from Lemma 12.4, the conjugate of a power is the power of the conjugate. Now, note that

$$x_k^2 - dy_k^2 = (x_k + y_k\sqrt{d})(x_k - y_k\sqrt{d})$$
$$= (x_1 + y_1\sqrt{d})^k(x_1 - y_1\sqrt{d})^k$$
$$= (x_1^2 - dy_1^2)^k$$
$$= 1.$$

Hence, x_k, y_k is a solution for $k = 1, 2, 3, \ldots$.

To show that every positive solution is equal to x_k, y_k for some positive integer k, assume that X, Y is a positive solution from x_k, y_k for $k = 1, 2, 3, \ldots$. Then there is an integer n such that

$$(x_1 + y_1\sqrt{d})^n < X + Y\sqrt{d} < (x_1 + y_1\sqrt{d})^{n+1}.$$

When we multiply this inequality by $(x_1 + y_1\sqrt{d})^{-n}$, we obtain

$$1 < (x_1 - y_1\sqrt{d})^n(X + Y\sqrt{d}) < x_1 + y_1\sqrt{d},$$

because $x_1^2 - dy_1^2 = 1$ implies that $x_1 - y_1\sqrt{d} = (x_1 + y_1\sqrt{d})^{-1}$.

Now, let

$$s + t\sqrt{d} = (x_1 - y_1\sqrt{d})^n(X + Y\sqrt{d})$$

and note that

$$s^2 - dt^2 = (s - t\sqrt{d})(s + t\sqrt{d})$$
$$= (x_1 + y_1\sqrt{d})^n(X - Y\sqrt{d})(x_1 - y_1\sqrt{d})^n(X + Y\sqrt{d})$$
$$= (x_1^2 - dy_1^2)^n(X^2 - dY^2)$$
$$= 1.$$

We see that s, t is a solution of $x^2 - dy^2 = 1$ and, furthermore, we know that $1 < s + t\sqrt{d} < x_1 + y_1\sqrt{d}$. Moreover, because we know that $s + t\sqrt{d} > 1$, we see that $0 < (s + t\sqrt{d})^{-1} < 1$. Hence,

$$s = \frac{1}{2}\left[(s + t\sqrt{d}) + (s - t\sqrt{d})\right] > 0$$

and

$$t = \frac{1}{2\sqrt{d}}\left[(s + t\sqrt{d}) - (s - t\sqrt{d})\right] > 0.$$

This means that s, t is a positive solution, so that $s \geq x_1$, and $t \geq y_1$, by the choice of x_1, y_1 as the smallest positive solution. But this contradicts the inequality $s + t\sqrt{d} < x_1 + y_1\sqrt{d}$. Therefore, X, Y must be x_k, y_k for some choice of k. ∎

The following example illustrates the use of Theorem 13.11.

Example 13.11. From Example 13.9, we know that the least positive solution of the diophantine equation $x^2 - 13y^2 = 1$ is $x_1 = 649$, $y = 180$. Hence, all positive solutions are given by x_k, y_k where

$$x_k + y_k\sqrt{13} = (649 + 180\sqrt{13})^k.$$

For instance, we have

$$x_2 + y_2\sqrt{13} = 842{,}401 + 233{,}640\sqrt{13}.$$

Hence, $x_2 = 842{,}401$, $y_2 = 233{,}640$ is the least positive solution of $x^2 - 13y^2 = 1$, other than $x_1 = 649$, $y_1 = 180$. ◄

13.4 Exercises

1. Find all of the solutions, where x and y are integers, of each of the following equations.

   a) $x^2 + 3y^2 = 4$ b) $x^2 + 5y^2 = 7$ c) $2x^2 + 7y^2 = 30$

2. Find all of the solutions, where x and y are integers, of each of the following equations.

   a) $x^2 - y^2 = 8$ b) $x^2 + 4y^2 = 40$ c) $4x^2 + 9y^2 = 100$

3. For which of the following values of n does the diophantine equation $x^2 - 31y^2 = n$ have a solution?

   a) 1   c) 2   e) 4

   b) -1 d) -3 f) -45

4. Find the least positive solution in integers of each of the following diophantine equations.

   a) $x^2 - 29y^2 = -1$ b) $x^2 - 29y^2 = 1$

5. Find the three smallest positive solutions of the diophantine equation $x^2 - 37y^2 = 1$.

6. For each of the following values of d, determine whether the diophantine equation $x^2 - dy^2 = -1$ has solutions in integers.

   a) 2   e) 17

   b) 3   f) 31

   c) 6   g) 41

   d) 13   h) 50

7. The least positive solution of the diophantine equation $x^2 - 61y^2 = 1$ is $x_1 = 1{,}766{,}319{,}049$, $y_1 = 226{,}153{,}980$. Find the least positive solution other than x_1, y_1.

 * 8. Show that if p_k/q_k is a convergent of the simple continued fraction expansion of $\sqrt{d}$, then $|p_k^2 - dq_k^2| < 1 + 2\sqrt{d}$.

9. Show that if d is a positive integer divisible by a prime of the form $4k + 3$, then the diophantine equation $x^2 - dy^2 = -1$ has no solutions.

10. Let d and n be positive integers.
    a) Show that if r, s is a solution of the diophantine equation $x^2 - dy^2 = 1$ and X, Y is a solution of the diophantine equation $x^2 - dy^2 = n$, then $Xr \pm dYs, Xs \pm Yr$ is also a solution of $x^2 - dy^2 = n$.
    b) Show that the diophantine equation $x^2 - dy^2 = n$ either has no solutions or has infinitely many solutions.

11. Find those right triangles having legs with lengths that are consecutive integers. (*Hint:* Use Theorem 13.1 to write the lengths of the legs as $x = s^2 - t^2$ and $y = 2st$, where s and t are positive integers such that $(s, t) = 1, s > t$ and s and t have opposite parity. Then $x - y = \pm 1$ implies that $(s - t)^2 - 2t^2 = \pm 1$.)

12. Show that the diophantine equation $x^4 - 2y^4 = 1$ has no nontrivial solutions.

13. Show that the diophantine equation $x^4 - 2y^4 = -1$ has no nontrivial solutions.

14. Show that if t_n, the nth triangular number, equals the mth square, so that $n(n + 1)/2 = m^2$, then $x = 2n + 1$ and $y = m$ are solutions of the diophantine equation $x^2 - 8y^2 = 1$. Find the first five solutions of this diophantine equation in terms of increasing values of the positive integer x and the corresponding pairs of triangular and square numbers.

13.4 Computational and Programming Exercises

Computations and Explorations

Using a computation program such as Maple or *Mathematica,* or programs you have written, carry out the following computations and explorations.

1. Find the least positive solution of the diophantine equation $x^2 - 109y^2 = 1$. (This problem was posed by Fermat to English mathematicians in the mid-1600s.)

2. Find the least positive solution of the diophantine equation $x^2 - 991y^2 = 1$.

3. Find the least positive solution of the diophantine equation $x^2 - 1,000,099y^2 = 1$.

Programming Projects

Write programs using Maple, *Mathematica,* or a language of your choice to do the following.

1. Find those integers n with $| n | < \sqrt{d}$ such that the diophantine equation $x^2 - dy^2 = n$ has no solutions.

2. Find the least positive solutions of the diophantine equations $x^2 - dy^2 = 1$ and $x^2 - dy^2 = -1$.

3. Find the solutions of Pell's equation from the least positive solution (see Theorem 13.12).

14

The Gaussian Integers

Introduction

In previous chapters we studied properties of the set of integers. A particularly appealing aspect of number theory is that many basic properties of the integers relating to divisibility, primality, and factorization can be carried over to other sets of numbers. In this chapter we study the set of Gaussian integers, numbers of the form $a + bi$, where a and b are integers and $i = \sqrt{-1}$. We will introduce the concept of divisibility for Gaussian integers and establish a version of the division algorithm for these numbers. We will describe what it means for a Gaussian integer to be prime. We will develop the notion of greatest common divisors for pairs of Gaussian integers and show that Gaussian integers can be written uniquely as the product of Gaussian primes (taking into account a few minor details). Finally, we will show how the Gaussian integers can be used to determine how many ways a positive integer can be written as the sum of two squares. The material in this chapter is a small step into the world of algebraic number theory, a major branch of number theory devoted to the study of algebraic numbers and their properties. Students continuing their study of number theory will find this fairly concrete treatment of the Gaussian integers a useful bridge to more advanced studies. Excellent references for the study of algebraic number theory include [AlWi03], [Mo96], [Mo99], [Po99], and [Ri01].

14.1 Gaussian Integers and Gaussian Primes

In this chapter we extend our study of number theory into the realm of complex numbers. We begin with a brief review of the basic properties of the complex numbers for those who have either never seen this material or need a brief refresher.

The complex numbers are the numbers of the form $x + yi$, where $i = \sqrt{-1}$. Complex numbers can be added, subtracted, multiplied, and divided, according to the following rule.

$$(a + bi) + (c + di) = (a + c) + (b + d)i$$
$$(a + bi) - (c + di) = (a - c) + (b - d)i$$
$$(a + bi)(c + di) = ac + adi + bci + bdi^2 = (ac - bd) + (ad + bc)i$$
$$\frac{a + bi}{c + di} = \frac{a + bi}{c + di} \cdot \frac{c - di}{c - di} = \frac{ac + bd}{c^2 + d^2} + \frac{(-ad + bc)i}{c^2 + d^2}$$

Note that addition and multiplication of complex numbers are commutative. We use the absolute value of an integer to describe the size of this integer. For complex numbers, there are several commonly used ways to describe the size of numbers.

Definition.   If $z = x + iy$ is a complex number, then $|z|$, the *absolute value* of z, equals

$$|z| = \sqrt{x^2 + y^2},$$

and $N(z)$, the *norm* of z, equals

$$|z|^2 = x^2 + y^2.$$

Given a complex number, we can form another complex number with the same absolute value and norm by changing the sign of the imaginary part of the number.

Definition.   The *conjugate* of the complex number $z = a + bi$, denoted by $\bar{z}$, is the complex number $x - iy$.

Note that if w and z are two complex numbers, then the conjugate of wz is the product of the conjugates of w and z. That is, $\overline{(wz)} = (\overline{w})(\overline{z})$. Also note that if $z = x + iy$ is a complex number, then

$$z\bar{z} = (x + iy)(x - iy) = x^2 + y^2 = N(z).$$

We will now prove some useful properties of norms.

Theorem 14.1.   The norm function N from the set of complex numbers to the set of nonnegative real numbers satisfies the following properties.

(i) $N(z)$ is a nonnegative real number for all complex numbers z.

(ii) $N(zw) = N(z)N(w)$ for all complex numbers z and w.

(iii) $N(z) = 0$ if and only if $z = 0$.

To prove (i), suppose that z is a complex number. Then $z = x + iy$, where x and y are real numbers. It follows that $N(z) = x^2 + y^2$ is a nonnegative real number because both x^2 and y^2 are nonnegative real numbers.

To prove (ii), note that

$$N(zw) = (zw)\overline{(zw)} = (zw)(\overline{z}\,\overline{w}) = (z\overline{z})(w\overline{w}) = N(z)N(w),$$

whenever z and w are complex numbers.

To prove (iii), note that $0 = 0 + 0i$, so that $N(0) = 0^2 + 0^2 = 0$. Conversely, suppose that $N(x + iy) = 0$, where x and y are integers. Then $x^2 + y^2 = 0$, which implies that $x = 0$ and $y = 0$ because both x^2 and y^2 are nonnegative. Hence, $x + iy = 0 + i0 = 0$.
∎

Gaussian Integers

In previous chapters we generally restricted ourselves to the rational numbers and integers. An important branch of number theory, called *algebraic number theory*, extends the theory we have developed for the integers to particular sets of algebraic integers. By an algebraic integer, we mean a root of a monic polynomial (that is, with leading coefficient 1) with integer coefficients. We now introduce the particular set of algebraic integers we will study in this chapter.

Definition.   Complex numbers of the form $a + bi$, where a and b are integers, are called *Gaussian integers*. The set of all Gaussian integers is denoted by $\mathbf{Z}[i]$.

Note that if $\gamma = a + bi$ is a Gaussian integer, then it is an algebraic integer satisfying the equation

$$\gamma^2 - 2a\gamma + (a^2 + b^2) = 0,$$

as the reader should verify. Because γ satisfies a monic polynomial with integer coefficients of degree two, it is called a *quadratic irrational*. Conversely, note that if α is a number of the form $r + si$, where r and s are rational numbers and α is a root of a monic quadratic polynomial with integer coefficients, then α is a Gaussian integer (see Exercise 20.) The Gaussian integers are named after the great German mathematician Carl Friedrich Gauss, who was the first to extensively study their properties.

The usual convention is to use Greek letters, such as α, β, γ, and δ to denote Gaussian integers. Note that if n is an integer, then $n = n + 0i$ is also a Gaussian integer. We call an integer n a *rational integer* when we are discussing Gaussian integers.

The Gaussian integers are closed under addition, subtraction, and multiplication, as the following theorem shows.

Theorem 14.2.   Suppose that $\alpha = x + iy$ and $\beta = w + iz$ are Gaussian integers, where x, y, w, and z are rational integers. Then $\alpha + \beta, \alpha - \beta$, and $\alpha\beta$ are all Gaussian integers.

Proof.   We have $\alpha + \beta = (x + iy) + (w + iz) = (x + w) + i(y + z)$, $\alpha - \beta = (x + iy) - (w + iz) = (x - w) + i(y - z)$, and $\alpha\beta = (x + iy)(w + iz) = xw + iyw + ixz + i^2yz = (xw - yz) + i(yw + xz)$. Because the rational integers are closed under addition, subtraction, and multiplication, it follows that each of $\alpha + \beta, \alpha - \beta$, and $\alpha\beta$ are Gaussian integers.
∎

Although the Gaussian integers are closed under addition, subtraction, and multiplication, they are not closed under division, which is also the case for the rational integers. Also, note that if $\alpha = a + bi$ is a Gaussian integer, then $N(\alpha) = a^2 + b^2$ is a nonnegative rational integer.

Divisibility of Gaussian Integers

We can study the set of Gaussian integers much as we have studied the set of rational integers. There are straightforward analogies to many of the basic properties of the integers for the Gaussian integers. To develop these properties for the Gaussian integers, we need to introduce some concepts for the Gaussian integers analogous to those for the ordinary integers. In particular, we need to define what it means for one Gaussian integer to divide another. Later, we will define Gaussian primes, greatest common divisors of pairs of Gaussian integers, and other important notions.

Definition.   Suppose that α and β are Gaussian integers. We say that α *divides* β if there exists a Gaussian integer γ such that $\beta = \alpha\gamma$. If α divides β, we write $\alpha \mid \beta$, whereas if α does not divide β, we write $\alpha \nmid \beta$.

Example 14.1.   We see that $2 - i \mid 13 + i$ because

$$(2 - i)(5 + 3i) = 13 + i.$$

However, $3 + 2i \nmid 6 + 5i$ because

$$\frac{6 + 5i}{3 + 2i} = \frac{(6 + 5i)(3 - 2i)}{(3 + 2i)(3 - 2i)} = \frac{28 + 3i}{13} = \frac{28}{13} + \frac{3i}{13},$$

which is not a Gaussian integer.   ◀

Example 14.2.   We see that $-i \mid (a + bi)$ for all Gaussian integers $a + bi$ because $a + bi = -i(-b + ai)$, whenever a and b are integers. The only other Gaussian integers that divide all other Gaussian integers are 1, -1, and i. We will see why this is true later in this section.   ◀

Example 14.3.   The Gaussian integers divisible by the Gaussian integer $3 + 2i$ are the numbers $(3 + 2i)(a + ib)$, where a and b are integers. Note that $(3 + 2i)(a + ib) = 3a + 2ia + 3ib + 2i^2 b = (3a - 2b) + i(2a + 3b)$. We display these Gaussian integers in Figure 14.1.   ◀

Divisibility in the Gaussian integers satisfies many of the same properties satisfied by divisibility of rational integers. For example, if α, β, and γ are Gaussian integers and $\alpha \mid \beta$ and $\beta \mid \gamma$, then $\alpha \mid \gamma$. Furthermore, if α, β, γ, ν, and μ are Gaussian integers and $\gamma \mid \alpha$ and $\gamma \mid \beta$, then $\gamma \mid (\mu\alpha + \nu\beta)$. We leave it to the reader to verify that these properties hold.

In the integers, there are exactly two integers that are divisors of the integer 1, namely 1 and -1. We now determine which Gaussian integers are divisors of 1. We begin with a definition.

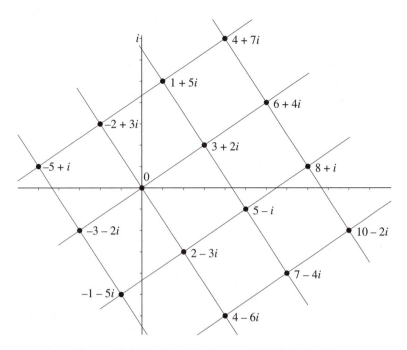

Figure 14.1 *The Gaussian integers divisible by* $3 + 2i$.

Definition.   A Gaussian integer ϵ is called a *unit* if ϵ divides 1. When ϵ is a unit, $\epsilon\alpha$ is an *associate* of the Gaussian integer α

We now characterize which Gaussian integers are units in a way that will make them easy to find.

Theorem 14.3.   A Gaussian integer ϵ is a unit if and only if $N(\epsilon) = 1$.

Proof.   First suppose that ϵ is a unit. Then there a Gaussian integer ν such that $\epsilon\nu = 1$. By part (ii) of Theorem 14.1, it follows that $N(\epsilon\nu) = N(\epsilon)N(\nu) = 1$. Because ϵ and ν are Gaussian integers, both $N(\epsilon)$ and $N(\nu)$ are positive integers. It follows that $N(\epsilon) = N(\nu) = 1$.

Conversely, suppose that $N(\epsilon) = 1$. Then $\epsilon\bar{\epsilon} = N(\epsilon) = 1$. It follows that $\epsilon \mid 1$ and ϵ is a unit.   ∎

We now determine which Gaussian integers are units.

Theorem 14.4.   The Gaussian integers that are units are $1, -1, i$, and $-i$.

Proof.   By Theorem 14.3, the Gaussian integer $\epsilon = a + bi$ is a unit if and only if $N(\epsilon) = 1$. Because $N(\epsilon) = N(a + bi) = a^2 + b^2$, ϵ is a unit if and only if $a^2 + b^2 = 1$. Because a and b are rational integers, we can conclude that $\epsilon = a + bi$ is a unit if and only if $(a, b) = (1, 0), (-1, 0), (0, 1)$, or $(0, -1)$. It follows that ϵ is a unit if and only if $\epsilon = 1, -1, i$, or $-i$.   ∎

Now that we know which Gaussian integers are units, we see that the associates of a Gaussian integer β are the four Gaussian integers $\beta, -\beta, i\beta$, and $-i\beta$.

Example 14.4. The associates of the Gaussian integer $-2 + 3i$ are $-2 + 3i$, $-(-2 + 3i) = 2 - 3i$, $i(-2 + 3i) = -2i + 3i^2 = -3 - 2i$, and $-i(-2 + 3i) = 2i - 3i^2 = 3 + 2i$. ◄

Gaussian Primes

Note that a rational integer is prime if and only if it is not divisible by an integer other than $1, -1$, itself, or its negative. To define Gaussian primes, we want to ignore divisibility by units and associates.

Definition. A nonzero Gaussian integer π is a Gaussian *prime* if it is not a unit and is divisible only by units and its associates.

It follows from the definition of a Gaussian prime that a Gaussian integer π is prime if and only if it has exactly eight divisors, the four units and its four associates, namely $1, -1, i, -i, \pi, -\pi, i\pi$, and $-i\pi$. (Units in the Gaussian integers have exactly four divisors, namely the four units. Gaussian integers that are not prime and are not units have more than eight different divisors.)

An integer that is prime in the set of integers is called a *rational prime*. Later we will see that some rational primes are Gaussian primes, but some are not. Prior to providing examples of Gaussian primes, we prove a useful result which we can use to help determine whether a Gaussian integer is prime.

Theorem 14.5. If π is a Gaussian integer and $N(\pi) = p$, where p is a rational prime, then π and $\overline{\pi}$ are Gaussian primes, but p is not a Gaussian prime.

Proof. Suppose that $\pi = \alpha\beta$, where α and β are Gaussian integers. Then $N(\pi) = N(\alpha\beta) = N(\alpha)N(\beta)$, so that $p = N(\alpha)N(\beta)$. Because $N(\alpha)$ and $N(\beta)$ are positive integers, it follows that $N(\alpha) = 1$ and $N(\beta) = p$ or $N(\alpha) = p$ and $N(\beta) = 1$. We conclude by Theorem 14.3 that either α is a unit or β is a unit. This means that π cannot be factored into two Gaussian integers neither of which is a unit, so it must be a Gaussian prime.

Note that $N(\pi) = \pi \cdot \overline{\pi}$. Because $N(\pi) = p$, it follows that $p = \pi\overline{\pi}$, which means that p is not a Gaussian prime. Note that because $N(\overline{\pi}) = p, \overline{\pi}$ is also a Gaussian prime. ∎

We now give some examples of Gaussian primes.

Example 14.5. We can use Theorem 14.5 to show that $2 - i$ is a Gaussian prime because $N(2 - i) = 2^2 + 1^2 = 5$ and 5 is a rational prime. Also, note that $5 = (2 + i)(2 - i)$, so that 5 is not a Gaussian prime. Similarly, $2 + 3i$ is a Gaussian prime because $N(2 + 3i) = 2^2 + 3^2 = 13$ and 13 is a rational prime. Moreover, 13 is not a Gaussian prime because $13 = (2 + 3i)(2 - 3i)$. ◄

The converse of Theorem 14.5 is not true. It is possible for a Gaussian prime to have a norm that is not a rational prime, as we will see in Example 14.6.

Example 14.6. The integer 3 is a Gaussian prime, as we will show, but $N(3) = N(3 + 0i) = 3^2 + 0^2 = 9$ is not a rational prime. To see that 3 is a Gaussian prime, suppose that $3 = (a + bi)(c + di)$, where $a + bi$ and $c + di$ are not units. By taking norms of both sides of this equation, we find that

$$N(3) = N((a + bi) \cdot (c + di)).$$

It follows that

$$9 = N(a + ib)N(c + id),$$

using part (ii) of Theorem 14.1. Because neither $a + ib$ nor $c + id$ is a unit, $N(a + ib) \neq 1$ and $N(c + id) \neq 1$. Consequently, $N(a + ib) = N(c + id) = 3$. This means that $N(a + ib) = a^2 + b^2 = 3$, which is impossible because 3 is not the sum of two squares. It follows that 3 is a Gaussian prime.   ◄

We now determine whether the rational prime 2 is also a Gaussian prime.

Example 14.7. To determine whether 2 is a Gaussian prime, we determine whether there are Gaussian integers α and β neither a unit such that $2 = \alpha\beta$, where $\alpha = a + ib$ and $\beta = c + id$. If $2 = \alpha\beta$, by taking norms, we see that

$$N(2) = N(\alpha)N(\beta).$$

Because $N(2) = N(2 + 0i) = 2^2 + 0^2 = 4$, this means that

$$N(\alpha)N(\beta) = (a^2 + b^2)(c^2 + d^2) = 4.$$

Because neither α nor β is a unit, we know that $N(\alpha) \neq 1$ and $N(\beta) \neq 1$. It follows that $a^2 + b^2 = 2$ and $c^2 + d^2 = 2$ so that each of a, b, c, and d equals 1 or -1. Consequently, α and β must take on one of the values $1 + i$, $-1 + i$, $1 - i$, or $-1 - i$. On inspection, we find that when $\alpha = 1 + i$ and $\beta = 1 - i$, we have $\alpha\beta = 2$. We conclude that 2 is not a Gaussian prime and $2 = (1 + i)(1 - i)$.

However, $1 + i$ and $1 - i$ are both Gaussian primes, because $N(1 + i) = N(1 - i) = 2$ and 2 is prime, so that Theorem 14.5 applies.   ◄

Looking at Examples 14.5, 14.6, and 14.7, we see that some rational primes are also Gaussian primes, such as 3, while other rational primes, such as $2 = (1 - i)(1 + i)$ and $5 = (2 + i)(2 - i)$ are not Gaussian primes. In Section 14.3 we will determine which rational primes are also Gaussian primes and which are not.

The Division Algorithm for Gaussian Integers

In the first chapter of this book we introduced the division algorithm for rational integers, which shows that when we divide an integer a by a positive integer divisor b, we obtain a nonnegative remainder r less than b. Furthermore, the quotient and remainder we obtain are unique. We would like an analogous result for the Gaussian integers, but in the Gaussian integers it does not make sense to say that a remainder of a division is smaller than the divisor. We overcome this difficulty by developing a division algorithm where the remainder of a division has norm less than the norm of the divisor. However, unlike the situation for rational integers, the quotient and remainder we compute are not unique, as we will illustrate with a subsequent example.

Theorem 14.6. *The Division Algorithm for Gaussian Integers.* Let α and β be Gaussian integers with $\beta \neq 0$. Then there exist Gaussian integers γ and ρ such that

$$\alpha = \beta\gamma + \rho$$

and $0 \leq N(\rho) < N(\beta)$. Here γ is called the *quotient* and ρ is called the *remainder* of this division.

Proof. Suppose that $\alpha/\beta = x + iy$. Then $x + iy$ is a complex number which is a Gaussian integer if and only if β divides α. Suppose that $s = [x + \frac{1}{2}]$ and $t = [y + \frac{1}{2}]$ (these are the integers closest to x and y, respectively, rounded up if the fractional part of x or y equals 1/2; see Figure 14.2).

With these choices for s and t, we find that

$$x + iy = (s + f) + i(t + g),$$

where f and g are real numbers with $|f| \leq 1/2$ and $|g| \leq 1/2$. Now, let $\gamma = s + ti$ and $\rho = \alpha - \beta\gamma$. By Theorem 14.1, we know that $N(\rho) \geq 0$.

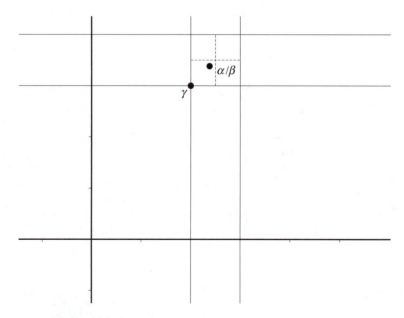

Figure 14.2 *Determining the quotient γ when α is divided by β.*

To show that $N(\rho) < N(\beta)$, recalling that $\alpha/\beta = x + iy$ and using Theorem 14.1 (ii), we see that

$$N(\rho) = N(\alpha - \beta\gamma) = N(((\alpha/\beta) - \gamma)\beta) = N((x + iy) - \gamma)\beta)$$
$$= N((x + iy) - \gamma)N(\beta).$$

Because $\gamma = s + ti$, $x - s = f$, and $y - t = g$, we find that

$$N(\rho) = N((x + iy) - (s + ti))N(\beta) = N(f + ig)N(\beta).$$

Finally, because $|f| \leq 1/2$ and $|g| \leq 1/2$, we conclude that

$$N(\rho) = N(f + ig)N(\beta) \leq ((1/2)^2 + (1/2)^2)N(\beta) \leq N(\beta)/2 < N(\beta).$$

This completes the proof.   ∎

Remark.   In the proof of Theorem 14.6 when we divide a Gaussian integer α by a nonzero Gaussian integer β, we construct a remainder ρ such that $0 \leq N(\rho) \leq N(\beta)/2$. That is, the norm of the remainder does not exceed 1/2 of the norm of the divisor. This will be a useful fact to remember.

Example 14.8 illustrates how to find the quotient and remainder computed in the proof of Theorem 14.6. This example also illustrates that these values are not unique, in the sense that there are other possible values that satisfy the conclusions of the theorem.

Example 14.8.   Let $\alpha = 13 + 20i$ and $\beta = -3 + 5i$. We can follow the steps in the proof of Theorem 14.6 to find γ and ρ such that $\alpha = \beta\gamma + \rho$ and $N(\rho) < N(\beta)$, that is, with $13 + 20i = (-3 + 5i)\gamma + \rho$ and $0 \leq N(\rho) < N(-3 + 5i) = 34$. We first divide α by β to obtain

$$\frac{13 + 20i}{-3 + 5i} = \frac{61}{34} - \frac{125}{34}i.$$

Next, we find the integers closest to $\frac{61}{34}$ and $\frac{-125}{34}$, namely 2 and -4, respectively. Consequently, we take $\gamma = 2 - 4i$ as the quotient. The corresponding remainder is $\rho = \alpha - \beta\gamma = (13 + 20i) - (-3 + 5i)\gamma = (13 + 20i) - (-3 + 5i)(2 - 4i) = -1 - 2i$. We verify that $N(\rho) < N(\beta)$ by noting that $N(-1 - 2i) = 5 < N(-3 + 5i) = 34$, as expected.

Other choices for γ and ρ besides those produced by the construction in the proof of Theorem 14.6 satisfy the consequences of the division algorithm. For example, we can take $\gamma = 2 - 3i$ and $\rho = 4 + i$, because $13 + 20i = (-3 + 5i)(2 - 3i) + (4 + i)$ and $N(4 + i) = 17 < N(-3 + 5i) = 34$. (See Exercise 19.)   ◄

14.1 Exercises

1. Simplify each of the following expressions, expressing your answer in the form of a Gaussian integer $a + bi$.

   a) $(2 + i)^2(3 + i)$ b) $(2 - 3i)^3$ c) $-i(-i + 3)^3$

2. Simplify each of the following expressions, expressing your answer in the form of a Gaussian integer $a + bi$.

   a) $(-1+i)^3(1+i)^3$ b) $(3+2i)(3-i)^2$ c) $(2+i)^2(5-i)^3$

3. Determine whether the Gaussian integer α divides the Gaussian integer β if

   a) $\alpha = 2 - i$, $\beta = 5 + 5i$.   c) $\alpha = 5$, $\beta = 2 + 3i$.

   b) $\alpha = 1 - i$, $\beta = 8$.   d) $\alpha = 3 + 2i$, $\beta = 26$.

4. Determine whether the Gaussian integer α divides the Gaussian integer β, where

   a) $\alpha = 3$, $\beta = 4 + 7i$.   c) $\alpha = 5 + 3i$, $\beta = 30 + 6i$.

   b) $\alpha = 2 + i$, $\beta = 15$.   d) $\alpha = 11 + 4i$, $\beta = 274$.

5. Give a formula for all Gaussian integers divisible by $4 + 3i$ and display the set of all such Gaussian integers in the plane.

6. Give a formula for all Gaussian integers divisible by $4 - i$ and display the set of all such Gaussian integers in the plane.

7. Show that if α, β, and γ are Gaussian integers and $\alpha \mid \beta$ and $\beta \mid \gamma$, then $\alpha \mid \gamma$.

8. Show that if α, β, γ, μ, and ν are Gaussian integers and $\gamma \mid \alpha$ and $\gamma \mid \beta$, then $\gamma \mid (\mu\alpha + \nu\beta)$.

9. Show that if ϵ is a unit for the Gaussian integers, then $\epsilon^5 = \epsilon$.

10. Find all Gaussian integers $\alpha = a + bi$ such that $\overline{\alpha} = a - bi$, the conjugate of α, is an associate of α.

11. Show that the Gaussian integers α and β are associates if $\alpha \mid \beta$ and $\beta \mid \alpha$.

12. Show that if α and β are Gaussian integers and $\alpha \mid \beta$, then $N(\alpha) \mid N(\beta)$.

13. Suppose that $N(\alpha) \mid N(\beta)$, where α and β are Gaussian integers. Does it necessarily follow that $\alpha \mid \beta$? Supply either a proof or a counterexample.

14. Show that if α divides β, where α and β are Gaussian integers, then $\overline{\alpha}$ divides $\overline{\beta}$.

15. Show that if $\alpha = a + bi$ is a nonzero Gaussian integer, then α has exactly one associate $c + di$ (including α itself), where $c > 0$ and $d \geq 0$.

16. For each pair of values for α and β, find the quotient γ and the remainder ρ when α is divided by β computed following the construction in the proof of Theorem 14.6, and verify that $N(\rho) < N(\beta)$.

    a) $\alpha = 14 + 17i$, $\beta = 2 + 3i$ c) $\alpha = 33$, $\beta = 5 + i$

    b) $\alpha = 7 - 19i$, $\beta = 3 - 4i$

17. For each pair of values for α and β, find the quotient γ and the remainder ρ when α is divided by β computed following the construction in the proof of Theorem 14.6, and verify that $N(\rho) < N(\beta)$.

    a) $\alpha = 24 - 9i$, $\beta = 3 + 3i$ c) $\alpha = 87i$, $\beta = 11 - 2i$

    b) $\alpha = 18 + 15i$, $\beta = 3 + 4i$

18. For each pair of values for α and β in Exercise 16, find a pair of Gaussian integers γ and ρ such that $\alpha = \beta\gamma + \rho$ and $N(\rho) < N(\beta)$ different from that computed following the construction in Theorem 14.6.

19. For each pair of values for α and β in Exercise 17, find a pair of Gaussian integers γ and ρ such that $\alpha = \beta\gamma + \rho$ and $N(\rho) < N(\beta)$ different from that computed following the construction in Theorem 14.6.

20. Show that for every pair of Gaussian integers α and β with $\beta \neq 0$ and $\beta \nmid \alpha$, there are at least two different pairs of Gaussian integers γ and ρ such that $\alpha = \beta\gamma + \rho$ and $N(\rho) < N(\beta)$.

* 21. Determine all possible values for the number of pairs of Gaussian integers γ and ρ such that $\alpha = \beta\gamma + \rho$ and $N(\rho) < N(\beta)$ when α and β are Gaussian integers and $\beta \neq 0$. (*Hint:* Analyze this geometrically by looking at the position of α/β in the square containing it and with four lattice points as its corners.)

22. Show that if a number of the form $r + si$, where r and s are rational numbers, is an algebraic integer, then r and s are integers.

23. Show that $1 + i$ divides a Gaussian integer $a + ib$ if and only if a and b are both even or both odd.

24. Show that if π is a Gaussian prime, then $N(\pi) = 2$ or $N(\pi) \equiv 1 \pmod 4$.

25. Find all Gaussian primes of the form $\alpha^2 + 1$, where α is a Gaussian integer.

26. Show that if $a + bi$ is a Gaussian prime, then $b + ai$ is also a Gaussian prime.

27. Show that the rational prime 7 is also a Gaussian prime by adapting the argument given in Example 14.6 that shows 3 is a Gaussian prime.

28. Show that every rational prime p of the form $4k + 3$ is also a Gaussian prime.

29. Suppose that α is a nonzero Gaussian integer which is neither a unit nor a prime. Show that a Gaussian integer β exists such that $\beta \mid \alpha$ and $1 < N(\beta) \leq \sqrt{N(\alpha)}$.

30. Explain how to adapt the sieve of Eratosthenes to find all the Gaussian primes with norm less than a specified limit.

31. Find all the Gaussian primes with norm less than 100.

32. Display all the Gaussian primes with norm less than 200 as lattice points in the plane.

We can define the notion of congruence for Gaussian integers. Suppose that α, β, and γ are Gaussian integers and that $\gamma \neq 0$. We say that α is *congruent* to β modulo γ and we write $\alpha \equiv \beta \pmod \gamma$ if $\gamma \mid (\alpha - \beta)$.

33. Suppose that μ is a nonzero Gaussian integer. Show that each of the following properties holds.
    a) If α is a Gaussian integer, then $\alpha \equiv \alpha \pmod \mu$.
    b) If $\alpha \equiv \beta \pmod \mu$, then $\beta \equiv \alpha \pmod \mu$.
    c) If $\alpha \equiv \beta \pmod \mu$ and $\beta \equiv \gamma \pmod \mu$, then $\alpha \equiv \gamma \pmod \mu$.

34. Suppose that $\alpha \equiv \beta \pmod \mu$ and $\gamma \equiv \delta \pmod \mu$, where $\alpha, \beta, \gamma, \delta$, and μ are Gaussian integers and $\mu \neq 0$. Show that each of these properties holds.

    a) $\alpha + \gamma \equiv \beta + \delta \pmod \mu$ c) $\alpha\gamma \equiv \beta\delta \pmod \mu$
    b) $\alpha - \gamma \equiv \beta - \delta \pmod \mu$

35. Show that two Gaussian integers $\alpha = a_1 + ib_1$ and $\beta = a_2 + ib_2$ can multiplied using only three multiplications of rational integers, rather than the four in the equation shown

in the text, together with five additions and subtractions. (*Hint:* One way to do this uses the product $(a_1 + b_1)(a_2 + b_2)$. A second way uses the product $b_2(a_1 + b_1)$.)

36. When a and b are real numbers, let $\{a + bi\} = \{a\} + \{b\}i$, where $\{x\}$ is the closest integer to the real number x, rounding up in the case of a tie. Show that if z is a complex number, no Gaussian integer is closer to z than $\{z\}$ and $N(z - \{z\}) \leq 1/2$.

Let k be a nonnegative integer. The *Gaussian Fibonacci number* G_k is defined in terms of the Fibonacci numbers with $G_k = f_k + i f_{k+1}$. Exercises 37–39 involve Gaussian Fibonacci numbers.

37. a) List the terms of the Gaussian Fibonacci sequence for $k = 0, 1, 2, 3, 4, 5$. (Recall that $f_0 = 0$.)

   b) Show that $G_k = G_{k-1} + G_{k-2}$ for $k = 2, 3, \ldots$.

38. Show that $N(G_k) = f_{2k+1}$ for all nonnegative integers k.

39. Show that $G_{n+2}G_{n+1} - G_{n+3}G_n = (-1)^n(2 + i)$, whenever n is a positive integer.

40. Show that every Gaussian integer can be written in the form $a_n(-1 + i)^n + a_{n-1}(-1 + i)^{n-1} + \cdots + a_1(-1 + i) + a_0$, where $a_j = 0$ or 1 for $j = 0, 1, \ldots, n - 1, n$.

41. Show that if α is a number of the form $r + si$, where r and s are rational numbers and α is a root of a monic quadratic polynomial with integer coefficients, then α is a Gaussian integer.

42. What can you conclude if $\pi = a + bi$ is a Gaussian prime and one of the Gaussian integers $(a + 1) + bi$, $(a - 1) + bi$, $a + (b + 1)i$, and $a + (b - 1)i$ is also a Gaussian prime?

43. Show that if $\pi_1 = a - 1 + bi$, $\pi_2 = a + 1 + bi$, $\pi_3 = a + (b - 1)i$, and $\pi_4 = a + (b + 1)i$ are all Gaussian primes and $|a| + |b| > 5$, then 5 divides both a and b and neither a nor b is zero.

44. Describe the block of Gaussian integers containing no Gaussian primes that can be constructed by first forming the product of all Gaussian integers $a + bi$ with a and b rational integers, $0 \leq a \leq m$, and $0 \leq b \leq n$.

45. Find all Gaussian integers α, β, and γ such that $\alpha\beta\gamma = \alpha + \beta + \gamma = 1$.

46. Show that if π is a Gaussian prime with $N(\pi) \neq 2$, then exactly one of the associates of π is congruent to either 1 or $3 + 2i$ modulo 4.

14.1   Computational and Programming Exercises

Computations and Explorations

Using a computation program such as Maple or *Mathematica,* or programs you have written, carry out the following computations and explorations.

1. Find all pairs of Gaussian integers γ and ρ such that $180 - 181i = (12 + 13i)\gamma + \rho$ and $N(\rho) < N(12 + 13i)$.

2. Use a version of the sieve of Eratosthenes to find all Gaussian primes with norm less than 1000.

3. Find as many different pairs of Gaussian primes that differ by 2 as you can.

4. Find as many triples of Gaussian primes that form an arithmetic progression with a common difference of 2 as you can.

5. Find as many Gaussian primes of the form $\alpha^2 + \alpha + (9 + 4i)$ as you can.

6. Estimate the probability that two randomly chosen Gaussian integers are relatively prime by testing whether a large number of randomly chosen pairs of Gaussian integers are relatively prime.

Programming Projects

Write programs using Maple, *Mathematica,* or a language of your choice to do the following.

1. Given two Gaussian integers α and β, find all pairs of Gaussian integers γ and ρ such that $\alpha = \gamma\beta + \rho$.

2. Implement a version of the sieve of Eratosthenes to find all Gaussian primes with norm less than a specified integer.

3. Given a positive real number k and a positive integer n, find all Gaussian primes with norm less than n that can be reached, starting with a Gaussian prime with norm not exceeding five moving from one Gaussian prime to the next in steps not exceeding k.

4. Display a graph of the Gaussian primes that can be reached as described in the preceding programming project.

∗∗ **5.** Given a positive real number k, search for *Gaussian moats*, which are regions of width k in the complex plane surrounding the origin that contain no Gaussian integers. (See [GeWaWi98] for more information about Gaussian moats.)

14.2 Greatest Common Divisors and Unique Factorization

In Chapter 3 we showed that every pair of rational integers not both zero has a greatest common divisor. Using properties of the greatest common divisor, we showed that if a prime divides the product of two integers, it must divide one of these integers. We used this fact to show that every integer can be uniquely written as the product of the powers of primes when these primes are written in increasing order. In this section we will establish analogous results for the Gaussian integers. We first develop the concept of greatest common divisors for Gaussian integers. We will show that every pair of Gaussian integers, not both zero, has a greatest common divisor. Then we will show that if a Gaussian prime divides the product of two Gaussian integers, it must divide one of these integers. We will use this result to develop a unique factorization theorem for the Gaussian integers.

Greatest Common Divisors

We cannot adapt the original definition we gave for greatest common divisors of integers because it does not make sense to say that one Gaussian integer is larger than another one. However, we will be able to define the notion of a greatest common divisor for a pair of Gaussian integers by adapting the characterization of the greatest common divisor of two rational integers that does not use the ordering of the integers given in Theorem 3.10.

Definition.   Let α and β be Gaussian integers. A *greatest common divisor* of α and β is a Gaussian integer γ with these two properties:

(i) $\gamma \mid \alpha$ and $\gamma \mid \beta$;

and

(ii)   if $\delta \mid \alpha$ and $\delta \mid \beta$, then $\delta \mid \gamma$.

If γ is a greatest common divisor of the Gaussian integers α and β, then it is straightforward to show that all associates of γ are also greatest common divisors of α and β (see Exercise 5). Consequently, if γ is a greatest common divisor of α and β, then $-\gamma, i\gamma,$ and $-i\gamma$ are also greatest common divisors of α and β. The converse is also true, that is, any two greatest common divisors of two Gaussian integers are associates, as we will prove later in this section. First, we will show that a greatest common divisor exists for every two Gaussian integers.

Theorem 14.7.   If α and β are Gaussian integers, not both zero, then

(i)   there exists a greatest common divisor γ of α and β;

and

(ii)   if γ is a greatest common divisor of α and β, then there exist Gaussian integers μ and ν such that $\gamma = \mu\alpha + \nu\beta$.

Proof.   Let S be the set of norms of nonzero Gaussian integers of the form

$$\mu\alpha + \nu\beta,$$

where μ and ν are Gaussian integers. Because $\mu\alpha + \nu\beta$ is a Gaussian integer when μ and ν are Gaussian integers and the norm of a nonzero Gaussian integer is a positive integer, every element of S is a positive integer. S is nonempty, which can be seen because $N(1 \cdot \alpha + 0 \cdot \beta) = N(\alpha)$ and $N(0 \cdot \alpha + 1 \cdot \beta) = N(\beta)$ both belong to S and both cannot be 0.

Because S is a nonempty set of positive integers, by the well-ordering property, it contains a least element. Consequently, a Gaussian integer γ exists with

$$\gamma = \mu_0\alpha + \nu_0\beta,$$

where μ_0 and ν_0 are Gaussian integers and $N(\gamma) \leq N(\mu\alpha + \nu\beta)$ for all Gaussian integers μ and ν.

We will show that γ is a greatest common divisor of α and β. First, suppose that $\delta \mid \alpha$ and $\delta \mid \beta$. Then there exist Gaussian integers ρ and σ such that $\alpha = \delta\rho$ and $\beta = \delta\sigma$. It follows that

$$\gamma = \mu_0\alpha + \nu_0\beta = \mu_0\delta\rho + \nu_0\delta\sigma = \delta(\mu_0\rho + \nu_0\sigma).$$

We see that $\delta \mid \gamma$.

To show that $\gamma \mid \alpha$ and $\gamma \mid \beta$ we will show that γ divides every Gaussian integer of the form $\mu\alpha + \nu\beta$. So suppose that $\tau = \mu_1\alpha + \nu_1\beta$ for Gaussian integers μ_1 and ν_1. By

Theorem 14.6, the division algorithm for Gaussian integers, we see that

$$\tau = \gamma \eta + \zeta,$$

where η and ζ are Gaussian integers with $0 \leq N(\zeta) < N(\gamma)$. Furthermore, ζ is a Gaussian integer of the form $\mu\alpha + \nu\beta$. To see this note that

$$\zeta = \tau - \gamma\eta = (\mu_1\alpha + \nu_1\beta) - (\mu_0\alpha + \nu_0\beta)\eta = (\mu_1 - \mu_0\eta)\alpha + (\nu_1 - \nu_0\eta)\beta.$$

Recall that γ was chosen as an element with smallest possible norm among the nonzero Gaussian integers of the form $\mu\alpha + \nu\beta$. Consequently, because ζ has this form and $0 \leq N(\zeta) < N(\gamma)$, we know that $N(\zeta) = 0$. By Theorem 14.1, we see that $\zeta = 0$. Consequently, $\tau = \gamma\eta$. We conclude that every element Gaussian integer of the form $\mu\alpha + \nu\beta$ is divisible by γ. ∎

We now show that any two greatest common divisors of two Gaussian integers must be associates.

Theorem 14.8. If both γ_1 and γ_2 are greatest common divisors of the Gaussian integers α and β, not both zero, then γ_1 and γ_2 are associates of each other.

Proof. Suppose that γ_1 and γ_2 are both greatest common divisors of α and β. By part (ii) of the definition of greatest common divisor, it follows that $\gamma_1 \mid \gamma_2$ and $\gamma_2 \mid \gamma_1$. This means there are Gaussian integers ϵ and θ such that $\gamma_2 = \epsilon\gamma_1$ and $\gamma_1 = \theta\gamma_2$. Combining these two equations, we see that

$$\gamma_1 = \theta\epsilon\gamma_1.$$

Divide both sides by γ_1 (which does not equal 0 because 0 is not a common divisor of two Gaussian integers if they are not both zero) to see that

$$\theta\epsilon = 1.$$

We conclude that θ and ϵ are both units. Because $\gamma_1 = \theta\gamma_2$, we see that γ_1 and γ_2 are associates. ∎

The demonstration that the converse of Theorem 14.8 is also true is left as Exercise 5 at the end of this section.

Definition. The Gaussian integers α and β are *relatively prime* if 1 is a greatest common divisor of α and β.

Note that 1 is a greatest common divisor of α and β if and only if the associates of 1, namely $-1, i$, and $-i$, are also greatest common divisors of α and β. For example, if we know that i is a greatest common divisor of α and β, then these two Gaussian integers are relatively prime.

We can adapt the Euclidean algorithm (Theorem 3.11) to find a greatest common divisor of two Gaussian integers.

Theorem 14.9. *A Euclidean Algorithm for Gaussian Integers.* Let $\rho_0 = \alpha$ and $\rho_1 = \beta$ be nonzero Gaussian integers. If the division algorithm for Gaussian integers is successively applied to obtain $\rho_j = \rho_{j+1}\gamma_{j+1} + r_{j+2}$, with $N(\rho_{j+2}) < N(\rho_{j+1})$ for $j = 0, 1, 2, \ldots, n-2$ and $\rho_{n+1} = 0$, then ρ_n, the last nonzero remainder, is a greatest common divisor of α and β.

We leave the proof of Theorem 14.9 to the reader; it is a straightforward adaption of the proof of Theorem 3.11. Note that we can also work backward through the steps of the Euclidean algorithm for Gaussian integers to express the greatest common divisor found by the algorithm as a linear combination of the two Gaussian integers provided as input to the algorithm. We illustrate this in the following example.

Example 14.9. Suppose that $\alpha = 97 + 210i$ and $\beta = 123 + 16i$. The version of the Euclidean algorithm based on the version of the division algorithm in the proof of Theorem 4.6 can be used to find the greatest common divisors of α and β with the following steps.

$$97 + 210i = (123 + 16i)(1 + 2i) + (6 - 52i)$$
$$123 + 16i = (6 - 52i)(2i) + (19 + 4i)$$
$$6 - 52i = (19 + 4i)(-3i) + (-6 + 5i)$$
$$19 + 4i = (-6 + 5i)(-2 - 2i) + (-3 + 2i)$$
$$-6 + 5i = (-3 + 2i)2 + i$$
$$-3 + 2i = i(2 + 3i) + 0$$

We conclude that i is a greatest common divisor of $97 + 210i$ and $123 + 16i$. Consequently, all greatest common divisors of these two Gaussian integers are the associates of i, namely 1, -1, i, and $-i$. It follows that $97 + 210i$ and $123 + 6i$ are relatively prime.

Because $97 + 210i$ and $123 + 16i$ are relatively prime, we can express 1 as a linear combination of these Gaussian integers. We can find Gaussian integers μ and ν such that $1 = \mu\alpha + \nu\beta$ by working backward through these steps and then multiplying both sides by $-i$ to obtain 1. These computations, which we leave to the reader, show that

$$(97 + 210i)(-24 + 21i) + (123 + 16i)(57 + 17i) = 1. \qquad \blacktriangleleft$$

Unique Factorization for Gaussian Integers

The fundamental theorem of arithmetic states that every rational integer has a unique factorization into primes. Its proof depends on the fact that if the rational prime p divides the product of two rational integers ab, then p divides either a or b. We now prove an analogous fact about the Gaussian integers which will play the crucial role in proving unique factorization for the Gaussian integers.

Lemma 14.1. If π is a Gaussian prime and α and β are Gaussian integers such that $\pi \mid \alpha\beta$, then $\pi \mid \alpha$ or $\pi \mid \beta$.

Proof. Suppose that π does not divide α. We will show that π must then divide β. If $\pi \nmid \alpha$, then we also know that $\epsilon\pi \nmid \alpha$ when ϵ is a unit. Because the only divisors of π are $1, -1, i, -i, \pi, -\pi, i\pi$, and $-i\pi$, it follows that a greatest common divisor of π and α must be a unit. This means that 1 is a greatest common divisor of π and α. By Theorem 14.7, we know that there exist Gaussian integers μ and ν such that

$$1 = \mu\pi + \nu\alpha.$$

Multiplying both sides of this equation by β, we see that

$$\beta = \pi(\mu\beta) + \nu(\alpha\beta).$$

By the hypotheses of the theorem, we know that $\pi \mid \alpha\beta$ so that $\pi \mid \nu(\alpha\beta)$. Because $\beta = \pi(\mu\beta) + \nu(\alpha\beta)$, it follows (using Exercise 8 of Section 14.1) that $\pi \mid \beta$. ∎

Lemma 14.1 is a key ingredient in proving that the Gaussian integers enjoy the unique factorization property. Other sets of algebraic integers, such as $Z[\sqrt{-5}]$, the set of quadratic integers of the form $a + b\sqrt{-5}$, do not enjoy a property analogous to Lemma 14.1 and do not enjoy unique factorization.

We can extend Lemma 14.1 to products with more than two terms.

Lemma 14.2. If π is a Gaussian prime and $\alpha_1, \alpha_2, \cdots, \alpha_m$ are Gaussian integers such that $\pi \mid \alpha_1\alpha_2 \cdots \alpha_m$, then there is an integer j such that $\pi \mid \alpha_j$, where $1 \leq j \leq m$.

Proof. We can prove this result using mathematical induction. When $m = 1$, the result is trivial. Now suppose that the result is true for $m = k$, where k is a positive integer. That is, suppose that if

$$\pi \mid \alpha_1\alpha_2 \cdots \alpha_k,$$

where α_i is a Gaussian integer for $i = 1, 2, \ldots, k$, then $\pi \mid \alpha_i$ for some integer i with $1 \leq i \leq k$. Now suppose that

$$\pi \mid \alpha_1\alpha_2 \cdots \alpha_k\alpha_{k+1},$$

where $\alpha_i, i = 1, 2, \ldots, k + 1$ are Gaussian integers. Then $\pi \mid \alpha_1(\alpha_2 \cdots \alpha_k\alpha_{k+1})$, so that by Lemma 14.1, we know that $\pi \mid \alpha_1$ or $\pi \mid \alpha_2 \cdots \alpha_k\alpha_{k+1}$. If $\pi \mid \alpha_2 \cdots \alpha_k\alpha_{k+1}$, we can use the induction hypothesis to conclude that $\pi \mid \alpha_j$ for some integer j with $2 \leq j \leq k + 1$. It follows that $\pi \mid \alpha_j$ for some integer j with $1 \leq j \leq k + 1$, completing the proof. ∎

We can now state and prove the unique factorization theorem for Gaussian integers. Not surprising, Carl Friedrich Gauss was the first to prove this theorem.

Theorem 14.10. *The Unique Factorization Theorem for Gaussian Integers.* Suppose that γ is a nonzero Gaussian integer which is not a unit. Then

(i) γ can be written as the product of Gaussian primes; and

(ii) this factorization is unique in the sense that if

$$\gamma = \pi_1\pi_2 \cdots \pi_s = \rho_1\rho_2 \cdots \rho_t,$$

where $\pi_1, \pi_2, \ldots, \pi_s, \rho_1, \rho_2, \ldots, \rho_t$ are all Gaussian primes, then $s = t$, and after renumbering the terms, if necessary, π_i and ρ_i are associates for $i = 1, 2, \ldots, s$.

Proof. We will prove part (i) using the second principle of mathematical induction where the variable is $N(\gamma)$, the norm of γ. First note that because $\gamma \neq 0$ and γ is not a unit, by Theorem 14.3, we know that $N(\gamma) \neq 1$. It follows that $N(\gamma) \geq 2$.

When $N(\gamma) = 2$, by Theorem 14.5, we know that γ is a Gaussian prime. Consequently, in this case, γ is the product of exactly one Gaussian prime, itself.

Now assume that $N(\gamma) > 2$. We assume that every Gaussian integer δ with $N(\delta) < N(\gamma)$ can be written as the product of Gaussian primes; this is the induction hypothesis. If γ is a Gaussian prime, it can be written as the product of exactly one Gaussian prime, itself. Otherwise, $\gamma = \eta\theta$, where η and θ are Gaussian integers which are not units. Because η and θ are not units, by Theorem 14.3, we know that $N(\eta) > 1$ and $N(\theta) > 1$. Furthermore, because $N(\gamma) = N(\eta)N(\theta)$, we know that $2 \leq N(\eta) < N(\gamma)$ and $2 \leq N(\theta) < N(\gamma)$. Using the induction hypothesis, we know that both η and θ are products of Gaussian primes. That is, $\eta = \pi_1\pi_2 \cdots \pi_s$, where $\pi_1, \pi_2, \ldots, \pi_k$ are Gaussian primes and $\theta = \rho_1\rho_2 \cdots \rho_t$, where $\rho_1, \rho_2, \ldots, \rho_t$ are Gaussian primes. Consequently,

$$\gamma = \theta\eta = \pi_1\pi_2 \cdots \pi_s\rho_1\rho_2 \cdots \rho_t$$

is the product of Gaussian primes. This finishes the proof that every Gaussian integer can be written as the product of Gaussian primes.

We will also use the second principle of mathematical induction to prove part (ii) of the theorem, the uniqueness of the factorization in the sense described in the statement of the theorem. Suppose that γ is a nonzero Gaussian integer which is not a unit. By Theorem 14.3, we know that $N(\gamma) \geq 2$. To begin the proof by mathematical induction, note that when $N(\gamma) = 2$, γ is a Gaussian prime, so γ can only be written in one way as the product of Gaussian primes, namely the product with one term, γ.

Now assume that part (ii) of the statement of the theorem is true when δ is a Gaussian integer with $N(\delta) < N(\gamma)$. Assume that γ can be written as the product of Gaussian primes in two ways, that is,

$$\gamma = \pi_1\pi_2 \cdots \pi_s = \rho_1\rho_2 \cdots \rho_t,$$

where $\pi_1, \pi_2, \ldots, \pi_s, \rho_1, \rho_2, \ldots, \rho_t$ are all Gaussian primes. Note that $s > 1$; otherwise, γ is a Gaussian prime which already can be written uniquely as the product of Gaussian primes.

Because $\pi_1 \mid \pi_1\pi_2 \cdots \pi_s$ and $\pi_1\pi_2 \cdots \pi_s = \rho_1\rho_2 \cdots \rho_t$, we see that $\pi_1 \mid \rho_1\rho_2 \cdots \rho_t$. By Lemma 14.2, we know that $\pi_1 \mid \rho_k$ for some integer k with $1 \leq k \leq t$. We can reorder the primes $\rho_1, \rho_2, \ldots, \rho_k$, if necessary, so that $\pi_1 \mid \rho_1$. Because ρ_1 is a Gaussian prime, it is only divisible by units and associates, so that π_1 and ρ_1 must be associates. It follows that $\rho_1 = \epsilon \pi_1$, where ϵ is a unit. This implies that

$$\pi_1 \pi_2 \cdots \pi_s = \rho_1 \rho_2 \cdots \rho_t = \epsilon \pi_1 \rho_2 \cdots \rho_t.$$

We now divide both sides of this last equation by π_1 to obtain

$$\pi_2 \pi_3 \cdots \pi_s = (\epsilon \rho_2) \rho_3 \cdots \rho_t.$$

Because π_1 is a Gaussian prime, we know that $N(\pi_1) \geq 2$. Consequently,

$$1 \leq N(\pi_2 \pi_3 \cdots \pi_s) < N(\pi_1 \pi_2 \cdots \pi_s) = N(\gamma).$$

By the induction hypothesis and the fact that $\pi_2 \pi_3 \cdots \pi_s = (\epsilon \rho_2) \rho_3 \cdots \rho_t$, we can conclude that $s - 1 = t - 1$, and that after reordering of terms, if necessary, ρ_i is an associate of π_i for $i = 1, 2, \ldots, s - 1$. This completes the proof of part (ii).   ■

Factoring a Gaussian integer into a product of Gaussian primes can be done by computing its norm. For each prime in the factorization of this norm as a rational integer, we look for possible Gaussian prime divisors of the Gaussian integer with this norm. We can perform trial division by each possible Gaussian prime divisor to see whether it divides the Gaussian integer.

Example 14.10.   To find the factorization of 20 into Gaussian integers, we note that $N(20) = 20^2 = 400$. It follows that the possible Gaussian prime divisors of 20 have norm 2 or 5. We find that we can divide 20 by $1 + i$ four times, leaving a quotient of -5. Because $5 = (1 + 2i)(1 - 2i)$, we see that

$$20 = -(1 + i)^4 (1 + 2i)(1 - 2i). \qquad \blacktriangleleft$$

14.2 Exercises

1. Use the definition of the greatest common divisor of two Gaussian integers to show that if π_1 and π_2 are Gaussian primes that are not associates, then 1 is a greatest common divisor of π_1 and π_2.

2. Use the definition of the greatest common divisor of two Gaussian integers to show that if ϵ is a unit and α is a Gaussian integer, then 1 is a greatest common divisor of α and ϵ.

3. Show that if γ is a greatest common divisor of the Gaussian integers α and β, then $\overline{\gamma}$ is a greatest common divisor of $\overline{\alpha}$ and $\overline{\beta}$.

4. a) By extending the definition of a greatest common divisor of two Gaussian integers, define the greatest common divisor of a set of more than two Gaussian integers.

   b) Show from your definition that a greatest common divisor of three Gaussian integers α, β, and γ is a greatest common divisor of γ and a greatest common divisor of α and β.

5. Show that if α and β are Gaussian integers and γ is a greatest common divisor of α and β, then all associates of γ are also greatest common divisors of α and β.

6. Show that if α and β are Gaussian integers and $N(\alpha)$ and $N(\beta)$ are relatively prime rational integers, then α and β are relatively prime Gaussian integers.

7. Show that the converse of the statement in Exercise 6 is not necessarily true, that is, find Gaussian integers α and β such that α and β are relatively prime Gaussian integers, but $N(\alpha)$ and $N(\beta)$ are not relatively prime positive integers.

8. Show that if α and β are Gaussian integers and γ is a greatest common divisor of α and β, then $N(\gamma)$ divides $((N(\alpha), N(\beta)))$.

9. Show if a and b are relatively prime rational integers, then they are also relatively prime Gaussian integers.

10. Show that if α, β, and γ are Gaussian integers and n is a positive integer such that $\alpha\beta = \gamma^n$ and α and β are relatively prime, then $\alpha = \epsilon\delta^n$, where ϵ is a unit and δ is a Gaussian integer.

11. a) Show all steps of the version of the Euclidean algorithm for the Gaussian integers described in the text to find a greatest common divisor of $\alpha = 44 + 18i$ and $\beta = 12 - 16i$.

    b) Use the steps in part (a) to find Gaussian integers μ and ν such that $\mu(44 + 18i) + \nu(12 - 16)$ equals the greatest common divisor found in part (a).

12. a) Show all steps of the version of the Euclidean algorithm for the Gaussian integers described in the text to show that $2 - 11i$ and $7 + 8i$ are relatively prime.

    b) Use the steps in part (a) to find Gaussian integers μ and ν such that $\mu(2 - 11i) + \nu(7 + 8i) = 1$.

13. Show that two consecutive Gaussian Fibonacci numbers G_k and G_{k+1} (defined in the preamble to Exercise 37 of Section 14.1), where k is a positive integer, are relatively prime Gaussian integers.

14. How many divisions are used to find a greatest common divisor of two consecutive Gaussian Fibonacci numbers G_k and G_{k+1} (defined in Exercise 37 of Section 14.1), where k is a positive integer? Justify your answer.

15. Derive a big-O estimate for the number of bit operations required to find a greatest common divisor of two nonzero Gaussian integers α and β, where $N(\alpha) \leq N(\beta)$. (*Hint:* Use the remark following the proof of Theorem 14.6.)

16. For each of these Gaussian integers, find its factorization into Gaussian primes and a unit where each Gaussian prime has a positive real part and a nonnegative imaginary part.

    a) $9 + i$ b) 4 c) $22 + 7i$ d) $210 + 2100i$

17. For each of these Gaussian integers, find its factorization into Gaussian primes and a unit where each Gaussian prime has a positive real part and a nonnegative imaginary part.

    a) $7 + 6i$ b) $3 - 13i$ c) 28 d) $400i$

18. Find the factorization into Gaussian primes of each of the Gaussian integers $k + (7 - k)i$ for $k = 1, 2, 3, 4, 5, 6, 7$, where each Gaussian prime has a positive real part and a nonnegative imaginary part.

19. Determine the number of different Gaussian integers, counting associates separately, that divide

    a) 10.   c) 27000.
    b) $256 + 128i$.   d) $5040 + 40320i$.

20. Determine the number of different Gaussian integers, counting associates separately, that divide

a) 198.   b) $128 + 256i$.   c) 169000.   d) $4004 + 8008i$.

21. Suppose that $a + ib$ is a Gaussian integer and n is a rational integer. Show that n and $a + ib$ are relatively prime if and only if n and $b + ai$ are relatively prime.

22. Use the unique factorization theorem for Gaussian integers (Theorem 14.10) and Exercise 13 in Section 10.1 to show that every nonzero Gaussian integer can be written uniquely, except for the order of terms, as $\epsilon \pi_1^{e_1} \pi_2^{e_2} \cdots \pi_k^{e_k}$, where ϵ is a unit and for $j = 1, 2, \ldots, k$, $\pi_j = a_j + ib_j$ is a Gaussian prime with $a_j > 0$ and $b_j \geq 0$, and e_j is a positive integer.

23. Adapt Euclid's proof that there are infinitely many primes (Theorem 3.1) to show that there are infinitely many Gaussian primes.

Exercises 24–41 rely on the notion of a congruence for Gaussian integers defined in the preamble to Exercise 33 in Section 14.1.

24. a) Define what it means for β to be an inverse of the α modulo μ, where α, β, and μ are Gaussian integers.

b) Show that if α and μ are relatively prime Gaussian integers, then there exists a Gaussian integer β which is an inverse of α modulo μ.

25. Find an inverse of $1 + 2i$ modulo $2 + 3i$.

26. Find an inverse of 4 modulo $5 + 2i$.

27. Explain how a linear congruence of the form $\alpha x \equiv \beta \pmod{\mu}$ can be solved, where α, β, and μ are Gaussian integers and α and μ are relatively prime.

28. Solve each of these linear congruences in Gaussian integers.

a) $(2 + i)x \equiv 3 \pmod{4 - i}$ c) $2x \equiv 5 \pmod{3 - 2i}$

b) $4x \equiv -3 + 4i \pmod{5 + 2i}$

29. Solve each of these linear congruences in Gaussian integers.

a) $3x \equiv 2 + i \pmod{13}$ c) $(3 + i)x \equiv 4 \pmod{2 + 3i}$

b) $5x \equiv 3 - 2i \pmod{4 + i}$

30. Solve each of these linear congruences in Gaussian integers.

a) $5x \equiv 2 - 3i \pmod{11}$ c) $(2 + 5i)x \equiv 3 \pmod{4 - 7i}$

b) $4x \equiv 7 + i \pmod{3 + 2i}$

31. Develop and prove a version of the Chinese remainder theorem for systems of congruences for Gaussian integers.

32. Find the simultaneous solutions in Gaussian integers of the system of congruences
$$x \equiv 2 \pmod{2 + 3i}$$
$$x \equiv 3 \pmod{1 + 4i}.$$

33. Find the simultaneous solutions in Gaussian integers of the system of congruences
$$x \equiv 1 + 3i \pmod{2 + 5i}$$
$$x \equiv 2 - i \pmod{3 - 4i}.$$

34. Find a Gaussian integer congruent to 1 modulo 11, to 2 modulo $4 + 3i$, and to 3 modulo $1 + 7i$.

A *complete residue system* modulo γ, where γ is a Gaussian integer, is a set of Gaussian integers such that every Gaussian integer is congruent modulo γ to exactly one element of this set.

35. Find a complete residue system modulo

    a) $1 - i$.   b) 2.   c) $2 + 3i$.

36. Find a complete residue system modulo

    a) $1 + 2i$.   b) 3.   c) $4 - i$.

37. Prove that a complete residue system of α, where α is a Gaussian integer, has $N(\alpha)$ elements.

A *reduced residue system* modulo γ, where γ is a Gaussian integer, is a set of Gaussian integers such that every Gaussian integer that is relatively prime to γ is congruent to exactly one element of this set.

38. Find a reduced residue system modulo

    a) $-1 + 3i$.   b) 2.   c) $5 - i$.

39. Find a reduced residue system modulo

    a) $2 + 2i$.   b) 4.   c) $4 + 2i$.

40. Suppose that π is a Gaussian prime. Determine the number of elements in a reduced residue system modulo π.

41. Suppose that π is a Gaussian prime. Determine the number of elements in a reduced residue system modulo π^e, where e is a positive integer.

42. a) Show that the algebraic integers of the form $r + s\sqrt{-3}$, where r and s are rational numbers, are the numbers of the form $a + b\omega$, where a and b are integers and where $\omega = (-1 + \sqrt{-3})/2$. Numbers of this form are called *Eisenstein integers* after Max Eisenstein who studied them in the mid-nineteenth century. (They are also sometimes called *Eisenstein-Jacobi integers* because they were also studied by Carl Jacobi.) The set of Eisenstein integers is denoted by $Z[\omega]$.

    b) Show that the sum, difference, and product of two Eisenstein integers is also an Eisenstein integer.

    c) Show that if α is an Eisenstein integer, then $\overline{\alpha}$, the complex conjugate of α, is also an Eisenstein integer. (*Hint*: First show that $\overline{\omega} = \omega^2$.)

    d) If α is an Eisenstein integer, we define the *norm* of this integer by $N(\alpha) = a^2 - ab + b^2$ if $\alpha = a + b\omega$, where a and b are integers. Show that $N(\alpha) = \alpha\overline{\alpha}$ whenever α is an Eisenstein integer.

    e) If α and β are Eisenstein integers, we say that α divides β if there exists an element γ in $Z[\omega]$ such that $\beta = \alpha\gamma$. Determine whether $1 + 2\omega$ divides $1 + 5\omega$ and whether $3 + \omega$ divides $9 + 8\omega$.

    f) An Eisenstein integer ϵ is a *unit* if ϵ divides 1. Find all the Eisenstein integers that are units.

g) An *Eisenstein prime* π in $Z[\omega]$ is an element divisible only by a unit or an associate of π. (An associate of an Eisenstein integer is the product of that integer and a unit.) Determine whether each of the following elements are Eisenstein primes: $1 + 2\omega$, $3 - 2\omega$, $5 + 4\omega$, and $-7 - 2w$.

*h) Show that if α and $\beta \neq 0$ belong to $Z[\omega]$, there are numbers γ and ρ such that $\alpha = \beta\gamma + \rho$ and $N(\rho) < N(\beta)$. That is, establish a version of the division algorithm for the Eisenstein integers.

i) Using part (h), show that Eisenstein integers can be uniquely written as the product of Eisenstein primes, with the appropriate considerations about associated primes taken into account.

j) Find the factorization into Eisenstein primes of each of the following Eisenstein integers: 6, $5 + 9\omega$, 114, $37 + 74\omega$.

43. a) Show that the algebraic integers of the form $r + s\sqrt{-5}$, where r and s are rational numbers, are the numbers of the form $a + b\sqrt{-5}$, where a and b are rational integers. (Recall that we briefly studied such numbers in Chapter 3. In this exercise, we look at these numbers in more detail.)

b) Show that the sum, difference, and product of numbers of the form $a + b\sqrt{-5}$, where a and b are rational integers, is again of this form.

c) We denote the set of numbers $a + b\sqrt{-5}$ by $Z[\sqrt{-5}]$. Suppose that α and β belong to $Z[\sqrt{-5}]$. We say that α *divides* β if there exists a number γ in $Z[\sqrt{-5}]$ such that $\beta = \alpha\gamma$. Determine whether $-9 + 11\sqrt{-5}$ is divisible by $2 + 3\sqrt{-5}$ and whether $8 + 13\sqrt{-5}$ is divisible by $1 + 4\sqrt{-5}$.

d) We define the *norm* of a number $\alpha = a + b\sqrt{-5}$ to be $N(\alpha) = a^2 + 5b^2$. Show that $N(\alpha\beta) = N(\alpha)N(\beta)$ whenever α and β belong to $Z[\sqrt{-5}]$.

e) We say ϵ is a *unit* of $Z[\sqrt{-5}]$ if ϵ divides 1. Show that the units in $Z[\sqrt{-5}]$ are 1 and -1.

f) We say that an element α in $Z[\sqrt{-5}]$ is *prime* if its only divisors in $Z(\sqrt{-5})$ are 1, -1, α, and $-\alpha$. Show that 2, 3, $1 + \sqrt{-5}$, and $1 - \sqrt{-5}$ are all primes, that 2 does not divide either $1 + \sqrt{-5}$ or $1 - \sqrt{-5}$. Conclude that $6 = 2 \cdot 3 = (1 + \sqrt{-5})(1 - \sqrt{-5})$ can be written as the product of primes in two different ways. This means that $Z[\sqrt{-5}]$ does not have unique factorization into primes.

g) Show that there do not exist elements γ and ρ in $Z[\sqrt{-5}]$ such that $7 - 2\sqrt{-5} = (1 + \sqrt{-5})\gamma + \rho$, where $N(\rho) < N(1 + \sqrt{-5}) = 6$. Conclude that there is no analog for the division algorithm in $Z[\sqrt{-5}]$.

h) Show that if $\alpha = 3$ and $\beta = 1 + \sqrt{-5}$, there do not exist numbers μ and ν in $Z[\sqrt{-5}]$ such that $\alpha\mu + \beta\nu = 1$, even though α and β are both primes, neither of which divides the other.

14.2 Computational and Programming Exercises

Computations and Explorations

Using a computation program such as Maple or *Mathematica*, or programs you have written, carry out the following computations and explorations.

1. Find the unique factorization into a unit and a product of Gaussian primes, where each Gaussian prime has a positive real part and a nonnegative imaginary part of $(2007 - k) + (2008 - k)i$ for all positive integers k with $k \leq 8$.

2. Find a prime factor of smallest norm of each of the Gaussian integers formed by adding 1 to the product of all Gaussian primes with norm less than n for as many n as possible. Do you think that infinitely many of these numbers are Gaussian primes?

3. Determine whether two randomly selected Gaussian integers are relatively prime, and by doing this repeatedly, estimate the probability that two randomly selected Gaussian integers are relatively prime.

Programming Projects

Write programs using Maple, *Mathematica*, or a language of your choice to do the following.

1. Find a greatest common divisor of two Gaussian integers using a version of the Euclidean algorithm for Gaussian integers.

2. Express a greatest common divisor of two Gaussian integers as a linear combination of these Gaussian integers.

3. Keep track of the number of steps used by the version of the Euclidean algorithm for Gaussian integers that uses the construction in the proof of the division algorithm for Gaussian integers to find quotients and remainders.

4. Find the unique factorization of a Gaussian integer into a unit times Gaussian primes, where each Gaussian prime in the factorization is in the first quadrant.

14.3 Gaussian Integers and Sums of Squares

In Section 13.3 we determined which positive integers are the sum of two squares. In this section we will show that we can prove this result using what we have learned about Gaussian primes. We will also be able to determine the number of different ways that a positive integer can be written as the sum of two squares using Gaussian primes.

In Section 13.3 we proved that every prime of the form $4k + 1$ is the sum of two squares. We can prove this fact in a different way using Gaussian primes.

Theorem 14.11.   If p is a rational prime of the form $4k + 1$, where k is a positive integer, then p is the sum of two squares.

Proof.   Suppose that p is of the form $4k + 1$, where k is a positive integer. To prove that p can be written as the sum of two squares, we show that p is not a Gaussian prime. By Theorem 11.5, we know that -1 is a quadratic residue of p. Consequently, we know that there is a rational integer t such that $t^2 \equiv -1 \pmod{p}$. It follows that $p \mid (t^2 + 1)$. We can use this divisibility relation for rational integers to conclude that $p \mid (t + i)(t - i)$. If p is a Gaussian prime, then by Lemma 14.1, it follows that $p \mid t + i$ or $p \mid t - i$. Both of these cases are impossible because the Gaussian integers divisible by p have the form $p(a + bi) = pa + pbi$, where a and b are rational integers. Neither $t + i$ nor $t - i$ has this form. We can conclude that p is not a Gaussian prime.

Because p is not a Gaussian prime, there are Gaussian integers α and β, neither a unit, such that $p = \alpha\beta$. Taking norms of both sides of this equation, we find that

$$N(p) = p^2 = N(\alpha\beta) = N(\alpha)N(\beta).$$

Because neither α nor β is a unit, $N(\alpha) \neq 1$ and $N(\beta) \neq 1$. This implies that $N(\alpha) = N(\beta) = p$. Consequently, if $\alpha = a + bi$ and $\beta = c + di$, we know that

$$p = N(\alpha) = a^2 + b^2 \quad \text{and} \quad p = N(\beta) = c^2 + d^2.$$

It follows that p is the sum of two squares. ∎

To find which rational integers are the sum of two squares, we will need to determine which rational integers are Gaussian primes and which factor into Gaussian primes. To accomplish that task, we will need the following lemma.

Lemma 14.3. If π is a Gaussian prime, then there is exactly one rational prime p such that π divides p.

Proof. We first factor the rational integer $N(\pi)$ into prime factors, say $N(\pi) = p_1 p_2 \cdots p_t$, where p_j is prime for $j = 1, 2, \ldots, t$. Because $N(\pi) = \pi\overline{\pi}$, it follows that $\pi \mid N(\pi)$, so that $\pi \mid p_1 p_2 \cdots p_t$. By Lemma 14.2, it follows that $\pi \mid p_j$ for some integer j with $1 \leq j \leq t$. We have shown that π divides a rational prime.

To complete the proof, we must show that π cannot divide two different rational primes. So suppose that $\pi \mid p_1$ and $\pi \mid p_2$, where p_1 and p_2 are different rational primes. Because p_1 and p_2 are relatively prime, by Corollary 3.8.1, there are rational integers m and n such that $mp_1 + np_2 = 1$. Moreover, because $\pi \mid p_1$ and $\pi \mid p_2$ we see that $\pi \mid 1$ (using the divisibility property in Exercise 8 of Section 14.1.) But this implies that π is a unit, which is impossible, so π does not divide two different rational primes. ∎

We can now determine which rational primes are also Gaussian primes and the factorization into Gaussian primes of those that are not.

Theorem 14.12. If p is a rational prime, then p factors as a Gaussian integer according to these rules.

(i) If $p = 2$, then $p = -i(1+i)^2 = i(1-i)^2$, where $1+i$ and $1-i$ are both Gaussian primes with norm 2.

(ii) If $p \equiv 3 \pmod 4$, then $p = \pi$ is a Gaussian prime with $N(\pi) = p^2$.

(iii) If $p \equiv 1 \pmod 4$, then $p = \pi\pi'$, where π and π' are Gaussian primes which are not associates with $N(\pi) = N(\pi') = p$.

Proof. To prove (i), we note that $2 = -i(1+i)^2 = i(1-i)^2$, where the factors $-i$ and i are units. Furthermore, $N(1+i) = N(1-i) = 1^2 + 1^2 = 2$. Since $N(1+i) = N(1-i)$ is a rational prime by Theorem 14.3, it follows that $1+i$ and $1-i$ are Gaussian primes.

To prove (ii), let p be a rational prime with $p \equiv 3 \pmod 4$. Suppose that $p = \alpha\beta$, where α and β are Gaussian integers with $\alpha = a + bi$ and $\beta = c + di$ and neither α nor β is a unit. By part (ii) of Theorem 14.1, it follows that $N(p) = N(\alpha\beta) = N(\alpha)N(\beta)$. Because $N(p) = p^2$, $N(\alpha) = a^2 + b^2$, and $N(\beta) = c^2 + d^2$, we see that $p^2 = (a^2 + b^2)(c^2 + d^2)$. Neither α nor β is a unit, so neither has norm 1. It follows that $N(\alpha) = a^2 + b^2 = p$ and $N(\beta) = c^2 + d^2 = p$. However, this is impossible because $p \equiv 3 \pmod 4$, so that p is not the sum of two squares.

To prove (iii), let p be a rational prime with $p \equiv 1 \pmod 4$. By Theorem 14.11, there are integers a and b such that $p = a^2 + b^2$. If $\pi_1 = a - bi$ and $\pi_2 = a + bi$, then $p^2 = N(p) = N(\pi_1)N(\pi_2)$, so that $N(\pi_1) = N(\pi_2) = p$. It follows by Theorem 14.5 that π_1 and π_2 are Gaussian primes.

Next, we show that π_1 and π_2 are not associates. Suppose that $\pi_1 = \epsilon \pi_2$, where ϵ is a unit. Because ϵ is a unit, $\epsilon = 1, -1, i$, or $-i$.

If $\epsilon = 1$, then $\pi_1 = \pi_2$. This means that $x + yi = x - yi$, so that $y = 0$. This implies that $p = x^2 + y^2 = x^2$, which is impossible because p is prime. Similarly, when $\epsilon = -1$, then $\pi_1 = -\pi_2$. This implies that $x + yi = -x + yi$, which makes $x = 0$. This implies that $y^2 = p$, which is also impossible. If $\epsilon = i$, then $x + iy = i(x - iy) = y + ix$, so that $x = y$. Similarly, if $\epsilon = -i$, then $x + iy = -i(x - iy)$, so that $x = -y$. In both of these cases, $p = x^2 + y^2 = 2x^2$, which is impossible because p is an odd prime. We have shown that all four possible values of ϵ are impossible. It follows that π_1 and π_2 are not associates, completing the proof of (iii). ∎

We have all the ingredients we need to determine the number of representations of a positive integer as the sum of two squares using the unique factorization theorem for the Gaussian integers. Recall that we determined which positive integers can be written as the sum of two squares in Section 13.6.

Theorem 14.13.   Suppose that n is a positive integer with prime power factorization

$$n = 2^m p_1^{e_1} p_2^{e_2} \cdots p_s^{e_s} q_1^{f_1} q_2^{f_2} \cdots q_t^{f_t},$$

where m is a nonnegative integer, $p_1, p_2, \ldots, p_s$ are primes of the form $4k + 1$, q_1, $q_2, \ldots, q_t$ are primes of the form $4k + 3$, $e_1, e_2 \ldots, e_s$ are nonnegative integers, and $f_1, f_2, \ldots, f_t$ are even nonnegative integers. Then there are

$$4(e_1 + 1)(e_2 + 1) \cdots (e_s + 1)$$

ways to express n as the sum of two squares. (Here the order in which squares appear in the sum and the sign of the integer being squared both matter.)

Proof.   To count the number of ways to write n as the sum of the squares, that is, the number of solutions of $n = a^2 + b^2$, we can count the number of ways to factor n into Gaussian integers $a + ib$ and $a - ib$, that is, to write $n = (u + iv)(u - iv)$.

We will use the factorization of n to count the number of ways we can factor n as the product of two conjugates, that is, $n = (u + iv)(u - iv)$. First, note that by Theorem 14.11, for each prime p_k of the form $4k + 1$ that divides n, there are integers a_k and b_k such that $p_k = a_k^2 + b_k^2$. Also, note that because $1 + i = i(1 - i)$, we have $2^m = (1 + i)^m(1 - i)^m = (i(1 - i))^m(1 - i)^m = i^m(1 - i)^{2m}$.

Consequently, we have

$$n = i^m(1-i)^{2m}(a_1+b_1i)^{e_1}(a_1-b_1i)^{e_1}(a_2+b_2i)^{e_2}(a_2-b_2i)^{e_2}$$
$$\cdots(a_s-b_si)^{e_s}(a_s+b_si)^{e_s}q_1{}^{f_1}q_2{}^{f_2}\cdots q_t{}^{f_t}.$$

Next, note that $\epsilon = i^m$ is a unit because it takes on one of the values $1, -1, i,$ or $-i$. This means that a factorization of n into the product of a unit and Gaussian primes is

$$n = \epsilon(1-i)^{2m}(a_1+b_1i)^{e_1}(a_1-b_1i)^{e_1}(a_2+b_2i)^{e_2}(a_2-b_2i)^{e_2}$$
$$\cdots(a_s-b_si)^{e_s}(a_s+b_si)^{e_s}q_1{}^{f_1}q_2{}^{f_2}\cdots q_t{}^{f_t}.$$

Because the Gaussian integer $u+iv$ divides n, its factorization into a unit and Gaussian primes must have the form

$$u+iv = \epsilon_0(1-i)^w(a_1+b_1i)^{g_1}(a_1-b_1i)^{h_1}(a_2+b_2i)^{g_2}(a_2-b_2i)^{h_2}$$
$$\cdots(a_t-b_si)^{g_s}(a_s-b_si)^{h_s}q_1{}^{k_1}q_2{}^{k_2}\cdots q_t{}^{k_t},$$

where ϵ_0 is a unit, $w, g_1, \ldots, g_s, h_1, \ldots, h_s,$ and $k_1, \ldots, k_t$ are nonnegative integers with $0 \le w \le 2m, 0 \le g_i \le e_i, 0 \le h_i \le e_i$ for $i = 1, \ldots, s,$ and $0 \le k_j \le f_j$ for $j = 1, \ldots, t$.

Forming the conjugate of $u+iv$, we find

$$u-iv = \overline{\epsilon_0}(1+i)^w(a_1-b_1i)^{g_1}(a_1+b_1i)^{h_1}(a_2-b_2i)^{g_2}(a_2+b_2i)^{h_2}$$
$$\cdots(a_s-b_si)^{g_s}(a_s+b_si)^{h_s}q_1{}^{k_1}q_2{}^{k_2}\cdots q_t{}^{k_t}.$$

We can now rewrite the equation $n = (u+iv)(u-iv)$ as

$$n = 2^w p_1{}^{g_1+h_1}\cdots p_s{}^{g_s+h_s}q_1{}^{2k_1}\cdots q_t{}^{2k_t}.$$

Comparing this with the factorization of n into a unit and Gaussian primes, we see that $w = m$, $g_i + h_i = e_j$ for $i = 1, \ldots, s,$ and $2k_j = f_j$ for $j = 1, \ldots, t$. We see that the values of w and k_j for $j = 1, \ldots, t$ are determined, but we have $e_i + 1$ choices for g_i, namely $g_i = 0, 1, 2, \ldots, e_i,$ and that once g_i is determined, so is $h_i = e_i - g_i$. Furthermore, we have four choices for the unit ϵ_0. We conclude that there are $4(e_1 + 1)(e_2 + 1) \cdots (e_s + 1)$ choices for the factor $u+iv$ and for the number of ways to write n as the sum of two squares. $\blacksquare$

Example 14.11. Suppose that $n = 25 = 5^2$. Then by Theorem 14.13, there are $4 \cdot 3 = 12$ ways to write 25 as the sum of two squares. (These are $(\pm 3)^2 + (\pm 4)^2, (\pm 4)^2 + (\pm 3)^2,$ $(\pm 5)^2 + 0^2,$ and $0 + (\pm 5)^2$. Note that the order in which terms appear matters when we count these representations.)

Suppose that $n = 90 = 2 \cdot 5 \cdot 3^2$. Then by Theorem 14.13, there are $4 \cdot 2 = 8$ ways to write 90 as the sum of two squares. (These are $(\pm 3)^2 + (\pm 9)^2$ and $(\pm 9)^2 + (\pm 3)^2$. Note that the order in which terms appear matters when we count these representations.)

Let $n = 16{,}200 = 2^3 \cdot 5^2 \cdot 3^4$. By Theorem 14.13, there are $4 \cdot 3 = 12$ ways to write 16,200 as the sum of two squares. We leave it to the reader to find these representations. ◄

Conclusion

In this section we used the Gaussian integers to study the solutions of the diophantine equation $x^2 + y^2 = n$, where n is a positive integer. The Gaussian integers are useful in studying a variety of other types of diophantine equations. For example, we can find Pythagorean triples using the Gaussian integers (Exercise 7), and we can find the solutions in rational integers of the diophantine equation $x^2 + y^2 = z^3$ (Exercise 8).

14.3 Exercises

1. Determine the number of ways to write each of the following rational integers as the sum of squares of two rational integers.

   a) 5   b) 20   c) 120   d) 1000

2. Determine the number of ways to write each of the following rational integers as the sum of squares of two rational integers.

   a) 16   b) 99   c) 650   d) 1001000

3. Explain how to solve a linear diophantine equation of the form $\alpha x + \beta y = \gamma$, where α, β, and γ are Gaussian integers so that the solution (x, y) is a pair of Gaussian integers.

4. Find all solutions in Gaussian integers of each of these linear diophantine equations.

   a) $(3 + 2i)x + 5y = 7i$ b) $5x + (2 - i)y = 3$

5. Find all solutions in Gaussian integers of each of the following linear diophantine equations.

   a) $(3 + 4i)x + (3 - i)y = 7i$ b) $(7 + i)x + (7 - i)y = 1$

6. In this exercise we will use the Gaussian integers to find the solutions in rational integers of the diophantine equation $x^2 + 1 = y^3$.
   a) Show that if x and y are integers such that $x^2 + 1 = y^3$, then $x - i$ and $x + i$ are relatively prime.
   b) Show that there are integers r and s such that $x = r^3 - 3rs^2$ and $3r^2s - s^3 = 1$. (*Hint:* Use part (a) and Exercise 10 in Section 14.2 to show that there is a unit ϵ and a Gaussian integer δ such that $x + i = (\epsilon\delta)^3$.)
   c) Find all solutions in integers $x^2 + 1 = y^3$ by analyzing the equations for r and s in part (b).

7. Use the Gaussian integers to prove Theorem 13.1 in Section 13.1, which gives primitive Pythagorean triples, that is, solutions of the equation $x^2 + y^2 = z^2$ in integers x, y, and z, where x, y, and z are pairwise relatively prime. (*Hint:* Begin with the factorization $x^2 + y^2 = (x + iy)(x - iy)$. Show that $x + iy$ and $x - iy$ are relatively prime Gaussian integers and then use Exercise 10 in Section 14.1.)

* 8. Use the Gaussian integers to find all solutions of the diophantine equation $x^2 + y^2 = z^3$ in rational integers x, y, and z.

* 9. Prove the analog of Fermat's little theorem for the Gaussian integers, which states that if α and π are relatively prime, then $\alpha^{N(\pi)-1} \equiv 1 \pmod{\pi}$. (*Hint:* Suppose that p is the

unique rational prime with $\pi \mid p$. Consider separately the cases where $p \equiv 1 \pmod 4$, $p \equiv 2 \pmod 4$, and $p \equiv 3 \pmod 4$.

10. Define $\phi(\gamma)$, where γ is a Gaussian integer, to be the number of elements in a reduced residue system modulo γ. Prove the analog of Euler's theorem for the Gaussian integers, which states that if γ is a Gaussian integer and α is a Gaussian integer that is relatively prime to γ, then

$$\alpha^{\phi(\gamma)} \equiv 1 \pmod \gamma.$$

11. Prove the analog of Wilson's theorem for the Gaussian integers, which states that if π is a Gaussian prime and $\{\alpha_1, \alpha_2, \ldots, \alpha_r\}$ is a reduced system of residues modulo π, then

$$\alpha_1 \alpha_2 \cdots \alpha_r \equiv -1 \pmod \pi.$$

12. Show that in the Eisenstein integers (defined in Exercise 42 in Section 14.2)
    a) the rational prime 2 is an Eisenstein prime.
    b) a rational prime of the form $3k + 2$, where k is a positive integer, is an Eisenstein prime.
    c) a rational prime of the form $3k + 1$, where k is a positive integer, factors into the product of two primes that are not associates of one another.

14.3 Computational and Programming Exercises

Computations and Explorations

Using a computation program such as Maple or *Mathematica,* or programs you have written, carry out the following computations and explorations.

1. In Chapter 13 we mentioned that Catalan's conjecture has been settled, showing that 2^3 and 3^2 are the only powers of rational integers that differ by 1. An open question for Gaussian integers is to find all powers of Gaussian integers that differ by a unit. Show that $(11 + 11i)^2$ and $(3i)^5$, $(1 - i)^5$ and $(1 + 2i)^2$, and $(78 + 78i)^2$ and $(23i)^3$ are such pairs of powers. Can you find other such pairs?

2. Show that $(3 + 13i)^3 + (7 + i)^3 = (3 + 10i)^3 + (1 + 10i)^3$, $(6 + 3i)^4 + (2 + 6i)^4 = (4 + 2i)^4 + (2 + i)^4$, $(2 + 3i)^5 + (2 - 3i)^5 = 3^5 + 1$, $(1 + 6i)^5 + (3 - 2i)^5 = (6 + i)^5 + (-2 + 3i)^5$, $(9 + 6i)^5 + (3 - 10i)^5 = (6 + i)^5 + (6 - 5i)^5$, and $(15 + 14i)^5 + (5 - 18i)^5 = (18 - 7i)^5 + (2 + 3i)^5$. Can you find other solutions of the equation $x^n + y^n = w^n + z^n$, where x, y, z, and w are Gaussian integers and n is a positive integer?

3. Show that Beal's conjecture, which asserts that there are no nontrivial solutions of the diophantine equation $x^a + y^b = z^c$, where a, b, and c are integers with $a \geq 3$, $b \geq 3$, and $c \geq 3$, does not hold when x, y, and z are allowed to be pairwise relatively prime Gaussian integers by showing that $(-2 + i)^3 + (-2 - i)^3 = (1 + i)^4$. Can you find other counterexamples?

Programming Projects

Write programs using Maple, *Mathematica,* or a language of your choice to do the following.

1. Find the number of ways to write a positive integer n as the sum of two squares.

2. Find all representations of a positive integer n as the sum of two squares.

A

Axioms for the Set of Integers

In this appendix, we state a collection of fundamental properties for the set of *integers* $\{\ldots, -2, -1, 0, 1, 2, \ldots\}$ that we have taken as axioms in the main body of the text. These properties provide the foundations for proving results in number theory. We begin with properties dealing with addition and multiplication. As usual, we denote the sum and product of a and b by $a + b$ and $a \cdot b$, respectively. Following convention, we write ab for $a \cdot b$.

- *Closure:* $a + b$ and $a \cdot b$ are integers whenever a and b are integers.
- *Commutative laws:* $a + b = b + a$ and $a \cdot b = b \cdot a$ for all integers a and b.
- *Associative laws:* $(a + b) + c = a + (b + c)$ and $(a \cdot b) \cdot c = a \cdot (b \cdot c)$ for all integers a, b, and c.
- *Distributive law:* $(a + b) \cdot c = a \cdot c + b \cdot c$ for all integers a, b, and c.
- *Identity elements:* $a + 0 = a$ and $a \cdot 1 = a$ for all integers a.
- *Additive inverse:* For every integer a there is an integer solution x to the equation $a + x = 0$; this integer x is called the *additive inverse* of a and is denoted by $-a$. By $b - a$ we mean $b + (-a)$.
- *Cancellation law:* If a, b, and c are integers with $a \cdot c = b \cdot c, c \neq 0$, then $a = b$.

We can use these axioms and the usual properties of equality to establish additional properties of integers. An example illustrating how this is done follows. In the main body of the text, results that are easily proved from these axioms are used without comment.

Example A.1. To show that $0 \cdot a = 0$, begin with the equation $0 + 0 = 0$; this holds because 0 is an identity element for addition. Next, multiply both sides by a to obtain $(0 + 0) \cdot a = 0 \cdot a$. By the distributive law, the left-hand side of this equation equals $(0 + 0) \cdot a = 0 \cdot a + 0 \cdot a$. Hence, $0 \cdot a + 0 \cdot a = 0 \cdot a$. Next subtract $0 \cdot a$ from both sides (which is the same as adding the inverse of $0 \cdot a$). Using the associative law for addition and the fact that 0 is an additive identity element, the left-hand side becomes $0 \cdot a + (0 \cdot a - 0 \cdot a) = 0 \cdot a + 0 = 0 \cdot a$. The right-hand side becomes $0 \cdot a - 0 \cdot a = 0$. We conclude that $0 \cdot a = 0$. ◄

Ordering of integers is defined using the set of *positive integers* $\{1, 2, 3, \ldots\}$. We have the following definition.

Definition.   If a and b are integers, then $a < b$ if $b - a$ is a positive integer. If $a < b$, we also write $b > a$.

Note that a is a positive integer if and only if $a > 0$.

The fundamental properties of ordering of integers follow.

- *Closure for the Positive Integers:* $a + b$ and $a \cdot b$ are positive integers whenever a and b are positive integers.
- *Trichotomy law:* For every integer a, exactly one of the statements $a > 0$, $a = 0$, and $a < 0$ is true.

The set of integers is said to be an *ordered set* because it has a subset that is closed under addition and multiplication and because the trichotomy law holds for every integer.

Basic properties of ordering of integers can now be proved using our axioms, as the following example shows. Throughout the text we have used without proof properties of ordering that easily follow from our axioms.

Example A.2.   Suppose that a, b, and c are integers with $a < b$ and $c > 0$. We can show that $ac < bc$. First, note that by the definition of $a < b$ we have $b - a > 0$. Because the set of positive integers is closed under multiplication, $c(b - a) > 0$. Because $c(b - a) = cb - ca$, it follows that $ca < cb$. ◄

We need one more property to complete our set of axioms.

- *The Well-Ordering Property:* Every nonempty set of positive integers has a least element.

We say that the set of positive integers is *well ordered.* On the other hand, the set of all integers is not well ordered, because there are sets of integers that do not have a smallest element (as the reader should verify). Note that the principle of mathematical induction discussed in Section 1.3 is a consequence of the set of axioms listed in this appendix. Sometimes, the principle of mathematical induction is taken as an axiom replacing the well-ordering property. When this is done, the well-ordering property follows as a consequence.

Exercises

1. Use the axioms for the set of integers to prove the following statements for all integers a, b, and c.

   a) $a \cdot (b + c) = a \cdot b + a \cdot c$ c) $a + (b + c) = (c + a) + b$

   b) $(a + b)^2 = a^2 + 2ab + b^2$ d) $(b - a) + (c - b) + (a - c) = 0$

2. Use the axioms for the set of integers to prove the following statements for all integers a and b.

   a) $(-1) \cdot a = -a$ c) $(-a) \cdot (-1) = ab$

   b) $-(a \cdot b) = a \cdot (-b)$ d) $-(a + b) = (-a) + (-b)$

3. What is the value of -0? Give a reason for your answer.

4. Use the axioms for the set of integers to show that if a and b are integers with $ab = 0$, then $a = 0$ or $b = 0$.

5. Show that an integer a is positive if and only if $a > 0$.

6. Use the definition of the ordering of integers, and the properties of the set of positive integers, to prove the following statements for integers a, b, and c with $a < b$ and $c < 0$.

   a) $a + c < b + c$ c) $ac > bc$

   b) $a^2 \geq 0$ d) $c^3 < 0$

7. Show that if a, b, and c are integers with $a > b$ and $b > c$, then $a > c$.

* 8. Show that there is no positive integer that is less than 1.

B

Binomial Coefficients

Sums of two terms are called *binomial expressions.* Powers of binomial expressions are used throughout number theory and throughout mathematics. In this section we will define the *binomial coefficients* and show that these are precisely the coefficients that arise in expansions of powers of binomial expressions.

Definition. Let m and k be nonnegative integers with $k \leq m$. The *binomial coefficient* $\binom{m}{k}$ is defined by

$$\binom{m}{k} = \frac{m!}{k!(m-k)!}.$$

When k and m are positive integers with $k > m$, we define $\binom{m}{k} = 0$.

In computing $\binom{m}{k}$, we see that there is a good deal of cancellation, because

$$\binom{m}{k} = \frac{m!}{k!(m-k)!} = \frac{1 \cdot 2 \cdot 3 \cdots (m-k)(m-k+1) \cdots (m-1)m}{k! \; 1 \cdot 2 \cdot 3 \cdots (m-k)}$$

$$= \frac{(m-k+1) \cdots (m-1)m}{k!}.$$

Example B.1. To evaluate the binomial coefficient $\binom{7}{3}$, we note that

$$\binom{7}{3} = \frac{7!}{3!4!} = \frac{1 \cdot 2 \cdot 3 \cdot 4 \cdot 5 \cdot 6 \cdot 7}{1 \cdot 2 \cdot 3 \cdot 1 \cdot 2 \cdot 3 \cdot 4} = \frac{5 \cdot 6 \cdot 7}{1 \cdot 2 \cdot 3} = 35. \qquad \blacktriangleleft$$

We now prove some simple properties of binomial coefficients.

Theorem B.1. Let n and k be nonnegative integers with $k \leq n$. Then

(i) $\dbinom{n}{0} = \dbinom{n}{n} = 1$, and

(ii) $\dbinom{n}{k} = \dbinom{n}{n-k}$.

Proof. To see that (i) is true, note that

$$\binom{n}{0} = \frac{n!}{0!n!} = \frac{n!}{n!} = 1$$

and

$$\binom{n}{n} = \frac{n!}{n!0!} = \frac{n!}{n!} = 1.$$

To verify (ii), we see that

$$\binom{n}{k} = \frac{n!}{k!(n-k)!} = \frac{n!}{(n-k)!(n-(n-k))!} = \binom{n}{n-k}.$$ ∎

An important property of binomial coefficients is the following identity.

Theorem B.2. *Pascal's Identity.* Let n and k be positive integers with $n \geq k$. Then

$$\binom{n}{k} + \binom{n}{k-1} = \binom{n+1}{k}.$$

Proof. We perform the addition

$$\binom{n}{k} + \binom{n}{k-1} = \frac{n!}{k!(n-k)!} + \frac{n!}{(k-1)!(n-k+1)!}$$

by using the common denominator $k!(n-k+1)!$. This gives

$$\binom{n}{k} + \binom{n}{k-1} = \frac{n!(n-k+1)}{k!(n-k+1)!} + \frac{n!k}{k!(n-k+1)!}$$

$$= \frac{n!((n-k+1)+k)}{k!(n-k+1)!}$$

$$= \frac{n!(n+1)}{k!(n-k+1)!}$$

$$= \frac{(n+1)!}{k!(n-k+1)!}$$

$$= \binom{n+1}{k}$$ ∎

Using Theorem B.2, we can construct *Pascal's triangle*, named after French mathematician *Blaise Pascal* who used the binomial coefficients in his analysis of gambling games. In Pascal's triangle, the binomial coefficient $\binom{n}{k}$ is the $(k+1)$st number in the

$(n + 1)$st row. The first nine rows of Pascal's triangle are displayed in Figure B.1. Pascal's triangle appeared in Indian and Islamic mathematics several hundred years before it was studied by Pascal.

$$
\begin{array}{c}
1 \\
1\ 2\ 1 \\
1\ 3\ 3\ 1 \\
1\ 4\ 6\ 4\ 1 \\
1\ 5\ 10\ 10\ 5\ 1 \\
1\ 6\ 15\ 20\ 15\ 6\ 1 \\
1\ 7\ 21\ 35\ 35\ 21\ 7\ 1 \\
1\ 8\ 28\ 56\ 70\ 56\ 28\ 8\ 1
\end{array}
$$

Figure B.1 *Pascal's triangle.*

We see that the exterior numbers in the triangle are all 1. To find an interior number, we simply add the two numbers in the positions above, and to either side, of the position being filled. From Theorem B.2, this yields the correct integer.

Binomial coefficients occur in the expansion of powers of sums. Exactly how they occur is described by the *binomial theorem.*

Theorem B.3. *The Binomial Theorem.* Let x and y be variable, and n be a positive integer. Then,

$$
(x + y)^n = \binom{n}{0}x^n + \binom{n}{1}x^{n-1}y + \binom{n}{2}x^{n-2}y^2 + \cdots
$$

$$
+ \binom{n}{n-2}x^2y^{n-2} + \binom{n}{n-1}xy^{n-1} + \binom{n}{n}y^n,
$$

or, using summation notation,

BLAISE PASCAL (1623–1662) exhibited his mathematical talents early even though his father, who had made discoveries in analytic geometry, kept mathematical books from him to encourage his other interests. At 16, Pascal discovered an important result concerning conic sections. At 18, he designed a calculating machine, which he had built and successfully sold. Later, Pascal made substantial contributions to hydrostatics. Pascal, together with Fermat, laid the foundations for the modern theory of probability. It was in his work on probability that Pascal made new discoveries concerning what is now called Pascal's triangle, and gave what is considered to be the first lucid description of the principle of mathematical induction. In 1654, catalyzed by an intense religious experience, Pascal abandoned his mathematical and scientific pursuits to devote himself to theology. He returned to mathematics only once: one night, he had insomnia caused by the discomfort of a toothache and, as a distraction, he studied the mathematical properties of the cycloid. Miraculously, his pain subsided, which he took as a signal of divine approval of the study of mathematics.

$$(x + y)^n = \sum_{j=0}^{n} \binom{n}{j} x^{n-j} y^j.$$

Proof. We use mathematical induction. When $n = 1$, according to the binomial theorem, the formula becomes

$$(x + y)^1 = \binom{1}{0} x^1 y^0 + \binom{1}{1} x^0 y^1.$$

But because $\binom{1}{0} = \binom{1}{1} = 1$, this states that $(x + y)^1 = x + y$, which is obviously true.

We now assume that the theorem is true for the positive integer n, that is, we assume that

$$(x + y)^n = \sum_{j=0}^{n} \binom{n}{j} x^{n-j} y^j.$$

We must now verify that the corresponding formula holds with n replaced by $n + 1$, assuming the result holds for n. Hence, we have

$$(x + y)^{n+1} = (x + y)^n (x + y)$$

$$= \left[\sum_{j=0}^{n} \binom{n}{j} x^{n-j} y^j \right] (x + y)$$

$$= \sum_{j=0}^{n} \binom{n}{j} x^{n-j+1} y^j + \sum_{j=0}^{n} \binom{n}{j} x^{n-j} y^{j+1}.$$

We see by removing terms from the sums and subsequently shifting indices, that

$$\sum_{j=0}^{n} \binom{n}{j} x^{n-j+1} y^j = x^{n+1} + \sum_{j=1}^{n} \binom{n}{j} x^{n-j+1} y^j$$

and

$$\sum_{j=0}^{n} \binom{n}{j} x^{n-j} y^{j+1} = \sum_{j=0}^{n-1} \binom{n}{j} x^{n-j} y^{j+1} + y^{n+1}$$

$$= \sum_{j=1}^{n} \binom{n}{j-1} x^{n-j+1} y^j + y^{n+1}.$$

Hence, we find that

$$(x + y)^{n+1} = x^{n+1} + \sum_{j=1}^{n} \left[\binom{n}{j} + \binom{n}{j-1} \right] x^{n-j+1} y^j + y^{n+1}.$$

By Pascal's identity, we have

$$\binom{n}{j} + \binom{n}{j-1} = \binom{n+1}{j},$$

so we conclude that

$$(x + y)^{n+1} = x^{n+1} + \sum_{j=1}^{n} \binom{n+1}{j} x^{n-j+1} y^j + y^{n+1}.$$

$$= \sum_{j=0}^{n+1} \binom{n+1}{j} x^{n+1-j} y^j.$$

This establishes the theorem.   ∎

The binomial theorem shows that the coefficients of $(x + y)^n$ are the numbers in the $(n + 1)$st row of Pascal's triangle.

We now illustrate one use of the binomial theorem.

Corollary B.1.   Let n be a nonnegative integer. Then

$$2^n = (1 + 1)^n = \sum_{j=0}^{n} \binom{n}{j} 1^{n-j} 1^j = \sum_{j=0}^{n} \binom{n}{j}.$$

Proof.   Let $x = 1$ and $y = 1$ in the binomial theorem.   ∎

Corollary B.1 shows that if we add all elements of the $(n + 1)$st row of Pascal's triangle, we get 2^n. For instance, for the fifth row, we find that

$$\binom{4}{0} + \binom{4}{1} + \binom{4}{2} + \binom{4}{3} + \binom{4}{4} = 1 + 4 + 6 + 4 + 1 = 16 = 2^4.$$

Exercises

1. Find the value of each of the following binomial coefficients.

   a) $\binom{100}{0}$ c) $\binom{20}{3}$ e) $\binom{10}{7}$

   b) $\binom{50}{1}$ d) $\binom{11}{5}$ f) $\binom{70}{70}$

2. Find the binomial coefficients $\binom{9}{3}$, $\binom{9}{4}$, and $\binom{10}{4}$, and verify that $\binom{9}{3} + \binom{9}{4} = \binom{10}{4}$.

3. Use the binomial theorem to write out all terms in the expansions of the following expressions.

   a) $(a + b)^5$ c) $(m - n)^7$ e) $(3x - 4y)^5$

   b) $(x + y)^{10}$ d) $(2a + 3b)^4$ f) $(5x + 7)^8$

4. What is the coefficient of $x^{99} y^{101}$ in $(2x + 3y)^{200}$?

5. Let n be a positive integer. Using the binomial theorem to expand $(1 + (-1))^n$, show that

$$\sum_{k=0}^{n} (-1)^k \binom{n}{k} = 0.$$

6. Use Corollary B.1 and Exercise 5 to find

$$\binom{n}{0} + \binom{n}{2} + \binom{n}{4} + \cdots$$

and

$$\binom{n}{1} + \binom{n}{3} + \binom{n}{5} + \cdots.$$

7. Show that if $n, r,$ and k are integers with $0 \le k \le r \le n$, then

$$\binom{n}{r}\binom{r}{k} = \binom{n}{k}\binom{n-k}{r-k}.$$

* **8.** What is the largest value of $\binom{m}{n}$, where m is a positive integer and n is an integer such that $0 \le n \le m$? Justify your answer.

9. Show that

$$\binom{r}{r} + \binom{r+1}{r} + \cdots + \binom{n}{r} = \binom{n+1}{r+1},$$

where n and r are integers with $1 \le r \le n$.

The binomial coefficients $\binom{x}{n}$, where x is a real number and n is a positive integer, can be defined recursively by the equations $\binom{x}{1} = x$ and

$$\binom{x}{n+1} = \frac{x-n}{n+1}\binom{x}{n}.$$

10. Show from the recursive definition that if x is a positive integer, then $\binom{x}{k} = \frac{x!}{k!(x-k)!}$, where k is a integer with $1 \le k \le x$.

11. Show from the recursive definition that if x is a positive integer, then $\binom{x}{n} + \binom{x}{n+1} = \binom{x+1}{n+1}$, whenever n is a positive integer.

12. Show that the binomial coefficient $\binom{n}{k}$, where n and k are integers with $0 \le k \le n$, gives the number of subsets with k elements of a set with n elements.

13. Use Exercise 12 to give an alternate proof of the binomial theorem.

14. Let S be a set with n elements and let P_1 and P_2 be two properties that an element of S may have. Show that the number of elements of S possessing neither property P_1 nor property P_2 is

$$n - [n(P_1) + n(P_2) - n(P_1, P_2)],$$

where $n(P_1), n(P_2),$ and $n(P_1, P_2)$ are the number of elements of S with property P_1, with property P_2, and both properties P_1 and P_2, respectively.

15. Let S be a set with n elements and let P_1, P_2, and P_3 be three properties that an element S may have. Show that the number of elements of S possessing none of the properties $P_1, P_2,$ and P_3 is

$$n - [n(P_1) + n(P_2) + n(P_3)]$$
$$- n(P_1, P_2) - n(P_1, P_3) - n(P_2, P_3) + n(P_1, P_2, P_3)],$$

where $n(P_{i_1}, \ldots, P_{i_k})$ is the number of elements of S with properties $P_{i_1} \ldots, P_{i_k}$.

* **16.** In this exercise we develop the *principle of inclusion-exclusion.* Suppose that S is a set with n elements and let $P_1, P_2, \ldots, P_t$ be t different properties that an element of S may

have. Show that the number of elements of S possessing *none* of the t properties is

$$n - [n(P_1) + n(P_2) + \cdots + n(P_t)]$$
$$+ [n(P_1, P_2) + n(P_1, P_3) + \cdots + n(P_{t-1}, P_t)]$$
$$- [n(P_1, P_2, P_3) + n(P_1, P_2, P_4) + \cdots + n(P_{t-2}, P_{t-1}, P_t)]$$
$$+ \cdots + (-1)^t n(P_1, P_2, \ldots, P_t),$$

where $n(P_{i_1}, P_{i_2}, \ldots, P_{i_j})$ is the number of elements of S possessing all of the properties $P_{i_1}, P_{i_2}, \ldots, P_{i_j}$. The first expression in brackets contains a term for each property, the second expression in brackets contains terms for all combinations of two properties, the third expression contains terms for all combinations of three properties, and so forth. (*Hint:* For each element of S, determine the number of times it is counted in the above expression. If an element has k of the properties, show that it is counted $1 - \binom{k}{1} + \binom{k}{2} - \cdots + (-1)^k \binom{k}{k}$ times; this is 0 when $k > 0$, by Exercise 5.)

* **17.** What are the coefficients of $(x_1 + x_2 + \cdots + x_m)^n$? These coefficients are called *multinomial coefficients*.

18. Write out all terms in the expansion of $(x + y + z)^7$.

19. What is the coefficient of $x^3 y^4 z^5$ in the expansion of $(2x - 3y + 5z)^{12}$?

Computational and Programming Exercises

Computations and Explorations

Using a computation program such as Maple or *Mathematica,* or programs you have written, carry out the following computations and explorations.

1. Find the least integer n such that there is a binomial coefficient $\binom{n}{k}$, where k is a positive integer greater than 1,000,000.

Programming Projects

Write computer programs using Maple, *Mathematica,* or a language of your choice to do the following.

1. Evaluate binomial coefficients.

2. Given a positive integer n, print out the first n rows of Pascal's triangle.

3. Expand $(x + y)^n$, given a positive integer n, using the binomial theorem.

C

Using Maple and *Mathematica* for Number Theory

Investigating questions in number theory often requires computations with large integers. Fortunately, there are many tools available today that can be used for such computations. This appendix describes how two of the most popular of these tools, Maple and *Mathematica*, can be used to perform computations in number theory. We will concentrate on existing commands in these two systems, both of which support extensive programming environments that can be used to create useful programs for studying number theory. We will not describe these programming environments here.

C.1 Using Maple for Number Theory

The Maple system is a comprehensive environment for numerical and symbolic computations. It can also be used to develop additional functionality. We will briefly describe some of the existing support for number theory in Maple. For additional information about Maple, consult the Maple Web site at `http://www.maplesoft.com`.

In Maple, commands for computations in number theory can be found in the `numtheory` package. Some useful commands for number theory are included in the standard set of Maple commands, and a few are found in other packages, such as the `combinat` package of combinatorics commands. You need to let Maple know when you want to use one or more commands from a package. This can be done in two ways: You can either load the package and then use any of its commands, or you can prepend the name of the package to a particular command. For example, after running the command `with(numtheory)`, you can use commands from the `numtheory` package as you would standard commands. You can also run commands from this package by simply prepending the name of the package before the command. You will need to do this every time you use a command from the package, unless you run the `with(numtheory)` command.

Additional Maple commands for number theory can be found in the Maple V Share Library, which can be accessed at `http://www.cybermath.com/share_home.html`.

A useful reference for using Maple to explore number theory (and other topics in discrete mathematics) is *Exploring Discrete Mathematics with Maple* [Ro97]. This book explains how to use Maple to find greatest common divisors and least common multiples, apply the Chinese remainder theorem, factor integers, run primality tests, find base *b* expansions, encrypt and decrypt using classical ciphers and the RSA cryptosystem, and perform other number theoretic computations. Also, Maple worksheets for number theory and cryptography, written by John Cosgrave for a course at St. Patrick's College in Dublin, Ireland, can be found at `http://www.spd.dcu.ie/johnbcos` `/Maple_3rd_year.htm`.

Maple Number Theory Commands

The Maple commands relevant to material in this text are presented according to the chapter in which that material is covered. These commands are useful for checking computations in the text, for working or checking some exercises, and for the computations and explorations at the end of each section. Furthermore, programs in Maple can be written for many of the explorations and programming projects listed at the end of each section. Consult the appropriate Maple reference materials, such as the *Maple V Programming Guide* [Mo96], for information about writing programs in Maple.

Chapter 1

`combinat[fibonacci](n)` computes the nth Fibonacci number.

`iquo($int_1, int_2$)` computes the quotient when int_1 is divided by int_2.

`irem($int_1, int_2$)` computes the remainder when int_1 is divided by int_2.

`floor(`*expr*`)` computes the largest integer less than or equal to the real expression *expr*.

`numtheory[divisors]($n$)` computes the positive divisors of the integer n.

Maple code for investigating the Collatz $3x + 1$ problem has been written by Gaston Gonnet, and is available in the Maple V Release 5 Share Library.

Chapter 2

`convert(`*int*`, base, `*posint*`)` converts the integer *int* in decimal notation to a list representing its digits base *posint*.

`convert(`*int*`,binary)` converts the integer *int* in decimal notation to its binary equivalent.

`convert(`*int*`,hex)` converts the integer *int* in decimal notation to its hexadecimal equivalent.

`convert(`*bin*`,decimal,binary)` converts the integer *bin* in binary notation to its decimal equivalent.

`convert(`*oct*`,decimal,octal)` converts the integer *oct* in octal notation to its decimal equivalent.

`convert`(*hex*,`decimal`,`octal`) converts the integer *hex* in hexadecimal notation to its decimal equivalent.

Chapter 3

`isprime`(*n*) tests whether *n* is prime.

`ithprime`(*n*) calculates the *n*th prime number where *n* is a positive integer.

`prevprime`(*n*) calculates the largest prime smaller than the integer *n*.

`numbertheory[fermat]`(*n*) calculates the *n*th Fermat number.

`ifactor`(*n*) finds the prime-power factorization of an integer *n*.

`ifactors`(*n*) finds the prime integer factors of an integer *n*.

`igcd`($int_1, \ldots, int_n$) computes the greatest common divisor of integers $int_1, \ldots, int_n$.

`igcdex`(int_1, int_2) computes the greatest common divisor of the integers int_1 and int_2 using the extended Euclidean algorithm, which also expresses the greatest common divisor as a linear combination of int_1 and int_2.

`ilcm`($int_1, \ldots, int_n$) computes the least common multiple of the integers $int_1, \ldots, int_n$.

Chapter 4

The operator `mod` can be used in Maple; for example, 17 mod 4 tells Maple to reduce 17 to its least residue modulo 4.

`msolve`(*eqn*,*m*) finds the integer solutions modulo *m* of the equation *eqn*.

`chrem`($[n_1 \ldots, n_r], [m_1, \ldots, m_r]$) computes the unique positive integer *int* such that *int* **mod** $m_i = n_i$ for $i = 1, \ldots, r$.

Chapter 6

`numtheory[phi]`(*n*) computes the value of the Euler phi function at *n*.

Chapter 7

`numtheory[invphi]`(*n*) computes the positive integers *m* with $\phi(m) = n$.

`numtheory[sigma]`(*n*) computes the sum of the positive divisors of the integer *n*.

`numtheory[tau]`(*n*) computes the number of positive divisors of the integer *n*.

`numbertheory[bigomega]`(*n*) computes the value of $\Omega(n)$, the number of prime factors of *n*.

`numtheory[mersenne]`(*n*) determines whether the *n*th Mersenne number $M_n = 2^n - 1$ is prime.

`numtheory[mobius]`(*n*) computes the value of the Möbius function at the integer *n*.

Chapter 9

`numtheory[order]`(n_1, n_2) computes the order of n_1 modulo n_2.

`numtheory[primroot]`(*n*) computes the smallest primitive root modulo *n*.

numtheory[mlog] (n_1, n_2, n_3) computes the index, or discrete logarithmn, of n_1 to the base n_2 modulo n_3. (The function numtheory[index] (n_1, n_2, n_3) is identical to this function.)

numtheory[lambda] (n) computes the minimal universal exponent of n.

Chapter 11

numtheory[quadres] (int_1, int_2) determines whether int_1 is a quadratic residue modulo int_2.

numtheory[legendre] (n_1, n_2) computes the value of the Legendre symbol $\left(\frac{n_1}{n_2}\right)$.

numtheory[jacobi] (n_1, n_2) computes the value of the Jacobi symbol $\left(\frac{n_1}{n_2}\right)$.

numtheory[msqrt] (n_1, n_2) computes the square root of n_1 modulo n_2.

Chapter 12

numtheory[pdexpand] (rat) computes the periodic decimal expansion of the rational number *rat*.

numtheory[cfrac] (rat) computes the continued fraction of the rational number *rat*.

numtheory[invcfrac] (cf) converts a periodic continued fraction *cf* to a quadratic irrational number.

Chapter 13

numtheory[sum2sqr] (n) computes all sums of two squares that sum to n.

Chapter 14

Maple supports a special package for working with Gaussian integers. To use the commands in this package, first run the command

```
with(GaussInt);
```

After running this command you can add, subtract, multiply, and form powers of Gaussian integers using the same operators as you ordinarily do. Maple requires that you enter the Gaussian integer $a + ib$ as a + b*I. (That is, you must include the * operator between b and the letter I, which Maple uses to represent the imaginary number i.)

GaussInt[GInearest] (c) returns the Gaussian integer closest to the complex number c, where the Gaussian integer of smallest norm is chosen in the case of ties.

GaussInt[GIquo] (m, n) finds the Gaussian integer quotient when m is divided by n.

GaussInt[GIrem] (m, n) finds the remainder Gaussian integer divisor when m is divided by n.

GaussInt[GInorm] (m) gives the norm of the complex number m.

GaussInt[GIprime] (m) returns true when m is a Gaussian prime and false otherwise.

GaussInt[GIfactor] (m) returns a factorization of m into a unit and Gaussian primes.

GaussInt[GIfactors] (*m*) finds a unit and Gaussian prime factors and their multiplicities in a factorization of the Gaussian integer *m*.

GaussInt[GIsieve] (*m*), where *m* is a positive integer, generates a list of Gauss primes $a + ib$ with $0 \leq a \leq b$ and norm not exceeding m^2.

GaussInt[GIdivisor] (*m*) finds the set of divisors of the Gaussian integer *m* in the first quadrant.

GaussInt[GInodiv] (*m*) computes the number of nonassociated divisors of *m*.

GaussInt[GIgcd] ($m_1, m_2, \ldots, m_r$) finds the greatest common divisor in the first quadrant of the Gaussian integers $m_1, m_2, \ldots, m_r$.

GaussInt[GIgcdex] ($a, b, 's', 't'$) finds the greatest common divisor in the first quadrant of the Gaussian integers *a* and *b* and finds integers *s* and *t* such that as $as + bt$ equals this greatest common divisor.

GaussInt[GIchrem] ([$a_0, a_1, \ldots, a_r$], [$u_0, u_1, \ldots, u_r$]) computes the unique Gaussian integer *m* such that *m* is congruent to a_i modulo u_i for $i = 1, 2, \ldots, r$.

GaussInt[GIlcm] ($a_1, \ldots, a_r$) finds the least common multiple in the first quadrant (that is, with positive real part and nonnegative part), in terms of norm, of the Gaussian integers $a_1, \ldots, a_r$.

GaussInt[GIphi] (*n*) returns the number of Gaussian integers in a reduced residue set modulo *n*, where *n* is a Gaussian integer.

GaussInt[GIquadres] (a, b) returns 1 if the Gaussian integer *a* is a quadratic residue of the Gaussian integer *b* and -1 if *a* is a quadratic nonresidue of *b*.

Appendices

binomial(n, r) computes the binomial coefficient *n* choose *r*.

C.2 Using *Mathematica* for Number Theory

The *Mathematica* system provides a comprehensive environment for numerical and symbol computations. It can also be used to develop additional functionality. We will describe the existing *Mathematica* support for computations relating to the number theory covered in this text. For additional information on *Mathematica*, consult the *Mathematica* Web site at http://www.mathematica.com.

Mathematica supports many number theory commands as part of its basic system. Additional number theory commands can be found in *Mathematica* packages that are collections of programs implementing functions in particular areas. The *Mathematica* system bundles some add-on packages, called standard packages, with its basic offerings. These standard packages include a group supporting commands for functions from number theory, including ContinuedFractions, FactorIntegerECM, NumberTheoryFunctions, and PrimeQ. There are other *Mathematica* packages that can be obtained using the Internet; access them at http://www.mathsource.com. Consult the *Mathematica Book* [Wo03] to learn how to load and use them.

You cannot use a command form package without having first told *Mathematica* that you want to run commands from this package, which is done by loading it. For example, to load the package NumberTheoryFunctions, use the command: In[1]:=NumberTheory'NumberTheoryFunctions'

Another resource for using *Mathematica* for number theory computations is *Mathematica in Action* by Stan Wagon [Wa99]. This book contains useful discussions of how to use *Mathematica* to investigate large primes, run extended versions of the Euclidean algorithm, solve linear diophantine equations, use the Chinese remainder theorem, work with continued fractions, and generate prime certificates.

Number Theory Commands in *Mathematica*

The *Mathematica* commands relevant to material covered in this book are presented here according to the chapter in which that material is covered. (The command for loading these functions if they are part of add-on packages is also provided.) These commands are useful for checking computations in the text, for working or checking some of the exercises, and for the computations and explorations at the end of each section. Furthermore, it is possible to write programs in *Mathematica* for many of the explorations and programming projects listed at the end of each section. Consult *Mathematica* reference materials, such as the *Mathematica Book* [Wo03], for information about writing programs in *Mathematica*.

Chapter 1

Fibonacci[n] gives the nth Fibonacci number f_n.

Quotient[m, n] gives the integer quotient when m is divided by n.

Mod[m,n] gives the remainder when m is divided by n.

The Collatz ($3x + 1$) problem has been implemented in *Mathematica* by Ilan Vardi. You can access this *Mathematica* package at http://www.mathsource.com /Content/Applications/Mathematics/0200-305.

Chapter 2

IntegerDigits[n, b] gives a list of the base b digits of n.

Chapter 3

PrimeQ[n] produces output True if n is prime and False if n is not prime.

Prime[n] gives the nth prime number.

PrimePi[x] gives the number of primes less than or equal to x.

In[1]:=NumberTheory'NumberTheoryFunctions'
NextPrime[n] gives the smallest prime larger than n.

GCD[$n_1, n_2, \ldots, n_k$] gives the greatest common divisor of the integers $n_1, n_2, \ldots, n_k$.

ExtendedGCD[n, m] gives the extended greatest common divisor of the integers n and m.

LCM $[n_1, n_2, \ldots, n_k]$ gives the least common multiple of the integers $n_1, n_2, \ldots, n_k$.

FactorInteger $[n]$ produces a list of the prime factors of n and their exponents.

Divisors $[n]$ gives a list of the integers that divide n.

IntegerExponent $[n, b]$ gives the highest power of b that divides n.

In[1]:=NumberTheory'NumberTheoryFunctions'
SquareFreeQ $[n]$ returns True if n contains a squared factor and False otherwise.

In[1]:=NumberTheory'FactorIntegerECM'
FactorIntegerECM $[n]$ gives a factor of a composite integer n produced using Lenstra's elliptic curve factorization method.

Chapter 4

Mod $[k, n]$ gives the least nonnegative residue of k modulo n.

Mod $[k, n, 1]$ gives the least positive residue of k modulo n.

Mod $[k, n, -n/2]$ gives the absolute least residue of k modulo n.

PowerMod $[a, b, n]$ gives the value of a^b **mod** n. Taking $b = -1$ gives the inverse of a modulo n, if it exists.

In[1]:=NumberTheory'NumberTheoryFunctions'
ChineseRemainder $[list_1, \ list_2]$ gives the smallest nonnegative integer r such that Mod $[r, list_2]$ equals $list_1$. (For example, ChineseRemainder $[\{r_1, r_2\}, \{m_1 m_2\}]$ produces the solution of the simultaneous congruence $x \equiv r_1$ **mod** m_1 and $x \equiv r_2$ **mod** m_2.)

Chapter 6

EulerPhi $[n]$ gives the value of the Euler phi function at n.

Chapter 7

DivisorSigma $[k, n]$ gives the value of the sum of the kth powers of divisors function at n. Taking $k = 1$ gives the sum of divisors function at n. Taking $k = 0$ gives the number of divisors of n.

MoebiusMu $[n]$ gives the value of $\mu(n)$.

Chapter 8

The RSA Public Key Cryptosystem has been implemented in *Mathematica* by Stephan Kaufmann. You can obtain the *Mathematica* package, instructions for how to use it, and a *Mathematica* notebook from the Mathsource Web site at http://www.mathsource.com/Content/Applications/ComputerScience/0204-130.

Chapter 9

MultiplicativeOrder $[k, n]$ gives the order of k modulo n.

PrimitiveRoot $[n]$ gives a primitive root of n when n has a primitive root, and does not evaluate when it does not.

```
In[1]:=NumberTheory'PrimeQ'
```
`PrimeQCertificate[n]` produces a certificate verifying that n is prime or composite.

`CarmichaelLambda[n]` gives the minimal universal exponent $\lambda(n)$.

Chapter 11

`JacobiSymbol[n, m]` gives the value of the Jacobi symbol $\left(\frac{n}{m}\right)$.

`SqrtMod[d, n]` gives a square root of d modulo n for odd n.

Chapter 12

`RealDigits[x]` gives a list of the digits in the decimal expansion of x.

`RealDigits[x, b]` gives a list of the digits in the base b expansion of x.

The following functions dealing with decimal expansions are part of the Number Theory'ContinuedFractions' package. Load this package using `In[1]:=Number Theory'Continued Fractions'` before using them.

`PeriodicForm[{a_0, ..., {a_m, ...}}, exp]` presents a repeated decimal expansion in terms of a preperiodic and a periodic part.

`PeriodicForm[{a_0, ..., {a_m, ...}}, expr, b]` represents a base b expansion.

`Normal[PeriodicForm[args]]` gives the rational number corresponding to a decimal expansion.

The following functions dealing with continued fractions are part of the Number Theory'Continued Fractions' package. Load this package using `In[1]:=Number Theory'Continued Fractions'` before using them.

`ContinuedFraction[x, n]` gives the first n terms of the continued fraction expansion of x.

`ContinuedFraction[x]` gives the complete continued fraction expansion of a quadratic irrational number.

`FromContinued Fraction[list]` finds a number from its continued fraction expansion.

`ContinuedFractionForm[{a_0, a_1, ...}]` represents the continued fraction with partial quotients $a_0, a_1 ...$

`ContinuedFractionForm[{a_0, a_1, ..., {p_0, p_1, ...}}]` represents the continued fraction with partial quotients $a_0, a_1 ...$ and additional quotients $p_1, p_2, ...$.

`Normal[ContinuedFractionForm[quotients]]` gives the rational or quadratic irrational number corresponding to the given continued fraction.

`Convergents[rat]` gives the convergents for all terms of the continued fraction of a rational or quadratic irrational x.

`Convergents[num, terms]` gives the convergents for the given number of terms of the continued fraction expansion of *num*.

`Convergents[cf]` gives the convergents for the particular continued fraction cf returned from `ContinuedFraction` or `ContinuedFractionForm`.

QuadraticIrrationalQ [*expr*] tests whether *expr* is a quadratic irrational.

Chapter 14

Divisors [*n*, GaussianIntegers -> True] lists all Gaussian integer divisors of the Gaussian integer *n*.

DivisorSigma [*k*, *n*, GaussianIntegers -> True] gives the sum of the *k*th powers of the Gaussian integer divisors of the Gaussian integer *n*.

FactorInteger [*n*, GaussianIntegers -> True] produces a list of the Gaussian prime factors of the Gaussian integer *n* with positive real parts, and nonnegative imaginary parts, their exponents, and a unit.

PrimeQ [*n*, GaussianIntegers -> True] returns the value of True if *n* is a Gaussian prime and False otherwise.

Appendices

Binomial [*n*, *m*] gives the values of the binomial coefficient $\binom{n}{m}$.

D

Number Theory Web Links

In this appendix we provide an annotated list of key Web sites for number theory. These sites are excellent starting points for an exploration of number theory resources on the Web. At the time of publication of this book, these sites could be found at the URLs listed here. However, with the ephemeral nature of the Web, the addresses of these sites may change, they may cease to exist, or their content may change, and neither the author nor the publisher of this book is able to vouch for the contents of these sites. If you have trouble locating these sites, you may want to try using a search engine to see whether they can be found at a new URL. You will also want to consult the comprehensive guide to all the Web references for this book at `http://www.awlonline.com/rosen`. This guide will help you locate some of the more difficult-to-find sites relevant to number theory and to cryptography.

The Fibonacci Numbers and the Golden Section (`http://www.mcs.surrey.ac.uk /Personal/R.Knott/Fibonacci/fib.html`)

An amazing collection of information about the Fibonacci numbers, including their history, where they arise in nature, puzzles involving the Fibonacci numbers, and their mathematical properties can be found on this site. Additional material addresses the golden section. An extensive collection of links to other sites makes this an excellent place to start your exploration for information about Fibonacci numbers.

The Prime Pages (`http://www.utm.edu/research/primes/`)

This is the premier site for information about prime numbers. You can find a glossary, primers, articles, the Prime FAQ, current records, conjectures, extensive lists of primes and prime factorizations, as well as links to other sites, including those that provide useful software. This is a great site for exploring the world of primes!

The Great Internet Prime Search (http://www.mersenne.org)

Find the latest discoveries about Mersenne primes at this site. You can download software from this site to search for Mersenne primes, as well as primes of other special forms. Links to other sites related to searching for primes and factoring are provided. This is the site to visit to sign up for the communal search for a new prime of world-record size!

The MacTutor History of Mathematics Archives (http://www-groups.dcs .st-and.ac.uk/history/index.html)

This is the main site to visit for biographies of mathematicians. Hundreds of important mathematicians from ancient to modern times are covered. You can also find essays on the history of important mathematical topics, including the prime numbers and Fermat's last theorem.

Frequently Asked Questions in Mathematics (http://db.uwaterloo.ca /alopez-o/math-faq/math-faq.html)

This is a compilation of the frequently asked questions from the USENET newsgroup sci.math. It contains several sections of questions relating to number theory, including primes and Fermat's last theorem, as well as a potpourri of historical information and mathematical trivia.

The Number Theory Web (http://www.numbertheory.org/ntw/web.html)

This site provides an amazing collection to links to sites containing information relevant to number theory. You can find links to sites providing software for number theory calculations, course notes, articles, online theses, historical and biographical information, conference information, job postings, and everything else on the Web related to number theory.

RSA Labs—Cryptography FAQ (http://www.rsasecurity.com/rsalabs/faq/)

This site provides an excellent overview of modern cryptography. You can find descriptions of cryptographic applications, cryptographic protocols, public and private key cryptosystems, and the mathematics behind them.

The Mathematics of Fermat's Last Theorem (http://www.best.com/~cgd /home/flt/flt01.htm)

This site provides an excellent introduction to Fermat's last theorem. It provides discussions of each of the important topics involved in the proof of the theorem.

NOVA Online—The Proof (http://www.pbs.org/wgbh/nova/proof)

This site provides material relating to a television program on the proof of Fermat's last theorem. Included are transcripts of the program and of an interview with Andrew Wiles, and links to other sites on Fermat's last theorem.

E

Tables

Table E.1 gives the least prime factor of each odd positive integer less than 10,000 and not divisible by 5. The initial digits of the integer are listed to the side and the last digit is at the top of the column. Primes are indicated with a dash. The table is reprinted with permission from U. Dudley, *Elementary Number Theory*, Second Edition, Copyright © 1969 and 1978 by W. H. Freeman and Company. All rights reserved.

Table E.3 gives the least primitive root r modulo p for each prime p, $p < 1000$.

Table E.4 is reprinted with permission from J. V. Uspensky and M. A. Heaslet, *Elementary Number Theory*, McGraw-Hill Book Company. Copyright © 1939.

	1	3	7	9		1	3	7	9		1	3	7	9		1	3	7	9
0	—	—	—	3	40	—	13	11	—	80	3	11	3	—	120	—	3	17	3
1	—	—	—	—	41	3	7	3	—	81	—	3	19	3	121	7	—	—	23
2	3	—	3	—	42	—	3	7	3	82	—	—	—	—	122	3	—	3	—
3	—	3	—	3	43	—	—	19	—	83	3	7	3	—	123	—	3	—	3
4	—	—	—	7	44	3	—	3	—	84	29	3	7	3	124	17	11	29	—
5	3	—	3	—	45	11	3	—	3	85	23	—	—	—	125	3	7	3	—
6	—	3	—	3	46	—	—	—	7	86	3	—	3	11	126	13	3	7	3
7	—	—	7	—	47	3	11	3	—	87	13	3	—	3	127	31	19	—	—
8	3	—	3	—	48	13	3	—	3	88	—	—	—	7	128	3	—	3	—
9	7	3	—	3	49	—	17	7	—	89	3	19	3	29	129	—	3	—	3
10	—	—	—	—	50	3	—	3	—	90	17	3	—	3	130	—	—	—	7
11	3	—	3	7	51	7	3	11	3	91	—	11	7	—	131	3	13	3	—
12	11	3	—	3	52	—	—	17	23	92	3	13	3	—	132	—	3	—	3
13	—	7	—	—	53	3	13	3	7	93	7	3	—	3	133	11	31	7	13
14	3	11	3	—	54	—	3	—	3	94	—	23	—	13	134	3	17	3	19
15	—	3	—	3	55	19	7	—	13	95	3	8	3	7	135	7	3	23	3
16	7	—	—	13	56	3	—	3	—	96	31	3	—	3	136	—	29	—	37
17	3	—	3	—	57	—	3	—	3	97	—	7	—	11	137	3	—	3	7
18	—	3	11	3	58	7	11	—	19	98	3	—	3	23	138	—	3	19	3
19	—	—	—	—	59	3	—	3	—	99	—	3	—	3	139	13	7	11	—
20	3	7	3	11	60	—	3	—	3	100	7	17	19	—	140	3	23	3	—
21	—	3	7	3	61	13	—	—	—	101	3	—	3	—	141	17	3	13	3
22	13	—	—	—	62	3	7	3	17	102	—	3	13	3	142	7	—	—	—
23	3	—	3	—	63	—	3	7	3	103	—	—	17	—	143	3	—	3	—
24	—	3	13	3	64	—	—	—	11	104	3	7	3	—	144	11	3	—	3
25	—	11	—	7	65	3	—	3	—	105	—	3	7	3	145	—	—	31	—
26	3	—	3	—	66	—	3	23	3	106	—	—	11	—	146	3	7	3	13
27	—	3	—	3	67	11	—	—	7	107	3	29	3	13	147	—	3	7	3
28	—	—	7	17	68	3	—	3	13	108	23	3	—	3	148	—	—	—	—
29	3	—	3	13	69	—	3	17	3	109	—	—	—	7	149	3	—	3	—
30	7	3	—	3	70	—	19	7	—	110	3	—	3	—	150	19	3	11	3
31	—	—	—	11	71	3	23	3	—	111	11	3	—	3	151	—	17	37	7
32	3	17	3	7	72	7	3	—	3	112	19	—	7	—	152	3	—	3	11
33	—	3	—	3	73	17	—	11	—	113	3	11	3	17	153	—	3	29	3
34	11	7	—	—	74	3	—	3	7	114	7	3	31	3	154	23	—	7	—
35	3	—	3	—	75	—	3	—	3	115	—	—	13	19	155	3	—	3	—
36	19	3	—	3	76	—	7	13	—	116	3	—	3	7	156	7	3	—	3
37	7	—	13	—	77	3	—	3	19	117	—	3	11	3	157	—	11	19	—
38	7	—	3	—	78	11	3	—	3	118	—	7	—	29	158	3	—	3	7
39	17	3	—	3	79	7	13	—	17	119	3	—	3	11	159	37	3	—	3

Table E.1 *Factor table.*

	1	3	7	9		1	3	7	9		1	3	7	9		1	3	7	9
160	—	7	—	—	200	3	—	3	7	240	7	3	29	3	280	—	—	7	53
161	3	—	3	—	201	—	3	—	3	241	—	19	—	41	281	3	29	3	—
162	—	3	—	3	202	43	7	—	—	242	3	—	3	7	282	7	3	11	3
163	7	23	—	11	203	3	19	3	—	243	11	3	—	3	283	19	—	—	17
164	3	31	3	17	204	13	3	23	3	244	—	7	—	31	284	3	—	3	7
165	13	3	—	3	205	7	—	11	29	245	3	11	3	—	285	—	3	—	3
166	11	—	—	—	206	3	—	3	—	246	23	3	—	3	286	—	7	47	19
167	3	7	3	23	207	19	3	31	3	247	7	—	—	37	287	3	13	3	—
168	41	3	7	3	208	—	—	—	—	248	3	13	3	19	288	43	3	—	3
169	19	—	—	—	209	3	7	3	—	249	47	3	11	3	289	7	11	—	13
170	3	13	3	—	210	11	3	7	3	250	41	—	23	13	290	3	—	3	—
171	29	3	17	3	211	—	—	29	13	251	3	7	3	11	291	41	3	—	3
172	—	—	11	7	212	3	11	3	—	252	—	3	7	3	292	23	37	—	29
173	3	—	3	37	213	—	3	—	3	253	—	17	43	—	293	3	7	3	—
174	—	3	—	3	214	—	—	19	7	254	3	—	3	—	294	17	3	7	3
175	17	—	7	—	215	3	—	3	17	255	—	3	—	3	295	13	—	—	11
176	3	41	3	29	216	—	3	11	3	256	13	11	17	7	296	3	—	3	—
177	7	3	—	3	217	13	41	7	—	257	3	31	3	—	297	—	3	13	3
178	13	—	—	—	218	3	37	3	11	258	29	3	13	3	298	11	19	29	7
179	3	11	3	7	219	7	3	13	3	259	—	—	7	23	299	3	41	3	—
180	—	3	13	3	220	31	—	—	47	260	3	19	3	—	300	—	3	31	3
181	—	7	23	17	221	3	—	3	7	261	7	3	—	3	301	—	23	7	—
182	3	—	3	31	222	—	3	17	3	262	—	43	37	11	302	3	—	3	13
183	—	3	11	3	223	23	7	—	—	263	3	—	3	7	303	7	3	—	3
184	7	19	—	43	224	3	—	3	13	264	19	3	—	3	304	—	17	11	—
185	3	17	3	11	225	—	3	37	3	265	11	7	—	—	305	3	43	3	7
186	—	3	—	3	226	7	31	—	—	266	3	—	3	17	306	—	3	—	3
187	—	—	—	—	227	3	—	3	43	267	—	3	—	3	307	37	7	17	—
188	3	7	3	—	228	—	3	—	3	268	7	—	—	—	308	3	—	3	—
189	31	3	7	3	229	29	—	—	11	269	3	—	3	—	309	11	3	19	3
190	—	11	—	23	230	3	7	3	—	270	37	3	—	3	310	7	29	13	—
191	3	—	3	19	231	—	3	7	3	271	—	—	11	—	311	3	11	3	—
192	17	3	41	3	232	11	23	13	17	272	3	7	3	—	312	—	3	53	3
193	—	—	13	7	233	3	—	3	—	273	—	3	7	3	313	31	13	—	43
194	3	29	3	—	234	—	3	—	3	274	—	13	41	—	314	3	7	3	47
195	—	3	19	3	235	—	13	—	7	275	3	—	3	31	315	23	3	7	3
196	37	13	7	11	236	3	17	3	23	276	11	3	—	3	316	29	—	—	—
197	3	—	3	—	237	—	3	—	3	277	17	47	—	7	317	3	19	3	11
198	7	3	—	3	238	—	—	7	—	278	3	11	3	—	318	—	3	—	3
199	11	—	—	—	239	3	—	3	—	279	—	3	—	3	319	—	31	23	7

Table E.1 (continued)

	1	3	7	9		1	3	7	9		1	3	7	9		1	3	7	9
320	3	—	3	—	360	13	3	—	3	400	—	—	—	19	440	3	7	3	—
321	13	3	—	3	361	23	—	—	7	401	3	—	3	—	441	11	3	7	3
322	—	11	7	—	362	3	—	3	19	402	—	3	—	3	442	—	—	19	43
323	3	53	3	41	363	—	3	—	3	403	29	37	11	7	443	3	11	3	23
324	7	3	17	3	364	11	—	7	41	404	3	13	3	—	444	—	3	—	3
325	—	—	—	—	365	3	13	3	—	405	—	3	—	3	445	—	61	—	7
326	3	13	3	7	366	7	3	19	3	406	31	17	7	13	446	3	—	3	41
327	—	3	29	3	367	—	—	—	13	407	3	—	3	—	447	17	3	11	3
328	17	7	19	11	368	3	29	3	7	408	7	3	61	3	448	—	—	7	67
329	3	37	3	—	369	—	3	—	3	409	—	—	17	—	449	3	—	3	11
330	—	3	—	3	370	—	7	11	—	410	3	11	3	7	450	7	3	—	3
331	7	—	31	—	371	3	47	3	—	411	—	3	23	3	451	13	—	—	—
332	3	—	3	—	372	61	3	—	3	412	13	7	—	—	452	3	—	3	7
333	—	3	47	3	373	7	—	37	—	413	3	—	3	—	453	23	3	13	3
334	13	—	—	17	374	3	19	3	23	414	41	3	11	3	454	19	7	—	—
335	3	7	3	—	375	11	3	13	3	415	7	—	—	—	455	3	29	3	47
336	—	3	7	3	376	—	53	—	—	416	3	23	3	11	456	—	3	—	3
337	—	—	11	31	377	3	7	3	—	417	43	3	—	3	457	7	17	23	19
338	3	17	3	—	378	19	3	7	3	418	37	47	53	59	458	3	—	3	13
339	—	3	43	3	379	17	—	—	29	419	3	7	3	13	459	—	3	—	3
340	19	41	—	7	380	3	—	3	31	420	—	3	7	3	460	43	—	17	11
341	3	—	3	13	381	37	3	11	3	421	—	11	—	—	461	3	7	3	31
342	11	3	23	3	382	—	—	43	7	422	3	41	3	—	462	—	3	7	3
343	47	—	7	19	383	3	—	3	11	423	—	3	19	3	463	11	41	—	—
344	3	11	3	—	384	23	3	—	3	424	—	—	31	7	464	3	—	3	—
345	7	3	—	3	385	—	—	7	17	425	3	—	3	—	465	—	3	—	3
346	—	—	—	—	386	3	—	3	53	426	—	3	17	3	466	59	—	13	7
347	3	23	3	7	387	7	3	—	3	427	—	—	7	11	467	3	—	3	—
348	59	3	11	3	388	—	11	13	—	428	3	—	3	—	468	31	3	43	3
349	—	7	13	—	389	3	17	3	7	429	7	3	—	3	469	—	13	7	37
350	3	31	3	11	390	47	3	—	3	430	11	13	59	31	470	3	—	3	17
351	—	3	—	3	391	—	7	—	—	431	3	19	3	7	471	7	3	53	3
352	7	13	—	—	392	3	—	3	—	432	29	3	—	3	472	—	—	29	—
353	3	—	3	—	393	—	3	31	3	433	61	7	—	—	473	3	—	3	7
354	—	3	—	3	394	7	—	—	11	434	3	43	3	—	474	11	3	47	3
355	53	11	—	—	395	3	59	37	3	435	19	3	—	3	475	—	7	67	—
356	3	7	3	43	396	17	3	—	3	436	7	—	11	17	476	3	11	3	19
357	—	3	7	3	397	11	29	41	23	437	3	—	3	29	477	13	3	17	3
358	—	—	17	37	398	3	7	3	—	438	13	3	41	3	478	7	—	—	—
359	3	—	3	59	399	13	3	7	3	439	—	23	—	53	479	3	—	3	—

Table E.1 (continued)

	1	3	7	9		1	3	7	9		1	3	7	9		1	3	7	9
480	—	3	11	3	520	7	11	41	—	560	3	13	3	71	600	17	3	—	3
481	17	—	—	61	521	3	13	3	17	561	31	3	41	3	601	—	7	11	13
482	3	7	3	11	522	23	3	—	3	562	7	—	17	13	602	3	19	3	—
483	—	3	7	3	523	—	—	—	13	563	3	43	3	—	603	37	3	—	3
484	47	29	37	13	524	3	7	3	29	564	—	3	—	3	604	7	—	—	23
485	3	23	3	43	525	59	3	7	3	565	—	—	—	—	605	3	—	3	73
486	—	3	31	3	526	—	19	23	11	566	3	7	3	—	606	11	3	—	3
487	—	11	—	7	527	3	—	3	—	567	53	3	7	3	607	13	—	59	—
488	3	19	3	—	528	—	3	17	3	568	13	—	11	—	608	3	7	3	—
489	67	3	59	3	529	11	67	—	7	569	3	—	3	41	609	—	3	7	3
490	13	—	7	—	530	3	—	3	—	570	—	3	13	3	610	—	17	31	41
491	3	17	3	—	531	47	3	13	3	571	—	29	—	7	611	3	—	311	29
492	7	3	13	3	532	17	—	7	73	572	3	59	3	17	612	—	3	11	3
493	—	—	—	11	533	3	—	3	19	573	11	3	—	3	613	—	—	17	7
494	3	—	3	7	534	7	3	—	3	574	—	—	7	—	614	3	—	3	11
495	—	3	—	3	535	—	53	11	23	575	3	11	3	13	615	—	3	47	3
496	11	7	—	—	536	3	31	3	7	576	7	3	73	3	616	61	—	7	31
497	3	—	3	13	537	41	3	19	3	577	29	23	53	—	617	3	—	3	37
498	17	3	—	3	538	—	7	—	17	578	3	—	3	7	618	7	3	23	3
499	7	—	19	—	539	3	—	3	—	579	—	3	11	3	619	41	11	—	—
500	3	—	3	—	540	11	3	—	3	580	—	7	—	37	620	3	—	3	7
501	—	3	29	3	541	7	—	—	—	581	3	—	3	11	621	—	3	—	3
502	—	—	11	47	542	3	11	3	61	582	—	3	—	3	622	—	7	13	—
503	3	7	3	—	543	—	3	—	3	583	7	19	13	—	623	3	23	3	17
504	71	3	7	3	544	—	—	13	—	584	3	—	3	—	624	79	3	—	3
505	—	31	13	—	545	3	7	3	53	585	—	3	—	3	625	7	13	—	11
506	3	61	3	37	546	43	3	7	3	586	—	11	—	—	626	3	—	3	—
507	11	3	—	3	547	—	13	—	—	587	3	7	3	—	627	—	3	—	3
508	—	13	—	7	548	3	—	3	11	588	—	3	7	3	628	11	61	—	19
509	3	11	3	—	549	17	3	23	3	589	43	71	—	17	629	3	7	3	—
510	—	3	—	3	550	—	—	—	7	590	3	—	3	19	630	—	3	7	3
511	19	—	7	—	551	3	37	3	—	591	23	3	61	3	631	—	59	—	71
512	3	47	3	23	552	—	3	—	3	592	31	—	—	7	632	3	—	3	—
513	7	3	11	3	553	—	11	7	29	593	3	17	3	—	633	13	3	—	3
514	53	37	—	19	554	3	23	3	31	594	13	3	19	3	634	17	—	11	7
515	3	—	3	7	555	7	3	—	3	595	11	—	7	59	635	3	—	3	—
516	13	3	—	3	556	67	—	19	—	596	3	67	3	47	636	—	3	—	3
517	—	7	31	—	557	3	—	3	7	597	7	3	43	3	637	23	—	7	—
518	3	71	3	—	558	—	3	37	3	598	—	31	—	53	638	3	13	3	—
519	29	3	—	3	559	—	7	29	11	599	3	13	3	7	639	7	3	—	3

Table E.1 (continued)

	1	3	7	9		1	3	7	9		1	3	7	9		1	3	7	9
640	37	19	43	13	680	3	—	3	11	720	19	3	—	3	760	11	—	—	7
641	3	11	3	7	681	7	3	17	3	721	—	—	7	—	761	3	23	3	19
642	—	3	—	3	682	19	—	—	—	722	3	31	3	—	762	—	3	29	3
643	59	7	41	47	683	3	—	3	7	723	7	3	—	3	763	13	17	7	—
644	3	19	3	—	684	—	3	41	3	724	13	—	—	11	764	3	—	3	—
645	—	3	11	3	685	13	7	—	19	725	3	—	3	7	765	7	3	13	3
646	7	23	29	—	686	3	—	3	—	726	53	3	13	3	766	47	79	11	—
647	3	—	3	11	687	—	3	13	3	727	11	7	19	29	767	3	—	3	7
648	—	3	13	3	688	7	—	71	83	728	3	—	3	37	768	—	3	—	3
649	—	43	73	67	689	3	61	3	—	729	23	3	—	3	769	—	7	43	—
650	3	7	3	23	690	67	3	—	3	730	7	67	—	—	770	3	—	3	13
651	17	3	7	3	691	—	31	—	11	731	3	71	3	13	771	11	3	—	3
652	—	11	61	—	692	3	7	3	13	732	—	3	17	3	772	7	—	—	59
653	3	47	3	13	693	29	3	7	3	733	—	—	11	41	773	3	11	3	71
654	31	3	—	3	694	11	53	—	—	734	3	7	3	—	774	—	3	61	3
655	—	—	79	7	695	3	17	3	—	735	—	3	7	3	775	23	—	—	—
656	3	—	3	—	696	—	3	—	3	736	17	37	53	—	776	3	7	3	17
657	—	3	—	3	697	—	19	—	7	737	3	73	3	47	777	19	3	7	3
658	—	29	7	11	698	3	—	3	29	738	11	3	83	3	778	31	43	13	—
659	3	19	3	—	699	—	3	—	3	739	19	—	13	7	779	3	—	3	11
660	7	3	—	3	700	—	47	7	43	740	3	11	3	31	780	29	3	37	3
661	11	17	13	—	701	3	—	3	—	741	—	3	—	3	781	73	13	—	7
662	3	37	3	7	702	7	3	—	3	742	41	13	7	17	782	3	—	3	—
663	19	3	—	3	703	79	13	31	—	743	3	—	3	43	783	41	3	17	3
664	29	7	17	61	704	3	—	3	7	744	7	3	11	3	784	—	11	7	47
665	3	—	3	—	705	11	3	—	3	745	—	29	—	—	785	3	—	3	29
666	—	3	59	3	706	23	7	37	—	746	3	17	3	7	786	7	3	—	3
667	7	—	11	—	707	3	11	3	—	747	31	3	—	3	787	17	—	—	—
668	3	41	3	—	708	73	3	19	3	748	—	7	—	—	788	3	—	3	7
669	—	3	37	3	709	7	41	47	31	749	3	59	3	—	789	13	3	53	3
670	—	—	19	—	710	3	—	3	—	750	13	3	—	3	790	—	7	—	11
671	3	7	3	—	711	13	3	11	3	751	7	11	—	73	791	3	41	3	—
672	11	3	7	3	712	—	17	—	—	752	3	—	3	—	792	89	3	—	3
673	53	—	—	23	713	3	7	3	11	753	17	3	—	3	793	7	—	—	17
674	3	11	3	17	714	37	3	7	3	754	—	19	—	—	794	3	13	3	—
675	43	3	29	3	715	—	23	17	—	755	3	7	3	—	795	—	3	73	3
676	—	—	67	7	716	3	13	3	67	756	—	3	7	3	796	19	—	31	13
677	3	13	3	—	717	71	3	—	3	757	67	—	—	11	797	3	7	3	79
678	—	3	11	3	718	43	11	—	7	758	3	—	3	—	798	23	3	7	3
679	—	—	7	13	719	3	—	3	23	759	—	3	71	3	799	61	—	11	19

Table E.1 (continued)

	1	3	7	9		1	3	7	9		1	3	7	9		1	3	7	9
800	3	53	3	—	840	31	3	7	3	880	13	—	—	23	920	3	—	3	—
801	—	3	—	3	841	13	47	19	—	881	3	7	3	—	921	61	3	13	3
802	13	71	23	7	842	3	—	3	—	882	—	3	7	3	922	—	23	—	11
803	3	29	3	—	843	—	3	11	3	883	—	11	—	—	923	3	7	3	—
804	11	3	13	3	844	23	—	—	7	884	3	37	3	—	924	—	3	7	3
805	83	—	7	—	845	3	79	3	11	885	53	3	17	3	925	11	19	—	47
806	3	11	3	—	846	—	3	—	3	886	—	—	—	7	926	3	59	3	13
807	7	3	41	3	847	43	37	7	61	887	3	19	3	13	927	73	3	—	3
808	—	59	—	—	848	3	17	3	13	888	83	3	—	3	928	—	—	37	7
809	3	—	3	7	849	7	3	29	3	889	17	—	7	11	929	3	—	3	17
810	—	3	11	3	850	—	11	47	67	890	3	29	3	59	930	71	3	41	3
811	—	7	—	23	851	3	—	3	7	891	7	3	37	3	931	—	67	7	—
812	3	—	3	11	852	—	3	—	3	892	11	—	79	—	932	3	—	3	19
813	47	3	79	3	853	19	7	—	—	893	3	—	3	7	933	7	3	—	3
814	7	17	—	29	854	3	—	3	83	894	—	3	23	3	934	—	—	13	—
815	3	31	3	41	855	17	3	43	3	895	—	7	13	17	935	3	47	3	7
816	—	3	—	3	856	7	—	13	11	896	3	—	3	—	936	11	3	14	3
817	—	11	13	—	857	3	—	3	23	897	—	3	47	3	937	—	7	—	83
818	3	7	3	19	858	—	3	31	3	898	7	13	11	89	938	3	11	3	41
819	—	3	7	3	859	11	13	—	—	899	3	17	3	—	939	—	3	—	3
820	59	13	29	—	860	3	7	3	—	900	—	3	—	3	940	7	—	23	97
821	3	43	3	—	861	79	3	7	3	901	—	—	71	29	941	3	—	3	—
822	—	3	19	3	862	37	—	—	—	902	3	7	3	—	942	—	3	11	3
823	—	—	—	7	863	3	89	3	53	903	11	3	7	3	943	—	—	—	—
824	3	—	3	73	864	—	3	—	3	904	—	—	83	—	944	3	7	3	11
825	37	3	23	3	865	41	17	11	7	905	3	11	3	—	945	13	3	7	3
826	11	—	7	—	866	3	—	3	—	906	13	3	—	3	946	—	—	—	17
827	3	—	3	17	867	13	3	—	3	907	47	43	29	7	947	3	—	3	—
828	7	3	—	3	868	—	19	7	—	908	3	31	3	61	948	19	3	53	3
829	—	—	—	43	869	3	—	3	—	909	—	3	11	3	949	—	11	—	7
830	3	19	3	7	870	7	3	—	3	910	19	—	7	—	950	3	13	3	37
831	—	3	—	3	871	31	—	23	—	911	3	31	3	11	951	—	3	31	3
832	53	7	11	—	872	3	11	3	7	912	7	3	—	3	952	—	89	7	13
833	3	13	3	31	873	—	3	—	3	913	23	—	—	13	953	3	—	3	—
834	19	3	17	3	874	—	7	—	13	914	3	41	3	7	954	7	3	—	3
835	7	—	61	13	875	3	—	3	193	915	—	3	—	3	955	—	41	19	11
836	3	—	3	—	876	—	3	11	3	916	—	7	89	53	956	3	73	3	7
837	11	3	—	3	877	7	31	67	—	917	3	—	3	67	957	17	3	61	3
838	17	83	—	—	878	3	—	3	11	918	—	3	—	3	958	11	7	—	43
839	3	7	3	37	879	59	3	19	3	919	7	29	17	—	959	3	53	3	29

Table E.1 (continued)

	1	3	7	9		1	3	7	9		1	3	7	9		1	3	7	9
960	—	3	13	3	970	89	31	18	7	980	3	—	3	17	990	—	3	—	3
961	7	—	59	—	971	3	11	3	—	981	—	3	—	3	991	11	23	47	7
962	3	—	3	—	972	—	3	71	3	982	7	11	31	—	992	3	—	3	—
963	—	3	23	3	973	37	—	7	—	983	3	—	3	—	993	—	3	19	3
964	31	—	11	—	974	3	—	3	—	984	13	3	43	3	994	—	61	7	—
965	3	7	3	13	975	7	3	11	3	985	—	59	—	—	995	3	37	3	23
966	—	3	7	3	976	43	13	—	—	986	3	7	3	71	996	7	3	—	3
967	19	17	—	—	977	3	29	3	7	987	—	3	7	3	997	13	—	11	17
968	3	23	3	—	978	—	3	—	3	988	41	—	—	11	998	3	67	3	7
969	11	3	—	3	979	—	7	97	41	989	3	13	3	19	999	97	3	13	3

Table E.1 (continued)

n	$\phi(n)$	$\tau(n)$	$\sigma(n)$
1	1	1	1
2	1	2	3
3	2	2	4
4	2	3	7
5	4	2	6
6	2	4	12
7	6	2	8
8	4	4	15
9	6	3	13
10	4	4	18
11	10	2	12
12	4	6	28
13	12	2	14
14	6	4	24
15	8	4	24
16	8	5	31
17	16	2	18
18	6	6	39
19	18	2	20
20	8	6	42
21	12	4	32
22	10	4	36
23	22	2	24
24	8	8	60
25	20	3	31
26	12	4	42
27	18	4	40
28	12	6	56
29	28	2	30
30	8	8	72
31	30	2	32
32	16	6	63
33	20	4	48
34	16	4	54
35	24	4	48
36	12	9	91
37	36	2	38
38	18	4	60
39	24	4	56
40	16	8	90
41	40	2	42
42	12	8	96
43	42	2	44
44	20	6	84
45	24	6	78
46	22	6	72
47	46	2	48
48	16	10	124
49	42	3	57

Table E.2 *Values of some arithmetic functions.*

n	$\phi(n)$	$\tau(n)$	$\sigma(n)$
50	20	6	93
51	32	4	72
52	24	6	98
53	52	2	54
54	18	8	120
55	40	4	72
56	24	8	120
57	36	4	80
58	28	4	90
59	58	2	60
60	16	12	168
61	60	2	62
62	30	4	96
63	36	6	104
64	32	7	127
65	48	4	84
66	20	8	144
67	66	2	68
68	32	6	126
69	44	4	96
70	24	8	144
71	70	2	72
72	24	12	195
73	72	2	74
74	36	4	114
75	40	6	124
76	36	6	140
77	60	4	96
78	24	8	168
79	78	2	80
80	32	10	186
81	54	5	121
82	40	4	126
83	82	2	84
84	24	12	224
85	64	4	108
86	42	4	132
87	56	4	120
88	40	8	180
89	88	2	90
90	24	12	234
91	72	4	112
92	44	6	168
93	60	4	128
94	46	4	144
95	72	4	120
96	32	12	252
97	96	2	98
98	42	6	171
99	60	6	156
100	40	9	217

Table E.2 (continued)

p	r	p	r	p	r	p	r
2	1	191	19	439	15	709	2
3	2	193	5	443	2	719	11
5	2	197	2	449	3	727	5
7	3	199	3	457	13	733	6
11	2	211	2	461	2	739	3
13	2	223	3	463	3	743	5
17	3	227	2	467	2	751	3
19	2	229	6	479	13	757	2
23	5	233	3	487	3	761	6
29	2	239	7	491	2	769	11
31	3	241	7	499	7	773	2
37	2	251	6	503	5	787	2
41	6	257	3	509	2	797	2
43	3	263	5	521	3	809	3
47	5	269	2	523	2	811	3
53	2	271	6	541	2	821	2
59	2	277	5	547	2	823	3
61	2	281	3	557	2	827	2
67	2	283	3	563	2	829	2
71	7	293	2	569	3	839	11
73	5	307	5	571	3	853	2
79	3	311	17	577	5	857	3
83	2	313	10	587	2	859	2
89	3	317	2	593	3	863	5
97	5	331	3	599	7	877	2
101	2	337	10	601	7	881	3
103	5	347	2	607	3	883	2
107	2	349	2	613	2	887	5
109	6	353	3	617	3	907	2
113	3	359	7	619	2	911	17
127	3	367	6	631	3	919	7
131	2	373	2	641	3	929	3
137	3	379	2	643	11	937	5
139	2	383	5	647	5	941	2
149	2	389	2	653	2	947	2
151	6	397	5	659	2	953	3
157	5	401	3	601	2	967	5
163	2	409	21	673	5	971	6
167	5	419	2	677	2	977	3
173	2	421	2	683	5	983	5
179	2	431	7	691	3	991	6
181	2	433	5	701	2	997	7

Table E.3 *Primitive roots modulo primes.*

	Numbers															
p	1	2	3	4	5	6	7	8	9	10	11	12	13	14	15	16
3	2	1											Indices			
5	4	1	3	2												
7	6	2	1	4	5	3										
11	10	1	8	2	4	9	7	3	6	5						
13	12	1	4	2	9	5	11	3	8	10	7	6				
17	16	14	1	12	5	15	11	10	2	3	7	13	4	9	6	8
19	18	1	13	2	16	14	6	3	8	17	12	15	5	7	11	4
23	22	2	16	4	1	18	19	6	10	3	9	20	14	21	17	8
29	28	1	5	2	22	6	12	3	10	23	25	7	18	13	27	4
31	30	24	1	18	20	25	28	12	2	14	23	19	11	22	21	0
37	36	1	26	2	23	27	32	3	16	24	30	28	11	33	13	4
41	40	26	15	12	22	1	39	38	30	8	3	27	31	25	37	24
43	42	27	1	12	25	28	35	39	2	10	30	13	32	20	26	24
47	46	18	20	36	1	38	32	8	40	19	7	10	11	4	21	26
53	52	1	17	2	47	18	14	3	34	48	6	19	24	15	12	4
59	58	1	50	2	6	51	18	3	42	7	25	52	45	19	56	4
61	60	1	6	2	22	7	49	3	12	23	15	8	40	50	28	4
67	66	1	39	2	15	40	23	3	12	16	59	41	19	24	54	4
71	70	6	26	12	28	32	1	18	52	34	31	38	39	7	54	24
73	72	8	6	16	1	14	33	24	12	9	55	22	59	41	7	32
79	78	4	1	8	62	5	53	12	2	66	68	9	34	57	63	16
83	82	1	72	2	27	73	8	3	62	28	24	74	77	9	17	4
89	88	16	1	32	70	17	81	48	2	86	84	33	23	9	71	64
97	96	34	70	68	1	8	31	6	44	35	86	42	25	65	71	40

	Numbers																
p	17	18	19	20	21	22	23	24	25	26	27	28	29	30	31	32	33
19	10	9												Indices			
23	7	12	15	5	13	11											
29	21	11	9	24	17	26	20	8	16	19	15	14					
31	7	26	4	8	29	17	27	13	10	5	3	16	9	15			
37	7	17	35	25	22	31	15	29	10	12	6	34	21	14	9	5	20
41	33	16	9	34	14	29	36	13	4	17	5	11	7	23	28	10	18
43	38	29	19	37	36	15	16	40	8	17	3	5	41	11	34	9	31
47	16	12	45	37	6	25	5	28	2	29	14	22	35	39	3	44	27
53	10	35	37	49	31	7	39	20	42	25	51	16	46	13	33	5	23
59	40	43	38	8	10	26	15	53	12	46	34	20	28	57	49	5	17
61	47	13	26	24	55	16	57	9	44	41	18	51	35	29	59	5	21
67	64	13	10	17	62	60	28	42	30	20	51	25	44	55	47	5	32
71	49	58	16	40	27	37	15	44	56	45	8	13	68	60	11	30	57
73	21	20	62	17	39	63	46	30	2	67	18	49	35	15	11	40	61
79	21	6	32	70	54	72	26	13	46	38	3	61	11	67	56	20	69
83	56	63	47	29	80	25	60	75	56	78	52	10	12	18	38	5	14
89	6	18	35	14	82	12	57	49	52	39	3	25	59	87	31	80	85
97	89	78	81	69	5	24	77	76	2	59	18	3	13	9	46	74	60

Table E.4 *Indices.*

	Numbers															
p	34	35	36	37	38	39	40	41	42	43	44	45	46	47	48	49
37	8	19	18													
41	19	21	2	32	35	6	20					Indices				
43	23	18	14	7	4	33	22	6	21							
47	34	33	30	42	17	31	9	15	24	13	43	41	23			
53	11	9	36	30	38	41	50	45	32	22	8	29	40	44	21	23
59	41	24	44	55	39	37	9	14	11	33	27	48	16	23	54	36
61	48	11	14	39	27	46	25	54	56	43	17	34	58	20	10	38
67	65	38	14	22	11	58	18	53	63	9	61	27	29	50	43	46
71	55	29	64	20	22	65	46	25	33	48	43	10	21	9	50	2
78	29	34	28	64	70	65	25	4	47	51	71	13	54	31	38	66
79	25	37	10	19	36	35	74	75	58	49	76	64	30	59	17	28
83	57	35	64	20	48	67	30	40	81	71	26	7	61	23	76	16
89	22	63	34	11	51	24	30	21	10	29	28	72	73	54	65	74
97	27	32	16	91	19	95	7	85	39	4	58	45	15	84	14	62

	Numbers															
p	50	51	52	53	54	55	56	57	58	59	60	61	62	63	64	65
53	43	27	26													
59	13	32	47	22	35	31	21	30	29			Indices				
61	45	53	42	33	19	37	52	32	36	31	30					
67	31	37	21	57	52	8	26	49	45	36	56	7	48	35	6	34
71	62	5	51	23	14	59	19	42	4	3	66	69	17	53	36	67
73	10	27	3	53	26	56	57	68	43	5	23	58	19	45	48	60
79	50	22	42	77	7	52	65	33	15	31	71	45	60	55	24	18
83	55	46	79	59	53	51	11	37	13	34	19	66	39	70	6	22
89	68	7	55	78	19	66	41	36	75	43	15	69	47	83	8	5
97	36	63	93	10	52	87	37	55	47	67	43	64	80	75	12	26

	Numbers															
p	66	67	68	69	70	71	72	73	74	75	76	77	78	79	80	81
67	33															
71	63	47	61	41	35						Indices					
78	69	50	37	52	42	44	36									
79	73	48	29	27	41	51	14	44	23	47	40	43	39			
83	15	45	58	50	36	33	65	69	21	44	49	32	68	43	31	42
89	13	56	38	58	79	62	50	20	27	53	67	77	40	42	46	4
97	94	57	61	51	66	11	50	28	29	72	53	21	33	30	41	88

	Numbers														
p	82	83	84	85	86	87	88	89	90	91	92	93	94	95	96
83	41														
89	37	61	26	76	45	60	44					Indices			
97	23	17	73	90	38	83	92	54	79	56	49	20	22	82	48

Table E.4 (continued)

Indices

p	1	2	3	4	5	6	7	8	9	10	11	12	13	14	15	16
3	2	1														
5	2	4	3	1												
7	3	2	6	4	5	1						Numbers				
11	2	4	8	5	10	9	7	3	6	1						
13	2	4	8	3	6	12	11	9	5	10	7	1				
17	3	9	10	13	5	15	11	16	14	8	7	4	12	2	6	1
19	2	4	8	16	13	7	14	9	18	17	15	11	3	6	12	5
23	5	2	10	4	20	8	17	16	11	9	22	18	21	13	19	3
29	2	4	8	16	3	6	12	24	19	9	18	7	14	28	27	25
31	3	9	27	19	26	16	17	20	29	25	13	8	24	10	30	28
37	2	4	8	16	32	27	17	34	31	25	13	26	15	30	23	9
41	6	36	11	25	27	39	29	10	19	32	28	4	24	21	3	18
43	3	9	27	38	28	41	37	25	32	10	30	4	12	36	22	23
47	5	25	31	14	23	21	11	8	40	12	13	18	43	27	41	17
53	2	4	8	16	32	11	22	44	35	17	34	15	30	7	14	28
59	2	4	8	16	32	5	10	20	40	21	42	25	50	41	23	46
61	2	4	8	16	32	3	6	12	24	48	35	9	18	36	11	22
67	2	4	8	16	32	64	61	55	43	19	38	9	18	36	5	10
71	7	49	59	58	51	2	14	27	47	45	31	4	28	54	23	19
73	5	25	52	41	59	3	15	2	10	50	31	9	45	6	30	4
79	3	9	27	2	6	18	54	4	12	36	29	8	24	72	58	16
83	2	4	8	16	32	64	45	7	14	28	56	29	58	33	66	49
89	3	9	27	81	65	17	51	64	14	42	37	22	66	20	60	2
97	2	25	28	43	21	8	40	6	30	53	71	64	29	48	46	36

Indices

p	17	18	19	20	21	22	23	24	25	26	27	28	29	30	31	32	33
19	10	1															
23	15	6	7	12	14	1							Numbers				
29	21	13	26	23	17	5	10	20	11	22	15	1					
31	22	4	12	5	15	14	11	2	6	18	23	7	21	1			
37	18	36	35	33	29	21	5	10	20	3	6	12	24	11	22	7	14
41	26	33	34	40	35	5	30	16	14	2	12	31	22	9	13	37	17
43	26	35	19	14	42	40	34	16	5	15	2	6	18	11	33	13	39
47	38	2	10	3	15	28	46	42	22	16	33	24	26	36	39	7	35
53	3	6	12	24	48	43	33	13	26	52	51	49	45	37	21	42	31
59	33	7	14	28	56	53	47	35	11	22	44	29	58	57	55	51	43
61	44	27	54	47	33	5	10	20	40	19	38	15	30	60	59	57	53
67	20	40	13	26	52	37	7	14	28	56	45	23	46	25	50	33	66
71	62	8	56	37	46	38	53	16	41	3	21	5	35	32	11	6	42
73	20	27	62	18	17	12	60	8	40	54	51	36	34	24	47	16	7
79	48	65	37	32	17	51	74	64	34	23	69	49	68	46	59	19	57
83	15	30	60	37	74	65	47	11	22	44	5	10	20	40	80	77	71
89	6	18	54	73	41	34	13	39	28	84	74	44	43	40	31	4	12
97	83	27	38	93	77	94	82	22	13	65	34	73	74	79	7	35	78

Table E.4 (continued)

Indices

p	34	35	36	37	38	39	40	41	42	43	44	45	46	47	48	49
37	28	19	1													
41	20	38	23	15	8	7	1					Numbers				
43	31	7	21	20	17	8	24	29	1							
47	34	29	4	20	6	30	9	45	37	44	32	19	1			
53	9	18	36	19	38	23	46	39	25	50	47	41	29	5	10	20
59	27	54	49	39	19	38	17	34	9	18	36	13	26	52	45	31
61	45	29	58	55	49	37	13	26	52	43	25	50	39	17	34	7
67	65	63	59	51	35	3	6	12	24	48	29	58	49	31	62	57
71	10	70	64	22	12	13	20	69	57	44	24	26	40	67	43	17
73	35	29	72	68	48	21	32	14	70	58	71	63	23	42	64	28
79	13	39	38	35	26	78	76	70	52	77	73	61	25	75	67	43
83	59	35	70	57	31	62	41	82	81	79	75	67	51	19	38	76
89	36	19	57	82	68	26	78	56	79	59	88	86	80	62	8	24
97	2	10	50	56	86	42	16	80	12	60	9	45	31	58	96	92

Indices

p	50	51	52	53	54	55	56	57	58	59	60	61	62	63	64	65
53	40	27	1									Numbers				
59	3	6	12	24	48	37	15	30	1							
61	14	28	56	51	41	21	42	23	46	31	1					
67	47	27	54	41	15	30	60	53	39	11	22	44	21	42	17	34
71	48	52	9	63	15	34	25	33	18	55	30	68	50	66	36	39
73	67	43	69	53	46	11	55	56	61	13	65	33	19	22	37	39
79	50	71	55	7	21	63	31	14	42	47	62	28	5	15	45	56
83	69	55	27	54	25	50	17	34	68	53	23	46	9	18	36	72
89	72	38	25	75	47	52	67	23	69	29	87	83	71	35	16	48
97	72	69	54	76	89	57	91	67	44	26	33	68	49	51	61	14

Indices

p	66	67	68	69	70	71	72	73	74	75	76	77	78	79	80	81
67	1															
71	60	65	29	61	1											
73	49	26	57	66	38	44	1									
79	10	30	11	33	20	60	22	66	40	41	44	53	1			
83	61	39	78	73	63	43	3	6	12	24	48	13	26	52	21	42
89	55	76	50	61	5	15	45	46	49	58	85	77	53	70	32	7
97	70	59	4	20	3	15	75	84	32	63	24	23	18	90	62	19

Indices

p	82	83	84	85	86	87	88	89	90	91	92	93	94	95	96
83	1										Numbers				
89	21	63	11	33	10	30	1								
97	95	87	47	41	11	55	81	17	85	37	88	52	66	39	1

Table E.4 (continued)

d	$\sqrt{d}$	d	$\sqrt{d}$
2	$[1;\overline{2}]$	53	$[7;\overline{3,1,1,3,14}]$
3	$[1;\overline{1,2}]$	54	$[7;\overline{2,1,6,2,14}]$
5	$[2;\overline{4}]$	55	$[7;\overline{2,2,2,14}]$
6	$[2;\overline{2,4}]$	56	$[7;\overline{2,14}]$
7	$[2;\overline{1,1,1,4}]$	57	$[7;\overline{1,1,4,1,1,14}]$
8	$[2;\overline{1,4}]$	58	$[7;\overline{1,1,1,1,1,1,14}]$
10	$[3;\overline{6}]$	59	$[7;\overline{1,2,7,2,1,14}]$
11	$[3;\overline{3,6}]$	60	$[7;\overline{1,2,1,14}]$
12	$[3;\overline{2,6}]$	61	$[7;\overline{1,4,3,1,2,2,1,3,4,1,14}]$
13	$[3;\overline{1,1,1,1,6}]$	62	$[7;\overline{1,6,1,14}]$
14	$[3;\overline{1,2,1,6}]$	63	$[7;\overline{1,14}]$
15	$[3;\overline{1,6}]$	65	$[8;\overline{16}]$
17	$[4;\overline{8}]$	66	$[8;\overline{8,16}]$
18	$[4;\overline{4,8}]$	67	$[8;\overline{5,2,1,1,7,1,1,2,5,16}]$
19	$[4;\overline{2,1,3,1,2,8}]$	68	$[8;\overline{4,16}]$
20	$[4;\overline{2,8}]$	69	$[8;\overline{3,3,1,4,1,3,3,16}]$
21	$[4;\overline{1,1,2,1,1,8}]$	70	$[8;\overline{2,1,2,1,2,16}]$
22	$[4;\overline{1,2,4,2,1,8}]$	71	$[8;\overline{2,2,1,7,1,2,2,16}]$
23	$[4;\overline{1,3,1,8}]$	72	$[8;\overline{2,16}]$
24	$[4;\overline{1,8}]$	73	$[8;\overline{1,1,5,5,1,1,16}]$
26	$[5;\overline{10}]$	74	$[8;\overline{1,1,1,1,16}]$
27	$[5;\overline{5,10}]$	75	$[8;\overline{1,1,1,16}]$
28	$[5;\overline{3,2,3,10}]$	76	$[8;\overline{1,2,1,1,5,4,5,1,1,2,1,16}]$
29	$[5;\overline{2,1,1,2,10}]$	77	$[8;\overline{1,3,2,3,1,16}]$
30	$[5;\overline{2,10}]$	78	$[8;\overline{1,4,1,16}]$
31	$[5;\overline{1,1,3,5,3,1,1,10}]$	79	$[8;\overline{1,7,1,16}]$
32	$[5;\overline{1,1,1,10}]$	80	$[8;\overline{1,16}]$
33	$[5;\overline{1,2,1,10}]$	82	$[9;\overline{18}]$
34	$[5;\overline{1,4,1,10}]$	83	$[9;\overline{9,18}]$
35	$[5;\overline{1,10}]$	84	$[9;\overline{6,18}]$
37	$[6;\overline{12}]$	85	$[9;\overline{4,1,1,4,18}]$
38	$[6;\overline{6,12}]$	86	$[9;\overline{3,1,1,1,8,1,1,1,3,18}]$
39	$[6;\overline{4,12}]$	87	$[9;\overline{3,18}]$
40	$[6;\overline{3,12}]$	88	$[9;\overline{2,1,1,1,2,18}]$
41	$[6;\overline{2,2,12}]$	89	$[9;\overline{2,3,3,2,18}]$
42	$[6;\overline{2,12}]$	90	$[9;\overline{2,18}]$
43	$[6;\overline{1,1,3,1,5,1,3,1,1,12}]$	91	$[9;\overline{1,1,5,1,5,1,1,18}]$
44	$[6;\overline{1,1,1,2,1,1,1,12}]$	92	$[9;\overline{1,1,2,4,2,1,1,18}]$
45	$[6;\overline{1,2,2,2,1,12}]$	93	$[9;\overline{1,1,1,4,6,4,1,1,1,18}]$
46	$[6;\overline{1,3,1,1,2,6,2,1,1,3,1,12}]$	94	$[9;\overline{1,2,3,1,1,5,1,8,1,5,1,1,3,2,1,18}]$
47	$[6;\overline{1,5,1,12}]$	95	$[9;\overline{1,2,1,18}]$
48	$[6;\overline{1,12}]$	96	$[9;\overline{1,3,1,18}]$
50	$[7;\overline{14}]$	97	$[9;\overline{1,5,1,1,1,1,1,1,5,1,18}]$
51	$[7;\overline{7,14}]$	98	$[9;\overline{1,8,1,18}]$
52	$[7;\overline{4,1,2,1,4,14}]$	99	$[9;\overline{1,18}]$

Table E.5 *Simple continued fractions for square roots of positive integers.*

Answers to Odd-Numbered Exercises

Section 1.1

1. a. well-ordered; every subset of this set is also a subset of the set of positive integers and hence must have a least element
b. well-ordered; every subset of this set is also a subset of the set of positive integers and hence must have a least element
c. not well-ordered; the set of positive rational numbers
d. well-ordered; the set of numerators of the numbers in any subset is a subset of the set of positive integers, so it must have a least element b, and then $b/2$ is the least element of the subset
e. not well-ordered; the set of positive rational numbers

3. Let a/b and c/d be the given rational numbers, where a, b, c, and d are integers with b and d nonzero. The sum of the two rational numbers is $(ad + bc)/(bd)$, which is a rational number since the numerator and denominator are integers and the denominator is not 0. Similarly, their product is $(ac)/(bd)$, which is a rational number for the same reasons.

5. Suppose that $\sqrt{3} = a/b$, with a and b positive integers. Then the set $S = \{k\sqrt{3} \mid k$ and $k\sqrt{3}$ are positive integers$\}$ is nonempty since it contains $a = b\sqrt{3}$. By the well-ordering property, S has a smallest element, say $s = t\sqrt{3}$. Consider $s' = s\sqrt{3} - s = 3t - s$. Since $3t$ and s are both integers, s' must also be an integer. Note that $s' = s(\sqrt{3} - 1)$, so s' is positive since $\sqrt{3} > 1$, and s' is less than s since $\sqrt{3} < 2$. This contradicts the choice of s, so our original assumption that $\sqrt{3}$ is rational is wrong.

7. a. 0   **b.** −1   **c.** 3   **d.** −2   **e.** 0   **f.** −4

9. a. $\{8/5\} = 3/5$ **b.** $\{1/7\} = 1/7$ **c.** $\{-11/4\} = 1/4$ **d.** $\{7\} = 0$

11. 0 if x is an integer; −1 otherwise

13. We have $[x] \leq x$ and $[y] \leq y$. Adding these two inequalities gives $[x] + [y] \leq x + y$. Hence, $[x + y] \geq [[x] + [y]] = [x] + [y]$.

15. Let $x = a + r$ and $y = b + s$, where a and b are integers and r and s are real numbers such that $0 \leq r, s < 1$. Then $[xy] = [ab + as + br + sr] = ab + [as + br + sr]$, whereas $[x][y] = ab$. Thus

$[xy] \geq [x][y]$. If x and y are both negative, then $[xy] \leq [x][y]$. If one of x and y is positive and the other negative, then either $[xy]$ or $[x][y]$ could be larger.

17. Let $x = [x] + r$. Since $0 \leq r < 1$, $x + \frac{1}{2} = [x] + r + \frac{1}{2}$. If $r < \frac{1}{2}$, then $[x]$ is the integer nearest to x, and $[x + \frac{1}{2}] = [x]$ since $[x] \leq x + \frac{1}{2} = [x] + r + \frac{1}{2} < [x] + 1$. If $r \geq \frac{1}{2}$, then $[x] + 1$ is the integer nearest to x (choosing this integer if x is midway between $[x]$ and $[x + 1]$), and $[x + \frac{1}{2}] = [x] + 1$ since $[x] + 1 \leq x + r + \frac{1}{2} < [x] + 2$.

19. If x is a positive integer, then the two sides are identical. So suppose that $x = n^2 + m + \epsilon$, where n is the largest perfect square integer less than x, m is a nonnegative integer, and $0 < \epsilon < 1$. Then both $\sqrt{x}$ and $\sqrt{[x]} = \sqrt{n^2 + m}$ are between n and $n + 1$. Therefore, both sides of the equation equal n.

21. a. $8n - 5$ **b.** $2^n + 3$ **c.** $[[\sqrt{n}]/\sqrt{n}]$ **d.** $a_1 = 1$, $a_2 = 3$, and $a_n = a_{n-1} + a_{n-2}$ for $n \geq 3$

23. $a_n = 2^{n-1}$; $a_n = (n^2 - n + 2)/2$; $a_1 = 1$, $a_2 = 2$, and $a_n = a_{n-1} + 2a_{n-2}$ for $n \geq 3$

25. This set is exactly the sequence $a_n = n - 100$ and hence is countable.

27. The function $f(a + b\sqrt{2}) = 2^a 3^b$ is a one-to-one map of this set into the set of positive integers, which is countable.

29. Suppose that $\{A_i\}$ is a countable collection of countable sets. Then each A_i can be represented by a sequence: $A_1 = \{a_{11}, a_{12}, a_{13}, \ldots\}$, $A_2 = \{a_{21}, a_{22}, a_{23}, \ldots\}$, $A_3 = \{a_{31}, a_{32}, a_{33}, \ldots\}$, $\ldots$ Consider the listing $a_{11}, a_{12}, a_{21}, a_{13}, a_{22}, a_{31}, \ldots$, in which we first list the elements with subscripts adding to 2, then the elements with subscripts adding to 3, and so on. Further, we order the elements with subscripts adding to k in order of the first subscript. Form a new sequence c_i as follows. Let $c_1 = a_1$. Given that c_n is determined, let c_{n+1} be the next element in the listing that is different from each c_i with $i = 1, 2, \ldots, n$. Then this sequence is exactly the elements of $\bigcup_{i=1}^{\infty} A_i$, which is therefore countable.

31. a. $a = 4, b = 7$ **b.** $a = 7, b = 10$ **c.** $a = 7, b = 69$ **d.** $a = 1, b = 20$

33. The number α lies in an interval of the form $r/k \leq \alpha < (r + 1)/k$, where $0 \leq r \leq k - 1$. If we divide this interval into equal halves, then α must lie in one of the halves, so either $r/k \leq \alpha < (2r + 1)/2k$ or $(2r + 1)/2k \leq \alpha < (r + 1)/k$. In the first case, because $|\alpha - r/k| < 1/2k$, we can take $u = r$. In the second case, we can take $u = r + 1$, because $|\alpha - (r + 1)/k| < 1/2k$.

35. First we have $|\sqrt{2} - 1/1| = 0.414\ldots < 1/1^2$. Second, by Exercise 30 (a), we have $|\sqrt{2} - 7/5| < 1/50 < 1/5^2$. Third, observing that $3/7 = 0.428\ldots$ leads us to try $|\sqrt{2} - 10/7| = 0.014\ldots < 1/7^2 = 0.0204\ldots$. Fourth, observing that $5/12 = 0.4166\ldots$ leads us to try $|\sqrt{2} - 17/12| = 0.00245\ldots < 1/12^2 = 0.00694\ldots$.

37. Assume that $b > 0$ and $q > 0$. Note that if $q > b$, then $|p/q - a/b| = |pb - aq|/qb \geq 1/qb > 1/q^2$. Therefore, solutions to the inequality have $1 \leq q \leq b$. For a given q, there can be only finitely many p such that the distance between the rational numbers a/b and p/q is less than $1/q^2$ (indeed, there is at most one). Therefore, there are only finitely many p/q satisfying the inequality.

39. a. 3, 6, 9, 12, 15, 18, 21, 24, 27, 30 **b.** 1, 3, 5, 6, 8, 10, 12, 13, 15, 17 **c.** 2, 4, 7, 9, 11, 14, 16, 18, 21, 23 **d.** 3, 6, 9, 12, 15, 18, 21, 25, 28, 31

41. Assume that $1/\alpha + 1/\beta = 1$. First, show that the sequences $m\alpha$ and $n\beta$ are disjoint. Then, for an integer k, define $N(k)$ to be the number of elements of the sequences $m\alpha$ and $n\beta$ less than k. Then $N(k) = [k/\alpha] + [k/\beta]$. By definition of the greatest integer function, $k/\alpha - 1 < [k/\alpha] < k/\alpha$ and $k/\beta - 1 < [k/\beta] < k/\beta$. Add these inequalities to deduce that $k - 2 < N(k) < k$. Hence, $N(k) = k - 1$, and the conclusion follows. To prove the converse, note that if $1/\alpha + 1/\beta \neq 1$, then the spectrum sequences cannot partition the integers.

43. Assume that there are only finitely many Ulam numbers. Let the two largest Ulam numbers be u_{n-1} and u_n. Then the integer $u_{n-1} + u_n$ is an Ulam number larger than u_n. It is the unique sum of two distinct Ulam numbers u_i and u_j with $i < j$, since $u_i + u_j < u_{n-1} + u_n$ if $j < n$ or if $j = n$ and $i < n - 1$.

Section 1.2

1. a. 55   **b.** -15 **c.** 29/20

3. a. 510   **b.** 24600   **c.** $-255/256$

5. The sum $\sum_{k=1}^{n} [\sqrt{k}]$ counts for every value of k with $\sqrt{k} \geq 1$. There are n such values of k in the range $k = 1, 2, 3, \ldots, n$. It counts another 1 for every value of k with $\sqrt{k} \geq 2$. There are $n - 3$ such values in the range. The sum counts another 1 for each value of k with $\sqrt{k} \geq 3$. There are $n - 8$ such values in the range. In general, for $m = 1, 2, 3, \ldots, [\sqrt{n}]$ the sum counts a 1 for each value of k with $\sqrt{k} \geq m$, and there are $n - (m^2 - 1)$ values in the range. Therefore, $\sum_{k=1}^{n} [\sqrt{k}] = \sum_{m=1}^{[\sqrt{n}]} n - (m^2 - 1) = [\sqrt{n}](n + 1) - \sum_{m=1}^{[\sqrt{n}]} m^2 = [\sqrt{n}](n + 1) - ([\sqrt{n}]([\sqrt{n}] + 1)(2[\sqrt{n}] + 1))/6$.

7. The total number of dots in the n by $n + 1$ rectangle, namely $n(n + 1)$, is $2t_n$ since the rectangle is made from two triangular arrays. Dividing both sides by 2 gives the desired formula.

9. From Exercise 8 we have $p_n = \sum_{k=1}^{n} (3k - 2) = 3 \sum_{k=1}^{n} k - 2 \sum_{k=1}^{n} 1 = 3n(n + 1)/2 - 2n = (3n^2 - n)/2$. On the other hand, $t_{n-1} + n^2 = n(n - 1)/2 + n^2 = (3n^2 - n)/2$ as well.

11. a. Consider a regular heptagon which we border successively by heptagons with 3, 4, 5, $\ldots$ dots on each side. Define the heptagonal number s_k to be the number of dots contained in the k nested heptagons.
b. $(5k^2 - 3k)/2$

13. By Exercise 12 we have $T_n = \sum_{k=1}^{n} t_k = \sum_{k=1}^{n} k(k + 1)/2$. Note that $(k + 1)^3 - k^3 = 3k^2 + 3k + 1 = 3(k^2 + k) + 1$, so that $k^2 + k = (k + 1)^3 = k^3/3 - 1/3$. It follows that $T_n = (1/2) \sum_{k=1}^{n} k(k + 1) = (1/6) \sum_{k=1}^{n} ((k + 1)^3 - k^3) - (1/6) \sum_{k=1}^{n} 1$. Because the first sum telescopes, we conclude that $T_n = (1/6)((n + 1)^3 - 1^3) - 1/6 = (n^3 + 3n^2 + 2n)/6$.

15. Each of these four quantities is the product of 100 integers. The largest product is 100^{100}, since it is the product of 100 factors of 100. The second largest is 100! which is the product of the integers $1, 2, \ldots, 100$, and each of these terms is less than or equal to 100. The third largest is $(50!)^2$ which is the product of $1^2, 2^2, \ldots, 50^2$, and each of these factors j^2 is less than $j(50 + j)$. The smallest is 2^{100} as is easily seen.

17. We have $\sum_{k=1}^{n} 1/(k(k + 1)) = \sum_{k=1}^{n} (1/k - 1/(k + 1))$. Let $a_j = 1/(j + 1)$. Notice that this is a telescoping sum. Using the notation in the text preceding Example 1.19, we have $\sum_{k=1}^{n} (1/k - 1/(k + 1)) = \sum_{j=1}^{n} (a_{j-1} - a_j) = -(a_n - a_0) = 1 - 1/(n + 1)$.

19. We sum both sides of the identity $(k + 1)^3 - k^3 = 3k^2 + 3k + 1$ from $k = 1$ to $k = n$. Then the sum telescopes, as in Example 1.19, yielding $\sum_{k=1}^{n} ((k + 1)^3 - k^3) = (n + 1)^3 - 1$. Also $\sum_{k=1}^{n} (3k^2 + 3k + 1) = 3(\sum_{k=1}^{n} k^2) + 3(\sum_{k=1}^{n} k) + \sum_{k=1}^{n} 1 = 3(\sum_{k=1}^{n} k^2) + 3n(n + 1)/2 + n$. As these two expressions are equal, solving for $\sum_{k=1}^{n} k^2$ yields, after several steps of algebra, $\sum_{k=1}^{n} k^2 = n(n + 1)(2n + 1)/6$.

21. a. $10! = (7!)(8 \cdot 9 \cdot 10) = (7!)(720) = (7!)(6!)$
b. $10! = (7!)(6!) = (7!)(5!) \cdot 6 = (7!)(5!)(3!)$
c. $16! = (14!)(15 \cdot 16) = (14!)(240) = (14!)(5!)(2!)$
d. $9! = (7!)(8 \cdot 9) = (7!)(6 \cdot 6 \cdot 2) = (7!)(3!)(3!)(2!)$

23. $x = y = 1, z = 2$

Section 1.3

1. For $n = 1$ we have $1 < 2^1 = 2$. Now assume $n < 2^n$. Then $n + 1 < 2^n + 1 < 2^n + 2^n = 2^{n+1}$.

3. For the basis step, $\sum_{k=1}^{1} 1/k^2 = 1 \leq 2 - \frac{1}{1}$. For the inductive step, we assume that $\sum_{k=1}^{n} 1/k^2 \leq 2 - 1/n$. Then $\sum_{k=1}^{n+1} 1/k^2 = \sum_{k=1}^{n} 1/k^2 + 1/(n+1)^2 \leq 2 - 1/n + 1/(n+1)^2 = 2 - ((n+1)^2 - n)/(n(n+1)^2) = 2 - (n^2 + n + 1)/(n(n+1)^2) \leq 2 - (n^2 + n)/(n(n+1)^2) = 2 - 1/(n+1)$, as desired.

5. $\mathbf{A}^n = \begin{bmatrix} 1 & n \\ 0 & 1 \end{bmatrix}$. The basis step is trivial. For the inductive step, assume that $\mathbf{A}^n = \begin{bmatrix} 1 & n \\ 0 & 1 \end{bmatrix}$. Then $\mathbf{A}^{n+1} = \mathbf{A}^n \mathbf{A} = \begin{bmatrix} 1 & n \\ 0 & 1 \end{bmatrix} \begin{bmatrix} 1 & 1 \\ 0 & 1 \end{bmatrix} = \begin{bmatrix} 1 & n+1 \\ 0 & 1 \end{bmatrix}$.

7. For the basis step, $\sum_{j=1}^{1} j^2 = 1 = 1(1+1)(2 \cdot 1 + 1)/6$. For the inductive step, we assume that $\sum_{j=1}^{n} j^2 = n(n+1)(2n+1)/6$. Then $\sum_{j=1}^{n+1} j^2 = \sum_{j=1}^{n} j^2 + (n+1)^2 = n(n+1)(2n+1)/6 + (n+1)^2 = (n+1)((n+1)+1)(2(n+1)+1)/6$.

9. For the basis step, $\sum_{j=1}^{1} j(j+1) = 2 = 1 \cdot 2 \cdot 3/3$. Assume it is true for n. Then $\sum_{j=1}^{n+1} j(j+1) = n(n+1)(n+2)/3 + (n+1)(n+2) = (n+1)(n+2)(n/3+1) = (n+1)(n+2)(n+3)/3$.

11. $2^{n(n+1)/2}$

13. Proof using mathematical induction. We see that $12 = 4 \cdot 3$. Now assume that postage of n cents can be formed, with $n = 4a + 5b$, where a and b are nonnegative integers. To form $n + 1$ cents postage, if $a > 0$ we can replace a 4-cent stamp with a 5-cent stamp; that is, $n + 1 = 4(a - 1) + 5(b + 1)$. If no 4-cent stamps are present, then all 5-cent stamps were used. It follows that there must be at least three 5-cent stamps and these can be replaced by four 4-cent stamps; that is, $n + 1 = 4(a + 4) + 5(b - 3)$.

15. We use mathematical induction. The inequality is true for $n = 0$ since $H_{2^0} = H_1 = 1 \geq 1 = 1 + 0/2$. Now assume that the inequality is true for n, that is, $H_{2^n} \geq 1 + n/2$. Then $H_{2^{n+1}} = \sum_{j=1}^{2^n} 1/j + \sum_{j=2^n+1}^{2^{n+1}} 1/j \geq H_{2^n} + \sum_{j=2^n+1}^{2^{n+1}} 1/2^{n+1} \geq 1 + n/2 + 2^n/2^{n+1} = 1 + n/2 + 1/2 = 1 + (n+1)/2$.

17. For the basis step, $(2 \cdot 1)! = 2 < 2^{2 \cdot 1}(1!)^2 = 4$. For the inductive step, we assume that $(2n)! < 2^{2n}(n!)^2$. Then $(2(n+1))! = (2n)!(2n+1)(2n+2) < 2^{2n}(n!)^2(2n+1)(2n+2) < 2^{2n}(n!)^2(2n+2)^2 = 2^{2(n+1)}((n+1)!)^2$.

19. Let A be such a set. Let $B = \{x - k + 1 \mid x \in A \text{ and } x \geq k\}$, which is clearly a set of positive integers. Since $k \in A$ and $k \geq k$, we know that $k - k + 1 = 1$ is in B. Since $n + 1$ is in A whenever n is, $n + 1 - k + 1$ is in B whenever $n - k + 1$ is. Thus B satisfies the hypothesis for mathematical induction, i.e., B is the set of positive integers. Mapping B back to A in the natural manner, we find that A contains the set of integers greater than or equal to k.

21. For the basis step, $4^2 = 16 < 24 = 4!$. For the inductive step, we assume that $n^2 < n!$. Then $(n+1)^2 = n^2 + 2n + 1 < n! + 2n + 1 < n \cdot n! + n! = (n+1)n! = (n+1)!$.

23. We use the second principle of mathematical induction to prove that $n - 1$ moves are necessary and sufficient to assemble a puzzle with n pieces. For the basis step $(n = 1)$, if the puzzle has only one piece, then it clearly may be assembled with no moves. For the inductive step, assume that a puzzle of k pieces takes $k - 1$ moves, for all $k \leq n$. To assemble a puzzle of $n + 1$ pieces, first assemble n pieces, using $n - 1$ moves. This leaves two blocks—the assembled n pieces and the last piece. Now make the move consisting of putting these two blocks together. Thus assembling a puzzle of $n + 1$ pieces can be done in $n = (n + 1) - 1$ moves. To see that it cannot be done

in fewer moves, look at the situation just before making any last move, where there are two blocks, say of sizes i and $n + 1 - i$. By the inductive hypothesis, it required $i - 1$ moves to put together the first block and $n + 1 - i - 1 = n - i$ moves to put together the second block. Thus $i - 1 + n - i = n - 1$ moves have been required thus far. These, together with the final move, account for n moves, as desired.

25. Suppose that $f(n)$ is defined recursively by specifying the value of $f(1)$ and a rule for finding $f(n + 1)$ from $f(n)$. We will prove by mathematical induction that such a function is well-defined. First note that $f(1)$ is well-defined since this value is explicitly stated. Now assume that $f(n)$ is well-defined. Then $f(n + 1)$ also is well-defined since a rule is given for determining this value from $f(n)$.

27. 65,536

29. We use the second principle of mathematical induction. The basis step consists of verifying the formula for $n = 1$ and $n = 2$. For $n = 1$ we have $f(1) = 1 = 2^1 + (-1)^1$, and for $n = 2$ we have $f(2) = 5 = 2^2 + (-1)^2$. Now assume that $f(k) = 2^k + (-1)^k$ for all positive integers k with $k < n$, where $n > 2$. By the inductive hypothesis, $f(n) = f(n - 1) + 2f(n - 2) = (2^{n-1} + (-1)^{n-1}) + 2(2^{n-2} + (-1)^{n-2}) = (2^{n-1} + 2^{n-1}) + (-1)^{n-2}(-1 + 2) = 2^n + (-1)^n$.

31. We use the second principle of mathematical induction. We see that $a_0 = 1 \le 3^0 = 1$, $a_1 = 3 \le 3^i = 3$, and $a_2 = 9 \le 3^2 = 9$. These are the basis cases. Now assume that $a_k \le 3^k$ for all integers k with $0 \le k < n$. It follows that $a_n = a_{n-1} + a_{n-2} + a_{n-3} \le 3^{n-1} + 3^{n-2} + 3^{n-3} = 3^{n-3}(1 + 3 + 9) = 13 \cdot 3^{n-3} < 27 \cdot 3^{n-3} = 3^n$.

33. Let P_n be the statement for n. Then P_2 is true, since we have $((a_1 + a_2)/2)^2 - a_1a_2 = ((a_1 - a_2)/2)^2 \ge 0$. Assume that P_n is true. Then by P_2, for $2n$ positive real numbers $a_1, \ldots, a_{2n}$ we have $a_1 + \cdots + a_{2n} \ge 2(\sqrt{a_1a_2} + \sqrt{a_3a_4} + \cdots + \sqrt{a_{2n-1}a_{2n}})$. Apply P_n to this last expression to get $a_1 + \cdots + a_{2n} \ge 2n(a_1a_2 \cdots a_{2n})^{1/(2n)}$. This establishes P_n for $n = 2^k$ for all k. Again, assume P_n is true. Let $g = (a_1a_2 \cdots a_{n-1})^{1/(n-1)}$. Applying P_n, we have $a_1 + a_2 + \cdots + a_{n-1} + g \ge n(a_1a_2 \cdots a_{n-1}g)^{1/n} = n(g^{n-1}g)^{1/n} = ng$. Therefore $a_1 + a_2 + \cdots + a_{n-1} \ge (n - 1)g$, which establishes P_{n-1}. Thus P_n implies P_{n-1}. Putting these two pieces together establishes P_n for all n.

35. We follow the hint. The basis step follows immediately because the algorithm stops after 1 step when applied to a fraction of the form $1/q$. To carry out the induction step, assume that the algorithm terminates for all fractions with numerator less than p. Given a fraction p/q, apply the algorithm and find the unit fraction $1/s$ such that $1/(s - 1) > p/q > 1/s$. When we subtract $1/s$ from p/q, the remainder is $p/q - 1/s = (ps - q)/qs$. On the other hand, when we multiply the inequality $1/(s - 1) > p/q > 1/s$ by $q(s - 1)$, we see that $q > p(s - 1)$. This implies that $p > ps - q$, showing that the numerator of $(ps - q)/q$ is less than the numerator of the fraction p/q. Applying the induction hypothesis finishes the proof.

Section 1.4

1. **a.** 55   **b.** 233   **c.** 610   **d.** 2584   **e.** 6765   **f.** 75,025

3. Note that $2f_{n+2} - f_n = f_{n+2} + (f_{n+2} - f_n) = f_{n+2} + f_{n+1} = f_{n+3}$. Add f_n to both sides.

5. For the basis step (when $n = 1$ and $n = 2$), note that $f_2 = f_1^2 + 2f_0f_1$ because $1 = 1^2 + 2 \cdot 0 \cdot 1$ and $f_4 = f_2^2 + 2f_1f_2$ because $3 = 1^2 + 2 \cdot 1 \cdot 1$. For the induction step, assume that $f_{2k} = f_k^2 + 2f_{k-1}f_k$ for $k = 1, 2, \ldots, n$, where $n \ge 3$. Using the induction hypothesis, we have $f_{2n-4} = f_{n-2}^2 + 2f_{n-3}f_{n-2}$ and $f_{2n-2} = f_{n-1}^2 + 2f_{n-2}f_{n-1}$. We have $f_{2n} = f_{2n-1} + f_{2n-2} = 2f_{2n-2} + f_{2n-3} = 3f_{2n-2} - f_{2n-4}$ because $f_{2n-1} = f_{2n-2} + f_{2n-3}$ and $f_{2n-2} = f_{2n-3} + f_{2n-4}$. Using the inductive hypothesis, this last expression equals $3f_{n-1}^2 + 6f_{n-2}f_{n-1} - f_{n-2}^2 -$

$2f_{n-3}f_{n-2} = 3f_{n-1}^2 + 6(f_n - f_{n-1})f_{n-1} - (f_n - f_{n-1})^2 - 2(f_{n-1} - f_{n-2})(f_n - f_{n-1}) = -2f_{n-1}^2 + 6f_n f_{n-1} - f_n^2 + 2f_n(f_n - f_{n-1}) - 2f_{n-1}(f_n - f_{n-1}) = f_n^2 + 2f_{n-1}f_n$, which completes the induction step.

7. $\sum_{j=1}^n f_{2j-1} = f_{2n}$. Basis case $n = 1$ is trivial. Now assume that $\sum_{j=1}^n f_{2j-1} = f_{2n}$. Using this inductive hypothesis, we have $\sum_{j=1}^{n+1} f_{2j-1} = (\sum_{j=1}^n f_{2j-1}) + f_{2n+1} = f_{2n} + f_{2n+1} = f_{2n+2}$.

9. The sum is $f_{n-1} - (-1)^n$. To see this in the case that $n = 2k$ is even, note that the sum is $(f_2 + f_4 + \cdots + f_{2k}) - (f_1 + f_3 + \cdots + f_{2k-1})$, which equals $(f_{2k+1} - 1) - f_{2k} = f_{2k-1} - 1 = f_{n-1} - 1 = f_{n-1} - (-1)^n$ by Exercises 8 and 9. Similarly, when $n = 2k + 1$ is odd, we have $(f_1 + f_3 + \cdots + f_{2k+1}) - (f_2 + f_4 + \cdots + f_{2k}) = f_{2k+2} - (f_{2k+1} - 1) = f_{2k} + 1 = f_{n-1} + 1 = f_{n-1} - (-1)^n$.

11. By Exercise 5, we have $f_{2n} = f_n^2 + 2f_{n-1}f_n = f_n(f_n + f_{n-1} + f_{n-1}) = (f_{n+1} - f_{n-1})(f_{n+1} + f_{n-1}) = f_{n+1}^2 - f_{n-1}^2$.

13. We use mathematical induction. To complete the basis step, note that $\sum_{j=1}^1 f_j^2 = f_1 f_2$ because the left-hand side is $f_1^2 = 1^2 = 1$ and the right-hand side is $f_1 f_2 = 1 \cdot 1 = 1$. For the induction step, assume that $\sum_{j=1}^n f_j^2 = f_n f_{n+1}$. It follows that $\sum_{j=1}^{n+1} f_j^2 = \sum_{j=1}^n f_j^2 + f_{n+1}^2 = f_n f_{n+1} + f_{n+1}^2 = f_{n+1}(f_n + f_{n+1}) = f_{n+1}f_{n+2}$, completing the proof.

15. We use mathematical induction and the recursive definition $f_n = f_{n-1} + f_{n-2}$, with $f_0 = 0$ and $f_1 = 1$. For $n = 1$, we have $f_2 f_0 - f_1^2 = 1 \cdot 0 - 1^2 = -1 = (-1)^1$. Hence, the basis step holds. Now assume that $f_{n+1}f_{n-1} - f_n^2 = (-1)^n$. Then $f_{n+2}f_n - f_{n+1}^2 = (f_{n+1} + f_n)f_n - f_{n+1}(f_n + f_{n-1}) = f_n^2 - f_{n+1}f_{n-1} = -(-1)^n = (-1)^{n+1}$.

17. For fixed m, we proceed by induction on n. For the basis step, note that when $n = 1$, the identity holds because $f_{m+1} = f_m f_2 + f_1 f_{m-1} = f_m + f_{m-1}$. When $n = 2$, the identity holds because $f_{m+2} = f_m f_3 + f_2 f_{m-1} = 2f_m + f_{m-1} = f_m + (f_m + f_{m-1}) = f_m + f_{m+1}$. For the induction step, assume that the identity holds for $1, 2, \ldots, k$, where $k \geq 3$. Then $f_{m+k} = f_m f_{k+1} + f_{m-1}f_k$ and $f_{m+k-1} = f_m f_k + f_{m-1}f_{k-1}$. Adding these equations gives us $f_{m+k} + f_{m+k-1} = f_m(f_{k+1} + f_k) + f_{m-1}(f_k + f_{k-1})$. This simplifies to $f_{m+k+1} = f_m f_{k+2} + f_{m-1}f_{k+1}$.

19. $\sum_{i=1}^n L_i = L_{n+2} - 3$. We prove this by induction. The basis step is $L_1 = 1 = L_3 - 3$. Assume that the formula holds for n and compute $\sum_{i=1}^{n+1} L_i = \sum_{i=1}^n L_i + L_{n+1} = L_{n+2} - 3 + L_{n+1} = L_{n+3} - 3$.

21. $\sum_{i=1}^n L_{2i} = L_{2n+1} - 1$. We prove this by induction. The basis step is $L_2 = 3 = L_3 - 1$. Assume that the formula holds for n and compute $\sum_{i=1}^{n+1} L_{2i} = \sum_{i=1}^n L_{2i} + L_{2n+2} = L_{2n+1} - 1 + L_{2n+2} = L_{2n+3} - 1$.

23. We proceed by induction. The basis step is $L_1^2 = 1 = L_1 L_2 - 2$. Assume that the formula holds for n. Then $\sum_{i=1}^{n+1} L_i^2 = \sum_{i=1}^n L_i^2 + L_{n+1}^2 = L_n L_{n+1} - 2 + L_{n+1}^2 = L_{n+1}(L_n + L_{n+1}) - 2 = L_{n+1}L_{n+2} - 2$.

25. For the basis step we check that $L_1 f_1 = 1 = f_2$ and $L_2 f_2 = 3 = f_4$. Assume that the identity is true for all positive integers up to n. Then $f_{n+1}L_{n+1} = (f_{n+2} - f_n)(f_{n+2} - f_n)$ from Exercise 24. This equals $f_{n+2}^2 - f_n^2 = (f_{n+1} + f_n)^2 - (f_{n-1} + f_{n-2})^2 = f_{n+1}^2 + 2f_{n+1}f_n + f_n^2 - f_{n-1}^2 - 2f_{n-1}f_{n-2} - f_{n-2}^2 = (f_{n+1}^2 - f_{n-1}^2) + (f_n^2 - f_{n-2}^2) + 2(f_{n+1}f_n - f_{n-1}f_{n-2}) = (f_{n+1} - f_{n-1})(f_{n+1} + f_{n-1}) + (f_n - f_{n-2})(f_n + f_{n-2}) + 2f_{2n-1}$, where the last term is obtained from Exercise 16. This equals $f_n L_n + f_{n-1}L_{n-1} + 2f_{2n-1}$. Applying the induction hypothesis yields $f_{2n} + f_{2n-2} + 2f_{2n-1} = (f_{2n} + f_{2n-1}) + (f_{2n-1} + f_{2n-2}) = f_{2n+1} + f_{2n} = f_{2n+2}$, which completes the induction.

27. We prove this by induction on n. If $n = 2$, use induction on m to establish the basis step for n. For the induction step on n, note that $L_{m+n+1} = L_{m+n} + L_{m+n-1} = (f_{m+1}L_n + f_m L_{n-1}) + (f_{m+1}L_{n-1} + f_m L_{n-2}) = f_{m+1}(L_n + L_{n-1}) + f_m(L_{n-1} + L_{n-2}) = f_{m+1}L_{n+1} + f_m L_n$.

29. $50 = f_9 + f_7 + f_4, 85 = f_{10} + f_8 + f_6 + f_2, 110 = f_{11} + f_8, 200 = f_{12} + f_{10} + f_2$

31. We proceed by mathematical induction. The basis steps ($n = 2, 3$) are easily seen to hold. For the inductive step, we assume that $f_n \le \alpha^{n-1}$ and $f_{n-1} \le \alpha^{n-2}$. Now $f_{n+1} = f_n + f_{n-1} \le \alpha^{n-1} + \alpha^{n-2} = \alpha^n$, since α satisfies $\alpha^n = \alpha^{n-1} + \alpha^{n-2}$.

33. We use Theorem 1.3. Note that $\alpha^2 = \alpha + 1$ and $\beta^2 = \beta + 1$, since α and β are the roots of $x^2 - x - 1 = 0$. Then $f_{2n} = (\alpha^{2n} - \beta^{2n})/\sqrt{5} = (1/\sqrt{5})((\alpha + 1)^n - (\beta + 1)^n) = (1/\sqrt{5}) \left(\sum_{j=0}^{n} \binom{n}{j} \alpha^j - \sum_{j=0}^{n} \binom{n}{j} \beta^j \right) = (1/\sqrt{5}) \sum_{j=0}^{n} \binom{n}{j}(\alpha^j - \beta^j) = \sum_{j=1}^{n} \binom{n}{j} f_j$ since the first term is 0 in the second-to-last sum.

35. We have $\det(\mathbf{F}^n) = \det(\mathbf{F})^n = (-1)^n$ and $\det \begin{bmatrix} f_{n+1} & f_n \\ f_n & f_{n-1} \end{bmatrix} = f_{n+1}f_{n-1} - f_n^2$.

37. $f_n = f_{n+2} - f_{n+1}$; $f_{-1} = 1$, $f_{-2} = -1$, $f_{-3} = 2$, $f_{-4} = -3$, $f_{-5} = 5$, $f_{-6} = -8$, $f_{-7} = 13$, $f_{-8} = -21$, $f_{-9} = 34$, $f_{-10} = -55$

39. The square has area 64 square units, while the rectangle has area 65 square units. This corresponds to the identity in Exercise 14, which tells us that $f_7 f_5 - f_6^2 = 1$. Notice that the slope of the hypotenuse of the triangular piece is $\frac{3}{8}$, while the slope of the top of the trapezoidal piece is $\frac{2}{5}$. We have $\frac{2}{5} - \frac{3}{8} = \frac{1}{40}$. Thus the "diagonal" of the rectangle is really a very skinny parallelogram of area 1, hidden visually by the fact that the two slopes are nearly equal.

41. We solve the equation $r^2 - r - 1 = 0$ to discover the roots $r_1 = (1 + \sqrt{5})/2$ and $r_2 = (1 - \sqrt{5})/2$. Then according to the theory in the preamble, $f_n = C_1 r_1^n + C_2 r_2^n$. For $n = 0$ we have $0 = C_1 r_1^0 + C_2 r_2^0 = C_1 + C_2$, and for $n = 1$ we have $1 = C_1 r_1 + C_2 r_2 = C_1(1 + \sqrt{5})/2 + C_2(1 - \sqrt{5})/2$. Solving these two equations simultaneously yields $C_1 = 1/\sqrt{5}$ and $C_2 = -1/\sqrt{5}$. So the explicit formula is $f_n = (1/\sqrt{5})r_1^n - (1/\sqrt{5})r_2^n = (r_1^n - r_2^n)/\sqrt{5}$.

43. We seek to solve the recurrence relation $L_n = L_{n-1} + L_{n-1}$ subject to the initial conditions $L_1 = 1$ and $L_2 = 3$. We solve the equation $r^2 - r - 1 = 0$ to discover the roots $\alpha = (1 + \sqrt{5})/2$ and $\beta = (1 - \sqrt{5})/2$. Then according to the theory in the preamble to Exercise 41, $L_n = C_1\alpha^n + C_2\beta^n$. For $n = 1$ we have $L_1 = 1 = C_1\alpha + C_2\beta$, and for $n = 2$ we have $3 = C_1\alpha^2 + C_2\beta^2$. Solving these two equations simultaneously yields $C_1 = 1$ and $C_2 = 1$. So the explicit formula is $L_n = \alpha^n + \beta^n$.

45. First check that $\alpha^2 = \alpha + 1$ and $\beta^2 = \beta + 1$. We proceed by induction. The basis steps are $(\alpha - \beta)/\sqrt{5} = \sqrt{5}/\sqrt{5} = 1 = f_1$ and $(\alpha^2 - \beta^2)/\sqrt{5} = ((1 + \alpha) - (1 + \beta))/\sqrt{5} = (\alpha - \beta)/\sqrt{5} = 1 = f_2$. Assume that the identity is true for all positive integers up to n. Then $f_{n+1} = f_n + f_{n-1} = (\alpha^n - \beta^n)/\sqrt{5} + (\alpha^{n-1} - \beta^{n-1})/\sqrt{5} = (\alpha^{n-1}(\alpha + 1) - \beta^{n-1}(\beta + 1))/\sqrt{5} = (\alpha^{n-1}(\alpha^2) - \beta^{n-1}(\beta^2))/\sqrt{5} = (\alpha^{n+1} - \beta^{n+1})/\sqrt{5}$, which completes the induction.

Section 1.5

1. $3 \mid 99$ since $99 = 3 \cdot 33$; $5 \mid 145$ since $145 = 5 \cdot 29$; $7 \mid 343$ since $343 = 7 \cdot 49$; $888 \mid 0$ since $0 = 888 \cdot 0$

3. a. yes **b.** yes **c.** no **d.** no **e.** no **f.** no

5. a. $q = 5, r = 15$ **b.** $q = 17, r = 0$ **c.** $q = -3, r = 7$ **d.** $q = -6, r = 2$

7. By the hypothesis, $b = ra$ and $d = sc$ for some r and s. Thus $bd = rs(ac)$, so $ac \mid bd$.

9. If $a \mid b$, then $b = na$ and $bc = n(ca)$, i.e., $ac \mid bc$. Now suppose that $ac \mid bc$. Then $bc = nac$, and, as $c \ne 0$, $b = na$, i.e., $a \mid b$.

11. The statement is trivially true for $k = 1$. Assume therefore that $a \mid b$ and $a^k \mid b^k$; we want to show that $a^{k+1} \mid b^{k+1}$. But this follows directly from Exercise 7.

13. Let $a = 2x + 1$ and $b = 2y + 1$ be odd, and $c = 2z$ even. Then $ab = (2x + 1)(2y + 1) = 4xy + 2x + 2y + 1 = 2(2xy + x + y) + 1$, so ab is odd. On the other hand, if w is any integer, then $cw = (2z)w = 2(zw)$ is even.

15. By the division algorithm $a = bq + r$ with $0 \le r < b$. Thus $-a = -bq - r = -(q + 1)b + b - r$. If $0 \le b - r < b$, then we are done. Otherwise, $b - r = b$, or $r = 0$ and $-a = -qb + 0$.

17. a. Note that if $a = bq + r$, then $a = (-b)(-q) + r$. Therefore, to divide by a negative number, just divide by the corresponding positive number and take the negative of the quotient. The remainder stays the same.
   b. 3

19. By the division algorithm, let $m = qn + r$, with $0 \le r \le n - 1$ and $q = [m/n]$. Then $[(m + 1)/n] = [(qn + r + 1)/n] = [q + (r + 1)/n] = q + [(r + 1)/n]$. If $r = 0, 1, 2, \ldots, n - 2$, then $m \ne kn - 1$ for any integer k and $1/n \le (r + 1)/n < 1$, and so $[(r + 1)/n] = 0$. In this case, we have $[(m + 1)/n] = q + 0 = [m/n]$. On the other hand, if $r = n - 1$, then $m = qn + n - 1 = n(q + 1) - 1 = nk - 1$, and $[(r + 1)/n] = 1$. In this case, we have $[(m + 1)/n] = q + 1 = [m/n] + 1$.

21. The positive integers divisible by the positive integer d are those integers of the form kd, where k is a positive integer. The number of these that are less than x is the number of positive integers k with $kd \le x$, or equivalently with $k \le x/d$. There are $[x/d]$ such integers.

23. 128, 18

25. 457

27. It costs $11 - 22[-x]$ cents to mail a letter weighing x ounces. It cannot cost $1.45; a 10-ounce letter costs $2.31.

29. Multiplying two integers of this form gives us $(4n + 1)(4m + 1) = 16mn + 4m + 4n + 1 = 4(4mn + m + n) + 1$. Similarly, $(4n + 3)(4m + 3) = 16mn + 12m + 12n + 9 = 4(4mn + 3m + 3n + 2) + 1$.

31. Every odd integer can be written in the form $4k + 1$ or $4k + 3$. Observe that $(4k + 1)^4 = 16^2 k^4 + 4(4k)^3 + 6(4k)^2 + 4(4k) + 1 = 16(16k^4 + 16k^3 + 6k^2 + k) + 1$. Proceeding further, $(4k + 3)^4 = (4k)^4 + 12(4k)^3 + 54(4k)^2 + 108(4k) + 3^4 = 16(16k^4 + 48k^3 + 54k^2 + 27k + 5) + 1$.

33. Of any three consecutive integers, one is a multiple of 3. Also, at least one is even. Therefore, the product is a multiple of $2 \cdot 3 = 6$.

35. The basis case is true: $1^3 + 2^3 + 3^3 = 36$ is divisible by 9. Assume the inductive hypothesis that $n^3 + (n + 1)^3 + (n + 2)^3$ is divisible by 9. Then $(n + 1)^3 + (n + 2)^3 + (n + 3)^3 = (n^3 + (n + 1)^3 + (n + 2)^3) + ((n + 3)^3 - n^3) = (n^3 + (n + 1)^3 + (n + 2)^3) + (9n^2 + 27n + 27)$. In this last expression, the first summand is divisible by 9 by the inductive hypothesis, and the second is clearly divisible by 9, so we are done.

37. We proceed by mathematical induction. The basis step is clear. Assume that $3 \mid f_i$ if and only if $4 \mid i$, for all $i \le 4k$. Since $f_{4k+1} = f_{4k} + f_{4k-1}$, knowing that $3 \mid f_{4k}$ and $3 \nmid f_{4k-1}$ tells us that $3 \nmid f_{4k+1}$. Similarly, $3 \mid f_{4k}$ and $3 \nmid f_{4k+1}$ imply $3 \nmid f_{4k+2}$. Also $f_{4k+3} = 2f_{4k+1} + f_{4k}$, and since $3 \mid f_{4k}$ but $3 \nmid 2f_{4k+1}$, we have $3 \nmid f_{4k+3}$. Finally, as $f_{4k+4} = 3f_{4k+1} + 2f_{4k}$ and $3 \mid 2f_{4k}$ and $3 \mid 3f_{4k+1}$, we see that $3 \mid f_{4k+4}$. This has taken us up to the next value of k, as required.

39. The basis cases ($n = 6$ and $n = 7$) state that $f_6 = 5f_2 + 3f_1$ and $f_7 = 5f_3 + 3f_2$, which are true, since $8 = 5 + 3$ and $13 = 5 \cdot 2 + 3 \cdot 1$. Assume the inductive hypothesis (second principle). Then $f_{n+1} = f_n + f_{n-1} = 5f_{n-4} + 3f_{n-5} + 5f_{n-5} + 3f_{n-6} = 5(f_{n-4} + f_{n-5}) + 3(f_{n-5} + f_{n-6}) =$

$5f_{n-3} + 3f_{n-4}$, as desired. For the second statement, the basis case is the true statement that $f_5 = 5$ is divisible by 5. Assuming the inductive hypothesis that f_n is divisible by 5, we see that f_{n+5} ($n + 5$ being the next multiple of 5 after n) is the sum of two multiples of 5, namely $5f_{n+1}$ and $3f_n$.

41. 39, 59, 89, 134, 67, 101, 152, 76, 38, 19, 29, 44, 22, 11, 17, 26, 13, 20, 10, 5, 8, 4, 2, 1

43. We prove this using the second principle of mathematical induction. Since $T(2) = 1$, the Collatz conjecture is true for $n = 2$. Now assume that the conjecture holds for all integers less than n. By assumption, there is an integer k such that k iterations of the transformation T, starting at n, produces an integer m less than n. By the inductive hypothesis, there is an integer l such that iterating T l times starting at m produces the integer 1. Hence, iterating T $k + l$ times starting with n leads to 1.

45. We first show that $(2 + \sqrt{3})^n + (2 - \sqrt{3})^n$ is an even integer. From the binomial theorem it follows that $(2 + \sqrt{3})^n + (2 - \sqrt{3})^n = \sum_{j=0}^{n} \binom{n}{j} 2^{n-j} \sqrt{3}^j + \sum_{j=0}^{n} \binom{n}{j} 2^{n-j} (-1)^j \sqrt{3}^j = 2(2^n + 3\binom{n}{2}2^{n-2} + 3^2\binom{n}{4}2^{n-4} + \cdots) = 2l$, where l is an integer. Next, note that $(2 - \sqrt{3})^n < 1$. We see that $[(2 + \sqrt{3})^n] = (2 + \sqrt{3})^n + (2 - \sqrt{3})^n - 1$. It follows that $[(2 + \sqrt{3})^2]$ is odd.

Section 2.1

1. $(5554)_7$, $(2112)_{10}$

3. $(175)_{10}$, $(1111100111)_2$

5. $(8F5)_{16}$, $(74E)_{16}$

7. The reason is that we are using the blocks of three digits as one "digit," which has 1000 possible values

9. -39, 26

11. If m is any integer weight less than 2^k, then by Theorem 2.1, m has a base 2 expansion $m = a_{k-1}2^{k-1} + a_{k-2}2^{k-2} + \cdots + a_1 2^1 + a_0 2^0$, where each a_i is 0 or 1. The 2^i weight is used if and only if $a_i = 1$.

13. Let w be the weight to be measured. By Exercise 12, w has a unique balanced ternary expansion. Place the object in pan 1. If $e_i = 1$, then place a weight of 3^i in pan 2. If $e_i = -1$, then place a weight of 3^i in pan 1. If $e_i = 0$, then do not use the weight of 3^i. Now the pans will be balanced.

15. To convert a number from base r to base r^n, take the number in blocks of size n. To go the other way, convert each digit of a base r^n number to base r, and concatenate the results.

17. $(a_k a_{k-1} \ldots a_1 a_0 00 \ldots 00)_b$, where we have placed m zeros at the end of the base b expansion of n

19. a. -6 **b.** 13   **c.** -14 **d.** 0

21. If m is positive, then $a_{n-1} = 0$ and $a_{n-2}a_{n-3} \ldots a_0$ is the binary expansion of m. Hence, $m = \sum_{i=0}^{n-2} a_i 2^i$ as desired. If m is negative, then the one's complement expansion for m has its leading bit equal to 1. By the definition of one's complement, we can think of obtaining the remaining $n - 1$ bits by subtracting $-m$, written in binary, from $111 \ldots 1$ (with $n - 1$ 1s), since subtracting a bit from 1 is the same thing as complementing it. Equivalently, if we view the bit string $(a_{n-2}a_{n-1} \ldots a_0)$ as a binary number, then it represents $(2^{n-1} - 1) - (-m)$. In symbols, this says that $(2^{n-1} - 1) - (-m) = \sum_{i=0}^{n-2} a_i 2^i$. Solving for m gives us the equation we are trying to prove (since $a_{n-1} = 1$).

23. a. -7 **b.** 13   **c.** -15 **d.** -1

25. Complement each of the digits in the two's complement representation for m and then add 1.

27. $4n$

29. We first show that every positive integer has a Cantor expansion. To find a Cantor expansion of the positive integer n, let m be the unique positive integer such that $m! \le n < (m+1)!$. By the division algorithm, there is an integer a_m such that $n = m! \cdot a_m + r_m$, where $0 \le a_m \le m$ and $0 \le r_m < m!$. We iterate, finding that $r_m = (m-1)! \cdot a_{m-1} + r_{m-1}$, where $0 \le a_{m-1} \le m-1$ and $0 \le r_{m-1} < (m-1)!$. We iterate $m-2$ more times, obtaining $r_i = (i-1)! \cdot a_{i-1} + r_{i-1}$, where $0 \le a_{i-1} \le i-1$ and $0 \le r_{i-1} < (i-1)!$ for $i = m+1, m, m-1, \dots, 2$, with $r_{m+1} = n$. At the last stage, we have $r_2 = 1! \cdot a_1 + 0$, where $r_2 = 0$ or 1 and $r_2 = a_1$.

31. Call a position *good* if the number of ones in each column is even, and *bad* otherwise. Since a player can affect only one row, he or she must change some column sums. Thus any move from a good position produces a bad position. To find a move from a bad position to a good one, construct a binary number by putting a 1 in the place of each column with odd sum, and a 0 in the place of each column with even sum. Subtracting this number of matches from the largest pile will produce a good position.

33. a. First show that the result of the operation must yield a multiple of 9. Then it suffices to check only multiples of 9 with decreasing digits. There are only 79 of these. If we perform the operation on each of these 79 numbers and reorder the digits, we will have one of the following 23 numbers: 7551, 9954, 5553, 9990, 9981, 8820, 9810, 9620, 8532, 8550, 9720, 9972, 7731, 6543, 8730, 8640, 8721, 7443, 9963, 7632, 6552, 6642, or 6174. It will suffice to check only 9810, 7551, 9990, 8550, 9720, 8640, and 7632.
b. 8

35. Consider $a_0 = (1234)_6$. We find that T_6 repeats with period 6. Therefore it never goes to a Kaprekar's constant for the base 6. Hence there is no Kaprekar's constant for the base 6.

Section 2.2

1. $(10010110110)_2$

3. $(1011101100)_2$

5. $(10110001101)_2$

7. $q = (11111)_2, r = (1100)_2$

9. $(3314430)_5$

11. $(4320023)_5$

13. $(16665)_{16}$

15. $(B705736)_{16}$

17. Represent $(18235187)_{10}$ using three words, $((018)(235)(187))_{1000}$, and $(22135674)_{10}$ using three words, $((022)(135)(674))_{1000}$, where each base 1000 digit is represented by three base 10 digits in parentheses. To find the sum, difference, and product of these integers from their base 1000 representations we carry out the algorithms for such computations for base 1000.

19. We must assume that the sum actually represents a number in the appropriate range. Assume that n bits are being used, so that numbers strictly between -2^{n-1} and 2^{n-1} can be represented. The answer is almost, but not quite, that to obtain the one's complement representation of the sum of two numbers, we simply add the two strings representing these numbers using the usual grade-school right-to-left algorithm, as in Example 2.4. Instead, after performing this operation, there may be a carry out of the left-most column; in such a case, we then add 1 more to the answer.

21. Let $a = (a_m a_{m-1} \dots a_2 a_1)_!$ and $b = (b_m b_{m-1} \dots b_2 b_1)_!$. Then $a + b = (d_{m+1} d_m d_{m-1} \dots d_2 d_1)_!$ is obtained by adding the digits from right to left with the following rule for producing carries.

If $a_j + b_j + c_{j-1}$, where c_{j-1} is the carry from adding a_{j-1} and b_{j-1}, is greater than j, then $c_j = 1$, and the resulting jth digit is $d_j = a_j + b_j + c_{j-1} - j - 1$. Otherwise, the resulting digit is $d_j = a_j + b_j + c_{j-1}$, and $c_j = 0$. To subtract b from a, assuming $a > b$, we let $d_j = a_j - b_j + c_{j-1}$ and set $c_j = 0$ if $a_j - b_j + c_{j-1}$ is between 0 and j (inclusive). Otherwise, $d_j = a_j - b_j + c_{j-1} + j + 1$ and $c_j = -1$. In this manner, $a - b = (d_m d_{m-1} \ldots d_2 d_1)_!$.

23. We have $(a_n \ldots a_1 5)_{10}^2 = (10(a_n \ldots a_1)_{10} + 5)^2 = 100(a_n \ldots a_1)_{10}^2 + 100(a_n \ldots a_1)_{10} + 25 = 100(a_n \ldots a_1)_{10} \cdot ((a_n \ldots a_1)_{10} + 1) + 25$. The decimal digits of this number consist of the decimal digits of $(a_n \ldots a_1)_{10} \cdot ((a_n \ldots a_1)_{10} + 1)$ followed by 25 since this first product is multiplied by 100, which shifts its decimal expansion two digits.

Section 2.3

1. a. yes   **b.** no   **c.** yes   **d.** yes   **e.** yes   **f.** yes

3. First note that $(n^3 + 4n^2 \log n + 101n^2)$ is $O(n^3)$ and that $(14n \log n + 8n)$ is $O(n \log n)$ as in Example 2.11. Now applying Theorem 2.3 yields the result.

5. Use Exercise 4 and follow Example 2.9, noting that $(\log n)^3 \le n^3$ whenever n is a positive integer.

7. What we want to show is equivalent to the statement that $\log(n^n)$ is at most a constant times $\log(n!)$, which in turn is equivalent to the statement that n^n is at most a constant power of $n!$ (because of the fact that $C \log A = \log(A^C)$). We will show that in fact $n^n \le (n!)^2$ for all $n > 1$. To do this, let us write $(n!)^2$ as $(n \cdot 1) \cdot ((n-1) \cdot 2) \cdot ((n-2) \cdot 3) \cdots (2 \cdot (n-1)) \cdot (1 \cdot n)$. Now clearly each product pair $(i+1) \cdot (n-i)$ is at least as big as n (indeed, the ones near the middle are significantly bigger than n). Therefore, the entire product is at least as big as n^n, as desired.

9. Suppose that f is $O(g)$, where $f(n)$ and $g(n)$ are positive integers for every integer n. Then there is an integer C such that $f(n) < Cg(n)$ for all $x \in S$. Then $f^k(n) < C^k g^k(n)$ for all $x \in S$. Hence, f^k is $O(g^k)$.

11. The number of digits in the base b expansion of n is $1 + k$, where k is such that $b^k \le n < b^{k+1}$, since there is a digit for each of the powers of $b^0, b^1, \ldots, b^k$. Note that this inequality is equivalent to $k \le \log_b n < k + 1$, so $k = [\log_b n]$. Hence, there are $[\log_b n] + 1$ digits in the base b expansion of n.

13. To multiply an n-digit integer by an m-digit integer in the conventional manner, one must multiply every digit of the first number by every digit of the second number (with carries), to produce an array of about mn digits. Then we need to add these partial products, but this requires only about mn additions as we proceed column by column. In all, the number of operations is at most a constant multiple of mn, as desired.

15. a. $O((n \log n)^{1+\epsilon})$ for every $\epsilon > 0$
   b. $O((n \log n)^{1+\epsilon})$ for every $\epsilon > 0$

17. $(1100011)_2$

19. a. $ab = (10^{2n} + 10^n)A_1 B_1 + 10^n(A_1 - A_0)(B_0 - B_1) + (10^n + 1)A_0 B_0$, where A_i and B_i are defined as in identity (2.2)
   b. 6351 **c.** 11522328

21. That the given equation is an identity may be seen by direct calculation. The seven multiplications necessary to use this identity are $a_{11}b_{11}$, $a_{12}b_{21}$, $(a_{11} - a_{21} - a_{22})(b_{11} - b_{12} - b_{22})$, $(a_{21} + a_{22})(b_{12} - b_{11})$, $(a_{11} + a_{12} - a_{21} - a_{22})b_{22}$, $(a_{11} - a_{21})(b_{22} - b_{12})$, and $a_{22}(b_{11} - b_{21} - b_{12} + b_{22})$.

23. Let $k = [\log_2 n] + 1$. Then the number of multiplications for $2^k \times 2^k$ matrices is $O(7^k)$. But $7^k = 2^{(\log_2 7)([\log_2 n]+1)}$, which is $O(2^{\log_2 n \log_2 7} 2^{\log_2 7}) = O(n^{\log_2 7})$. The other bit operations are absorbed into this term.

Section 3.1

1. a. yes   **b.** yes   **c.** yes   **d.** no   **e.** yes   **f.** no

3. 2, 3, 5, 7, 11, 13, 17, 19, 23, 29, 31, 37, 41, 43, 47, 53, 59, 61, 67, 71, 73, 79, 83, 89, 97, 101, 103, 107, 109, 113, 127, 131, 137, 139, 149

5. none

7. If n is not prime, then let $n = kl$, where $1 < k < n$. From the hint (which follows from algebra) we see that $a^k - 1$ is a factor of $a^n - 1$. This means that $a^k - 1 = 1$, whence $a^k = 2$. But this is impossible. Therefore, n is prime. Clearly, $a \neq 1$. Note that n cannot equal 2 unless $a = 2$, since otherwise $a^n - 1$ factors nontrivially as the difference of two squares; so we can assume that n is an odd prime. But then $a^n - 1$ factors algebraically as $(a - 1)(a^{n-1} + \cdots + a + 1)$, and this is nontrivial if $a > 2$. We conclude that $a = 2$.

9. We can show that there are infinitely many primes by showing that given an integer n, there is a prime p with $p > n$. We assume that $n \geq 3$. By Lemma 3.1, $S_n = n! - 1$ has a prime divisor p. If $p \leq n$ then $p \mid n!$, and so $p \mid n! - S_n = 1$, a contradiction. It follows that $p > n$.

11. 3, 7, 31, 211, 2311, 59

13. If n is prime, then we are done. Otherwise, $n/p < (\sqrt[3]{n})^2$. If n/p is prime, then we are done. Otherwise, by Theorem 3.2, n/p has a prime factor less than $\sqrt{n/p} < \sqrt[3]{n}$, a contradiction.

15. a. 7   **b.** 19   **c.** 71

17. If n is prime, then the statement is true for n. Otherwise, n is composite, so n is the product of two integers a and b such that $1 < a \leq b < n$. Since $n = ab$ and by the inductive hypothesis both a and b are the product of primes, we conclude that n is also the product of primes.

19. 53

21. For $n = 0, 1, 2, \ldots, 10$, the values of the function are 11, 13, 19, 29, 43, 61, 83, 109, 139, 173, 211, each of which is prime. But $2 \cdot 11^2 + 11 = 11(2 \cdot 11 + 1) = 11 \cdot 23$.

23. Assume not. Let x_0 be a positive integer. It follows that $f(x_0) = p$, where p is prime. Let k be an integer. We have $f(x_0 + kp) = a_n(x_0 + kp)^n + \cdots + a_1(x_0 + kp) + a_0$. Note that by the binomial theorem $(x_0 + kp)^j = \sum_{i=0}^{j} \binom{j}{i} x_0^{j-i}(kp)^i$. It follows that $f(x_0 + kp) = \sum_{j=0}^{n} a_j x_0^j + Np = f(x_0) + Np$, for some integer N. Since $p \mid f(x_0)$ it follows that $p \mid (f(x_0) + Np) = f(x_0 + kp)$. Since $f(x_0 + kp)$ is supposed to be prime, it follows that $f(x_0 + kp) = p$ for all integers k. This contradicts the fact that a polynomial of degree n takes on each value no more than n times. Hence, $f(y)$ is composite for at least one integer y.

25. At each stage of the procedure for generating the lucky numbers, the smallest number left, say k, is designated to be a lucky number and infinitely many numbers are left after the deletion of every kth integer left. It follows that there are infinitely many steps, and at each step a new lucky number is added to the sequence. Hence, there are infinitely many lucky numbers.

Section 3.2

1. 24, 25, 26, 27, 28

3. Suppose that p, $p + 2$, and $p + 4$ are all prime. We consider three cases. First, suppose that p is of the form $3k$. Then p cannot be prime unless $k = 1$, and then the prime triplet is 3, 5, 7. Next, suppose that p is of the form $3k + 1$. Then $p + 2 = 3k + 3 = 3(k + 1)$ is not prime. We obtain no prime triplets in this case. Finally, suppose that p is of the form $3k + 2$. Then $p + 4 = 3k + 6 = 3(k + 2)$ is not prime. We obtain no prime triplet in this case either.

5. $(7, 11, 13), (13, 17, 19), (37, 41, 43), (67, 71, 73)$.

7. **a.** 5   **b.** 7   **c.** 29   **d.** 53

9. 127, 149, 173, 197, 227, 257, 293, 331, 367, 401

11. **a.** $7 = 3 + 2 + 2$ **b.** $17 = 11 + 3 + 3$ **c.** $27 = 23 + 2 + 2$
    d. $97 = 89 + 5 + 3$ **e.** $101 = 97 + 2 + 2$ **f.** $199 = 191 + 5 + 3$

13. Suppose that $n > 5$ and that Goldbach's conjecture is true. Apply it to $n - 2$ if n is even or to $n - 3$ if n is odd. Conversely, suppose that every integer greater than 5 is the sum of three primes. Let $n > 2$ be an even integer. Then $n + 2$ is also even and the sum of three primes, not all odd.

15. Let $p < n$ be prime. Using the division algorithm, divide each of the first $p + 1$ integers in the sequence by p to get $a = q_0 p + r_0, a + k = q_1 p + r_1, \ldots, a + pk = q_p p + r_p$, with $0 \le r_i < p$ for each i. By the pigeonhole principle, at least two of the remainders are equal, say $r_i = r_j$. Subtract the corresponding equations to get $a + ik - a - jk = q_i p + r_i - q_j p - r_j$, which reduces to $(i - j)k = (q_i - q_j)p$. Therefore, $p \mid (i - j)k$, and because p is prime, it must divide one of the factors. But since $(i - j) < p$, we must have $p \mid k$.

17. The difference is 6, achieved with 5, 11, 17, 23.

19. The difference is 30, achieved with 7, 37, 67, 97, 127, 157.

21. If $p^\alpha - q^\beta = 1$, with p and q primes, then p or q is even, so that p or q is 2. If $p = 2$, there are several cases: we have $2^\alpha - q^\beta = 1$. If α is even, say $\alpha = 2k$, then $(2^{2k} - 1) = (2^k - 1)(2^k + 1) = q^\beta$. So $q \mid (2^k - 1)$ and $q \mid (2^k + 1)$; hence $q = 1$, a contradiction. If α is odd and β is odd, then $2^\alpha = 1 + q^\beta = (1 + q)(q^{\beta-1} - q^{\beta-2} + \cdots \pm 1)$. Thus $1 + q = 2^n$ for some n. Then $2^\alpha = (2^n - 1)^\beta + 1 = 2^n \cdot (\text{odd number})$, since β is odd. So $2^{\alpha-n}$ is odd and therefore $\alpha = n$. Thus $2^\alpha = 1 + (2^\alpha - 1)^\beta$, and so $\beta = 1$, which is not allowed. If $\alpha = 2k + 1$ and $\beta = 2n$, then $2^{2k+1} = 1 + q^{2n}$. Since q is odd, q^2 is of the form $4m + 1$, and by the binomial theorem, so is q^{2n}. Thus the right-hand side of the last equation is of the form $4m + 2$, but this forces $k = 0$, a contradiction. If $q = 2$, then $p^\alpha - 2^\beta = 1$, whence $2^\beta = (p - 1)(p^{\alpha-1} + p^{\alpha-2} + \cdots + p + 1)$, where the last factor is the sum of α odd terms but must be a power of 2; therefore, $\alpha = 2k$ for some k. Then $2^\beta = (p^k - 1)(p^k + 1)$. These last two factors are powers of 2 that differ by 2; this forces $k = 1$, $\alpha = 2$, $\beta = 3$, $p = 3$, and $q = 2$ as the only solution: $3^2 - 2^3 = 9 - 8 = 1$.

23. Since $3p > 2n$, we see that p and $2p$ are the only multiples of p that appear as factors in $(2n)!$. Thus p divides $(2n)!$ exactly twice. Since $2p > n$, we know that p is the only multiple of p that appears as a factor in $n!$. Thus p divides $n!$ exactly once. Then since $\binom{2n}{n} = (2n)!/(n!n!)$, the two factors of p in the numerator are canceled by the two in the denominator. and therefore p does not divide the quotient.

25. By Bertrand's postulate, there must be a prime in each interval of the form $(2^{k-1}, 2^k)$, for $k = 2, 3, 4, \ldots$. Thus there are at least $k - 1$ primes less than 2^k. Since the prime 2 is not counted here, we have at least k primes less than 2^k.

27. First suppose that $m < n$. Then $1/n + 1/(n + 1) + \cdots + 1/(n + m) \le 1/n + 1/(n + 1) + \cdots + 1/(2n - 1) < 1/n + 1/n + \cdots + 1/n \le n(1/n) = 1$, so the sum cannot be an integer. Now suppose $m \ge n$. By Bertrand's postulate, there is a prime p such that $n < p < n + m$. Let p be the largest such prime. Then $n + m < 2p$. Suppose that $1/n + 1/(n + 1) + \cdots + 1/p + \cdots + 1/(n + m) = a$, where a is an integer. Note that p occurs as a factor in only one denominator, since $2p > n + m$. Let $Q = \prod_{j=n}^{n+m} j$, and let $Q_i = Q/i$, for $i = n, n + 1, \ldots, n + m$. Multiply the equation by Q to get $Q_n + Q_{n+1} + \cdots + Q_p + \cdots + Q_{n+m} = Qa$. Except for the term Q_p on the left-hand side of the equation, every term on both sides is divisible by p. When we solve the equation for Q_p and factor p out, we obtain an equation of the form $Q_p = pN$, where N is an integer. But this implies that p divides Q_p, a contradiction.

29. Suppose that n has the stated property and $n \geq p^2$ for some prime p. Since p^2 is not prime, there must be a prime dividing both p^2 and n, and the only possibility is p itself, that is, $p \mid n$. Now if $n \geq 7^2$, then n is greater than 2^2, 3^2, and 5^2 and hence divisible by 2, 3, 5, and 7. This is the basic step for induction. Now assume n is divisible by $p_1, p_2, \ldots, p_k$. By Bonse's inequality, $p_{k+1}^2 < p_1 p_2 \cdots p_k < n$, so $p_{k+1} \mid n$ also. This induction implies that every prime divides n, which is absurd. Therefore, if n has the stated property, it must be less than $7^2 = 49$. To finish, check the remaining cases.

31. First suppose $n \geq 8$. By Bertrand's postulate, we have $p_{n-1} < p_n < 2p_{n-1}$ and $p_{n-2} < p_{n-1} < 2p_{n-2}$. Therefore, $p_n^2 < (2p_{n-1})(2p_{n-1}) < (2p_{n-1})(4p_{n-2}) < p_{n-1}p_{n-2}8 < p_{n-1}p_{n-2}p_5 \leq p_{n-1}p_{n-2}p_{n-3}$, since $n \geq 8$. Now check the cases $n = 6$ and 7.

Section 3.3

1. **a.** 5   **b.** 111   **c.** 6   **d.** 1   **e.** 11   **f.** 2

3. a

5. 1

7. By Theorem 3.8, $(ca, cb) = cma + cnb = |c| \cdot |ma + nb|$, where $cma + cnb$ is as small as possible. Therefore, $|ma + nb|$ is as small a positive integer as possible, i.e., equal to (a, b).

9. 1 or 2

11. Let $a = 2k$. Since $(a, b) \mid b$ and b is odd, (a, b) is odd. But $(a, b) \mid a = 2k$. Thus $(a, b) \mid k$. Therefore $(a, b) = (k, b) = (a/2, b)$.

13. Let $d = (a, b)$. Then $(a/d, b/d) = 1$, so if $g \mid (a/d)$, then $(g, b/d) = 1$. In particular, if we let $e = (a/d, bc/d)$, then $e \mid (a/d)$, so $(e, b/d) = 1$; therefore $e \mid c$. Since $e \mid (a/d)$, we know that $e \mid a$, so $e \mid (a, c)$. Conversely, if $f = (a, c)$, then $(f, b) = 1$, so $(d, f) = 1$; therefore $f \mid (a/d)$, and trivially $f \mid (bc/d)$. Therefore, $f \mid e$, whence $e = f$. Then $(a, b)(a, c) = de = d(a/d, bc/d) = (a, bc)$.

15. 10, 26, 65

17. **a.** 2   **b.** 5   **c.** 99   **d.** 3   **e.** 7   **f.** 1001

19. We proceed by induction, the basis case $n = 2$ being Exercise 7. Then using the basis case and Lemma 3.2, we have $(ca_1, ca_2, \ldots, ca_n) = (ca_1, ca_2, \ldots, ca_{n-2}, (ca_{n-1}, ca_n)) = (ca_1, ca_2, \ldots, ca_{n-2}, c(a_{n-1}, a_n)) = c(a_1, a_2, \ldots, a_{n-2}, (a_{n-1}, a_n))$ by the inductive hypothesis. But this last expression equals $c(a_1, a_2, \ldots, a_n)$ by Lemma 3.2.

21. Suppose that $(6k + a, 6k + b) = d$. Then $d \mid b - a$. We have $a, b \in \{-1, 1, 2, 3, 5\}$, so if $a < b$ it follows that $b - a \in \{1, 2, 3, 4, 6\}$. Hence, $d \in \{1, 2, 3, 4, 6\}$. To show that $d = 1$ it is sufficient to show that neither 2 nor 3 divides $(6k + a, 6k + b)$. If $p = 2$ or 3 and $p \mid (6k + a, 6k + b)$, then $p \mid a$ and $p \mid b$. However, there are no such pairs a, b in the set $\{-1, 1, 2, 3, 5\}$.

23. We proceed with the Euclidean algorithm: $8a + 3 = 1(5a + 2) + (3a + 1)$, $5a + 2 = 1(3a + 1) + (2a + 1)$, $3a + 1 = 1(2a + 1) + (a)$, $2a + 1 = 2(a) + (1)$. Therefore $(8a + 3, 5a + 2) = 1$.

25. From Exercise 21, we know that $6k - 1$, $6k + 1$, $6k + 2$, $6k + 3$, and $6k + 5$ are pairwise relatively prime. To represent n as the sum of two relatively prime integers greater than 1, let $n = 12k + h$, $0 \leq h < 12$. We now examine the twelve cases, one for each possible value of h: $h = 0, n = (6k - 1) + (6k + 1)$; $h = 1, n = (6k - 1) + (6k + 2)$; $h = 2, n = (6k - 1) + (6k + 3)$; $h = 3, n = (6k + 1) + (6k + 2)$; $h = 4, n = (6k + 1) + (6k + 3)$; $h = 5, n = (6k + 2) + (6k + 3)$;

$h = 6, n = (6k + 1) + (6k + 5); h = 7, n = (6k + 2) + (6k + 5); h = 8, n = (6k + 3) + (6k + 5);$
$h = 9, n = (12k + 7) + 2; h = 10, n = (12k + 7) + 3; h = 11, n = (12k + 9) + 2.$

27. Let S be the set of all fractions $P/Q = (xa + ye)/(xb + yf)$, where x and y are relatively prime positive integers. Then every element of S lies between a/b and e/f and is in lowest terms. The first element of S to appear in a Farey series will have the smallest Q, i.e., $x = y = 1$. This fraction must be c/d by hypothesis.

29. Since $a/b < (a + c)/(b + d) < c/d$, we have $b + d > n$, or a/b and c/d would not be consecutive, since otherwise $(a + c)/(b + d)$ would have appeared in the Farey series of order n.

31. Since $(a/b) + (c/d) = (ad + bc)/(bd)$ is an integer, $bd \mid ad + bc$. Certainly, then, $bd \mid d(ad + bc) = ad^2 + cbd$. Now since $bd \mid cbd$, we have $bd \mid ad^2$. From this, $bdn = ad^2$ for some integer n, and it follows that $bn = ad$, or $b \mid ad$. Since $(a, b) = 1$, we must have $b \mid d$. Similarly, we can show that $d \mid b$; hence $b = d$.

33. Note that a lattice point lies on the diagonal from $(0, 0)$ to (a, b) if and only if $[bx/a]$ is an integer. Let $d = (a, b)$ and $a = cd$, so that $(c, b) = 1$. There are exactly x multiples of c less than or equal to a since $cd = a$, so there are exactly $d + 1$ lattice points on the diagonal. One way to count the lattice points is to consider the rectangle, which has $(a + 1)(b + 1)$ points, and divide by 2. But we need to add back in half the points on the diagonal, which gives us $(a + 1)(b + 1)/2 + ((a, b) + 1)/2$. Another way is to count each column above the horizontal axis, starting with $i = 1, 2, \ldots, a - 1$. The equation of the diagonal is $y = (ba)x$, so for a given i, the number of points on or below the diagonal is $[bi/a]$. So the total number of interior points in the triangle plus the points on the the diagonal is $\sum_{i=1}^{a-1}[bi/a]$. Then the right-hand boundary has b points and the lower boundary has $a + 1$ points. So in all, we have $\sum_{i=1}^{a-1}[bi/a] + a + b + 1$ points. Equating the two expressions and simplifying gives the identity.

35. Assume there are exactly r primes and consider the $r + 1$ numbers $(r + 1)! + 1$. From Lemma 3.1, each of these numbers has a prime divisor, but from Exercise 34, these numbers are pairwise relatively prime, so these prime divisors must be unique, so we must have at least $r + 1$ different prime divisors, a contradiction.

Section 3.4

1. a. 15 **b.** 6 **c.** 2 **d.** 5

3. a. $-1 \cdot 75 + 2 \cdot 45$ **b.** $6 \cdot 222 + (-13) \cdot 102$ **c.** $-138 \cdot 666 + 65 \cdot 1414$ **d.** $-1707 \cdot 20785 + 800 \cdot 44350$

5. a. 1 **b.** 7 **c.** 5

7. a. $1 \cdot 6 + 1 \cdot 10 + (-1) \cdot 15$ **b.** $0 \cdot 70 + (-1) \cdot 98 + 1 \cdot 105$ **c.** $-13 \cdot 280 + 0 \cdot 330 + 9 \cdot 405 + 0 \cdot 490$

9. 2

11. $2n - 2$

13. Suppose that we have the balanced ternary expansions for integers $a \geq b$. If both expansions end in 0, then both are divisible by 3, and we can divide this factor of 3 out by deleting the trailing 0s (a shift), in which case $(a, b) = 3(a/3, b/3)$. If exactly one expansion ends in 0, then we can divide the factor of 3 out by shifting, and we have $(a, b) = (a/3, b)$, say. If both expansions end in 1 or in -1, then we can subtract the larger from the smaller to get $(a, b) = (a - b, b)$, say, and then the expansion for $a - b$ ends in 0. Finally, if one expansion ends in 1 and the other in -1, then we can add the two to get $(a + b, b)$, where the expansion of $a + b$ now ends in 0. Since $a + b$ is no larger than $2a$ and since we can now divide $a + b$ by 3, the larger term is reduced by a factor of at least 2/3 after two steps. Therefore, this algorithm will terminate in a finite number of steps, when we finally have $a = b = 1$.

15. *Lemma:* If c and d are integers and $c = dq \pm r$, where q and r are integers, then $(c, d) = (d, r)$. [*Proof of lemma:* If an integer e divides both c and d, then since $r = \pm(c - dq)$, Theorem 1.8 shows that $e \mid r$. If $e \mid d$ and $e \mid r$, then since $c = dq + r$, from Theorem 1.8 we see that $e \mid c$. Since the common divisors of c and d are the same as the common divisors of d and r, we see that $(c, d) = (d, r)$.] Let $r_0 = a$ and $r_1 = b$ be positive integers with $a \geq b$. By successively applying the least-remainder division algorithm, we find that $r_0 = r_1 q_1 + e_2 r_2$, $-r_1/2 < e_2 r_2 \leq r_1/2$; $\ldots$; $r_{n-2} = r_{n-1} q_{n-1} + e_n r_n$, $-r_{n-1}/2 < e_n r_n \leq r_{n-1}/2$; $r_{n-1} = r_n q_n$. We eventually obtain a remainder of 0 since the sequence of remainders $a = r_0 > r_1 > r_2 > \cdots \geq 0$ cannot contain more than a terms. By the lemma we see that $(a, b) = (r_0, r_1) = (r_1, r_2) = \cdots = (r_{n-2}, r_{n-1}) = (r_{n-1}, r_n) = (r_n, 0) = r_n$. Hence $(a, b) = r_n$, the last nonzero remainder.

17. Let $v_2 = v_3 = 2$ and $v_i = 2v_{i-1} + v_{i-2}$ for $i \geq 4$.

19. Performing the Euclidean algorithm with $r_0 = m$ and $r_1 = n$, we find that $r_0 = r_1 q_1 + r_2$, $0 \leq r_2 < r_1$, $r_1 = r_2 q_2 + r_3$, $0 \leq r_3 < r_2$, $\ldots$, $r_{k-3} = r_{k-2} q_{k-2} + r_{k-1}$, $0 \leq r_{k-1} < r_{k-2}$, and $r_{k-2} = r_{k-1} q_{k-1}$. We have $(m, n) = r_{k-1}$. We will use these steps to find the greatest common divisor of $a^m - 1$ and $a^n - 1$. First, we show that if u and v are positive integers, then the least positive residue of $a^u - 1$ modulo $a^v - 1$ is $a^r - 1$, where r is the least positive residue of u modulo v. To see this, note that $u = vq + r$, where r is the least positive residue of u modulo v. It follows that $a^u - 1 = a^{vq+r} - 1 = (a^v - 1)(a^{v(q-1)+r} + \cdots + a^{v+r} + a^r) + (a^r - 1)$. This shows that the remainder is $a^r - 1$ when $a^u - 1$ is divided by $a^v - 1$. Now let $R_0 = a^m - 1$ and $R_1 = a^n - 1$. When we perform the Euclidean algorithm starting with R_0 and R_1 we obtain $R_0 = R_1 Q_1 + R_2$, where $R_2 = a^{r_2} - 1$, $R_1 = R_2 Q_2 + R_3$, where $R_3 = a^{r_3} - 1, \ldots, R_{k-3} = R_{k-2} Q_{k-2} + R_{k-1}$, where $R_{k-1} = a^{r_{k-1}-1}$. Hence the last nonzero remainder, $R_{k-1} = a^{r_{k-1}} - 1 = a^{(m,n)} - 1$ is the greatest common divisor of $a^m - 1$ and $a^n - 1$.

21. Note that $(x, y) = (x - ty, y)$, as every divisor of x and y is also a divisor of $x - ty$. So, every move in the game of Euclid preserves the g.c.d. of the two numbers. Since $(a, 0) = a$, if the game beginning with $\{a, b\}$ terminates, then it must do so at $\{(a, b), 0)\}$. Since the sum of the two numbers is always decreasing and positive, the game must terminate.

23. Choose m so that d has no more than m bits and q has $2m$ bits, if necessary appending initial zeros to q. By Theorems 2.5 and 2.7, q can be divided by d using $O(m^2) = O(\log_2 q \log_2 d)$ bit operations. Suppose that n is the number of steps used by the Euclidean algorithm to find (a, b). Then by Theorem 3.13, $n = O(\log_2 a)$. The total number of bit operations for divisions in the Euclidean algorithm is $\sum_{i=1}^{n} O(\log_2 q_i \log_2 r_i) = \sum_{i=1}^{n} O(\log_2 q_i \log_2 b) = O(\log_2 b \sum_{i=1}^{n} \log_2 q_i) = O(\log_2 b \log_2 \prod_{i=1}^{n} q_i)$, where q_i and r_i are as in the proof of Theorem 3.13. Dropping the remainder in each step of the Euclidean algorithm, we have inequalities $r_i \geq r_{i+1} q_{i+1}$, for $i = 0, 1, \ldots, n-1$. Multiplying these inequalities together yields $\prod_{i=0}^{n-1} r_i \geq \prod_{i=1}^{n} r_i q_i$. Cancelling common factors reduces this to $a = r_0 \geq r_n \prod_{i=1}^{n} q_i$. Therefore, the total number of bit operations is $O(\log_2 b \log_2 \prod_{i=1}^{n} q_i) = O(\log_2 b \log_2 a) = O((\log_2 a)^2)$.

25. We apply the Q_i's one at a time. When we multiply $\begin{pmatrix} q_n & 1 \\ 1 & 0 \end{pmatrix} \begin{pmatrix} r_n \\ 0 \end{pmatrix} = \begin{pmatrix} q_n r_n \\ r_n \end{pmatrix} = \begin{pmatrix} r_{n-1} \\ r_n \end{pmatrix}$, the top component is the last equation in the series of equations in the proof of Lemma 3.3. When we multiply this result on the left by the next matrix, we get $\begin{pmatrix} q_{n-1} & 1 \\ 1 & 0 \end{pmatrix} \begin{pmatrix} r_{n-1} \\ r_n \end{pmatrix} = \begin{pmatrix} q_{n-1} r_{n-1} + r_n \\ r_{n-1} \end{pmatrix} = \begin{pmatrix} r_{n-2} \\ r_{n-1} \end{pmatrix}$, which is the matrix version of the last two equations in the proof of Lemma 3.3. In general, at the ith step we have $\begin{pmatrix} q_{n-1} & 1 \\ 1 & 0 \end{pmatrix} \begin{pmatrix} r_{n-i-1} \\ r_{n-i} \end{pmatrix} =$

$$\begin{pmatrix} q_{n-i}r_{n-i-1} + r_{n-i} \\ r_{n-i-1} \end{pmatrix} = \begin{pmatrix} r_{n-i-2} \\ r_{n-i-1} \end{pmatrix},$$ so that we inductively work our way up the equations

in the proof of Lemma 3.3, until finally we have $\begin{pmatrix} r_0 \\ r_1 \end{pmatrix} = \begin{pmatrix} a \\ b \end{pmatrix}$.

Section 3.5

1. a. $2^2 \cdot 3^2$ **b.** $3 \cdot 13$ **c.** $2^2 \cdot 5^2$ **d.** 17^2 **e.** $2 \cdot 3 \cdot 37$ **f.** 2^8 **g.** $5 \cdot 103$ **h.** $23 \cdot 43$
   i. $2^4 \cdot 3^2 \cdot 5 \cdot 7$ **j.** $2^6 \cdot 5^3$ **k.** $3 \cdot 5 \cdot 7^2 \cdot 13$ **l.** $3^2 \cdot 11 \cdot 101$

3. $3 \cdot 5 \cdot 7 \cdot 11 \cdot 13 \cdot 17 \cdot 19$

5. a. 2, 3   **b.** 2, 3, 5   **c.** 2, 3, 5, 7, 11, 13, 17, 19   **d.** 2, 3, 7, 13, 29, 31, 37, 41, 43, 47

7. integers of the form p^2, where p is prime; integers of the form pq or p^3, where p and q are distinct primes

9. Let $n = p_1^{2a_1} p_2^{2a_2} \cdots p_k^{2a_k} q_1^{2b_1+3} q_2^{2b_2+3} \cdots q_l^{2b_l+3}$ be the factorization of a powerful number. Then $n = (p_1^{a_1} p_2^{a_2} \cdots p_k^{a_k} q_1^{b_1} q_2^{b_2} \cdots q_l^{b_l})^2 (q_1 q_2 \cdots q_l)^3$ is a product of a square and a cube.

11. Suppose that $p^a \| m$ and $p^b \| n$. Then $m = p^a Q$ and $n = p^b R$, where both Q and R are products of primes other than p. Hence, $mn = (p^a Q)(p^b R) = p^{a+b} QR$. It follows that $p^{a+b} \| mn$ since p does not divide QR.

13. Suppose that $p^a \| m$ and $p^b \| n$ with $a \neq b$. Then $m = p^a Q$ and $n = p^b R$, where both Q and R are products of primes other than p. Suppose, without loss of generality, that $a = \min(a, b)$. Then $m + n = p^a Q + p^b R = p^{\min(a,b)}(Q + p^{b-a} R)$. Then $p \nmid (Q + p^{b-a} R)$ because $p \nmid Q$ but $p \mid p^{b-a} R$. It follows that $p^{\min(a,b)} \| (m + n)$.

15. $2^{18} \cdot 3^8 \cdot 5^4 \cdot 7^2 \cdot 11 \cdot 13 \cdot 17 \cdot 19$

17. 300, 301, 302, 303, 304

19. We compute $\alpha\beta = (ac - 5bd) + (ad + bc)\sqrt{-5}$. Thus $N(\alpha\beta) = (ac - 5bd)^2 + 5(ad + bc)^2 = a^2c^2 - 10acbd + 25b^2d^2 + 5a^2d^2 + 10adbc + 5b^2c^2 = a^2(c^2 + 5d^2) + 5b^2(5d^2 + c^2) = (a^2 + 5b^2)(c^2 + 5d^2) = N(\alpha)N(\beta)$.

21. Suppose $3 = \alpha\beta$. Then from Exercise 19 we know that $9 = N(3) = N(\alpha)N(\beta)$. Then $N(\alpha) = 1$, 3, or 9. Let $\alpha = a + b\sqrt{-5}$. Then we must have $a^2 + 5b^2 = 1$, 3, or 9. So either $b = 0$ and $a = \pm 1$ or ± 3, or $b = \pm 1$ and $a = \pm 2$. Since $a = \pm 1$, $b = 0$ is excluded, and since $a = \pm 3$ forces $\beta = \pm 1$, we must have $b = \pm 1$. That is, $\alpha = \pm 2 \pm \sqrt{-5}$. But then $N(\alpha) = 9$, and hence $N(\beta) = 1$, which forces $\beta = \pm 1$.

23. Note that $21 = 3 \cdot 7 = (1 + 2\sqrt{-5})(1 - 2\sqrt{-5})$. We know that 3 is prime from Exercise 21. Similarly, if we seek $\alpha = a + b\sqrt{-5}$ such that $N(\alpha) = a^2 + 5b^2 = 7$, then we find there are no solutions. Indeed, $b = 0$ implies $a^2 = 7$, $|b| = 1$ implies $a^2 = 2$, and $|b| > 1$ implies $a^2 < 0$, and in each case there is no such a. Hence, if $\alpha\beta = 7$, then $N(\alpha\beta) = N(\alpha)N(\beta) = N(7) = 49$. So one of $N(\alpha)$ and $N(\beta)$ must be equal to 49 and the other equal to 1. Hence, 7 is also prime. We have shown that there are no numbers of the form $a + b\sqrt{-5}$ with norm 3 or 7. So in a similar fashion to the argument above, if $\alpha\beta = 1 \pm 2\sqrt{-5}$, then $N(\alpha\beta) = N(\alpha)N(\beta) = N(1 \pm 2\sqrt{-5}) = 21$. Since there are no numbers with norm 3 or 7, one of α and β has norm 21 and the other has norm 1. Hence, $1 \pm 2\sqrt{-5}$ is also prime.

25. The product of $4k + 1$ and $4l + 1$ is $(4k + 1)(4l + 1) = 16kl + 4k + 4l + 1 = 4(4kl + k + l) + 1 = 4m + 1$, where $m = 4kl + k + l$. Hence, the product of two integers of the form $4k + 1$ is also of this form.

27. We proceed by mathematical induction on the elements of H. The first Hilbert number greater than 1, 5, is a Hilbert prime because it is an integer prime. This completes the basis step. For the inductive step, we assume that all numbers in H less than or equal to n can be factored into Hilbert primes. The next greatest number in H is $n + 4$. If $n + 4$ is a Hilbert prime, then we are done. Otherwise, $n = hk$, where h and k are less than n and in H. By the inductive hypothesis, h and k can be factored into Hilbert primes. Thus $n + 4$ can be written as the product of Hilbert primes.

29. $1, 2, 3, 4, 6, 8, 12, 24$

31. a. 77 **b.** 36 **c.** 150 **d.** 33633 **e.** 605605 **f.** 277200

33. a. $2^2 3^3 5^3 7^2$, $2^7 3^5 5^5 7^7$ **b.** $1, 2 \cdot 3 \cdot 5 \cdot 7 \cdot 11 \cdot 13 \cdot 17 \cdot 19 \cdot 23 \cdot 29$ **c.** $2 \cdot 5 \cdot 11, 2^3 \cdot 3 \cdot 5^7 \cdot 7 \cdot 11^{13} \cdot 13$
d. 101^{1000}, $41^{11} 47^{11} 79^{111} 83^{111} 101^{1001}$

35. the year 2121

37. Let $a = p_1^{r_1} p_2^{r_2} \cdots p_k^{r_k}$ and $b = p_1^{s_1} p_2^{s_2} \cdots p_k^{s_k}$, where each p_i is a prime and r_i and s_i are nonnegative. Then $(a, b) = p_1^{\min(r_1, s_1)} \cdots p_k^{\min(r_k, s_k)}$ and $[a, b] = p_1^{\max(r_1, s_1)} \cdots p_k^{\max(r_k, s_k)}$. So $[a, b] = (a, b) \cdot p_1^{\max(r_1, s_1) - \min(r_1, s_1)} \cdots p_k^{\max(r_k, s_k) - \min(r_k, s_k)}$. Since $\max(r_i, s_i) - \min(r_i, s_i)$ is clearly nonnegative, we now see that $(a, b) \mid [a, b]$. Clearly, $(a, b) = [a, b]$ if and only if each $r_i = s_i$, which means $a = b$.

39. If $[a, b] \mid c$, then since $a \mid [a, b]$, we have $a \mid c$. Similarly, $b \mid c$. Conversely, suppose that $a = p_1^{a_1} p_2^{a_2} \cdots p_n^{a_n}$, $b = p_1^{b_1} p_2^{b_2} \cdots p_n^{b_n}$, and $c = p_1^{c_1} p_2^{c_2} \cdots p_n^{c_n}$. If $a \mid c$ and $b \mid c$, then $\max(a_i, b_i) \leq c_i$ for $i = 1, 2, \ldots, n$. Hence $[a, b] \mid c$.

41. Assume that $p \mid a^n = \pm |a| \cdot |a| \cdots |a|$. Then by Lemma 3.5, $p \mid |a|$ and so $p \mid a$.

43. a. Suppose that $(a, b) = 1$ and $p \mid (a^n, b^n)$, where p is a prime. It follows that $p \mid a^n$ and $p \mid b^n$. By Exercise 43, $p \mid a$ and $p \mid b$. But then $p \mid (a, b) = 1$, which is a contradiction.
b. Suppose that a does not divide b, but $a^n \mid b^n$. Then there is some prime power, say p^r, that divides a but does not divide b (else $a \mid b$ by the fundamental theorem of arithmetic). Thus $a = p^r Q$, where Q is an integer. Now $a^n = (p^r Q)^n = p^{rn} Q^n$, so $p^{rn} \mid a^n$ and it follows that $p^{rn} \mid b^n$. Then $b^n = mp^{rn}$, from which it follows that each of the n b's must by symmetry contain at least r p's. But this is a contradiction.

45. Let $x = \sqrt{2} + \sqrt{3}$. Then $x^2 = 2 + 2\sqrt{2}\sqrt{3} + 3 = 5 + 2\sqrt{6}$. Hence, $x^2 - 5 = 2\sqrt{6}$. It follows that $x^4 - 10x^2 + 25 = 24$. Consequently, $x^4 - 10x^2 + 1 = 0$. From Theorem 3.18 it follows that $\sqrt{2} + \sqrt{3}$ is irrational, since it is not an integer (we can see this since $3 < \sqrt{2} + \sqrt{3} < 4$).

47. Suppose that $m/n = \log_p b$. This implies that $p^{m/n} = b$, from which it follows that $p^m = b^n$. Since b is not a power of p, there must be another prime, say q, such that $q \mid b$. But then $q \mid b^n = p^m = p \cdot p \cdots p$. By Lemma 3.5, $q \mid p$, which is impossible since p is a prime number.

49. Let p be a prime. Define s and t by $p^s \| a$ and $p^t \| b$, say $a = xp^s$ and $b = yp^t$, where $p \nmid xy$. Without loss of generality, suppose that $s \leq t$. Then $a + b = p^s(x + p^{t-s})$, so $p^s \| a + b$. Also, $p^{\max(s,t)} \| [a, b]$. But $\max(s, t) = t$, so $p^t \| [a, b]$. Therefore $p^{\min(s,t)} \| (a + b, [a, b])$. But $p^{\min(s,t)} \| (a, b)$, so the same power of p divides both sides of the equation. Since this holds for each p, the two sides must be equal.

51. It suffices to prove this "one prime at a time"; to this end, let r, s, and t be the exponents on the prime p in the prime factorizations of a, b, and c, respectively. We know that the exponent on p in $[a, b]$ is $\max(r, s)$, and so the exponent on p in $([a, b], c)$ is $\min(t, \max(r, s))$. We also know that the exponent on p in (a, c) is $\min(r, t)$ and the exponent on p in (b, c) is $\min(s, t)$, so the exponent on p in $[(a, c), (b, c)]$ is $\max(\min(r, t), \min(s, t))$. But it is not hard to see that $\min(t, \max(r, s))$ and $\max(\min(r, t), \min(s, t))$ always represent the same value. It follows that the exponent on p in the prime factorizations of $([a, b], c)$ and $[(a, c), (b, c)]$ are the same for each prime p, and we conclude therefore that $([a, b], c) = [(a, c), (b, c)]$. In a similar manner, we find that $[(a, b), c] = ([a, c], [b, c])$.

53. Let $c = [a_1, \ldots, a_n]$, $d = [[a_1, \ldots, a_{n-1}], a_n]$, and $e = [a_1, \ldots, a_{n-1}]$. If $c \mid m$, then all a_i's divide m; hence $e \mid m$ and $a_n \mid m$, so $d \mid m$. Conversely, if $d \mid m$, then $e \mid m$ and $a_n \mid m$, so all a_i's divide m; thus $c \mid m$. Since c and d divide all the same numbers, they must be equal.

55. a. There are six cases, all handled the same way. So without loss of generality, suppose that $a \le b \le c$. Then $\max(a, b, c) = c$, $\min(a, b) = a$, $\min(a, c) = a$, $\min(b, c) = b$, and $\min(a, b, c) = a$. Hence $c = \max(a, b, c) = a + b + c - \min(a, b) - \min(a, c) - \min(b, c) + \min(a, b, c) = a + b + c - a - a - b + a$.

b. The exponent on a prime p that occurs in the prime factorization of $[a, b, c]$ is $\max(x, y, z)$, where x, y, and z are the exponents on this prime in the factorizations of a, b, and c, respectively. Also $x + y + z$ is the exponent on p in abc, $\min(x, y, z)$ is the exponent on p in (a, b, c), $\min(x, y)$ is the exponent on p in (a, b), $\min(x, z)$ is the exponent on p in (a, c), $\min(y, z)$ is the exponent on p in (b, c), and $\min(x, y, z)$ is the exponent on p in (a, b, c). It follows that $x + y + z + \min(x, y, z) - \min(x, y) - \min(x, z) - \min(y, z)$ is the exponent on p in $abc(a, b, c)/((a, b)(a, c)(b, c))$. Hence from part (a), $[a, b, c] = abc(a, b, c)/((a, b)(a, c)(b, c))$.

57. Let $a = p_1^{r_1} p_2^{r_2} \cdots p_k^{r_k}$, $b = p_1^{s_1} p_2^{s_2} \cdots p_k^{s_k}$, and $c = p_1^{t_1} p_2^{t_2} \cdots p_k^{t_k}$, with p_i prime and r_i, s_i, and t_i nonnegative. Then $p_i^{r_i + s_i + t_i} \parallel abc$, but $p_i^{\min(r_i, s_i, t_i)} \parallel (a, b, c)$ and $p_i^{r_i + s_i + t_i - \min(r_i, s_i, t_i)} \parallel [ab, ac, ab]$, and $p_i^{\min(r_i, s_i, t_i)} \cdot p_i^{r_i + s_i + t_i - \min(r_i, s_i, t_i)} = p_i^{r_i + s_i + t_i}$.

59. It suffices to prove this "one prime at a time"; to this end, let r, s, and t be the exponents on the prime p in the prime factorizations of a, b, and c, respectively. Then, using the facts that the exponent on p in (a, b, c) is $\min(r, s, t)$, and the exponent on p in $[a, b, c]$ is $\max(r, s, t)$, we see that the exponent on p in $([a, b], [a, c], [b, c])$ is $\min(\max(r, s), \max(r, t), \max(s, t))$, whereas the exponent on p in $[(a, b), (a, c), (b, c)]$ is $\max(\min(r, s), \min(r, t), \min(s, t))$. But these two are equal (examine the six orderings $r \ge s \ge t, \ldots$).

61. First note that there are arbitrarily long sequences of composites in the integers. For example, $(n + 2)! + 2, (n + 2)! + 3, \ldots, (n + 2)! + (n + 2)$ is a sequence of n consecutive composites. To find a sequence of n composites in the sequence $a, a + b, a + 2b, \ldots$, look at the integers in $a, a + b, a + 2b, \ldots$ with absolute values between $(nb + 2)! + 2$ and $(nb + 2)! + (nb + 2)$. There are clearly n or $n + 1$ such integers, and all are composite.

63. 103

65. 701

67. Let $a = \prod_{i=1}^{s} p_i^{\alpha_i}$ and $b = \prod_{i=1}^{t} p_i^{\beta_i}$. The condition $(a, b) = 1$ is equivalent to $\min(\alpha_i, \beta_i) = 0$ for all i, and the condition $ab = c^n$ is equivalent to $n \mid \alpha_i + \beta_i$ for all i. Hence $n \mid \alpha_i$ and $\beta_i = 0$, or $n \mid \beta_i$ and $\alpha_i = 0$. Let d be the product of $p_i^{\alpha_i/n}$ over all i of the first kind, and let e be the product of $p_i^{\beta_i/n}$ over all i of the second kind. Then $d^n = a$ and $e^n = b$.

69. Partition the set of integers $S = \{1, 2, 3, \ldots, 2n\}$ into n subsets in the following way. Let $S_1 = \{1, 2, 4, 8, \ldots\} \cap S$; let $S_2 = \{3, 6, 12, 24, \ldots\} \cap S$; let $S_3 = \{5, 10, 20, 40, \ldots\} \cap S$; and so on, with the last set being $S_n = \{2n - 1\}$. In other words, S_i is the set of elements in S whose "odd part" (the number with all factors of 2 divided out) is $2i - 1$, for $i = 1, 2, \ldots, n$. By the pigeonhole principle, at least two of the $n + 1$ given numbers must lie in the same S_i, and clearly the smaller of the two will divide the larger.

71. $m = n$ or $\{m, n\} = \{2, 4\}$

73. For $j \ne i$, $p_i \mid Q_j$, since it is one of the factors. So p_1 must divide $S - \sum_{j \ne i} Q_j = Q_i = p_1 \cdots p_{i-1} p_{i+1} \cdots p_r$, but by the fundamental theorem of arithmetic, p_i must be equal to one of these last factors, a contradiction.

75. Let p be the largest prime less than or equal to n. If $2p < n$, then Bertrand's postulate guarantees another prime q such that $p < q < 2p \leq n$, contradicting the choice of p. Therefore, we know that $n < 2p$. Therefore, in the prime factorization of the product $n! = 1 \cdot 2 \cdot 3 \cdots n$, only one multiple of p appears.

77. a. Uniqueness follows from the fundmental theorem of arithmetic. Because $e_i \geq 0$, we have
$$p_1^{e_1} = p_1^{e_1} p_2^0 \cdots p_r^0 \leq p_1^{e_1} p_2^{e_2} \cdots p_r^{e_r} = m.$$
b. Because $p_1^{e_i} < p_i^{e_i} \leq m \leq Q = p_r^n$, taking logarithms it follows that $e_i \log p_1 = n \log p_r$. Dividing by $\log p_1$ gives the first inequality. If $1 \leq m \leq Q$, then m has a prime-power factorization of the form in part (a), so the r-tuples of exponents count the number of integers in the range $1 \leq m \leq Q$.
c. To bound the number of r-tuples, by part (b) there are at most $C(n+1)^r$ r-tuples. By part (b) we have $p_r^n \leq (Cn + 1)^r = (n(C + 1/n))^r \leq n^r (C + 1)^r$.
d. Taking logarithms of both sides of the inequality in part (c), we obtain $n \leq r(\log n + \log(C + 1))/\log p_r$, but because n grows much faster than $\log n$, the left-hand side must be larger than the right-hand side for large values of n.

79. $S(40) = 5$, $(41) = 41$, $S(43) = 43$

81. $a(n) = 1, 2, 3, 4, 5, 9, 7, 32, 27, 25, 11, \ldots$

83. By Exercise 80, we have $S(p) = p$ whenever p is prime. If $m < p$ and $m \mid S(p)! = p!$, then $m \mid (p-1)!$, so $S(p)$ must be the first time that $S(n)$ takes on the value p.

85. Let n be square-free. Then no prime can appear to a power greater than one in the prime-power factorization of n. So $n = p_1 p_2 \cdots p_r$ for some distinct primes p_i, which implies that $\mathrm{rad}(n) = p_1 p_2 \cdots p_r = n$. Conversely, if n is not square-free, then $d^2 \mid n$ for some integer d with $d > 2$, and some prime factor p_1 of d appears to an even power in the prime-power factorization of n. So $n = p_1^{2a} p_2^{b_2} \cdots p_r^{b_r}$. This implies that $\mathrm{rad}(n) = p_1 p_2 \cdots p_r \neq n$.

87. Since every prime occurring in the prime-power factorization of mn occurs in either the factorization of m or n, every factor in $\mathrm{rad}(mn)$ occurs at least once in the product $\mathrm{rad}(m)\mathrm{rad}(n)$, yielding the inequality. If $m = p_1^{a_1} \cdots p_r^{a_r}$ and $n = q_1^{b_1} \cdots q_s^{b_s}$ are relatively prime, then $\mathrm{rad}(mn) = p_1 \cdots p_r q_1 \cdots q_s = \mathrm{rad}(m)\mathrm{rad}(n)$.

89. First note that if $p \mid \binom{2n}{n}$, then $p \leq 2n$. This is true because every factor of the numerator of $\binom{2n}{n} = (2n)!/(n!)^2$ is less than or equal to $2n$. Let $\binom{2n}{n} = p_1^{r_1} p_2^{r_2} \cdots p_k^{r_k}$ be the factorization of $\binom{2n}{n}$ into distinct prime powers. By the definition of the function $\pi(x)$, we have $k \leq \pi(2n)$. By Exercise 88, $p_i^{r_i} \leq 2n$. It now follows that $\binom{2n}{n} = p_1^{r_1} p_2^{r_2} \cdots p_k^{r_k} \leq (2n)(2n) \cdots (2n) \leq (2n)^{\pi(2n)}$.

91. Note that $\binom{2n}{n} \leq \sum_{a=0}^{2n} \binom{2n}{a} = (1+1)^{2n} = 2^{2n}$. Then from Exercise 90, $n^{\pi(2n)-\pi(n)} < \binom{2n}{n} \leq 2^{2n}$. Taking logarithms gives $(\pi(2n) - \pi(n)) \log n < \log(2^{2n}) = n \log 4$. Now divide by $\log n$.

93. Note that $2^n = \prod_{a=1}^n 2 \leq \prod_{a=1}^n (n+a)/a = \binom{2n}{n}$. Then by Exercise 89, $2^n \leq (2n)^{\pi(2n)}$. Taking logarithms gives $\pi(2n) \geq n \log 2/(\log 2n)$. Hence for a real number x we have $\pi(x) \geq [x/2] \log 2/\log[x] > c_1 x/\log x$. For the other half, Exercise 91 gives $\pi(x) - \pi(x/2) < ax/\log x$, where a is a constant. Then $\log(x/2^m)\pi(x/2^m) - \log(x/2^{m+1})\pi(x/2^{m+1}) < ax/2^m$ for every positive integer m. Thus $(\log x)\pi(x) = \sum_{m=0}^v \left(\log(x/2^m)\pi(x/2^m) - \log(x/2^{m+1})\pi(x/2^{m+1})\right) < ax \sum_{m=0}^v 1/2^m < c_2 x$, where v is the largest integer such that $2^{v+1} \leq x$. Then $\pi(x) < c_2 x/\log x$.

Section 3.6

1. a. $3 \cdot 5^2 \cdot 7^3 \cdot 13 \cdot 101$ **b.** $11^3 \cdot 13 \cdot 19 \cdot 641$ **c.** $13 \cdot 17 \cdot 19 \cdot 47 \cdot 71 \cdot 97$
3. a. $143 = 12^2 - 1 = (12 + 1)(12 - 1) = 13 \cdot 11$
b. $2279 = 48^2 - 5^2 = (48 + 5)(48 - 5) = 53 \cdot 43$

c. 43 is prime.

d. $11413 = 107^2 - 6^2 = (107 + 6)(107 - 6) = 113 \cdot 101$

5. Note that $(50 + n)^2 = 2500 + 100n + n^2$ and $(50 - n)^2 = 2500 - 100n + n^2$. The first equation shows that the possible final two digits of squares can be found by examining the squares of the integers $0, 1, \ldots, 49$, and the second equation shows that these final two digits can be found by examining the squares of the integers $0, 1, \ldots, 25$. We find that $0^2 = 0, 1^2 = 1, 2^2 = 4, 3^2 = 9, 4^2 = 16, 5^2 = 25, 6^2 = 36, 7^2 = 49, 8^2 = 64, 9^2 = 81, 10^2 = 100, 11^2 = 121, 12^2 = 144, 13^2 = 169, 14^2 = 196, 15^2 = 225, 16^2 = 256, 17^2 = 289, 18^2 = 324, 19^2 = 361, 20^2 = 400, 21^2 = 441, 22^2 = 484, 23^2 = 529, 24^2 = 576, 25^2 = 625$. It follows that the last two digits of a square are $00, e1, e4, 25, o6, e9$, where e represents an even digit and o represents an odd digit.

7. Suppose that $x^2 - n$ is a perfect square with $x > (n + p^2)/(2p)$, say a^2. Now, $a^2 = x^2 - n > ((n + p^2)/(2p))^2 - n = ((n - p^2)/(2p))^2$. It follows that $a > (n + p^2)/(2p)$. From these inequalities for x and a, we see that $x + a > n/p$, or $n < p(x + a)$. Also, $a^2 = x^2 - n$ tells us that $(x - a)(x + a) = n$. Now, $(x - a)(x + a) = n < p(x + a)$. Canceling, we find that $x - a < p$. But since $x - a$ is a divisor of n less than p, the smallest prime divisor of n, it follows that $x - a = 1$. In this case, $x = (n + 1)/2$.

9. From the identity in Exercise 8, it is clear that if $n = n_1$ is a multiple of $2k + 1$, then so is n_k, since it is the difference of two multiples of $2k + 1$. If $2k + 1 \mid n_k$, then $2k + 1 \mid r_k$, and it follows from $r_k < 2k + 1$ that $r_k = 0$. Thus $n_k = (2k + 1)q_k$. Continuing, we see that $n = n + 2n_k - 2(2k + 1)q_k = (2k + 1)n + 2(n_k - kn) - 2(2k + 1)q_k$. It follows from Exercise 8 that $n = (2k + 1)n - 2(2k + 1)\sum_{i=1}^{k-1} q_i - 2(2k + 1)q_k = (2k + 1)n - 2(2k + 1)\sum_{i=1}^{k} q_i$. Using Exercise 8 again, we conclude that $n = (2k + 1)(n - 2\sum_{i=1}^{k} q_i) = (2k + 1)m_{k+1}$.

11. To see that u is even, note that $a - c$ is the difference of odd numbers and that $b - d$ is the difference of even numbers. Thus $a - c$ and $b - d$ are even, and u must be as well. That $(r, s) = 1$ follows trivially from Theorem 3.6. To continue, $a^2 + b^2 = c^2 + d^2$ implies that $(a + c)(a - c) = (d - b)(d + b)$. Dividing both sides of this equation by u, we find that $r(a + c) = s(d + b)$. From this it is clear that $s \mid r(a + c)$. But since $(r, s) = 1$, we have $s \mid a + c$.

13. To factor n, observe that $((u/2)^2 + (v/2)^2)(r^2 + s^2) = \frac{1}{4}(u^2r^2 + u^2s^2 + v^2r^2 + v^2s^2)$. Substituting $a - c, d - b, a + c$, and $d + b$ for ru, su, sv, and rv, respectively, will allow everything to be simplified down to n. As u and v are both even, both of the factors are integers.

15. We have $2^{4n+2} + 1 = 4(2^n)^4 + 1 = (2 \cdot 2^{2n} + 2 \cdot 2^n + 1)(2 \cdot 2^{2n} - 2 \cdot 2^n + 1)$. Using this identity we have the factorization $2^{18} + 1 = 4(2^4)^4 + 1 = (2 \cdot 2^8 + 2 \cdot 2^4 + 1)(2 \cdot 2^8 - 2 \cdot 2^4 + 1) = (2^9 + 2^5 + 1)(2^9 - 2^5 + 1) = 545 \cdot 481$.

17. We can prove that the last digit in the decimal expansion of F_n is 7 for $n \geq 2$ by proving that the last digit in the decimal expansion of 2^{2^n} is 6 for $n \geq 2$. This can be done using mathematical induction. We have $2^{2^2} = 16$ so the result is true for $n = 2$. Now assume that the last decimal digit of 2^{2^n} is 6, i.e., $2^{2^n} \equiv 6 \pmod{10}$. It follows that $2^{2^{n+1}} = (2^{2^n})^{2^{n+1}-2^n} \equiv 6^{2^{n+1}-2^n} \equiv 6 \pmod{10}$. This completes the proof.

19. Since every prime factor of $F_5 = 2^{2^5} + 1 = 4,294,967,297$ is of the form $2^7k + 1 = 128k + 1$, we attempt to factor F_5 by trial division by primes of this form. We find that $128 \cdot 1 + 1 = 129$ is not prime, $128 \cdot 2 + 1 = 257$ is prime but does not divide $4,294,967,297$, $128 \cdot 3 + 1 = 385$ is not prime, $128 \cdot 4 + 1 = 513$ is not prime, and $128 \cdot 5 + 1 = 641$ is prime and does divide $4,294,967,297$, with $4,294,967,297 = 641 \cdot 6,700,417$. Every factor of $6,700,417$ is also a factor of $4,294,967,297$. We attempt to factor $6,700,417$ by trial division by primes of the form $128k + 1$ beginning with 641. We first note that 641 does not divide $6,700,417$. Among the other integers of the form $128k + 1$ less than $\sqrt{6700417}$, namely the integers $769, 897, 1025, 1153, 1281, 1409, 1537, 1665, 1793, 1921, 2049, 2177, 2305, 2433$, and 2561, only $769, 1153$, and 1409 are prime,

and none of them divides 6,700,417. Hence 6,700,417 is prime and the prime factorization of F_5 is $641 \cdot 6{,}700{,}417$.

21. $2^n \log_{10} 2$

23. See Exercise 21 of Section 3.2.

Section 3.7

1. a. $x = 33 - 5t$, $y = -11 + 2t$
   b. $x = -300 + 13t$, $y = 400 - 17t$
   c. $x = 21 - 2t$, $y = -21 + 3t$
   d. no solutions
   e. $x = 889 - 1969t$, $y = -633 + 1402t$

3. 39 US\$ and 94 Can\$, or 95 US\$ and 33 Can\$

5. Solving $111e + 169p = 11798$ yields $e = 53$, $p = 35$.

7. 17 apples and 23 oranges

9. a. $(1, 16), (4, 14), (7, 12), \ldots, (22, 2), (25, 0)$
   b. no solutions
   c. $(0, 37), (3, 35), (6, 33), \ldots, (51, 3), (54, 1)$

11. a. $x = -5 + 3s - 2t$, $y = 5 - 2s$, $z = t$
   b. no solutions
   c. $x = -1 + 102s + t$, $y = 1 - 101s - 2t$, $z = t$

13. $(9, 9, 0), (19, 8, 0), \ldots, (99, 0, 0); (4, 7, 1), (14, 6, 1), \ldots, (74, 0, 1); (9, 4, 2), (19, 3, 2), \ldots, (49, 0, 2); (4, 2, 3), (14, 1, 3), (24, 0, 3)$

15. a. $x = 92 + 6t$, $y = 8 - 7t$, $z = t$
   b. no solutions
   c. $x = 50 - t$, $y = -100 + 3t$, $z = 150 - 3t$, $w = t$

17. $9, 19, 41$

19. The quadrilateral with vertices $(b, 0)$, $(0, a)$, $(b - 1, -1)$, and $(-1, a - 1)$ has area $a + b$. Pick's theorem, from elementary geometry, states that the area of a simple polygon whose vertices are lattice points (points with integer coordinates) is given by $\frac{1}{2}x + y - 1$, where x is the number of lattice points on the boundary and y is the number of lattice points inside the polygon. Since $(a, b) = 1$, we have $x = 4$, and therefore by Pick's theorem the quadrilateral contains $a + b - 1$ lattice points. Every point corresponds to a different value of n in the range $ab - a - b < n < ab$. Therefore, every n in the range must get hit, so the equation is solvable.

21. (See the solution for Exercise 19.) The line $ax + by = ab - a - b$ bisects the rectangle with vertices $(-1, a - 1)$, $(-1, -1)$, $(b - 1, a - 1)$, and $(b - 1, -1)$ but contains no lattice points. Hence half the interior points are below the line and half are above. The half below correspond to $n < ab - a - b$, and there are $(a - 1)(b - 1)/2$ of them.

23. $(0, 25, 75), (4, 18, 78), (8, 11, 81), (12, 4, 84)$

Section 4.1

1. a. $2 \mid (13 - 1) = 12$ **b.** $5 \mid (22 - 7) = 15$ **c.** $13 \mid (91 - 0) = 91$ **d.** $7 \mid (69 - 62) = 7$
   e. $3 \mid (-2 - 1) = -3$ **f.** $11 \mid (-3 - 30) = -33$ **g.** $40 \mid (111 - (-9)) = 120$
   h. $37 \mid (666 - 0) = 666$

3. a. $1, 2, 11, 22$ **b.** $1, 3, 9, 27, 37, 111, 333, 999$ **c.** $1, 11, 121, 1331$

5. Suppose that a is odd. Then $a = 2k + 1$ for some integer k. Then $a^2 = (2k+1)^2 = 4k^2 + 4k + 1 = 4k(k+1) + 1$. If k is even, then $k = 2l$, where l is an integer. Then $a^2 = 8l(2l+1) + 1$. Hence, $a^2 \equiv 1 \pmod 8$. If k is odd, then $k = 2l + 1$ when l is an integer. Then $a^2 = 4(2l+1)(2l+2) + 1 = 8(2l+1)(l+1)$. Again, $a^2 \equiv 1 \pmod 8$.

7. **a.** 1   **b.** 5   **c.** 9   **d.** 13

9. Since $a \equiv b \pmod m$, there exists an integer k such that $a = b + km$. Thus $ac = (b + km)c = bc + k(mc)$. By Theorem 4.1, $ac \equiv bc \pmod{mc}$.

11. **a.** We proceed by induction on n. It is clearly true for $n = 1$. For the inductive step, we assume that $\sum_{j=1}^{n} a_j \equiv \sum_{j=1}^{n} b_j \pmod m$ and that $a_{n+1} \equiv b_{n+1} \pmod m$. Now $\sum_{j=1}^{n+1} a_j = (\sum_{j=1}^{n} a_j) + a_{n+1} \equiv (\sum_{j=1}^{n} b_j) + b_{n+1} = \sum_{j=1}^{n+1} b_j \pmod m$ by Theorem 4.5(i).
b. We use induction on n. For $n = 1$, the identity clearly holds. For the inductive step, we assume that $\prod_{j=1}^{n} a_j \equiv \prod_{j=1}^{n} b_j \pmod m$ and $a_{n+1} \equiv b_{n+1} \pmod m$. Then $\prod_{j=1}^{n+1} a_j = a_{n+1}(\prod_{j=1}^{n} a_j) \equiv b_{n+1}(\prod_{j=1}^{n} b_j) = \prod_{j=1}^{n+1} b_j \pmod m$ by Theorem 4.5(iii).

13. $0 - 0 = 0, 0 - 1 = 5, 0 - 2 = 4, 0 - 3 = 3, 0 - 4 = 2, 0 - 5 = 1; 1 - 0 = 1, 1 - 1 = 0, 1 - 2 = 5, 1 - 3 = 4, 1 - 4 = 3, 1 - 5 = 2; 2 - 0 = 2, 2 - 1 = 1, 2 - 2 = 0, 2 - 3 = 5, 2 - 4 = 4, 2 - 5 = 3; 3 - 0 = 3, 3 - 1 = 2, 3 - 2 = 1, 3 - 3 = 0, 3 - 4 = 5, 3 - 5 = 4; 4 - 0 = 4, 4 - 1 = 3, 4 - 2 = 2, 4 - 3 = 1, 4 - 4 = 0, 4 - 5 = 5; 5 - 0 = 5, 5 - 1 = 4, 5 - 2 = 3, 5 - 3 = 2, 5 - 4 = 1, 5 - 5 = 0$

15. **a.** 4 o'clock   **b.** 6 o'clock   **c.** 4 o'clock

17. $a \equiv \pm b \pmod p$

19. Note that $1 + 2 + 3 + \cdots + (n-1) = (n-1)n/2$. If n is odd, then $(n-1)$ is even, so $(n-1)/2$ is an integer. Hence, $n \mid 1 + 2 + 3 + \cdots + (n-1)$ if n is odd, and so $1 + 2 + 3 + \cdots + (n-1) \equiv 0 \pmod n$. If n is even, then $n = 2k$, where k is an integer. Then $(n-1)n/2 = (n-1)k$. We can easily see that n does not divide $(n-1)k$ since $(n, n-1) = 1$ and $k < n$. It follows that $1 + 2 + \cdots + (n-1) \not\equiv 0 \pmod n$ if n is even.

21. those n relatively prime to 6

23. If $n = 1$, then $5 = 5^1 \equiv 1 + 4(1) \pmod{16}$, so the basis step holds. For the inductive step, we assume that $5^n \equiv 1 + 4n \pmod{16}$. Now $5^{n+1} \equiv 5^n 5 \equiv (1 + 4n)5 \pmod{16}$ by Theorem 4.3(iii). Further, $(1 + 4n)5 = 5 + 20n \equiv 5 + 4n \pmod{16}$. Finally, $5 + 4n = 1 + 4(n+1) \pmod{16}$. Thus $5^{n+1} \equiv 1 + 4(n+1) \pmod{16}$.

25. Note that if $x \equiv 0 \pmod 4$ then $x^2 \equiv 0 \pmod 4$, if $x \equiv 1 \pmod 4$ then $x^2 \equiv 1 \pmod 4$, if $x \equiv 2 \pmod 4$ then $x^2 \equiv 4 \equiv 0 \pmod 4$, and if $x \equiv 3 \pmod 4$ then $x^2 \equiv 9 \equiv 1 \pmod 4$. Hence, $x^2 \equiv 0$ or $1 \pmod 4$ whenever x is an integer. It follows that $x^2 + y^2 \equiv 0, 1$, or $2 \pmod 4$ whenever x and y are integers. Therefore, n is not the sum of two squares when $n \equiv 3 \pmod 4$.

27. By Theorem 4.1, $ap^k = x^2 - x = x(x-1)$ for some integer a. By the fundamental theorem of arithmetic, p^k is a factor of $x(x-1)$. Since p cannot divide both x and $x - 1$, we know that $p^k \mid x$ or $p^k \mid x - 1$. Thus $x \equiv 0$ or $x \equiv 1 \pmod{p^k}$.

29. First note that there are m_1 possibilities for a_1, m_2 possibilities for a_2, and in general m_i possibilities for a_i. Thus there are $m_1 m_2 \cdots m_k$ expressions of the form $M_1 a_1 + M_2 a_2 + \cdots + M_k a_k$, where $a_1, a_2, \ldots, a_k$ run through complete systems of residues modulo $m_1, m_2, \ldots, m_k$, respectively. Since this is exactly the size of a complete system of residues modulo M, the result will follow if we can show that each of these expressions is distinct modulo M. Suppose, by way of contradiction, that $M_1 a_1 + M_2 a_2 + \cdots + M_k a_k \equiv M_1 a_1' + M_2 a_2' + \cdots + M_k a_k' \pmod M$. Then $M_1 a_1 \equiv M_1 a_1' \pmod{m_1}$, and therefore $a_1 \equiv a_1' \pmod{m_1}$ since $(M_1, m_1) = 1$. Similarly, $a_i \equiv a_i' \pmod{m_i}$. Thus a_i' is in the same congruence class modulo m_i as a_i, for all i.

31. a. Let $\sqrt{n} = a + r$, where a is an integer, and $0 \le r < 1$. We now consider two cases: when $0 \le r < \frac{1}{2}$ and when $\frac{1}{2} \le r < 1$. For the first case, $T = [\sqrt{n} + \frac{1}{2}] = a$, and so $t = T^2 - n = -(2ar + r^2)$. Thus $|t| = 2ar + r^2 < 2a(\frac{1}{2}) + (\frac{1}{2})^2 = a + \frac{1}{4}$. Since both T and n are integers, t is also an integer. It follows that $|t| \le [a + \frac{1}{4}] = a = T$. For the second case, when $\frac{1}{2} \le r < 1$, we find that $T = [\sqrt{n} + \frac{1}{2}] = a + 1$ and $t = 2a(1 - r) + (1 - r^2)$. Since $\frac{1}{2} \le r < 1$, we have $0 < 1 - r \le \frac{1}{2}$ and $0 < 1 - r^2 < 1$. It follows that $t \le 2a(\frac{1}{2}) + (1 - r^2)$. Because t is an integer, we can say that $t \le [a + (1 - r^2)] = a < T$.
b. By the division algorithm, if we divide x by T we get $x = aT + b$, where $0 \le b < T$. If a were negative, then $x = aT + b \le (-1)T + b < 0$; but we assumed x to be nonnegative. This shows that $0 \le a$. Suppose now that $a > T$. Then $x = aT + b \ge (T + 1)T = T^2 + T \ge (\sqrt{n} - \frac{1}{2})^2 + (\sqrt{n} - \frac{1}{2}) = n - \frac{1}{4}$ and, as x and n are integers, $x \ge n$. This is a contradiction, which shows that $a \le T$. Similarly, $0 \le c \le T$ and $0 \le d < T$.
c. $xy = (aT + b)(cT + d) = acT^2 + (ad + bc)T + bd \equiv ac(T^2 - n) + zT + bd \equiv act + zT + bd \pmod{n}$
d. Use part (c), substituting $eT + f$ for ac.
e. The first half is identical to part (b); the second half follows by substituting $gT + h$ for $z + et$ and noting that $T^2 \equiv t \pmod{n}$.
f. Certainly ft and gt can be computed since all three numbers are less than T, which is less than $\sqrt{n} + 1$. So $(f + g)t$ is less than $2n < w$. Similarly, we can compute $j + bd$ without exceeding the word size. Using the same arguments, we can compute $hT + k$ without exceeding the word size.

33. a. 1   **b.** 1   **c.** 1   **d.** 1   **e.** Fermat's little theorem (Section 6.1)

35. Since $f_{n-2} + f_{n-1} \equiv f_n \pmod{m}$, if two consecutive numbers recur in the same order, then the sequence must be repeating both as n increases and as it decreases. But there are only m residues and therefore only m^2 ordered sequence of two residues. As the sequence is infinite, some two elements of the sequence must recur by the pigeonhole principle. Thus the sequence of least positive residues of the Fibonacci numbers repeats. It follows that if m divides some Fibonacci number, that is, if $f_n \equiv 0 \pmod{m}$, then m divides infinitely many Fibonacci numbers. To see that m does divide some Fibonacci number, note that $f_0 = 0$.

37. Let a and b be positive integers less than m. Then they have $O(\log m)$ digits (bits). Therefore by Theorem 2.4 we can multiply them using $O(\log^2 m)$ operations. Division by m takes $O(\log^2 m)$ operations by Theorem 2.7. Thus in all we have $O(\log^2 m)$ operations.

39. Let N_i be the number of coconuts the ith man leaves for the next man, with $N_0 = N$. At each stage, the ith man finds N_{i-1} coconuts, gives k coconuts to the monkeys, takes $(1/n)(N_{i-1} - k)$ coconuts for himself, and leaves the rest for the next man. This yields the recursive formula $N_i = (N_{i-1} - k)(n - 1)/n$. For convenience, let $w = (n - 1)/n$. If we iterate this formula a few times, we get $N_1 = (N_0 - k)w$, $N_2 = (N_1 - k)w = ((N_0 - k)w - k)w = N_0w^2 - kw^2 - kw$, $N_3 = N_0w^3 - kw^3 - kw^2 - kw$, ... The general pattern $N_i = N_0w^i - kw^i - kw^{i-1} - \cdots - kw = N_0w^i - kw(w^i - 1)/(w - 1)$ may be proved by induction. When the men rise in the morning they find $N_n = N_0w^n - kw(w^n - 1)/(w - 1)$ coconuts, and we must have $N_n \equiv k \pmod{n}$, that is, $N_n = N_0w^n - kw(w^n - 1)/(w - 1) = k + tn$ for some integer t. Substituting $w = (n - 1)/n$ back in for w, solving for N_0, and simplifying yields $N = N_0 = n^{n+1}(t + k)/(n - 1)^n - kn + k$. For N to be an integer, since $(n, n - 1) = 1$, we must have $(t + k)/(n - 1)^n$ an integer. Since we seek the smallest positive value for N, we take $t + k = (n - 1)^n$, so $t = (n - 1)^n - k$. Substituting this value back into the formula for N yields $N = n^{n+1} - kn + k$.

41. a. Let $f_1(x) = \sum_{i=0}^n a_ix^i$, $f_2(x) = \sum_{i=1}^n b_ix^i$, $g_1(x) = \sum_{i=1}^n c_ix^i$, and $g_2(x) = \sum_{i=1}^n d_ix^i$, where the leading coefficients may be zero to keep the limits of summation the same for all

polynomials. Then $a_i \equiv c_i \pmod{n}$ and $b_i \equiv d_i \pmod{n}$. Therefore, $a_i + b_1 \equiv c_i + d_i \pmod{n}$ for $i = 0, 1, \ldots, n$. Because $(f_1 + f_2)(x) = \sum_{i=1}^{n}(a_i + b_i)x^i$ and $(g_1 + g_2)(x) = \sum_{i=1}^{n}(c_i + d_i)x^i$, showing the sums of the polynomials are congruent modulo n.
b. The coefficient of x^k in $(f_1 f_2)(x)$ is $a_0 b_k + a_1 b_{k-1} + \cdots + a_k b_0$, and the corresponding coefficient in $(g_1 g_2)(x)$ is $c_0 d_k + c_1 d_{k-1} + \cdots + c_k d_0$. Because $a_i \equiv c_i \pmod{n}$ and $b_i \equiv d_i \pmod{n}$ for all i, the two expressions are congruent modulo n, and so, therefore, are the polynomials.

43. The basis step for induction on k is Exercise 42. Assume that $f(x) \equiv h(x) \pmod{p}$ and $f(x) = (x - a_1) \cdots (x - a_{k-1})h(x)$. Substituting a_k for x in this equation makes both sides 0, and none of the factors $a_k - a_i$ can be congruent to 0 modulo p, so we must have $h(a_k) \equiv 0 \pmod{p}$. Apply Exercise 42 to $h(x)$ and a_k to get $h(x) \equiv (x - a_k)g(x) \pmod{p}$ and substitute this in the congruence for $f(x)$.

Section 4.2

1. a. $x \equiv 6 \pmod 7$ **b.** $x \equiv 2,\ 5,$ or $8 \pmod 9$ **c.** $x \equiv 10 \pmod{40}$ **d.** $x \equiv 20 \pmod{25}$
e. $x \equiv 111 \pmod{999}$ **f.** $x \equiv 75 + 80k \pmod{1600}$, where k is an integer

3. $x \equiv 1074 + 3157k \pmod{28{,}927{,}591}$

5. 19 hours

7. 77 solutions when c is a multiple of 77

9. a. 13   **b.** 7   **c.** 5   **d.** 16

11. a. 1, 7, 11, 13, 17, 19, 23, 29
b. 1, 11, 19, and 29 are their own inverses; 7 and 13 are inverses of each other, as are 23 and 17

13. If $ax + by \equiv c \pmod{m}$, then there exists an integer k such that $ax + by - mk = c$. Because $d = (a, b, m) \mid ax + by - mk$, it follows that $d \mid c$, which shows that there are no solutions when $d \nmid c$. So suppose that $d \mid c$. Let $a = da'$, $b = db'$, $c = dc'$, and $m = dm'$, so that $(a', b', m') = 1$. When we divide both sides of the original congruence by d, we obtain $a'x + b'y \equiv c' \pmod{m'}$, or $a' \equiv c' - b'y \pmod{m'}$. This congruence has solutions if and only if $g = (a', m') \mid c - b'y$, or equivalently, if and only if $b'y \equiv c' \pmod{g}$ has solutions. Because $(a', b', m') = 1$, and $(a', m') = g$, it follows that $(b', g) = 1$. This means that the last congruence has only one incongruent solution y_0 modulo g. But the m'/g solutions, $y_0, y_0 + g, y_0 + 2g, \ldots, y_0 + (m'/g + 1)g$ are incongruent modulo m'. Each of these yields g incongruent values of x in the congruence $a'x \equiv c' - b'y \pmod{m'}$. Therefore, there are $g(m'/g) = m'$ incongruent solutions of $a'x \equiv c' - b'y \pmod{m'}$. Now let (x_1, y_1) be one solution of the original congruence. Then the d values $x_1, x_1 + m', x_1 + 2m', \ldots, x_1 + (d - 1)m'$ are congruent modulo m' but incongruent modulo m. Likewise, the d values $y_1, y_1 + m', y_1 + 2m', \ldots, y_1 + (d - 1)m'$ are congruent modulo m' but incongruent modulo m. So for each solution of $a'x \equiv c' - b'y \pmod{m'}$, there are d^2 solutions of the original congruence. This means that there are $d^2 m' = dm$ solutions to the original congruence.

15. Suppose that $x^2 \equiv 1 \pmod{p^k}$, where p is an odd prime and k is a positive integer. Then $x^2 - 1 \equiv (x + 1)(x - 1) \equiv 0 \pmod{p^k}$. Hence $p^k \mid (x + 1)(x - 1)$. Since $(x + 1) - (x - 1) = 2$ and p is an odd prime, p divides at most one of $x - 1$ and $x + 1$. It follows that either $p^k \mid x + 1$ or $p^k \mid x - 1$, so $p \equiv \pm 1 \pmod{p^k}$.

17. To find the inverse of a modulo m, we must solve the diophantine equation $ax + my = 1$, which can be done using the Euclidean algorithm. Using Corollary 3.13.1 we can find the greatest common divisor in $O(\log^3 m)$ bit operations. The back substitution to find x and y will take no more than $O(\log m)$ multiplications, each taking $O(\log^2 m)$ operations. Therefore, the total number of operations is $O(\log^3 m) + O(\log m)O(\log^2 m) = O(\log^3 m)$.

Section 4.3

1. $x \equiv 1 \pmod 6$

3. $32 + 60k$

5. $x \equiv 1523 \pmod{2310}$

7. 204

9. 1023

11. 2101

13. We can construct a sequence of k consecutive integers each divisible by a square as follows. Consider the system of congruences $x \equiv 0 \pmod{p_1^2}$, $x \equiv -1 \pmod{p_2^2}$, $x \equiv -2 \pmod{p_3^2}$, ..., $x \equiv -k + 1 \pmod{p_k^2}$, where p_k is the kth prime. By the Chinese remainder theorem there is a solution to this simultaneous system of congruences since the moduli are relatively prime. It follows that there is a positive integer N that satisfies each of these congruences. Each of the k integers $N, N+1, \ldots, N+k-1$ is divisible by a square since p_j^2 divides $N + j - 1$ for $j = 1, 2, \ldots, k$.

15. Suppose that x is a solution to the system of congruences. Then, $x \equiv a_1 \pmod{m_1}$, so that $x = a_1 + km_1$ for some integer k. Substituting this into the second congruence gives $a_1 + km_1 \equiv a_2 \pmod{m_2}$ or $km_1 = (a_2 - a_1) \pmod{m_2}$, which has a solution in k if and only if $(m_1, m_2) \mid (a_1, a_2)$. Now assume such a solution k_0 exists. Then all incongruent solutions are given by $k = k_0 + m_2 t/(m_1, m_2)$, where t is an integer. Then $x = a_1 + km_1 = a_1 + \left(k_0 + \frac{m_2 t}{(m_1, m_2)}\right)m_1 = a_1 + k_0 m_1 + \frac{m_1 m_2}{(m_1, m_2)}t$. Note that $m_1 m_2/(m_1, m_2) = [m_1, m_2]$, so that if we set $x_1 = a_1 + k_0 m_1$, we have $x = x_1 + [m_1, m_2]t \equiv x_1 \pmod{[m_1, m_2]}$, and so the solution is unique modulo $[m_1, m_2]$.

17. a. $x = 430 + 2100j$ **b.** $x = 9102 + 10{,}010j$

19. The basis step $r = 2$ is given by Exercise 15. Suppose that the system of the first k congruences has a unique solution A modulo $M = [m_1, \ldots, m_k]$ and $(m_i, m_j) \mid a_j - a_i$ for $1 \le i < j \le k$. Consider the system $x \equiv A \pmod M$, $x \equiv a_{r+1} \pmod{m_{r+1}}$. First suppose that it has a solution B modulo $[[m_1, m_2, \ldots, m_k], m_{k+1}]$. Then by Exercise 15, $([m_1, m_2, \ldots, m_k], m_{r+1}) \mid B - a_{k+1}$. Since $m_i \mid [m_1, m_2, \ldots, m_k]$ for $1 \le i \le k$, we have $(m_i, m_{k+1}) \mid B - a_{kr+1}$. That is, there exists an integer n_i such that $(m_i, m_{k+1})n_i = B - a_{kr+1}$. If we reduce this equation modulo m_i, for $1 \le i \le k$, then $(0, m_{k+1})n_i \equiv m_{k+1} \equiv a_i - a_{k+1} \pmod{m_i}$. If we reduce modulo m_{k+1}, then $(m_i, 0)n_i = m_i n_i \equiv 0 \pmod{m_{k+1}}$. In either case, we have that $(m_i, m_j) \mid a_j - a_i$ for $1 \le i < j \le k + 1$. Conversely, suppose that $(m_i, m_j) \mid a_j - a_i$ for $1 \le i < j \le k + 1$. Then as we have just shown, $([m_1, m_2, \ldots, m_k], m_{k+1}) \mid A - a_{k+1}$. Therefore by Exercise 15 there is a unique solution B to the first $k + 1$ congruences. This completes the induction step.

21. 2101

23. 73800 grams

25. 0000, 0001, 0625, 9376

27. none

29. every 85008 quarter-days, starting at 0

31. If the set of distinct congruences cover the integers modulo the least common multiple of the moduli, then that set will cover all integers. Examine the integers modulo 210, the l.c.m. of the moduli in this set of congruences. The first four congruences take care of all numbers containing a prime divisor of 2, 3, 5, or 7. The remaining numbers can be examined one at a time, and each can be seen to satisfy one (or more) of the congruences.

33. most likely 318 inches

35. $x = 225a_1 + 1000a_2 + 576a_3 + 1800k$, where k is an integer and a_1 is 3 or 7, a_2 is 2 or 7, and a_3 is 14 or 18

Section 4.4

1. a. 1 or 2 (mod 7)   **b.** 8 or 37 (mod 49)   **c.** 106 or 233 (mod 343)

3. 785 or 1615 (mod 2401)

5. 184, 373, 562, 751, 940, 1129, or 1318 (mod 1323)

7. 279 or 3404 (mod 4375)

9. two

11. Since $(a, p) = 1$, we know that a has an inverse b modulo p. Let $f(x) = ax - 1$. Then $x \equiv b$ (mod p) is the unique solution to $f(x) \equiv 0$ (mod p). Since $f'(x) = a \not\equiv 0$ (mod p), we know that $r \equiv b$ lifts uniquely to solutions modulo p^k for all natural numbers k. By Corollary 4.14.1, $r_k = r_{k-1} - f(r_{k-1})\overline{f'(b)} = r_{k-1} - (ar_{k-1} - 1)\overline{a} = r_{k-1} - (ar_{k-1} - 1)b = r_{k-1}(1 - ab) + b$. This gives a recursive formula for lifting b to a solution modulo p^k for every k.

13. There are 1, 3, 3, 9, and 18 solutions for $n = 1, 2, 3, 4$, and $n \geq 5$, respectively.

Section 4.5

1. a. $x \equiv 2$ (mod 5) and $y \equiv 2$ (mod 5)
b. no solutions
c. $x \equiv 3$ (mod 5) and $y \equiv 0$ (mod 5); $x \equiv 4$ (mod 5) and $y \equiv 1$ (mod 5); $x \equiv 0$ (mod 5) and $y \equiv 2$ (mod 5); $x \equiv 1$ (mod 5) and $y \equiv 3$ (mod 5); $x \equiv 2$ (mod 5) and $y \equiv 4$ (mod 5)

3. 0, 1, p, or p^2

5. The basis step, where $k = 1$, is clear by assumption. For the inductive hypothesis assume that $\mathbf{A} \equiv \mathbf{B}$ (mod m) and $\mathbf{A}^k \equiv \mathbf{B}^k$ (mod m). Then using Theorem 4.10 we have $\mathbf{A}^{k+1} = \mathbf{A} \cdot \mathbf{A}^k \equiv \mathbf{A} \cdot \mathbf{B}^k \equiv \mathbf{B} \cdot \mathbf{B}^k = \mathbf{B}^{k+1}$ (mod m).

7. false; take $m = 8$ and $\mathbf{A} = \begin{bmatrix} 3 & 0 \\ 0 & 1 \end{bmatrix}$

9. a. $\begin{bmatrix} 4 & 4 & 3 \\ 4 & 3 & 4 \\ 3 & 4 & 4 \end{bmatrix}$ **b.** $\begin{bmatrix} 2 & 0 & 6 \\ 2 & 1 & 4 \\ 3 & 4 & 0 \end{bmatrix}$ **c.** $\begin{bmatrix} 5 & 5 & 5 & 4 \\ 5 & 5 & 4 & 5 \\ 5 & 4 & 5 & 5 \\ 4 & 5 & 5 & 5 \end{bmatrix}$

11. a. 5   **b.** 5   **c.** 5   **d.** 1

13. In Gaussian elimination, the chief operation is to subtract a multiple of one equation or row from another, in order to put a 0 in a desirable place. Given that an entry a must be changed to 0 by subtracting a multiple of b, we proceed as follows: Let $\overline{b}$ be the inverse of b (mod k). Then $a - (a\overline{b})b \equiv 0$ (mod k), and elimination proceeds as for real numbers. If $\overline{b}$ doesn't exist, and one cannot swap rows to get an invertible b, then the system is underdetermined.

15. Consider summing the ith row. Let $k = xn + y$, where $0 \leq y < n$. Then x and y must satisfy the diophantine equation $i \equiv a + cy + ex$ (mod n), if k is in the ith row. Then $x - ct$ and $y + et$ is also a solution for every integer t. By Exercise 14 there must be n positive solutions, which yield n numbers k between 0 and n^2. Let $s, s + 1, \ldots, s + n - 1$ be the values of t that give these solutions. Then the sum of the ith row is $\sum_{r=0}^{n-1}(n(x - c(s + r)) + y + e(s + r)) = n(n + 1)$, which is independent of i.

Section 4.6

1. **a.** $7 \cdot 19$ **b.** $29 \cdot 41$ **c.** $41 \cdot 47$ **d.** $47 \cdot 173$ **e.** $131 \cdot 277$ **f.** $29 \cdot 1663$

3. Numbers generated by linear functions where $a > 1$ will not be random in the sense that $x_{2s} - x_k = ax_{2s-1} + b - (ax_{s-1} + b) = a(x_{2s-1} - x_{s-1})$ is a multiple of a for all s. If $a = 1$, then $x_{2s} - x_s = x_0 + sb$. In this case, if $x_0 \neq 0$, then we will not notice if a factor of b that is not a factor of x_0 is a divisor of n.

Section 5.1

1. **a.** $2^8 = 256$ **b.** $2^4 = 16$ **c.** $2^{10} = 1024$ **d.** $2^1 = 2$

3. **a.** by 3 but not by 9   **b.** by both 3 and 9   **c.** by both 3 and 9   **d.** by neither 3 nor 9

5. **a.** $2^1 = 2$ **b.** $2^0 = 1$ **c.** $2^6 = 64$ **d.** $2^0 = 1$

7. **a.** no   **b.** no   **c.** yes   **d.** yes

9. **a.** by neither 3 nor 5   **b.** by both 3 and 5   **c.** by neither 3 nor 5   **d.** by 5 but not by 3

11. if and only if the number of digits is a multiple of 3 (respectively, 9)

13. if and only if the number of digits is a multiple of 6 in each case

15. if and only if the number of digits is a multiple of d, where $d \mid b - 1$

17. A palindromic integer with $2k$ digits has the form $(a_k a_{k-1} \ldots a_1 a_1 a_2 \ldots a_k)_{10}$. Using the test for divisibility by 11 developed in this section, we find that $a_k - a_{k-1} + \cdots \pm a_1 \mp a_1 \pm a_2 \mp \cdots -a_k = 0$, and so $(a_k a_{k-1} \ldots a_1 a_1 a_2 \ldots a_k)_{10}$ is divisible by 11.

19. an integer $a_k a_{k-1} \ldots a_1 a_0$ is divisible by 37 if and only if $a_0 a_1 a_2 + a_3 a_4 a_5 + a_6 a_7 a_8 + \cdots$ is; $37 \nmid 443{,}692$; $37 \mid 11{,}092{,}785$

21. **a.** no   **b.** by 5 but not by 2   **c.** by neither 5 nor 13   **d.** yes

23. 6

25. no

Section 5.2

1. answer is person-dependent

3. once

5. $W \equiv k + [2.6m - 0.2] - 2C + Y + [Y/4] + [C/4] - [N/4000] \pmod 7$

7. answer is person-dependent

9. 2500

11. If the 13th falls on the same day of the week on two consecutive months, then the number of days in the first month is congruent to 0 modulo 7, and the only such month is February during a nonleap year. If February 13th is a Friday, then January 1st is a Thursday.

13. Let $W = 5$ and $k = 13$ in the formula for the day of the week to obtain $5 \equiv 13 + [2.6m - 0.2] - 2C + Y + [Y/4] + [C/4] \pmod 7$. This implies that $[2.6m - 0.2] \equiv 6 + 2C - Y - [Y/4] - [C/4] \pmod 7$. For every pair of values of C and Y, there is an m satisfying this congruence because $[2.6m - 0.2]$ takes on all possible remainders modulo 7 as the month varies from March to December and m takes on the values from 0 to 10.

15. Months with 31 days are March, May, July, August, October, December, and January (considered to be in the previous year); the corresponding values of m are $1, 3, 5, 6, 8, 10$, and 11, respectively. Given Y and C, let $k = 31$ to obtain $W \equiv 31 + [2.6m - 0.2] - 2C + Y + [Y/4] + [C/4] \equiv 3 + [2.6m - 0.2] - 2C + Y + [Y/4] + [C/4] \pmod 7$. To determine the days of the week the 31st falls on, first let m equal $1, 3, 5, 6, 8, 10$, and reduce modulo 7. Finally, decrease the year by one and find the new values of Y and C, and let $m = 11$ and reduce modulo 7. The values of W we find tell us the days of the week on which the 31st falls.

Section 5.3

1. a. Teams i and j are paired in round k if and only if $i + j \equiv k \pmod 7$, with team i drawing a bye if $2i \equiv k \pmod 7$. Round 1: 1-7, 2-6, 3-5, 4-bye; round 2: 2-7, 3-6, 4-5, 1-bye; round 3: 1-2, 3-7, 4-6, 5-bye; round 4: 1-3, 4-7, 5-6, 2-bye; round 5: 1-4, 2-3, 5-7, 6-bye; round 6: 1-5, 2-4, 6-7, 3-bye; round 7: 1-6, 2-5, 3-4, 7-bye.
b. Teams i and j are paired in round k if and only if $i + j \equiv k \pmod 7$, $i, j \neq 8$; team i plays team 8 if $2i \equiv k \pmod 7$.
c. Teams i and j are paired in round k if and only if $i + j \equiv k \pmod 9$, with team i drawing a bye if $2i \equiv k \pmod 9$.
d. Teams i and j are paired in round k if and only if $i + j \equiv k \pmod 9$, $i, j \neq 10$; team i plays team 10 if $2i \equiv k \pmod 9$.

3. a. home teams in round 1: 4 and 5; round 2: 2 and 3; round 3: 1 and 5; round 4: 3 and 4; round 5: 1 and 2
b. home teams in round 1: 5, 6, and 7; round 2: 2, 3, and 4; round 3: 1, 6, and 7; round 4: 3, 4, and 5; round 5: 1, 2, and 7; round 6: 4, 5, and 6; round 7: 1, 2, and 3
c. home teams in round 1: 6, 7, 8, and 9; round 2: 2, 3, 4, and 5; round 3: 1, 7, 8, and 9; round 4: 3, 4, 5, and 6; round 5: 1, 2, 8, and 9; round 6: 4, 5, 6, and 7; round 7: 1, 2, 3, and 9; round 8: 5, 6, 7, and 8; round 9: 1, 2, 3, and 4

Section 5.4

1. Let k be the six-digit number on the license plate of a car. We can assign this car the space numbered $h(k) = k \bmod 101$. When a car is assigned the same space as another car we can assign it to the space $(h(k) + g(k)) \bmod 101$, where $g(k) = (k \bmod 99) + 1$. When this space is occupied we next try $(h(k) + 2g(k)) \bmod 101$, then $(h(k) + 3g(k)) \bmod 101$, and so on. All spaces are examined since $(g(k), 101) = 1$.

3. a. It is clear that m memory locations will be probed as $j = 0, 1, 2, \ldots, m - 1$. To see that they are all distinct, and hence that every memory location is probed, assume that $h_i(K) \equiv h_j(K) \pmod m$. Then $h(K) + iq \equiv h(K) + jq \pmod m$. From this it follows that $iq \equiv jq \pmod m$, and, as $(q, m) = 1$, we have $i \equiv j \pmod m$ by Corollary 4.4.1. Therefore $i = j$ since i and j are both less than m.
b. It is clear that m memory locations will be probed as $j = 0, 1, 2, \ldots, m - 1$. To see that they are all distinct, and hence that every memory location is probed, assume that $h_i(K) \equiv h_j(K) \pmod m$. Then $h(K) + iq \equiv h(K) + jq \pmod m$. From this it follows that $iq \equiv jq \pmod m$, and, as $(q, m) = 1$, we have $i \equiv j \pmod m$ by Corollary 4.4.1. Therefore $i = j$ since i and j are both less than m.

5. 558, 1002, 2174, 4035

Section 5.5

1. a. 0   **b.** 0   **c.** 1   **d.** 1   **e.** 0   **f.** 1

3. a. 0   **b.** 1   **c.** 0

5. a. 7   **b.** 1   **c.** 4

7. Transposition means that adjacent digits are in the wrong order. Suppose, first, that the first two digits, x_1 and x_2, or, equivalently, the fourth and fifth digits, are exchanged, and the error is not detected. Then $x_7 \equiv 7x_1 + 3x_2 + x_3 + 7x_4 + 3x_5 + x_6 \equiv 7x_2 + 3x_1 + x_3 + 7x_4 + 3x_5 + x_6$ (mod 10). It follows that $7x_1 + 3x_2 \equiv 7x_2 + 3x_1$ (mod 10) or $4x_1 \equiv 4x_2$ (mod 10). By Corollary 4.4.1, $x_1 \equiv x_2$ (mod 5). This is equivalent to $|x_1 - x_2| = 5$, as x_1 and x_2 are distinct single digits. Similarly, if the second and third (or fifth and sixth) digits are transposed, then $2x_2 \equiv 2x_3$ (mod 10), which again reduces to $x_2 \equiv x_3$ (mod 5) by Corollary 4.4.1. Also, if the third and fourth digits are transposed, then $6x_3 \equiv 6x_4$ (mod 10) and $x_3 \equiv x_4$ (mod 5), similarly as before. The reverse argument will complete the proof.

9. a. 0   **b.** 3   **c.** 4   **d.** X

11. a. valid   **b.** not valid   **c.** valid   **d.** valid   **e.** not valid

13. 0-07-289905-0

15. a. no   **b.** yes   **c.** yes   **d.** no

17. It can.

19. a. yes   **b.** no

21. a. 94

b. If x_i is misentered as y_i, then if the congruence defining x_{10} holds, we see that $ax_i \equiv ay_i$ (mod 11) by setting the two definitions of x_{10} congruent. From this, it follows from Corollary 4.4.1 that $x_i \equiv y_i$ (mod 11), and so $x_i = y_i$. If the last digit, x_{11}, is misentered as y_{11}, then the congruence defining x_{11} will hold if and only if $x_{11} = y_{11}$.

c. Suppose that x_i is misentered as y_i and x_j is misentered as y_j, with $i < j < 10$. Suppose that both of the congruences defining x_{10} and x_{11} hold. Then by setting the two versions of each congruence congruent to each other we obtain $ax_i + bx_j \equiv ay_i + by_j$ (mod 11) and $cx_i + dx_j \equiv cy_i + dy_j$ (mod 11), where $a \neq b$ and $c \neq d$. If $ad - bc \not\equiv 0$ (mod 11), then the coefficient matrix is invertible and we can multiply both sides of this system of congruences by the inverse to obtain $x_i = y_i$ and $x_j = y_j$. Indeed, after (tediously) checking each possible choice of a, b, c, and d, we find that all the matrices are invertible modulo 11.

23. a. 1   **b.** 1   **c.** 6

25. Errors involving a difference of 7 cannot be detected: 0 for 7, 1 for 8, 2 for 9, or vice versa. All others can be detected.

27. a. 1   **b.** X   **c.** 2   **d.** 8

29. Yes. Assume not and compare the expressions modulo 11, to get a congruence of the form $ad_i + bd_j \equiv ad_j + bd_i$ (mod 11), which reduces to $(a - b)d_i \equiv (a - b)d_j$ (mod 11). Because $0 < a - b < 11$ and 11 is prime, it follows that $d_i \equiv d_j$ (mod 11). Because these are digits between 0 and X, they must be equal.

Section 6.1

1. We have $10! + 1 = 1(2 \cdot 6)(3 \cdot 4)(5 \cdot 9)(7 \cdot 8)10 + 1 = 1 \cdot 12 \cdot 12 \cdot 45 \cdot 56 \cdot 10 + 1 \equiv 1 \cdot 1 \cdot 1 \cdot 1 \cdot 1 \cdot (-1) + 1 \equiv 0$ (mod 11). Therefore, 11 divides $10! + 1$.

3. 9

5. 6

7. 436

9. 2

11. 6

13. $(3^5)^2 \equiv 243^2 \equiv 1^2 \equiv 1 \pmod{11^2}$

15. a. $x \equiv 9 \pmod{17}$ **b.** $x \equiv 17 \pmod{19}$

17. Suppose that p is an odd prime. Then Wilson's theorem tells us that $(p-1)! \equiv -1 \pmod{p}$. Since $(p-1)! = (p-3)!(p-1)(p-2) \equiv (p-3)!(-1)(-2) \equiv 2 \cdot (p-3)! \pmod{p}$, this implies that $2 \cdot (p-3)! \equiv -1 \pmod{p}$.

19. Since $(a, 35) = 1$, we have $(a, 7) = (a, 5) = 1$, so we may apply Fermat's little theorem to get $a^{12} - 1 = (a^6)^2 - 1 \equiv 1^2 - 1 = 0 \pmod{7}$, and $a^{12} - 1 = (a^4)^3 - 1 \equiv 1^3 - 1 = 0 \pmod{5}$. Both 5 and 7 divide $a^{12} - 1$, so 35 must also divide it.

21. When n is even, so is n^7, and when n is odd, so is n^7. It follows that $n^7 \equiv n \pmod{2}$. Similarly, since $n^3 \equiv n \pmod{3}$, it follows that $n^7 = (n^3)^2 \cdot n \equiv n^2 \cdot n \equiv n^3 \equiv n \pmod{3}$. We also know by Fermat's little theorem that $n^7 \equiv n \pmod{7}$. Since $42 = 2 \cdot 3 \cdot 7$, it follows that $n^7 \equiv n \pmod{42}$.

23. By Fermat's little theorem, $\sum_{k=1}^{p-1} k^{p-1} \equiv \sum_{k=1}^{p-1} 1 = p - 1 \equiv -1 \pmod{p}$.

25. By Fermat's little theorem, $a \equiv a^p \equiv b^p \equiv b \pmod{p}$; hence $b = a + kp$ for some integer k. Then by the binomial theorem, $b^p = (a + kp)^p = a^p + \binom{p}{1}a^{p-1}kp + p^2N$, where N is some integer. Then $b^p \equiv a^p + p^2a^{p-1}k + p^2N \equiv a^p \pmod{p^2}$.

27. 641 (at $k = 8$)

29. Suppose that p is prime. Then by Fermat's little theorem, $a^p \equiv a \pmod{p}$ for every integer a. Also, by Wilson's theorem, $(p-1)! \equiv -1 \pmod{p}$, so $a(p-1)! \equiv -a \pmod{p}$. It follows that $a^p + (p-1)!a \equiv a + (-a) \equiv 0 \pmod{p}$. Consequently, $p \mid a^p + (p-1)!a$.

31. Since $p - 1 \equiv -1$, $p - 2 \equiv -2$, $\ldots$, $(p+1)/2 \equiv -(p-1)/2 \pmod{p}$, we have $((p-1)/2)!^2 \equiv -(p-1)! \equiv 1 \pmod{p}$. (Because $p \equiv 3 \pmod{4}$ the minus signs work out.) If $x^2 \equiv 1 \pmod{p}$, then $p \mid x^2 - 1 = (x-1)(x+1)$, so $x \equiv \pm 1 \pmod{p}$.

33. Suppose that $p \equiv 1 \pmod{4}$. Let $y = \pm((p-1)/2)!$. Then $y^2 \equiv ((p-1)/2)!^2 \equiv ((p-1)/2)!^2(-1)^{(p-1)/2} \equiv (1 \cdot 2 \cdot 3 \cdots (p-1)/2)((-(p-1)/2) \cdots (-3) \cdot (-2) \cdot (-1)) \equiv 1 \cdot 2 \cdot 3 \cdots (p-1)/2 \cdot (p+1)/2 \cdots (p-3)(p-2)(p-1) = (p-1)! \equiv -1 \pmod{p}$, where we have used Wilson's theorem. Now suppose that $x^2 \equiv -1 \pmod{p}$. Then $x^2 \equiv y^2 \pmod{p}$. Hence $(x^2 - y^2) = (x-y)(x+y) \equiv 0 \pmod{p}$. It follows that $p \mid x - y$ or $p \mid x + y$, so $x \equiv \pm y \pmod{p}$.

35. If n is composite and $n \neq 4$, then Exercise 16 shows that $(n-1)!/n$ is an integer, so $[((n-1)! + 1)/n - [(n-1)!/n]] = [(n-1)!/n + 1/n - (n-1)!/n] = [1/n] = 0$. If $n = 4$, then the same expression is also equal to 0. But if n is prime, then by Wilson's theorem, $(n-1)! = Kn - 1$ for some integer K. So $[((n-1)! + 1)/n - [(n-1)!/n]] = [(Kn - 1 + 1)/n - [(Kn-1)/n]] = [K - (K-1)] = 1$. Therefore, the sum increases by 1 exactly when n is prime, so it equals $\pi(n)$.

37. Suppose that n and $n + 2$ are twin primes. Then by Wilson's theorem, $(n-1)! \equiv -1 \pmod{n}$. Hence, $4((n-1)! + 1) + n \equiv 4 \cdot 0 + n \equiv 0 \pmod{n}$. Also, since $n + 2$ is prime it follows from Wilson's theorem that $(n+1)! \equiv -1 \pmod{n+2}$, so $(n+1)n \cdot (n-1)! \equiv (-1)(-2)(n-1)! \equiv 2(n-1)! \equiv -1 \pmod{n+2}$. Hence $4((n-1)! + 1) + n \equiv 2(2 \cdot (n-1)!) + 4 + n \equiv 2 \cdot (-1) + 4 + n = n + 2 \equiv 0 \pmod{n+2}$. Since $(n, n+2) = 1$ it follows that $4((n-1)! + 1) + n \equiv 0 \pmod{n(n+2)}$.

39. Note that $1 \cdot 2 \cdots (p-1) \equiv (p+1)(p+2) \cdots (2p-1) \pmod{p}$. Each factor is relatively prime to p, so $1 \equiv (p+1)(p+2) \cdots (2p-1)/(1 \cdot 2 \cdots (p-1)) \pmod{p}$. Thus $2 \equiv (p+1)(p+2) \cdots (2p-1)2p/(1 \cdot 2 \cdots (p-1)p) = \binom{2p}{p} \pmod{p}$.

41. We first note that $1^p \equiv 1 \pmod{p}$. Now suppose that $a^p \equiv a \pmod{p}$. Then by Exercise 40 we see that $(a+1)^p \equiv a^p + 1 \pmod{p}$. But by the inductive hypothesis $a^p \equiv a \pmod{p}$, so $a^p + 1 \equiv a + 1 \pmod{p}$. Hence, $(a+1)^p \equiv a + 1 \pmod{p}$.

43. a. If $c < 26$, then c extra cards are put into the deck above the card, so it ends up in the $(2c)$th position; $2c < 52$, so $b = 2c$. If $c \geq 26$, then $c - 26 - 1$ extra cards are put into the deck above the card, but 26 cards are taken away above it, so it ends up in the $b = (c - 26 - 1 + c - 26) = (2c - 53)$th place. Then $b = 2c - 53 \equiv 2c \pmod{53}$.
b. 52

45. There are two cases. Assume first that $a_k \equiv 0 \pmod{p}$ and $b_j \equiv 0 \pmod{p}$, where $k \neq j$. Then two of the products $a_i b_i \equiv 0 \pmod{p}$, and this would contradict Wilson's theorem if the $a_i b_i$ formed a complete system, since the product of all but one of them must be $-1 \pmod{p}$, not 0. For the second case, assume without loss of generality that $a_p \equiv b_p \equiv 0 \pmod{p}$. Then by Wilson's theorem, $a_1 a_2 \cdots a_{p-1} \equiv b_1 b_2 \cdots b_{p-1} \equiv -1 \pmod{p}$. Then $a_1 b_1 \cdots a_{p-1} b_{p-1} \equiv (-1)^2 = 1 \pmod{p}$. If the set were a complete system, the last product would have been $\equiv -1 \pmod{p}$.

47. The basis step for induction is Wilson's theorem. Assume $(p-1)!^{p^{k-1}} \equiv -1 \pmod{p^k}$. Then $(p-1)!^{p^k} \equiv ((p-1)!^{p^{k-1}})^p \equiv (-1 + mp^k)^p \equiv -1 + \binom{p}{1}mp^k + \cdots + (mp^k)^p \equiv -1 \pmod{p^{k+1}}$, where we have used the fact that $p \mid \binom{p}{j}$ for $j \neq 0$ or p.

49. If n is prime, then n divides the binomial coefficient $\binom{n}{r}$, $r = 1, 2, \ldots, n-1$. It follows that $(x - a)^n$ and $x^n - a^n$ are congruent modulo n as polynomials because the coefficient of x^r in $(x - a)^n$ is congruent to 0 modulo n for $r = 1, 2, \ldots, n-1$. By Fermat's little theorem, it follows that $a^n \equiv a \pmod{n}$, so that $(x - a)^n$ and $x^n - a$ are congruent modulo n as polynomials. Now suppose that n is a composite integer with $n > 1$. We know that n has a prime divisor q. Suppose that q^k is the greatest power of q dividing n. A short argument shows that $q^k \nmid \binom{n}{q} = n(n-1) \cdots (n - q + 1)/q!$ and $(q^k, a^{n-q}) = 1$. It follows that the coefficient of x^q in $(x - a)^n$ is not congruent to 0 modulo p, but the coefficient of $x^n - a$ is 0. It follows that $(x - a)^n$ and $x^n - a$ are not congruent as polynomials modulo n.

Section 6.2

1. $3^{90} \equiv 1 \pmod{91}$, but $91 = 7 \cdot 13$

3. Either computation by hand or a computational program shows that $2^{161038} \equiv 2 \pmod{161038}$.

5. $(n - a)^n \equiv (-a)^n \equiv -(a^n) \equiv -a \equiv (n - a) \pmod{n}$

7. Raise the congruence $2^{2^m}; \equiv; -1 \pmod{F_m}$ to the $(2^{2^m - m})$th power, to obtain $2^{2^{2^m}}; \equiv; 1 \pmod{2^{2^m}; +; 1}$, which says that $2^{F_m - 1} \equiv 1 \pmod{F_m}$.

9. Suppose that n is a pseudoprime to the bases a and b. Then $b^n \equiv b \pmod{n}$ and $a^n \equiv a \pmod{n}$. It follows that $(ab)^n \equiv a^n b^n \equiv ab \pmod{n}$. Hence, n is a pseudoprime to the base ab.

11. a. If n is a pseudoprime to the base ab, then $(ab)^{n-1} \equiv 1 \pmod{n}$, so $1 \equiv a^{n-1}b^{n-1} \equiv 1 \cdot b^{n-1} \pmod{n}$, which implies that n is a pseudoprime to the base b, a contradiction.
b. Let $a_1, a_2, \ldots, a_r$ be the bases to which n is a pseudoprime and for which $(a_i, n) = 1$ for each i. Then by part (a) we know that, for each i, n is not a pseudoprime to the base ba_i. Thus we have $2r$ different elements relatively prime to n. Then by the definition of $\phi(n)$, we have $r \leq \phi(n)/2$.

13. A computation shows that $2^{1387} \equiv 2 \pmod{1387}$, so 1387 is a pseudoprime. But $1387 - 1 = 2 \cdot 693$ and $2^{693} \equiv 512 \pmod{1387}$, which is all that must be checked, since $s = 1$. Thus 1387 fails Miller's test and hence is not a strong pseudoprime.

15. Note that $25{,}326{,}001 - 1 = 2^4 \cdot 1{,}582{,}875 = 2^s t$. With this value of t, we see with the help of computational software that $2^t \equiv -1 \pmod{25{,}326{,}001}$, $3^t \equiv -1 \pmod{25{,}326{,}001}$, and $5^t \equiv 1 \pmod{25{,}326{,}001}$.

17. Suppose that $c = 7 \cdot 23 \cdot q$, with q an odd prime, is a Carmichael number. Then by Theorem 6.7 we must have $7 - 1 \mid c - 1$, so $c = 7 \cdot 23 \cdot q \equiv 1 \pmod 6$. Solving this yields $q \equiv 5 \pmod 6$. Similarly, $23 - 1 \mid c - 1$, so $7 \cdot 23 \cdot q \equiv 1 \pmod{22}$. Solving this yields $q \equiv 19 \pmod{22}$. If we apply the Chinese remainder theorem to these two congruences, we obtain $q \equiv 41 \pmod{66}$, that is, $q = 41 + 66k$ for some k. Then we must have $q - 1 \mid c - 1$, which is $40 + 66k \mid 7 \cdot 23 \cdot (41 + 66k) - 1$. So there is an integer m such that $m(40 + 66k) = 6600 + 10626k = 160 + 6440 + 10626k = 160 + 161(40 + 66k)$. Therefore, 160 must be a multiple of $40 + 66k$, which happens only when $k = 0$. Therefore, $q = 41$ is the only such prime.

19. We have $32{,}111{,}197{,}185 - 1 = 321{,}197{,}184 = 4 \cdot 80{,}299{,}296 = 18 \cdot 17{,}844{,}288 = 22 \cdot 14{,}599{,}872 = 28 \cdot 11{,}471{,}328 = 36 \cdot 8{,}922{,}144 = 136 \cdot 2{,}361{,}744$, so $p - 1 \mid 321{,}197{,}185 - 1$ for every prime p that divides $321{,}197{,}185$. Therefore, by Theorem 5.7, $321{,}197{,}185$ is a Carmichael number.

21. We can assume that $b < n$. Then b has fewer than $\log_2 n$ bits. Also, $t < n$, so t has fewer than $\log_2 n$ bits. It takes at most $\log_2 n$ multiplications to calculate b^{2^s}, so it takes $O(\log_2 n)$ multiplications to calculate $b^{2^{\log_2 t}} = b^t$. Each multiplication is of two $(\log_2 n)$-bit numbers and so takes $O((\log_2 n)^2)$ operations. In all, we have $O((\log_2 n)^3)$ operations.

Section 6.3

1. a. $\{1, 5\}$ **b.** $\{1, 2, 4, 5, 7, 8\}$ **c.** $\{1, 3, 7, 9\}$ **d.** $\{1, 3, 5, 9, 11, 13\}$
   e. $\{1, 3, 5, 7, 9, 11, 13, 15\}$ **f.** $\{1, 2, 3, 4, 5, 6, 7, 8, 9, 10, 11, 12, 13, 14, 15, 16\}$

3. If $(a, m) = 1$, then $(-a, m) = 1$, so $-c_i$ must appear among the c_j. Also $c_i \not\equiv -c_i \pmod m$, else $2c_i \equiv 0 \pmod m$ and so $(c_i, m) \neq 1$. Hence, the elements in the sum can be paired so that each pair sums to $0 \pmod m$, and thus the entire sum is $0 \pmod m$.

5. 1

7. 11

9. Since $a^2 \equiv 1 \pmod 8$ whenever a is odd, $a^{12} = (a^2)^6 \equiv 1 \pmod 8$ whenever $(a, 32{,}760) = 1$. Euler's theorem tells us that $a^{\phi(9)} = a^6 \equiv 1 \pmod 9$ whenever $(a, 9) = 1$, so $a^{12} = (a^6)^2 \equiv 1 \pmod 9$ whenever $(a, 32{,}760) = 1$. Furthermore, Fermat's little theorem tells us that $a^4 \equiv 1 \pmod 5$ whenever $(a, 5) = 1$, $a^6 \equiv 1 \pmod 7$ whenever $(a, 7) = 1$, and $a^{12} \equiv 1 \pmod{13}$ whenever $(a, 13) = 1$. It follows that $a^{12} \equiv (a^4)^3 \equiv 1 \pmod 5$, $a^{12} \equiv (a^6)^2 \equiv 1 \pmod 7$, and $a^{12} \equiv 1 \pmod{13}$ whenever $(a, 32{,}760) = 1$. Since $32{,}760 = 2^3 \cdot 3^2 \cdot 5 \cdot 7 \cdot 13$ and the moduli 8, 9, 5, 7, and 13 are pairwise relatively prime, we see that $a^{12} \equiv 1 \pmod{32{,}760}$.

11. a. $x \equiv 9 \pmod{14}$ **b.** $x \equiv 13 \pmod{15}$ **c.** $x \equiv 7 \pmod{16}$

13. a. $x \equiv 37 \pmod{187}$ **b.** $x \equiv 23 \pmod{30}$ **c.** $x \equiv 6 \pmod{210}$ **d.** $x \equiv 150{,}999 \pmod{554{,}268}$

15. 1

17. $\phi(13) = 12$, $\phi(14) = 6$, $\phi(15) = 8$, $\phi(16) = 8$, $\phi(17) = 16$, $\phi(18) = 6$, $\phi(19) = 18$, $\phi(20) = 8$

19. If $(a, b) = 1$ and $(a, b - 1) = 1$, then $a \mid (b^{k\phi(a)} - 1)/(b - 1)$, which is a base b repunit. If $(a, b - 1) = d > 1$, then d divides every repunit of length $k(b - 1)$ and $(a/d) \mid (b^{k\phi(a/d)} - 1)/(b - 1)$, and these sets intersect infinitely often.

Section 7.1

1. a. f is completely multiplicative since for all positive integers m and n, $f(mn) = 0 = 0 \cdot 0 = f(m) \cdot f(n)$.

b. f is not completely multiplicative since $f(6) = 2$, but $f(2) \cdot f(3) = 2 \cdot 2 = 4$.

c. f is not completely multiplicative since $f(6) = 3$, but $f(2) \cdot f(3) = \frac{2}{2} \cdot \frac{3}{2} = \frac{3}{2}$.

d. f is not completely multiplicative since $f(4) = \log(4) > 1$, but $f(2) \cdot f(2) = \log(2) \cdot \log(2) < 1$.

e. f is completely multiplicative since for every positive integer m and n, $f(mn) = (mn)^2 = m^2 n^2 = f(m) \cdot f(n)$.

f. f is not completely multiplicative since $f(4) = 4! = 24$, but $f(2) \cdot f(2) = 2!2! = 4$.

g. f is not completely multiplicative since $f(6) = 7$, but $f(2) \cdot f(3) = 3 \cdot 4 = 12$.

h. f is not completely multiplicative since $f(4) = 4^4 = 256$, but $f(2) \cdot f(2) = 2^2 2^2 = 16$.

i. f is completely multiplicative since for every positive integer m and n, $f(mn) = \sqrt{mn} = \sqrt{m}\sqrt{n} = f(m) \cdot f(n)$.

3. We have the following prime factorizations of 5186, 5187, and 5188: $5186 = 2 \cdot 2593$, $5187 = 3 \cdot 7 \cdot 13 \cdot 19$, and $5188 = 2^2 1297$. Hence, $\phi(5186) = \phi(2)\phi(2593) = 1 \cdot 2592 = 2592$, $\phi(5187) = \phi(3)\phi(7)\phi(13)\phi(19) = 2 \cdot 6 \cdot 12 \cdot 18 = 2592$, and $\phi(5188) = \phi(2^2)\phi(1297) = 2 \cdot 1296 = 2592$. It follows that $\phi(5186) = \phi(5187) = \phi(5188)$.

5. 7, 9, 14, 18

7. 35, 39, 45, 52, 56, 70, 72, 78, 84, 90

9. The nth term of this sequence is $\phi(2n)$.

11. multiples of 3

13. powers of 2 greater than 1

15. If n is odd, then $(2, n) = 1$ and $\phi(2n) = \phi(2)\phi(n) = 1 \cdot \phi(n) = \phi(n)$. If n is even, say $n = 2^s t$ with t odd, then $\phi(2n) = \phi(2^{s+1}t) = \phi(2^{s+1})\phi(t) = 2^s \phi(t) = 2(2^{s-1}\phi(t)) = 2(\phi(2^s)\phi(t)) = 2(\phi(2^s t)) = 2\phi(n)$.

17. $n = 2^k p_1 p_2 \cdots p_r$, where each p_i is a distinct Fermat prime

19. Let $n = p_1^{a_1} \cdots p_r^{a_r}$ be the prime-power factorization for n. If $n = 2\phi(n)$, then $p_1^{a_1} \cdots p_r^{a_r} = 2 \prod_{j=1}^r p_j^{a_j - 1}(p_j - 1)$. This implies that $p_1 \cdots p_r = 2 \prod_{j=1}^r (p_j - 1)$. If any p_j is an odd prime, then the factor $(p_j - 1)$ is even and must divide the product on the left-hand side. But there can be at most one factor of 2 on the left-hand side and it is accounted for by the factor of 2 in front of the product on the right-hand side. Therefore, no odd primes appear in the product. That is, $n = 2^j$ for some j.

21. Since $(m, n) = p$, p divides one of the terms, say n, exactly once, so $n = kp$ with $(m, k) = 1 = (n, k)$. Then $\phi(n) = \phi(kp) = \phi(k)\phi(p) = \phi(k)(p - 1)$, and $\phi(mp) = p\phi(m)$ by the formula in Example 7.7. Consequently, $\phi(mn) = \phi(mkp) = \phi(mp)\phi(k) = (p\phi(m))(\phi(n)/(p - 1))$.

23. Let $p_1, \ldots, p_r$ be those primes dividing a but not b. Let $q_1, \ldots, q_s$ be those primes dividing b but not a. Let $r_1, \ldots, r_t$ be those primes dividing a and b. Let $P = \prod(1 - (1/p_i))$, $Q = \prod(1 - (1/q_i))$, and $R = \prod(1 - (1/r_i))$. Then $\phi(ab) = abPQR = aPRbQR/R = \phi(a)\phi(b)/R$. But $\phi((a, b)) = (a, b)R$, so $R = \phi((a, b))/(a, b)$ and we have $\phi(ab) = \phi(a)\phi(b)/R = (a, b)\phi(a)\phi(b)/\phi((a, b))$, as desired. The final conclusion now follows immediately from the obvious fact that $\phi((a, b)) < (a, b)$ when $(a, b) > 1$.

25. Assume that there are only finitely many primes, $2, 3, 5, \ldots, p$. Let $N = 2 \cdot 3 \cdot 5 \cdots p$. Then $\phi(N) = 1$ since there is exactly one positive integer less than N that is relatively prime to N, namely 1, because every prime is a factor of N. However, $\phi(N) = \phi(2)\phi(3)\phi(5)\cdots\phi(p) = 1\cdot 2\cdot 4\cdots(p-1) > 1$. This contradiction shows that there must be infinitely many primes.

27. From the formula for the ϕ function, we see that if $p \mid n$, then $p-1 \mid \phi(n)$. Since $\phi(n)$ is finite, there are only finitely many possibilities for prime divisors of n. Further, if p is prime and $p^a \mid n$, then $p^{a-1} \mid \phi(n)$. Hence, $a \leq \log_p \phi(n) + 1$. Therefore, each of the finitely many primes that might divide n may appear to only finitely many exponents. This gives only finitely many possibilities for n.

29. As suggested, we take $k = 2\cdot 3^{6j+1}$ with $j \geq 1$ and suppose that $\phi(n) = k$. From the formula for $\phi(n)$ we see that $\phi(n)$ has a factor of $p-1$, which is even, for every odd prime that divides n. Since there is only one factor of 2 in k, there is at most one odd prime divisor of n. Further, since $2 \parallel k$ and $n > 4$, we know that $4 \nmid n$. Since k is not a power of 2, we know that an odd prime p must divide n. So n is of the form p^a or $2p^a$. Recall that $\phi(p^a) = \phi(2p^a)$. It remains to discover the values of p and a. If $a = 1$, then $\phi(p^a) = p-1 = 2\cdot 3^{6j+1}$. But then $p = 2\cdot 3^{6j+1} + 1 = 6\cdot(3^6)^j + 1 \equiv (-1)(1)^j + 1 \equiv 0 \pmod 7$. Hence, $p = 7$. But $\phi(7) = 6 = 2\cdot 3^{6j+1}$ implies that $j = 0$, contrary to hypothesis, so this is not a solution. Therefore, $a > 1$ and we have $\phi(p^a) = (p-1)p^{a-1} = 2\cdot 3^{6j+1}$, from which we conclude that $p = 3$ and $a = 6j+2$. Therefore, the only solutions are $n = 3^{6j+2}$ and $n = 2\cdot 3^{6j+2}$.

31. If $n = p^r m$, where $p \nmid m$, then $\phi(p^r m) = (p^r - p^{r-1})\phi(m) \mid p^r m - 1$; hence, $p \mid 1$ or $r = 1$. So n is square-free. If $n = pq$, then $\phi(pq) = (p-1)(q-1) \mid pq - 1$. Then $p-1 \mid (pq-1) - (p-1)q = q-1$. Similarly, $q-1 \mid p-1$, a contradiction.

33. Let $n = p_1^{a_1} p_2^{a_2} \cdots p_k^{a_k}$. Let P_i be the property that an integer is divisible by p_i. Let S be the set $\{1, 2, \ldots, n-1\}$. To compute $\phi(n)$ we need to count the elements of S with none of the properties $P_1, P_2, \ldots, P_k$. Let $n(P_{i_1}, P_{i_2}, \ldots, P_{i_m})$ be the number of elements of S with all of properties $P_{i_1}, P_{i_2}, \ldots, P_{i_m}$. Then $n(P_{i_1}, \ldots, P_{i_m}) = n/(p_{i_1} p_{i_2} \cdots p_{i_m})$. By the principle of inclusion–exclusion, we have $\phi(n) = n - (n/p_1 + n/p_2 + \cdots + n/p_k) + (n/(p_1 p_2) + n/(p_1 p_3) + \cdots + n/(p_{k-1} p_k)) - \cdots + (-1)^k \cdot n/(p_1 \cdots p_k) = n(1 - \sum_{p_i \mid n} 1/p_i + \sum_{p_{i_1} p_{i_2} \mid n} 1/(p_{i_1} p_{i_2}) - \sum_{p_{i_1} p_{i_2} p_{i_3} \mid n} 1/(p_{i_1} p_{i_2} p_{i_3}) + \cdots + (-1)^k \cdot n/(p_1 \cdots p_k))$. On the other hand, notice that each term in the expansion of $(1 - 1/p_1)(1 - 1/p_2)\cdots(1 - 1/p_k)$ is obtained by choosing either 1 or $-1/p_i$ from each factor and multiplying the choices together. This gives each term the form $(-1)^m/(p_{i_1} p_{i_2} \cdots p_{i_m})$. Note that each term can occur in only one way. Thus $n(1 - 1/p_1)(1 - 1/p_2)\cdots(1 - 1/p_k) = n(1 - \sum_{p_i \mid n} 1/p_i + \sum_{p_{i_1} p_{i_2} \mid n} 1/(p_{i_1} p_{i_2}) - \cdots + (-1)^k n/(p_1 \cdots p_k)) = \phi(n)$.

35. Note that $1 \leq \phi(m) \leq m-1$ for $m > 1$. Hence, if $n \geq 2$, then $n > n_1 > n_2 > \cdots \geq 1$, where $n_1 = \phi(n)$ and $n_i = \phi(n_{i-1})$ for $i > 1$. Since n_i, $i = 1, 2, 3, \ldots$, is a decreasing sequence of positive integers, there must be a positive integer r such that $n_r = 1$.

37. Note that the definition of $f * g$ can also be expressed as $(f * g)(n) = \sum_{a \cdot b = n} f(a)g(b)$. Then the fact that $f * g = g * f$ is evident.

39. a. If either $m > 1$ or $n > 1$, then $mn > 1$ and one of $\iota(m)$ or $\iota(n)$ is equal to 0. Then $\iota(mn) = 0 = \iota(m)\iota(n)$. Otherwise, $m = n = 1$ and we have $\iota(mn) = 1 = 1\cdot 1 = \iota(m)\iota(n)$. Therefore, $\iota(n)$ is multiplicative.
b. $(\iota * f)(n) = \sum_{d \mid n} \iota(d)f(n/d) = \iota(1)f(n/1) = f(n)$ since $\iota(d) = 0$ except when $d = 1$; $(f * \iota)(n) = (\iota * f)(n) = f(n)$ by Exercise 37.

41. Let $h = f * g$ and let $(m, n) = 1$. Then $h(mn) = \sum_{d \mid mn} f(d)g(mn/d)$. Since $(m, n) = 1$, each divisor d of mn can be expressed in exactly one way as $d = ab$, where $a \mid m$ and $b \mid n$. Then $(a, b) = 1$ and $(m/a, n/b) = 1$. Thus there is a one-to-one correspondence between the divisors d of mn

and the pairs of products ab, where $a \mid m$ and $b \mid n$. Then $h(mn) = \sum_{a \mid m, \, b \mid n} f(ab)g(mn/(ab)) = \sum_{a \mid m, \, b \mid n} f(a)f(b)g(m/a)g(n/b) = \sum_{a \mid m} f(a)g(m/a) \cdot \sum_{b \mid n} f(b)g(n/b) = h(m)h(n)$.

43. a. -1 **b.** -1 **c.** 1 **d.** 1 **e.** -1 **f.** -1 **g.** 1

45. Let $f(n) = \sum_{d \mid n} \lambda(d)$. Then by Theorem 7.8, f is multiplicative. Now $f(p^t) = \lambda(1) + \lambda(p) + \lambda(p^2) + \cdots + \lambda(p^t) = 1 - 1 + 1 - \cdots + (-1)^t = 0$ if t is odd and $= 1$ if t is even. Then $f(p_1^{a_1} p_2^{a_2} \cdots p_r^{a_r}) = \prod f(p_i^{a_i}) = 0$ if any a_i is odd (n is not a square) and $= 1$ if all a_i are even (n is a square).

47. If f and g are completely multiplicative and m and n are positive integers, then $(fg)(mn) = f(mn) \cdot g(mn) = f(m) \cdot f(n) \cdot g(m) \cdot g(n) = f(m) \cdot g(m) \cdot f(n) \cdot g(n) = (fg)(m) \cdot (fg)(n)$, so fg is also completely multiplicative.

49. $f(mn) = \log mn = \log m + \log n = f(m) + f(n)$

51. a. 2 **b.** 3 **c.** 1 **d.** 4 **e.** 8 **f.** 15

53. Let $(m, n) = 1$. Then by the additivity of f we have $f(mn) = f(m) + f(n)$. Thus $g(mn) = 2^{f(mn)} = 2^{f(m)+f(n)} = 2^{f(m)}2^{f(n)} = g(m)g(n)$.

Section 7.2

1. a. 48 **b.** 399 **c.** 2340 **d.** $2^{101} - 1$ **e.** 6912 **f.** $813{,}404{,}592$ **g.** $15{,}334{,}088$
h. $13{,}891{,}399{,}238{,}731{,}734{,}720$

3. perfect squares

5. a. $6, 11$ **b.** $10, 17$ **c.** $14, 15, 23$ **d.** $33, 35, 47$ **e.** none   **f.** $44, 65, 83$

7. Note that $\tau(p^{k-1}) = k$ whenever p is prime and k is an integer greater than 1. Hence the equation $\tau(n) = k$ has infinitely many solutions.

9. squares of primes

11. $n^{\tau(n)/2}$

13. a. The nth term is $\sigma(2n)$.
   b. The nth term is $\sigma(n) - \tau(n)$.
   c. The nth term is the least positive integer m with $\tau(m) = n$.
   d. The nth term is the number of solutions to the equation $\sigma(x) = n$.

15. $2, 4, 6, 12, 24, 36$

17. Let a be the largest highly composite integer less than or equal to n. Note that $2a$ is less than or equal to $2n$ and has more divisors than a, and hence $\tau(2a) > \tau(a)$. By Exercise 16 there must be a highly composite integer b with $a < b \leq 2a$. If $b \leq n$, this contradicts the choice of a. Therefore, $n < b \leq 2n$. It follows that there must be a highly composite integer k with $2^m < k \leq 2^{m+1}$ for every nonnegative integer m. Therefore, there are at least m highly composite integers less than or equal to 2^m. Thus the mth highly composite integer is less than or equal to 2^m.

19. $1, 2, 4, 6, 12, 24, 36, 48$

21. $1 + p^k$

23. Suppose that a and b are positive integers with $(a, b) = 1$. Then $\sum_{d \mid ab} d^k = \sum_{d_1 \mid a, \, d_2 \mid b} (d_1 d_2)^k = \sum_{d_1 \mid a} d_1^k \cdot \sum_{d_2 \mid b} d_2^k = \sigma_k(a)\sigma_k(b)$.

25. prime numbers

27. Let $n = p_1^{a_1} p_2^{a_2} \cdots p_r^{a_r}$, and let x and y be integers such that $[x, y] = n$. Then $x \mid n$ and $y \mid n$, so $x = p_1^{b_1} p_2^{b_2} \cdots p_r^{b_r}$ and $y = p_1^{c_1} p_2^{c_2} \cdots p_r^{c_r}$, where b_i and c_i are between 0 and a_i. Since $[x, y] = n$, we must have $\max(b_i, c_i) = a_i$ for each i. Then one of b_i and c_i must be equal to

a_i and the other can range over $0, 1, 2, \ldots, a_i$. Therefore, we have $2a_i + 1$ ways to choose the pair (b_i, c_i) for each i. Thus in total we can choose the exponents $b_1, b_2, \ldots, b_r, c_1, c_2, \ldots, c_r$ in $(2a_1 + 1)(2a_2 + 1) \cdots (2a_r + 1) = \tau(n^2)$ ways.

29. Suppose that n is composite. Then $n = ab$, where a and b are integers with $1 < a \le b < n$. It follows that either $a \ge \sqrt{n}$ or $b \ge \sqrt{n}$. Consequently, $\sigma(n) \ge 1 + a + b + n > 1 + \sqrt{n} + n > n + \sqrt{n}$. Conversely, suppose that n is prime. Then $\sigma(n) = n + 1$, so $\sigma(n) \le n + \sqrt{n}$. Hence, $\sigma(n) > n + \sqrt{n}$ implies that n is composite.

31. For $n = 1$, the statement is true. Suppose that $\sum_{j=1}^{n-1} \tau(j) = 2 \sum_{j=1}^{[\sqrt{n-1}]} [(n-1)/j] - [\sqrt{n-1}]^2$. For the induction step, if n is not a perfect square, it suffices to show that $\tau(n) = 2 \sum_{j=1}^{[\sqrt{n-1}]} ([n/j] - [(n-1)/j]) = 2 \sum_{j \le [\sqrt{n-1}], j|n} 1$, which is true by the definition of $\tau(n)$, since there is one factor less than $\sqrt{n}$ for every factor greater than $\sqrt{n}$. Note that if n is a perfect square, we must add the term $2\sqrt{n} - (2\sqrt{n} - 1) = 1$ to the last two sums. For $n = 100$, we have $\sum_{j=1}^{100} \tau(j) = 2 \sum_{j=1}^{10} [100/j] - 100 = 482$.

33. Let $a = \prod p_i^{a_i}$ and $b = \prod p_i^{b_i}$, and let $c_i = \min(a_i, b_i)$ for each i. We first prove that the product $\prod_{p_i} \sum_{j=0}^{c_i} p_i^j \sigma(p_i^{a_i + b_i - 2j}) = \sum_{d|(a,b)} d\sigma(ab/d^2)$. To see this, let d be any divisor of (a, b), say $d = \prod p_i^{d_i}$. Then $d_i \le c_i$ for each i, so each of the terms $p_i^{d_i} \sigma(p_i^{a_i + b_i - 2d_i})$ appears in exactly one of the sums in the product. Therefore, if we expand the product, we will find, exactly once, the term $\prod_{p_i} p_i^{d_i} \sigma(p_i^{a_i + b_i - 2d_i}) = d\sigma\left(\prod_{p_i} p_i^{a_i + b_i - 2d_i}\right) = d\sigma\left(\prod_{p_i} (p_i^{a_i}/p_i^{d_i})(p_i^{b_i}/p_i^{d_i})\right) = d\sigma((a/d)(b/d))$. This proves the first identity. Next, consider the sum $\sum_{j=0}^{c} (p^{a+b-j} + p^{a+b-j-1} + \cdots + p^j)$, where $c = \min(a, b)$. The term p^k appears in this sum once each time that $k = a + b - j$, which happens exactly when $a + b - c \le k \le a + b$, that is, $c + 1$ times. On the other hand, in the expansion of the product $(p^a + p^{a-1} + \cdots + 1)(p^b + p^{b-1} + \cdots + 1) = \sigma(p^a)\sigma(p^b)$, the same term p^k appears whenever $k = (a - m) + (b - n)$, where $0 \le m \le a$ and $0 \le n \le b$. Each of m and n determines the other, so p^k appears exactly $\min(a + 1, b + 1) = c + 1$ times. Given this identity, we have $\sigma(a)\sigma(b) = \prod_{p_i} (p_i^{a_i} + p_i^{a_i - 1} + \cdots + 1)(p_i^{b_i} + p_i^{b_i - 1} + \cdots + 1) = \prod_{p_i} \sum_{j=0}^{c_i} (p_i^{a_i + b_i - j} + p_i^{a_i + b_i - j - 1} + \cdots + p_i^j)$, which is the right side of the identity, as we proved above.

35. From Exercises 52 and 53 in Section 7.1, we know that the arithmetic function $f(n) = 2^{\omega(n)}$ is multiplicative. Therefore, the Dirichlet product $h(n) = \sum_{d|n} 2^{\omega(d)} = f * g(n)$, where $g(n) = 1$, is also multiplicative (see Exercise 41 in Section 7.1). Since $\tau(n)$ and n^2 are multiplicative, so is $\tau(n^2)$. Therefore, it is sufficient to prove the identity for n equal to a prime power, p^a. We have $\tau(p^{2a}) = 2a + 1$. On the other hand, $\sum_{d|p^a} 2^{\omega(d)} = \sum_{i=0}^{a} 2^{\omega(p^i)} = 1 + \sum_{i=1}^{a} 2^1 = 2a + 1$.

37. $\phi(1)\phi(2) \cdots \phi(n)$

39. If p and $p + 2$ are prime, then $\phi(p + 2) = p + 1 = \sigma(p)$. If $2^p - 1$ is prime, then $\phi(2^{p+1}) = 2^p = \sigma(2^p - 1)$.

Section 7.3

1. $6, 28, 496, 8128, 33{,}550{,}336, 8{,}589{,}869{,}056$

3. a. 31 **b.** 127 **c.** 127

5. $12, 18, 20, 24, 30, 36$

7. Suppose that $n = p^k$, where p is prime and k is a positive integer. Then $\sigma(p^k) = (p^{k+1} - 1)/(p - 1)$. Note that $2p^k - 1 < p^{k+1}$ since $p \ge 2$. It follows that $p^{k+1} - 1 < 2(p^{k+1} - p^k) = 2p^k(p - 1)$, so $(p^{k+1} - 1)/(p - 1) < 2p^k = 2n$. It follows that $n = p^k$ is deficient.

9. Suppose that n is abundant or perfect. Then $\sigma(n) \geq 2n$. Suppose that $m = nk$ for some integer $k > 1$. The divisors of m include the integers kd whenever $d \mid n$, as well as the number 1. Hence, $\sigma(m) \geq 1 + \sum_{d|n} kd = 1 + k \sum_{d|n} d = 1 + k\sigma(n) \geq 1 + 2nk > 2kn = 2m$. Hence, m is abundant.

11. If p is any prime, then $\sigma(p) = p + 1 < 2p$, so p is deficient; and we know that there are infinitely many primes.

13. (See Exercises 6 and 9.) For a positive integer a, let $n = 3^a \cdot 5 \cdot 7$, and compute $\sigma(n) = \sigma(3^a \cdot 5 \cdot 7) = ((3^{a+1} - 1)/(3 - 1))(5 + 1)(7 + 1) = (3^{a+1} - 1)24 = 3^{a+1}24 - 24 = 2 \cdot 3^a(36) - 24 = 2 \cdot 3^a(35) + 2 \cdot 3^a - 24 = 2n + 2 \cdot 3^a - 24$, which will be greater than $2n$ whenever $a \geq 3$. This demonstrates infinitely many odd abundant integers.

15. a. The prime factorizations of 220 and 284 are $220 = 2^2 \cdot 5 \cdot 11$ and $284 = 2^2 \cdot 71$. Hence, $\sigma(220) = \sigma(2^2)\sigma(5)\sigma(11) = 7 \cdot 6 \cdot 12 = 504$ and $\sigma(284) = \sigma(2^2)\sigma(71) = 7 \cdot 72 = 504$. Since $\sigma(220) = \sigma(284) = 220 + 284 = 504$, it follows that 220 and 284 form an amicable pair.
b. The prime factorizations of 1184 and 1210 are $1184 = 2^5 \cdot 37$ and $1210 = 2 \cdot 5 \cdot 11^2$. Hence, $\sigma(1184) = \sigma(2^5)\sigma(37) = 63 \cdot 38 = 2394$ and $\sigma(1210) = \sigma(2)\sigma(5)\sigma(11^2) = 3 \cdot 6 \cdot 133 = 2394$. Since $\sigma(1184) = \sigma(1210) = 1184 + 1210 = 2394$, 1184 and 1210 form an amicable pair.
c. The prime factorizations of 79,750 and 88,730 are $79{,}750 = 2 \cdot 5^3 \cdot 11 \cdot 29$ and $88{,}730 = 2 \cdot 5 \cdot 19 \cdot 467$. Hence, $\sigma(79{,}750) = \sigma(2)\sigma(5^3)\sigma(11)\sigma(29) = 3 \cdot 156 \cdot 12 \cdot 30 = 168{,}480$ and similarly $\sigma(88{,}730) = \sigma(2)\sigma(5)\sigma(19)\sigma(467) = 3 \cdot 6 \cdot 20 \cdot 468 = 168{,}480$. Since $\sigma(79{,}750) = \sigma(88{,}730) = 79{,}750 + 88{,}730 = 168{,}480$, it follows that 79,750 and 88,730 form an amicable pair.

17. $\sigma(120) = \sigma(2^3 \cdot 3 \cdot 5) = \sigma(2^3)\sigma(3)\sigma(5) = 15 \cdot 4 \cdot 6 = 360 = 3 \cdot 120$

19. $\sigma(2^7 \cdot 3^4 \cdot 5 \cdot 7 \cdot 11^2 \cdot 17 \cdot 19) = (2^8 - 1) \cdot (3^5 - 1)/2 \cdot (5 + 1) \cdot (7 + 1) \cdot (11^3 - 1)/10 \cdot (17 + 1) \cdot (19 + 1) = 255 \cdot 121 \cdot 6 \cdot 8 \cdot 133 \cdot 18 \cdot 20 = 70{,}912{,}195{,}200 = 5 \cdot 14{,}182{,}439{,}040$

21. Suppose that n is 3-perfect and 3 does not divide n. Then $\sigma(3n) = \sigma(3)\sigma(n) = 4 \cdot 3n = 12n = 4 \cdot 3n$. Hence, $3n$ is 4-perfect.

23. 908,107,200

25. $\sigma(\sigma(16)) = \sigma(31) = 32 = 2 \cdot 16$

27. Certainly if r and s are integers greater than 1, then $\sigma(rs) \geq rs + s + 1$. Suppose $n = 2^q t$ is superperfect with t odd and $t > 1$. Then $2n = 2^{q+1}t = \sigma(\sigma(2^q t)) = \sigma\left((2^{q+1} - 1)\sigma(t)\right) \geq (2^{q+1} - 1)\sigma(t) + \sigma(t) + 1 > 2^{q+1}\sigma(t) \geq 2^{q+t}(t + 1)$. Then $t > t + 1$, a contradiction. Therefore, we must have $n = 2^q$, in which case we have $2n = 2^{q+1} = \sigma(\sigma(2^q)) = \sigma\left(2^{q+1} - 1\right) = \sigma(2n - 1)$. Therefore, $2n - 1 = 2^{q+1} - 1$ is prime.

29. a. yes **b.** no **c.** yes **d.** no

31. a. Note that $M_n(M_n + 2) = (2^n - 1)(2^n + 1) = 2^{2n} - 1$. If $2n + 1$ is prime, then $\phi(2n + 1) = 2n$ and $2^{2n} \equiv 1 \pmod{2n + 1}$. Thus $2n + 1 \mid 2^{2n} - 1 = M_n(M_n + 2)$. Therefore, $2n + 1 \mid M_n$ or $2n + 1 \mid M_n + 2$.
b. $23 \nmid 2049$, so $23 \mid 2047 = M_{11}$; $47 \nmid 8{,}388{,}609$ so $47 \mid 8{,}388{,}607 = M_{23}$

33. Since m is odd, $m^2 \equiv 1 \pmod 8$, so $n = p^a m^2 \equiv p^a \pmod 8$. By Exercise 32(a), $a \equiv 1 \pmod 4$, so $p^a \equiv p^{4k}p \equiv p \pmod 8$, since p^{4k} is an odd square. Therefore, $n \equiv p \pmod 8$.

35. First suppose that $n = p^a$, where p is an odd prime and a is a positive integer. Then $\sigma(n) = (p^{a+1} - 1)/(p - 1) < p^{a+1}/(p - 1) = np/(p - 1) = n/(1 - (1/p)) \leq n/(\frac{2}{3}) = 3n/2$, so $\sigma(n) \neq 2n$ and n is not perfect. Next suppose that $n = p^a q^b$, where p and q are primes and a and b are positive integers. Then $\sigma(n) = (p^{a+1} - 1)/(p - 1) \cdot (q^{b+1} - 1)/(q - 1) < p^{a+1}q^{b+1}/((p - 1)(q - 1)) = npq/((p - 1)(q - 1)) = n/((1 - (1/p))(1 - (1/q))) \leq n/(\frac{2}{3} \cdot \frac{4}{5}) = 15n/8 < 2n$. Hence, $\sigma(n) \neq 2n$ and n is not perfect.

37. integers of the form p^5 and p^2q, where p and q are primes

39. Suppose that $M_n = 2^n - 1 = a^k$, with n and k integers greater than 1. Then a must be odd. If $k = 2j$, then $2^n - 1 = (a^j)^2$. Since $n > 1$ and the square of an odd integer is congruent to 1 modulo 4, reduction of the last equation modulo 4 yields the contradiction $-1 \equiv 1 \pmod{4}$. Therefore, k must be odd. Then $2^n = a^k + 1 = (a+1)(a^{k-1} - a^{k-2} + \cdots + 1)$. So $a + 1 = 2^m$ for some integer m. Then $2^n - 1 = (2^m - 1)^k \geq 2^{mk} - k2^{m(k-1)} \geq 2^{m(k-1)} \geq 2^{2m}$, so $n \geq 2m$. Then reduction modulo 2^{2m} gives $-1 \equiv k2^m - 1 \pmod{2^{2m}}$, or, since k is odd, $2^m \equiv 0 \pmod{2^{2m}}$, a contradiction.

Section 7.4

1. a. 0 **b.** 1 **c.** -1 **d.** 0 **e.** -1 **f.** 1 **g.** 0

3. $0, -1, -1, -1, 0, -1, 1, -1, 0, -1, -1$, respectively

5. 1, 6, 10, 14, 15, 21, 22, 26, 33, 34, 35, 38, 39, 46, 51, 55, 57, 58, 62, 65, 69, 74, 77, 82, 85, 86, 87, 91, 93, 94, 95

7. $1, 0, -1, -1, -2, -1, -2, -2, -2, -1$, respectively

9. Since $\mu(n)$ is 0 for non-square-free n, is 1 for n a product of an even number of distinct primes, and is -1 for n a product of an odd number of distinct primes, the sum $M(n) = \sum_{i=1}^{n} \mu(i)$ is unaffected by the non-square-free numbers, but counts 1 for every even product and -1 for every odd product. Thus $M(n)$ counts how many more even products than odd products there are.

11. For any nonnegative integer k, the numbers $n = 36k + 8$ and $n + 1 = 36k + 9$ are consecutive and divisible by $4 = 2^2$ and $9 = 3^2$, respectively. Therefore, $\mu(36k + 8) + \mu(36k + 9) = 0 + 0 = 0$.

13. 3

15. Let $h(n) = n$ be the identity function. Then from Theorem 7.7 we have $h(n) = n = \sum_{d|n} \phi(n)$. Then by the Möbius inversion formula $\phi(n) = \sum_{d|n} \mu(d)h(n/d) = \sum_{d|n} \mu(d)(n/d) = n \sum_{d|n} \mu(d)/d$.

17. Since μ and f are multiplicative, so is their product μf, by Exercise 46 of Section 7.1. The summatory function $\sum_{d|n} \mu(d)f(d)$ is also multiplicative by Theorem 7.17. Therefore, it suffices to prove the proposition for n a prime power. We compute $\sum_{d|p^a} \mu(d)f(d) = \mu(p^a)f(p^a) + \mu(p^{a-1})f(p^{a-1}) + \cdots + \mu(p)f(p) + \mu(1)f(1)$. But for exponents j greater than 1, $\mu(p^j) = 0$, so this sum equals $\mu(p)f(p) + \mu(1)f(1) = -f(p) + 1$.

19. $\phi(n)/n$

21. $(-1)^k \prod_{i=1}^{k} p_i$

23. Since both sides of the equation are known to be multiplicative (see the solution to Exercise 35 in Section 7.2, Exercise 46 in Section 7.1, Theorem 7.17, and Theorem 7.14), it suffices to prove the identity for $n = p^a$, a prime power. On one hand, $\sum_{d|p^a} \mu^2(d) = \mu^2(p) + \mu^2(1) = 1 + 1 = 2$. On the other hand, $\omega(p^a) = 1$, so the right side is $2^1 = 2$.

25. Let λ play the role of f in the identity of Exercise 17. Then the left side equals $\prod_{j=1}^{k}(1 - \lambda(p_j)) = \prod_{j=1}^{k}(1 - (-1)) = 2^k = 2^{\omega(n)}$.

27. By Theorem 7.15, $\mu * v(n) = \sum_{d|n} \mu(d)v(n/d) = \sum_{d|n} \mu(d) = \iota(n)$.

29. Since $v(n)$ is identically 1, we have $F(n) = \sum_{d|n} f(d) = \sum_{d|n} f(d)v(n/d) = f * v(n)$. If we Dirichlet multiply both sides on the right by μ, then $F * \mu = f * v * \mu = f * \iota = f$.

31. From the Möbius inversion formula, Exercise 30, and Theorem 7.15, $\Lambda(n) = \sum_{d|n} \mu(d) \log(n/d) = \sum_{d|n} \mu(d)(\log n - \log d) = \sum_{d|n} \mu(d) \log(n) - \sum_{d|n} \mu(d) \log(d) = \log n \sum_{d|n} \mu(d) - $

$\sum_{d|n} \mu(d) \log(d) = (\log n)\nu(n) - \sum_{d|n} \mu(d) \log(d) = -\sum_{d|n} \mu(d) \log(d)$, since $\nu(n) = 0$ if $n \neq 1$ and $\log n = 0$ if $n = 1$.

Section 8.1

1. DWWDF NDWGD ZQ

3. IEXXK FZKXC UUKZC STKJW

5. READ MY LIPS

7. 12

9. AN IDEA IS LIKE A CHILD NONE IS BETTER THAN YOUR OWN FROM CHINESE FORTUNE COOKIE

11. 9, 12

13. THIS MESSAGE WAS ENCIPHERED USING AN AFFINE TRANSFORMATION

15. $C \equiv 7P + 16 \pmod{26}$

Section 8.2

1. VSPFXH HIPKLB KIPMIE GTG

3. TJEVT EESPZ TJIAN IARAB GSHWQ HASBU BJGAO XYACF XPHML AWVMO XANLB GABMS HNEIA TIEZV VWNQF TLEZF HJWPB WKEAG AENOF UACIH LATPR RDADR GKTJR XJDWA XXENB KA

5. Let n be the key length, and suppose that $k_1, k_2, \ldots, k_n$ are the numerical equivalents of the letters of the keyword. If $p_i = p_j$ are two plaintext characters separated by a multiple of the key length, when we break the plaintext into blocks of length n, p_i and p_j will be in the same position in their respective blocks, say the mth position. So when we encrypt them, we get $c_i \equiv p_i + k_m \equiv p_j + k_m \equiv c_j \pmod{26}$.

7. The key is YES, and the plaintext is MISTA KESAR EAPAR TOFBE INGHU MANAP PRECI ATEYO URMIS TAKES FORWH ATTHE YAREP RECIO USLIF ELESS ONSTH ATCAN ONLYB ELEAR NEDTH EHARD WAYUN LESSI TISAF ATALM ISTAK EWHIC HATLE ASTOT HERSC ANLEA RNFRO M.

9. The key is BIRD, and the plaintext is IONCE HADAS PARRO WALIG HTUPO NMYSH OULDE RFORA MOMEN TWHIL EIWAS HOEIN GINAV ILLAG EGARD ENAND IFELT THATI WASMO REDIS TINGU ISHED BYTHA TCIRC UMSTA NCETH ATISH OULDH AVEBE ENBYA NYEPA ULETI COULD HAVEW ORN.

11. The key is SAGAN, and the plaintext is BUTTH EFACT THATS OMEGE NIUSE SWERE LAUGH EDATD OESNO TIMPL YTHAT ALLWH OAREL AUGHE DATAR EGENI USEST HEYLA UGHED ATCOL UMBUS THEYL AUGHE DATFU LTONT HEYLA UGHED ATTHE WRIGH TBROT HERSB UTTHE YALSO LAUGH EDATB OZOTH ECLOW N.

13. RL OQ NZ OF XM CQ KG QI VD AZ

15. TO SLEEP PERCHANCE TO DREAM

17. 3, 24, 24, 25

19. $C \equiv AP \pmod{26}$. Multiplying both sides on the left by A gives $AC \equiv A^2P \equiv IP \equiv P \pmod{26}$. The congruence $A^2 \equiv I \pmod{26}$ follows since A is involutory. It follows that A is also a decrypting matrix.

21. $C \equiv \begin{bmatrix} 11 & 6 \\ 2 & 13 \end{bmatrix} P \pmod{26}$

23. If the plaintext is grouped into blocks of size m, we may take $[m, n]/m$ of these blocks to form a super-block of size $[m, n]$. If $\mathbf{A}$ is the $m \times m$ encrypting matrix, form the $[m, n] \times [m, n]$ matrix $\mathbf{B}$ with $[m, n]/m$ copies of $\mathbf{A}$ on the diagonal and zeros elsewhere. Then $\mathbf{B}$ will encrypt $[m, n]/m$ blocks of size m at once. Similarly, if $\mathbf{C}$ is the $n \times n$ encrypting matrix, form the corresponding $[m, n] \times [m, n]$ matrix $\mathbf{D}$. Then by Exercise 22, $\mathbf{BD}$ is an $[m, n] \times [m, n]$ encrypting matrix which does everything at once.

25. Multiplication of $[\, 0 \quad \ldots \quad 0 \quad 1 \quad 0 \quad \ldots \quad 0 \,]$, with the 1 in the ith place, by $[\, P_1 \ P_2 \ \ldots$ $P_n]^T$ yields the 1×1 matrix $[\, P_i \,]$ (T denotes transpose). If the jth row of a matrix $\mathbf{A}$ is $[\, 0 \quad \ldots \quad 0 \quad 1 \quad 0 \quad \ldots \quad 0 \,]$, then $\mathbf{A}\,[\, P_1 \quad P_2 \quad \ldots \quad P_n \,]^T = [\, C_1 \quad C_2 \quad \ldots \quad C_n \,]$ gives $C_j = P_i$. So if every row of $\mathbf{A}$ has its 1 in a different column, then each C_j is equal to a different P_i. Hence, $\mathbf{A}$ is a "permutation" matrix.

27. $P \equiv \begin{bmatrix} 17 & 4 \\ 1 & 7 \end{bmatrix} C + \begin{bmatrix} 22 \\ 15 \end{bmatrix} \pmod{26}$

29. TOXIC WASTE

31. Make a frequency count of the trigraphs and use a published English language count of frequencies of trigraphs. Then proceed as in Exercise 30. There are 12 variables to determine, so four guesses are needed.

33. yes

35. 01 1101 1010

37. RENDEZVOUZ

39. Let $p_1 p_2 \cdots p_m$ and $q_1 q_2 \cdots q_m$ be two different plaintext bit streams. Let $k_1, k_2, \ldots, k_m$ be the keystream by which these two plaintexts are encrypted. Note that for $i = 1, 2, \ldots, m$, $E_{k_i}(p_i) + E_{k_i}(q_i) = k_i + p_i + k_i + q_i = 2k_i + p_i + q_i \equiv p_i + q_i \pmod{2}$. Therefore, by adding corresponding bits of the ciphertext streams, we get the sums of the corresponding bits of the plaintext streams. This partial information may lead to successful cryptanalysis of encrypted messages.

Section 8.3

1. 14 17 17 27 11 17 65 76 07 76 14

3. BEAM ME UP

5. We encrypt messages using the transformation $C \equiv P^{11} \pmod{31}$. The decrypting exponent is the inverse of 11 modulo 30 since $\phi(31) = 30$. But 11 is its own inverse modulo 30 since $11 \cdot 11 \equiv 121 \equiv 1 \pmod{30}$. It follows that 11 is both the encrypting and decrypting exponent.

Section 8.4

1. 151, 97

3. Since a block of ciphertext p is less than n, we must have $(p, n) = p$ or q. Therefore, the cryptanalyst has a factor of n.

5. 1215 1224 1471 0023 0116

7. GREETINGS

9. 2145 0672 0724 1404 1630

11. No. This is the same as using the RSA cryptosystem with encryption key $(e_1 e_2, n)$. It is no easier, or more difficult, to discover the inverse of $e = e_1 e_2$ than it is to discover the inverse of either of the factors modulo $\phi(n)$.

13. Suppose that P is a plaintext message and the two encrypting exponents are e_1 and e_2. Let $a = (e_1, e_2)$. Then there exist integers x and y such that $e_1 x + e_2 y = a$. Let $C_1 \equiv P^{e_1} \pmod{n}$ and $C_2 \equiv P^{e_2} \pmod{n}$ be the two ciphertexts. Since C_1, C_2, e_1, and e_2 are known to the decipherer, and since x and y are relatively easy to compute, it is also easy to compute $C_1^x C_2^y \equiv P^{e_1 x} P^{e_2 y} = P^{e_1 x + e_2 y} = P^a \pmod{n}$. If $a = 1$, then P has been recovered. If a is fairly small, then it may not be too difficult to compute ath roots of P^a and thereby recover P.

15. Encryption works the same as for the two-prime case. For decryption, we must compute an inverse d for e modulo $\phi(n) = (p-1)(q-1)(r-1)$, where $n = pqr$ the product of three primes. Then we proceed as in the case with two primes.

Section 8.5

1. a. yes   **b.** no   **c.** yes   **d.** no

3. Proceed by induction. Certainly, $a_1 < 2a_1 < a_2$. Suppose $\sum_{j=1}^{n-1} a_j < a_n$. Then $\sum_{j=1}^{n} a_j = \sum_{j=1}^{n-1} a_j + a_n < a_n + a_n = 2a_n < a_{n+1}$.

5. $(17, 51, 85, 7, 14, 45, 73)$

7. NUTS

9. If the multipliers and moduli are $(w_1, m_1), (w_2, m_2), \ldots, (w_r, m_r)$, then the inverses $\overline{w_1}, \overline{w_2}, \ldots, \overline{w_r}$ can be computed with respect to their corresponding moduli. Then we multiply and reduce successively by $(\overline{w_r}, m_r), (\overline{w_{r-1}}, m_{r-1}), \ldots, (\overline{w_1}, m_1)$. The result will be the plaintext sequence of easy knapsack problems.

11. $8 \cdot 21 \cdot 95$

13. For $i = 1, 2, 3, \ldots, n$, we have $b^{\alpha_i} \equiv a_i \pmod{m}$. Then $b^S \equiv P \equiv (b^{\alpha_1})^{x_1}(b^{\alpha_2})^{x_2} \cdots (b^{\alpha_n})^{x_n} \equiv b^{\alpha_1 x_1 + \cdots + \alpha_n x_n} \pmod{m}$. Then $S \equiv \alpha_1 x_1 + \cdots + \alpha_n x_n \pmod{\phi(m)}$. Since $S + k\phi(m)$ is also a logarithm of P to the base b, we may take the congruence to be an equation. Since each x_i is 0 or 1, this becomes an additive knapsack problem on the sequence $(\alpha_1, \alpha_2, \ldots, \alpha_n)$.

Section 8.6

1. 90

3. 476

5. Let $k_1, k_2, \ldots, k_n$ be the private keys for parties 1 through n, respectively. There are n steps in this protocol. The first step is for each party i to compute the least positive residue of $r^{k_i} \pmod{p}$ and send this value y_i to the $(i+1)$th party. (The nth party sends his value to the first party.) Now the ith party has the value y_{i-1} (where we take y_0 to be y_n). The second step is for each party i to compute the least positive residue of $y_{i-1}^{k_i} \pmod{p}$ and send this value to the $(i+1)$th party. Now the ith party has the least positive residue of $r^{k_{i-1}+k_{i-2}} \pmod{p}$. This process is continued for a total of n steps. However, at the nth step, the computed value is not sent on to the next party. Then the ith party will have the least positive residue of $r^{k_{i-1}+k_{i-2}+\cdots+k_1+k_n+k_{n-1}+\cdots+k_{i+1}+k_i} \pmod{p}$, which is exactly the value of K desired.

7. a. 0371 0354 0858 0858 0087 1369 0354 0000 0087 1543 1797 0535
   b. 0833 0475 0074 0323 0621 0105 0621 0865 0421 0000 0746 0803 0105 0621 0421

9. a. If $n_i < n_j$, the block sizes are chosen small enough so that each block is unique modulo n_i. Since $n_i < n_j$, each block will be unique modulo n_j after applying the transformation D_{k_j}. Therefore, we can apply E_{k_j} to $D_{k_i}(P)$ and retain uniqueness of blocks. The argument is similar when $n_i > n_j$.

b. If $n_i < n_j$, individual j receives $E_{k_j}(D_{k_i}(P))$ and knows an inverse for e_j modulo $\phi(n_j)$. So he can apply $D_{k_j}(E_{k_j}(D_{k_i}(P))) = D_{k_i}(P)$. Since he also knows e_i, he can apply $E_{k_i}(D_{k_i}(P)) = P$ and discover the plaintext P. If $n_i > n_j$, individual j receives $D_{k_i}(E_{k_j}(P))$. Since he knows e_i he can apply $E_{k_i}(D_{k_i}(E_{k_j}(P))) = E_{k_j}(P)$. Since he also knows $\overline{e_j}$ he can apply $D_{k_j}(E_{k_j}(P)) = P$ and discover the plaintext P.

c. Since only individual i knows $\overline{e_i}$, only he can apply the transfomation D_{k_i} and thereby make $E_{k_i}(D_{k_i}(P))$ intelligible.

d. We have $n_i = 2867 > n_j = 2537$, so we compute $D_{k_i}(E_{k_j}(P))$. Both n_i and n_j are greater than 2525, so we use blocks of 4. REGARDS FRED becomes 1704 0600 1703 1805 1704 0323 (adding an X to fill out the last block). Now $e_i = 11$ and $\phi(n_i) = 2760$, so $\overline{e_i} = 251$. We apply $E_{k_j}(P) \equiv P^{e_j} \equiv P^{13} \pmod{2537}$ to each block and get 1943 0279 0847 0171 1943 0088. Then we apply $D_{k_i}(E) \equiv E^{251} \pmod{2867}$ and get 0479 2564 0518 1571 0479 1064. Now since $n_j < n_i$ individual j must send $E_{k_j}(D_{k_i}(P))$, $e_j = 13$, $\phi(2537) = 2436$, and $\overline{e_j} = 937$. Then $D_{k_j}(P) \equiv P^{937} \pmod{2537}$ and $E_{k_i}(D) \equiv D^{11} \pmod{2867}$. The ciphertext for REGARDS ZELDA is 1609 1802 0790 2508 1949 0267.

11. $k_1 \equiv 4 \pmod 8$, $k_2 \equiv 5 \pmod 9$, $k_3 \equiv 2 \pmod{11}$

13. The three shadows from Exercise 11 are $k_1 = 4$, $k_2 = 5$, and $k_3 = 2$. We solve the system $K_0 \equiv 4 \pmod 8$, $K_0 \equiv 5 \pmod 9$, and $K_0 \equiv 2 \pmod{11}$, where the moduli are the m_i's. The Chinese remainder theorem yields $K_0 \equiv 68 \pmod{8 \cdot 9 \cdot 11}$. Then $K = K_0 - tp = 68 - 13 \cdot 5 = 3$.

Section 9.1

1. a. 4   **b.** 4   **c.** 6   **d.** 4

3. a. $\phi(6) = 2$, and $5^2 \equiv 1 \pmod 6$
    b. $\phi(11) = 10$, and $2^2 \equiv 4$, $2^5 \equiv -1$, $2^{10} \equiv 1 \pmod{11}$

5. Only 1, 5, 7, 11 are relatively prime to 12. Each one squared is congruent to 1, but $\phi(12) = 4$.

7. There are 2: 3 and 5.

9. That $\mathrm{ord}_n a = \mathrm{ord}_n \overline{a}$ follows from the fact that $a^t \equiv 1 \pmod n$ if and only if $\overline{a}^t \equiv 1 \pmod n$. To see this, suppose that $a^t \equiv 1 \pmod n$. Then $\overline{a}^t \equiv \overline{a}^t a^t a^t \equiv (\overline{a}^t a^t) a^t \equiv (a\overline{a})^t a^t \equiv 1^t \cdot 1 \equiv 1 \pmod n$. The converse is shown in a similar manner.

11. $[r, s]/(r, s) \le \mathrm{ord}_n ab \le [r, s]$, where $r = \mathrm{ord}_n a$ and $s = \mathrm{ord}_n b$

13. Let $r = \mathrm{ord}_m a^t$. Then $a^{tr} \equiv 1 \pmod m$; hence, $tr \ge ts$ and $r \ge s$. Since $1 \equiv a^{st} \equiv (a^t)^s \pmod m$, we have $s \ge r$.

15. Suppose that r is a primitive root modulo the odd prime p. Then $r^{(p-1)/q} \not\equiv 1 \pmod p$ for all prime divisors q of $p - 1$ since no smaller power than the $(p - 1)$th of r is congruent to 1 modulo p. Conversely, suppose that $r^{(p-1)/q} \not\equiv 1 \pmod p$ for all prime divisors of $p - 1$. Suppose that r is not a primitive root of p. Then there is an integer t such that $r^t \equiv 1 \pmod p$ with $t < p - 1$. Since t must divide $p - 1$, we have $p - 1 = st$ for some positive integer s greater than 1. Then $(p - 1)/s = t$. Let q be a prime divisor of s. Then $(p - 1)/q = ts/q$, so $r^{(p-1)/q} = r^{ts/q} = (r^t)^{s/q} \equiv 1 \pmod p$. This contradicts the original assumption, so r is a primitive root modulo p.

17. Since $2^{2^n} + 1 \equiv 0 \pmod{F_n}$, we have $2^{2^n} \equiv -1 \pmod{F_n}$. Squaring gives $(2^{2^n})^2 = 2^{2^n \cdot 2} \equiv 1 \pmod{F_n}$. Thus $\mathrm{ord}_{F_n} 2 \le 2^n \cdot 2 = 2^{n+1}$.

19. Note that $a^t < m = a^n - 1$ whenever $1 \le t < n$. Hence, a^t cannot be congruent to 1 modulo m when t is a positive integer less than n. However, $a^n \equiv 1 \pmod m$ since $m = a^n - 1 \mid a^n - 1$. It follows that $\mathrm{ord}_m a = n$. Since $\mathrm{ord}_m a \mid \phi(m)$, we see that $n \mid \phi(m)$.

21. First suppose that pq is a pseudoprime to the base 2. By Fermat's little theorem, $2^p \equiv 2 \pmod{p}$, so there exists an integer k such that $2^p - 2 = kp$. Then $2^{M_p-1} - 1 = 2^{2^p-2} - 1 = 2^{kp} - 1$. This last expression is divisible by $2^p - 1 = M_p$ by Lemma 6.1. Hence, $2^{M_p-1} \equiv 1 \pmod{M_p}$, or $2^{M_p} \equiv 2 \pmod{M_p}$. Since pq is a pseudoprime to the base 2, we have $2^{pq} \equiv 2 \pmod{pq}$, so $2^{pq} \equiv 2 \pmod{p}$. But $2^{pq} \equiv (2^p)^q \equiv 2^q \pmod{p}$. Therefore, $2^q \equiv 2 \pmod{p}$. Thus there exists an integer l such that $M_q - 1 = 2^q - 2 = lp$. Then $2^{M_q-1} - 1 = 2^{2^q-2} = 2^{lp} - 1$, so $2^p - 1 = M_p$ divides $2^{M_q-1} - 1$. Therefore, $2^{M_q} \equiv 2 \pmod{M_p}$. Then we have $2^{M_pM_q} \equiv (2^{M_p})^{M_q} \equiv 2^{M_q} \equiv 2 \pmod{M_p}$. Similarly, $2^{M_pM_q} \equiv 2 \pmod{M_q}$. By the Chinese remainder theorem, noting that M_p and M_q are relatively prime, we have $2^{M_pM_q} \equiv 2 \pmod{M_pM_q}$. Therefore, M_pM_q is a pseudoprime to the base 2. Conversely, suppose that M_pM_q is a pseudoprime to the base 2. From the reasoning in the proof of Theorem 6.6, $2^{M_p} \equiv 2 \pmod{p}$. Therefore, $2^{M_pM_q} \equiv 2^{(M_p-1)M_q+M_q} \equiv 2^{M_q} \equiv 2 \pmod{p}$. But since $M_p = 2^p - 1 \equiv 0 \pmod{M_p}$, the order of 2 modulo M_p is p. Therefore, $p \mid M_q - 1$. In other words, $2^q \equiv 2 \pmod{p}$. Then $2^{pq} \equiv 2^q \equiv 2 \pmod{p}$. Similarly, $2^{pq} \equiv 2 \pmod{q}$. Therefore, by the Chinese remainder theorem, $2^{pq} \equiv 2 \pmod{pq}$. Since pq is composite, it is a pseudoprime to the base 2.

23. Let $j = \text{ord}_{\phi(n)}\, e$. Then $e^j \equiv 1 \pmod{\phi(n)}$. Since $\text{ord}_n P \mid \phi(n)$, we have $e^j \equiv 1 \pmod{\text{ord}_n P}$. Then by Theorem 9.2, $P^{e^j} \equiv P \pmod{n}$, so $C^{e^{j-1}} \equiv (P^e)^{e^{j-1}} \equiv P^{e^j} \equiv P \pmod{n}$ and $C^{e^j} \equiv P^e \equiv C \pmod{n}$.

Section 9.2

1. a. 2   **b.** 2   **c.** 3   **d.** 0

3. a. 2   **b.** 4   **c.** 8   **d.** 6   **e.** 12   **f.** 22

5. 2, 6, 7, 11

7. 2, 3, 10, 13, 14, 15

9. By Lagrange's theorem there are at most two solutions to $x^2 \equiv 1 \pmod{p}$, and we know that $x \equiv \pm 1$ are the two solutions. Since $p \equiv 1 \pmod{4}$, we have $4 \mid p - 1 = \phi(p)$, so by Theorem 9.8 there is an element x of order 4 modulo p. Then $x^4 = (x^2)^2 \equiv 1 \pmod{p}$, so $x^2 \equiv \pm 1 \pmod{p}$. If $x^2 \equiv 1 \pmod{p}$, then x does not have order 4. Therefore, $x^2 \equiv -1 \pmod{p}$.

11. a. Let $f(x) = a_n x^n + a_{n-1}x^{n-1} + \cdots + a_0$, and let k be the largest integer such that p does not divide a_k. Let $g(x) = a_k x^k + a_{k-1}x^{k-1} + \cdots + a_0$. Then $f(x) \equiv g(x) \pmod{p}$ for every value of x. In particular, $g(x)$ has the same set of roots modulo p as $f(x)$. Since the number of roots is greater than $n > k$, this contradicts Lagrange's theorem. Therefore, no such k exists, and p must divide every coefficient of $f(x)$.
b. Note that the degree of $f(x)$ is $p - 2$ (the x^{p-1} terms cancel). By Fermat's little theorem we have that $x^{p-1} - 1 \equiv 0 \pmod{p}$ for $x = 1, 2, \ldots, p - 1$. Further, each x in the same range is patently a zero for $(x - 1)(x - 2) \cdots (x - p + 1)$. Therefore, each such x is a root of $f(x)$. Since $f(x)$ has degree $p - 2$ and $p - 1$ roots, part (a) tells us that all the coefficients of $f(x)$ are divisible by p.
c. From part (b) we know that the constant term of $f(x)$ is divisible by p. The constant term is $f(0)$. Thus $f(0) = (-1)(-2) \cdots (-p + 1) + 1 = (-1)^{p-1}(p - 1)! + 1 = (p - 1)! + 1 \equiv 0 \pmod{p}$, which is Wilson's theorem.

13. a. Since $q_i^{t_i} \mid \phi(p) = p - 1$, by Theorem 9.8 there exist $\phi(q_i^{t_i})$ elements of order $q_i^{t_i}$ for each $i = 1, 2, \ldots, r$. Let a_i be a fixed element of this order.
b. Using mathematical induction and Exercise 10 of Section 9.1, we have $\text{ord}_p(a) = \text{ord}_p(a_1 \cdots a_r) = \text{ord}_p(a_1 \cdots a_{r-1}) \cdot \text{ord}_p(a_r) = \cdots = \text{ord}_p(a_1) \cdots \text{ord}_p(a_r)$ since

$\{\text{ord}_p(a_1), \ldots, \text{ord}_p(a_r)\} = \{q_1^{t_1}, \ldots, q_r^{t_r}\}$ are pairwise relatively prime.
c. 18

15. If n is odd, composite, and not a power of 3, then the product in Exercise 14 is $\prod_{j=1}^r(n-1, p_j - 1) \geq (n-1, 3-1)(n-1, 5-1) \geq 2 \cdot 2 = 4$. Therefore, there must be two bases other than -1 and $+1$.

17. a. Suppose that $f(x)$ is a polynomial of degree $n-1$ with integer coefficients. Suppose that $x_1, x_2, \ldots, x_n$ are incongruent modulo p, where p is prime. Consider the polynomial $g(x) = f(x) - \sum_{j=1}^n (f(x_j) \cdot \prod_{i \neq j}(x - x_i)\overline{(x_j - x_i)})$. Note that x_j, $j = 1, 2, \ldots, n$, is a root of this polynomial modulo p since its value at x_j is $f(x_j) - (0 + 0 + \cdots + f(x_j)\prod_{i \neq j}(x_j - x_i)\overline{(x_j - x_i)} + \cdots + 0) \equiv f(x_j) - f(x_j) \cdot 1 \equiv 0 \pmod{p}$. Since $g(x)$ has n incongruent roots modulo p and since it is of degree $n-1$ or less, we can easily use Lagrange's theorem to see that $g(x) \equiv 0 \pmod{p}$ for every integer x.
b. 10

19. By Exercise 23 of Section 9.1, $j \mid \text{ord}_{\phi(n)} e$. Here $\phi(n) = \phi(pq) = 4p'q'$, so $j \mid \phi(4p'q') = 2(p'-1)(q'-1)$. Choose e to be a primitive root modulo p'. Then $p'-1 = \phi(p') \mid \phi(\phi(n))$, so $p'-1 \mid \text{ord}_{\phi(n)} e$. The decrypter needs $e^j \equiv 1 \pmod{n}$, but this choice of e forces $j = p'-1$, which will take quite some time to find.

Section 9.3

1. 4, 10, 22

3. a. 2   **b.** 2   **c.** 5   **d.** 2

5. a. 2   **b.** 2   **c.** 2   **d.** 3

7. a. 3   **b.** 3   **c.** 3   **d.** 3

9. 7, 13, 17, 19

11. 3, 13, 15, 21, 29, 33

13. Suppose that r is a primitive root of m and suppose further that $x^2 \equiv 1 \pmod{m}$. Let $x \equiv r^t \pmod{m}$, where $0 \leq t \leq p-1$. Then $r^{2t} \equiv 1 \pmod{m}$. Since r is a primitive root, $\phi(m) \mid 2t$, so $2t = k\phi(m)$ and $t = k\phi(m)/2$ for some integer k. Then $x \equiv r^t = r^{k\phi(m)/2} = r^{(\phi(m)/2)k} \equiv (-1)^k \equiv \pm 1 \pmod{m}$. Conversely, suppose that m has no primitive root. Then m is not of one of the forms 2, 4, p^a, or $2p^a$, with p an odd prime. So either two distinct odd primes divide m, or $m = 2^b p^a$ with p an odd prime and $b > 1$, or $m = 2^b$ with $b > 2$. Let p be an odd prime dividing m, say $p^\alpha \| m$. Then the solution to the system $x \equiv 1 \pmod{p^\alpha}$, $x \equiv -1 \pmod{m/p^\alpha}$ cannot be congruent to $\pm 1 \pmod{m}$. But $x^2 \equiv 1 \pmod{p^\alpha}$ and $x^2 \equiv 1 \pmod{m/p^\alpha}$, so by the Chinese remainder theorem $x^2 \equiv 1 \pmod{m}$. If no odd prime divides m, then $m = 2^b$ with $b \geq 3$. From Theorem 9.12, we know there are at least three solutions y_1, y_2, and y_3 to $y^2 \equiv 1 \pmod{2^b}$. So in each case, there is at least one solution that is not congruent to $\pm 1 \pmod{m}$.

15. By Theorem 9.12 we know that $\text{ord}_{2^k} 5 = \phi(2^k)/2 = 2^{k-2}$. Hence, the 2^{k-2} integers 5^j, $j = 0, 1, \ldots, 2^{k-2} - 1$, are incongruent modulo 2^k. Similarly, the 2^{k-2} integers -5^j, $j = 0, 1, \ldots, 2^{k-2} - 1$, are incongruent modulo 2^k. Note that 5^j cannot be congruent to -5^i modulo 2^k since $5^j \equiv 1 \pmod{4}$ but $-5^i \equiv 3 \pmod{4}$. It follows that the integers $1, 5, \ldots, 5^{2^{k-2}-1}, -1, -5, \ldots, -5^{2^{k-2}-1}$ are 2^{k-1} incongruent integers modulo 2^k. Since $\phi(2^k) = 2^{k-1}$ and every integer of the form $(-1)^\alpha 5^\beta$ is relatively prime to 2^k, it follows that every odd integer is congruent to an integer of this form with $\alpha = 0$ or 1 and $0 \leq \beta \leq 2^{k-2} - 1$.

Section 9.4

1. The values of $\text{ind}_5 i$, $i = 1, 2, \ldots, 22$, are 22, 2, 16, 4, 1, 18, 19, 6, 10, 3, 9, 20, 14, 21, 17, 8, 7, 12, 15, 5, 13, 11, respectively.

3. a. 7, 18   **b.** none

5. 8, 9, 20, 21, 29 (mod 29)

7. All positive integers $x \equiv 0$ 1, 12, 23, 24, 45, 46, 47, 67, 69, 70, 78, 89, 91, 92, 93, 100, 111, 115, 116, 133, 137, 138, 139, 144, 155, 161, 162, 177, 183, 184, 185, 188, 199, 207, 208, 210, 221, 229, 230, 231, 232, 243, 253, 254, 265, 275, 276, 277, 287, 299, 300, 309, 321, 322, 323, 331, 345, 346, 353, 367, 368, 369, 375, 386, 391, 392, 397, 413, 414, 415, 419, 430, 437, 438, 441, 459, 460, 461, 463, 483, 484, 485, 496, or 505 (mod 506)

9. Let r be a primitive root of p. Suppose that $x^4 \equiv -1 \pmod{p}$, and let $y = \text{ind}_r x$. Then $-x$ is also a solution, and by Exercise 8, $\text{ind}_r(-x) \equiv \text{ind}_r(-1) + \text{ind}_r(x) \equiv (p-1)/2 + y \pmod{p-1}$. So without loss of generality we may take $0 < y < (p-1)/2$, or $0 < 4y < 2(p-1)$. Taking indices of both sides of the congruence yields $4y \equiv \text{ind}_r(-1) \equiv (p-1)/2 \pmod{p-1}$, again using Exercise 8. So $4y = (p-1)/2 + m(p-1)$ for some m. But $4y < 2(p-1)$, so either $4y = (p-1)/2$ and so $p = 8y + 1$, or $4y = 3(p-1)/2$. In the latter case, 3 must divide y, so we have $p = 8(y/3) + 1$. In either case, p is of the desired form. Conversely, suppose that $p = 8k + 1$, and let r be a primitive root of p. Take $x = r^k$. Then $x^4 \equiv r^{4k} \equiv r^{(p-1)/2} \equiv -1$ $\pmod{p}$ by Exercise 8. So this x is a solution.

11. $(1, 2)$; $(0, 2)$

13. $x \equiv 29 \pmod{32}$; $x \equiv 4 \pmod 8$

15. $(0, 0, 1, 1)$; $(0, 0, 1, 4)$

17. $x \equiv 17 \pmod{60}$

19. We seek a solution to $x^k \equiv a \pmod{2^e}$. We take indices as described before Exercise 11. Suppose that $a \equiv (-1)^\alpha 5^\beta$ and $x \equiv (-1)^\gamma 5^\delta$. Then the index system of x^k is $(k\gamma, k\delta)$, and the index system of a is (α, β), so $k\gamma \equiv \alpha \pmod 2$ and $k\delta \equiv \beta \pmod{2^{e-2}}$. Since k is odd, both congruences are solvable for γ and δ, which determine x.

21. First we show that $\text{ord}_{2^e} 5 = 2^{e-2}$. Indeed, $\phi(2^e) = 2^{e-1}$, so it suffices to show that the highest power of 2 dividing $5^{2^{e-2}} - 1$ is 2^e. We proceed by induction. The basis step is the case $e = 2$, which is true. Note that $5^{2^{e-2}} - 1 = (5^{2^{e-3}} - 1)(5^{2^{e-3}} + 1)$. The first factor is exactly divisible by 2^{e-1} by the induction hypothesis. The second factor differs from the first by 2, so it is exactly divisible by 2; therefore, $5^{2^{e-2}} - 1$ is exactly divisible by 2^e, as desired. Hence if k is odd, then the numbers $(\pm 5)^k$, $(\pm 5)^{2k}$, $\ldots$, $(\pm 5)^{2^{e-2}k}$ are 2^{e-1} incongruent kth power residues, which is the number given by the formula. If 2^m exactly divides k, then $5^k \equiv (-5)^k \pmod{2^e}$, so the formula must be divided by 2, hence the factor $(k, 2)$ in the denominator. Further, 5^{2^m} has order $2^{e-2}/2^m$ if $m \le e - 2$ and order 1 if $m > e - 2$, so the list must repeat modulo 2^e every $\text{ord}_{2^e} 5^{2^m}$ terms, whence the other factor in the denominator.

23. a. From the first inequality in the proof, if n is not square-free, then the probability is strictly less than $2n/9$, which is substantially smaller than $(n-1)/4$ for large n. If n is square-free, the argument following (9.6) shows that if n has 4 or more factors, then the probability is less than $n/8$. The next inequality shows that the worst case for $n = p_1 p_2$ is when $s_1 = s_2$ and s_1 is as small as possible, which is the case stated in the exercise.
b. $0.24999249\ldots$

Section 9.5

1. We have $2^2 \equiv 4 \pmod{101}$, $2^5 \equiv 32 \pmod{101}$, $2^{10} \equiv (2^5)^2 \equiv 32^2 \equiv 14 \pmod{101}$, $2^{20} \equiv (2^{10})^2 \equiv 14^2 \equiv 95 \pmod{101}$, $2^{25} \equiv (2^5)^5 \equiv 32^5 \equiv (32^2)^2 \cdot 32 \equiv 1024^2 \cdot 32 \equiv 14^2 \cdot 32 \equiv 196 \cdot 32 \equiv -6 \cdot 32 \equiv -192 \equiv 10 \pmod{101}$, $2^{50} \equiv (2^{25})^2 \equiv 10^2 \equiv 100 \equiv -1 \pmod{101}$, $2^{100} \equiv (2^{50})^2 \equiv (-1)^2 \equiv 1 \pmod{101}$. Since $2^{(101-1)/q} \not\equiv 1 \pmod{101}$ for every proper divisor q of 100 but $2^{101-1} \equiv 1 \pmod{101}$, it follows that 101 is prime.

3. $233 - 1 = 2^3 \cdot 29$, $3^{116} \equiv -1 \pmod{233}$, $3^8 \equiv 37 \not\equiv 1 \pmod{233}$

5. The first condition implies that $x^{F_n-1} \equiv 1 \pmod{F_n}$. The only prime dividing $F_n - 1 = 2^{2^n}$ is 2, and $(F_n - 1)/2 = 2^{2^n-1}$, so the second condition implies that $2^{(F_n-1)/2} \not\equiv 1 \pmod{F_n}$. Then by Theorem 9.18, F_n is prime.

7. See [Le80].

9. Since $n - 1 = 9928 = 2^3 \cdot 17 \cdot 73$, we take $F = 2^3 \cdot 17 = 136$ and $R = 73$, noting that $F > R$. We apply Pocklington's test with $a = 3$. We check (using a calculator or computational software) that $3^{9928} \equiv 1 \pmod{9929}$ and that $(3^{9928/2} - 1, 9929) = 1$ and $(3^{9928/17} - 1, 9929) = 1$, since 2 and 17 are the only primes dividing F. Therefore, n passes Pocklington's test and so is prime.

11. Note that $3329 = 2^8 \cdot 13 + 1$ and $13 < 2^8$, so it is of the form that can be tested by Proth's test. With the help of computational software, we compute $3^{(3329-1)/2} \equiv -1 \pmod{3329}$, which shows that 3329 is prime.

13. We apply Pocklington's test to this situation. Note that $n - 1 = hq^k$, so we let $F = q^k$ and $R = h$ and observe that by hypothesis $F > R$. Since q is the only prime dividing F, we need only check that there is an integer a such that $a^{n-1} \equiv 1 \pmod{n}$ and $(a^{(n-1)/q} - 1, n) = 1$. But both of these conditions are hypotheses.

Section 9.6

1. **a.** 20   **b.** 12   **c.** 36   **d.** 48   **e.** 180   **f.** 388,080   **g.** 8640   **h.** 125,411,328,000

3. 65,520

5. Suppose that $m = 2^{t_0} p_1^{t_1} \cdots p_s^{t_s}$. Then $\lambda(m) = [\lambda(2^{t_0}), \phi(p_1^{t_1}), \ldots, \phi(p_s^{t_s})]$. Furthermore, $\phi(m) = \phi(2^{t_0})\phi(p_1^{t_1}) \cdots \phi(p_s^{t_s})$. Since $\lambda(2^{t_0}) = 1, 2$, or 2^{t_0-2} when $t_0 = 1, 2$, or $t_0 \geq 3$, respectively, we have $\lambda(2^{t_0}) \mid \phi(2^{t_0}) = 2^{t_0-1}$. Since the least common multiple of a set of numbers divides the product of these numbers, or their multiples, we see that $\lambda(m) \mid \phi(m)$.

7. For any integer x with $(x, n) = (x, m) = 1$ we have $x^a \equiv 1 \pmod{n}$ and $x^a \equiv 1 \pmod{m}$. Then the Chinese remainder theorem gives us $x^a \equiv 1 \pmod{[n, m]}$. But since n is the largest integer with this property, we must have $[n, m] = n$, so $m \mid n$.

9. Suppose that $ax \equiv b \pmod{m}$. Multiplying both sides of this congruence by $a^{\lambda(m)-1}$ gives $a^{\lambda(m)}x \equiv a^{\lambda(m)-1}b \pmod{m}$. Since $a^{\lambda(m)} \equiv 1 \pmod{m}$, it follows that $x \equiv a^{\lambda(m)-1}b \pmod{m}$. Conversely, let $x_0 \equiv a^{\lambda(m)-1}b \pmod{m}$. Then $ax_0 \equiv aa^{\lambda(m)-1}b = a^{\lambda(m)}b \equiv b \pmod{m}$, so x_0 is a solution.

11. **a.** First, suppose that $m = p^a$. Then $x(x^{c-1} - 1) \equiv 0 \pmod{p^a}$. Let r be a primitive root for p^a. Then the solutions to $x^{c-1} \equiv 1 \pmod{p^a}$ are the powers r^k with $(c - 1)k \equiv 1 \pmod{\phi(p^a)}$; there are $(c - 1, \phi(p^a))$ of these. Since 0 is a solution, there are $1 + (c - 1, \phi(p^a))$ solutions. If $m = p_1^{a_1} \cdots p_t^{a_t}$, the result follows by the Chinese remainder theorem since there is a one-to-one correspondence between solutions modulo m and the set of t-tuples of solutions to the system of congruences modulo each of the prime powers.
b. If $(c - 1, \phi(m)) = 2$, then $c - 1$ is even. Since $\phi(p^a)$ is even for all prime powers, except 2, $(c - 1, \phi(p_i^{a_i})) = 2$ for each i. By (a) the number of solutions is 3^r. If 2^1 is a prime-power factor,

then $\phi(m) = \phi(m/2)$, and since x^c and x have the same parity, x is a solution modulo m if and only if it is a solution modulo $m/2$, so the result still holds.

13. Let $n = 3pq$, with $p < q$ odd primes, be a Carmichael number. Then by Theorem 9.23, $p - 1 \mid 3pq - 1 = 3(p - 1)q + 3q - 1$, so $p - 1 \mid 3q - 1$, say $(p - 1)a = 3q - 1$. Since $q > p$, we must have $a \geq 4$. Similarly, there is an integer b such that $(q - 1)b = 3p - 1$. Solving these two equations for p and q yields $q = (2a + ab - 3)/(ab - 9)$ and $p = (2b + ab - 3)/(ab - 9) = 1 + (2b + 6)/(ab - 9)$. Then since p is an odd prime greater than 3, we must have $4(ab - 9) \leq 2b + 6$, which reduces to $b(2a - 1) \leq 21$. Since $a \geq 4$, this implies that $b \leq 3$. Then $4(ab - 9) \leq 2b + 6 \leq 12$, so $ab \leq 21/4$, so $a \leq 5$. Therefore, $a = 4$ or 5. If $b = 3$, then the denominator in the expression for q is a multiple of 3, so the numerator must be a multiple of 3, but that is impossible since there is no choice for a that is divisible by 3. Thus $b = 1$ or 2. The denominator of q must be positive, so $ab > 9$, which eliminates all remaining possibilities except $a = 5$, $b = 2$, in which case $p = 11$ and $q = 17$. So the only Carmichael number of this form is $561 = 3 \cdot 11 \cdot 17$.

15. Assume that $q < r$. By Theorem 9.23, $q - 1 \mid pqr - 1 = (q - 1)pr + pr - 1$. Therefore, $q - 1 \mid pr - 1$, say $a(q - 1) = pr - 1$. Similarly, $b(r - 1) = pq - 1$. Since $q < r$, we must have $a > b$. Solving these two equations for q and r yields $r = (p(a - 1) + a(b - 1))/(ab - p^2)$ and $q = (p(b - 1) + b(a - 1))/(ab - p^2) = 1 + (p^2 + pb - p - b)/(ab - p^2)$. Since this last fraction must be an integer, $ab - p^2 \leq p^2 + pb - p - b$, which reduces to $a(b - 1) \leq 2p^2 + p(b - 1)$ or $a - 1 \leq (2p^2/b) + (p(b - 1)/b) \leq 2p^2 + p$. So there are only finitely many values for a. Likewise, the same inequality gives us $b(a - 1) \leq 2p^2 + pb - p$ or $b(a - 1 - p) \leq 2p^2 - p$. Since $a > b$ and the denominator of the expression for q must be positive, $a \geq p + 1$. If $a = p + 1$, then $(p + 1)(q - 1) = pq - p + q - 1 = pr - 1$, which implies that $p \mid q$, a contradiction. Therefore, $a > p + 1$, and so $a - 1 - p$ is a positive integer. The last inequality gives us $b \leq b(a - 1 - p) \leq 2p^2 - p$. Therefore, there are only finitely many values for b. Since a and b determine q and r, there can be only finitely many Carmichael numbers of this form.

17. We have $q_n(ab) \equiv ((ab)^{\lambda(n)} - 1)/n = (a^{\lambda(n)}b^{\lambda(n)} - a^{\lambda(n)} - b^{\lambda(n)} + 1 + a^{\lambda(n)} + b^{\lambda(n)} - 2)/n = (a^{\lambda(n)} - 1)(b^{\lambda(n)} - 1)/n + ((a^{\lambda(n)} - 1) + (b^{\lambda(n)} - 1))/n \equiv q_n(a) + q_n(b) \pmod{n}$. At the last step, we use the fact that n^2 must divide $(a^{\lambda(n)} - 1)(b^{\lambda(n)} - 1)$, since $\lambda(n)$ is the universal exponent.

Section 10.1

1. 69, 76, 77, 92, 46, 11, 12, 14, 19, 36, 29, 84, 05, 02, 00, 00, 00, $\ldots$

3. 10

5. **a.** $a \equiv 1 \pmod{20}$ **b.** $a \equiv 1 \pmod{30{,}030}$ **c.** $a \equiv 1 \pmod{111111}$ **d.** $a \equiv 1 \pmod{2^{25} - 1}$

7. **a.** 31   **b.** 715,827,882   **c.** 31   **d.** 195,225,786   **e.** 1,073,741,823   **f.** 1,073,741,823

9. 8, 64, 15, 71, 36, 64, 15, 71, 36, $\ldots$

11. First we find that $\text{ord}_{77} 8$ is 10 so that $t = 1$ and $s = 5$. Since $\text{ord}_5 2 = 4$, the period length is 4.

13. Using the notation of Theorem 10.4, we have $\phi(77) = 60$, so $\text{ord}_{77} x_0$ is a divisor of $60 = 2^2 \cdot 3 \cdot 5$. Thus the only possible values for s are the odd divisors of 60, which are 3, 5, and 15. But $\text{ord}_3 2 = 2$, $\text{ord}_5 2 = 4$, and $\text{ord}_{15} 2 = 4$. Hence, by Theorem 10.4 the maximum period length is 4.

15. 24, 25, 18, 12, 30, 11, 10, 21

17. Check that 7 has maximal order 1800 modulo $2^{25} - 1$. To make a large enough multiplier, raise 7 to a power relatively prime to $\phi(2^{25} - 1) = 32{,}400{,}000$, for example, to the 11th power.

19. 665

21. **a.** 8, 2, 8, 2, 8, 2, $\ldots$ **b.** 9, 12, 6, 13, 8, 18, 2, 4, 16, 3, 9, 12, 6, 13, 8, 18, 2, 4, 16, 3, $\ldots$

Section 10.2

1. We select $k = 1234$ for our random integer. Converting the plaintext into numerical equivalents results in 0700 1515 2401 0817 1907 0300 2423, where we filled out the last block with an X. With the help of a computational program, we find $\gamma \equiv r^k = 6^{1234} \equiv 517 \pmod{2551}$. Then for each block P we compute $\delta \equiv P \cdot b^k = P \cdot 33^{1234} \equiv P \cdot 651 \pmod{2551}$. The resulting blocks are $0700 \cdot 651 \equiv 1622 \pmod{2551}$, $1515 \cdot 651 \equiv 1579 \pmod{2551}$, $2401 \cdot 651 \equiv 1839 \pmod{2551}$, $0817 \cdot 651 \equiv 1259 \pmod{2551}$, $1907 \cdot 651 \equiv 1671 \pmod{2551}$, $0300 \cdot 651 \equiv 1424 \pmod{2551}$, and $2423 \cdot 651 \equiv 855 \pmod{2551}$. Therefore, the ciphertext is $(517, 1622)$, $(517, 1579)$, $(517, 1839)$, $(517, 1259)$, $(517, 1671)$, $(517, 1424)$, $(517, 855)$. To decrypt this ciphertext, we compute $\gamma^{p-1-a} = 517^{2551-1-13} = 517^{2537} \equiv 337 \pmod{2551}$. Then for each block of the ciphertext we compute $P \equiv 337 \cdot \delta \pmod{2551}$. For the first block we have $337 \cdot 1622 \equiv 0700 \pmod{2551}$, which was the first block of the plaintext; and so on.

3. RABBIT

5. $(\gamma, s) = (2022, 833)$; to verify this signature, we use a computational program to find that $V_1 \equiv 2022^{833} \cdot 801^{2022} \equiv 1014 \equiv 3^{823} \equiv V_2 \pmod{2657}$

7. Let $\delta_1 = P_1 b^k$ and $\delta_2 = P_2 b^k$ in the ElGamal cryptosystem. If P_1 is known, then it is easy to compute an inverse for P_1 modulo p. Then $b^k \equiv \overline{P_1}\delta_1 \pmod p$. Then it is also easy to compute an inverse for $b^k \pmod p$. Then $P_2 \equiv \overline{b^k}\delta_2 \pmod p$. Hence the plaintext P_2 is recovered.

Section 10.3

1. a. 8 **b.** 5 **c.** 2 **d.** 6 **e.** 30 **f.** 20

3. a. At each stage of the splicing, the kth wire of one section is connected to the $S(k)$th wire, where $S(k)$ is the least positive residue of $3k - 2 \pmod{50}$.
b. At each stage of the splicing, the kth wire of one section is connected to the $S(k)$th wire, where $S(k)$ is the least positive residue of $21k + 56 \pmod{76}$.
c. At each stage of the splicing, the kth wire of one section is connected to the $S(k)$th wire, where $S(k)$ is the least positive residue of $2k - 1 \pmod{125}$.

Section 11.1

1. a. 1 **b.** 1, 4 **c.** 1, 3, 4, 9, 10, 12 **d.** 1, 4, 5, 6, 7, 9, 11, 16, 17

3. $1, -1, -1, 1$

5. a. $\left(\frac{7}{11}\right) \equiv 7^{\frac{11-1}{2}} \equiv 7^5 \equiv 49^2 \cdot 7 \equiv 5^2 \cdot 7 \equiv 3 \cdot 7 \equiv -1 \pmod{11}$
b. $(7, 14, 21, 28, 35) \equiv (7, 3, 10, 6, 2) \pmod{11}$ and three of these are greater than $\frac{11}{2}$, so $\left(\frac{7}{11}\right) = (-1)^3 = -1$

7. We have $\left(\frac{-2}{p}\right) = \left(\frac{-1}{p}\right)\left(\frac{2}{p}\right)$ by Theorem 11.4. Using Theorems 11.5 and 11.6, we have: If $p \equiv 1$ $\pmod 8$, then $\left(\frac{-2}{p}\right) = (1)(1) = 1$; if $p \equiv 3 \pmod 8$, then $\left(\frac{-2}{p}\right) = (-1)(-1) = 1$; if $p \equiv -1$ $\pmod 8$, then $\left(\frac{-2}{p}\right) = (-1)(1) = -1$; if $p \equiv -3 \pmod 8$, then $\left(\frac{-2}{p}\right) = (1)(-1) = -1$.

9. Since $p - 1 \equiv -1$, $p - 2 \equiv -2$, $\ldots$, $(p+1)/2 \equiv (p-1)/2 \pmod p$, we have $((p-1)/2)!^2 \equiv -(p-1)! \equiv 1 \pmod p$ by Wilson's theorem (since $p \equiv 3 \pmod 4$ the minus signs cancel). By Euler's criterion $((p-1)/2)!^{(p-1)/2} \equiv \left(\frac{1}{p}\right)\left(\frac{2}{p}\right)\cdots\left(\frac{(p-1)/2}{p}\right) \equiv (-1)^t \pmod p$, by definition of the Legendre symbol. Since $((p-1)/2)! \equiv \pm 1 \pmod p$ and $(p-1)/2$ is odd, we have the result.

11. If $p \equiv 1 \pmod 4$, then $\left(\frac{-a}{p}\right) = \left(\frac{-1}{p}\right)\left(\frac{a}{p}\right) = 1 \cdot 1 = 1$. If $p \equiv 3 \pmod 4$, then $\left(\frac{-a}{p}\right) = \left(\frac{-1}{p}\right)\left(\frac{a}{p}\right) = (-1) \cdot 1 = -1$.

13. **a.** $x \equiv 4$ or $2 \pmod 7$ **b.** $x \equiv 1 \pmod 7$ **c.** no solutions

15. Suppose that p is a prime greater than 6. At least one of the three incongruent integers 2, 3, and 6 is a quadratic residue of p, because if neither 2 nor 3 is a quadratic residue of p, then $2 \cdot 3 = 6$ is a quadratic residue of p. If 2 is a quadratic residue, then 2 and 4 are quadratic residues that differ by 2; if 3 is a quadratic residue, then 1 and 3 are quadratic residues that differ by 2; while if 6 is a quadratic residue, then 4 and 6 are quadratic residues that differ by 2.

17. **a.** Since $p = 4n + 3$, we have $2n + 2 = (p + 1)/2$. Then $x^2 \equiv (\pm a^{n+1})^2 \equiv a^{2n+2} \equiv a^{(p+1)/2} \equiv a^{(p-1)/2}a \equiv 1 \cdot a \equiv a \pmod p$ using the fact that $a^{(p-1)/2} \equiv 1 \pmod p$ since a is a quadratic residue of p. By Lemma 11.1, there are only these two solutions.
    b. By Lemma 11.1, there are exactly two solutions to $y^2 \equiv 1 \pmod p$, namely $y \equiv \pm 1 \pmod p$. Since $p \equiv 5 \pmod 8$, we know that -1 is a quadratic residue of p and 2 is a quadratic nonresidue of p. Since $p = 8n + 5$, we have $4n + 2 = (p - 1)/2$ and $2n + 2 = (p + 3)/4$. Then $(\pm a^{n+1})^2 \equiv a^{(p+3)/4} \pmod p$ and $(\pm 2^{2n+1}a^{n+1})^2 \equiv 2^{(p-1)/2}a^{(p+3)/4} \equiv -a^{(p+3)/4} \pmod p$ by Euler's criterion. We must show that one of $a^{(p+3)/4}$ or $-a^{(p+3)/4}$ is congruent to $a \pmod p$. Now a is a quadratic residue of p, so $a^{(p-1)/2} \equiv 1 \pmod p$, and therefore $a^{(p-1)/4}$ solves $x^2 \equiv 1 \pmod p$. But then $a^{(p-1)/4} \equiv \pm 1 \pmod p$, that is, $a^{(p+3)/4} \equiv \pm a \pmod p$ or $\pm a^{(p+3)/4} \equiv a \pmod p$, as desired.

19. $x \equiv 1$, 4, 11, or 14 $\pmod{15}$

21. 47, 96, 135, 278, 723, 866, 905, 954 $\pmod{1001}$

23. If $x_0^2 \equiv a \pmod{p^{e+1}}$, then $x_0^2 \equiv a \pmod{p^e}$. Conversely, if $x_0^2 \equiv a \pmod{p^e}$, then $x_0^2 = a + bp^e$ for some integer b. We can solve the linear congruence $2x_0y \equiv -b \pmod p$, say $y = y_0$. Let $x_1 = x_0 + y_0p^e$. Then $x_1^2 \equiv x_0^2 + 2x_0y_0p^e = a + p^e(b + 2x_0y_0) \equiv a \pmod{p^{e+1}}$ since $p \mid 2x_0y_0 + b$. This is the induction step in showing that $x^2 \equiv a \pmod{p^e}$ has solutions if and only if $\left(\frac{a}{p}\right) = 1$.

25. **a.** 4   **b.** 8   **c.** 0   **d.** 16

27. Suppose that $p_1, p_2, \ldots, p_n$ are the only primes of the form $4k + 1$. Let $N = 4(p_1p_2 \cdots p_n)^2 + 1$. Let q be an odd prime factor of N. Then $q \neq p_i$ for $i = 1, 2, \ldots, n$, but $N \equiv 0 \pmod q$, so $4(p_1p_2 \cdots p_n)^2 \equiv -1 \pmod q$. Therefore $\left(\frac{-1}{q}\right) = 1$, so $q \equiv 1 \pmod 4$ by Theorem 11.5.

29. Let b_1, b_2, b_3, and b_4 be the four modular square roots of a modulo pq. Then each b_i is a solution to exactly one of the four systems of congruences given in the text. For convenience let the subscripts correspond to the lowercase Roman numerals of the systems. Suppose that two of the b_i's were quadratic residues modulo pq. Without loss of generality, say $b_1 \equiv y_1^2 \pmod{pq}$ and $b_2 \equiv y_2^2 \pmod{pq}$. Then from systems (i) and (ii) we have that $y_1^2 \equiv b_1 \equiv x_2 \pmod q$ and $y_2^2 \equiv b_2 \equiv -x_2 \pmod q$. Therefore, both x_2 and $-x_2$ are quadratic residues modulo q, but this is impossible since $q \equiv 3 \pmod 4$. The other cases are identical.

31. Let r be a primitive root for p, and let $a \equiv r^s \pmod p$ and $b \equiv r^t \pmod p$ with $1 \leq s, t \leq p - 1$. If $a \equiv b \pmod p$, then $s = t$, and so s and t have the same parity; by Theorem 11.2, we have part (i). Further, $ab \equiv r^{s+t} \pmod p$. Thus the right-hand side of (ii) is 1 exactly when s and t have the same parity, which is exactly when the left-hand side is 1. This proves part (ii). Finally, since $a^2 \equiv r^{2s} \pmod p$ and $2s$ is even, a^2 must be a quadratic residue modulo p, proving part (iii).

33. If r is a primitive root of q, then the set of all primitive roots is given by $\{r^k \mid (k, \phi(q)) = (k, 2p) = 1\}$. Thus the $p - 1$ numbers $\{r^k \mid k$ is odd, $k \neq p$, $1 \leq k < 2p\}$ are all the primitive roots of q.

On the other hand, q has $(q-1)/2 = p$ quadratic residues, which are given by $\{r^2, r^4, \ldots, r^{2p}\}$. This set has no intersection with the first one.

35. First suppose that $p = 2^{2^n} + 1$ is a Fermat prime, and let r be a primitive root for p. Then $\phi(p) = 2^{2^n}$. An integer a is a nonresidue if and only if $a = r^k$ with k odd. But then $(k, \phi(p)) = 1$, so a is also a primitive root. Conversely, suppose that p is an odd prime and every quadratic nonresidue of p is also a primitive root of p. Let r be a particular primitive root of p. Then r^k is a quadratic nonresidue and hence a primitive root for p if and only if k is odd. But this implies that every odd number is relatively prime to $\phi(p)$, so $\phi(p)$ must be a power of 2. Thus $p = 2^b + 1$ for some b. If b had a nontrivial odd divisor, then we could factor p, contradicting the primality of p. Therefore, b is a power of 2, and so p is a Fermat prime.

37. a. We have $q = 2p + 1 = 2(4k + 3) + 1 = 8k + 7$, so $\left(\frac{2}{q}\right) = 1$ by Theorem 11.6. Then by Euler's criterion, $2^{(q-1)/2} \equiv 2^p \equiv 1 \pmod{q}$. Therefore, $q \mid 2^p - 1$.
b. $11 = 4(2) + 3$ and $23 = 2(11) + 1$, so $23 \mid 2^{11} - 1 = M_{11}$, by part (a); $23 = 4(5) + 3$ and $47 = 2(23) + 1$, so $47 \mid M_{23}$; $251 = 4(62) + 3$ and $503 = 2(251) + 1$, so $503 \mid M_{251}$

39. Let $q = 2k + 1$. Since q does not divide $2^p + 1$, we must have, by Exercise 38, that $k \equiv 0$ or 3 (mod 4). That is, $k \equiv 0, 3, 4,$ or 7 (mod 8). Then $q \equiv 2 \cdot (0, 3, 4,$ or $7) + 1 \equiv \pm 1 \pmod 8$.

41. Note that $\left(\frac{j(j+1)}{p}\right) = \left(\frac{j^2(1+\bar{j})}{p}\right) = \left(\frac{1+\bar{j}}{p}\right)$ since j^2 is a perfect square. Then $\sum_{j=1}^{p-2} \left(\frac{j(j+1)}{p}\right) = \sum_{j=1}^{p-2} \left(\frac{\bar{j}+1}{p}\right) = \sum_{j=2}^{p-1} \left(\frac{j}{p}\right) = \sum_{j=1}^{p-1} \left(\frac{j}{p}\right) - 1 = -1$. Here we have used the method in the solution to Exercise 10 to evaluate the last sum, and the fact that as j runs through the values 1 through $p - 2$, so does $\bar{j}$.

43. Let r be a primitive root of p. Then the congruence $x^2 \equiv a \pmod p$ has a solution in x if and only if the congruence $2 \cdot \text{ind}_r x \equiv \text{ind}_r a \pmod{p-1}$ has a solution in $\text{ind}_r x$. Since $p - 1$ is even, the last congruence is solvable if and only if $\text{ind}_r a$ is even, which happens when $a = r^2, r^4, \ldots, r^{p-1}$, i.e., $(p-1)/2$ times.

45. We have $q = 2(4k + 1) + 1 = 8k + 3$, so 2 is a quadratic nonresidue of q. By Exercise 33, 2 is a primitive root.

47. Check that $q \equiv 3 \pmod 4$, so -1 is a quadratic nonresidue of q. Since $4 = 2^2$, we have $\left(\frac{-4}{q}\right) = \left(\frac{-1}{q}\right)\left(\frac{2^2}{q}\right) = (-1)(1) = -1$. Therefore, -4 is a nonresidue of q. By Exercise 33, -4 is a primitive root.

49. a. By adding $(\bar{2}b)^2$ to both sides, we complete the square.
b. There are four solutions to $x^2 \equiv C + a \pmod{pq}$. From each, subtract $\bar{2}b$.
c. DETOUR

51. a. -1 **b.** -1 **c.** -1 **d.** -1 **e.** 1 **f.** 1

53. $1, 3, 4$

Section 11.2

1. a. -1 **b.** 1 **c.** 1 **d.** 1 **e.** 1 **f.** 1

3. If $p \equiv 1 \pmod 6$, then there are two cases: If $p \equiv 1 \pmod 4$, then $\left(\frac{-1}{p}\right) = 1$ and $\left(\frac{3}{p}\right) = \left(\frac{p}{3}\right) = \left(\frac{1}{3}\right) = 1$, so $\left(\frac{-3}{p}\right) = 1$. If $p \equiv 3 \pmod 4$, then $\left(\frac{-1}{p}\right) = -1$ and $\left(\frac{3}{p}\right) = -\left(\frac{p}{3}\right)$, so $\left(\frac{-3}{p}\right) = (-1)(-1) = 1$. If $p \equiv -1 \pmod 6$ and $p \equiv 1 \pmod 4$, then $\left(\frac{-3}{p}\right) = \left(\frac{-1}{p}\right)\left(\frac{3}{p}\right) = 1 \cdot \left(\frac{p}{3}\right) = \left(\frac{-1}{3}\right) = -1$. If $p \equiv 3 \pmod 4$, then $\left(\frac{-3}{p}\right) = \left(\frac{-1}{p}\right)\left(\frac{3}{p}\right) = (-1)\left(-\left(\frac{p}{3}\right)\right) = \left(\frac{p}{3}\right) = \left(\frac{-1}{3}\right) = -1$.

5. $p \equiv 1, 3, 9, 19, 25,$ or $27 \pmod{28}$

7. **a.** $F_1 = 2^{2^1} + 1 = 5.$ We find that $3^{(F_1-1)/2} = 3^{(5-1)/2} = 3^2 = 9 \equiv -1 \pmod{F_1}.$ Hence, by Pepin's test, $F_1 = 5$ is prime.
   b. $F_3 = 2^{2^3} + 1 = 257.$ We find that $3^{(F_3-1)/2} = 3^{(257-1)/2} = 3^{128} \equiv (3^8)^{16} \equiv 136^{16} \equiv (136^4)^4 \equiv 64^4 \equiv (64^2)^2 \equiv 241^2 \equiv 256 \equiv -1 \pmod{257}.$
   c. $3^{32,768} \equiv 3^{255 \cdot 128} \cdot 3^{128} \equiv 94^{128} \cdot 3^{128} \equiv -1 \pmod{F_4}$

9. The lattice points in the rectangle are the points (i, j), where $0 < i < p/2$ and $0 < j < q/2$. These are the lattice points (i, j) with $i = 1, 2, \ldots, (p-1)/2$ and $j = 1, 2, \ldots, (q-1)/2$. Consequently, there are $((p-1)/2)((q-1)/2)$ such lattice points.
   b. The points on the diagonal connecting **O** and **C** are the points (x, y), where $y = (q/p)x$. Suppose that x and y are integers with $y = (q/p)x$. Then $py = qx$. Since $(p, q) = 1$, it follows that $p \mid x$, which is impossible if $0 < x < p/2$. Hence, there are no lattice points on this diagonal.
   c. The number of lattice points in the triangle with vertices **O**, **A**, and **C** is the number of lattice points (i, j) with $i = 1, 2, \ldots, (p-1)/2$ and $1 \le j \le iq/p$. For a fixed value of i in the indicated range, there are $[iq/p]$ lattice points (i, j) in the triangle. Hence, the total number of lattice points in the triangle is $\sum_{i=1}^{(p-1)/2}[iq/p]$.
   d. The number of lattice points in the triangle with vertices **O**, **B**, and **C** is the number of lattice points (i, j) with $j = 1, 2, \ldots, (q-1)/2$ and $1 \le i < jp/q$. For a fixed value of j in the indicated range, there are $[jp/q]$ lattice points (i, j) in the triangle. Hence, the total number of lattice points in the triangle is $\sum_{j=1}^{(q-1)/2}[jp/q]$.
   e. Since there are $((p-1)/2)((q-1)/2)$ lattice points in the rectangle and no points on the diagonal **OC**, the sum of the numbers of lattice points in the triangles **OBC** and **OAC** is $((p-1)/2)((q-1)/2)$. From parts (b) and (c), it follows that $\sum_{j=1}^{(p-1)/2}[jq/p] + \sum_{j=1}^{(q-1)/2}[jp/q] = ((p-1)/2)((q-1)/2)$. From Lemma 11.3, it follows that $\left(\frac{p}{q}\right) = (-1)^{T(p,q)}$ and $\left(\frac{q}{p}\right) = (-1)^{T(q,p)}$, where $T(p, q) = \sum_{j=1}^{(p-1)/2}[jp/q]$ and $T(q, p) = \sum_{j=1}^{(q-1)/2}[jq/p]$. We conclude that $\left(\frac{p}{q}\right)\left(\frac{q}{p}\right) = (-1)^{((p-1)/2)((q-1)/2)}$. This is the law of quadratic reciprocity.

11. First suppose that $a = 2$. Then $p \equiv \pm q \pmod 8$, and so $\left(\frac{a}{p}\right) = \left(\frac{a}{q}\right)$ by Theorem 11.6. Now suppose that a is an odd prime. If $p \equiv q \pmod{4a}$, then $p \equiv q \pmod a$, and so $\left(\frac{q}{a}\right) = \left(\frac{p}{a}\right)$. Since $p \equiv q \pmod 4$, we have $(p-1)/2 \equiv (q-1)/2 \pmod 2$. Then by Theorem 11.7, $\left(\frac{a}{p}\right) = \left(\frac{p}{a}\right)(-1)^{((p-1)/2)((a-1)/2)} = \left(\frac{q}{a}\right)(-1)^{((q-1)/2)((a-1)/2)} = \left(\frac{a}{q}\right)$. But if $p \equiv -q \pmod{4a}$, then $p \equiv -q \pmod a$, and so $\left(\frac{-q}{a}\right) = \left(\frac{p}{a}\right)$. Since $p \equiv -q \pmod 4$, we have $(p-1)/2 \equiv ((q-1)/2) + 1 \pmod 2$. Then by Theorem 11.7, $\left(\frac{a}{p}\right) = \left(\frac{p}{a}\right)(-1)^{((p-1)/2)((a-1)/2)} = \left(\frac{-q}{a}\right)(-1)^{((q-1)/2+1)((a-1)/2)} = \left(\frac{-1}{a}\right)(-1)^{(a-1)/2}\left(\frac{a}{q}\right) = \left(\frac{a}{q}\right)$. The general case follows from the multiplicativity of the Legendre symbol.

13. **a.** Recall that $e^{xi} = 1$ if and only if x is a multiple of 2π. First we compute $(e^{(2\pi i/n)k})^n = e^{(2\pi i/n)nk} = (e^{2\pi i})^k = 1^k = 1$, so $e^{(2\pi i/n)k}$ is an nth root of unity. Now if $(k, n) = 1$, then $((2\pi i/n)k)a$ is a multiple of $2\pi i$ if and only if $n \mid a$. Therefore, $a = n$ is the least positive integer for which $(e^{(2\pi i/n)k})^a = 1$. Therefore, $e^{(2\pi i/n)k}$ is a primitive nth root of unity. Conversely, suppose that $(k, n) = d > 1$. Then $(e^{(2\pi i/n)k})^{n/d} = e^{(2\pi i)k/d} = 1$, since k/d is an integer, and so in this case, $e^{(2\pi i/n)k}$ is not a primitive nth root of unity.
   b. Let $m = l + kn$, where k is an integer. Then $\zeta^m = \zeta^{l+kn} = \zeta^l \zeta^{kn} = \zeta^l$. Now suppose that ζ is a primitive nth root of unity and that $\zeta^m = \zeta^l$, and without loss of generality, assume that $m \ge l$. From the first part of this exercise, we may take $0 \le l \le m < n$. Then $0 = \zeta^m - \zeta^l = \zeta^l(\zeta^{m-l} - 1)$.

Hence, $\zeta^{m-l} = 1$. Since n is the least positive integer such that $\zeta^n = 1$, we must have $m - l = 0$.

c. First, $f(z + 1) = e^{2\pi i(z+1)} - e^{-2\pi i(z+1)} = e^{2\pi i z}e^{2\pi i} - e^{-2\pi i z}e^{-2\pi i} = e^{2\pi i z} \cdot 1 - e^{-2\pi i z} \cdot 1 = f(z)$. Next, $f(-z) = e^{-2\pi i z} - e^{2\pi i z} = -(e^{2\pi i z} - e^{-2\pi i z}) = -f(z)$. Finally, suppose that $f(z) = 0$. Then $0 = e^{2\pi i z} - e^{-2\pi i z} = e^{-2\pi i z}(e^{4\pi i z} - 1)$, so $e^{4\pi i z} = 1$. Therefore, $4\pi i z = 2\pi i n$ for some integer n, and so $z = n/2$.

d. Fix y, and consider $g(x) = x^n - y^n$ and $h(x) = (x - y)(\zeta x - \zeta^{-1}y) \cdots (\zeta^{n-1}x - \zeta^{-(n-1)}y)$ as polynomials in x. Both polynomials have degree n. The leading coefficient in $h(x)$ is $\zeta^{1+2+\cdots+(n-1)} = \zeta^{n(n-1)/2} = (\zeta^n)^{(n-1)/2} = 1$, since $n - 1$ is even. So both polynomials are monic. Further, note that $g(\zeta^{-2k}y) = (\zeta^{-2k}y)^n - y^n = y^n - y^n = 0$ for $k = 0, 1, 2, \ldots, n - 1$. Also $h(\zeta^{-2k}y)$ has $\zeta^k \zeta^{-2k}y - \zeta^{-k}y = \zeta^{-k}y - \zeta^{-k}y = 0$ as one of its factors. So g and h are monic polynomials sharing these n distinct zeros (since $-2k$ runs through a complete set of residues modulo n). By the fundamental theorem of algebra, g and h are identical.

e. Let $x = e^{2\pi i z}$ and $y = e^{-2\pi i z}$ in the identity from part (d). Then the right-hand side becomes $\prod_{k=0}^{n-1} \left(\zeta^k e^{2\pi i z} - \zeta^{-k} e^{-2\pi i z} \right) = \prod_{k=0}^{n-1} \left(e^{2\pi i(z+k/n)} - e^{-2\pi i(z+k/n)} \right) = \prod_{k=0}^{n-1} f\,(z + k/n) = f(z) \cdot \prod_{k=1}^{(n-1)/2} f\,(z + k/n) \cdot \prod_{k=(n+1)/2}^{n-1} f\,(z + k/n)$. From part (c), this last product is equal to $\prod_{k=(n+1)/2}^{n-1} f\,(z + k/n) = \prod_{k=1}^{(n-1)/2} f\,(z + (n - k)/n) = \prod_{k=1}^{(n-1)/2} f\,(z + 1 - k/n) = \prod_{k=1}^{(n-1)/2} f\,(z - k/n)$. So the product above is equal to $f(z) \cdot \prod_{k=1}^{(n-1)/2} f\,(z + k/n) \cdot \prod_{k=1}^{(n-1)/2} f\,(z - k/n) = f(z) \cdot \prod_{k=1}^{(n-1)/2} f\,(z + k/n) \cdot f\,(z - k/n)$. Then noting that the left side of the identity in part (d) is $(e^{2\pi i z})^n - (e^{-2\pi i z})^n = e^{2\pi i n z} - e^{-2\pi i n z} = f(nz)$ finishes the proof.

f. For $l = 1, 2, \ldots, (p - 1)/2$, let k_l be the least positive residue of la modulo p. Then $\prod_{l=1}^{(p-1)/2} f\,(la/p) = \prod_{l=1}^{(p-1)/2} f\,(k_l/p)$ by the periodicity of f established in part (c). We break this product into two pieces: $\prod_{k_l < p/2} f\,(k_l/p) \cdot \prod_{k_l > p/2} f\,(k_l/p) = \prod_{k_l < p/2} f\,(k_l/p) \cdot \prod_{k_l > p/2} -f\,(-k_l/p) = \prod_{k_l < p/2} f\,(k_l/p) \cdot \prod_{k_l > p/2} -f\,((p - k_l)/p) = \prod_{l=1}^{(p-1)/2} f\,(l/p)\,(-1)^N$, where N is the number of k_l exceeding $p/2$. But by Gauss's lemma, $(-1)^N = \left(\frac{a}{p}\right)$. This establishes the identity.

g. Let $z = l/p$ and $n = q$ in the identities in parts (e) and (f). Then we have $\left(\frac{q}{p}\right) = \prod_{l=1}^{(p-1)/2} f\,(lq/p) / f\,(l/p) = \prod_{l=1}^{(p-1)/2} \prod_{k=1}^{(q-1)/2} f\,(l/p + k/q) \cdot f\,(l/p - k/q) = \prod_{l=1}^{(p-1)/2} \prod_{k=1}^{(q-1)/2} f\,(k/q + l/p) \cdot f\,(k/q - l/p) \cdot (-1)^{(p-1)/2 \cdot (q-1)/2}$, where we have used the fact that $f(-z) = -f(z)$ and the fact that there are exactly $((p - 1)/2) \cdot ((q - 1)/2)$ factors in the double product. But by symmetry, this is exactly the expression for $\left(\frac{q}{p}\right)(-1)^{((p-1)/2)\cdot((q-1)/2)}$.

15. Since $p \equiv 1 \pmod 4$, we have $\left(\frac{q}{p}\right) = \left(\frac{p}{q}\right)$. And since $p \equiv 1 \pmod q$ for all primes $q \le 23$, we have $\left(\frac{p}{q}\right) = \left(\frac{1}{q}\right) = 1$. If a is an integer with $0 < a < 29$ and prime factorization $a = p_1 p_2 \cdots p_k$, then each $p_i < 29$ and $\left(\frac{a}{p}\right) = \left(\frac{p_1}{p}\right) \cdots \left(\frac{p_k}{p}\right) = 1^k = 1$. So there are no quadratic nonresidues modulo p less than 29. Further, since a quadratic residue must be an even power of any primitive root r, we know that $r = r^1$ cannot be less than 29.

17. a. If $a \in T$, then $a = qk$ for some $k = 1, 2, \ldots (p - 1)/2$. So $1 \le a \le q(p - 1)/2 \le (pq - 1)/2$. Furthermore, $(p, k) = 1$ because $k \le (p - 1)/2$, and p is prime. Because $(q, p) = 1$, it follows that $(a, p) = (qk, p) = 1$, so that $a \in S$, and hence, $T \subset S$. Now suppose $a \in S - T$. Then $1 \le a \le (pq - 1)/2$ and $(a, p) = 1$. Because $a \notin T$, it follows that $a \ne qk$ for any k. We conclude that $(a, q) = 1$, which means that $(a, pq) = 1$, and so $a \in R$. Thus $S - T \subset R$. Conversely, if $a \in R$, then $1 \le a \le (pq - 1)/2$ and $(a, pa) = 1$. This implies that $(a, q) = 1$, and so a is not a multiple of q. Hence, $a \notin T$, so that $a \in S - T$. It follows that $R \subset S - T$. Therefore, $R = S - T$.

b. By part (a), $R = S - T$, so that by Euler's criterion, $\prod\limits_{a \in S} a = \prod\limits_{a \in R} a \prod\limits_{a \in T} a = A(q \cdot 2q \cdots$
$((p-1)/2)a)) = Aq^{(p-1)/2}((p-1)/2)! \equiv A\left(\frac{q}{p}\right)((p-1)/2)! \pmod{p}$. Note that $(pq-1)/2 = p(q-1)/2 + (p-1)/2$, so that we can evaluate $\prod\limits_{a \in S} a \equiv ((p-1)!)^{(q-1)/2}((p-1)/2!) \equiv (-1)^{(q-1)/2}((p-1)/2)! \pmod{p}$ by Wilson's theorem. When we set these two expressions congruent to each other modulo p and simplify, we obtain $A \equiv (-1)^{(q-1)/2}\left(\frac{q}{p}\right) \pmod{p}$ as desired.

c. Because the roles of p and q are identical in the hypotheses and in parts (a) and (b), the result follows by symmetry.

d. Assume that $(-1)^{(q-1)/2}\left(\frac{q}{p}\right) = (-1)^{(p-1)/2}\left(\frac{p}{q}\right)$. Then $A = \pm 1$, so that $A \equiv \pm 1 \pmod{pq}$. Conversely, suppose that $A \equiv 1 \pmod{pq}$. Then $A \equiv 1 \pmod{p}$ and $A \equiv 1 \pmod{q}$. By parts (b) and (c), we have $(-1)^{(q-1)/2}\left(\frac{q}{p}\right) = A = (-1)^{(p-1)/2}\left(\frac{p}{q}\right)$. The same argument works if $A \equiv -1 \pmod{pq}$.

e. If a is an integer in R, it satisfies $1 \le a \le (pq-1)/2$. Therefore, its additive inverse modulo pq is in the range $(pq+1)/2 \le -a \le pq - 1$ and in the set of reduced residue classes. By the Chinese remainder theorem, the congruence $a^2 \equiv 1 \pmod{pq}$ has exactly four solutions, $1, -1, b,$ and $-b$ $\pmod{pq}$ and the congruence $a^2 \equiv -1 \pmod{pq}$ has solutions if and only if $p \equiv q \equiv 1 \pmod 4$, and in this case, it has exactly four solutions, $i, -i, ib,$ and $-ib \pmod{pq}$. For each element $a \in R$, $(a, pq) = 1$, so a has a multiplicative inverse u. It follows that exactly one of $u, -u$ is in R. Let $U = \{a \in R \mid a^2 \equiv \pm 1 \pmod{pq}\}$. Then when we compute A, all other elements will be paired with an element that is either its inverse or the negative of its inverse. Thus, $A \equiv \prod\limits_{a \in R} a \equiv \prod\limits_{a \in U} a$ $\pmod{pq}$. So if $p \equiv q \equiv 1 \pmod{pq}$, then $A \equiv \prod\limits_{a \in U} a \equiv \pm(1 \cdot c \cdot i \cdot ic) \equiv c^2 i^2 \equiv \mp 1 \pmod{pq}$. Conversely, in the other case, $A \equiv \prod\limits_{a \in U} a \equiv \pm(1 \cdot c) \not\equiv \pm 1 \pmod{pq}$, which completes the proof.

f. By parts (d) and (e), we have $(-1)^{(q-1)/2}\left(\frac{q}{p}\right) = (-1)^{(p-1)/2}\left(\frac{p}{q}\right)$ if and only if $p \equiv q \equiv 1 \pmod 4$. So if $p \equiv q \equiv 1 \pmod 4$, we have $\left(\frac{q}{p}\right) = \left(\frac{p}{q}\right)$. But if $p \equiv 1 \pmod 4$ while $q \equiv 3 \pmod 4$, then we must have $-\left(\frac{q}{p}\right) \ne \left(\frac{p}{q}\right)$, which means we must change the sign, yielding $\left(\frac{q}{p}\right) = \left(\frac{p}{q}\right)$. The case where $p \equiv 3 \pmod 4$ but $q \equiv 1 \pmod 4$ is identical. If $p \equiv q \equiv 3 \pmod 4$, then we have $-\left(\frac{q}{p}\right) \ne -\left(\frac{p}{q}\right)$ so that we have $-\left(\frac{q}{p}\right) = \left(\frac{p}{q}\right)$, which concludes the proof.

Section 11.3

1. **a.** 1 **b.** −1 **c.** 1 **d.** 1 **e.** −1 **f.** 1

3. 1, 7, 13, 17, 19, 29, 37, 49, 71, 83, 91, 101, 103, 107, 113, or 119 (mod 120)

5. The pseudo-squares modulo 21 are 5, 17, and 20.

7. The pseudo-squares modulo 143 are 1, 3, 4, 9, 12, 14, 16, 23, 25, 27, 36, 38, 42, 48, 49, 52, 56, 64, 69, 75, 81, 82, 92, 100, 103, 108, 113, 114, 126, and 133.

9. Since n is odd and square-free, n has prime factorization $n = p_1 p_2 \cdots p_r$. Let b be one of the $(p-1)/2$ quadratic nonresidues of p_1, so that $\left(\frac{b}{p_1}\right) = -1$. By the Chinese remainder theorem, let a be a solution to the system of linear congruences $x \equiv b \pmod{p_1}$ and $x \equiv 1 \pmod{p_i}$, $2 \le i \le r$. Then $\left(\frac{a}{p_1}\right) = \left(\frac{b}{p_1}\right) = -1$, $\left(\frac{a}{p_2}\right) = \left(\frac{1}{p_2}\right) = 1$, $\dots$, $\left(\frac{a}{p_r}\right) = \left(\frac{1}{p_r}\right) = 1$. Therefore, $\left(\frac{a}{n}\right) = \left(\frac{a}{p_1}\right)\left(\frac{a}{p_2}\right) \cdots \left(\frac{a}{p_r}\right) = (-1) \cdot 1 \cdots 1 = -1$.

11. a. Note that $(a, b) = (b, r_1) = (r_1, r_2) = \cdots = (r_{n-1}, r_n) = 1$, and, since the q_i are even, the r_i are odd. Since $r_0 = b$ and $a \equiv \epsilon_1 r_1 \pmod{b}$, $\left(\frac{a}{b}\right) = \left(\frac{\epsilon_1 r_1}{r_0}\right) = \left(\frac{\epsilon_1}{r_0}\right)\left(\frac{r_1}{r_0}\right) = \left(\frac{\epsilon_1}{r_0}\right)\left(\frac{r_0}{r_1}\right) \cdot$ $(-1)^{((r_0-1)/2)((r_1-1)/2)}$ by Theorem 11.11. If $\epsilon_1 = 1$, then $\left(\frac{a}{b}\right) = (-1)^{((r_0-1)/2)((\epsilon_1 r_1 - 1)/2)}\left(\frac{r_0}{r_1}\right)$. If $\epsilon_1 = -1$, then $\left(\frac{\epsilon_1}{r_0}\right) = (-1)^{((r_0-1)/2)}$, and so we have $\left(\frac{a}{b}\right) = (-1)^{((r_0-1)/2)((r_1+1)/2)}\left(\frac{r_0}{r_1}\right) = (-1)^{((r_0-1)/2)((-r_1-1)/2)}\left(\frac{r_0}{r_1}\right) = (-1)^{((r_0-1)/2)((\epsilon_1 r_1 - 1)/2)}\left(\frac{r_0}{r_1}\right)$ since $(r_1 + 1)/2$ and $-(r_1 - 1)/2$ have the same parity. Similarly, $\left(\frac{r_0}{r_1}\right) = (-1)^{((r_1-1)/2)((\epsilon_2 r_2 - 1)/2)}\left(\frac{r_1}{r_2}\right)$, so that $\left(\frac{a}{b}\right) = (-1)^{((r_0-1)/2)((\epsilon_1 r_1 - 1)/2)+((r_1-1)/2)((\epsilon_2 r_2 - 1)/2)}\left(\frac{r_1}{r_2}\right)$. Proceed inductively until the last step, when $\left(\frac{r_n}{r_{n-1}}\right) = \left(\frac{1}{r_{n-1}}\right) = 1$.

b. If either $r_{i-1} \equiv 1 \pmod{4}$ or $\epsilon_i r_i \equiv 1 \pmod 4$, then $((r_{i-1} - 1)/2)((\epsilon_i r_i - 1)/2)$ is even. Otherwise, that is, if $r_{i-1} \equiv \epsilon_i r_i \equiv 3 \pmod 4$, then $((r_{i-1} - 1)/2)((\epsilon_i r_i - 1)/2)$ is odd. Then $((r_{n-1} - 1)/2)((\epsilon_n r_n - 1)/2)$, the exponent in part (a), is even or odd as T is even or odd.

13. a. -1 **b.** -1 **c.** -1

15. Let $n_1 = p_1^{a_1} p_2^{a_2} \cdots p_r^{a_r}$ and $n_2 = q_1^{b_1} q_2^{b_2} \cdots q_s^{b_s}$ be the prime factorizations of n_1 and n_2. Then by the definition of the Kronecker symbol, $\left(\frac{a}{n_1 n_2}\right) = \left(\frac{a}{p_1}\right)^{a_1} \cdots \left(\frac{a}{p_r}\right)^{a_r}\left(\frac{a}{q_1}\right)^{b_1} \cdots \left(\frac{a}{q_s}\right)^{b_s} = \left(\frac{a}{n_1}\right)\left(\frac{a}{n_2}\right)$.

17. If $a \equiv 1 \pmod 4$, then by Exercise 16, we have $\left(\frac{a}{n_1}\right) = \left(\frac{n_1}{|a|}\right)$. This last is equivalent to the Jacobi symbol, so by Theorem 11.10(i), we have $\left(\frac{n_1}{|a|}\right) = \left(\frac{n_2}{|a|}\right) = \left(\frac{a}{n_2}\right)$, using Exercise 16 again. If $a \equiv 0 \pmod 4$, say $a = 2^s t$ with t odd and $s \geq 2$, then Exercise 16 gives $\left(\frac{a}{n_1}\right) = \left(\frac{2}{n_1}\right)^s (-1)^{((t-1)/2)\cdot((n_1-1)/2)}\left(\frac{n_1}{|a|}\right)$ and $\left(\frac{a}{n_2}\right) = \left(\frac{2}{n_2}\right)^s (-1)^{((t-1)/2)\cdot((n_2-1)/2)}\left(\frac{n_2}{|t|}\right)$. Since $n_1 \equiv n_2 \pmod{|t|}$, we have $\left(\frac{n_1}{|t|}\right) = \left(\frac{n_2}{|t|}\right)$, and since $4 \mid a$, we have $n_1 \equiv n_2 \pmod 4$, and so $(-1)^{((t-1)/2)\cdot((n_1-1)/2)} = (-1)^{((t-1)/2)\cdot((n_2-1)/2)}$. If $s = 2$, then certainly $\left(\frac{2}{n_1}\right)^2 = \left(\frac{2}{n_2}\right)^2$. If $s > 2$, then $8 \mid a$ and $n_1 \equiv n_2 \pmod 8$, and hence $n_1^2 \equiv n_2^2 \pmod 8$. So $\left(\frac{2}{n_1}\right) = (-1)^{(n_1^2-1)/8} = (-1)^{(n_2^2-1)/8} = \left(\frac{2}{n_2}\right)$. Therefore, $\left(\frac{a}{n_1}\right) = \left(\frac{a}{n_2}\right)$.

19. If $a \equiv 1 \pmod 4$, then $|a| \equiv 1 \pmod 4$ if $a > 0$ and $|a| \equiv -1 \pmod 4$ if $a < 0$, so by Exercise 16, $\left(\frac{a}{|a|-1}\right) = \left(\frac{|a|-1}{|a|}\right) = \left(\frac{-1}{|a|}\right) = (-1)^{\frac{|a|-1}{2}} = 1$ if $a > 0$ and $= -1$ if $a < 0$. If $a \equiv 0 \pmod 4$, then $a = 2^s t$ with t odd and $t \geq 2$, so by Exercise 16, $\left(\frac{a}{|a|-1}\right) = \left(\frac{2}{|a|-1}\right)^s (-1)^{\frac{t-1}{2}}\left(\frac{|a|-1}{|t|}\right)$. Since $s \geq 2$, check that $\left(\frac{2}{|a|-1}\right)^s = 1$. (Indeed, $|a| - 1 \equiv 7 \pmod 8$ if $s > 2$.) Also $(-1)^{\frac{t-1}{2}}\left(\frac{|a|-1}{|t|}\right) = (-1)^{\frac{t-1}{2}}\left(\frac{-1}{|t|}\right) = (-1)^{\frac{t-1}{2}+\frac{|t|-1}{2}} = 1$ if $t > 0$ and $= -1$ if $t < 0$.

Section 11.4

1. We have $2^{(561-1)/2} = 2^{280} = (2^{10})^{28} \equiv (-98)^{28} \equiv ((-98)^2)^{14} \equiv 67^{14} \equiv (67^2)^7 \equiv 1^7 = 1 \pmod{561}$. Furthermore, we see that $\left(\frac{2}{561}\right) = 1$ since $561 \equiv 1 \pmod 8$. But $561 = 3 \cdot 11 \cdot 17$ is not prime.

3. Suppose that n is an Euler pseudoprime to both the bases a and b. Then $a^{(n-1)/2} \equiv \left(\frac{a}{n}\right)$ (mod n) and $b^{(n-1)/2} \equiv \left(\frac{b}{n}\right)$ (mod n). It follows that $(ab)^{(n-1)/2} \equiv \left(\frac{a}{n}\right)\left(\frac{b}{n}\right) = \left(\frac{ab}{n}\right)$. Hence, n is an Euler pseudoprime to the base ab.

5. Suppose that $n \equiv 5$ (mod 8) and n is an Euler pseudoprime to the base 2. Since $n \equiv 5$ (mod 8) we have $\left(\frac{2}{n}\right) = -1$. Since n is an Euler pseudoprime to the base 2, we have $2^{(n-1)/2} \equiv \left(\frac{2}{n}\right) = -1$ (mod n). Write $n - 1 = 2^t t$, where t is odd. Since $2^{(n-1)/2} \equiv 2^{2t} \equiv -1$ (mod n), n is a strong pseudoprime to the base 2.

7. $n \equiv 5$ (mod 40)

9. 80

Section 11.5

1. 1229

3. Since $p, q \equiv 3$ (mod 4), we know that -1 is not a quadratic residue modulo p or q. If the four square roots are found using the method in Example 11.19, then only one of each possibility for choosing $+$ or $-$ can yield a quadratic residue in each congruence, so there is only one system that results in a square.

5. If Paula chooses $c = 13$, then $v = 713$, which is a quadratic residue of 1411, and which has square root $u \equiv 837$ (mod 1411). Her random number is 822, so she computes $x \equiv 822^2 \equiv 1226$ (mod 1411) and $y \equiv v\overline{x} \equiv 713 \cdot 961 \equiv 858$ (mod 1411). She sends $x = 1226$, $y = 858$ to Vince. Vince checks that $xy \equiv 1226 \cdot 858 \equiv 713$ (mod 1411) and then sends the bit $b = 1$ to Paula, so she computes $\overline{r} \equiv \overline{822} \equiv 1193$ (mod 1411) and then $u\overline{r} \equiv 837 \cdot 1193 \equiv 964$ (mod 1411), which she sends to Vince. Since Vince sent $b = 1$, he computes $964^2 \equiv 858$ (mod 1411) and notes that it is indeed equal to y.

7. The prover sends $x = 1403^2 = 1968409 \equiv 519$ (mod 2491). The verifier sends $\{1, 5\}$. The prover sends $y = 1425$. The verifier computes $y^2 \cdot s_1 \cdot s_5 = 1425^2 \cdot 197 \cdot 494 \equiv 519 \equiv x$ (mod 2491).

9. **a.** 959, 1730, 2895, 441, 2900, 2684   **b.** 1074   **c.** $1074^2 \cdot 959 \cdot 1730 \cdot 441 \cdot 2684 \equiv 336 \equiv 403^2$ (mod 3953)

11. If Paula sends back a to Vince, then $a^2 \equiv w^2$ (mod n), with $a \not\equiv w$ (mod n). Then $a^2 - w^2 = (a - w)(a + w) \equiv 0$ (mod n). By computing $(a - w, n)$ and $(a + w, n)$, Vince can produce a nontrivial factor of n.

Section 12.1

1. **a.** $.4$ **b.** $.41\overline{6}$ **c.** $.\overline{923076}$ **d.** $.5\overline{3}$ **e.** $.00\overline{9}$ **f.** $.\overline{000999}$

3. **a.** 3/25   **b.** 11/90   **c.** 4/33

5. $b = 2^r 3^s 5^t 7^u$, with r, s, t, and u positive integers

7. **a.** pre-period 1, period 0   **b.** pre-period 2, period 0   **c.** pre-period 1, period 4
   d. pre-period 2, period 0   **e.** pre-period 11, period 1   **f.** pre-period 2, period 4

9. **a.** 3   **b.** 11   **c.** 37   **d.** 101   **e.** 41, 271   **f.** 7, 13

11. Using the construction from Theorem 12.2 and Example 12.1, we show by mathematical induction that $c_k = k - 1$ and $\gamma_k = (kb - k + 1)/(b - 1)^2$. The inductive step is as follows: $c_{k+1} = [b\gamma_k] = [(kb^2 - bk + b)/(b - 1)^2] = [(k(b - 1)^2 + b(k + 1) - k)/(b - 1)^2] = [k + (b(k + 1) - k)/(b - 1)^2] = k$, and $\gamma_{k+1} = (k + 1)b - k$, if $k \neq b - 2$. If $k = b - 2$, then

$c_{b-2} = b$, so we have determined $b - 1$ consecutive digits of the expansion. From the binomial theorem, $(x + 1)^a \equiv ax + 1 \pmod{x^2}$, so $\operatorname{ord}_{(b-1)^2} b = b - 1$, which is the period length. Therefore, we have determined the entire expansion.

13. The base b expansion is $(.100100001\ldots)_b$, which is nonrepeating and therefore, by Theorem 12.4, represents an irrational number.

15. Let γ be a real number. Set $c_0 = [\gamma]$ and $\gamma_1 = \gamma - c_0$. Then $0 \le \gamma_1 < 1$ and $\gamma = c_0 + \gamma_1$. From the condition that $c_k < k$ for $k = 1, 2, 3, \ldots$, we must have $c_1 = 0$. Let $c_2 = [2\gamma_1]$ and $\gamma_2 = 2\gamma_1 - c_2$. Then $\gamma_1 = (c_2 + \gamma_2)/2$, so $\gamma = c_0 + c_1/1! + c_2/2! + \gamma_2/2!$. Now let $c_3 = [3\gamma_2]$ and $\gamma_3 = 3\gamma_2 - c_3$. Then $\gamma_2 = (c_3 + \gamma_3)/3$ and so $\gamma = c_0 + c_1/1! + c_2/2! + c_3/3! + \gamma_3/3!$. Continuing in this fashion, for each $k = 2, 3, \ldots$, define $c_k = [k\gamma_{k-1}]$ and $\gamma_k = k\gamma_{k-1} - c_k$. Then $\gamma = c_0 + c_1/1! + c_2/2! + c_3/3! + \cdots + c_k/k! + \gamma_k/k!$. Since each $\gamma_k < 1$, we know that $\lim_{k\to\infty} \gamma_k/k! = 0$, so we conclude that $\gamma = c_0 + c_1/1! + c_2/2! + c_3/3! + \cdots + c_k/k! + \cdots$.

17. In the proof of Theorem 12.2, the numbers $p\gamma_n$ are the remainders of b^n upon division by p. The process recurs as soon as some γ_i repeats a value. Since $1/p = (.\overline{c_1 c_2 \ldots c_{p-1}})$ has period length $p - 1$, we have, by Theorem 12.4, that $\operatorname{ord}_p b = p - 1$, so there is an integer k such that $b^k \equiv m \pmod{p}$. So the remainders of mb^n upon division by p are the same as the remainders of $b^k b^n$ upon division by p. Hence, the nth digit of the expansion of m/p is determined by the remainder of b^{k+n} upon division by p. Therefore, it will be the same as the $(k + n)$th digit of $1/p$.

19. n must be prime with 2 a primitive root.

21. Let $\gamma b^{j-1} = a + \epsilon$, where a is an integer and $0 \le \epsilon < 1$. Then $[\gamma b^j] - b[\gamma b^{j-1}] = [(a + \epsilon)b] - b[a + \epsilon] = ab + [\epsilon b] - ab = [\epsilon b]$. Since $0 \le \epsilon < 1$, this last expression is an integer between 0 and $b - 1$. Therefore, $0 \le [\gamma b^j] - b[\gamma b^{j-1}] \le b - 1$. Now consider the sum $\sum_{j=1}^{N}([\gamma b^j] - b[\gamma b^{j-1}])/b^j$. Factor out $1/b^N$ to clear fractions and this becomes $(1/b^N)\sum_{j=1}^{N}(b^{N-j}[\gamma b^j] - b^{N-(j-1)}[\gamma b^{j-1}])$. This sum telescopes to $(-b^N[\gamma] + [\gamma b^N])/b^N = [\gamma b^N]/b^N$ since $[\gamma] = 0$. But $[\gamma b^N]/b^N = (\gamma b^N - \gamma b^N + [\gamma b^N])/b^N = \gamma - (\gamma b^N - [\gamma b^N])/b^N$. But $0 \le \gamma b^N - [\gamma b^N] < 1$, so taking limits as $N \to \infty$ of both sides of this equation yields $\gamma = \sum_{j=1}^{\infty}([\gamma b^j] - b[\gamma b^{j-1}])/b^j$. By the uniqueness of the base b expansion given in Theorem 12.1, we must have $c_j = [\gamma b^j] - b[\gamma b^{j-1}]$ for each j.

23. Let $\alpha = \sum_{i=1}^{\infty}(-1)^{a_i}/10^{i!}$ and $p_k/q_k = \sum_{i=1}^{k}(-1)^{a_i}/10^{i!}$. Then $\left|\alpha - p_k/q_k\right| = \left|\sum_{i=k+1}^{\infty}(-1)^{a_i}/10^{i!}\right| \le \sum_{i=k+1}^{\infty} 1/10^{i!}$. As in the proof of Corollary 12.5.1, it follows that $\left|\alpha - p_k/q_k\right| < 2/10^{(k+1)!}$, which shows that there can be no real number C, as in Theorem 12.5. Hence, α must be transcendental.

25. Suppose that $e = h/k$. Then $k!(e - 1 - 1/1! - 1/2! - \cdots - 1/k!)$ is an integer. But this is equal to $k!(1/(k + 1)! + 1/(k + 2)! + \cdots) = 1/(k + 1) + 1/((k + 1)(k + 2)) + \cdots < 1/(k + 1) + 1/(k + 1)^2 + \cdots = 1/k < 1$. But $k!(e - 1 - 1/1! - 1/2! - \cdots - 1/k!)$ is positive, and therefore cannot be an integer, a contradiction.

Section 12.2

1. **a.** 15/7   **b.** 10/7   **c.** 6/31   **d.** 355/113   **e.** 2   **f.** 3/2   **g.** 5/3   **h.** 8/5

3. **a.** $[1; 2, 1, 1, 2]$ **b.** $[1; 1, 7, 2]$ **c.** $[2; 9]$ **d.** $[3; 7, 1, 1, 1, 1, 2]$ **e.** $[-1; 13, 1, 1, 2, 1, 1, 2, 2]$
   f. $[0, 9, 1, 3, 6, 2, 4, 1, 2]$

5. **a.** 1, 3/2, 4/3, 7/5, 18/13   **b.** 1, 2, 15/8, 32/17   **c.** 2, 19/9
   d. 3, 22/7, 25/8, 47/15, 72/23, 119/38, 310/99
   e. -1, $-12/13$, $-13/14$, $-25/27$, $-63/68$, $-88/95$, $-151/163$, $-390/421$, $-931/1005$
   f. 0, 1/9, 1/10, 4/39, 25/244, 54/527, 241/2352, 295/2879, 831/8110

7. a. $3/2 > 7/5$ and $1 < 4/3 < 18/13$ **b.** $2 > 32/17$ and $1 < 15/8$ **c.** vacuous
d. $22/7 > 47/15 > 119/38$ and $3 < 25/8 < 72/23 < 310/99$
e. $-12/13 > -25/27 > -88/95 > -390/421$ and $-1 < -13/14 < -63/68 < -151/163 < -931/1005$
f. $1/9 > 4/39 > 54/527 > 295/2879$ and $0 < 1/10 < 25/244 < 241/2352 < 831/8110$

9. Let $\alpha = r/s$. The Euclidean algorithm for $1/\alpha = s/r < 1$ gives $s = 0(r) + s$, $r = a_0(s) + a_1$, and continues just as for r/s.

11. Proceed by induction. The basis case is trivial. Assume that $q_j \geq f_j$ for $j < k$. Then $q_k = a_k q_{k-1} + q_{k-2} \geq a_k f_{k-1} + f_{k-2} \geq f_{k-1} + f_{k-2} = f_k$, as desired.

13. By Exercise 10, we have $p_n/p_{n-1} = [a_n; a_{n-1}, \ldots, a_0] = [a_0; a_1, \ldots, a_n] = p_n/q_n = r/s$ if the continued fraction is symmetric. Then $q_n = p_{n-1} = s$ and $p_n = r$, so by Theorem 12.10, we have $p_n q_{n-1} - q_n p_{n-1} = r q_{n-1} - s^2 = (-1)^{n-1}$. Then $r q_{n-1} = s^2 + (-1)^{n-1}$, and so $r \mid s^2 + (-1)^{n-1}$. Conversely, if $r \mid s^2 + (-1)^{n-1}$, then $(-1)n - 1 = p_n q_{n-1} - q_n p_{n-1} = r q_{n-1} - p_{n-1} s$. Therefore, $r \mid p_{n-1} s + (-1)^{n-1}$, and hence, $r \mid (s^2 + (-1)^{n-1}) - (p_{n-1} s + (-1)^{n-1}) = s(s - p_{n-1})$. Since $s, p_{n-1} < r$ and $(r, s) = 1$, we have $s = p_{n-1}$. Then $[a_n; a_{n-1}, \ldots, a_0] = p_n/p_{n-1} = r/s = [a_0; a_1, \ldots, a_n]$.

15. Note that the notation $[a_0; a_1, \ldots, a_n]$ makes sense, even if the a_j are not integers. Use induction. Assume that the statement is true for k odd and prove it for $k + 2$. Define $a_k' = [a_k; a_{k+1}, a_{k+2}]$ and check that $a_k' < [a_k; a_{k+1}, a_{k+2} + x] = a_k' + x'$. Then $[a_0; a_1, \ldots, a_{k+2}] = [a_0; a_1, \ldots, a_k'] > [a_0; a_1, \ldots, a_k' + x'] = [a_0; a_1, \ldots, a_{k+2} + x]$. Proceed similarly for k even.

Section 12.3

1. a. $[1; 2, 2, 2, \ldots]$ **b.** $[1; 1, 2, 1, 2, \ldots]$ **c.** $[2; 4, 4, 4, \ldots]$ **d.** $[1; 1, 1, 1, \ldots]$

3. $312689/99532$

5. If $a_1 > 1$, let $A = [a_2; a_3, \ldots]$. Then $[a_0; a_1, \ldots] + [-a_0 - 1; 1, a_1 - 1, a_2, a_3, \ldots] = a_0 + 1/(a_1 + 1/A) + (-a_0 - 1 + 1/(1 + 1/(a_1 - 1 + 1/A))) = 0$. Similarly if $a_1 = 1$.

7. If $\alpha = [a_0; a_1, a_2, \ldots]$, then $1/\alpha = 1/[a_0; a_1, a_2, \ldots] = 0 + 1/(a_0 + 1/(a_1 + \cdots)) = [0; a_0, a_1, a_2, \ldots]$. Then the kth convergent of $1/\alpha$ is $[0; a_0, a_1, a_2, \ldots, a_{k-1}] = 1/[a_0; a_1, a_2, \ldots, a_{k-1}]$, which is the reciprocal of the $(k-1)$th convergent of α.

9. By Theorem 12.19, such a p/q is a convergent of α. Now $(\sqrt{5} + 1)/2 = [1; 1, 1, \ldots]$, so $q_n = f_n$ (Fibonacci) and $p_n = q_{n+1}$. Then $\lim_{n \to \infty} q_{n-1}/q_n = \lim_{n \to \infty} q_{n-1}/p_{n-1} = 2/(\sqrt{5} + 1) = (\sqrt{5} - 1)/2$. Therefore, $\lim_{n \to \infty}((\sqrt{5} + 1)/2 + q_{n-1}/q_n) = (\sqrt{5} + 1)/2 + (\sqrt{5} - 1)/2 = \sqrt{5}$. So $(\sqrt{5} + 1)/2 + q_{n-1}/q_n > c$ only finitely often, whence $1/(((\sqrt{5} + 1)/2 + q_{n-1}/q_n)q_n^2) < 1/(cq_n^2)$. The following identity finishes the proof. Note that $\alpha_n = \alpha$ for all n. Then $|\alpha - p_n/q_n| = |(\alpha_{n+1} p_n + p_{n-1})/(\alpha_{n+1} q_n + q_{n-1}) - p_n/q_n| = |-(p_n q_{n-1} - p_{n-1} q_n)/(q_n(\alpha q_n + q_{n-1}))| = 1/(q_n^2(\alpha + q_{n-1}/q_n))$.

11. If β is equivalent to α, then $\beta = (a\alpha + b)/(c\alpha + d)$. Solving for α gives $\alpha = (-d\beta + b)/(c\beta - a)$, so α is equivalent to β.

13. By symmetry and transitivity (Exercises 11 and 12), it suffices to show that every rational number $\alpha = m/n$ (which we can assume is in lowest terms) is equivalent to 1. By the Euclidean algorithm, we can find a and b such that $ma + nb = 1$. Let $d = m + b$ and $c = a - n$. Then $(a\alpha + b)/(c\alpha + d) = 1$.

15. Note that $p_{k,t} q_{k-1} - q_{k,t} p_{k-1} = t(p_{k-1} q_{k-1} - q_{k-1} p_{k-1}) + (p_{k-2} q_{k-1} - p_{k-1} q_{k-2}) = \pm 1$. Thus $p_{k,t}$ and $q_{k,t}$ are relatively prime.

17. See, for example, the classic work by O. Perron, *Die Lehre von den Kettenbrüchen*, Leipzig, Teubner (1929).

19. 179/57

21. Note first that if $b < d$, then $|a/b - c/d| < 1/2d^2$ implies that $|ad - bc| < b/2d < 1/2$. Because $b \neq d$, $|ad - bc|$ is a positive integer, and so is greater than $1/2$. Thus $b \geq d$. Now assume that c/d is not a convergent of the continued fraction for a/b. Since the denominators of the convergents increase to b, there must be two successive convergents p_n/q_n and p_{n+1}/q_{n+1} such that $q_n < d < q_{n+1}$. Next, by the triangle inequality, $\frac{1}{2d^2} > \left| \frac{a}{b} - \frac{c}{d} \right| = \left| \frac{c}{d} - \frac{p_n}{q_n} \right| - \left| \frac{a}{b} - \frac{p_n}{q_n} \right| \geq$ $\left| \frac{c}{d} - \frac{p_n}{q_n} \right| - \left| \frac{p_{n+1}}{q_{n+1}} - \frac{p_n}{q_n} \right|$, because the $(n+1)$st convergent is on the other side of a/b from the nth convergent. Because the numerator of the first difference is a nonzero integer, applying Corollary 12.10.2 to the second difference shows that the last expression is greater than or equal to $1/dq_n - 1/q_{n+1}q_n$. If we multiply through by d^2, we obtain $\frac{1}{2} > \frac{d}{q_n} \left(1 - \frac{d}{q_{n+1}} \right) > 1 - \frac{d}{q_{n+1}}$ because $d/q_n > 1$. We deduce that $1/2 < d/q_{n+1}$.

    The convergents p_n/q_n and p_{n+1}/q_{n+1} divide the line into three regions. As c/d could be in any of these, there are three cases.

Case 1: If c/d is between the convergents, then $\frac{1}{dq_n} \leq \left| \frac{c}{d} - \frac{p_n}{q_n} \right|$, because the numerator of the fraction is a positive integer and the denominators on both sides of the inequality are the same. This last term is less than or equal to $\left| \frac{p_{n+1}}{q_{n+1}} - \frac{p_n}{q_n} \right| = \frac{1}{q_{n+1}q_n}$, because the $n + 1$st convergent is further from the nth convergent than c/d, where we have applied Corollary 12.10.2. But this implies that $d \geq q_{n+1}$, a contradiction.

Case 2: If c/d is closer to p_n/q_n, then we also have $\frac{1}{dq_n} \leq \left| \frac{c}{d} - \frac{p_n}{q_n} \right| \leq \left| \frac{a}{b} - \frac{c}{d} \right|$, because a/b is on the other side of the nth convergent from c/d. But this last term is less than $1/2d^2$, and if we multiply through by d we have $1/q_n < 1/2d$, which implies that $q_n > d$, a contradiction.

Case 3: If c/d is closer to p_{n+1}/q_{n+1}, then with the same reasoning as in Case 2, we have $\frac{1}{dq_{n+1}} \leq \left| \frac{c}{d} - \frac{p_{n+1}}{q_{n+1}} \right| < \left| \frac{a}{b} - \frac{c}{d} \right| < 1/2d^2$. But this implies that $d/q_{n+1} < 1/2$, contradicting the inequality established above. Having exhausted all the cases, we must conclude that c/d must be a convergent of the continued fraction for a/b.

Section 12.4

1. **a.** $[2; \overline{1, 1, 1, 4}]$ **b.** $[3; \overline{3, 6}]$ **c.** $[4; \overline{1, 3, 1, 8}]$ **d.** $[6; \overline{1, 5, 1, 12}]$ **e.** $[7; \overline{1, 2, 7, 2, 1, 14}]$
   f. $[9; \overline{1, 2, 3, 1, 1, 5, 1, 8, 1, 5, 1, 1, 3, 2, 1, 18}]$

3. **a.** $[2; \overline{2}]$ **b.** $[1; \overline{2, 2, 2, 1, 12, 1}]$ **c.** $[0; 1, 1, \overline{2, 3, 10, 3}]$

5. **a.** $(23 + \sqrt{29})/10$ **b.** $(-1 + 3\sqrt{5})/2$ **c.** $(8 + \sqrt{82})/6$

7. **a.** $\sqrt{10}$ **b.** $\sqrt{17}$ **c.** $\sqrt{26}$ **d.** $\sqrt{37}$

9. **a.** We have $\alpha_0 = \sqrt{d^2 - 1}$, $a_0 = d - 1$, $P_0 = 0$, $Q_0 = 1$, $P_1 = d - 1$, $Q_1 = 2d - 2$, $\alpha_1 = \frac{1}{2} + \frac{1}{2}\sqrt{(d+1)/(d-1)}$, $a_1 = 1$, $P_2 = d - 1$, $Q_2 = 1$, $\alpha_2 = d - 1 + \sqrt{d^2 - 1}$, $a_2 = 2d - 2$, $P_3 = d - 1 = P_1$, $Q_3 = 2d - 2 = Q_1$, so $\alpha = [d - 1; \overline{1, 2(d-1)}]$.
   b. We have $\alpha_0 = \sqrt{d^2 - d}$, $a_0 = d - 1$ since $(d-1)^2 < d^2 - d < d^2$. Then $P_0 = 0$, $Q_0 = 1$, $P_1 = d - 1$, $Q_1 = d - 1$, $\alpha_1 = 1 + \sqrt{d/(d-1)}$, $a_1 = 2$, $P_2 = d - 1$, $Q_2 = 1$, $\alpha_2 = (d-1) + \sqrt{d^2 - d}$, $a_2 = 2(d - 1)$, $P_3 = P_1$, $Q_3 = Q_1$. Therefore, $\sqrt{d^2 - d} = [d - 1; \overline{2, 2(d-1)}]$.
   c. $[9; \overline{1, 18}]$, $[10; \overline{2, 20}]$, $[16; \overline{2, 32}]$, $[24; \overline{2, 48}]$

11. **a.** Note that $d < \sqrt{d^2 + 4} < d + 1$. Then $\alpha_0 = \sqrt{d^2 + 4}$, $a_0 = d$, $P_0 = 0$, $Q_0 = 1$, $P_1 = d$, $Q_1 = 4$, $\alpha_1 = (d + \sqrt{d^2 + 4})/4$, $a_1 = \lfloor 2d/4 \rfloor = (d - 1)/2$, since d is odd. Also, $P_2 = d - 2$,

$Q_2 = d$, $\alpha_2 = (d - 2 + \sqrt{d^2 + 4})/d$, $((d - 2) + d)/d < \alpha_2 < (d - 2 + d + 1)/d$, so $a_2 = 1$, $P_3 = 2$, $Q_3 = d$, $\alpha_3 = (2 + \sqrt{d^2 + 4})/d$, $a_3 = 1$, $P_4 = d - 2$, $Q_4 = 4$, $\alpha_4 = (d - 2 + \sqrt{d^2 + 4})/4$, $(d - 2 + d)/4 = (d - 1)/2 < \alpha_4 < (d - 2 + d + 1)/4$, so $a_4 = (d - 1)/2$, $P_5 = d$, $Q_5 = 1$, $\alpha_5 = d + \sqrt{d^2 + 4}$, $a_5 = 2d$, $P_6 = d = P_1$, $Q_6 = 4 = Q_1$. Thus $\alpha = [d; \overline{(d - 1)/2, 1, 1, (d - 1)/2, 2d}]$.
b. Note that $d - 1 < \sqrt{d^2 - 4} < d$. Then $\alpha_0 = \sqrt{d^2 - 4}$, $a_0 = d - 1$, $P_0 = 0$, $Q_0 = 1$, $P_1 = d - 1$, $Q_1 = 2d - 5$, $\alpha_1 = (d - 1 + \sqrt{d^2 - 4})/(2d - 5)$, $(d - 1 + d - 1)/(2d - 5) < \alpha_0 < (d - 1 + d)/(2d - 5)$ and $d > 3$, so $a_1 = 1$, $P_2 = d - 4$, $Q_2 = 4$, $a_2 = (d - 4 + \sqrt{d^2 - 4})/4$, $a_2 = (d - 3)/2$, $P_3 = d - 2$, $Q_3 = d - 2$, $\alpha_3 = (d - 2 + \sqrt{d^2 - 4})/(d - 2)$, $a_3 = 2$, $P_4 = d - 2$, $Q_4 = 4$, $\alpha_4 = (d - 2 + \sqrt{d^2 - 4})/4$, $a_4 = (d - 3)/2$, $P_5 = d - 4$, $Q_5 = 2d - 5$, $\alpha_5 = (d - 4 + \sqrt{d^2 - 4})/(2d - 5)$, $a_5 = 1$, $P_6 = d - 1$, $Q_6 = 1$, $\alpha_6 = d - 1 + \sqrt{d^2 - 4}$, $a_6 = 2d - 2$, $P_7 = d - 1 = P_1$, $Q_7 = 2d - 5 = Q_1$. Thus $\alpha = [d - 1; \overline{1, (d - 3)/2, 2, (d - 3)/2, 1, 2d - 2}]$.

13. Suppose that $\sqrt{d}$ has period length two. Then $\sqrt{d} = [a; \overline{c, 2a}]$ from the discussion preceding Example 12.15. Then $\sqrt{d} = [a; y]$ with $y = [\overline{c; 2a}] = [c; 2a, y] = c + 1/(2a + 1/y) = (2acy + c + y)/(2ay + 1)$. Then $2ay^2 - 2acy - c = 0$, and since y is positive, $y = (2ac + \sqrt{(2ac)^2 + 4(2a)c})/(4a) = (ac + \sqrt{(ac)^2 + 2ac})/(2a)$. Then $\sqrt{d} = [a; y] = a + 1/y = a + 2a/(ac + \sqrt{(ac)^2 + 2ac}) = \sqrt{a^2 + 2a/c}$, so $d = a^2 + 2a/c$, and $b = 2a/c$ is an integral divisor of $2a$. Conversely, let $\alpha = \sqrt{a^2 + b}$ and $b \mid 2a$, say $kb = 2a$. Then $a_0 = [\sqrt{a^2 + b}] = a$, since $a^2 < a^2 + b < (a + 1)^2$. Then $P_0 = 0$, $Q_0 = 1$, $P_1 = a$, $Q_1 = b$, $\alpha_1 = (a + \sqrt{a^2 + b})/b$, $a_1 = 4k$, $P_2 = a$, $Q_2 = 1$, $\alpha_2 = a + \sqrt{a^2 + b}$, $a_2 = 2a$, $P_3 = a = P_1$, $Q_3 = b = Q_1$, so $\alpha = [a; \overline{4k, 2a}]$, which has period length two.

15. **a.** no   **b.** yes   **c.** yes   **d.** no   **e.** yes   **f.** no

17. Let $\alpha = (a + \sqrt{b})/c$. Then $-1/\alpha' = -c/(a - \sqrt{b}) = (ca + \sqrt{bc^2})/(b - a^2) = (A + \sqrt{B})/C$, say. By Exercise 16, $0 < a < \sqrt{b}$ and $\sqrt{b} - a < c < \sqrt{b} + a < 2\sqrt{b}$. Multiplying by c gives $0 < ca < \sqrt{bc^2}$ and $\sqrt{bc^2} - ca < c^2 < \sqrt{bc^2} + ca < 2\sqrt{bc^2}$. That is, $0 < A < \sqrt{B}$ and $\sqrt{B} - A < c^2 < \sqrt{B} + A < 2\sqrt{B}$. Multiply $\sqrt{b} - a < c$ by $\sqrt{b} + a$ to get $C = b - a^2 < \sqrt{bc^2} + ca = A + \sqrt{B}$. Multiply $c < \sqrt{b} + a$ by $\sqrt{b} - a$ to get $\sqrt{B} - A = \sqrt{bc^2} - ac < b - a^2 = C$. So, $-1/\alpha'$ satisfies all the inequalities in Exercise 16 and therefore is reduced.

19. Start with $\alpha_0 = \sqrt{D_k} + 3^k + 1$ (this will have the same period since it differs from $\sqrt{D_k}$ by an integer) and use induction. Apply the continued fraction algorithm to show that $\alpha_{3i} = \sqrt{D_k} + 3^k - 2 \cdot 3^{k-i} + 2/(2 \cdot 3^{k-i})$ for $i = 1, 2, \ldots, k$, but $\alpha_{3k+3i} = \sqrt{D_k} + 3^k - 2/(2 \cdot 3^i)$ for $i = 1, 2, \ldots, k - 1$, and $\alpha_{6k} = \sqrt{D_k} + 3^k + 1 = \alpha_0$. Since $\alpha_i \neq \alpha_0$ for $i < 6k$, the period is $6k$.

Section 12.5

1. Note that $19^2 - 2^2 = (19 - 2)(19 + 2) \equiv 0 \pmod{119}$. Then $(19 - 2, 119) = (17, 119) = 17$ and $(19 + 2, 119) = (21, 119) = 7$ are factors of 119.

3. $3119 \cdot 4261$

5. We have $17^2 = 289 \equiv 3 \pmod{143}$ and $19^2 = 361 \equiv 3 \cdot 5^2 \pmod{143}$. Combining these, we have $(17 \cdot 19)^2 \equiv 3^2 5^2 \pmod{143}$. Hence, $323^2 \equiv 15^2 \pmod{143}$. It follows that $323^2 - 15^2 = (323 - 15)(323 + 15) \equiv 0 \pmod{143}$. This produces the two factors $(323 - 15, 143) = (308, 143) = 11$ and $(323 + 15, 143) = (338, 143) = 13$ of 143.

7. $3001 \cdot 4001$

Section 13.1

1. a. $(3, 4, 5)$, $(5, 12, 13)$, $(15, 8, 17)$, $(7, 24, 25)$, $(21, 20, 29)$, $(35, 12, 37)$
b. those in part (a) and $(6, 8, 10)$, $(9, 12, 15)$, $(12, 16, 20)$, $(15, 20, 25)$, $(18, 24, 30)$, $(21, 28, 35)$, $(24, 32, 40)$, $(10, 24, 26)$, $(15, 36, 39)$, $(30, 16, 34)$

3. By Lemma 13.1, 5 divides at most one of x, y, and z. If $5 \nmid x$ or y, then $x^2 \equiv \pm 1 \pmod 5$ and $y^2 \equiv \pm 1 \pmod 5$. Then $z^2 \equiv 0$, 2, or $-2 \pmod 5$. But ± 2 is not a quadratic residue modulo 5, so $z^2 \equiv 0 \pmod 5$, whence $5 \nmid z$.

5. Let k be an integer ≥ 3. If $k = 2n + 1$, let $m = n + 1$. Then m and n have opposite parity, $m > n$ and $m^2 - n^2 = 2n + 1 = k$, so m and n define the desired triple. If k has an odd divisor $d > 1$, then use the construction above for d and multiply the result by k/d. If k has no odd divisors, then $k = 2^j$ for some integer $j > 1$. Let $m = 2^{j-1}$ and $n = 1$. Then $k = 2mn$, $m > n$, and m and n have opposite parity, so m and n define the desired triple.

7. Substituting $y = x + 1$ into the Pythagorean equation gives us $2x^2 + 2x + 1 = z^2$, which is equivalent to $m^2 - 2z^2 = -1$, where $m = 2x + 1$. Dividing by z^2 yields $m^2/z^2 - 2 = -1/z^2$. Note that $m/z \geq 1$ and $1/z^2 = 2 - m^2/z^2 = (\sqrt 2 + m/z)(\sqrt 2 - m/z) < 2(\sqrt 2 - m/z)$. So by Theorem 12.19, m/z must be a convergent of the continued fraction expansion of $\sqrt 2$. Further, by the proof of Theorem 12.13, it must be one of even-subscripted convergents. Therefore, each solution is given by the recurrence $m_{n+1} = 3m_n + 2z_n$, $z_{n+1} = 2m_n + 3m_n$. (See, for example, Theorem 13.11.) Substituting x back in yields the recurrences of Exercise 6.

9. see Exercise 15 with $p = 3$

11. $(9, 12, 15)$, $(35, 12, 37)$, $(5, 12, 13)$, $(12, 16, 20)$

13. $x = 2m$, $y = m^2 - 1$, $z = m^2 + 1$, $m > 1$

15. primitive solutions given by $x = (m^2 - pn^2)/2$, $y = mn$, $z = (m^2 + pn^2)/2$, where $m > \sqrt{pn}$

17. Substituting $f_n = f_{n+2} - f_{n+1}$ and $f_{n+3} = f_{n+2} + f_{n+1}$ into $(f_n f_{n+3})^2 + (2f_{n+1}f_{n+2})^2$ yields $(f_{n+2} - f_{n+1})^2(f_{n+2} + f_{n+1})^2 + 4f_{n+1}^2 f_{n+2}^2 = (f_{n+2}^2 - f_{n+1}^2)^2 + 4f_{n+1}^2 f_{n+2}^2 = f_{n+2}^4 - 2f_{n+1}^2 f_{n+2}^2 + f_{n+1}^4 + 4f_{n+1}^2 f_{n+2}^2 = f_{n+2}^4 + 2f_{n+1}^2 f_{n+2}^2 + f_{n+1}^4 = (f_{n+2}^2 + f_{n+1}^2)^2$, proving the result.

Section 13.2

1. Assume without loss of generality that $x < y$. Then $x^n + y^n = x^2 x^{n-2} + y^2 y^{n-2} < (x^2 + y^2)y^{k-2} = z^2 y^{n-2} < z^2 z^{n-2} = z^n$.

3. a. If $p \mid x$, y, or z, then certainly $p \mid xyz$. If not, then by Fermat's little theorem $x^{p-1} \equiv y^{p-1} \equiv z^{p-1} \equiv 1 \pmod p$. Hence $1 + 1 \equiv 1 \pmod p$, which is impossible.
b. We know that $a^p \equiv a \pmod p$ for every integer a. Then $x^p + y^p \equiv z^p \pmod p$ implies $x + y \equiv z \pmod p$, so $p \mid x + y - z$.

5. Let x and y be the lengths of the legs and z be the hypotenuse. Then $x^2 + y^2 = z^2$. If the area is a perfect square, we have $A = \frac{1}{2}xy = r^2$. Then if $x = m^2 - n^2$ and $y = 2mn$, we have $r^2 = mn(m^2 - n^2)$. All of these factors are relatively prime, so $m = a^2$, $n = b^2$, and $m^2 - n^2 = c^2$, say. Then $a^4 - b^4 = c^2$, which contradicts Exercise 4.

7. We use the method of infinite descent. Assume that there is a nonzero solution where $|x|$ is minimal. Then $(x, y) = 1$. Also, x and z cannot both be even, because then y would be odd and then $z^2 \equiv 8 \pmod{16}$, but 8 is not a quadratic residue modulo 16. Therefore x and z are both odd, since $8y^4$ is even. From here it is easy to check that $(x, z) = 1$. We may also assume (by negating if necessary) that $x \equiv 1 \pmod 4$ and $z \equiv 3 \pmod 4$. Clearly $x^2 > |z|$. We have $8y^4 = x^4 - z^2 = (x^2 - z)(x^2 + z)$. Since $z \equiv 3 \pmod 4$, we have $x^2 - z \equiv 2 \pmod 4$, so

$m = (x^2 - z)/2$ is odd, and $n = (x^2 + z)/4$ is an integer. Since no odd prime can divide both m and n, we have $(m, n) = 1$, $m, n > 0$, and $mn = y^4$, whence $m = r^4$ and $n = s^4$, with $(r, s) = 1$. So now $r^4 + 2s^4 = m + 2n = x^2$. This implies $(x, r) = 1$, since no odd prime divides r and x but not s, and r and x are both odd. Also, $|x| > r^2 > 0$. Now consider $2s^4 = (x^2 - r^4) = (x - r^2)(x + r^2)$. Then s must be even since a difference of squares is not congruent to 2 (mod 4), so $s = 2t$ and $32t^4 = (x - r^2)(x + r^2)$. Recalling that $x \equiv 1 \pmod 4$ and r is odd, we know that $U = (x + r^2)/2$ is odd and $V = (x - r^2)/16$ is an integer. Again $(U, V) = 1$ and $UV = t^4$, but we don't know the sign of x. So $U = \pm u^4$ and $V = \pm v^4$, depending on the sign of x. Now $r^2 = \pm(u^4 - 8v^4)$. But since u is odd, the sign can't be $-$ (or else $r^2 \equiv 7 \pmod 8$). So the sign is $+$ (hence x is positive), and $u^4 - 8v^4 = r^2$. Finally, $|v| > 0$ because $|x + r^2| > 0$, so we have not reduced to a trivial case. Then $u^4 = U < |x + r^2|/2 < x$, so $|u| < x$, and so $|x|$ was not minimal. This contradiction shows that there are no nontrivial solutions.

9. Suppose that $x = a/b$, where a and b are relatively prime integers with $b \neq 0$. Then $y^2 = (a^4 + b^4)/b^4$, from which we can deduce that $y = z/b^2$ for some integer z. Then $z^2 = a^4 + b^4$, which has no nonzero solutions by Theorem 13.3. Because $b \neq 0$, it follows that $z \neq 0$. Therefore, $a = 0$, and hence $x = 0$, and consequently $y = \pm 1$. These are the only solutions.

11. If x were even, then $y^2 = x^3 + 23 \equiv 3 \pmod 4$, which is impossible, so x must be odd, making y even, say $y = 2v$. If $x \equiv 3 \pmod 4$, then $y^2 \equiv 3^3 + 23 \equiv 2 \pmod 4$, which is also impossible, so $x \equiv 1 \pmod 4$. Next, add 4 to both sides of the equation to get $y^2 + 4 = 4v^2 + 4 = x^3 + 27 = (x + 3)(x^2 - 3x + 9)$. Then $z = x^2 - 3x + 9 \equiv 1 - 3 + 9 \equiv 3 \pmod 4$, so a prime $p \equiv 3 \pmod 4$ must divide z. Then $4v^2 + 4 \equiv 0 \pmod p$ or $v^2 \equiv -1 \pmod p$. But this shows that a prime congruent to 3 modulo 4 has -1 as a quadratic residue, which contradicts Theorem 11.5. Therefore, the equation has no solutions.

13. This follows from Exercise 4 and Theorem 13.2.

15. Assume that $n \nmid xyz$ and $(x, y, z) = 1$. Now $(-x)^n = y^n + z^n = (y + z)(y^{n-1} - y^{n-2}z + \cdots + z^{n-1})$, and these factors are relatively prime, so they are nth powers, say $y + z = a^n$ and $y^{n-1} - y^{n-2}z + \cdots + z^{n-1} = \alpha^n$, whence $-x = a\alpha$. Similarly, $z + x = b^n$, $z^{n-1} - z^{n-2}x + \cdots + x^{n-1} = \beta^n$, $-y = b\beta$, $x + y = c^n$, $x^{n-1} - x^{n-2}y + \cdots + y^{n-1} = \gamma^n$, and $-z = c\gamma$. Since $x^n + y^n + z^n \equiv 0 \pmod p$, we have $p \mid xyz$, say $p \mid x$. Then $\gamma^n = x^{n-1} - x^{n-2}y + \cdots + y^{n-1} \equiv y^{n-1} \pmod p$. Also $2x \equiv b^n + c^n + (-a)^n \equiv 0 \pmod p$, so by the condition on p, we have $p \mid abc$. If $p \mid b$, then $y = -b\beta \equiv 0 \pmod p$, but then $p \mid x$ and y, a contradiction. Similarly, p cannot divide c. Therefore, $p \mid a$, so $y \equiv -z \pmod p$, and so $\alpha^n \equiv y^{n-1} - y^{n-2}z + \cdots + z^{n-1} \equiv ny^{n-1} \equiv n\gamma^n \pmod p$. Let g be the inverse of γ modulo p; then $(ag)^n \equiv n \pmod p$, which contradicts the condition that there is no solution to $w^n \equiv n \pmod p$.

17. $3, 4, 5, 6$

19. If $m \geq 3$, then modulo 8 we have $3^n \equiv -1 \pmod 8$, which is impossible, so $m = 1$ or 2. If $m = 1$, then $3^n = 2 - 1 = 1$, which implies that $n = 0$, which is not a positive integer, so that we have no solutions in this case. If $m = 2$, then $3^n = 2^2 - 2 = 3$, which implies that $n = 1$, and this is the only solution.

21. **a.** Substituting the expressions into the left-hand side of the equation yields $a^2 + b^2 + (3ab - c)^2 = a^2 + b^2 + 9a^2b^2 - 6abc + c^2 = (a^2 + b^2 + c^2) + 9a^2b^2 - 6abc$. Since (a, b, c) is a solution to Markov's equation, we substitute $a^2 + b^2 + c^2 = 3abc$ to get the last expression equal to $3abc + 9a^2b^2 - 6abc = 9a^2b^2 - 3abc = 3ab(3ab - c)$, which is the right-hand side of Markov's equation evaluated at these expressions.

b. *Case 1:* If $x = y = z$, then Markov's equation becomes $3x^2 = 3xyz$ so that $1 = yz$. Then $y = z = 1$ and then $x = 1$, so the only solution in this case is $(1, 1, 1)$.

Case 2: If $x = y \neq z$, then $2x^2 + z^2 = 3x^2z$, which implies that $x^2 \mid z^2$ or $x \mid z$, say $dx = z$. Then $2x^2 + d^2x^2 = 3dx^3$ or $2 + d^2 = 3dx$ or $2 = d(3x - d)$. So $d \mid 2$, but because $x \neq z$, we must have

$d = 2$. Then $3x - d = 1$ so that $x = 1 = y$ and $z = 2$. It follows that the only solution in this case is $(1, 1, 2)$.

Case 3: Assume that $x < y < z$. From $z^2 - 3xyz + x^2 = y^2 + z^2$ we apply the quadratic formula to get $2z = 3xy \pm \sqrt{9x^2y^2 - 4(x^2 + y^2)}$. Note that $8x^2y^2 - 4x^2 - 4y^2 = 4x^2(y^2 - 1) + 4y^2(x^2 - 1) > 0$, so in the "minus" case of the quadratic formula, we have $2z < 3xy - \sqrt{9x^2y^2 - 8x^2y^2} = 3xy - xy = 2xy$, or $z < xy$. But $3xyz = x^2 + y^2 + z^2 < 3z^2$ so that $xy < z$, a contradiction, therefore we must have the case corresponding to the plus sign in the quadratic formula and $2z = 3xy + \sqrt{9x^2y^2 - 4(x^2 + y^2)} > 3xy$, so that $z > 3xy - z$. This last expression is the formula for the generation of z in part (a). Therefore, by successive use of the formula in part (a), we will reduce the value of $x + y + z$ until it is one of the solutions in Case 1 and Case 2.

23. Let $\epsilon > 0$ be given. Then the abc conjecture implies that $\max(|a|, |b|, |c|) \leq K(\epsilon)\text{rad}(abc)^{1+\epsilon}$ for integers $(a, b) = 1$ and $a + b = c$. Set $M = \log K / \log 2 + (3 + 3\epsilon)$. Suppose x, y, z, a, b, c are positive integers with $(x, y) = 1$ and $x^a + y^b = c^z$, so that we have a solution to Beal's equation. Assume $\min(a, b, c) > M$. From the abc conjecture we have $\max(x^a, y^b, z^c) \leq K(\epsilon)\text{rad}(xyz)^{1+\epsilon} \leq K(\epsilon)(xyz)^{1+\epsilon}$. If $\max(x, y, z) = x$, then we would have $x^a \leq K(\epsilon)x^{3(1+\epsilon)}$. Taking algorithms of both sides yields $a \leq \log K / \log x + (3 + 3\epsilon) < \log K / \log 2 + (3 + 3\epsilon) = M$, a contradiction. A similar argument applies if the maximum is y or z. Therefore, if the *abc* conjecture is true, then there are no solutions to the Beal conjecture for sufficiently large exponents.

25. **a.** If 1 is a congruent number, then there exist rational numbers r, s, and t such that $r^2 + s^2 = t^2$ and $rs/2 = 1$. Let $r = a/d$, $s = b/d$, and $t = c/d$, where a, b, c, and d are integers and d is the least common denominator of the rational numbers r, s, and t. Then $a^2 + b^2 = (rd)^2 + (sd)^2 = d^2t^2 = c^2$, so that (a, b, c) is a Pythagorean triple, consisting of the lengths of the sides of a right triangle with area $ab/2 = (rd)(sd)/2 = (d^2)(rs/2) = d^2$, a perfect square. Conversely, if there is a right triangle with area a perfect square, d^2, then its side lengths form a Pythagorean triple (a, b, c), and $a^2 + b^2 + c^2$. We can divide through by d^2 to get $(a/d)^2 + (b/d)^2 = (c/d)^2$ and so this represents a right triangle with sides $(a/d, b/d, c/d)$ and area $1/2(a/d)(b/d) = (ab/2)(1/d^2) = d^2/d^2 = 1$.
b. Suppose that 1 is a congruent number. Then, by part (a), there exist integers a, b, c, and d, such that $a^2 + b^2 = c^2$ and $ab/2 = d^2$. If we add and subtract 4 times the second equation from the first we get $a^2 + 2ab + b^2 = (a + b)^2 = c^2 + (2d)^2$ and $a^2 - 2ab + b^2 = (a - b)^2 = c^2 - (2d)^2$. Since the right-hand sides of both equations are squares, then so is their product, and we have $(c^2 + (2d)^2)(c^2 - (2d)^2) = c^4 - (2d)^4 = (a + b)^2(a - b)^2$, but this is a solution to $x^4 - y^4 = z^2$, which contradicts Exercise 4. Therefore 1 is not a congruent number.

Section 13.3

1. **a.** $19^2 + 4^2$ **b.** $23^2 + 11^2$ **c.** $37^2 + 9^2$ **d.** $137^2 + 9^2$

3. **a.** $5^2 + 3^2$ **b.** $9^2 + 3^2$ **c.** $10^2 + 1^2$ **d.** $21^2 + 7^2$ **e.** $133^2 + 63^2$ **f.** $448^2 + 352^2$

5. **a.** $1^2 + 1^2 + 1^2$ **b.** not possible   **c.** $3^2 + 1^2 + 1^2$ **d.** $3^2 + 3^2 + 0^2$ **e.** not possible   **f.** not possible

7. Let $n = x^2 + y^2 + z^2 = 4^m(8k + 7)$. If $m = 0$, then see Exercise 6. If $m \geq 1$, then n is even, so zero or two of x, y, z are odd. If two are odd, then $x^2 + y^2 + z^2 \equiv 2$ or $6 \pmod 8$, but then $4 \nmid n$, a contradiction, so all of x, y, z are even. Then $4^{m-1}(8k + 7) = (x/2)^2 + (y/2)^2 + (z/2)^2$ is the sum of three squares. Repeat until $m = 0$ and use Exercise 6 to get a contradiction.

9. **a.** $10^2 + 1^2 + 0^2 + 2^2$ **b.** $22^2 + 4^2 + 1^2 + 3^2$ **c.** $14^2 + 4^2 + 1^2 + 5^2$ **d.** $56^2 + 12^2 + 17^2 + 1^2$

11. Let $m = n - 169$. Then m is the sum of four squares: $m = x^2 + y^2 + z^2 + w^2$. If, say, x, y, z are 0, then $n = w^2 + 169 = w^2 + 10^2 + 8^2 + 2^2 + 1^2$. If, say, x, y are 0, then $n = z^2 + w^2 + 169 =$

$z^2 + w^2 + 12^2 + 4^2 + 3^2$. If, say, x is 0, then $n = y^2 + z^2 + w^2 + 169 = y^2 + z^2 + w^2 + 12^2 + 5^2$. If none are 0, then $n = x^2 + y^2 + z^2 + w^2 + 13^2$.

13. If k is odd, then 2^k is not the sum of four positive squares. Suppose that $k \geq 3$ and $2^k = x^2 + y^2 + z^2 + w^2$. Modulo 8 we have $0 \equiv x^2 + y^2 + z^2 + w^2$, and since an odd square is congruent to 1 (mod 8), the only possibility is to have x, y, z, w all even. But then we can divide by 4 to get $2^{k-2} = (x/2)^2 + (y/2)^2 + (z/2)^2 + (w/2)^2$. Either $k - 2 \geq 3$ and we can repeat the argument, or $k - 2 = 1$, in which case we have 2 equal to the sum of four positive squares, a contradiction.

15. If $p = 2$, the theorem is obvious. Otherwise, $p = 4k + 1$, whence -1 is a quadratic residue modulo p, say $a^2 \equiv -1$ (mod p). Let x and y be as in Thue's lemma. Then $x^2 < p$ and $y^2 < p$ and $-x^2 \equiv (ax)^2 \equiv y^2$ (mod p). Thus $p \mid x^2 + y^2 < 2p$, so $p = x^2 + y^2$, as desired.

17. The left sum runs over all pairs of integers $i < j$ for $1 \leq i < j \leq 4$, so there are six terms. Each integer subscript 1, 2, 3, and 4 appears in exactly three pairs, so $\sum_{1 \leq i < j \leq 4}((x_i + x_j)^4 + (x_i - x_j)^4) = \sum_{1 \leq i < j \leq 4}(2x_i^4 + 12x_i^2 x_j^2 + 2x_j^4) = \sum_{k=1}^{4} 6x_k^4 + \sum_{1 \leq i < j \leq 4} 12x_i^2 x_j^2 = 6\left(\sum_{k=1}^{4} x_k^2\right)^2$.

19. If m is positive, then $m = \sum_{k=1}^{4} x_k^2$ for some x_k's. Then $6m = 6\sum_{k=1}^{4} x_k^2 = \sum_{k=1}^{4} 6x_k^2$. Each term of the last sum is the sum of 12 fourth powers by Exercise 18. Therefore $6m$ is the sum of 48 fourth powers.

21. For $n = 1, 2, \ldots, 50$ we have $n = \sum_1^n 1^4$. For $n = 51, 52, \ldots, 81$, we have $n - 48 = n - 3(2^4) = \sum_1^{n-48} 1^4$, so $n = 2^4 + 2^4 + 2^4 + \sum_1^{n-48} 1^4$ is the sum of $n - 45$ fourth powers, and $n - 45 \leq 36 \leq 50$. This result, coupled with the result from Exercise 20, shows that all positive integers can be written as the sum of 50 or fewer fourth powers. That is, $g(4) \leq 50$.

23. The only quartic residues modulo 16 are 0 and 1. Therefore, the sum of fewer than 15 fourth powers must have a least nonnegative residue between 0 and 14 (mod 16), which excludes any integer congruent to 15 (mod 16).

Section 13.4

1. a. $(\pm 2, 0)$, $(\pm 1, \pm 1)$ **b.** none   **c.** $(\pm 1, \pm 2)$

3. a. yes   **b.** no   **c.** yes   **d.** yes   **e.** yes   **f.** no

5. $(73, 12)$, $(10{,}657, 1752)$, $(1{,}555{,}849, 255{,}780)$

7. $(6{,}239{,}765{,}965{,}720{,}528{,}801, 798{,}920{,}165{,}762{,}330{,}040)$

9. Reduce modulo p to get $x^2 \equiv -1$ (mod p). Since -1 is a quadratic nonresidue modulo p if $p = 4k + 3$, there is no solution.

11. Let $p_0 = 0$, $p_1 = 3$, $p_k = 2p_{k-1} + 2p_{k-2}$, $q_0 = 1$, $q_1 = 1$, and $q_k = 2q_{k-1} + q_{k-2}$. Then the legs are $x = p_k^2 + 2p_k q_k$ and $y = 2p_k q_k + 2q_k^2$.

13. Suppose that (x, y) is a solution of $x^4 - 2y^2 = -1$. Note that x must be odd. Furthermore, note that $(x^2 + 1)^2 = x^4 + 2x^2 + 1 = 2y^2 + 2x^2$ and $(x^2 - 1)^2 = x^4 - 2x^2 + 1 = 2y^2 - 2x^2$. Multiplying these equations together yields $(x^4 - 1)^2 = 4(y^4 - x^4)$ so that $((x^4 - 1)/2)^2 = y^4 - x^4$. Since $(x^4 - 1)/2$ is an integer, this contradicts Exercise 4 in Section 13.2.

Section 14.1

1. a. $5 + 15i$ **b.** $-46 - 9i$ **c.** $-26 - 18i$

3. a. yes **b.** yes **c.** no **d.** yes

5. $(4a - 3b) + (3a + 4b)i$, where a and b are rational integers (see the *Student Solutions Manual* for the display of such integers)

7. Because $\alpha|\beta$ and $\beta|\gamma$, there are Gaussian integers μ and ν with $\mu\alpha = \beta$ and $\nu\beta = \gamma$, and hence, $\gamma = \nu\beta = \nu\mu\alpha$. Because the product of Gaussian integers is a Gaussian integer, $\nu\mu$ is also a Gaussian integer. It follows that $\alpha|\gamma$.

9. Note that $x^5 = x$ if and only if $x^5 - x = x(x-1)(x+1)(x-i)(x+i) = 0$. The solutions of this last equation are $0, 1, -1, i$, and $-i$. These are the four Gaussian integers that are units, together with 0.

11. Since $\alpha|\beta$ and $\beta|\alpha$, there are Gaussian integers μ and ν such that $\alpha\mu = \beta$ and $\beta\nu = \alpha$. It follows that $\alpha = \alpha\mu\nu$. By Theorem 14.1, this implies that $N(\alpha) = N(\alpha\mu\nu) = N(\alpha)N(\mu\nu)$. This means that $N(\mu)N(\nu) = 1$, and because the norm of a Gaussian integer is a nonnegative rational integer, $N(\mu) = N(\nu) = 1$. Consequently μ and ν are units, and hence, α and β are associates.

13. The pair $\alpha = 2 + i$, $\beta = 1 + 2i$ is a counterexample.

15. We first whow that such an associate exists. If $a > 0$ and $b \geq 0$, the desired inequalities are met; if $a < 0$ and $b > 0$, multiply by $-i$ to get $-i\alpha = b - ai = c + di$; if $a < 0$ and $b \leq 0$, multiply by -1 to get $-\alpha = -a - bi = c + di$; and if $a \geq 0$ and $b < 0$, multiply by i to get $i\alpha = -b + ai = c + di$. In all cases, $c > 0$ and $d \geq 0$. To prove uniqueness, note that when we multiply $c + di$ with $c > 0$ and $d \geq 0$ by a unit other than 1, we obtain $-c - di$, which has $-c < 0$, $-d + ci$, which has $-d \leq 0$, or $d - ci$, which has $-c < 0$.

17. a. $\gamma = 3 - 5i$, $\rho = -3i$, $N(\rho) = 3^2 + 0^2 = 9 < N(\beta) = 3^2 + 3^2 = 18$
 b. $\gamma = 5 - i$, $\rho = -1 - 2i$, $N(\rho) = 5 < N(\beta) = 25$
 c. $i\gamma = -1 + 8i$, $\rho = -5 - 3i$, $N(\rho) = 5^2 + 3^2 = 34 < N(\beta) = 11^2 + 2^2 = 125$

19. a. $\gamma = 2 - 5i$, $\rho = 3$ **b.** $\gamma = 4 - i$, $\rho = 2 + 2i$ **c.** $\gamma = -2 + 8i$, $\rho = 6 - 5i$

21. $1, 2$, and 4.

23. When a and b are both even, $2|a + ib$ because $a + ib = 2((a/2) + i(b/2))$, and $(a/2) + i(b/2)$ is a Gaussian integer. Because $1 + i | 2$, we conclude that $1 + i | a + ib$. When a and b are both odd, note that $a + bi = (1 + i) + (a - 1) + (b - 1)i$, where $a - 1$ and $b - 1$ are both even. Because $1 + i$ and $(a - 1) + (b - 1)i$ are both multiples of $1 + i$, so is their sum. On the other hand, if a is odd and b is even, then $(a - 1) + bi$ is a multiple of $1 + i$. Hence, if $a + bi$ is a multiple of $a + bi$, then $(a + bi) - (a - 1 + bi) = 1$ is a multiple of $1 + i$, a contradiction. A similar argument shows that if a is even and b is odd, then $1 + i$ does not divide $a + bi$.

25. $\pm 1 \pm 2i$

27. Suppose that $7 = (a + bi)(c + di)$, where $a + bi$ and $c + di$ are not units. Taking norms of both sides yields $49 = (a^2 + b^2)(c^2 + d^2)$. Because neither $a + bi$ nor $c + di$ is a unit, both factors on the right-hand side must equal 7. However, 7 is not the sum of 2 squares.

29. Because α is neither a unit nor a prime, there exist nonunit Gaussian integers α and β with $\alpha = \beta\gamma$, where neither β nor γ is a unit, so that $N(\alpha) = N(\beta)N(\gamma)$, $N(\beta) > 1$, and $N(\gamma) > 1$. If $N(\beta) > \sqrt{N(\alpha)}$, then $N(\gamma) = N(\alpha)/N(\beta) < N(\alpha)/\sqrt{N(\alpha)} = \sqrt{N(\alpha)}$. Consequently, either β or γ divides α and has norm not exceeding $\sqrt{N(\alpha)}$.

31. The Gaussian primes with norm less than 100 are $3, 7, 1 + i, 1 + 2i, 1 + 4i, 1 + 6i, 2 + 3i, 2 + 5i, 2 + 7i, 3 + 8i, 4 + 5i, 4 + 9i, 5 + 6i$, and $5 + 8i$, together with their associates and conjugates.

33. a. Note that $\alpha - \alpha = 0 = 0 \cdot \mu$, so that $\mu | \alpha - \alpha$. Thus, $\alpha \equiv \alpha \pmod{\mu}$.
b. Because $\alpha \equiv \beta \pmod{\mu}$, $\mu | \alpha - \beta$. Hence, there is a Gaussian integer γ with $\mu \gamma = \alpha - \beta$. This means that $\mu(-\gamma) = \beta - \alpha$, so that $\mu | \beta - \alpha$. Therefore, $\beta \equiv \alpha \pmod{\mu}$.
c. Because $\alpha \equiv \beta \pmod{\mu}$ and $\beta \equiv \gamma \pmod{\mu}$, there are Gaussian integers δ and ϵ such that $\mu \delta = \alpha - \beta$ and $\mu \epsilon = \beta - \gamma$. It follows that $\alpha - \gamma = \alpha - \beta + \beta - \gamma = \mu \delta + \mu \epsilon = \mu(\delta + \epsilon)$. Therefore, $\alpha \equiv \gamma \pmod{\mu}$.

35. Let $\alpha = a_1 + ib_1$, $\beta = a_2 + ib_2$, and $(a_1 + b_1)(a_2 + b_2) = R + Si$. We have $R = a_1a_2 - b_1b_2$ and $S = a_1b_2 + a_2b_1$. We compute $m_1 = b_2(a_1 + b_1)$, $m_2 = a_2(a_1 - b_1)$, and $m_3 = b_1(a_2 - b_2)$ using a total of three multiplications. We use these three products to find R and S using the equations $R = m_2 + m_3$ and $S = m_1 + m_3$.

37. a. $i, 1 + i, 1 + 2i, 2 + 3i, 3 + 5i, 5 + 8i$
b. Using the definition of G_k and the recursive definition of the Fibonacci sequence, we have $G_k = f_k + if_{k+1} = (f_{k-1} + f_{k-2}) + (f_k + f_{k-1})i = (f_{k-1} + f_k i) + (f_{k-2} + f_{k-1}i) = G_{k-1} + G_{k-2}$.

39. We proceed by induction. For the basis step, note that $G_2 G_1 - G_3 G_0 = (1 + 2i)(1 + i) - (2 + 3i)(i) = 2 + i$. Now assume the identity holds for values less than n. Using the identity in Exercise 37(b), we see that $G_{n+2}G_{n+1} - G_{n+3}G_n = (G_{n+1} + G_n)G_{n+1} - (G_{n+2} + G_{n+1})G_n = G_{n+1}^2 - G_{n+2}G_n = G_{n+1}^2 - (G_{n+1} + G_n)G_n = G_{n+1}^2 - G_n^2 - G_{n+1}G_n = (G_{n+1} + G_n)(G_{n+1} - G_n) - G_{n+1}G_n = G_{n+2}G_{n-1} - G_{n+1}G_n = -(-1)^{n-1}(2 + i) = (-1)^n(2 + i)$, which completes the induction step.

41. Suppose that $r + si$, where r and $s \neq 0$ are rational, is a root of the monic quadratic polynomial $z^2 + az + b$, where a and b are integers. The other root of the quadratic polynomial must be $r - si$; the polynomial must be $(z - (r + si))(z - (r - si)) = z^2 - 2rz + r^2 + s^2$. Hence, the coefficients $a = 2r$ and $b = r^2 + s^2$ are integers. Solving for r and s, we see that $r = a/2$ and $s^2 = (4b - r^2)/4$. This implies that $s = c/2$ for some integer c. Multiplying by 4, we find that $a^2 + c^2 \equiv 0 \pmod{4}$. This implies that both a and c are even. Hence, r and s are integers and $r + si$ is a Gaussian integer.

43. By the proof of the division algorithm in the text, given a Gaussian integer α, there are Gaussian integers γ and ρ such that $\alpha = \gamma(1 + 2i) + \rho$ such that $N(\rho) \leq N(1 + 2i)/2 = 5/2$. Therefore, the only possible values of ρ are $1 + 2i$ are $0, 1, i, 1 + i$ and their associates. Observing that $\alpha = (1 + 2i)\gamma + (1 + i) = (1 + 2i)(\gamma + 1) + (1 + i) - (1 + 2i) = (1 + 2i)(\gamma + 1) - i$, by modifying the quotient if necessary, all Gaussian integers can be written as a multiple of $1 + 2i$ plus a remainder equal to $0, 1, -1, i$, or $-i$. Now consider dividing each of the Gaussian primes $\pi_1, \ldots, \pi_4$, by $1 + 2i$. If two of these can be written as a multiple of $1 + 2i$ plus the same remainder, then $1 + 2i$ divides their difference. But these differences are either 2 or $1 \pm i$, which are not divisible by $1 + 2i$. Furthermore, none of these remainders are 0 because each of these four numbers is prime. Therefore, we may rule out 0 as a possible remainder. Now divide the Gaussian integer $a + bi$ by $1 + 2i$ so that the remainder is one of $0, 1, -1, i$, or $-i$; let the remainder be ρ. I this remainder is not 0, one of $\pi_1, \ldots, \pi_4$ leaves the same remainder when divided by $1 + 2i$, where the quotient is selected so that the remainder is one of $0, 1, -1, i$, or $-i$, say π_k. It follows that $1 + 2i$ divides $\pi_k - (a + bi)$, which is impossible because this difference equals $1, -1, i$, or -1. Therefore, $\rho = 0$, so that $1 + 2i \mid a + bi$. A similar argument shows that $1 - 2i \mid a + bi$. Therefore, the product of these primes $(1 - 2i)(1 + 2i) = 5$ also divides $a + bi$, which implies that $5 \mid a$ and $5 \mid b$. Note that b cannot be 0; if it were, $a - 1, a$, and $a + 1$ would all be prime, which is impossible and a cannot be zero. If $a = 0$ and if b is odd, then $(b - 1)i$ and $(b + 1)i$ are both divisible by 2, while if $a = 0$ and b is even, either $b - 1$ or $b + 1$ is congruent to 1 modulo 4. This implies that either $b - 1$ or $b + 1$ is not a Gaussian prime, and consequently, either $(b - 1)i$ or $(b + 1)i$ is not a Gaussian prime.

45. Because $\alpha\beta\gamma = 1$, $N(\alpha\beta\gamma) = N(\alpha)N(\beta)N(\gamma) = 1$. This implies that $N(\alpha) = N(\beta) = N(\gamma) = 1$, which shows that α, β, and γ are all units in the Gaussian integers. This means that the only possible values for these three values are 1, -1, i, and $-i$. Check all possible values for these three variables in the equation $\alpha + \beta + \gamma = 1$, show that the possible solutions, up to permutation are $(1, 1, -1)$ and $(1, i, -i)$, but the first solution does not satisfy $\alpha\beta\gamma = 1$. This shows that $(1, i, -i)$ and its permutations are the only six solutions.

Section 14.2

1. Certainly, $1 \mid \pi_1$ and $1 \mid \pi_2$. Suppose that $\delta \mid \pi_1$ and $\delta \mid \pi_2$. Because π_1 and π_2 are Gaussian primes, δ must be either a unit or an associate of both primes. But because π_1 and π_2 are not associates, no Gaussian integer can be an associate of both, so that δ must be a unit if $\delta \mid 1$. Therefore, 1 satisfies the definition of a greatest common divisor for π_1 and π_2.

3. Because γ is a greatest common divisor of α and β, $\gamma \mid \alpha$ and $\gamma \mid \beta$. Hence, there exist Gaussian integers μ and ν such that $\mu\gamma = \alpha$ and $\nu\gamma = \beta$. It follows that $\overline{\mu\gamma} = \overline{\mu} \cdot \overline{\gamma} = \overline{\alpha}$ and $\overline{\nu\gamma} = \overline{\nu} \cdot \overline{\gamma} = \overline{\beta}$, which implies that $\overline{\gamma}$ is a common divisor of $\overline{\alpha}$ and $\overline{\beta}$. Furthermore, if $\delta \mid \overline{\alpha}$ and $\delta \mid \overline{\beta}$, then $\overline{\delta} \mid \alpha$ and $\overline{\delta} \mid \beta$. Consequently, $\overline{\delta} \mid \gamma$ by the definition of greatest common divisor. But this implies that $\overline{\overline{\delta}} = \delta \mid \overline{\gamma}$, which shows that $\overline{\gamma}$ is a greatest common divisor for $\overline{\alpha}$ and $\overline{\beta}$.

5. Let $\epsilon\gamma$, where ϵ is a unit, be an associate of γ. Because $\gamma \mid \alpha$, there is a Gaussian integer μ such that $\mu\gamma = \alpha$. Because ϵ is a unit, $1/\epsilon$ is also a Gaussian integer. Then $(1/\epsilon)\mu(\epsilon\gamma) = \alpha$, so that $\epsilon\gamma \mid \alpha$. Similarly, $\epsilon\gamma \mid \beta$. If $\delta \mid \alpha$ and $\delta \mid \beta$, then $\delta \mid \gamma$ by the definition of greatest common divisor. Consequently, there exists a Gaussian integer ν such that $\nu\delta = \gamma$. This implies that $\epsilon\nu\delta = \epsilon\gamma$, and because $\epsilon\nu$ is a Gaussian integer, we have $\delta \mid \epsilon\gamma$. Thus, $\epsilon\gamma$ satisfies the definition of a greatest common divisor.

7. Take $3 - 2i$ and $3 + 2i$, for example.

9. Because a and b are relatively prime rational integers, there exist rational integers m and n such that $am + bn = 1$. Let δ be a greatest common divisor of the Gaussian integers a and b. Then δ divides $am + bn = 1$. Therefore, δ is a unit in the Gaussian integers. Hence, a and b are relatively prime Gaussian integers.

11. a. We have $44 + 18i = (12 - 16i)(1 + 2i) + 10i$; $12 - 16i = (10i)(-2 - i) + (2 + 4i)$; $10i = (2 + 4i)(2 + i) + 0$. The last nonzero remainder, $2 + 4i$, is a greatest common divisor.
b. By part (a), $2 + 4i = (12 - 16i) - (10i)(-2 - i) = (12 - 16i) - ((44 + 18i) - (12 - 16i)(1 + 2i))(-2 - i) = (2 + i)(44 + 18i) + (1 + (1 + 2i)(-2 - i))(12 - 16i) = (2 + i)(44 + 18i) + (1 - 5i)(12 - 16i)$. Take $\mu = 2 + i$ and $\nu = 1 - 5i$.

13. We proceed by induction. We have $G_0 = i$ and $G_1 = 1 + i$. Because G_0 is a unit, G_0 and G_1 are relatively prime, completing the basis step. For the induction step, assume that G_k and G_{k-1} are relatively prime. Suppose that $\delta \mid G_k$ and $\delta \mid G_{k+1}$. Then $\delta \mid (G_{k+1} - G_k) = (G_k + G_{k-1} - G_k) = G_{k-1}$, so that δ is a common divisor of G_k and G_{k-1}, which are relatively prime. Hence, $\delta \mid 1$. Hence, 1 is a greatest common divisor of G_{k+1} and G_k.

15. Because the norm of the remainder in each step of the Euclidean algorithm for Gaussian integers based on the division algorithm described in the text does not exceed half of the norm of the divisor and the norm of the remainder is a positive integer, the maximum number of steps used to find a greatest common divisor of α and β is the largest number of times we can divide $N(\alpha)$ by 2 and obtain a positive integer. It follows that there cannot be more than $[\log_2 N(\alpha)] + 1$ divisions. This means that the number of steps is at most $O(\log_2 N(\alpha))$.

17. a. $(-1)(1 - 2i)(1 - 4i)$ **b.** $(-1)(1 + i)(5 + 8i)$ **c.** $(-1)(1 + i)^4 7$
    d. $i(1 + i)^8(1 + 2i)^2(1 - 2i)^2$

19. a. 48   **b.** 120   **c.** 1792   **d.** 2592

21. Assume that n and $a + bi$ are relatively prime. Then, there exist Gaussian integers μ and ν such that $\mu n + \nu(a + bi) = 1$. Taking conjugates of both sides, and recalling that the conjugate of a rational integer is itself, we have $\overline{\mu} n + \overline{\nu}(a - bi) = 1$. This implies that n is also relatively prime to $a - bi$. Since $a - bi = -i(b + ai)$ is an associate of $b + ai$, n and $b + ai$ are relatively prime. The converse follows by symmetry.

23. Suppose that $\pi_1, \pi_2, \ldots, \pi_k$ are all the Gaussian primes. Form the Gaussian integer $Q = \pi_1 \pi_2 \ldots \pi_k + 1$. By Theorem 14.10, Q has a unique factorization into Gaussian primes, and hence, is divisible by some Gaussian prime ρ. Because we have assumed that $\pi_1, \ldots, \pi_k$ are all the Gaussian primes, ρ must be somewhere in this list. It follows that $\rho \mid Q$ and $\rho \mid \pi_1 \pi_2 \ldots \pi_k$, which implies that ρ divides $1 = Q - \pi_1 \pi_2 \ldots \pi_k$, a contradiction. It follows that $\pi_1, \pi_2, \ldots \pi_k$ cannot be a list of all the Gaussian primes. Consequently, there must be infinitely many Gaussian primes.

25. $-2i$

27. Because α and μ are relatively prime, there exist Gaussian integers σ and τ such that $\sigma\alpha + \tau\mu = 1$. When we multiply both sides of this equation by β, we obtain $\beta\sigma\alpha + \beta\tau\mu = \beta$, so that $\alpha(\beta\sigma) \equiv \beta$ (mod μ). Thus, $x \equiv \beta\sigma$ (mod μ) is the solution.

29. a. $x \equiv 5 - 4i$ (mod 13)   **b.** $x \equiv 1 + i$ (mod $4 + i$)   **c.** $x \equiv 3i$ (mod $2 + 3i$).

31. *Chinese Remainder Theorem for the Gaussian Integers.* Let $\mu_1, \mu_2, \ldots, \mu_r$ be pairwise relatively prime Gaussian integers and let $\alpha_1, \alpha_2, \ldots, \alpha_r$ be Gaussian integers. Then the system of congruences $x \equiv \alpha_i$ (mod μ_i), $i = 1, \ldots, r$ has a unique solution modulo $M = \mu_1 \mu_2 \cdots \mu_r$. *Proof:* To construct a solution, for each $k = 1, \ldots, r$, let $M_k = M/\mu_k$. Then M_k and μ_k are relatively prime, because μ_k is relatively prime to the factors of M_k. By Exercise 24, M_k has an inverse λ_k modulo μ_k, so that $M_k \lambda_k \equiv 1$ (mod μ_k). Now let $x = \alpha_1 M_1 \lambda_1 + \cdots + \alpha_r M_r \lambda_r$. We show that x is the solution to the system. Because $\mu_k \mid M_j$ whenever $j \neq k$, we have $\alpha_j M_j \lambda_k \equiv 0$ (mod μ_k) whenever $j \neq k$. Therefore, $x \equiv \alpha_k M_k \lambda_k$ (mod μ_k). Also, because λ_k is an inverse for M_k modulo μ_k, we have $x \equiv \alpha_k$ (mod μ_k) for every k, as desired. Now suppose there is another solution y to the system. Then $x \equiv \alpha_k \equiv y$ (mod μ_k) and so $\mu_k \mid x - y$ for every k. Because the μ_k are pairwise relatively prime, no Gaussian prime appears in more than one of their prime factorizations. Therefore, if a Gaussian prime power $\pi^e \mid x - y$, it divides exactly one μ_k. Therefore, the product M also divides $x - y$. Hence, $x \equiv y$ (mod M), which proves that x is unique modulo M.

33. $x \equiv 9 + 23i$ (mod $26 + 7i$)

35. a. $\{0, 1\}$ **b.** $\{0, 1, i, 1 + i\}$ **c.** $\{0, 1, -1, i, 2i, 3i, -i, -2i, -3i, 1 + i, 1 + 2i, 1 - i, -1 + i\}$

37. Let $\alpha = a + bi$ and $d = \gcd(a, b)$. We assert that the set $S = \{p + qi \mid 0 \leq p < N(\alpha)/d - 1, 0 \leq q < d - 1\}$ is a complete residue system. This set consists of lattice points inside a rectangle in the plane. To see this, first note that $N(\alpha)/d = \alpha(\overline{\alpha}/d)$ is a real number and is also a multiple of α. Second, note that there exist rational integers r and s such that $ra + sb = d$. All multiples of α are given by $v = (s + ir)\alpha = (s + ir)(a + bi) = (as - br) + di$, where s and r are integers. Any Gaussian integer is congruent modulo α to an integer in the rectangle S, because we can add or subtract multiples of v until the imaginary part is between 0 and $d - 1$, and we can add and subtract multiples of $N(\alpha)/d$ until the real part is between 0 and $N(\alpha)/d - 1$. It remains to show that the elements of S are incongruent to each other modulo α. Suppose that β and γ are in S and congruent modulo α. Then the imaginary part of $\beta - \gamma$ is divisible by d, but since β and γ must lie in the interval from 0 to $d - 1$, they are equal. Therefore the difference between β and γ is real and divisible by α, and hence, by $\overline{\alpha}$, and hence, by $\alpha\overline{\alpha}/d = N(\alpha)/d$, proving they are equal. Since S has $N(\alpha)$ elements, we are done.

39. a. $\{1, i, 2+i, 3\}$ **b.** $\{i, 3i, 1, 1+2i, 2+i, 2+3i, 3, 3+2i\}$ **c.** $\{i, 1, 3, 4+i, 6+i, 7, 8+i, 9\}$

41. By the properties of the norm of Gaussian integers and Exercise 37, there are $N(\pi^e) = N(\pi)^e$ residue classes modulo π^e. Let $\pi = r + si$, and $d = \gcd(r, s)$. Also, by Exercise 37, a complete residue system modulo π^e is given by the lattice points $S = \{p + qi \mid 0 \le p < N(\pi^e)/d - 1, 0 \le q < d - 1\}$, while a complete residue system modulo π is given by the set of lattice points $T = \{p + qi \mid 0 \le p < N(\pi)/d - 1, 0 \le q < d - 1\}$. Note that in T there is exactly one element not relatively prime to π, and that there are $N(\pi)^{e-1}$ copies of T, congruent modulo π, inside of S. Therefore, there are exactly $N(\pi)^{e-1}$ elements in S not relatively prime to π. Thus there are $N(\pi)^e - N(\pi)^{e-1}$ elements in a reduced residue system modulo π^e.

43. a. First note that because $r + s\sqrt{-5}$ is a root of a monic polynomial with integer coefficients, the other root must be $r - s\sqrt{-5}$ and the polynomial is $(x - (r + s\sqrt{-5}))(x - (r - s\sqrt{-5})) = x^2 - 2rx + (r^2 + 5s^2) = x^2 - ax + b$, where a and b are rational integers. Then $r = a/2$ and $5s^2 = (4b - a^2)/4$, so that $s = c/2$ for some integer c. (Note that 5 cannot appear in the denominator of s, else when we square it, the single factor of 5 in the expression leaves a remaining factor in the denominator, which does not appear on the right side of the equation.) Substituting these expressions for r and s, we have $(a/2)^2 + 5(c/2)^2 = b^2$, or, upon multiplication by 4, $a^2 + 5c^2 = 4b^2 \equiv 0 \bmod 4$, which has solutions only when a and c are even. Therefore, r and s are rational integers.
b. Let $\alpha = a + b\sqrt{-5}$ and $\beta = c + d\sqrt{-5}$, where a, b, c, and d are rational integers. Then $(a + b\sqrt{-5}) + (c + d\sqrt{-5}) = (a + c) + (b + d)\sqrt{-5}$, $(a + b\sqrt{-5}) - (c + d\sqrt{-5}) = (a - c) + (b - d)\sqrt{-5}$, and $(a + b\sqrt{-5})(c + d\sqrt{-5}) = ac + bc\sqrt{-5} + ad\sqrt{-5} - 5bd = (ac - 5bd) + (bc + ad)\sqrt{-5}$. Each of these results is again of desired form.
c. yes, no
d. Let $\alpha = a + b\sqrt{-5}$ and $\beta = c + d\sqrt{-5}$. Then $N(\alpha)N(\beta) = (a^2 + 5b^2)(c^2 + 5d^2) = a^2c^2 + 5a^2d^2 + 5b^2c^2 + 25b^2d^2$. On the other hand, $\alpha\beta = (ac - 5bd) + (ad + bc)\sqrt{-5}$, so that $N(\alpha\beta) = N((ac - 5bd) + (ad + bc)\sqrt{-5}) = (ac - 5bd)^2 + 5(ad + bc)^2 = a^2c^2 - 10acbd + 25b^2d^2 + 5(a^2d^2 + 2adbc + b^2c^2) = a^2c^2 + 5a^2d^2 + 5b^2c^2 + 25b^2d^2$. It follows that $N(\alpha\beta) = N(\alpha)N(\beta)$.
e. If ϵ is a unit in $\mathbf{Z}[\sqrt{-5}]$, then there exists an η such that $\epsilon\eta = 1$. From part (d), we have $N(\epsilon\eta) = N(\epsilon)N(\eta) = N(1) = 1$, so that $N(\epsilon) = 1$. Now suppose that $\epsilon = a + b\sqrt{-5}$. Then $N(\epsilon) = a^2 + 5b^2 = 1$, which implies that $b = 0$ and $a = \pm 1$. Therefore, the only units are 1 and -1.
f. If an integer α in $\mathbf{Z}[\sqrt{-5}]$ is not a unit and not prime, it must have two nonunit divisors of β and γ such that $\alpha = \beta\gamma$. This implies that $N(\beta)N(\gamma) = N(\alpha)$. To see that 2 is prime, suppose that $\beta \mid 2$, where $\beta = a + b\sqrt{-5}$. It follows that $N(\beta) = a^2 + 5b^2 \mid N(2) = 4$. This implies that $b = 0$, and because β is not a unit, we have $a = \pm 2$. However, if $a = \pm 2$, then γ is a unit, which is a contradiction. Hence, 2 is prime. To see that 3 is prime, we seek divisors of $N(3) = 9$ among integers of the form $a^2 + 5b^2$. We see that b can be only 0 or ± 1. If $b = \pm 1$, then $a = \pm 3$. But this implies that the remaining factor is a unit. If $b = 0$, then $a = \pm 3$, and so 3 is prime. To see that $1 \pm \sqrt{-5}$ is prime, note that its norm is 6. A divisor $a + bi$ can have $b = 0$ or $b = \pm 1$, else its norm is too large. If $b = 0$, then $a^2 \mid 6$, a contradiction, so $b = \pm 1$. But then $a^2 + 5 \mid 6$, which implies that $a = \pm 1$. But $N(\pm 1 \pm \sqrt{-5}) = 6$ so the other divisor is a unit, and so $1 \pm \sqrt{-5}$ is also prime. Note then that $2 \cdot 3 = 6$ and $(1 - \sqrt{-5})(1 + \sqrt{-5}) = 6$, so that we do not have unique factorization into primes in $\mathbf{Z}[\sqrt{-5}]$.
g. Suppose γ and ρ exist. Note first that $(7 - 2\sqrt{-5})/(1 + \sqrt{-5}) = -1/2 - 3/2\sqrt{-5}$, so $\rho \ne 0$. Let $\gamma = a + b\sqrt{-5}$ and $\rho = c + d\sqrt{-5}$. Then from $7 - 2\sqrt{-5} = (1 + \sqrt{-5})(a + b\sqrt{-5}) + (c + d\sqrt{-5}) = (a - 5b + c) + (a + b + d)\sqrt{-5}$, we get $7 = a - 5b + c$ and $-2 = a + b + d$. If we subtract the second equation from the first, we have $9 = -6b + c - d$ or $c - d = 6b + 9$. Therefore, $3 \mid c - d$, and since $\rho \ne 0$, $c - d \ne 0$, so $|c - d| \ge 3$. We consider $N(\rho) = c^2 + 5d^2$.

If $d = 0$, then $N(\rho) \geq c^2 \geq 3^2 > 6$. If $d = \pm 1$, then $|c| \geq 2$ and $N(\rho) = c^2 + 5d^2 \geq 4 + 5 > 6$. If $|d| \geq 2$, then $N(\rho) \geq 5d^2 \geq 5 \cdot 2^2 = 20 > 6$, so in every case the norm of ρ is greater than 6. So no such γ and ρ exist, and there is no analog for the division algorithm in $\mathbf{Z}[\sqrt{-5}]$.
h. Suppose $\mu = a + b\sqrt{-5}$ and $\nu = c + d\sqrt{-5}$ is a solution to the equation. Then $3(a + b\sqrt{-5}) + (1 + \sqrt{-5})(c + d\sqrt{-5}) = (3a + c - 5d) + (3b + c + d)\sqrt{-5} = 1$. So we must have $3a + c - 5d = 1$ and $3b + c + d = 0$. If we subtract the second equation from the first, we get $3a - 3b - 6d = 1$, which implies that $3|1$, a contradiction. Therefore, no such solution exists.

Section 14.3

1. a. 8   **b.** 8   **8.** 8   **d.** 16

3. We first check that a greatest common divisor δ of α and β divides γ, otherwise no solution exists. If a solution exists, use the Euclidean algorithm and back substitution to express δ as a linear combination of α and β as $\alpha\mu + \beta\nu = \delta$. Because $\delta \mid \gamma$, there is a Gaussian integer η such that $\delta\eta = \gamma$. When we multiply the equation $\alpha\mu + \beta\nu = 1$ by η, we have $\alpha\mu\eta + \beta\nu\eta = \delta\eta = \gamma$, so we may take $x_0 = \mu\eta$ and $y_0 = \nu\eta$ as a solution. The set of all solutions is given by $x = x_0 + \beta\tau/\delta$, $y = y_0 - \alpha\tau/\delta$, where τ ranges over the Gaussian integers.

5. a. no solutions   **b.** no solutions

7. Suppose x, y, z is a primitive Pythagorean triple with y even, so that x and z are odd. Then $z^2 = x^2 + y^2 = (x + iy)(x - iy)$. If a rational prime p divides $x + iy$, it divides both x and y, contradicting the fact that the triple is primitive. Therefore, the only Gaussian primes that divide $x + iy$ are of the form $m + in$ with $n \neq 0$. Also, if $1 + i \mid x + iy$, then $1 - i \mid x - iy$, which implies that $2 = (1 - i)(1 + i)$ divides z^2, which is odd, a contradiction. Therefore, we conclude that $1 + i$ does not divide $x + iy$, and hence, neither does 2. Suppose δ is a common divisor of $x + iy$ and $x - iy$. Then δ divides the sum $2x$ and the difference $2iy$. Because 2 is not a common factor, δ must divide both x and y, which are relatively prime. Hence, δ is a unit and $x + iy$ and $x - iy$ are also relatively prime. Every prime that divides $x + iy$ is of the form $\pi = u + iv$, and so $\bar{\pi} = u - iv$ divides $x - iy$. Because their product equals a square, each factor is a square. Thus, $x + iy = (m + in)^2$ and $x - iy = (m - in)^2$ for some Gaussian integer $m + in$. But then $x + iy = m^2 - n^2 + 2mni$ so $x = m^2 - n^2$ and $y = 2mn$. And if $z^2 = (m + ni)^2(m - ni)^2 = (m^2 + n^2)^2$, so $z = m^2 + n^2$. Further, if m and n were both odd or both even, we would have z even, a contradiction. It follows that m and n have opposite parity. Finally, having found m and n, if $m < n$, we can multiply by i and reverse their roles to get $m > n$. The converse is exactly as shown in Section 13.1.

9. By Lemma 14.3, there is a unique rational prime p such that $\pi \mid p$. Let $\alpha = a + bi$. We separately consider three cases: $p = 2$, $p \equiv 3 \pmod 4$, and $p \equiv 1 \pmod 4$.
Case 1: If $p = 2$, then π is an associate of $1 + i$ and $N(\pi) - 1 = 1$. Because there are only two congruence classes modulo $1 + i$ and because α and $1 + i$ are relatively prime, we have $\alpha^{N(\pi)-1} = \alpha \equiv 1 \pmod{1 + i}$.
Case 2: If $p \equiv 3 \pmod 4$, then $\pi = p$ and $N(\pi) - 1 = p^2 - 1$. Also, $i^p = -i$. By the binomial theorem, we can show that $\alpha^p = (a + bi)^p \equiv a^p + (bi)^p \equiv a^p - ib^p \equiv a - bi \equiv \bar{\alpha} \pmod p$, using Fermat's little theorem. Similarly, $\bar{\alpha}^p \equiv \alpha \pmod p$, so that $\alpha^{p^2} \equiv \bar{\alpha}^p \equiv \alpha \pmod p$, and because $p = \pi$ and α and π are relatively prime, we have $\alpha^{N(\pi)-1} \equiv 1 \pmod p$.
Case 3: If $p \equiv 1 \pmod 4$, then $\pi\bar{\pi} = p$, $i^p = i$, and $N(\pi) - 1 = p - 1$. By the binomial theorem, we can show that $\alpha^p = (a + bi)^p \equiv a^p + (bi)^p \equiv a + bi \equiv \alpha \pmod p$, using Fermat's little theorem. This implies that $\alpha^{p-1} \equiv 1 \pmod p$, and because $\pi \mid p$, we have $\alpha^{N(\pi)-1} \equiv 1 \pmod \pi$, concluding the proof.

11. Let π be a Gaussian prime. If $\alpha^2 \equiv 1 \pmod{\pi}$, then $\pi \mid \alpha^2 - 1 = (\alpha - 1)(\alpha + 1)$, so that either $\alpha \equiv 1$ or $\alpha \equiv -1 \pmod{\pi}$. Therefore, only 1 or -1 can be its own inverse modulo π. Now let $\alpha_1 = 1, \alpha_2, \ldots, \alpha_{r-1}, \alpha_r = -1$ be a reduced residue system modulo π. For each α_k, $k = 2, 3, \ldots, r - 1$, there is a multiplicative inverse modulo π, α'_k, such that $\alpha_k \alpha'_k \equiv 1 \pmod{\pi}$. If we group together the two numbers in all such pairs in the reduced residue system, then the product is easy to evaluate: $\alpha_1 \alpha_2 \ldots \alpha_r = 1 (\alpha_2 \alpha'_2)(\alpha_3 \alpha'_3) \ldots (\alpha_{r-1})(\alpha'_{r-1})(-1) \equiv -1 \pmod{\pi}$, which proves the theorem.

Appendix A

1. a. $a(b + c) = (b + c)a = ba + ca = ab + ac$
b. $(a + b)^2 = (a + b)(a + b) = a(a + b) + b(a + b) = a^2 + ab + ba + b^2 = a^2 + ab \cdot 1 + ab \cdot 1 + b^2 = a^2 + ab \cdot 2 + b^2 = a^2 + 2ab + b^2$.
c. $a + (b + c) = a + (c + b) = (a + c) + b = (c + a) + b$
d. $(b - a) + (c - b) + (a - c) = (-a + b) + (-b + c) + (-c + a) = -a + (b - b) + (c - c) + a = -a + 0 + 0 + a = -a + a = 0$

3. $-0 = 0 + -0 = 0$

5. Let a be a positive integer. Since $a = a - 0$ is positive, $a > 0$. Now let $a > 0$. Then $a - 0 = a$ is positive.

7. $a - c = a + (-b + b) - c = (a - b) + (b - c)$, which is positive from our hypothesis and the closure of the positive integers.

Appendix B

1. a. 1   **b.** 50   **c.** 1140   **d.** 462   **e.** 120   **f.** 1

3. a. $a^5 + 5a^4b + 10a^3b^2 + 10a^2b^3 + 5ab^4 + b^5$
b. $x^{10} + 10x^9y + 45x^8y^2 + 120x^7y^3 + 210x^6y^4 + 252x^5y^5 + 210x^4y^6 + 120x^3y^7 + 45x^2y^8 + 10xy^9 + y^{10}$
c. $m^7 - 7m^6n + 21m^5n^2 - 35m^4n^3 + 35m^3n^4 - 21m^2n^5 + 7mn^6 - n^7$
d. $16a^4 + 96a^3b + 216a^2b^2 + 216ab^3 + 81b^4$
e. $243x^5 - 1620x^4y + 4320x^3y^2 - 5760x^2y^3 + 3840xy^4 - 1024y^5$
f. $390,625x^8 + 4,375,000x^7 + 21,437,500x^6 + 60,025,000x^5 + 105,043,750x^4 + 117,649,000x^3 + 82,354,300x^2 + 32,941,720x + 5,764,801$

5. On the one hand, $(1 + (-1))^n = 0^n = 0$. On the other hand, by the binomial theorem, $(1 + (-1))^n = \sum_{k=0}^{n} (-1)^k \binom{n}{k}$.

7. $\binom{n}{r}\binom{r}{k} = \frac{n!}{r!(n-r)!} \cdot \frac{r!}{k!(r-k)!} = \frac{n!(n-k)!}{k!(n-k)!(n-r)!(n-k-n+r)!} = \binom{n}{k}\binom{n-k}{n-r}$

9. We proceed using the second principle of mathematical induction on the variable n. The basis step $n = r = 1$ is clear. For the inductive step, we assume that $\binom{r}{r} + \binom{r+1}{r} + \cdots + \binom{n}{r} = \binom{n+1}{r+1}$ is true whenever r is an integer with $1 \le r \le n$. We will now examine the formula with $n + 1$ in the place of n. If $r < n + 1$, then $\binom{r}{r} + \binom{r+1}{r} + \cdots + \binom{n}{r} + \binom{n+1}{r} = \binom{n+1}{r+1} + \binom{n+1}{r} = \binom{n+2}{r+1}$ by Theorem A.2, so the formula holds in this case. If $r = n + 1$, then $\binom{r}{r} + \cdots + \binom{n+1}{n+1} = \binom{n+1}{n+1} = 1 = \binom{n+2}{n+2}$.

11. Using Exercise 10, $\binom{x}{n} + \binom{x}{n+1} = \frac{x!}{n!(x-n)!} + \frac{x!}{(n+1)!(x-n-1)!} = \frac{x!(n+1)}{(n+1)!(x-n)!} + \frac{x!(x-n)}{(n+1)!(x-n)!} = \frac{x!(x-n+n+1)}{(n+1)!(x-n)!} = \frac{(x+1)!}{(n+1)!(x-n)!} = \binom{x+1}{n+1}$.

13. Let S be a set of n copies of $x + y$. Consider the coefficient of $x^k y^{n-k}$ in the expansion of $(x + y)^n$. Choosing the x from each element of a k-element subset of S, we notice that the coefficient of $x^k y^{n-k}$ is the number of k-element subsets of S, $\binom{n}{k}$.

15. By counting elements with exactly 0, 1, 2, and 3 properties, we see that only elements with 0 properties are counted in $n - (n(P_1) + n(P_2) + n(P_3)) + (n(P_1, P_2) + n(P_1, P_3) + n(P_2, P_3)) - n(P_1, P_2, P_3)$, and those only once.

17. A term of the sum is of the form $ax_1^{k_1}x_2^{k_2} \cdots x_m^{k_m}$, where $k_1 + k_2 + \cdots + k_m = n$ and $a = n!/(k_1!k_2! \cdots k_m!)$.

19. 56133000000

Bibliography

An extensive bibliography of printed resources on number theory and its applications is provided here. Materials listed include both books and articles. To learn more about number theory, you may want to consult other number theory textooks, such as [AdGo76], [An94], [Ar70], [Ba69], [Be66], [Bo70], [BoSh66], [Bu01], [Da99], [Di57], [Du78], [ErSu03], [Fl89], [Gi70], [Go98], [Gr82], [Gu80], [HaWr79], [Hu82], [IrRo95], [Ki74], [La58], [Le90], [Le96], [Lo95], [Ma–], [Na81], [NiZuMo91], [Or67], [Or88], [PeBy70], [Ra77], [Re96], [Ro77], [Sh85], [Sh83], [Sh67], [Si87], [Si64], [Si70], [St78], [St64], [UsHe39], [Va01], [Vi54], and [Wr39].

Printed resources listed in this bibliography also include books and articles covering particular aspects of number theory and its applications, including factorization and primality testing, the history of number theory, and cryptography.

Additional information on number theory, including the latest discoveries, can be found on the many Web sites containing relevant information. Appendix D lists the top number theory and cryptography Web sites. A comprehensive set of links to relevant Web sites can be found on the Web site for this book `www.awlonline.com/rosen`.

[AdGo76]   W.W. Adams and L.J. Goldstein, *Introduction to Number Theory,* Prentice Hall, Englewood Cliffs, New Jersey, 1976.

[Ad79]   L.M. Adleman, "A subexponential algorithm for the discrete logarithm problem with applications to cryptography," *Proceedings of the 20th Annual Symposium on the Foundations of Computer Science*, 1979, 55–60.

[AdPoRu83] L.M. Adleman, C. Pomerance, and R.S. Rumely, "On distinguishing prime numbers from composite numbers," *Annals of Mathematics*, Volume 117 (1983).

[AgKaSa02]   M.A. Agrawal, N. Kayal, N. Saxena, "PRIMES is in P," Department of Computer Science & Engineering, Indian Institute of Technology, Kanpur, India, August 6, 2002.

[AiZi03]   M. Aigner and G.M. Ziegler, *Proofs from THE BOOK*, 3rd ed., Springer-Verlag, Berlin, 2003.

[AlWi03]   S. Alaca and K. Williams, *Introductory Algebraic Number Theory*, Cambridge University Press, 2003.

[AlGrPo94]   W.R. Alford, A. Granville, and C. Pomerance, "There are infinitely many Carmichael Numbers," *Annals of Mathematics*, Volume 140 (1994), 703–722.

[An94]   G.E. Andrews, *Number Theory*, Dover, New York, 1994.

[Ap76]   T.A. Apostol, *Introduction to Analytic Number Theory*, Springer-Verlag, New York, 1976.

[Ar70]   R.G. Archibald, *An Introduction to the Theory of Numbers*, Merrill, Columbus, Ohio, 1970.

[BaSh96]   E. Bach and J. Shallit, *Algorithmic Number Theory*, MIT Press, Cambridge, Massachusetts, 1996.

[Ba94]   P. Bachmann, *Die Analytische Zahlentheorie*, Teubner, Leipzig, Germany, 1894.

[Ba03]   E.J. Barbeau, *Pell's Equation*, Springer-Verlag, New York, 2003.

[Ba69]   I.A. Barnett, *Elements of Number Theory*, Prindle, Weber, and Schmidt, Boston, 1969.

[Be66]   A.H. Beiler, *Recreations in the Theory of Numbers*, 2nd ed., Dover, New York, 1966.

[BePi82]   H. Beker and F. Piper, *Cipher Systems*, Wiley, New York, 1982.

[Be65]   E.T. Bell, *Men of Mathematics*, Simon & Schuster, New York, 1965.

[Bl82]   M. Blum, "Coin-flipping by telephone—a protocol for solving impossible problems," *IEEE Proceedings, Spring Compcon 82,* 133–137.

[Bo70]   E.D. Bolker, *Elementary Number Theory*, Benjamin, New York, 1970.

[Bo99]   D. Boneh, "Twenty years of attacks on the RSA cryptosystem," *American Mathematical Society Notices*, Volumn 46 (1999), 203–213.

[Bo82]   B. Bosworth, *Codes, Ciphers, and Computers*, Hayden, Rochelle Park, New Jersey, 1982.

[Bo91]   C.B. Boyer, *A History of Mathematics*, 2nd ed., Wiley, New York, 1991.

[BoSh66]   Z.I. Borevich and I.R. Shafarevich, *Number Theory*, Academic Press, New York, 1966.

[Br91]   R.P. Brent, "Improved techniques for lower bounds for odd perfect numbers," *Mathematics of Computation*, Volume 57 (1991), 857–868.

[Br00]   R.P. Brent, "Recent progress and prospects for integer factorization algorithms," *Proc. COCOON* 2000, LNCS 1858, pages 3–22, Springer-Verlag, 2000.

[BrCote93]   R.P. Brent, G.L. Cohen, and H.J.J. te Riele, "Improved techniques for lower bounds for odd perfect numbers," *Mathematics of Computation*, Volume 61 (1993), 857–868.

[Br89]   D.M. Bressoud, *Factorization and Primality Testing*, Springer-Verlag, New York, 1989.

[BrWa00]   D. Bressoud and S. Wagon, *A Course in Computational Number Theory*, Key College Publishing, Emeryville, California, 2000.

[Br81]   J. Brillhart, "Fermat's factoring method and its variants," *Congressus Numerantium*, Volume 32 (1981), 29–48.

[Br88]   J. Brillhart, D.H. Lehmer, J.L. Selfridge, B. Tuckerman, S.S. Wagstaff, Jr., *Factorizations of $b^n \pm 1$, $b = 2, 3, 5, 6, 7, 10, 11, 12$ up to high powers*, revised ed., American Mathematical Society, Providence, Rhode Island, 1988.

[Bu01]   D.M. Burton, *Elementary Number Theory*, 5th ed., McGraw-Hill, New York, 2001.

[Bu02]   D.M. Burton, *The History of Mathematics*, 5th ed., McGraw-Hill, New York, 2002.

[Ca59]   R.D. Carmichael, *The Theory of Numbers and Diophantine Analysis*, Dover, New York, 1959 (reprint of the original 1914 and 1915 editions).

[Ch83]   D. Chaum, ed., *Advances in Cryptology—Proceedings of Crypto 83*, Plenum, New York, 1984.

[ChRiSh83]   D. Chaum, R.L. Rivest, A.T. Sherman, eds., *Advances in Cryptology—Proceedings of Crypto 82*, Plenum, New York, 1983.

[Ci88]   B. Cipra, "PCs Factor a 'Most Wanted' Number," *Science*, Volume 242 (1988), 1634–1635.

[Ci90]   B. Cipra, "Big Number Breakdown," *Science*, Volume 248 (1990), 1608.

[Co87]   G.L. Cohen, "On the largest component of an odd perfect number," *Journal of the Australian Mathematical Society, (A)*, Volume 42 (1987), 280–286.

[CoWe91]   W.N. Colquitt and L. Welsh, Jr., "A New Mersenne Prime," *Mathematics of Computation*, Volume 56 (1991), 867–870.

[CoGu96]   R.H. Conway and R.K. Guy, *The Book of Numbers*, Copernicus Books, New York, 1996.

[Co97]   D. Coppersmith, "Small solutions to polynomial equations, and low exponent RSA vulnerabilities," *Journals of Cryptology*, Volume 10 (1997), 233–260.

[CoLeRi01]   T.H. Cormen, C.E. Leierson, R.L. Rivest, *Introduction to Algorithms*, 2nd ed., MIT Press, Cambridge, Massachusetts, 2001.

[CoSiSt97]   G. Cornell, J.H. Silverman, and G. Stevens, *Modular Forms and Fermat's Last Theorem,* Springer-Verlag, New York, 1997.

[Cr94]   R.E. Crandall, *Projects in Scientific Computation*, Springer-Verlag, New York, 1994.

[CrPo01]   R. Crandall and C. Pomerance, *Prime Numbers, A Computational Perspective*, Springer-Verlag, New York, 2001.

[Da99]   H. Davenport, *The Higher Arithmetic*, 7th ed., Cambridge University Press, Cambridge, England, 1999.

[De82]   D.E.R. Denning, *Cryptography and Data Security*, Addison-Wesley, Reading, Massachusetts, 1982.

[De03]   J. Derbyshire, *Prime Obsession*, Joseph Henry Press, Washington, D.C., 2003.

[Di57]   L.E. Dickson, *Introduction to the Theory of Numbers*, Dover, New York, 1957 (reprint of the original 1929 edition).

[Di71]   L.E. Dickson, *History of the Theory of Numbers*, three volumes, Chelsea, New York, 1971 (reprint of the 1919 original).

[Di70]   *Dictionary of Scientific Biography*, Scribners, New York, 1970.

[DiHe76]   W. Diffie and M. Hellman, "New directions in cryptography," *IEEE Transactions on Information Theory*, Volume 22 (1976), 644–655.

[Di84]   J.D. Dixon, "Factorization and primality tests," *American Mathematical Monthly*, Volume 91 (1984), 333–353.

[Du78]   U. Dudley, *Elementary Number Theory*, 2nd ed., Freeman, New York, 1978.

[Ed96]   H.M. Edwards, *Fermat's Last Theorem*, 5th ed., Springer-Verlag, New York, 1996.

[Ed01]   H.M. Edwards, *Riemann's Zeta Function*, Dover, New York, 2001.

[ErSu03]   P. Erdős and J. Surányi, *Topics in the History of Numbers*, Springer-Verlag, New York, 2003.

[Ev92]   H. Eves, *An Introduction to the History of Mathematics*, 6th ed., Elsevier, New York, 1992.

[Ew83]   J. Ewing, "$2^{86243} - 1$ is prime," *The Mathematical Intelligencer*, Volume 5 (1983), 60.

[Fl89]   D. Flath, *Introduction to Number Theory*, Wiley, New York, 1989.

[Fl83]   D.R. Floyd, "Annotated bibliographical in conventional and public key cryptography," *Cryptologia*, Volume 7 (1983), 12–24.

[Fr56]   J.E. Freund, "Round robin mathematics," *American Mathematical Monthly*, Volume 63 (1956), 112–114.

[Fr78]   W.F. Friedman, *Elements of Cryptanalysis*, Aegean Park Press, Laguna Hills, California, 1978.

[Ga91]   J. Gallian, "The mathematics of identification numbers," *College Mathematics Journal*, Volume 22 (1991), 194–202.

[Ga92]   J. Gallian, "Assigning drivers license numbers," *Mathematics Magazine*, Volume 64 (1992), 13–22.

[Ga96]   J. Gallian, "Error Detection Methods," *ACM Computing Surveys*, Volume 28 (1996), 504–517.

[GaWi88]   J. Gallian and S. Winters, "Modular arithmetic in the marketplace," *American Mathematical Monthly*, Volume 95 (1988), 584–551.

[Ga86]   C.F. Gauss, *Disquisitiones Arithmeticae*, revised English translation by W.C. Waterhouse, Springer-Verlag, New York, 1986.

[Ge63]   M. Gerstenhaber, "The 152nd proof of the law of quadratic reciprocity," *American Mathematical Monthly*, Volume 70 (1963), 397–398.

[Ge82]   A. Gersho, ed., *Advances in Cryptography*, Department of Electrical and Computer Engineering, University of California, Santa Barbara, 1982.

[GeWaWi98] E. Gethner, S. Wagon, and B. Wick, "A stroll through the Gaussian primes," *American Mathematical Monthly*, Volume 104 (1998), 216–225.

[Gi70]   A.A. Gioia, *The Theory of Numbers*, Markham, Chicago, 1970.

[Go98]   J.R. Goldman, *The Queen of Mathematics: An Historically Motivated Guide to Number Theory*, A.K. Peters, Wellesley, Massachusetts, 1998.

[Go80]   J. Gordon, "Use of intractable problems in cryptography," *Information Privacy*, Volume 2 (1980), 178–184.

[Gr04]   A. Granville, "It is easy to determine whether a given integer is prime," *Current Events in Mathematics*, American Mathematical Society, 2004.

[GrTu02]   A. Granville and T.J. Tucker, "It's as Easy as abc," *Notices of the American Mathematical Society*, Volume 49 (2002), 1224–1231.

[Gr82]   E. Grosswald, *Topics from the Theory of Numbers*, 2nd ed., Birkhauser, Boston, 1982.

[GrKnPa94] R.L. Graham, D.E. Knuth, and O. Patashnik, *Concrete Mathematics*, 2nd ed., Addison-Wesley, Reading, Massachusetts, 1994.

[Gu80]   H. Gupta, *Selected Topics in Number Theory*, Abacus Press, Kent, England, 1980.

[Gu75]   R.K. Guy, "How to factor a number," *Proceedings of the Fifth Manitoba Conference on Numerical Mathematics*, Utilitas, Winnepeg, Manitoba, 1975, 49–89.

[Gu94]   R.K. Guy, *Unsolved Problems in Number Theory*, 2nd ed., Springer-Verlag, New York, 1994.

[Ha83]   P. Hagis, Jr., "Sketch of a proof that an odd perfect number relatively prime to 3 has at least eleven prime factors," *Mathematics of Computations*, Volume 46 (1983), 399–404.

[HaWr79]   G.H. Hardy and E.M. Wright, *An Introduction to the Theory of Numbers*, 5th ed., Oxford University Press, Oxford, 1979.

[He80]   A.K. Head, "Multiplication modulo n," *BIT*, Volume 20 (1980), 115–116.

[He79]   M.E. Hellman, "The mathematics of public-key cryptography," *Scientific American*, Volume 241 (1979) 146–157.

[Hi31]   L.S. Hill, "Concerning certain linear transformation apparatus of cryptography," *American Mathematical Monthly*, Volume 38 (1931), 135–154.

[Hu82]   L. Hua, *Introduction to Number Theory*, Springer-Verlag, New York 1982.

[Hw79]   K. Hwang, *Computer Arithmetic: Principles, Architecture and Design*, Wiley, New York, 1979.

[IrRo95]   K.F. Ireland and M.I. Rosen, *A Classical Introduction to Modern Number Theory*, 2nd ed., Springer-Verlag, New York, 1995.

[Ka96]   D. Kahn, *The Codebreakers, the Story of Secret Writing*, 2nd ed., Scribners, New York, 1996.

[Ka98]   V. Katz, *A History of Mathematics: An Introduction*, 2nd ed., Addison-Wesley, Boston, 1998.

[Ki04]   S.V. Kim, "An Elementary Proof of the Quadratic Reciprocity Law," *American Mathematical Monthly*, Volume 111, Number 1 (2004), 45–50.

[Ki74]   A.M. Kirch, *Elementary Number Theory: A Computer Approach*, Intext, New York, 1974.

[Ki01]   J. Kirtland, *Identification Numbers and Check Digit Schemes*, Mathematical Association of America, Washington, D.C., 2001.

[Kl72]   M. Kline, *Mathematical Thought from Ancient to Modern Times*, Oxford University, New York, 1972.

[Kn97]   D.E. Knuth, *Art of Computer Programming: Semi-Numerical Algorithms*, Volume 2, 3rd ed., Addison-Wesley, Reading, Massachusetts, 1997.

[Kn97a]   D.E. Knuth, *Art of Computer Programming: Sorting and Searching*, Volume 3, 2nd ed., Addison-Wesley, Reading, Massachusetts, 1997.

[Ko96]   N. Koblitz, *Introduction to Elliptic Curves and Modular Forms*, 2nd ed., Springer-Verlag, New York, 1996.

[Ko94]   N. Koblitz, *A Course in Number Theory and Cryptography*, 2nd ed., Springer-Verlag, New York, 1994.

[Ko96a]   P. Kocher, "Timing attacks on implementations of Diffie-Hellman, RSA, DSS, and other systems," *Advances in Cryptology—CRYPTO '96*, LNCS 1109, Springer-Verlag, New York, 1996, 104–113.

[Ko83]   G. Kolata, "Factoring Gets Easier," *Science*, Volume 222 (1983), 999–1001.

[Ko81]   A.G. Konheim, *Cryptography: A Primer*, Wiley, New York, 1981.

[Kr86]   E. Kranakis, *Primality and Cryptography*, Wiley-Teubner, Stuttgart, Germany, 1986.

[Kr79]   L. Kronsjo, *Algorithms: Their Complexity and Efficiency*, Wiley, New York, 1979.

[Ku76]   S. Kullback, *Statistical Methods in Cryptanalysis*, Aegean Park Press, Laguna Hills, California, 1976.

[La90]   J.C. Lagarias, "Pseudo-random number generators in cryptography and number theory," pages 115–143 in *Cryptology and Computational Number Theory*, Volume 42 of Proceedings of Symposia in Advanced Mathematics, American Mathematical Society, Providence, Rhode Island, 1990.

[LaOd82]   J.C. Lagarias and A.M. Odlyzko, "New algorithms for computing $\pi(x)$," Bell Laboratories Technical Memorandum TM-82-11218-57.

[La58]   E. Landau, *Elementary Number Theory*, Chelsea, New York, 1958.

[La60]   E. Landau, *Foundations of Analysis*, 2nd ed., Chelsea, New York, 1960.

[La35]   H.P. Lawther, Jr., "An application of number theory to the splicing of telephone cables," *American Mathematical Monthly*, Volume 42 (1935), 81–91.

[LePo31]   D.H. Lehmer and R.E. Powers, "On factoring large numbers," *Bulletin of the American Mathematical Society*, Volume 37 (1931), 770–776.

[Le00]   F. Lemmermeyer, *Reciprocity Laws I,* Springer-Verlag, Berlin, 2000.

[Le79]   A. Lempel, "Cryptology in transition," *Computing Surveys*, Volume 11 (1979), 285–303.

[Le80]   H.W. Lenstra, Jr., "Primality testing," *Studieweek Getaltheorie en Computers*, 1–5 September 1980, Stichting Mathematisch Centrum, Amsterdam, Holland.

[Le90]   W.J. Leveque, *Elementary Theory of Numbers*, Dover, New York, 1990.

[Le96]   W.J. LeVeque, *Fundamentals of Number Theory*, Dover, New York, 1996.

[Le74]   W.J. LeVeque, editor, *Reviews in Number Theory* [1940–1972], and R.K. Guy, editor, *Reviews in Number Theory* [1973–1983], six volumes each, American Mathematical Society, Washington, D.C., 1974 and 1984, respectively.

[LiDu87]   Y. Li and S. Du, *Chinese Mathematics: A Concise History*, translated by J. Crossley and A. Lun, Clarendon Press, Oxford, England, 1987.

[Li73]   U. Libbrecht, *Chinese Mathematics in the Thirteenth Century, The Shu-shu chiu-chang of Ch'in Chiu-shao*, MIT Press, 1973.

[Li79]   R.J. Lipton, "How to cheat at mental poker," and "An improved power encryption method," unpublished reports, Department of Computer Science, University of California, Berkeley, 1979.

[Lo95]   C.T. Long, *Elementary Introduction to Number Theory*, 3rd ed., Waveland Press, Prospect Heights, Illinois, 1995.

[Lo90]   J.H. Loxton, editor, *Number Theory and Cryptography*, Cambridge University Press, Cambridge, England, 1990.

[Ma79]   D.G. Malm, *A Computer Laboratory Manual for Number Theory*, COM-Press, Wentworth, New Hampshire, 1979.

[McRa79]   J.H. McClellan and C.M. Rader, *Number Theory in Digital Signal Processing*, Prentice Hall, Englewood Cliffs, New Jersey, 1979.

[Ma–]   G.B. Matthews, *Theory of Numbers*, Chelsea, New York (no publication date provided).

[Ma94]   U. Maurer, "Towards the equivalence of breaking the Diffie-Hellman protocol and computing discrete logarithms," *Advances in Cryptology—CRYPTO '94*, LNCS 839, 1994, 271–281.

[Ma95]   U. Maurer, "Fast generation of prime numbers and secure public-key cryptographic parameters," *Journal of Cryptology*, Volume 8 (1995), 123–155.

[Ma00]   B. Mazur, "Questions about powers of numbers," *Notices of the American Mathematical Society,* Volume 47 (2000), 195–202.

[MevaVa97]   A.J. Menezes, P.C. van Oorschot, S.A. Vanstone, *Handbook of Applied Cryptography*, CRC Press, Boca Raton, Florida, 1997.

[Me82]   R.C. Merkle, *Secrecy, Authentication, and Public Key Systems*, UMI Research Press, Ann Arbor, Michigan, 1982.

[MeHe78]   R.C. Merkle and M.E. Hellman, "Hiding information and signatures in trapdoor knapsacks," *IEEE Transactions in Information Theory*, Volume 24 (1978), 525–530.

[MeMa82]   C.H. Meyer and S.M. Matyas, *Cryptography: A New Dimension in Computer Data Security*, Wiley, New York, 1982.

[Mi76]   G.L. Miller, "Riemann's hypothesis and tests for primality," *Journal of Computer and Systems Science*, Volume 13 (1976), 300–317.

[Mi47]   W.H. Mills, "A prime-representing function," *Bulletin of the American Mathematical Society*, Volume 53 (1947), 604.

[Mo96]   R.A. Mollin, *Quadratics*, CRC Press, Boca Raton, Florida, 1996.

[Mo99]   R.A. Mollin, *Algebraic Number Theory*, CRC Press, Boca Raton, Florida, 1999.

[Mo96]   M.B. Monagan, K.O. Geddes, K.M. Heal, G. Labahn, and S.M. Vorkoetter, *Maple V Programming Guide*, Springer-Verlag, New York, 1996.

[Mo80]   L. Monier, "Evaluation and comparison of two efficient probabilistic primality testing algorithms, *Theoretical Computer Science*, Volume 11 (1980), 97–108.

[Mo69]   L.J. Mordell, *Diophantine Equations*, Academic Press, New York, 1969.

[Na81]   T. Nagell, *Introduction to Number Theory*, Chelsea, New York, 1981.

[Ne69]   O.E. Neugebauer, *The Exact Sciences in Antiquity*, Dover, New York, 1969.

[NeSc99]   J. Neukirch and N. Schappacher, *Algebraic Number Theory*, Springer-Verlag, New York, 1999.

[NiZuMo91]   I. Niven, H.S. Zuckerman, and H.L. Montgomery, *An Introduction to the Theory of Numbers*, 5th ed., Wiley, New York, 1991.

[Odte85]   A.M. Odlyzko and H.J.J. te Riele, "Disproof of the Mertens conjecture," *Journal für die reine und angewandte Mathematik*, Volume 357 (1985), 138–160.

[Od90]   A.M. Odlyzko, "The rise and fall of knapsack cryptosystems," pages 75–88 in *Cryptology and Computational Number Theory*, Volume 42 of Proceedings of Symposia in Applied Mathematics, American Mathematical Society, Providence, Rhode Island, 1990.

[Od95]   A.M. Odlyzko, "The future of integer factorization," *RSA CrytoBytes*, Volume 2, Number 1, 1995, 5–12.

[Or67]   O. Ore, *An Invitation to Number Theory*, Random House, New York, 1967.

[Or88]   O. Ore, *Number Theory and its History*, Dover, New York, 1988.

[PaMi88]   S.K. Park and K.W. Miller, "Random Number Generators: Good Ones are Hard to Find," *Communications of the ACM*, Volume 31 (1988), 1192–1201.

[PeBy70]   A.J. Pettofrezzo and D.R. Byrkit, *Elements of Number Theory*, Prentice Hall, Englewood Cliffs, New Jersey, 1970.

[Pf89]   C.P. Pfleeger, *Security in Computing*, Prentice Hall, Englewood Cliffs, New Jersey, 1989.

[Po14]   H.C. Pocklington, "The determination of the prime or composite nature of large numbers by Fermat's theorem," *Proceedings of the Cambridge Philosophical Society*, Volume 18 (1914/6), 29–30.

[PoHe78]   S. Pohlig and M. Hellman, "An improved algorithm for computing logarithms over $GF(p)$ and its cryptographic significance," *IEEE Transactions on Information Theory*, Volume 24 (1978), 106–110.

[Po99]   H. Pollard and H. Diamond, *The Theory of Algebraic Numbers*, 3rd ed., Dover, New York, 1999.

[Po74]   J.M. Pollard, "Theorems on Factorization and Primality Testing," *Proceedings of the Cambridge Philosophical Society,* Volume 76 (1974), 521–528.

[Po75]   J.M. Pollard, "A Monte Carlo Method for Factorization," *Nordisk Tidskrift for Informationsbehandling (BIT)*, Volume 15 (1975), 331–334.

[Po81]   C. Pomerance, "Recent developments in primality testing," *The Mathematical Intelligencer*, Volume 3 (1981), 97–105.

[Po82]   C. Pomerance, "The search for prime numbers," *Scientific American*, Volume 247 (1982), 136–147.

[Po84]   C. Pomerance, *Lecture Notes on Primality Testing and Factoring*, Mathematical Association of America, Washington, D.C., 1984.

[Po90]   C. Pomerance, ed., *Cryptology and Computational Number Theory*, American Mathematical Society, Providence, Rhode Island, 1990.

[Po93]   C. Pomerance, "Carmichael Numbers," *Nieuw Arch. v. Wiskunde*, Volume 4, number 11 (1993), 199–209.

[Ra79]   M.O. Rabin, "Digitalized signatures and public-key functions as intractable as factorization," M.I.T. Laboratory for Computer Science Technical Report LCS/TR-212, Cambridge, Massachusetts, 1979.

[Ra80]   M.O. Rabin, "Probabilistic algorithms for testing primality," *Journal of Number Theory*, Volume 12 (1980), 128–138.

[Ra77]   H. Rademacher, *Lectures on Elementary Number Theory*, Krieger, 1977.

[Re96]   D. Redfern, *The Maple Handbook*, Springer-Verlag, New York, 1996.

[Re96]   D. Redmond, *Number Theory: An Introduction*, Marcel Dekker, Inc., New York, 1996

[Ri79]   P. Ribenboim, *13 Lectures on Fermat's Last Theorem*, Springer-Verlag, New York, 1979.

[Ri96]   P. Ribenboim, *The New Book of Prime Number Record*, Springer-Verlag, New York, 1996.

[Ri01]   P. Ribenboim, *Classical Theory of Algebraic Integers*, 2nd ed., Springer-Verlag, New York, 2001.

[Ri59]   B. Riemann, "Uber die Anzahl der Primzahlen unter einer gegeben Grösse," *Monatsberichte der Berliner Akademie,* November, 1859.

[Ri85a]   H. Riesel, "Modern factorization methods," *BIT* (1985), 205–222.

[Ri94]   H. Riesel, *Prime Numbers and Computer Methods for Factorization*, 2nd ed., Birkhauser, Boston, 1994.

[Ri78]   R.L. Rivest, "Remarks on a proposed cryptanalytic attack on the M.I.T. public-key cryptosystem," *Cryptologia*, Volume 2 (1978), 62–65.

[RiShAd78] R.L. Rivest, A. Shamir, and L.M. Adleman, "A method for obtaining digital signatures and public-key cryptosystems," *Communications of the ACM*, Volume 21 (1978), 120–126.

[RiShAd83] R.L. Rivest, A. Shamir, and L.M. Adleman, "Cryptographic communications system and method," United States Patent #4,405,8239, issued September 20, 1983.

[Ro77]   J. Roberts, *Elementary Number Theory*, MIT Press, Cambridge, Massachusetts, 1977.

[Ro97]   K. Rosen et. al., *Exploring Discrete Mathematics with Maple,* McGraw-Hill, New York, 1997.

[Ro99a]   K.H. Rosen, *Handbook of Discrete and Combinatorial Mathematics*, CRC Press, Boca Raton, Florida, 1999.

[Ro03]   K.H. Rosen, *Discrete Mathematics and its Applications*, 5th ed., McGraw-Hill, New York, 2003.

[Ru64]   W. Rudin, *Principles of Mathematical Analysis*, 2nd ed., McGraw-Hill, New York, 1964.

[Ru83]   R. Rumely, "Recent advances in primality testing," *Notices of the American Mathematical Society*, Volume 30 (1983), 475–477.

[Sa03a]   K. Sabbagh, *The Riemann Hypothesis*, Farrar, Strauss, and Giroux, New York, 2003.

[Sa90]   A. Salomaa, *Public-Key Cryptography*, Springer-Verlag, New York, 1990.

[Sa03b]   M. du Sautoy, *The Music of the Primes*, Harper Collins, New York, 2003.

[ScOp85]   W. Scharlau and H. Opolka, *From Fermat to Minkowski, Lectures on the Theory of Numbers and its Historical Development*, Springer-Verlag, New York, 1985.

[Sc86]   M.R. Schroeder, *Number Theory in Science and Communication*, 2nd ed., Springer-Verlag, Berlin, 1986.

[SePi89]   J. Seberry and J. Pieprzyk, *Cryptography: An Introduction to Computer Security*, Prentice Hall, New York, 1989.

[Sh79]   A. Shamir, "How to share a secret," *Communications of the ACM*, Volume 22 (1979), 612–613.

[Sh83]   A. Shamir, "A polynomial time algorithm for breaking the basic Merkle-Hellman cryptosystem," in *Advances in Cryptology—Proceedings of Crypto 82*, 279–288.

[Sh84]   A. Shamir, "A polynomial time algorithm for breaking the basic Merkle-Hellman cryptosystem," *IEEE Transactions on Information Theory*, Volume 30 (1984), 699–704. (This is an improved version of [Sh83].)

[ShRiAd81] A. Shamir, R.L. Rivest, and L.M. Adleman, "Mental Poker," *The Mathematical Gardner*, ed. D.A. Klarner, Wadsworth International, Belmont, California, 1981, 37–43.

[Sh85]   D. Shanks, *Solved and Unsolved Problems in Number Theory*, 3rd ed., Chelsea, New York, 1985.

[Sh83]   H.S. Shapiro, *Introduction to the Theory of Numbers*, Wiley, New York, 1983.

[Sh67]   J.E. Shockley, *Introduction to Number Theory*, Holt, Rinehart, and Winston, New York, 1967.

[Si64]   W. Sierpinski, *A Selection of Problems in the Theory of Numbers*, Pergamon Press, New York, 1964.

[Si70]   W. Sierpinski, *250 Problems in Elementary Number Theory*, Polish Scientific Publishers, Warsaw, 1970.

[Si87]   W. Sierpinski, *Elementary Theory of Numbers*, 2nd ed., North-Holland, Amsterdam, 1987.

[Si82]   G.J. Simmons, ed., *Secure Communications and Asymmetric Cryptosystems*, AAAS Selected Symposium Series Volume 69, Westview Press, Boulder, Colorado, 1982.

[Si97]   S. Singh, *Fermat's Enigma: The Epic Quest to Solve the World's Greatest Mathematical Problem*, Walker and Company, New York, 1997.

[Si66]   A. Sinkov, *Elementary Cryptanalysis*, Mathematical Association of America, Washington, D.C., 1966.

[SlPl95]   N.J.A. Sloane and S. Plouffe, *The Encyclopedia of Integer Sequences*, Academic Press, New York, 1995.

[Sl78]   D. Slowinski, "Searching for the 27th Mersenne prime," *Journal of Recreational Mathematics*, Volume 11 (1978/9), 258–261.

[SoSt77]   R. Solovay and V. Strassen, "A fast Monte Carlo test for primality," *SIAM Journal for Computing*, Volume 6 (1977), 84–85 and erratum, Volume 7 (1978), 118.

[So86]   M.A. Soderstrand et al., editors, *Residue Number System Arithmetic: Modern Applications in Digital Signal Processing*, IEEE Press, New York, 1986.

[Sp82]   D.D. Spencer, *Computers in Number Theory*, Computer Science Press, Rockville, Maryland, 1982.

[St78]   H.M. Stark, *An Introduction to Number Theory*, Markham, Chicago, 1970; reprint MIT Press, Cambridge, Massachusetts, 1978.

[St64]   B.M. Stewart, *The Theory of Numbers*, 2nd ed., Macmillan, New York, 1964.

[St02]   D.R. Stinson, *Cryptography, Theory and Practice*, 2nd ed., Chapman & Hall/CRC, Boca Raton, Florida, 2002.

[SzTa67]   N.S. Szabo and R.J. Tanaka, *Residue Arithmetic and its Applications to Computer Technology*, McGraw-Hill, 1967.

[TrWa02]   W. Trappe and L. Washington, *Introduction to Cryptography with Coding Theory*, Prentice Hall, Upper Saddle River, New Jersey, 2002.

[UsHe39]   J.V. Uspensky and M.A. Heaslet, *Elementary Number Theory*, McGraw-Hill, New York, 1939.

[Va89]   S. Vajda, *Fibonacci & Lucas Numbers and the Golden Section: Theory and Applications*, Ellis Horwood, Chichester, England, 1989.

[Va96]   A.J. van der Poorten, *Notes on Fermat's Last Theorem*, Wiley, New York, 1996.

[Va01]   C. VandenEynden, *Elementary Number Theory*, McGraw-Hill, New York, 2001.

[Vi54]   I.M. Vinogradov, *Elements of Number Theory*, Dover, New York, 1954.

[Wa86]   S. Wagon, "Primality testing," *The Mathematical Intelligencer*, Volume 8, Number 3 (1986), 58–61.

[Wa99]   S. Wagon, *Mathematica in Action*, 2nd ed. Telos, New York, 1999.

[Wa86]   S.S. Wagstaff, "Using computers to teach number theory," *SIAM News*, Volume 19 (1986), 14 and 18.

[Wa90]   S.S. Wagstaff, "Some uses of microcomputers in number theory research," *Computers and Mathematics with Applications*, Volume 19 (1990), 53–58.

[WaSm87]   S.S. Wagstaff and J.W. Smith, "Methods of factoring large integers," in *Number Theory, New York, 1984–1985*, LNM, Volume 1240, Springer-Verlag, Berlin, 1987, 281–303.

[We84]   A. Weil, *Number Theory: An approach through history from Hummurapi to Legendre*, Birkhauser, Boston, 1984.

[Wi90]   M.J. Wiener, "Cryptanalysis of short RSA secret exponents," *IEEE Transactions on Information Theory*, Volume 36 (1990), 553–558.

[Wi95]   A. Wiles, "Modular elliptic-curves and Fermat's last theorem," *Annals of Mathematics,* Volume 141 (1995), 443–551.

[Wi86]   H.C. Williams, ed., *Advances in Cryptology—CRYPTO '85*, Springer-Verlag, Berlin, 1986.

[Wi78]   H.C. Williams, "Primality testing on a computer," *Ars Combinatorica*, Volume 5 (1978), 127–185.

[Wi82]   H.C. Williams, "The influence of computers in the development of number theory," *Computers and Mathematics with Applications*, Volume 8 (1982), 75–93.

[Wi84]   H.C. Williams, "An overview of factoring," in *Advances in Cryptology, Proceedings of Crypto 83*, Plenum, New York, 1984, 87–102.

[Wo03]   S. Wolfram, *The Mathematica Book*, 5th ed., Cambridge University Press, New York, 2003.

[Wr39]   H.N. Wright, *First Course in Theory of Numbers*, Wiley, New York, 1939.

[Wu85]   M.C. Wunderlich, "Implementing the continued fraction algorithm on parallel machines," *Mathematics of Computation*, Volume 44 (1985), 251–260.

[WuKu90]   M.C. Wunderlich and J.M. Kubina, "Extending Waring's conjecture to 471,600,000," *Mathematics of Computation*, Volume 55 (1990), 815–820.

Index of Biographies

Index

Photo Credits

Courtesy of The MacTutor History of Mathematics Archive, University of St. Andrews, Scotland: Stanislaw M. Ulam, Fibonacci, Francois-Edouard-Anatole Lucas, Paul Gustav Heinrich Bachmann, Edmund Landau, Pafnuty Lvovich Chebyshev, Jacques Hadamard, Alte Selberg, Joseph Louis François Bertrand, G. Lejeune Dirichlet, Gabriel Lamé, David Hilbert, Karl Friedrich Gauss, Kurt Hensel, Joseph Louis LaGrange, Georg Friedrich Bernhard Reimann, Leonhard Euler, Joesph Liouville, Srinivasa Ramanujan, Marin Mersenne, August Ferdinand Möbius, Adrien-Marie Legendre, Ferdinand Gotthold Max Eisenstein, Carl Gustav Jacob Jacobi, Leopold Kroneker, Georg Cantor, Pythagoras, Sophie Germain, Ernst Eduard Kummer, Andrew Wiles, Claude Bachet, Edward Waring, Axel Thue, and Blaise Pascal; Eratosthenes © Culver Pictures, Inc.; Paul Erdos © 1985 Wlodzimierz Kuperberg; Euclid © The Granger Collection; Pierre de Fermat © Giraudon/Art Resource, N.Y.; Derrick H. Lehmer © 1993 the American Mathematical Society; Gilbert S. Vernam © Worcester Polytechnic Institute, Class of 1914; Adi Shamir © the Weizmann Institute of Science, Israel; Ronald Rivest © 1999 Ronald Rivest; Leonard Adleman © 1999 Eric Mankin, University of Southern California; John Von Neumann © Corbis; Emil Artin © The Hall of Great Mathematicians, Southern Illinois University, Edwardsville; Eugène Catalan © Collections artistiques de l'Université de Liège.